LASER PROCESSING OF ENGINEERING MATERIALS

PRINCIPLES, PROCEDURE AND INDUSTRIAL APPLICATION

JOHN C. ION Eur. Ing., CEng, FIMMM

ELSEVIER
BUTTERWORTH
HEINEMANN

AMSTERDAM • BOSTON • HEIDELBERG • LONDON • NEW YORK • OXFORD
PARIS • SAN DIEGO • SAN FRANCISCO • SINGAPORE • SYDNEY • TOKYO

Elsevier Butterworth-Heinemann
Linacre House, Jordan Hill, Oxford OX2 8DP
30 Corporate Drive, Burlington, MA 01803

First published 2005

Permissions may be sought directly from Elsevier's Science & Technology
Rights Department in Oxford, UK: phone: (+44) 1865 843830,
fax: (+44) 1865 853333, e-mail: permissions@elsevier.co.uk.
You may also complete your request on-line via the Elsevier homepage
(http://www.elsevier.com), by selecting 'Customer Support' and then
'Obtaining Permissions'

British Library Cataloguing in Publication Data
A catalogue record for this book is available from the British Library

Library of Congress Cataloguing in Publication Data
A catalogue record for this book is available from the Library of Congress

ISBN 0 7506 6079 1

For information on all Elsevier Butterworth-Heinemann
publications visit our website at http://books.elsevier.com

Typeset by Charon Tec Pvt. Ltd, Chennai, India
www.charontec.com
Printed and bound in Great Britain by Biddles Ltd, King's Lynn, Norfolk

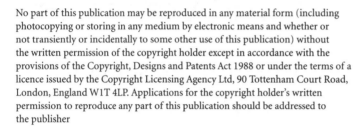

CONTENTS

13 CONDUCTION JOINING 327

14 CUTTING 347

15 MARKING 384

16 KEYHOLE WELDING 395

FOREWORD

Advances in technology frequently outstrip the ability to deploy them effectively in industry. Understandable concern for the risks associated with adopting new, untried methods of manufacture – particularly those requiring high up-front investment – hold users back; the prudent wait until someone else has demonstrated the viability and negotiated (or fallen into) the traps that lie along the way. The early bird may catch the first worm, but it is the second mouse that gets the cheese.

Considerations such as these held back the wide-scale adoption of lasers for welding, cutting and heat-treatment in the 1970s and early 80s, well after they had been demonstrated at a laboratory scale. But as experience accumulated, the pace of take-up increased, until today lasers are accepted as manufacturing tools. Capturing this experience, ordering it, relating it where possible to underlying fundamentals and predictive models is arguably the best way to accelerate this take-up further. This assembling of experience, understanding and knowledge is precisely what this book is about.

Few people could be better qualified to write a text on laser processing than the author, Dr John Ion. His professional life has spanned the time from which lasers first became powerful enough to permit material processing to the present day, a time when they are widely adopted for cutting, welding, marking, heat-treatment, cladding and much more. Throughout, Dr Ion has led research and development of lasers as processing tools and has sought to stimulate and enable their effective use.

One of Dr Ion's roles has been to act as an advisor and consultant, holding the hand, so to speak, of manufacturers striving to select, install and commission laser-based systems. Such a role requires experience and insight of exceptional breadth and depth. Both are rewardingly revealed in this book. The span is global, combining fundamental understanding of each process with a practical knowledge of where and how best to employ it – the 'dos' and 'don'ts' as well as the 'how it works'. Dr Ion pioneered the use of laser-processing maps, richly illustrated here, as a way of relating the process parameters of beam power, beam size and scan rate to specific manufacturing tasks and materials. But there is much more here than that. Dr Ion outlines his aims as those of providing understanding of underlying principles, documenting current best practice, developing predictive models for process control and clarifying the technical and economic criteria that guide process choice.

All of these he amply achieves. This is a truly comprehensive text in its coverage of the many diverse ways in which lasers are now used in manufacture, in the depth with which each of these is explored and in the vision for the future with which it concludes. It is a volume of lasting value.

M.F. Ashby
Cambridge, UK
November 2004

PREFACE

Lasers are now an integral part of modern society. The CD and DVD have revolutionized the way we listen to music and watch movies. The laser scanner at the supermarket checkout no longer raises an eyebrow. Corrective laser eye surgery has enabled many to discard their spectacles. Digital pictures from a colour laser printer are now of such high quality that photography has been transformed from a darkroom art into a computer science.

However, few appreciate the innovations that lasers have enabled in modern manufacturing. A new family automobile contains tens of metres of laser-welded bodywork, providing a stronger, safer passenger compartment. Control buttons, windows, mirrors and the vehicle identification number are neatly and indelibly laser marked. The automobile can be personalized by laser cutting myriad apertures and cavities for accessories and optional extras. Components in the engine and transmission are laser hardened so that they perform better and last longer. Hopefully the laser-cut material of the airbag will remain tucked inside the steering wheel, and its laser-welded sensor will never be actuated.

Given that this industry – known as laser processing of engineering materials – often achieves annual double-digit growth, and currently turns over more than one thousand million dollars each year, it is remarkable that it receives such little publicity. In education also, the subject is often only taught as a module as part of a course in engineering, the physical sciences, or materials, because of its multidisciplinary nature; a module that lacks the depth required to appreciate the opportunities for novel design and manufacturing procedures that the subject offers. Many in education and industry have identified a need for formal education and training in the subject if momentum is to be sustained. This book is intended to address that need.

The content is designed primarily for senior-level undergraduate students of Engineering and Materials Science, who have a basic understanding of the structure and properties of engineering materials. However, postgraduate research students will find reference data that provide an overview of the subject, and which suggest topics for further study. Scientifically inclined professionals will find practical information relevant to laser material processing in a wide range of industry sectors.

The book aims to provide answers to a number of fundamental questions. What are the differences between laser material processing and conventional methods of processing materials, and where do analogies exist? What opportunities does a clean, flexible and highly controllable source of laser power provide for novel processing? How can the economics of laser processing be assessed and compared with traditional costing models? Many of the answers lie in the underlying physical and chemical mechanisms of laser processing.

Chapters 1–6 set the stage. The breadth of current application is illustrated in Chapter 1 with a selection of familiar products used in the home, healthcare, manufacturing and the arts. Chapter 2 provides an account of the evolution of lasers, material processing systems, process development, and industrial application – highlighting the multidisciplinary nature of the subject, and illustrating the synergies that exist between the disciplines. The workings of lasers and systems for material processing are described in Chapters 3 and 4, respectively, to aid in practical selection of appropriate equipment. An overview of engineering materials is presented in Chapter 5, with emphasis on the interactions between laser beams and materials on the atomic scale and the effect of rapid laser thermal cycles on the structures and properties that can be produced, using material property charts as a guide to the selection of materials and processes. Chapter 6 examines the principal process variables of laser material processing; they indicate that two fundamental modes of processing are possible – thermal and athermal. The three fundamental mechanisms of thermal processing are described – heating, melting and vaporization – together with the basic mechanisms of athermal processing: making and breaking chemical bonds. An understanding of process variables and processing mechanisms leads to the concept of *Laser Processing Diagrams*, which are constructed for the most common processes using analytical mathematical modelling. These not only provide an overview of a process, enabling practical processing parameters to be extracted, but also highlight possibilities for novel processing techniques.

The chapters that follow cover the processes themselves. Since laser material processing originated as a thermal interaction between the beam and the material, the processes are presented in order of increasing temperature, starting with heating and ending with vaporization. But before thermal processes are examined, Chapter 7 describes athermal processing, which involves photoelectrical, photochemical and photophysical mechanisms – processes in which the variables span enormous ranges and which are enabling remarkable advances in process technology. The presentation framework created for these processes allows them to be placed in context with the more traditional thermal mode of laser material processing.

Chapters 8–10 consider processes in which classical heating is the principal processing mechanism: structural change driven by thermodynamic and kinetic reactions; surface hardening involving phase transformations; and deformation and controlled fracture. The format used for each chapter and process is designed to expedite data retrieval: process principles are explained in terms of the underlying materials science; process selection criteria are formulated by comparing technical and economic features with conventional manufacturing methods; industrial production techniques are described to show how principles are put into practice; the properties of commonly processed materials are characterized to aid comparison, identify trends, and provide a source of reference for further study; Laser Processing Diagrams are constructed for processes that can be modelled, and their uses illustrated; and manufacturing applications from different sectors of industry are examined to reveal the factors that influence decision making. The chapters conclude with recommendations for further reading, a comprehensive list of references, and a selection of exercises. Solutions to Exercises for recommending and adopting teachers, can be found at www.books.elsevier.com/manuals and follow the instructions on screen.

Chapters 11–14 use the same format to describe processes in which melting is the principal processing mechanism: surface melting (including alloying and particle injection); cladding; techniques of conduction joining; and cutting (which involves fusion, but introduces vaporization as a processing mechanism). Chapters 15–17 cover processes in which the dominant processing mechanism is vaporization (but which may also include melting): marking; keyhole (penetration) welding; and thermal machining.

The book concludes with Chapter 18 – an overview of the opportunities for sustainable growth in different economic sectors provided by laser material processing. The characteristics of laser processing are summarized to highlight the differences and similarities in comparison with conventional methods of material processing, and to help identify areas with potential for the development of innovative processes and the use of novel materials. The driving forces for change within the various manufacturing industry sectors are discussed, and compared with the characteristics of laser processing, to match potential cases of 'industry pull' and 'technology push'. The use of modelling to help in the development of new processes and materials is described, with particular reference to the growing use of non-metallic materials in laser processing. The spectacular growth of industrial laser material processing is quantified, and used to highlight the potential and need for education and practical training.

The appendices contain reference material. Appendix A is a glossary of terminology to help students of Engineering and Materials Science understand the concepts in each other's subject more easily. Appendix B contains the salient properties of lasers for material processing, which enable attributes such as wavelength, photon energy and interaction time to be compared with the properties of engineering materials, thus aiding in laser selection and predicting principal modes and mechanisms of processing.

Appendix C describes designation systems used for metals and alloys, allowing the common industrial codes mentioned throughout the book to be interpreted and the properties of processed materials to be compared. Appendix D contains the mechanical and thermal properties of popular engineering materials. These material data have many applications: they can be used in analytical formulae to make rapid calculations of the temperature profiles generated during laser processing; they are a means of characterizing and differentiating engineering materials, thus enabling their behaviour during laser processing to be understood, and their suitability for different laser-based processes to be assessed; and finally they are fundamental to designing novel materials suitable for innovative laser-based procedures. Appendix E contains useful analytical equations for heat flow and structural change, which illustrate the relationships between the principal processing variables, and allow rough calculations of processing parameters, temperature fields and processed material properties to be made. Finally, Appendix F lists standards relevant to laser material processing, from which practical requirements and guidance can be obtained for laboratory trials and industrial application.

The material is gathered from over 20 years' experience of laboratory investigations, university teaching, industrial workshops, and consultation with industrial users. The information provided helps the results of laser processing to be deduced. But more importantly, the underlying mechanisms of interaction are explained, in order to stimulate inductive reasoning, thereby opening new avenues for investigation of novel procedures and products. The book will have achieved its aim if the reader feels in a position to understand the principles, procedure and industrial application of laser material processing, and perhaps finds a means of developing a promising idea into a potentially profitable process.

John C. Ion
Adelaide
June 2004

SOLUTIONS MANUAL

Solutions to the exercises in this book are available for adopting and recommending teachers. To access this material visit www.books.elsevier.com/manuals and follow the instructions on screen.

Acknowledgements

It is a pleasure to acknowledge the contributions of colleagues, current and former, at: Imperial College, London University, Department of Materials, UK; Luleå University of Technology, Department of Engineering Materials, Sweden; Cambridge University, Department of Engineering, UK; the Centre for Industrial Research (now SINTEF), Oslo, Norway; TWI (The Welding Institute), Abington, UK; Lappeenranta University, Laser Processing Laboratory, Finland; and CSIRO Manufacturing & Infrastructure Technology, Adelaide, Australia.

When researching the evolution of lasers, systems and industrial material processing, I received considerable help from: Akira Matsunawa (Osaka University, Japan), David Belforte (Publisher/Editor-in-Chief, Industrial Laser Solutions, Sturbridge, MA, USA); Paul Hilton (Technology Manager – Lasers, TWI, Abington, UK); Sheldon Hochheiser (Corporate Historian, AT&T, Murray Hill, NJ, USA); Peter Houldcroft (formerly Director of Research, The Welding Institute, Abington, UK); Johnny Larsson (Advanced Body Engineering, Volvo Car Corporation, Gothenburg, Sweden); J. Ruary Muirhead (Marketing Manager, First Carton Thyne, Edinburgh, UK); Peter Sorokin (IBM Fellow, T.J. Watson Research Center, Yorktown Heights, NY, USA); William T. Walter (President, Laser Consultants, Inc., Huntington, NY, USA); and David Weeks (Communications Director, HRL Laboratories, Malibu, CA, USA). Their enthusiasm to delve into the past maintained my interest and motivated me to complete this project.

Manufacturing – particularly when it involves lasers – is shrouded in proprietary information. I am grateful to many who have been generous in providing examples and details. Their names appear in figure captions.

Three people deserve special mention. Henrikki Pantsar, a graduate student of Lappeenranta University of Technology, Finland, highlighted the most important information needed for students of Engineering and Materials Science. Steve Riches, formerly Head of Laser Processing at TWI, and now with Micro Circuit Engineering, Newmarket, UK, provided technical guidance in the organization and content of the manuscript. Michael Ashby, Royal Society Research Professor, Department of Engineering, University of Cambridge, UK, was a constant source of inspiration and encouragement throughout the venture.

PROLOGUE

Late in March 1964, principal photography on *Goldfinger* – Ian Fleming's seventh book to feature James Bond – began at Pinewood Studios in Buckinghamshire, England. It was Sean Connery's third outing as Britain's most famous fictional icon. On this occasion he was waking up strapped to a table confronted by a mysterious device with a pulsating spiral. In the book (published in 1959) Fleming had chosen an advancing circular saw as the *modus operandi* for Bond's demise. But the producers wanted something more breathtaking for the film.

Figure 0.1   A scene from the 1964 film *Goldfinger* – the first public demonstration of laser material processing. (Source: MGM, courtesy of the Directors Guild of America and the Screen Actors Guild of America)

Four years earlier, newspapers had described a new invention – the laser. The headlines had mainly been variations on a theme; a 'death ray'. This combination of mystique and terror was perfect for the film's producers. Production designer Ken Adam set about creating an industrial laser mounted overhead on a gantry. It would focus a beam of light that could cut through metal. Since such events were still the stuff of science fiction, a slot was cut in the table, which was filled with solder and painted over, after which Burt Luxford, a studio technician, knelt carefully under the table and proceeded to expel the solder from the slot using an oxyacetylene torch. The pyrotechnic effects were impressive.

Meanwhile, three-and-a-half thousand miles across the Atlantic Ocean, 24-year-old Kumar Patel and his team at the Bell Telephone Laboratories in Murray Hill, New Jersey, were working on a carbon dioxide gas laser. The power levels they were achieving were miniscule at first. But as they refined gas mixtures and devised methods of cooling their laser, the output power started to rise.

It would not take long for the paths of science fiction and industrial research to cross. In England, during the month of May 1967, Arthur Sullivan of the Services Electronics Research Laboratories in Harlow, and Peter Houldcroft of The Welding Institute in Abington, south of Cambridge, cut through tool steel one-tenth of an inch in thickness using the beam from a carbon dioxide laser.

The James Bond series went on to become the most successful franchise in the history of film making, earning over two thousand million dollars. Laser cutting became the most profitable facet of the industry now known as high power laser-based manufacturing, which is now worth over one thousand million dollars *each year*. Lasers continue to make regular appearances in James Bond films.

CHAPTER 1

INTRODUCTION

INTRODUCTION AND SYNOPSIS

Laser has become a catchword for precision, quality and speed. Desktop laser printers convert a colourful electronic image into a high quality permanent picture. The digital versatile disc (DVD) brings the experience of the cinema – high definition pictures and multi-channel sound – into the home. A laser scanner at the supermarket checkout translates a barcode into an item that appears on your receipt and disappears from the store's inventory. Lasers can be used to whiten your teeth, remove unwanted hair, and make your spectacles obsolete; the list grows every day.

Yet if any of these products had been demonstrated as recently as the early 1960s, the reaction would probably have been similar to that created by the first microwave oven – scepticism at best; a completely new way to solve a problem.

The breakthrough occurred in May 1960, when a working laser was constructed and demonstrated in a research laboratory in California. Only devotees would have noticed the announcement of what is now considered one of the ten most important inventions of the twentieth century. Many of today's lasers were invented in the mid-1960s, but they remained hidden in research laboratories and military establishments. Companies only started to manufacture lasers for commercial use towards the end of that decade. In the 1970s, enthusiasts – particularly in the automotive industry – started to realize their potential for *material processing*. Industrial applications began to appear, but the role of the laser remained hidden from the general public; it was simply a machine that performed a manufacturing task so well that it was hardly noticed. It was not until the early 1980s (when the compact disc was introduced and light shows experienced a step-change in sophistication) that the word 'laser' entered into general usage. Laser manufacturers, machine tool makers, and systems integrators interpreted this new familiarity as an opportunity to market flexible laser processing systems to enlightened manufacturing industries. The impact was immediate: sales of laser-fabricated products flourished from the late 1980s as the laser's high technology image was used in successful marketing campaigns for a wide range of products. A new industry had developed – *Laser Processing of Engineering Materials*.

This book is about this new industry. It examines how and why lasers are increasingly being used in manufacturing with a wide range of engineering materials. An enormous number of applications have been developed. The operation of a compact disc writer does not immediately appear to have much in common with the hardened steel components in an automobile, yet both depend on structural transformations in metallic alloys that are induced by the heating effect of a laser beam – albeit to very different degrees. To understand such connections, the emphasis of the book is on the underlying materials science of interactions between laser beams and engineering materials. The modes and mechanisms of processing, which link apparently unconnected techniques, are revealed. A framework is created in which the processes can be organized, and used as the basis for *Laser Processing Diagrams*, which are useful tools in education and industrial practice.

This introductory chapter examines why the laser is unique in the collection of modern machine tools: it has so many uses. The impact of laser material processing on modern life is illustrated through a snapshot of everyday products and procedures that involve a wide range of lasers, engineering materials and processing mechanisms.

THE LASER – AN INNOVATIVE MACHINE TOOL

Traditional machine tools are normally designed for a particular purpose: modern lathes can be programmed to machine with high precision; carburizing furnaces automatically harden to predetermined requirements; and automated cutting and welding gantries part and join materials quickly and accurately. They all perform to the most exacting standards. But they only perform a particular task.

The industrial laser is different. It is a *flexible* machine tool. It produces a beam of light with unique properties. Its light can be controlled accurately: it can be focused to a small spot, providing an intense source of energy that is ideal for penetrating materials; or spread into a diffuse heating pattern to treat surfaces. As well as processing materials by thermal modes, the interaction between the photons of the laser beam and atoms in materials enables processes to be performed athermally (without heat): bonds can be made and broken. The beam can be manipulated with optical components to perform a variety of operations simultaneously, or switched between locations for sequential processing.

The laser thus provides opportunities for *innovation* in material processing. Innovation: novel application of an existing idea; application of a novel process; or – in the most successful cases – novel application of a novel process. In some applications, the laser can simply be used in place of an existing machine tool. For example, when the beam is used to cut sheets of materials, the cost of a laser can be recouped in months because productivity can be increased and product quality can be improved. Equally, the laser can be the tool around which a new manufacturing process is developed – laser-based rapid manufacturing enables products to be made in novel ways with their design governed by the function of the product rather than the limitations imposed by traditional fabrication techniques. Both approaches have been implemented successfully to meet the constant demand for increasing competitiveness in manufacturing industry.

LASERS IN MATERIAL PROCESSING

The examples of lasers in material processing described in this chapter illustrate the widespread nature of laser processing. They represent a small fraction of a growing list of applications that take advantage of the unique properties of light produced from many different types of laser. Don't be concerned about unfamiliar terms; these will be explained later. Do, however, note:

- the different *types* of laser that are used – their active medium (gas, liquid or solid), wavelength, power, energy and mode of operation (continuous or pulsed);
- the variety of *materials* treated – metals and alloys, ceramics and glasses, polymers and composites;
- the *mode* of beam–material interaction – thermal (induced by heat transfer) and athermal (induced by changes on the atomic scale);
- the mechanisms of *thermal* processing – heating, melting and vaporization;
- the mechanisms of *athermal* processing – making and breaking chemical bonds; and
- the myriad *applications* that have been developed from a limited palette of processing mechanisms.

IN THE HOME

In 1895, King C. Gillette began work on a safety razor with disposable blades. A little over one hundred years later, the Gillette Corporation introduced the Sensor razor, which incorporated many unique components and novel approaches to design. One of these was the use of laser spot fusion welding. Each expensive thin platinum hardened stainless steel blade is joined to a cheaper rigid metal alloy support bar by 13 spot welds. The welds are made by using the energy in a pulsed beam of infrared light delivered from a solid Nd:YAG (neodymium yttrium aluminium garnet) laser with a power of about 250 W. Millions of blades are manufactured every day; each cartridge is welded in about one second. The MACH3 shaving system, Fig. 1.1, features cartridges with three blades that are welded using this technique.

Since 1994, Oras Oy (Rauma, Finland) has used lasers to mark its company and product logo on its bathroom products. An example is shown in Fig. 1.2. Polymer tap levers are electroplated with layers of chromium and nickel. The product logo is indelibly branded by thermal vaporization of part of this layer by a writing beam from a continuous 85 W Nd:YAG laser. It takes between two and five seconds to write the logo, depending on the product. Each year, the covers and levers of millions of single lever taps, thermostatic bath units and shower components are laser marked.

Nelko Oy (Lapinlahti, Finland), a manufacturer of kitchen furniture, produces a range of wooden finishes, some of which can be tailored to individual tastes. Wood is a natural polymer composite that comprises strong fibrous chains of cellulose embedded in a matrix of softer lignin. The panel shown in Fig. 1.3 was laser engraved from a digitized image. Different colours and textures are obtained by varying the pulse length of light from an infrared carbon dioxide gas laser (CO_2) and by using different gases to shield the interaction zone.

Modern kitchens provide industrial designers with opportunities to demonstrate their talents on a wealth of engineering materials: stainless steel sinks; wooden furniture; ceramic tiles and stove tops; tough resin mouldings; and granite worktops. When the talents of an industrial designer are combined with the knowledge of a laser processing engineer, the results can be spectacular. Or they can go unnoticed – few are likely to show off their laser-welded refrigerator door, washing machine drum, or, least likely of all, kitchen sink.

Figure 1.1   Laser spot welds on the three blades of the Gillette MACH3 shaving system. (Source: Mark Jahns, Gillette Consumer Services, Boston, MA, USA)

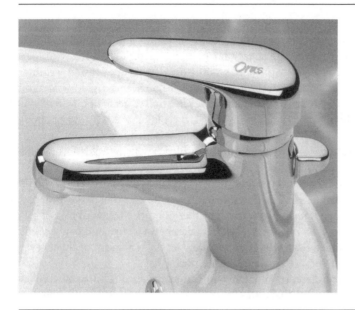

Figure 1.2   Laser-marked logo on an Oras Linea single lever tap. (Source: Jukka Koskinen, Oras Oy, Rauma, Finland)

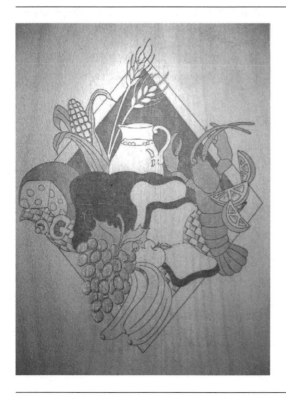

Figure 1.3   Laser-engraved wooden kitchen cupboard door. (Source: Henrikki Pantsar, Lappeenranta University of Technology, Finland)

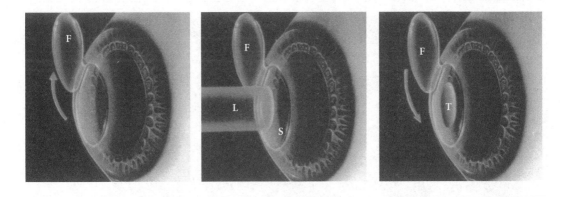

Figure 1.4 Laser *in-situ* keratomileusis (LASIK). (Source: Stephen Siebert, Laser Focus, Adelaide, Australia)

In Healthcare

The laser has made an outstanding contribution in many fields of medical science. Remarkable advances have been made in ophthalmology – the first user of medical lasers, and still the largest. Retinal attachment using a few watts of blue light from an argon ion gas laser brought the role of the laser to public attention. Today, laser-based cosmetic surgery probably commands the highest public profile. However, dentistry, gynaecology, urology, neurosurgery and many other disciplines have benefited from the minimally invasive nature of surgery that laser processing provides.

Laser *in-situ* keratomileusis (LASIK) is an ophthalmic procedure that combines laser processing and microsurgery. It is used for the treatment of nearsightedness (myopia) and farsightedness (hyperopia). A keratome (a device containing a high-speed rotating disc) is used to cut a layer of superficial corneal tissue (a natural polymer) about 0.15 mm in thickness, which is folded aside, Fig. 1.4. Corneal tissue is exposed to pulses of ultraviolet laser light (normally produced by an argon fluoride excimer laser), each of which removes a layer of tissue around one quarter of one micrometre (one thousandth of a millimetre) in thickness. Movement of the beam over the eye is computer controlled for precision profiling. The ablation process is *athermal*; the energy of the ultraviolet radiation is sufficient to break molecular bonds, which provides exceptional accuracy. The total amount of material removed is normally less than the thickness of a hair. Myopia is treated by removing tissue from the central region of the cornea, which decreases its curvature to reduce the focal length of the lens. Hyperopia is corrected by removing tissue from the periphery of the cornea. The flap, which acts as a natural bandage, is then replaced. The laser procedure is very brief – typically lasting for no more than 40 seconds – and is carried out under local anaesthetic. The complete treatment takes less than 15 minutes. The procedure is carried out in over 40 countries, and over one million people have been able to discard their spectacles.

The *thermal* effect of laser light is used in a many medical applications. Skin (another natural polymer) can be resurfaced by vaporizing superficial layers using pulses of infrared light from a CO_2 laser. This stimulates the growth of collagen in underlying tissue, which reduces the apparent depth of wrinkles. Hair can be removed by the heating action of red light from a solid ruby laser, or the infrared light of a semiconductor laser, which destroys the follicle. Laser light can be used to remove facial blemishes and tattoos; a laser that produces light similar in wavelength to the colour of the blemish is selected. Focused light from a CO_2 laser is absorbed well by the water in tissue and bone (a natural ceramic composite) – a laser light scalpel not only makes precision incisions, but also cauterizes the wound. The light from liquid dye lasers, whose wavelength can be tuned, is particularly suitable for precision processes involving athermal photochemical reactions in blood. Dental caries is removed

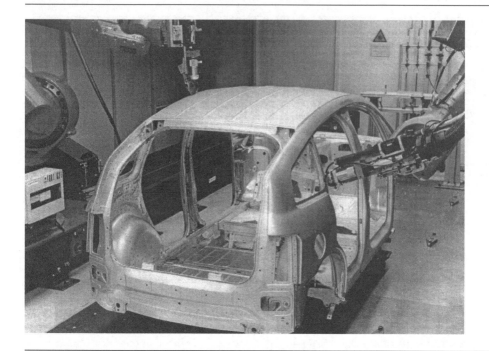

Figure 1.5   Audi A2 aluminium alloy spaceframe. (Source: Jens Christlein, Audi AG, Neckarsulm, Germany)

with pulsed laser light from a growing list of solid state lasers (based on active media comprising lanthanon ions in a YAG host) that are themselves excited using the light from a semiconductor laser – a more precise and less painful alternative to the conventional drill. 'White' tooth fillings (a man-made composite of inorganic particles in a polymer matrix) can be cured and hardened quickly with a few watts of blue light from an argon ion laser.

Laser light is used not only as a tool in medical procedures, but also as an energy source for the manufacture of tools found in medical devices. Catheters (tubes for introducing or removing fluid from the body) and stents (cylindrical scaffoldings that are inserted into arteries to restore blood flow) are among a large number of instruments made by laser-based micromachining and microjoining techniques.

IN MANUFACTURING

Few automobiles are made today without the use of lasers. Multikilowatt CO_2 and Nd:YAG laser systems are now common on production lines, where they weld bodywork, power train assemblies and accessories. Continuous seam welds in sheet materials not only improve the stiffness, handling, road noise and crashworthiness of an automobile, but they also enable new joint designs to be used, leading to cost reductions through savings in materials. In addition, light materials and novel designs are becoming more popular, notably the aluminium alloy space frame shown in Fig. 1.5. A hybrid system comprising a multikilowatt Nd:YAG laser beam and a metal–inert gas head is used for welding the spaceframe components combining the high penetration and productivity of laser welding with the tolerance for joint fitup and gap-filling ability of arc fusion welding.

The first commercial airbags appeared in automobiles in the 1980s. Airbags are constructed from tightly woven cloth of nylon or polyester. They are inflated with a gas produced from a contained chemical reaction that is triggered when a sensor detects a sharp deceleration. Airbag materials must

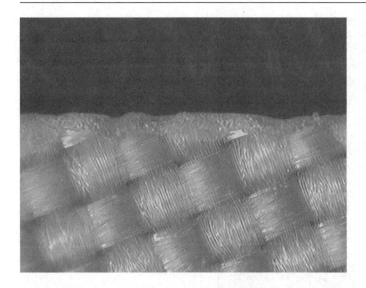

Figure 1.6   Laser-cut polymer airbag material. (Source: Geoff Shannon, Synrad Inc., Mukilteo, WA, USA)

Figure 1.7   Laser hardening of alloy steel gear wheel teeth. (Source: Tero Kallio, Lappeenranta University of Technology, Finland)

be tough but flexible, which causes difficulties for mechanical cutters. A sealed 240 W CO_2 laser is an excellent means of cutting airbag material, since the beam does not physically contact the material, and can be programmed to follow intricate contours easily. Nitrogen is used to assist the cutting process. The edge is simultaneously sealed, as shown in Fig. 1.6. Materials can be cut quickly in single or multiple layers.

Heat treatment has been used for centuries to improve the mechanical properties of materials. Manufacturers are now able to choose from a wide range of techniques to meet performance requirements. Laser hardening is a relative newcomer to manufacturing industry, offering precision, productivity, high quality, and opportunities for new design. Figure 1.7 shows the flanks of alloy steel

Figure 1.8 Component fabrication using Laser Engineered Net Shaping (LENS™) direct metal fabrication process. (Source: Clint Atwood, Sandia National Laboratories, Albuquerque, NM, USA and Sylvia Nasla, MTS, Eden Prarie, MN, USA.) LENS™ is a trademark of Sandia National Laboratories

gear wheel teeth being hardened by the shaped beam from a multikilowatt infrared laser. (Part of the mystery surrounding laser processing is that infrared light cannot be seen by the human eye.) The beam traverses the flank, heating the surface to a given depth, which then quenches to induce a phase transformation that produces a hardened surface. The benefits are four-fold: the life of the gear wheel is extended; the gear may now be used in high performance applications; new materials and gear designs can be introduced; and hardening patterns can be changed easily, resulting in rapid processing of a wide range of product geometries.

Powdered metallic alloys can be melted by a laser beam to create prearranged deposited patterns on surfaces. Laser cladding involves the production of surface layers with properties that are different from the substrate – they are tailored to the application. The technique has been extended to build up three-dimensional objects, and is one of a family of rapid manufacturing processes. These are novel techniques of fabricating components that are difficult to make by conventional means because of the material characteristics or the component geometry. Parts are constructed layer by layer from a CAD model, as shown in Fig. 1.8. The process is readily scaleable and easily integrated into a production line. Composite layers comprising different materials can also be produced. Thus mechanical properties can be customized in particular locations by varying the processing parameters or the material type. The method is particularly suitable for making prototype parts, which can be remodelled by making changes in the software. Stainless steel, titanium and nickel alloys are the most common materials used. (Titanium work hardens readily and so is difficult to forge.) In this example the laser is the basis of a novel means of manufacturing, rather than merely a replacement for an existing process. Laser-based near-net manufacturing processes can save 20% to 30% of the cost of the part by eliminating material waste and minimizing the use of expensive consumable cutting tools. Additional savings are possible because inventories can be reduced and manufacturing times shortened. The method can also be used to make moulds: conditions are created such that the deposit is easily removed from the substrate surface, thus creating an impression of the topography in a solid form suitable for the casting process.

Next time you step into an aircraft, consider the contributions that laser processing could have made: laser-welded aluminium fuselage skin stiffeners; laser-drilled nickel superalloy jet engine components; laser-marked polymer cabling insulation; and laser-cut titanium ducting. Ships, railway carriages, satellites, bicycles and Formula 1 racing cars also benefit from cost savings, new designs and reduced environmental impact that laser-based fabrication provides.

Industries have grown up around the short wavelength and rapid pulsing capabilities of laser light. A beam of short wavelength has a high energy, which is capable of vaporizing material. Vapour can be condensed to form thin films on materials; an ideal method of manufacturing sensitive electronic devices. An exciting development in nanotechnology is the ability to grow carbon nanotubes and other structures using a similar technique. Nanostructures have outstanding mechanical properties and can be used as the basis for atomic-scale electronic devices. Output from the Ti:sapphire laser can be obtained in pulses on the order of femtoseconds (10^{-15} s) in length. This is shorter than kinetic events in atoms and molecules, and provides means for high precision athermal micromachining of electronic components.

Environmental issues have a large impact on manufacturing: processes are required to generate fewer pollutants; materials that can be reused are favoured; and processes that use less energy are called for. The laser provides solutions to a number of challenges. Laser irradiation causes elements and compounds to fluoresce, enabling pollutants such as heavy metals and aromatic hydrocarbons in soil to be detected using spectroscopy. The same principle is used to identify specific alloys that can be recycled. Laser-based processes are normally clean in comparison with conventional manufacturing methods; processing is carried out under carefully controlled conditions and is relatively quiet. Modern diode lasers convert electrical energy into optical energy with up to 50% efficiency – a value that compares favourably with traditional manufacturing equipment.

In the Arts

The Finnish artist Reijo Hukkanen was commissioned to produce a piece of art in stainless steel, called *Tähtivyö* (Belt of Stars). It takes the form of an angel's wing, comprising five separate elements, shown in Fig. 1.9. The aesthetics belie the complexity of fabrication. The outlines of each element were first drawn and scanned to produce a digitized file. The file was used to programme the movement of a single multikilowatt CO_2 laser beam in *three* separate operations: cutting fixturing jigs to secure the parts; welding the periphery of each element; and cutting the outlines. The preforms created were inflated with high pressure nitrogen to create three-dimensional elements. The artwork was completed by attaching the elements to a curved beam.

Artwork is vulnerable to attack by atmospheric pollutants. Disfiguring encrustations are familiar sights on exposed buildings and statues. Systems have been developed that use a variety of laser sources and beam delivery techniques to remove the encrustation without damaging the underlying substrate. For example, an Nd:YAG laser beam can be carefully directed into intricate carvings by an articulated delivery system, Fig. 1.10. The encrustation typically absorbs about 90% of the incident light, whereas the absorptivity of the substrate is much lower. Many types of coating can be removed by a number of mechanisms: ejection through vaporization of materials at the interface with the substrate; fracture

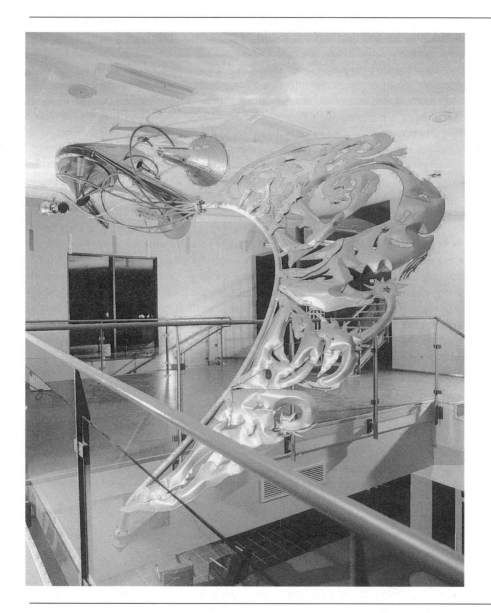

Figure 1.9   *Tähtivyö* (Belt of Stars), designed by the Finnish artist Reijo Hukkanen, and made by laser cutting and welding. (Source: Petri Metsola, Lappeenranta University of Technology, Finland)

through differential thermal expansion; and selective vaporization. Such techniques have been applied to materials as diverse as stained glass windows and painted aircraft surfaces. Conventional cleaning methods use chemicals or high pressure water jets and abrasives, which can erode the substrate and produce environmentally harmful waste.

Not only can lasers remove unwanted material, but they can also deposit material by both thermal means (as we have seen in the LENS™ process) and athermal mechanisms. Stereolithography is a technique for building three-dimensional models. By irradiating a liquid photopolymer with ultra-violet light from a gaseous helium–cadmium laser, a cured solid polymer surface can be produced by athermal chemical mechanisms. The model is built up by polymerizing layers sequentially. An

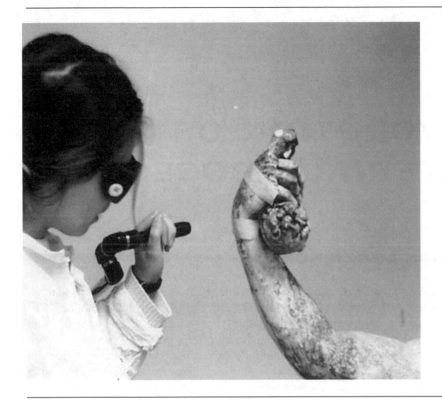

Figure 1.10   Restoration of a statue by pulsed Nd:YAG laser cleaning. (Source: Paolo Salvadeo, Quanta System SPA, Milan, Italy)

equally imaginative use of low power Nd:YAG laser light is to heat metals in gas atmospheres to grow microscopically thin films of coating compounds, which reflect and refract light to produce attractive interference colour distributions, which are ideal for jewellery. Titanium responds particularly well to such treatment.

SUMMARY AND CONCLUSIONS

The laser is a uniquely versatile tool for processing a remarkable range of metals and alloys, ceramics and glasses, and polymers and composites in both laboratory and industrial environments. Light can be produced in pulsed or continuous form, with a wavelength extending from subultraviolet through the visible to beyond the infrared, at power levels that span decades between milliwatt and multikilowatt. Laser-based fabrication is possible on both microscopic and macroscopic scales, in thermal and athermal modes, and uses a variety of mechanisms. The beam from a single laser can be manipulated for a multitude of tasks, giving unrivalled flexibility. Laser processing provides a competitive advantage over many traditional 'heat, beat and treat' methods of industrial fabrication. It can be a profitable replacement for an existing technique, or the basis for a completely new process. The scope of application is limited only by the imagination of designers and engineers. This book aims to stimulate that imagination.

To understand how laser processing has achieved such widespread acceptance in the manufacturing industry, it pays to look back over the evolution of lasers and systems for material processing. The following chapter provides an overview of the most important historical events.

CHAPTER 2

EVOLUTION OF LASER MATERIAL PROCESSING

INTRODUCTION AND SYNOPSIS

Throughout its evolution, laser processing has sustained times of 'technology push' (a solution looking for a problem), and periods of 'industry pull' (a problem looking for a solution). The invention of new laser sources, developments in automation, the production of new materials, and government policies (particularly those relating to potential military uses) push the technology forward into new fields. Enthusiastic champions from particular industry sectors appreciate the opportunities presented by laser-based fabrication, and pull the technology in directions that have enabled their companies to become more competitive. The evolution of laser material processing is a fascinating story of technological advance, government policies, and personalities.

This chapter describes the evolution of three discernible – but often overlapping – factors that influence the progress of laser material processing: the invention and commercialization of lasers; advances in automation, process monitoring and control, and system integration; and the attitudes of industries that enable processes to migrate from the applications laboratory onto the production line. New terms describing lasers and processes are introduced and explained briefly – they will become clearer in the chapters that follow. This chapter can be used as a chronology and overview for those new to the subject, or a source of historical information for experienced users of lasers in material processing.

FOUNDATIONS

THE NATURE OF LIGHT AND MATTER

Leonardo da Vinci was probably the first person to investigate the nature of light, using his considerable talents to study reflection, refraction, mirrors, lenses and the human eye – skills that he applied in painting the *Mona Lisa* between 1503 and 1505 (MacLeish and Amos, 1977). The Dutch physicist and astronomer Christiaan Huygens proposed a wave theory of light in his *Traité de la Lumière* of 1678, which stated that an expanding sphere of light behaves as if each point on the wavefront acts as a new source of radiation of the same frequency and phase (Huygens, 1690). The wave concept enabled the nature of light to be visualized and could be used to explain phenomena such as reflection and refraction. In 1801, Thomas Young (whose name is attached to relationships ranging from elasticity to medicinal doses for children) produced evidence that supported the wave theory. He described experiments undertaken in Cambridge in England between 1797 and 1799 in which he observed patterns of dark and light bands (interference or diffraction patterns) produced when light passed through two adjacent apertures. He also proposed that light was polarized, which suggested that the components of the wave possessed some form of directionality. In 1865, James Clerk Maxwell published a theory of electromagnetism, which concluded that light comprised electrical and magnetic vectors oscillating in orthogonal planes – an electromagnetic wave. This was demonstrated by Heinrich Hertz in 1886, when a high voltage induction coil was used to generate a spark discharge between two pieces of brass. Once the spark formed a conducting path between the two conductors, charge oscillated back and forth, and electromagnetic radiation was emitted. However, in 1887, Philipp von Lenard (working with Hertz) demonstrated that electrons of a fixed energy were emitted when light impinges on a surface, irrespective of the intensity of the incident light – the photoelectric effect. This suggested that light could also be considered in terms of particles of energy, which led to the idea of atoms comprising discrete particles.

The modern theory of atomic structure is founded on ideas proposed in the early nineteenth century. In 1808 John Dalton published four basic ideas of chemical atomic theory: each element is made up of tiny particles called atoms; the atoms of a given element are identical; chemical compounds are formed when atoms combine with each other; and chemical reactions involve the reorganization of atoms (Dalton, 1808). In 1897, Joseph John (J.J.) Thomson undertook experiments to gain information on cathode rays, and announced that they were negatively charged 'corpuscles' (electrons) (Thomson, 1897). Work in the fields of physics, chemistry and mathematics led to the notion that atoms, ions and molecules (species) exist in states characterized by discrete energy levels, and that they can interact with electromagnetic radiation. This led Max Planck to propose in 1901 that energy could only be emitted in discrete quanta (Planck, 1901). Three years later, Thomson suggested that electrons occupy geometrical orbitals (Thomson, 1904). In 1905, Albert Einstein proposed that light consists of bundles of wave energy, termed photons (Einstein, 1905). This enabled Ernest Rutherford to propose a model of the atom that comprised a positive nucleus and electrons (Rutherford, 1911), to which Niels Bohr applied quantum theory (Bohr, 1913).

It was then recognized that photons and species may interact either by absorption of a photon with a corresponding increase in energy, or by *spontaneous* emission of a photon by an atom originally in a higher energy state, leading to a reduction in energy. However, by considering the thermodynamics of photon emission, Einstein concluded in 1916 that a third process of interaction must be possible – *induced* or *stimulated* emission – in which an excited species could be stimulated to emit a photon by the arrival of another photon (Einstein, 1916, 1917). This is the basis of light amplification by stimulated emission of radiation, better known by the acronym *laser*.

In April 1924, Richard Tolman and Paul Ehrenfest, working at the California Institute of Technology, wrote a paper that discussed stimulated emission, under the title of 'negative absorption', or amplification (Ehrenfest and Tolman, 1924). They commented on the fact that such emission must be coherent (in step). A few years later, Rudolf Ladenburg demonstrated negative dispersion in a neon gas discharge tube, which was cited as evidence of stimulated light emission – a fundamental principle of laser emission (Ladenburg, 1928). However, it was a quarter of a century before a practical use for this observation was found.

DEVELOPMENT OF THE MASER

The conditions required to amplify radiation by stimulated emission were studied in the 1940s and early 1950s in both the former Soviet Union and the United States. In 1940, the Russian Valentin A. Fabricant described negative absorption, or 'negative temperature' in gas mixtures (Fabricant, 1940). This was later to become known as a *population inversion* – a prerequisite for stimulated emission and a fundamental principle of the laser. However, he was not able to demonstrate amplification of radiation, and the ideas were not pursued at the time.

Charles H. Townes tells of the events of 26 April 1951, when sitting on a bench in Franklin Park, Washington, D.C., he calculated how many molecules of ammonia it would take to make an oscillator capable of producing and amplifying radiation (Townes, 1999a). He committed these ideas to paper on 11 May 1951, but deliberately did not publish them openly. It was Joe Weber, working at the University of Maryland (College Park, Maryland, USA), who first publicly described microwave amplification by stimulated emission of radiation in June 1953 (Weber, 1953). In October 1954, Nicolai Basov and Alexander Prokhorov, working at the P.N. Lebedev Institute of Physics in Moscow, detailed the conditions required for the operation of a microwave amplifier (Basov and Prokhorov, 1954). However, a report of the first practical device had already appeared in an article by Townes and two of his students, James Gordon and Herb Zeiger, working at Columbia University (New York, NY, USA), which was published in July 1954 (Gordon *et al.*, 1954). After considering many cumbersome Greek and Latin names, Townes and co-workers coined the term *maser*, the acronym for microwave amplification by stimulated emission of radiation.

The first *solid state* maser, based on the gadolinium ion, was constructed in 1957 (Scovil *et al.*, 1957), swiftly followed by a device based on chromium ions (McWhorter and Meyer, 1958). Ruby was used as the host material for chromium ions in a maser developed at the University of Michigan (Ann Arbor, MI, USA) the following year (Makhov *et al.*, 1958) – a fact that was to take on great significance shortly afterwards. The potential of devices that could produce stimulated emission in the visible and infrared regions of the electromagnetic spectrum was soon realized.

PRINCIPLES OF THE LASER

In 1958, Townes co-authored a key theoretical paper with Arthur Schawlow, an optical spectroscopist (and by this time Townes's brother-in-law), in which they described the extension of maser techniques to the production of optical and infrared radiation (Schawlow and Townes, 1958). The paper also described the design principles of an infrared device that used potassium vapour as the active medium, excited by violet radiation from a potassium lamp. The requisite features of potential solid state devices were also considered. Schawlow had, in fact, been working with ruby crystals in the late 1950s, but it was believed that the pulsed output obtained was not suitable for communications technology, which was the driving force for the studies. Recognizing the potential importance of such a device in spectroscopy, they filed a patent application, dated 30 July 1958.

Around the same time, a graduate student of Columbia University, Gordon Gould, was looking into the conditions required for stimulated emission at visible wavelengths. After discussing his ideas with Townes, he committed his ideas to paper and had them notarized in mid-November 1957 (Townes, 1999b). However, a patent was not filed until 6 April 1959. Gould left Columbia University before finishing his degree to work for TRG, the Technical Research Group (Melville, NY, USA). In December 1958, TRG filed a proposal to the Pentagon's Advanced Research Projects Agency (ARPA) to study the properties of laser devices, requesting $300 000. The proposal was reviewed favourably (by Townes), and the company awarded $1 000 000. (The successful launch of the Sputnik satellite may have been a factor in the decision.) Gould challenged the patent held by Townes and Schawlow in the years that followed, claiming that it was limited in scope, and did not consider, for example, the use of high power levels for applications such as material processing. He eventually won the case 30 years later and collected royalties from laser manufacturers, benefiting financially from the delay in the award of the patent because of the increased number of lasers manufactured in the latter part of the twentieth century. It is Gould who claims the first use of the acronym *laser*, rather than *optical maser*.

Further theoretical analyses of the optical maser appeared in the early 1960s (Lamb, 1964; Scully and Lamb, 1967), but it was the potential for laser emission that captured the imagination.

THE 1960S

At the beginning of the 1960s, laser research and development activities were mainly sustained by an interest in their potential use in measurement and communications systems. Engineers at Bell Telephone Laboratories (Murray Hill, NJ, USA) were particularly active in this field.

THE FIRST LASER

The first working laser – the *ruby* laser – was demonstrated by Dr Theodore Maiman during the afternoon of 16 May 1960 (Maiman, 2000). A pink ruby cylinder 1 cm in diameter and 2 cm long was mounted on the axis of a helical xenon flashlamp, which was placed inside a polished aluminium cylinder. The ends of the crystal were ground and polished flat and parallel, and coated with silver. A hole 1 mm in diameter was made in one of the faces to allow light to escape. Pulses of visible red light were detected when the ruby was illuminated (Maiman, 1960a, b). The publicity photographs produced for the press conference in New York on 7 July 1960 showed a larger, more photogenic model, Fig. 2.1, which created a certain amount of confusion for others working in the field. A continuous

Figure 2.1   Dr Theodore H. Maiman with a photogenic model of the first laser. (Source: David L. Weeks, HRL Laboratories, Malibu, CA, USA)

wave ruby laser was subsequently constructed by replacing the flash lamp with a mercury–xenon arc lamp, which was focused onto the end of a trumpet-shaped crystal (Nelson and Boyle, 1962).

Maiman's report of optical *excitation* and *absorption* in ruby appears in the June 1960 issue of *Physical Review Letters* (Maiman, 1960a). However, his subsequent report of *stimulated* optical radiation was not deemed worthy of publication, since it was not felt that it represented a significant contribution to basic physics. As a result, Maiman wrote a second article, which appeared in *Nature* in August 1960 (Maiman, 1960b). Maiman had previously been working on ruby masers, deciding to study solid state devices because he believed that there were too many practical problems with gas lasers. In 1962, Maiman founded Korad Corporation (changing the 'C' in Coherent radiation to a 'K'), which was devoted to the research, development and manufacture of lasers. The first laboratory devices were the subject of intensive development work in the 1960s. However, the technology was not well understood, it was also difficult to obtain components, and there was a scarcity of instruments that could be used to characterize the output. Despite these obstacles, Korad and Raytheon marketed the first industrial ruby laser in 1963 (Klauminzer, 1984). It was initially used for drilling, but now find use in a wide range of medical procedures.

Following the demonstration of the ruby laser, research into lasers mushroomed. With hindsight, the characteristics of lasers that make them suitable for material processing can be appreciated, but at the time nobody knew which would be developed for such purposes. The emphasis in the following is placed on those lasers that did prove to be useful tools in material processing. However, their development owes much to the work done across the entire spectrum of laser research. More can be learned of the workings of industrial material processing lasers in Chapter 3.

THE LASER RUSH

The second laser was demonstrated two days before the Thanksgiving holiday in November 1960 by Peter Sorokin and Mirek Stevenson at Bell Labs (Sorokin and Stevenson, 1960). They used a flashlamp to excite a cryogenically cooled crystal of *uranium*-doped calcium fluoride, and obtained infrared light of wavelength 2500 nm. This was the first laser based on transitions between *four* levels, which theoretically (but not practically) could produce continuous wave radiation. (The ruby laser is based on three-level operation.) However, because of the need for cryogenic cooling, the uranium laser has never found many practical applications.

By mid-December 1960, Ali Javan and co-workers, also at Bell Labs, had obtained infrared radiation from the first *gas* laser, a helium–neon (He–Ne) device (Javan *et al.*, 1961). This laser was also the first to emit *continuous* wave (CW) radiation, and the first to be excited using an *electrical* discharge. Adjustments to the mirrors enabled visible red light to be produced (White and Rigden, 1962). This unique and spectacular piece of equipment was the prototype for the first *commercial* laser, a 1 mW He–Ne device, marketed for $7500 in 1962 by Spectra-Physics (Mountain View, CA, USA) (Klauminzer, 1984). The He–Ne Laser found use in measurement devices, and later became familiar as the source for product barcode readers.

Research into *semiconductors* as potential sources of coherent radiation had been initiated in the late 1950s in the United States, Japan, France and the Soviet Union. In 1961, a paper was published in the Soviet Union, based on work started in 1959, that proposed three different methods for obtaining a state of 'negative temperature' (a population inversion) in p–n junctions of degenerate semiconductors (Basov *et al.*, 1961). On 16 September 1962, Gunther Fenner, a member of Robert Hall's team at the General Electric Research Laboratory (Schenectady, NY, USA), operated the first semiconductor laser (Dupuis, 2004), which was made from gallium arsenide (Hall *et al.*, 1962). Within one month, laser light had been produced independently from the same material at three other sites: General Electric (Syracuse, NY, USA) (Holonyak Jr and Bevacqua, 1962); IBM Thomas J. Watson Research Center (Yorktown Heights, NY, USA) (Nathan *et al.*, 1962); and MIT Lincoln Laboratory

(Lexington, MA, USA) (Quist *et al.*, 1962). The output from such homostructure (single material) lasers was limited, and the operating temperatures were low; it was proposed that heterostructures (two dissimilar materials: an active layer and a cladding) would provide improved operation (Kroemer, 1963). Around five years later, the first room temperature semiconductor laser was demonstrated (Hayashi *et al.*, 1969). However, it would take over 30 years before power levels suitable for material processing were obtained at affordable prices from arrays of such devices.

Continuous laser action in pure *carbon dioxide* (CO_2) gas was reported early in 1964 by Kumar Patel, then aged only 24 years, working at Bell Labs (Patel *et al.*, 1964a, b). One milliwatt of continuous wave power was obtained initially (Patel, 1964c), rising to 200 mW with the addition of excited nitrogen (Patel, 1964d). Around 12 W was later obtained by direct excitation in a tube of length 2 metres containing a mixture of CO_2 and air (Patel, 1965a). Further increases in output were obtained by cooling the gas mixture to −60°C (Bridges and Patel, 1965), and by adding helium (Patel *et al.*, 1965b) and water vapour (Witteman, 1965). This *slow axial flow* design was the basis of the first commercial CO_2 laser, a 75 W Coherent General (Sturbridge, MA, USA) Model 40, produced in 1966, and delivered to Boeing (Seattle, WA, USA) in 1967. Further increases in power were obtained by increasing the cavity length: a 2.5 kW device 54 m in length was built in 1967 at the US Army Redstone Arsenal (Huntsville, AL, USA) (Patel, 1968), and by 1968 a folded cavity design with a total length of 229 m, capable of generating 8.8 kW, had been constructed.

In parallel with Patel's work in the United States, similar experiments were being carried out in France (Legay and Legay-Sommaire, 1964; Legay-Sommaire *et al.*, 1965). These resulted in the exhibition of the 'world's largest laser' in 1967 in the French pavilion at EXPO 67 in Montreal. This was a slow axial flow device 20 m in length, with tubes 15 cm in diameter, which generated several hundred watts. Around this time, engineers at SERL, the United Kingdom government Services Electronics Research Laboratories (Baldock, UK), were also working on a single tube device 10 m in length, rated at 250 W (Adams, 1970a, b; Houldcroft, 1972).

It was becoming clear that it would not be practical to continue to increase the power of CO_2 lasers by simply extending the length of the tube. Designs based on high gas pressures and convective cooling (Deutsch *et al.*, 1969), transverse gas flow (Tiffany *et al.*, 1969), transverse beam excitation (Beaulieu, 1970), transverse beam extraction (Brown, 1970), and large excited volumes (Hill, 1971) were therefore examined, and were commercialized in the following years. The first *gas dynamic* laser, capable of power levels on the order of tens of kilowatts, was developed in 1968 at the Avco Everett Research Laboratory (Everett, MA, USA) (Gerry, 1970). At the lower end of the power scale, *sealed* carbon dioxide devices were being developed with means for regenerating CO_2, which showed potential for compact designs (Carbone, 1967; Bridges *et al.*, 1972; Burkhardt *et al.*, 1972; Chester and Abrams, 1972; Jensen and Tobin, 1972). The first pulsed transverse excitation atmospheric (TEA) CO_2 lasers were reported by French and Canadian scientists in 1969 (Dumanchin and Rocca-Serra, 1969; Beaulieu, 1970). The CO_2 laser became a workhorse in material processing, and is still the source of choice in many applications.

In 1964, around the time of his discovery of laser action in CO_2, Patel also observed stimulated emission in *carbon monoxide* (CO) (Patel and Kerl, 1964e). The possibility of high power operation was proposed some years later (Bhaumik *et al.*, 1972), but although the shorter wavelength output from the CO laser has advantages for material processing, and commercial units have been built, practical requirements for efficient operation (notably cooling) have limited their industrial production.

Pulsed laser action in an active medium of *neodymium ions* in a host of $CaWO_4$ was reported in November 1961 at Bell Labs (Johnson and Nassau, 1961). The electronic transitions in the neodymium ion Nd^{3+} had been elucidated by Elias Snitzer in 1961 (Snitzer, 1961), who used barium crown glass as the host material. However, Geusic and co-workers at Bell Labs were the first to obtain infrared radiation (pulsed) from the neodymium yttrium aluminium garnet (Nd:YAG) crystal in 1964 (Geusic *et al.*, 1964), which was shown to be an ideal host. The device was referred to at the time as a YAlG:Nd^{3+} laser. Geusic used cylindrical rods 2.5 mm in diameter and 3 cm in length in the first Nd:YAG device, which was excited by a helical xenon flashlamp. A tungsten lamp was used to produce a continuous beam. The Nd:YAG laser is now manufactured with multikilowatt power levels, and is popular in many applications involving complex components since the light can be delivered via a fibreoptic cable that can be mounted on a robot.

The first report of generation of *ultraviolet* light from a laser was made in November 1963 by Heard working at Energy Systems Inc. (Palo Alto, CA, USA) (Heard, 1963). Nitrogen was excited at room temperature using a high voltage (100–150 kV) pulsed discharge. The short wavelength of ultraviolet light is ideal for micromachining, and is also suitable for exciting other types of laser. Avco introduced the first commercial nitrogen laser in 1969 (Klauminzer, 1984).

Investigations undertaken in 1963 by Bell at Spectra-Physics using pulsed operation in an ionized mercury–helium discharge revealed visible laser transitions that could not be obtained during continuous operation (Bell, 1964). This prompted the study of the ionic spectra of *atoms* for laser transitions, which helped Bridges at Hughes Research Laboratories (Malibu, CA, USA) to discover blue and green radiation in an *argon ion* medium on 14 February 1964 (Bridges, 1964). In 1965, Francis L'Esperance of the Columbia-Presbyterian Medical Center was using red ruby laser light to treat diabetic retinopathy (then the leading cause of blindness in people 20 to 64 years old), but noted that a blue–green laser would be a more efficient source for producing therapeutic lesions. The potential of this observation may have been one factor in the decision by Raytheon to commercialize the laser in 1966 (Klauminzer, 1984); such lasers are still used in medical procedures today.

In the early 1960s *metal vapours* were investigated for low-lying energy levels to provide efficient laser action. Green and yellow emission from the first copper vapour laser was demonstrated between 25 and 31 August 1965 by William Walter and co-workers (including Gordon Gould), working for TRG Inc. (Melville, NY, USA) (Walter *et al.*, 1966a). The relatively short wavelength light from such lasers can be focused to a small spot for accurate machining. However, because of the high temperatures involved, it would take around 15 years for Oxford Lasers (Abingdon, UK) to commercialize the copper vapour laser for material processing. Other metal vapours investigated at the time included manganese (Piltch *et al.*, 1965), lead (Fowles and Silfvast, 1965), and calcium (Walter *et al.*, 1966b). Another metal vapour laser, the helium–cadmium source, based on transitions in cadmium ions, was demonstrated in 1966 by William Silfvast at the University of Utah (Salt Lake City, UT, USA) (Silfvast *et al.*, 1966). Output at 442 nm was reported in the initial studies, with the ultraviolet 325 nm line being published in 1969 by James Goldsborough working at Bell Labs (Goldsborough, 1969). The laser was commercialized in 1970 by Spectra-Physics (Klauminzer, 1984), and has found niche markets in material processing because of its ability to cure photosensitive polymers – stereolithography being probably the best known application.

The concept of light emission during *chemical reactions* was first suggested in 1961 by John Polanyi (Polanyi, 1961), but it was not until 1965 that the first chemical laser, a device based on hydrogen chloride, was demonstrated at the University of California (Berkeley, CA, USA) (Kasper and Pimentel, 1965). Lasers based on chemical reactions involving deuterium fluoride and hydrogen fluoride were developed in the decades that followed, the aim being to produce sources of high peak pulse power thought suitable for disabling missiles and satellites in space.

Output from the first *liquid* laser – a dye – was demonstrated on 4 February 1966 by Peter Sorokin and John Lankard at the IBM T.J. Watson Research Center (Yorktown Heights, NY, USA) (Sorokin and Lankard, 1966). A ruby laser was used to excite chloro-aluminium phthalocyanine. The operation of a flashlamp-pumped dye laser is shown in Fig. 2.2. The popularity of organic dyes as active media for the production of tunable broadband radiation was soon realized (Schäfer *et al.*, 1966). Continuous wave emission from a dye laser was reported in 1970 (Peterson *et al.*, 1970), and the introduction of the telescope-grating oscillator provided a means to increase the power output (Hänsch, 1972). The first commercial dye laser was excited using a nitrogen laser, and was produced by Avco in 1969 (Klauminzer, 1984). Dye lasers are now used in a variety of medical applications.

Colour centre, or F-centre lasers, are based on crystalline active media that contain defects, which absorb intensely. Lithium fluoride, which has good thermal and optical properties, was investigated

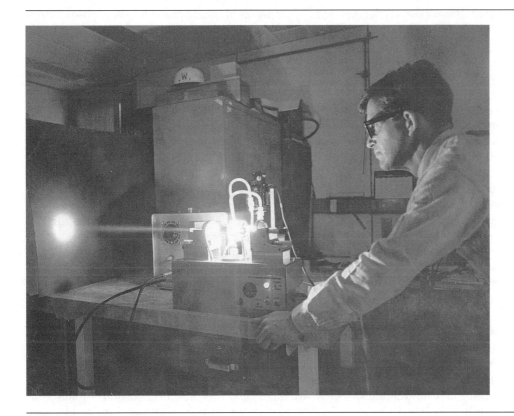

Figure 2.2   Peter Sorokin operating a flashlamp-pumped dye laser. (Source: Peter Sorokin, IBM, Yorktown Heights, NY, USA)

in 1966 (Nahum and Wiegand, 1966), and the potential of colour centres for making efficient broadly tunable CW lasers for the infrared portion of the electromagnetic spectrum was investigated in the mid-1970s (Mollenauer and Olson, 1975). Such lasers are also promising candidates for producing visible laser action in the range of the spectrum between green and red; they may be suitable for the fabrication of miniaturized optical devices (Kurobori *et al.*, 2003).

The first tunable lead–salt semiconductor lasers were developed by E. David Hinkley and Theodore C. Harman at the MIT Lincoln Laboratory in 1964. The devices, which were commercialized in 1975 by Laser Analytics (now part of Spectra-Physics) (Klauminzer, 1984), are used in a variety of civil and military applications including isotope analysis, pollution monitoring, toxin detection, communications, remote sensing, semiconductor processing, combustion diagnostics and explosives detection.

Theoretical and practical work was also being carried out into the practical construction of lasers. The nature of an optical cavity formed with spherical mirrors was analysed (Kogelnik and Li, 1966), and the results led to the concept of stable and unstable resonators. The subject is reviewed comprehensively by Siegman (2000a). More on laser cavity design can be found in Chapter 3.

Techniques of modulating the output of lasers were also developed in the 1960s. The principles of *Q-switching* (McClung and Hellwarth, 1962; Wagner and Lengyel, 1963) and *mode locking* (Hargrove *et al.*, 1964; Kuizenga and Siegman, 1970a, b) to obtain short pulses of high energy are discussed in Chapter 3, where details of frequency conversion by *harmonic generation* (Franken *et al.*, 1961; Armstrong *et al.*, 1962; Giordmaine and Miller, 1965; Ward and New, 1969) can also be found.

These techniques enable the wavelength and pulse properties of laser beams to be tailored to particular applications.

THE FIRST PROCESSES

It was the unique properties of laser light – spatial and temporal coherence, stability, monochromaticity and low divergence – that were exploited in the first applications. They included spectroscopy and measurement. As power levels began to rise, the potential of lasers for material processing began to be appreciated. Many of the first processes were developed with ruby lasers. Around the same time, digital technology, introduced in the 1950s to control the axes of milling machines, was maturing and the term 'numerical control' (NC) was coined. The first numerical controllers used vacuum tubes, later to be replaced by transistors. Numerical control and the possibility of using a laser beam in the manner of a milling tool provided the impetus for the first material processing systems. Chapter 4 provides more information on systems used today in laser material processing.

The first public 'demonstration' of laser material processing appeared in the 1964 film *Goldfinger*. More details of this innovative piece of cinematography can be found in the Prologue. The scene is remarkable – the laser had only been invented four years earlier, and most applications were still laboratory curiosities.

It did not take long for industrialists to realize that ruby laser pulses might be suitable for *drilling*. Engineers at Raytheon Research Division (Waltham, MA, USA) pioneered the field. At that time laser output was measured in 'Gillettes', which referred to the number of Gillette razor blades that the beam could penetrate. (Interestingly, laser spot welding of disposable razor blades made by Gillette is now one of the highest volume applications of laser welding – see Chapter 1.) Around the same time, ruby lasers were being developed in the former Soviet Union, and they were being used in an industrial capacity to drill holes in stainless steel. Drilling of diamond wire-drawing dies using a ruby laser was the first commercial application of laser material processing in the United States, introduced by Western Electric (Buffalo, NY, USA) in 1965 (Anon., 1966). (The laser, which was eventually replaced by an Nd:YAG unit, can now be seen in the Smithsonian Institute.) Drilling, and other techniques of laser machining are covered in Chapter 17.

The first laser *welds* were made around 1963, and involved butt and edge joints in 0.25 mm stainless steel foils, processed with a pulsed ruby laser (Platte and Smith, 1963). Welds were also made between gold connecting leads and aluminium-coated silicon surfaces in microelectronic devices. The laser beam was merely being used to fuse material, albeit very accurately, in a conduction-limited mode. Other reports of research into conduction-limited welding in metals were published (Fairbanks Sr and Adams Jr, 1964) and applied to joining wires, sheets and circuit boards shortly afterwards (Anderson and Jackson, 1965; Cohen *et al.*, 1969; Conti, 1969), when 0.5 mm was considered to be an upper limit of penetration. The first industrial application appeared around 1965, when a pulsed Nd:YAG laser was used to repair broken connectors inside assembled television tubes (Belforte, 1993). Ruby laser welding was used in fabrication of the first Apollo lunar sample return containers in 1969 (Moorhead, 1971). Various laser-based conduction joining techniques have been developed, including soldering and brazing, partly in response to the needs of the microelectronics industries. The first CNC laser soldering machine, based on a 50 W CO_2 laser was produced commercially in 1976 (Loeffer, 1977). More information on these conduction joining processes can be found in Chapter 13. The keyhole welding mode, already exploited in electron beam processes, was known at that time, but it was realized that considerable improvements to the characteristics of the laser beam would have to be made for the laser to be used in a similar manner.

The first gas-assisted carbon dioxide laser beam *cut* was made by Arthur Sullivan and co-workers at SERL (Baldock, UK) in May 1967 (Sullivan and Houldcroft, 1967). A 300 W laser beam coaxial with an oxygen jet was used to cut 1 mm thick steel using a potassium chloride lens, an aluminized beam

Figure 2.3   The first oxygen-assisted laser cut being made in May 1967. (Source: Paul Hilton, TWI, Abington, UK)

turning mirror, and parameters that were very close to those used today. The first oxygen-assisted laser cut is shown in Fig. 2.3.

It was thought at the time that gases other than oxygen might be beneficial for cutting those metals that produced refractory oxides, e.g. aluminium. Chlorine was therefore tried, which was vented out through the laboratory window. The wind took the fumes towards a small tree, which expired a few weeks later! Gas-assisted laser cutting was also demonstrated around the same time in Germany and Japan (Arata, 1987). The patent rights to the process have never been clarified satisfactorily.

In 1967, the first commercial application of continuous wave CO_2 lasers – *scribing* of ceramics – was demonstrated by Western Electric (Allentown, PA, USA). In August 1969, Boeing (Seattle, WA, USA) demonstrated laser cutting of non-ferrous alloys and controlled fracture of ceramic and glasses for aerospace applications (Bod *et al.*, 1969). Scribing is one mechanism of cutting, along with the melt fusion process demonstrated in May 1967 – these and other mechanisms of laser cutting are discussed further in Chapter 14.

First reports of the use of lasers for *heat treating* metals appeared in Germany in the early 1960s (DeMichelis, 1970). Similar work was reported some years later in the Russian literature (Gregson, 1983). Much of the early data were obtained as a by-product of investigations into the interaction between materials and the focused beams of pulsed ruby and Nd:glass lasers (Knecht, 1966). Early investigations in the United States of interactions between pulsed ruby laser radiation and a graphite-coated metal surface gave some indication of the potential for metal hardening (Lichtman and Ready, 1963). In the mid-1960s, researchers at US Steel (Speich and Szirmae, 1969) and in Japan (Namba *et al.*, 1965) studied laser-induced austenitization and hardening of ferrous alloys. The first mathematical models of heat flow provided greater insight into the role of the process variables in determining the thermal cycles induced (Rubanova and Sokolov, 1968), and the geometry of the hardened region (Veiko *et al.*, 1968). The automotive industry soon recognized the potential of the process, and transformation hardening became one of the first methods of laser-based material engineering to be industrialized; case studies of transformation hardening in the automotive and other industries can be found in Chapter 9.

Laser surface *melting* and the possibilities for surface *alloying* were investigated in 1963 (Lichtman and Ready, 1963; Cunningham, 1964). Reviews summarize early work performed mainly with pulsed solid state lasers, in which shallow surface alloys were produced (Draper, 1981, 1982; Draper and Ewing, 1984; Draper and Poate, 1985). Trials into laser surface melting were also carried out in 1965 in Sweden using a Q-switched ruby laser to irradiate the surface of various metals (Vogel and Backlund, 1965). In contrast to surface hardening, surface melting and alloying have received limited acceptance in modern laser material processing, but examples can be found in Chapter 11.

The mechanism of laser-induced *vaporization* was also studied in 1963 (Ready, 1963). This informed work on many methods of surface treatment, including *shock hardening*, which was developed at Battelle Memorial Institute (Columbus, OH, USA) in the late 1960s (Fairand *et al.*, 1972, 1974, 1977). At that time, lasers were only able to produce pulses with rise times and lengths on the order of nanoseconds (10^{-9} s), and as a result the development of industrial applications was hindered. When ultrafast lasers with pulse lengths on the order of femtoseconds (10^{-15} s) were developed in the 1980s, interest was renewed by the automotive and aerospace industries (Peyre *et al.*, 1998). This type of structural change induced by a laser beam is described further in Chapter 8.

In 1964, Charles Townes, Nicolai Basov and Aleksander Prokhorov shared the Nobel Prize in Physics for their fundamental work in the field of quantum electronics, which led to the construction of oscillators and amplifiers based on the maser–laser. On 7 February 1968 the Laser Institute of America (LIA) was founded, dedicated to fostering lasers, laser applications and safety worldwide, and serving the industrial, medical, research and government communities.

THE 1970S

Developments made in the early 1970s led to lasers being considered as machine tools, rather than simply light sources (Anon., 1971). Improvements were made in the mechanical construction of beam delivery systems as well as controllers. Application engineers were now able to take a number of processes out of the laboratory to be tested on the production line. This provided a stimulus to investigate new uses for the laser beam in industrial material processing.

LASERS

The slow axial flow laser design dominated CO_2 laser technology of the 1970s. However, the Welding Institute (Abington, UK) made modifications to the design (Crafer, 1974), which led to the construction of a fast axial flow laser rated at 2 kW. The device incorporated aluminium heat exchangers from the nearby Lotus Formula 1 racing team. Electrox (Letchworth, UK) commercialized the fast axial flow design in 1975. Studies of cavity design (Wisner *et al.*, 1973) pointed towards the use of unstable resonators for high power output. In the United States, the Avco Everett Research Laboratory (Everett, MA, USA) produced a commercial 15 kW machine, based on electron beam excitation and transverse gas flow technology (Hoag *et al.*, 1974). A new excitation technique for industrial CO_2 lasers, based on the use of radio frequency (RF) waves, was developed by Mitsubishi (Japan) and DFVLR (Germany) as an alternative to conventional direct current (DC) methods (Schock *et al.*, 1980).

The early 1970s saw the commercialization of low power pulsed output Nd:YAG systems. Holobeam Lasers first coupled modules of an Nd:YAG laser together to produce a 1 kW industrial unit in 1972. In 1970, Charles Kao and George Hockham working at Standard Telecommunication Laboratories in Harlow, UK, discovered that silica glass fibres could transmit laser light efficiently, thus opening the door to fibreoptic beam delivery – later to revolutionize beam delivery for processing of complex geometry components with the Nd:YAG laser. In 1976, reliable industrial Nd:YAG lasers were marketed by Quanta-Ray, which later became a part of Spectra-Physics.

In 1977, laser emission from the crystal alexandrite was obtained by John C. Walling working at the Allied Chemical Corporation (Morristown, NJ, USA) (Pryor and Frost, 1999). Alexandrite, named after Czar Alexander II, was discovered in the Ural mountains of Russia in the 1830s (Battis, 2001), and was synthesized as chromium-doped chrysoberyl in the early 1970s. The alexandrite laser was the first practical *tuneable* solid-state laser, and was commercialized in 1981 (Klauminzer, 1984). Tuneability means that a suitable wavelength can be selected for a particular application in scientific, material processing and medical procedures.

Developments in diode lasers concentrated on techniques for growing multilayer heterostructures based on gallium arsenide, which led to the first quantum well laser (van der Ziel *et al.*, 1975), and to improving output to obtain continuous wave operation at room temperature (Hsieh *et al.*, 1976).

The first *excimer* laser was demonstrated in 1970 – liquid xenon was excited with a pulsed electron beam (Basov *et al.*, 1970). Output around a wavelength of 170 nm was demonstrated in high pressure xenon gas shortly afterwards (Koehler *et al.*, 1972; Gerardo and Johnson, 1973). Stimulated ultraviolet emission was observed from high pressure krypton and mixtures of argon and xenon (Hoff *et al.*, 1973), and emission in the visible and near ultraviolet regions was observed in excited mixtures of diatomic xenon halides (Velazco and Setser, 1975). Laser action in an excited dimer was also reported at the Naval Research Laboratory (Washington, DC, USA) using xenon bromide pumped by an electron beam (Searles and Hart, 1975). In 1975, emission from excited xenon fluoride (Brau and Ewing, 1975; Burnham *et al.*, 1976), krypton fluoride and xenon chloride (Ewing and Brau, 1975) was obtained. Emission from excited argon fluoride was obtained in 1976 at Sandia National Laboratories (Albuquerque, NM, USA) (Hoffman *et al.*, 1976). Emissions from mixtures of fluorine and helium excited using an electron beam were demonstrated in 1977 (Rice *et al.*, 1977), and output from a fluorine laser obtained in 1979 (Pummer *et al.*, 1979). Lambda Physik (Göttingen, Germany) produced the first commercial excimer laser in 1977. However, the lack of industrial designs, and a poor understanding of the differences between ultraviolet and existing infrared lasers, meant that industrial application of excimer lasers would not become widespread for another decade.

Continuous wave light generation from a transition between two electronic states in *iodine* was demonstrated in 1978 at the Air Force Weapons Laboratory (Kirtland, NM, USA) (McDermott *et al.*, 1978). An output power of more than 4 mW was generated entirely from a chemical reaction, using no external power source. The laser is often referred to as a COIL (chemical oxygen iodine laser). There has been much interest in this laser for material processing, but until now its main use has been as a missile-destroying weapon, described later.

The emissions from lasers based on atomic or molecular media are discrete. In contrast, emission from *free electrons* in a periodic magnetic field can be tuned over a range of wavelengths. This is the principle of the free electron laser, details of which were published in 1971 (Madey, 1971). The first free electron laser was demonstrated at Stanford University (Stanford, CA, USA) in 1977, using an electron beam in a spatially periodic transverse magnetic field (Deacon *et al.*, 1977). The laser has potential applications material processing, but economic considerations have limited its application to date.

SYSTEMS AND PROCESSES

The 1970s saw a movement away from controller units to computers for directing the beam and manipulating the workpiece, and the term computer numerical control (CNC) became familiar. Systems were rigid, with simple geometry and few optical elements. Around 1971, Laser Work AG (now part of the Prima Industrie group) produced the first industrial flying optic two-axis cutting machine.

This combination accounted for the exponential growth in flat bed cutting of sheet materials, led by European industry, during the 1970s. In September 1978, the first *hybrid* laser cutting system was demonstrated in the United States – a combination laser/turret punch developed by Strippit and Photon Sources.

The first successful industrial application of laser cutting was dieboard slotting, which was demonstrated in August 1970 at William Thyne Ltd (Edinburgh, UK), using a BOC Falcon 200 W CO_2 unit. The namesake and son of the company founder had seen a television programme in which laser cutting had been demonstrated, and contacted BOC to find out if the laser could replace fretsawing with sufficient accuracy for making dies. Initially the process was difficult to control because inhomogeneities in the wood caused variations in beam absorptivity, and the workpiece could not be controlled with sufficient accuracy, but eventually these problems were overcome. In 1975, Ford published a photo showing three-dimensional auto body parts that had been cut with a 400 W CO_2 laser. In 1978, the first automobile to contain laser-cut parts was launched – the Ford Capri II.

In April 1970, Martin Adams working at the Welding Institute presented sections of welds made using a CO_2 laser beam in 1.5 mm metal that were indistinguishable from electron beam welds. In contrast to the conduction-limited welding mode demonstrated in the 1960s, the implication was that a laser beam could induce *keyholing* (the production of a penetrating vapour cavity) in a similar way to an electron beam (Houldcroft, 1972). In practice, a double weld run was made; the first oxidized the surface, which improved the absorption of the beam, and the second produced the weld. The technique was demonstrated in the USA in 1971 (Brown and Banas, 1971), where CO_2 lasers with 20 kW of power were being developed (Locke *et al.*, 1972). In 1977, full penetration, single pass welds were made in 50 mm thick steel plate, using a 100 kW gas dynamic CO_2 laser (Anon., 1977). Although it was realized that such a laser would not be suitable for industrial production, the feasibility of very thick section welding had been demonstrated. The principles, procedure and application of keyhole laser welding are described in Chapter 16.

In 1973, the Ford Motor Company (Dearborn, MI, USA) started to examine ways of industrializing laser welding in the automotive sector (Baardsen *et al.*, 1973; Yessik and Schmatz, 1975). Initially, manufacturers used existing tolerances for parts and there was no pressure to reduce production cycle times – two factors vital in ensuring both technical and economic viability of laser processing. Lasers were expensive and not reliable enough for production line work, and as a result industrial laser welding was treated with scepticism in the beginning. However, in 1975, Fiat (Orbassano, Italy) installed a transverse flow CO_2 laser for welding synchronous gears. Today, joining of bodywork and powertrain components are two of the largest applications of laser welding in automobile manufacturing.

Pioneering work was undertaken in 1972 by researchers from the University of California into the use of lasers for selective removal of undesirable coatings from artwork. In the first trials of laser *cleaning*, pulsed ruby laser light was used to remove black encrustations from white marble, utilizing the greater absorptivity of the encrustation in comparison with the substrate (Asmus *et al.*, 1973; Lazzarini and Asmus, 1973). Subsequent work showed that laser cleaning could be applied to stained glass, frescos, stone, leather and vellum, and that the more flexible Nd:YAG laser gave similar results with a greater coverage rate. An example of laser cleaning is given in Chapter 1, and further details can be found in Chapter 17.

In 1973, General Electric applied solid state laser drilling to aerospace applications (Heglin, 1979). Turbine blade hole drilling became a prime application for pulsed Nd:YAG lasers in the mid-1970s, later to be joined by shorter wavelength sources, notably the copper vapour laser.

In September 1974, the Saginaw Steering Gear Division of General Motors (Saginaw, MI, USA) started to laser harden its steering gear housings on a production basis (Wick, 1976; Miller and Wineman, 1977). This is considered to be the first industrial application of laser hardening. By 1977, 30 000 housings per day were being hardened using 15 CO_2 lasers with power levels of 0.5 and 1 kW.

The first patent referring to laser *cladding* was published on 20 April 1976, by the Avco Everett Research Laboratory (Avco, 1976). It describes a method of applying a metal coating by fusing metallic rods or wires through the action of a laser beam. On 3 August 1977 a patent was published in the UK by the Caterpillar Tractor Company describing a process for bonding a predeposited coating to a substrate by the use of a laser beam (Caterpillar, 1977). Cladding is now the most popular laser surfacing technique, although most cladding is carried out in a different manner to that described in these patents (see below). Further details of the process can be found in Chapter 12.

In 1976, Charles Townes was inducted into the Inventors Hall of Fame.

THE 1980S

The 1980s were notable for the development of integrated laser systems, which comprised a laser source, beam handling optics, and work handling equipment. User-friendly interfaces were developed to provide information and instant control to the operator. Flexible systems were introduced with Cartesian, cylindrical, spherical polar and articulated (anthropomorphic) geometries. Such steps facilitated further industrial application of developed processes. In addition, researchers turned to novel methods of using the laser beam for material processing, rather than as a direct replacement for a conventional process. During the late 1980s, annual growth rates in laser production of 10–12% per year were common.

LASERS

The early 1980s produced a generation of industrial CO_2 lasers that featured higher powers, greater reliability and more compact designs. Multikilowatt RF-excited units were introduced in 1985. The development of the radial impeller to circulate gas in the fast axial flow design enabled a more compact design to be engineered (Beck, 1987). The output power of DC-excited units rose to 30 kW, and the first European transverse flow laser, based on a design originated by the United Kingdom Atomic Energy Authority (Culham, UK), was licensed for manufacture in 1983. Not for the first time did parts from a high performance vehicle find their way into a laser design – fans from the Vulcan bomber were used to circulate the gas mixture. The success of CO_2 lasers in a variety of machining applications provided encouragement to broaden their use in industrial manufacturing (Walker, 1982).

The dominant Nd:YAG lasers available in the early 1980s were pulsed units. Until 1988 the maximum average power available from a commercial unit was 500 W. In that year a commercial 1 kW Nd:YAG laser was produced. This was preceded by the development of fibreoptic cable that could transmit a suprakilowatt near infrared beam, which meant that cumbersome mirror systems associated with CO_2 laser beam delivery could now be replaced with flexible optics mounted on an industrial robot. Complex geometry three-dimensional components were now able to be treated economically. The industrial Nd:YAG lasers of this time were based on active media comprising rods of crystals and lamp pumping, which results in a low efficiency of energy conversion (less than 5%) and a poor beam quality in comparison with gas lasers. As a result, researchers had begun to investigate novel forms of crystal geometry, such as slabs (Eggleston *et al.*, 1984) and rings (Kane and Byer, 1985) to improve performance; these are the basis of many of today's solid state laser designs.

The highlights of diode laser research in the 1980s were the development of molecular-beam epitaxy for growing heterostructures, and the construction of surface-emitting devices. The former meant that the threshold value for operation could be reduced (Tsang, 1981), which led to improved output. The latter enabled a very short cavity (of length of 5.5 μm), grown by chemical vapour deposition, to generate useful output (Koyama *et al.*, 1989; Jewell *et al.*, 1989).

In 1980, Geoffrey Pert's group in the Department of Applied Physics at Hull University (Hull, UK) reported emission from an *X-ray* laser (Jacoby *et al.*, 1981). In 1984, a soft X-ray laser was built by the group of Dennis Matthews in the Lawrence Livermore National Laboratory of the University of

California (Livermore, CA, USA) (Rosen *et al.*, 1985; Matthews *et al.*, 1985; Matthews and Rosen, 1988). This type of laser is still a laboratory device, but the short wavelength produced (about 20 nm) is useful for fine lithography and for manufacturing nanometre-scale structures.

In 1982, Peter Moulton demonstrated continuous laser output from titanium-doped sapphire at the Lincoln Laboratory of MIT (Moulton, 1982; Moulton, 1986). The laser was commercialized in 1988, and is now used in micromachining as a source of femtosecond-scale pulses.

Theoretical studies of laser beam propagation enabled expressions for Hermite Gaussian (Carter, 1980) and Laguerre Gaussian (Phillips and Andrews, 1983) beam intensity profiles to be generated, which provided means of characterizing the modes of a laser beam. Other developments in the field of laser beams and resonators at this time are reviewed by Siegman (2000b).

FLEXIBLE MANUFACTURING SYSTEMS

It became clear that the laser would need to be integrated into a flexible manufacturing system for it to be economically attractive to manufacturers who were looking to reduce costs, increase productivity, reliability and flexibility, save labour costs and improve working conditions, and improve quality. As a result, a variety of beam and workpiece manipulation systems became available in the 1980s.

The mechanisms of laser cutting were elucidated in the 1980s, and a marked improvement in edge quality resulted from improved laser performance, new nozzle designs, a better understanding of the role of process gases, and closer control of process parameters. The influence of beam polarization on cutting performance was first identified in 1980 (Olsen, 1980). By the mid-1980s the maximum cut thickness had increased to about 15 mm. Profitable job shop activities became more common, and production line applications increased rapidly. The first five-axis cutting system to be installed in the UK was commissioned in 1983 at the Swindon plant of Austin Rover, and was used for trimming of pre-production car body panels.

In the field of laser welding, progress was made with new welded joint designs, novel material combinations, and thick section welding, which led to improvements in quality, productivity and environmental friendliness. The automotive industry was quick to capitalize on such benefits, championed by a new generation of engineers. Progress was made in understanding the physics of keyhole formation and stability (Arata *et al.*, 1985), which provided greater confidence in the thick section keyhole welding process. At the same time, reliable high power industrial lasers were becoming available. This combination led to the appearance of integrated laser welding systems on industrial production lines. Thin section joining of pressed low alloy steel sheet became common, motivated by the need to increase manufacturing flexibility and to reduce automobile weight. Initially the laser beam simply replaced an electron beam, without consideration of the design opportunities afforded by the process. However, in August 1985, Thyssen Stahl AG (Duisburg, Germany) began laser welding two pieces of hot-dipped galvanized steel sheet to make a blank wide enough for the floorpan of an Audi automobile, leading the way for laser-welded tailored blanks (sections of different grade, thickness or coating welded into a single blank for pressing). (Although in 1981 British Leyland (Coventry, UK) laser welded sheets of differing thickness for production cars, but did not publicize the process.) In the mid-1980s, General Motors (Linden, NJ, USA) became the first automobile plant to replace spot welds with laser welds, using an integrated vision, clamping and welding system to join galvanized steel quarter panels to the uncoated roof of the Chevrolet Corsica/Baretta (Vaccari, 1994). Others were quick to follow: BMW (300 and 800 series), Volvo (850, S70 and V70), Mercedes (S class), Peugeot (406) and Audi (A6). The application shown in Fig. 2.4 is a modern example of this procedure. Progress was also made in conduction joining techniques, with the first soldering machine based on an Nd:YAG laser appearing in 1982 (Lea, 1989).

Rolls-Royce Ltd (Derby, UK) was the applicant of a patent published on 26 May 1981 that describes a method of applying a metallic coating to a metallic substrate by the interaction of a directed laser

Figure 2.4   Roof edge laser welding of the Volvo XC90. (Source: Johnny Larsson, Volvo Car Corporation, Gothenburg, Sweden)

beam and a gas stream containing entrained particles of the coating material (Rolls-Royce, 1981). Rolls-Royce (Macintyre, 1983a, b, 1986) and Pratt & Whitney (Furst, 1980) commercialized this blown powder process for producing hardfaced regions on expensive aeroengine turbine blades. Blown powder cladding is now the most popular laser-based surfacing technique, finding uses in the aerospace, automotive, power generation and machine tool industries, as well as forming the basis of a rapid manufacturing technique described in Chapter 1.

Laser *forming* (deliberate distortion through laser heating) was investigated in the early 1980s at Osaka University (Osaka, Japan), and Massachusetts Institute of Technology (Cambridge, MA, USA). Initial interest was tempered by doubts concerning the accuracy and cycle time that could be achieved for industrial application, but refinements to automatic process monitoring and control have enabled the process to enter the production line. Chapter 10 contains more information on these processes.

Much research was undertaken in the 1980s into monitoring signals emitted during laser processing, and using them in open-loop and closed-loop feedback control systems. A product of this work is commercial stereolithography apparatus (SLA), the first deliveries of which were made by 3D Systems (Valencia, CA, USA) on 3 January 1988, to Baxter Healthcare, Pratt & Whitney, and Eastman Kodak for rapid prototyping. This three-dimensional process used lasers, photopolymers, and optical scanning computers to turn CAD files into physical objects and is described in Chapter 7.

Ultraviolet excimer laser cleaning was reported in 1981 by the Canadian Conservation Institute (Cooper, 1998). The Nd:YAG laser was used in both normal and Q-switched pulse modes to clean marble, silver and stained glass. A portable system was developed for on-site work. In another machining application, in 1983 Stephen Trokel proposed the use of pulsed excimer laser to ablate regions of the corneal epithelium to correct vision defects (Trokel *et al.*, 1983). The process – photorefractive keratectomy (PRK) – laid the foundations for modern laser eye surgery; only two years later the first

Figure 2.5   Airborne laser: high energy chemical oxygen iodine laser (COIL) carried aboard a modified Boeing 747-400F freighter. (Source: Mary Kane, Boeing, Seattle, WA, USA)

excimer PRK procedure was demonstrated on humans in Germany, and in October 1995 the US Federal Drug Administration approved Summit Technology's excimer laser for the correction of myopia (Slade, 2002). (The LASIK laser eye surgery procedure is described in Chapter 1.)

No account of the development of laser technology in the 1980s would be complete without mention of the *Strategic Defense Initiative*, more commonly known as *Star Wars*. It was announced by President Reagan in March 1983, as a defensive shield designed to intercept incoming missiles, using, among other weaponry, laser beams. X-ray lasers held a particular fascination, particularly for the media. Although many of the original ideas were never developed, some are still being pursued. Considerable effort has been put into an airborne megawatt-class iodine laser mounted in a Boeing 747-400F, designed to intercept incoming missiles, and which is now flying, Fig. 2.5.

In 1981, Arthur Schawlow shared the Nobel Prize in Physics with Nicolaas Bloembergen and Kai Siegbahn for their contributions to the development of laser spectroscopy. In 1984, Ted Maiman was inducted into the Inventors Hall of Fame. In 1986, John Polanyi shared the Nobel Prize in Chemistry for his contribution to understanding the dynamics of chemical elementary processes.

THE 1990S

The recession of the early 1990s temporarily slowed growth in sales of laser processing systems. However, laser processing made the transition from the application laboratory to the production line in a large number of industrial sectors. This can be attributed to the considerable efforts of manufacturers to improve the reliability and ease of operation of lasers, and the availability of sophisticated automation and control systems for beam and material handling. Laser processing became simply another means of manufacturing. Turnkey systems for processes were marketed – often the laser itself was not even mentioned in promotional material. Steps were also taken to incorporate the opportunities provided by laser-based fabrication into the complete manufacturing process, from design through to fabrication. Developments continued in material processing lasers, but as the flexibility of the laser became appreciated more widely it became more common to talk of applications development in industry sectors, rather than in individual lasers and processes.

LASERS

The focus of developments in CO_2 laser technology in the early 1990s was on machines of higher power, better beam quality, greater reliability, reduced maintenance, improved ease of use, and compactness. A 2 kW diffusion-cooled CO_2 machine, of high beam quality and very low gas consumption, produced by Rofin-Sinar (Hamburg, Germany) was typical of this compact generation of machines (Bolwerk et al., 1997). Sealed CO_2 lasers with an output power under 100 W were incorporated into desktop manufacturing systems, while the small size of units with an output power up to 600 W enabled them to be mounted directly on articulated robots, and used in cost-effective manufacturing (Clarke, 1996). Fast axial flow technology was adopted in units up to 20 kW in output power, and modular transverse flow designs enabled units up to 60 kW to be produced. The CO_2 laser beam continued to be the source of choice for welding and cutting of long linear and rotationally symmetric parts. At the end of the decade CO_2 lasers represented about 55% of the industrial lasers sold for material processing, the majority being in the range 3–5 kW.

In 1997, Trumpf (Ditzingen, Germany) and Lumonics (Rugby, UK) marketed 4 kW CW Nd:YAG lasers, which provided direct competition, technically and economically, with CO_2 lasers. In addition, the availability of fibreoptic beam delivery opened up new fields of application involving parts of increasingly complex geometry. The automotive industry led the way in introducing Nd:YAG lasers on the production line, where they began to replace CO_2 lasers for complex geometry cutting and welding operations. A 10 kW continuous wave Nd:YAG unit was available commercially at the end of that decade.

Since around 1984, the available power from semiconductor diode lasers has doubled every year, and simultaneously the price has decreased by a factor of about 1.6. Because of their compact size, and the higher absorptivity of the shorter wavelength diode laser beam by materials, diode lasers were actively investigated in the 1990s as replacements for CO_2 and Nd:YAG sources in material processing. A major problem was the thermal load, coupled with the requirement to operate the lasers near room temperature, which required efficient cooling. To overcome this, various designs were developed: two-dimensional arrays were soldered directly onto heatsinks (Mott and Macomber, 1989); diode bars were attached to chip carrier bars, which were then attached to a heatsink (Mundinger et al., 1990); and diode bars were attached to their own heatsinks (Beach et al., 1992). Developments such as these meant that at the end of the 1990s, 6 kW infrared diode lasers were commercially available for material processing. The combination of light, compact multikilowatt diode lasers and industrial robots provided economic solutions to many challenges involving complex geometry processing. Diode lasers emitting in the shorter wavelength blue–green portion of the spectrum were also developed (Haase et al., 1991; Jeon et al., 1992), followed by even shorter wavelength blue–violet diode lasers based on gallium nitride (Nakamura et al., 1996), which have become popular in applications requiring a small focused spot of light, such as optical disc data storage (Gunshor and Nurmikko, 1996).

The replacement of flashlamps by more efficient diode lasers as pumping sources for solid state active media in the 1990s led to three major opportunities: active media could now be made in novel crystal geometries, such as discs, tubes and fibres; the media could be pumped and cooled through a variety of surfaces; and new active media based on a range of crystal hosts doped with lanthanon ions other than neodymium could now be used (King, 1990). In addition to YAG and glass, host materials used in such diode-pumped solid state lasers (DPSS) included yttrium lithium fluoride ($YLiF_4$, known as YLF) and perovskite ($YAlO_3$, YAP). Lanthanides such as holmium (Ho), erbium (Er) and thulium (Tm) were also used as dopants. Diode-pumped solid state lasers are efficient, reliable, long lasting sources able to produce a beam with a quality that is superior to conventional lamp-pumped units. When combined with non-linear frequency conversion, these lasers can produce output that spans the spectrum from the ultraviolet to the mid-infrared. Operation can also range from subpicosecond pulses to continuous. They are small, facilitating incorporation into moving laser processing systems. Their wall plug efficiency can be up to three times that of lamp-pumped lasers, with maintenance intervals about 20 times longer. These factors provide DPSS lasers with a competitive advantage in applications including material processing, medicine, metrology and remote sensing.

Ultrafast lasers, i.e. those producing pulses of light in the picosecond (10^{-12} s) and femtosecond (10^{-15} s) ranges, were researched extensively in the 1990s. Such short material interaction times lead to techniques of athermal processing, such as micromachining, discussed further in Chapters 7 and 17.

TURNKEY PROCESSING SYSTEMS

In the 1990s the availability of integrated circuit chips allowed controllers to be made smaller, with greater reliability and lower cost. Computer numerical control (CNC) became a self-contained NC system for a single machine tool. Direct numerical control (DNC) allowed a central computer to control several machine tools.

Since lasers and work handling could then be interfaced with CAD/CAM systems, component design, processing and inspection became integrated into turnkey manufacturing systems. This type of flexibility shortens the time taken to transfer a product from the 'drawing board' to the marketplace. The definition of the tool centre point, which is a point that usually lies along the last wrist axis of a robot at a specified distance from the wrist, plays the same role in laser processing as the cutting edge plays in machining. Open-loop control systems could then be developed to control actions automatically, but without reference to the state of the process or the position of the machine. Closed-loop control systems, which make use of process information gathered by sensors were also developed, and were to form the basis of completely automated processing systems. User-friendly machine controllers with interactive multimedia and on-line support further enhanced the attraction of turnkey systems.

High pressure inert gas cutting was introduced around 1990, enabling high quality edges, suitable for butt joints in laser welding, to be produced. The early 1990s saw maximum penetration of active gas laser cutting of structural steels increase to 25 mm. Complex three-dimensional cutting became feasible – in 1998 Volkswagen (Braunschweig, Germany) introduced laser cutting of hydroformed parts, which enabled the number of components required for substructures to be reduced. Flat sheet cutting still accounted for by far the largest portion of the total market for high power (>500 W) industrial laser material processing systems. Increased competition among suppliers of systems became based on the ease of use of integrated turnkey systems. An increased emphasis was placed on developing standards for classifying edge quality, including the design of new test pieces; a major incentive being the ability to laser weld pieces that had been laser cut. Lasers and conventional tooling systems were being combined for process optimization, notably the turret punch–laser hybrid cell: punching is effective for producing repeatable shapes in sheet materials; whereas the programmable laser is more suitable for complicated profiles that are frequently varied. Ships with superstructures constructed from 100% laser-cut plates were being launched, Fig. 2.6. (The plates are also laser marked, Chapter 15.)

The 1990s saw rapid growth in the use of laser welding of sheet components, motivated by the needs of the automotive industry, with a particular emphasis on the use of laser-welded blanks made from sheets with differing properties. Ford (Cologne, Germany) introduced fibreoptic Nd:YAG laser

Figure 2.6   Corvette constructed from laser-cut plates. (Source: Ian Perryman, Vosper Thorneycroft, Southampton, UK)

welding of car bodywork in 1993. February 1997 saw the first industrial application of diode laser-welded polymers – a key system for Mercedes Benz (Larsson, 1999). In 1998, Audi (Neckarsulm, Germany) started to weld the A2 series aluminium spaceframe with Nd:YAG lasers (see Chapter 1). The aerospace industry also recognized the potential of laser processing. In 1993, Air France introduce laser welding and cutting for repairing aircraft engine casings (Kunzig, 1994). In 1998, EADS Airbus delivered the first laser-welded aluminium alloy fuselage shell for A318 series aircraft (see Chapter 16). Much work was performed by research institutes, industries and classification societies to develop guidelines for laser welding thick section ship hull steels.

The availability of short wavelength and ultrafast lasers provided opportunities in many fields of machining. Drilling of fine holes using copper vapour lasers with an average power in excess of 2 kW was demonstrated in the laboratory. With the introduction of ultrafast lasers, femtosecond pulse machining became possible. In contrast to conventional pulsed laser machining, material is removed without the generation of heat. In the field of cleaning, work expanded to include various pulsed lasers: ultraviolet excimer, visible dye, near infrared Nd:YAG, and far infrared carbon dioxide. All were shown to remove dirt effectively, but the Nd:YAG was found to be the most selective (Cooper *et al.*, 1992).

Gordon Gould was inducted into the Inventors Hall of Fame in 1991. In 1997, in a courtroom in Orlando, Florida, Gould proved to the United States Patent Office that he had invented the laser. Arthur Schawlow was inducted into National Inventors Hall of Fame in 1996. He died at Stanford Hospital on 28 April 1999.

THE NEW MILLENNIUM

Electronics revolutionized the nature of manufacturing in the 1960s and 1970s. In the new millennium, photonics has the potential to influence manufacturing to an even greater extent. The lasers of the new millennium are the smallest, cheapest, most efficient and most flexible ever. The same can be said for the systems into which they are integrated. In addition to the traditional thermal means of processing, athermal fabrication on the nanoscale is now possible, with time scales on the order of femtoseconds.

LASERS

Carbon dioxide laser development continues in the direction of cost reduction and size, although the economic and technical competition from solid state devices for large-scale material processing continues to grow. Around the turn of the millennium, compact, energy efficient *multikilowatt* DPSS lasers started to appear. Their potential for material processing is enormous: energy efficiency is high compared with their lamp-pumped counterparts, increasing performance and reliability; they are small enough to mount directly onto robots for automated processing; the beam quality is high enough for accurate large-scale penetration and surface processing; and they are flexible enough to be attached to applicators used for intricate medical procedures. Ultrafast femtosecond lasers are available, many integrated into micromachining centres used in the production of microscopic parts for the medical and electronics industries (Leong, 2002). Blue output from argon ion lasers is used to etch semiconductors for optical waveguides. Blue–violet output from gallium nitride diode lasers is replacing the infrared output of gallium arsenide diode lasers in optical storage devices such as DVD recorders, described below. Ultraviolet output from excimer lasers is ideal for preparing thin films by chemical vapour deposition, and producing nanometre-scale devices. In summary, lasers that were once thought to have too little power for material processing are now finding niche uses in industries based on photonics, and the search continues for affordable lasers able to generate even shorter wavelengths.

PROCESSES

In 1961, Richard Feynman suggested that ordinary machines could build smaller machines that could build still smaller machines, working step by step down toward the molecular level. He also suggested that particle beams could be used to define two-dimensional patterns. Present microtechnology (exemplified by integrated circuits) has realized some of the potential envisioned by Feynman by following the same basic approach: working down from the macroscopic level to the microscopic. Lasers are now used in nanotechnology to create carbon nanotubes and other geometrical structures. More can be learned about this in Chapter 8.

In the world of large-scale material processing, cutting and welding still dominate.

Flat sheet cutting accounts for by far the largest portion of the total market for high power (>500 W) industrial laser material processing systems. Cutting is now a routine, and highly profitable, job shop activity. It is used for metals and alloys, a wide range of fabrics such as wool, cotton, leather, vinyl, polyester, velour, neoprene, polyaramide (Kevlar) and low density polyethylene, as well as composites. Around one quarter of industrial lasers sold today are used for cutting, mainly steel sheet of thickness below 6 mm. Japan is currently the leading user of laser sheet metal cutting systems, much being performed in job shops for large electronics corporations. Europe lies in second place in terms of usage.

Research into welding is concentrated in two main areas: the development of procedures for welding thick section structural steels and light alloys of aluminium, magnesium and titanium; and the use of hybrid processes that combine a laser beam and a secondary power source. The Airbus A380 is designed around techniques for laser welding aluminium fuselage stiffeners. The demand for lightweight metal casings for mobile phones, laptop computers and electronic packages provides material and processing challenges for the laser applications engineer. As lightweight materials encroach into traditional steel applications, steelmakers respond with high strength low alloy steels: the Volkswagen Golf V contains

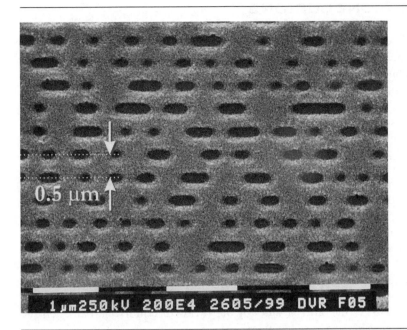

Figure 2.7   Pits (dark amorphous regions) in the crystalline recording medium of a Blu-ray Disc^{TM}. (Source: Ingrid Bal, Philips, Eindhoven, Holland.) Blu-ray Disc is a trademark of Sony Corporation

70 m of laser-welded joints in such materials. In 2002, Volkswagen AG (Wolfsburg, Germany) installed the first hybrid Nd:YAG laser/MIG welder for aluminium door parts. Welding applications account for about one quarter of CO_2 laser systems over 750 W, and more than half of the installed higher power (above 150 W) Nd:YAG systems. In comparison with cutting, little welding is performed on a job shop basis – it is still more likely to be found in specialized in-house installations. The use of lasers for welding is substantially higher in the USA than the rest of the world, because of the high volume of automobile production, although European manufacturers have made significant investments in recent years.

In the field of small-scale material processing, applications such as semiconductor etching, thin film deposition, nanoscale device fabrication, thin section forming, polymer ablation, and micromachining are creating the types of industry that multikilowatt lasers generated in the 1980s and 1990s. A new generation of medical devices is being produced for the healthcare industries. Laser treatment of biomaterials improves their properties, and reduces the chances of rejection by the body – precision machining of ceramic dental prostheses being just one example. Possibilities for the production of novel materials by ablation using short wavelength and ultrafast processing techniques are being investigated. The DVD has been so successful that even higher amounts of data storage are demanded, which has led to the development of blue–violet diode lasers capable of writing even more information onto optical discs, Fig. 2.7. Ironically, the growth in electronic applications, the relatively short lifetime of modern electronic devices, and environmental legislation are bound to create new applications for lasers in disassembly and recycling of electronic components. The list will continue to grow at an increasing rate in the new millennium.

The reader is directed to Chapter 18 to learn of more opportunities that modern laser processing affords in the new millennium.

Herbert Kroemer shared the Nobel Prize in Physics in 2000 for his work on semiconductor lasers. Nicolai Basov died at the age of 78 on 2 July 2001.

SUMMARY AND CONCLUSIONS

Throughout its evolution, laser processing of engineering materials has experienced periods of 'technology push' and 'industry pull'. As engineers in the applications laboratories of private companies and government organizations invented new lasers and found novel applications for them (often remote from those intended), companies realized their potential and decided to manufacture and market them, and end users profited from applying them in a wide range of industry sectors. Developments in automation, process monitoring, and control systems have facilitated system integration, providing user-friendliness – a much underrated factor in decision making. In some cases, existing lasers have been adapted for particular processes, whereas in others the requirements of the application have dictated the properties of the laser beam, and suitable sources have been developed. One factor in the evolution of the industry stands out: the enthusiasm of individuals from different industry sectors and their belief in laser processing (often against the advice of senior figures more comfortable with conventional methods of material processing) have been vital in the development of the industry.

New laser sources will continue to be developed. Existing sources will be made more compact, efficient and cheaper. Sophistication of process monitoring and control systems will continue to facilitate the migration of laboratory applications onto the production line. Government policies will continue to influence the industry, whether they are related to potential military uses or the increasing demands for a cleaner, more sustainable environment. One cannot help but consider the quantum changes that electronics brought to the latter decades of the twentieth century, and envision the potential of laser-based photonics in the new millennium.

FURTHER READING

Bertolotti, M. (1983). *Masers and Lasers*. Bristol: Adam Hilger.

Bromberg, J.L. (1991). *The Laser in America: 1950–1970*. Cambridge: The MIT Press.

Hecht, J. (1992). *Laser Pioneers*. Revised ed. Boston: Academic Press.

Klauminzer, G.K. (1984). Twenty years of commercial lasers – a capsule history. *Laser Focus/Electro-optics*, **20**, (December), 54–79.

Lengyel, B. (1966). Evolution of masers and lasers. *American Journal of Physics*, **34**, 903–913.

Maiman, T. (2001). *The Laser Odyssey*. Blaine: Laser Press.

Millman, S. ed. (1984). Chapter 5. Quantum electronics – the laser. In: *A History of Engineering and Science in the Bell System: Physical Sciences (1925–1980)*. Indianapolis: AT&T Bell Laboratories. pp. 151–210.

Myers, R.A. and Dixon, R.W. (2003). Who invented the laser: an analysis of the early patents. *Historical Studies in the Physical and Biological Sciences*, **34**, (1), 115–149.

Silfvast, W.T. ed. (1993). *Selected Papers on Fundamentals of Lasers*. Bellingham: SPIE.

Taylor, N. (2000). *Laser: The Inventor, the Nobel Laureate, the Thirty-year Patent War*. New York: Simon & Schuster.

ter Haar, D. (1967). *The Old Quantum Theory*. Oxford: Pergamon Press.

Townes, C.H. (1999). *How the Laser Happened*. New York: Oxford University Press.

Twenty Five Years of the Laser (1985). London: Taylor & Francis.

BIBLIOGRAPHY

Adams, M.J. (1970a). Gas jet laser cutting. In: Needham, J.C. ed. *Proceedings of the Conference Advances in Welding Processes*, 14–16 April 1970, Harrogate, UK. Abington: The Welding Institute. pp. 140–146.

Adams, M.J. (1970b). Introduction to gas jet laser cutting. *Metal Construction and British Welding Journal*, **2**, 1–8.

Anderson, J.E. and Jackson, J.E. (1965). Theory and application of pulsed laser welding. *Welding Journal*, **44**, (12), 1018–1026.

Anon. (1966). Laser 'punches' holes in diamond wire-drawing dies. *Laser Focus*, **2**, (1), 4–7.

Anon. (1971). New laser developments. *Welding and Metal Fabrication*, **39**, 268–270.

Anon. (1977). Laser welding at 100 kilowatts. *Laser Focus*, **13**, (March), 14–20.

Arata, Y. (1987). Challenge to laser advanced materials processing. In: Arata, Y. ed. *Proceedings of the Conference Laser Advanced Materials Processing (LAMP '87)*, 21–23 May 1987, Osaka, Japan. Osaka: High Temperature Society of Japan and Japan Laser Processing Society. pp. 3–12.

Arata, Y., Abe, N. and Oda, T. (1985). Fundamental phenomena in high power CO_2 laser welding (report 1). *Transactions of JWRI*, **14**, (1), 5–11.

Armstrong, J.A., Bloembergen, N., Ducuino, J. and Pershan, P.S. (1962). Interactions between light waves in a nonlinear dielectric. *Physical Review*, **127**, 1918–1939.

Asmus, J.F., Murphy, C.G. and Munk, W.H. (1973). Studies on the interaction of laser radiation with art artefacts. *Proceedings of SPIE*, **41**, 19–27.

Avco Everett Research Laboratory, Inc. (1976). Cladding, 20 April 1976. US Patent no. 3 952 180.

Baardsen, E.L., Schmatz, D.J. and Bisaro, R.E. (1973). High speed welding of sheet steel with a CO_2 laser. *Welding Journal*, **52**, (4), 227–229.

Basov, N.G. and Prokhorov, A.M. (1954). The use of molecular bundles in radiospectroscopic studies of rotating molecular spectra. *Zhurnal Eksperimental'noi i Teoreticheskoi Fiziki*, **27**, 431–438. (In Russian.)

Basov, N.G., Kroklin, O.N. and Popov, Y.M. (1961). Production of negative temperature state p-n junctions of degenerate semiconductors. *Zhurnal Eksperimental'noi i Teoreticheskoi Fiziki*, **40**, 1320. (In Russian.)

Basov, N.G., Danilychev, V.A., Popov, Yu.M. and Khodkevich, D.D. (1970). Laser in the vacuum spectral region during the electron-beam excitation of liquid xenon. *Pis'ma v Zhurnal Eksperimental'noi i Teoreticheskoi Fiziki*, **12**, (10), 473–474. (In Russian.)

Battis, R.D. (2001). Alexandrite lasers emerge. *Industrial Laser Solutions*, **16**, (12), 25–26.

Beach, R., Bennett, W.J., Freitas, B.L., Mundinger, D., Comaskey, B.J., Solarz, R.W. and Emanuel, M.A. (1992). Modular microchannel cooled heatsinks for high average power laser diode arrays. *IEEE Journal of Quantum Electronics*, **28**, (4), 966–976.

Beaulieu, A. (1970). Transversely excited atmospheric pressure CO_2 laser. *Applied Physics Letters*, **16**, 504–505.

Beck, R. (1987). Fast-axial-flow CO_2-laser with integrated turbo-blower. *Applied Physics B*, **42**, 233–236.

Belforte, D. (1993). Laser welding: a technology in transition. *Industrial Laser Review Buyers Guide*. Nashua: PennWell. p. 5.

Bell, W.E. (1964). Visible laser transitions in Hg^+. *Applied Physics Letters*, **4**, (2), 34–35.

Bhaumik, M.L., Lacina, W.B. and Mann, M.M. (1972). Characteristics of a CO_2 laser. *IEEE Journal of Quantum Electronics*, **QE-8**, (2), 150–160.

Bod, D., Brasier, R.E. and Parks, J. (1969). A powerful CO_2 laser cutting tool. *Laser Focus*, (August), 36–38.

Bohr, N. (1913). On the constitution of atoms and molecules. *Philosophical Magazine*, **26**, (6), 1–25.

Bolwerk, M., Jense, F. and Frauenpreiss, T. (1997). Slab laser advances boost welding quality. *EuroPhotonics*, **2**, (1), 30–31.

Brau, C.A. and Ewing, J.J. (1975). 354-nm laser action on XeF. *Applied Physics Letters*, **27**, (8), 435–437.

Bridges, T.J., Burkhardt, E.G. and Smith, P.W. (1972). CO_2 waveguide lasers. *Applied Physics Letters*, **20**, (10), 403–405.

Bridges, T.J. and Patel, C.K.N. (1965). High-power Brewster window laser at 10.6 microns. *Applied Physics Letters*, **7**, (9), 244–245.

Bridges, W.B. (1964). Laser oscillation in singly ionized argon in the visible spectrum. *Applied Physics Letters*, **4**, (7), 128–130. (Erratum: *Applied Physics Letters*, **5**, (3), 39.)

Brown, C.O. (1970). High-power CO_2 electric discharge mixing laser. *Applied Physics Letters*, **17**, (9), 388–391.

Brown, C.O. and Banas, C.M. (1971). Deep penetration laser welding. In: *American Welding Society 52nd Annual Meeting*, 26–29 April 1971, San Francisco, CA, USA.

Burkhardt, E.G., Bridges, T.J. and Smith, P.W. (1972). BeO capillary CO_2 waveguide laser. *Optics Communications*, **6**, (2), 193–195.

Burnham, R., Harris, N.W. and Djeu, N. (1976). Xenon fluoride laser excitation by transverse electric discharge. *Applied Physics Letters*, **28**, (2), 86–87.

Carbone, R.J. (1967). Long-term operation of a sealed CO_2 laser. *IEEE Journal of Quantum Electronics*, **QE-3**, (9), 373–375.

Carter, W.H. (1980). Spot size and divergence for Hermite Gaussian beams of any order. *Applied Optics*, **19**, (7), 1027–1029.

Caterpillar Tractor Co. (1977). Method and apparatus for fusibly bonding a coating material to a metal article, 3 August 1977. GB Patent no. 1 482 044.

Chester, A.N. and Abrams, R.L. (1972). Mode losses in hollow-waveguide lasers. *Applied Physics Letters*, **21**, (12), 576–578.

Clarke, D. (1996). Sealed CO_2 lasers – from technology to tools. *Industrial Laser Review*, **11**, (10), 9–14.

Cohen, M.I., Mainwaring, F.J. and Melone, T.G. (1969). Laser interconnection of wires. *Welding Journal*, **48**, (3), 191–197.

Conti, R.J. (1969). Carbon dioxide laser welding. *Welding Journal*, **48**, (10), 800–806.

Cooper, M.I., Emmony, D.C. and Larson, J.H. (1992). The use of laser energy to clean polluted stone sculpture. *Journal of Photographic Science*, **40**, 55–57.

Cooper, M.I. (1998). *Laser Cleaning in Conservation.* Oxford: Butterworth-Heinemann. p. 12.

Crafer, R.C. (1974). A 2 kW CO_2 laser system for welding sheet material. In: Needham, J.C. ed. *Proceedings of the Conference Advances in Welding Processes*, 7–9 May 1974, Harrogate, UK. Abington: The Welding Institute. pp. 178–184.

Cunningham, F.E. (1964). *The Use of Lasers for the Production of Surface Alloys.* MS thesis, Massachusetts Institute of Technology.

Dalton, J. (1808). *A New System of Chemical Philosophy.* London: S. Russell for R. Bickerstaff.

Deacon, D.A.G., Elias, L.R., Madey, J.M.J., Ramian, G.J., Schwettman, H.A. and Smith, T.I. (1977). First operation of a free-electron laser. *Physical Review Letters*, **38**, (16), 892–894.

DeMichelis, C. (1970). Laser interaction with solids – a bibliographical review. *IEEE Journal of Quantum Electronics*, **QE-6**, (10), 630–641.

Deutsch, T.F., Horrigan, F.A. and Rudko, R.I. (1969). CW operation of high-pressure flowing CO_2 lasers. *Applied Physics Letters*, **15**, (3), 88–91.

Draper, C.W. (1981). Laser surface alloying: the state of the art. In: Mukherjee, K. and Mazumder, J. eds *Proceedings of the Conference Lasers in Metallurgy*, 22–26 February 1981, Chicago, IL, USA. Warrendale: The Metallurgical Society of the AIME. pp. 67–92.

Draper, C.W. (1982). Laser surface alloying: the state of the art. *Journal of Metals*, **34**, (6), 24–32.

Draper, C.W. and Ewing, C.A. (1984). Review. Laser surface alloying: a bibliography. *Journal of Materials Science*, **19**, 3815–3825.

Draper, C.W. and Poate, J.M. (1985). Laser surface alloying. *International Metals Reviews*, **30**, (2), 85–108.

Dumanchin, R. and Rocca-Serra, J. (1969). Augmentation of the energy and power of a CO_2 laser in pulsed operation. *Comptes Rendus Hebdomadaires des Séances de l'Académie des Sciences*, **269**, (B), 916–917. (In French.)

Dupuis, R.D. (2004). The diode laser. The first 30 days, 40 years ago. *Optics and Photonics News*, **15**, (4), 30–35.

Eggleston, J., Kane, T., Kuhn, K., Unternahrer, J. and Byer, R. (1984). The slab geometry laser – Part I: Theory. *IEEE Journal of Quantum Electronics*, **20**, (3), 289–301.

Ehrenfest, P. and Tolman, R.C. (1924). Weak quantization. *Physical Review*, **24**, 287–295.

Einstein, A. (1905). On the electrodynamics of moving bodies. *Annalen der Physik*, **17**, 891–921. (In German.)

Einstein, A. (1916). On the quantum theory of radiation. *Physikalische Gesellschaft Zurich. Mitteilungen*, **16**, 47–52. (In German.)

Einstein, A. (1917). Quantum theory of radiation. *Physikalische Zeitschrift*, **18**, 220–233. (In German.)

Ewing, J.J. and Brau, C.A. (1975). Laser action on the $^2\Sigma^+_{1/2} \to {}^2\Sigma^+_{1/2}$ bands of KrF and XeCl. *Applied Physics Letters*, **27**, (6), 350–352.

Fabrikant, V.A. (1940). PhD thesis. In: Butayeva, F.A. and Fabrikant, V.A., *Investigations in Experimental and Theoretical Physics. A memorial to G.S. Landsberg.* Moscow: USSR Academy of Sciences, 1959. pp. 62–70.

Fairand, B.P., Clauer, A.H., Jung, R.G. and Wilcox, B.A. (1974). Quantitative assessment of laser-induced stress waves generated at confined surfaces. *Applied Physics Letters*, **25**, 431–433.

Fairand, B.P., Clauer, A.H. and Wilcox, B.A. (1977). Pulsed laser induced deformation in an Fe-3 wt pct Si alloy. *Metallurgical Transactions*, **8A**, 119–125.

Fairand, B.P., Wilcox, B.A., Gallagher, W.J. and Williams, D.N. (1972). Laser shock-induced microstructural and mechanical property changes in 7075 aluminum. *Journal of Applied Physics*, **43**, 3893–3896.

Fairbanks, R.H. Sr and Adams, C.M. Jr (1964). Laser beam fusion welding. *Welding Journal Research Supplement*, **43**, (3), 97s–102s.

Fowles, G.R. and Silfvast, W.T. (1965). High-gain transition in lead vapor. *Applied Physics Letters*, **6**, 236–237.

Franken, P.A., Hill, A.E., Peters, C.W. and Weinreich, G. (1961). Generation of optical harmonics. *Physical Review Letters*, **7**, 118–119.

Furst, A. (1980). American Market/Metalworking News, 17 November 1980.

Gerardo, J.B. and Johnson, A.W. (1973). High-pressure xenon laser at 1730 Angstroms. *IEEE Journal of Quantum Electronics*, **9**, (7), 748–754.

Gerry, E.T. (1970). Gasdynamic lasers. *IEEE Spectrum*, **7**, (November), 51–58.

Geusic, J.E., Marcos, H.M. and van Uitert, L.G. (1964). Laser oscillations in Nd-doped yttrium aluminum, yttrium gallium and gadolinium garnets. *Applied Physics Letters*, **4**, (10), 182–184.

Giordmaine, J.A. and Miller, R.C. (1965). Tunable coherent parametric oscillation in $LiNbO_3$ at optical frequencies. *Physical Review Letters*, **14**, (24), 973–976.

Goldsborough, J. (1969). Stable long life cw excitation of helium-cadmium lasers by DC cataphoresis. *Applied Physics Letters*, **15**, (6), 159–161.

Gordon, J.P., Zeiger, H.J. and Townes, C.H. (1954). Molecular microwave oscillator and new hyperfine structure in the microwave spectrum of NH_3. *Physical Review*, **95**, 282–284.

Gregson, V.G. (1983). Laser heat treatment. In: Bass, M. ed. *Laser Materials Processing.* Amsterdam: North-Holland Publishing Company. pp. 201–233.

Gunshor, R.L. and Nurmikko, A.V. (1996). Blue-Laser CD technology. *Scientific American*, **275**, (July), 48–51.

Haase, M.A., Qiu, J., DePuydt, J.M. and Cheng, H. (1991). Blue–green laser diodes. *Applied Physics Letters*, **59**, (11), 1272–1274.

Hall, R.N., Fenner, G.E., Kingsley, J.D., Soltys, T.J. and Carlson, R.O. (1962). Coherent light emission from GaAs junctions. *Physical Review Letters*, **9**, (9), 366–368.

Hänsch, T.W. (1972). Repetitively pulsed tunable dye laser for high resolution spectroscopy. *Applied Optics*, **11**, 895–898.

Hargrove, L.E., Fork, R.L. and Pollack, M.A. (1964). Locking of He–Ne laser modes induced by synchronous intracavity modulation. *Applied Physics Letters*, **5**, 4–5.

Hayashi, I., Panish, M. and Foy, P. (1969). A low-threshold room-temperature injection laser. *IEEE Journal of Quantum Electronics*, **5**, (4), 211–212.

Heard, H.G. (1963). Ultra-violet gas laser at room temperature. *Nature*, **200**, 667.

Heglin, L.M. (1979). Laser drilling for materials fabrication. In: Metzbower, E.A. ed. *Proceedings of the Conference Applications of Lasers in Materials Processing*. Metals Park, OH: American Society for Metals. pp. 129–147.

Hill, A.E. (1971). Uniform electrical excitation of large-volume high-pressure near-sonic CO_2–N_2–He flowstream. *Applied Physics Letters*, **18**, (5), 194–197.

Hoag, E., Pease, H., Staal, J. and Zar, J. (1974). Performance characteristics of a 10-kW industrial CO_2 laser system. *Applied Optics*, **13**, (8), 1959–1964.

Hoff, P.W., Swingle, J.C. and Rhodes, C.K. (1973). Observations of stimulated emission from high-pressure krypton and argon/xenon mixtures. *Applied Physics Letters*, **23**, (5), 245–246.

Hoffman, J.M., Hays, A.K. and Tisone, G.C. (1976). High-power uv noble-gas–halide lasers. *Applied Physics Letters*, **28**, (9), 538–539.

Holonyak, N. Jr and Bevacqua, S.F. (1962). Coherent (visible) light emission from GaAs$_{1-x}$P$_x$ junctions. *Applied Physics Letters*, **1**, (December), 82–83.

Houldcroft, P.T. (1972). The importance of the laser for cutting and welding. *Welding and Metal Fabrication*, **40**, (2), 42–46.

Hsieh, J.J., Rossi, J.A. and Donnelly, J.P. (1976). Room-temperature cw operation of GaInAsP/InP double-heterostructure diode lasers emitting at 1.1 μm. *Applied Physics Letters*, **28**, (12), 709–711.

Huygens, C. (1690). *Traité de la Lumière*. Leiden: Pierre van der Aa. (In French.)

Jacoby, D., Pert, G.J., Ramsden, S.A., Shorrock, L.D. and Tallents, G.J. (1981). Observation of gain in a possible extreme ultraviolet lasing system. *Optics Communications*, **37**, (3), 193–196.

Javan, A., Bennett, W.R. Jr and Herriott, D.R. (1961). Population inversion and continuous optical maser oscillation in a gas discharge containing a He–Ne mixture. *Physical Review Letters*, **6**, (3), 106–110.

Jensen, R.E. and Tobin, M.S. (1972). An investigation of gases for stark modulating the CO_2 laser. *IEEE Journal of Quantum Electronics*, **QE-8**, (2), 34–38.

Jeon, H., Ding, J., Nurmikko, A.V., Xie, W., Grillo, D.C., Kobayashi, M., Gunshor, R.L., Hua, G.C. and Otsuka, N. (1992). Blue and green diode lasers in ZnSe-based quantum wells. *Applied Physics Letters*, **60**, (17), 2045–2047.

Jewell, J.L., Scherer, A., McCall, S.L., Lee, Y.H., Walker, S., Harbison, J.P. and Florez, L.T. (1989). Low-threshold electrically pumped vertical-cavity surface-emitting microlasers. *Electronics Letters*, **25**, (17), 1123–1124.

Johnson, L.F. and Nassau, K. (1961). Infrared fluorescence and stimulated emission of Nd^{+3} in $CaWO_4$. *Proceedings of the IRE*, **49**, 1704–1706.

Kane, T.J. and Byer, R.L. (1985). Monolithic, unidirectional single-mode Nd:YAG ring laser. *Optics Letters*, **10**, (2), 65–67.

Kasper, J.V.V. and Pimentel, G.C. (1965). HCl chemical laser. *Physical Review Letters*, **14**, (10), 352–354.

King, T. (1990). Infrared solid-state lasers. In: Ireland, C.L.M. ed. *Proceedings of the Conference High-Power Solid State Lasers and Applications (ECO3)*. Vol. 1277. Bellingham: SPIE. pp. 2–13.

Klauminzer, G.K. (1984). Twenty years of commercial lasers – a capsule history. *Laser Focus/Electro-optics*, **20**, (December), 54–79.

Knecht, W.L. (1966). Surface temperature of laser heated metal. *Proceedings of the IEEE*, **54**, 692–693.

Koehler, H.A., Ferderber, L.J., Redhead, D.L. and Ebert, P.J. (1972). Stimulated VUV emission in high-pressure xenon excited by high-current relativistic electron beams. *Applied Physics Letters*, **21**, (5), 198–200.

Kogelnik, H. and Li, T. (1966). Laser beams and resonators. *Proceedings of the IEEE*, **54**, (10), 1312–1329.

Koyama, F., Kinoshita, S. and Iga, K. (1989). Room-temperature continuous wave lasing characteristics of a GaAs vertical cavity surface-emitting laser. *Applied Physics Letters*, **55**, (3), 221–222.

Kroemer, H. (1963). A proposed class of heterojunction laser. *Proceedings of the IEEE*, **51**, 1782–1784.

Kuizenga, D. and Siegman, A. (1970a). FM and AM mode locking of the homogeneous laser – Part I: theory. *IEEE Journal of Quantum Electronics*, **6**, (11), 694–708.

Kuizenga, D. and Siegman, A. (1970b). FM and AM mode locking of the homogeneous laser – Part II: experimental results in a Nd:YAG laser with internal FM modulation. *IEEE Journal of Quantum Electronics*, **6**, (11), 709–715.

Kunzig, L. (1994). Laser welding of automotive and aero components. *Welding and Metal Fabrication*, **62**, (1), 14–16.

Kurobori, T., Kawamura, K.-I., Hirano, M. and Hosono, H. (2003). Simultaneous fabrication of laser-active colour centres and permanent microgratings in lithium fluoride by a single femtosecond pulse. *Journal of Physics: Condensed Matter*, **15**, L399–L405.

Ladenburg, R. (1928). Research on the anomalous dispersion of gases. *Zeitschrift für Physik*, **48**, 15–25. (In German.)

Lamb, W.E. Jr (1964). Theory of an optical maser. *Physical Review*, **134**, A1429–A1450.

Larsson, J.K. (1999). Lasers for various materials processing. A review of the latest applications in automotive manufacturing. In: Kujanpää, V. and Ion, J.C. eds *Proceedings of the 7th Nordic Conference on Laser Materials Processing NOLAMP-7*. Lappeenranta: Acta Universitatis Lappeenrantaensis. pp. 26–37.

Lazzarini, L. and Asmus, J.F. (1973). The application of laser radiation to the cleaning of statuary. *Bulletin of the AIC*, **13**, (2), 39–49.

Lea, C. (1989). Laser soldering – production and microstructural benefits for SMT. *Soldering and Surface Mount Technology*, No. 2, (June), 13–21.

Legay, F. and Legay-Sommaire, N. (1964). On the possibilities of realizing an optical maser using the vibrational energy of gases excited by activated nitrogen. *Comptes Rendus Hebdomadaires des Séances de l'Académie des Sciences*, **259**, 99–102. (In French.)

Legay-Sommaire, N., Henry, L. and Legay, F. (1965). Development of a laser using the vibrational energy of gases excited by activated nitrogen (CO, CO_2 and N_2O). *Comptes Rendus Hebdomadaires des Séances de l'Académie des Sciences*, **260**, 3339–3342. (In French.)

Leong, K.H. (2002). Femtosecond lasers for manufacturing. *Industrial Laser Solutions*, **17**, (5), 11–15.

Lichtman, D. and Ready, J.F. (1963). Laser beam induced electron emission. *Physical Review Letters*, **10**, (8), 342–345.

Locke, E., Hoag, E. and Hella, R. (1972). Deep penetration welding with high power CO_2 lasers. *Welding Journal Research Supplement*, **51**, (5), 245s–249s.

Loeffer, J.R. (1977). Numerically controlled laser soldering – fast, low cost, no rejects. *Assembly Engineering*, **20**, (3), 32–34.

Macintyre, R.M. (1983a). Laser hardfacing of gas turbine blade shroud interlocks. In: Kimmitt, M.F. ed. *Proceedings of the 1st International Conference Lasers in Manufacturing*. Bedford: IFS Publications. pp. 253–261.

Macintyre, R.M. (1983b). Laser hard-surfacing of turbine blade shroud interlocks. In: Metzbower, E.A. ed. *Proceedings of the 2nd International Conference on Applications of Lasers in Materials Processing*. Metals Park: American Society for Metals. pp. 230–239.

Macintyre, R.M. (1986). The use of lasers in Rolls-Royce. In: Draper, C.W. and Mazzoldi, P. eds *Proceedings of the NATO ASI Laser Surface Treatment of Metals*. Dordrecht: Martinus Nijhoff. pp. 545–549.

MacLeish, K. and Amos, J.L. (1977). Leonardo da Vinci: a man for all ages. *National Geographic*, **152**, (3), 296–329.

Madey, J.M.J. (1971). Stimulated emission of Bremsstrahlung in a periodic magnetic field. *Journal of Applied Physics*, **42**, (5), 1906–1913.

Maiman, T.H. (1960a). Optical and microwave-optical experiments in ruby. *Physical Review Letters*, **4**, (11), 564–566.

Maiman, T.H. (1960b). Stimulated optical radiation in ruby. *Nature*, **187**, (6 August), 493–494.

Maiman, T.H. (2000). *The Laser Odyssey*. Blaine: Laser Press. p. 103.

Makhov, G., Kikuchi, C., Lambe, J. and Terhune, R.W. (1958). Maser action in ruby. *Physical Review*, **108**, 1399–1400.

Matthews, D.L., Hagelstein, P.L., Rosen, M.D., Eckart, M.J., Ceglio, N.M., Hazi, A.U., Medecki, H., MacGowan, B.J., Trebes, J.E., Whitten, B.L., Campbell, E.M., Hatcher, C.W., Hawryluk, A.M., Kauffman, R.L., Pleasance, L.D., Rambach, G., Schofield, J.H., Stone, G. and Weaver, T.A. (1985). Demonstration of a soft X-ray amplifier. *Physical Review Letters*, **54**, (2), 110–113.

Matthews, D.L. and Rosen, M.D. (1988). Soft-X-ray lasers. *Scientific American*, **264**, (December), 60–65.

McClung, F.J. and Hellwarth, R.W. (1962). Giant optical pulsations from ruby. *Journal of Applied Physics*, **33**, 838–841.

McDermott, W.E., Pchelkin, N.R., Benard, D.J. and Bousek, R.R. (1978). An electronic transition chemical laser. *Applied Physics Letters*, **32**, (8), 469–470.

McWhorter, A.L. and Meyer, J.W. (1958). Solid-state maser amplifier. *Physical Review*, **109**, (2), 312–318.

Miller, J.E. and Wineman, J.A. (1977). Laser hardening at Saginaw steering gear. *Metal Progress*, **111**, (5), 38–43.

Mollenauer, L.F. and Olson, D.H. (1975). Broadly tunable lasers using color centers. *Journal of Applied Physics*, **46**, (7), 3109–3118.

Moorhead, A.J. (1971). Laser welding and drilling applications. *Welding Journal*, **50**, (2), 97–106.

Mott, J.S. and Macomber, S.H. (1989). Two-dimensional surface emitting distributed feedback laser arrays. *IEEE Photonics Technology Letters*, **1**, (8), 202–204.

Moulton, P. (1982). Ti-doped sapphire: tunable solid-state laser. *Optics News*, **8**, 9.

Moulton, P.F. (1986). Spectroscopic and laser characteristics of Ti:Al_2O_3. *Journal of the Optical*

Society of America B (*Optical Physics*), **3**, (1), 125–133.

Mundinger, D., Beach, R., Benett, W., Solarz, R., Sperry, V. and Ciarlo, D. (1990). High average power edge emitting laser diode arrays on silicon microchannel coolers. *Applied Physics Letters*, **57**, (21), 2172–2174.

Nahum, J. and Wiegand, D. (1966). Optical properties of some F-aggregate centers in LiF. *Physical Review*, **154**, 817–830.

Nakamura, S., Senoh, M., Nagahama, S., Iwasa, N., Yamada, T., Matsushita, T., Kiyoku, H. and Sugimoto, Y. (1996). InGaN-based multi-quantum-well-structure laser diodes. *Japanese Journal of Applied Physics*, **35**, (2), L74–L76.

Namba, S., Kim, P.H., Nakayama, S. and Ida, I. (1965). The surface temperature of metals heated with laser. *Japanese Journal of Applied Physics*, **4**, 153–154.

Nathan, M.I., Dumke, W.P., Burns, G., Dill, F.H. Jr and Lasher, G. (1962). Stimulated emission of radiation from GaAs *p-n* junctions. *Applied Physics Letters*, **1**, (3), 62–64.

Nelson, D.F. and Boyle, W.S. (1962). A continuously operating ruby optical maser. *Applied Optics*, **1**, (2), 181–183.

Olsen, F.O. (1980). Cutting with polarised laser beams. *DVS Berichte*, **63**, 197–200.

Patel, C.K.N., Faust, W.L. and McFarlane, R.A. (1964a). CW laser action on rotational transitions of the Σ_u^+–Σ_g^+ vibrational band of CO_2. *Bulletin of the American Physical Society*, **9**, 500.

Patel, C.K.N. (1964b). Interpretation of CO_2 optical maser experiments. *Physical Review Letters*, **12**, (21), 588–590.

Patel, C.K.N. (1964c). Continuous-wave laser action on vibrational–rotational transitions of CO_2. *Physical Review*, **136**, (5A), 1187–1193.

Patel, C.K.N. (1964d). Selective excitation through vibrational energy transfer and optical maser action in N_2–CO_2. *Physical Review Letters*, **13**, (21), 617–619.

Patel, C.K.N. and Kerl, R.J. (1964e). Laser oscillation on $X^1\Sigma^+$ vibrational–rotational transitions of CO. *Applied Physics Letters*, **5**, (4), 81–83.

Patel, C.K.N. (1965a). CW high power in N_2–CO_2 laser. *Applied Physics Letters*, **7**, (1), 15–17.

Patel, C.K.N., Tien, P.K. and McFee, J.H. (1965b). CW high power CO_2–N_2–He laser. *Applied Physics Letters*, **7**, (11), 290–292.

Patel, C.K.N. (1968). High-power carbon dioxide lasers. *Scientific American*, **214**, (August), 23–33.

Peterson, O.G., Tuccio, S.A. and Snavely, B.B. (1970). CW operation of an organic dye solution laser. *Applied Physics Letters*, **17**, 245–247.

Peyre, P., Scherpereel, X., Berthe, L. and Fabbro, R. (1998). Current trends in laser shock processing. *Surface Engineering*, **14**, 377–380.

Phillips, R.L. and Andrews, L.C. (1983). Spot size and divergence for Laguerre Gaussian beams of any order. *Applied Optics*, **22**, (5), 643–644.

Piltch, M., Walter, W.T., Solimene, N., Gould, G. and Bennett, W.R. Jr (1965). Pulsed laser transitions in manganese vapor. *Applied Physics Letters*, **7**, 309–310.

Planck, M. (1901). On the energy distribution law in the normal spectrum radiation. *Annalen der Physik. Leipzig*, **4**, 553–563. (In German.)

Platte, W.N. and Smith, J.F. (1963). Laser techniques for metals joining. *Welding Journal Research Supplement*, **42**, (11), 481s–489s.

Polanyi, J.C. (1961). Proposal for an infrared maser dependent on vibrational excitation. *Journal of Chemical Physics*, **34**, 347–348.

Pryor, B. and Frost, R. (1999). Tunable alexandrite lasers find ultraviolet applications. *Laser Focus World*, **35**, (10), 73–76.

Pummer, H., Hohla, K., Diegelmann, M. and Reilly, J.P. (1979). Discharge pumped F_2 laser at 1580 Å. *Optics Communications*, **28**, (1), 104–106.

Quist, T.M., Rediker, R.H., Keyes, R.J., Krag, W.E., Lax, B., McWhorter, A.L. and Zeigler, H.J. (1962). Semiconductor maser of GaAs. *Applied Physics Letters*, **1**, (4), 91–92.

Ready, J.F. (1963). Development of plume of material vaporized by giant-pulse laser. *Applied Physics Letters*, **3**, (1), 11–13.

Rice, J.K., Hays, A.K. and Woodworth, J.R. (1977). vuv emissions from mixtures of F_2 and the noble gases – a molecular F_2 laser at 1575 Å. *Applied Physics Letters*, **31**, (1), 31–33.

Rolls-Royce (1981). Application of metallic coatings to metallic substrates, 26 May 1981. GB Patent no. 2 052 566.

Rosen, M.D., Hagelstein, P.L., Matthews, D.L., Campbell, E.M., Hazi, A.U., Whitten, B.L., MacGowan, B., Turner, R.E., Lee, R.W., Charatis, G., Busch, Gar.E., Shepard, C.L. and Rockett, P.D. (1985). Exploding-foil technique for achieving a soft X-ray laser. *Physical Review Letters*, **54**, (2), 106–109.

Rubanova, G.M. and Sokolov, A.P. (1968). Metal heating by laser radiation. *Soviet Physics – Technical Physics*, **12**, (9), 1226–1228.

Rutherford, E. (1911). The scattering of α and β particles by matter and the structure of the atom. *Philosophical Magazine*, **21**, (6), 669–688.

Schäfer, F.P., Schmidt, W. and Volze, J. (1966). Organic dye solution laser. *Applied Physics Letters*, **9**, 306–309.

Schawlow, A.L. and Townes, C.H. (1958). Infrared and optical masers. *Physical Review*, **112**, (6), 1940–1949.

Schock, W., Schall, W., Hügel, H. and Hoffmann, P. (1980). CW carbon monoxide laser with RF excitation in the supersonic flow. *Applied Physics Letters*, **36**, (10), 793–794.

Scovil, H.E.D., Feher, G. and Seidel, H. (1957). Operation of a solid state maser. *Physical Review*, **105**, 762–763.

Scully, M.O. and Lamb, W.E. Jr (1967). Quantum theory of an optical maser. *Physical Review*, **159**, 208–226.

Searles, S.K. and Hart, G.A. (1975). Stimulated emission at 281.8 nm from XeBr. *Applied Physics Letters*, **27**, (4), 243–245.

Siegman, A.E. (2000a). Laser beams and resonators: the 1960s. *IEEE Journal of Special Topics in Quantum Electronics*, **6**, (6), 1380–1388.

Siegman, A.E. (2000b). Laser beams and resonators: beyond the 1960s. *IEEE Journal of Special Topics in Quantum Electronics*, **6**, (6), 1389–1399.

Silfvast, W.T., Fowles, G.R. and Hopkins, B.D. (1966). Laser action in singly ionized Ge, Sn, Pb, In, Cd and Zn. *Applied Physics Letters*, **8**, (12), 318–319.

Slade, S.G. (2002). *The Complete Book of Laser Eye Surgery*. New York: Bantam. p. 188.

Snitzer, E. (1961). Optical maser action of Nd^{+3} in a barium crown glass. *Physical Review Letters*, **7**, (12), 444–446.

Sorokin, P.P. and Lankard, J.R. (1966). Stimulated emission observed from an organic dye, chloro-aluminum phthalocyanine. *IBM Journal*, **10**, 162–163.

Sorokin, P.P. and Stevenson, M.J. (1960). Stimulated infrared emission from trivalent uranium. *Physical Review Letters*, **5**, (12), 557–559.

Speich, G.R. and Szirmae, A. (1969). Formation of austenite from ferrite and ferrite–carbide aggregates. *Transactions of the Metallurgical Society of AIME*, **245**, 1063–1074.

Sullivan, A.B.J. and Houldcroft, P.T. (1967). Gas-jet laser cutting. *British Welding Journal*, **14**, (8), 443–445.

Thomson, J.J. (1897). Cathode rays. *Philosophical Magazine*, **44**, 293–316.

Thomson, J.J. (1904). On the structure of the atom: an investigation of the stability and periods of oscillation of a number of corpuscles arranged at equal intervals around the circumference of a circle; with application of the results to the theory of atomic structure. *Philosophical Magazine*, **7** (6), (39), 237–265.

Tiffany, W.B., Targ, R. and Foster, J.D. (1969). Kilowatt CO_2 gas transport laser. *Applied Physics Letters*, **15**, (3), 91–93.

Townes, C.H. (1999a). *How the Laser Happened: Adventures of a Scientist*. New York: Oxford University Press. p. 57.

Townes, C.H. (1999b). *How the Laser Happened: Adventures of a Scientist*. New York: Oxford University Press. p. 96.

Trokel, S.L., Srinivasan, R. and Braren, B.A. (1983). Excimer laser surgery of the cornea. *American Journal of Ophthalmology*, **96**, 710–715.

Tsang, W.T. (1981). Extremely low threshold (AlGa)As modified multiquantum well heterostructure lasers grown by molecular-beam epitaxy. *Applied Physics Letters*, **39**, (10), 786–788.

Vaccari, J.A. (1994). Lasers zapping spot welders off the auto line. *American Machinist*, March, 31–34.

van der Ziel, J.P., Dingle, R., Miller, R.C., Wiegmann, W. and Nordland, W.A. Jr (1975). Laser oscillation from quantum states in very thin $GaAs–Al_{0.2}Ga_{0.8}As$ multilayer structures. *Applied Physics Letters*, **26**, (8), 463–465.

Veiko, V.P., Kokora, A.N. and Libenson, M.P. (1968). Experimental verification of the temperature distribution in the area of laser-radiation action on a metal. *Soviet Physics – Technical Physics*, **13**, (3), 231–233.

Velazco, J.E. and Setser, D.W. (1975). Bound-free emission spectra of diatomic xenon halides. *Journal of Chemical Physics*, **62**, (5), 1990–1991.

Vogel, K. and Backlund, P. (1965). Application of optical microscopy in studying laser-irradiated metal surfaces. *Journal of Applied Physics*, **36**, (12), 3697–3701.

Wagner, W.G. and Lengyel, B.A. (1963). Evolution of the giant pulse in a laser. *Journal of Applied Physics*, **42**, 2040–2046.

Walker, R. (1982). CO_2 lasers: a machining success story. *Photonics Spectra*, **16**, (September), 65–72.

Walter, W.T., Piltch, M., Solimene, N. and Gould, G. (1966a). Pulsed-laser action in atomic copper vapor. *Bulletin of the American Physical Society*, **11**, (II), (1), 113.

Walter, W.T., Solimene, N., Piltch, M. and Gould, G. (1966b). 6C3 – Efficient pulsed gas discharge lasers. *IEEE Journal of Quantum Electronics*, **QE-2**, (9), 474–479.

Ward, J.F. and New, G.H.C. (1969). Optical third-harmonic generation in gases by a focused laser beam. *Physical Review A*, **185**, 57.

Weber, J. (1953). Amplification of microwave radiation by substances not in thermal equilibrium. *Transactions Institute of Radio Engineers Professional Group on Electron Devices PGED-3*, June, pp. 1–4.

White, A.D. and Rigden, J.D. (1962). Continuous gas maser operation in the visible. *Proceedings of IRE*, **50**, (July), 1697.

Wick, C. (1976). Laser hardening. *Manufacturing Engineering*, **76**, (6), 35–37.

Wisner, G.R., Foster, M.C. and Blaszuk, P.R. (1973). Unstable resonators for CO_2 electric-discharge convection lasers. *Applied Physics Letters*, **22**, (1), 14–15.

Witteman, W.J. (1965). Increasing continuous laser-action on CO_2 rotational vibrational transitions through selective depopulation of the lower laser level by means of water vapour. *Physics Letters*, **18**, (2), 125–127.

Yessik, M. and Schmatz, D.J. (1975). Laser processing at Ford. *Metal Progress*, **107**, (5), 61–66.

CHAPTER 3

LASERS

INTRODUCTION AND SYNOPSIS

Laser action has been demonstrated in the laboratory in hundreds of materials – some of the more surprising being whisky, jelly and Chinese tea. In 1996, the Hubble space telescope discovered a gas cloud that acts as a natural ultraviolet laser near the huge, unstable star *Eta Carinae*. In total, over 15 000 energy transitions that result in the production of laser light have been reported. However, commercial lasers make use of about 40 active media: many are too expensive to be commercially viable; the efficiency of light production is often very low; or it might not be possible to construct a unit for practical reasons.

One feature that all lasers share is the unique nature of the light that they produce – a coherent, monochromatic beam of low divergence and high brightness. These properties form the basis of applications in fields as diverse as measurement, holography, data storage and communications. Material processing makes use of the *thermal* and *photonic* effects associated with the interaction of a laser beam with various engineering materials.

This chapter concentrates on lasers used for material processing. The principles of laser light generation from transitions between energy levels in active media comprising gases, solids and liquids are explained. Practical means of exciting media to achieve stimulated emission of light are described. The conditions for amplification of light in different types of resonator are considered. The nature of laser output from various combinations of active media, excitation methods and resonators is compared. These principles are then applied to commercial lasers for processing engineering materials. Further information on the output from individual lasers can be found in Appendix B and in the reading lists at the end of the chapter.

GENERATION OF LASER LIGHT

After Hertz had generated electromagnetic waves using a high voltage induction coil (Chapter 2), it was realized that radiation could be produced over a continuous range – the electromagnetic spectrum – shown in Fig. 3.1.

The spectrum may be divided into portions. Radio waves occupy the low frequency, low energy, long wavelength range, and are produced by antennae. Microwaves are generated by electrical oscillators. Infrared radiation originates from electronic transitions and molecular vibrations in materials. Visible light (radiation in the wavelength range 390–780 nm) is characterized in order of increasing wavelength as violet (390–430 nm), indigo (430–455 nm), blue (455–492 nm), green (492–577 nm), yellow (577–597 nm), orange (597–622 nm), and red (622–780 nm), and is produced from transitions between energy states in the valence electrons of atoms. Ultraviolet light is emitted from corresponding high energy electronic transitions. X-rays result from deep electronic transitions. High frequency, high

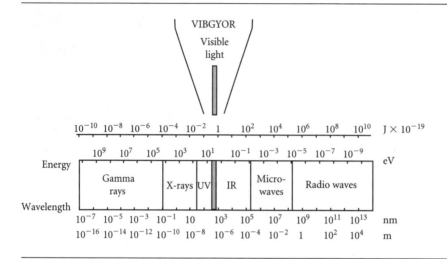

Figure 3.1 The electromagnetic spectrum: UV – ultraviolet; IR – infrared

energy, short wavelength gamma rays are produced by radioactive decay. In material processing applications, we are mainly interested in the infrared, visible and ultraviolet portions of the electromagnetic spectrum.

Laser light is generated by transitions between high and low states of energy in species (atoms, ions and molecules) in various media. Sustainable light generation depends on a suitable combination of fundamental physical phenomena to generate the light and an appropriate mechanical design to maintain and amplify the emission. We consider the fundamentals of laser light generation first.

ENERGY LEVELS

Atoms, ions and molecules, collectively known as *species*, exist in states characterized by discrete energy levels, also referred to as states.

The simplest forms of energy levels are those available to an isolated *atom*, such as hydrogen. The rules of quantum mechanics state that all particles have discrete energy states, which relate to different periodic motions of the constituent nuclei and electrons. The lowest possible energy level is the *ground state*, while other states are termed *excited states*.

When *molecules* in gases, liquids and solids are considered, the energy levels are no longer those of the individual atoms. Interactions with neighbours result in modifications to the energy levels. In condensed materials (liquids and solids), the atoms are packed together and the interactions are strong. The energy levels of the individual atoms then broaden and merge into an almost continuous band of closely spaced states.

In addition to electronic energy levels, molecules with more than one atom can possess quantized *vibrational* or *rotational* energy levels. Simple molecules such as nitrogen have only one vibrational mode. Complex molecules normally have many vibrational modes, all of which can interact. The carbon dioxide molecule, for example, can be visualized as a linear arrangement of two oxygen atoms and a carbon atom, with the carbon atom in the centre. The vibrational energy levels correspond to the motion of the oxygen atoms relative to the carbon atom. Vibrational transitions involve energy changes of about 0.1 eV – around an order of magnitude lower than the electronic transitions that produce visible light – and so such radiation normally lies in the mid-infrared region shown in Fig. 3.1. Rotational energy levels correspond to rotational motion of asymmetrical molecules, since angular momentum is quantized. These transitions involve energy changes around an order of magnitude smaller than vibrational transitions, and so are associated with far infrared radiation.

In addition to the values of the energy levels, the *lifetime* spent in those levels affects the temporal nature of radiation. The lifetime of a state depends on the ease with which the state can be depopulated. States that exist for time scales on the order of microseconds and milliseconds, which are long in terms of laser transitions, are known as *metastable* states, and are important means of storing energy in laser systems.

Energy Level Notations

An atom can be idealized as a positively charged nucleus surrounded by negatively charged electrons that are arranged in *quantum shells*. Each shell is described by using a principal quantum number, n. The shell is able to hold a certain number of electrons, given by $2n^2$. Thus the lowest energy quantum shell ($n = 1$) can hold 2 electrons. Successive quantum shells of higher energy hold greater numbers of electrons: 8 ($n = 2$), 18 ($n = 3$), 32 ($n = 4$) .... Electrons in a given shell have a similar energy, but no two are identical. Subshells, or *orbitals*, differentiate the probability that pairs of electrons occupy a given orbit relative to the nucleus. Electrons in a pair have identical energy, but opposite magnetic spin (the Pauli exclusion principle). The azimuthal quantum number, l, denotes states s, p, d and f, which have values of l of 0, 1, 2 and 3, respectively. The first quantum shell ($n = 1$) can contain only two electrons, both of which occupy the s orbital, which have a spherical probability distribution around the nucleus. The second quantum shell ($n = 2$) can contain eight electrons; two in the s orbital, and six in the p orbital, which has a slightly higher energy. The p orbitals have probability distributions shaped like dumbbells aligned with orthogonal axes. The third quantum shell ($n = 3$) can contain 18 electrons; two in the s orbital, six in the p orbital, and 10 in the higher energy d orbital. The fourth and fifth shells contain f orbitals, which can accommodate up to 14 electrons. The total number of electrons is equal to the atomic number of the element.

The Paschen notation describes the electrons in terms of particular shells and orbitals. The electrons in oxygen, of atomic number eight, are denoted $1s^2 2s^2 2p^4$: two electrons in the s orbital of the first shell, two electrons in the s orbital of the second shell, and four electrons in the p orbital of the second shell. As the number of shells and orbitals increases, the difference in energy between orbitals decreases, and some overlap in energy occurs. An inner orbital of an outer shell may have a lower energy than an outer orbital of an inner shell. The 26 electrons in iron, for example, are denoted $1s^2 2s^2 2p^6 3s^2 3p^6 3d^6 4s^2$. Electrons occupy the $4s$ subshell before the $3d$ subshell is filled. Transitions between adjacent states of angular momentum are denoted by lines in series: the *sharp* series lines denote transitions from p to higher s states; the *principal* series s to higher p; the diffuse series p to higher d; and the *fundamental* series d to higher f.

Energy levels involved in laser transitions are often named according to the Moore Sitterly convention. Each level is defined by an inner quantum number, J. Groups of related levels – *terms* – have multiplicities that are exclusively odd or even in a given spectrum. For terms of odd multiplicity the values of J are integers (0, 1, 2 ...). Values of J for terms of even multiplicity are odd multiples of the fraction ½ (½, 1½, 2½ ...). Terms are further defined by azimuthal quantum numbers, L, that have the values 0, 1, 2, 3, 4, 5, 6, 7 ... for terms labelled S, P, D, F, G, H, I, K..., respectively (the hyperfine quantum number). A term of a given type and multiplicity comprises a finite number of energy levels whose inner quantum number is stipulated by quantum theory. For example, an S term of multiplicity three has only one level with a value of J equal to 1 – it is designated $^3S_1$. A D term of multiplicity four comprises four levels whose values of J are 3½, 2½, 1½ and ½ – designated $^4D_{7/2}$, $^4D_{5/2}$, $^4D_{3/2}$ and $^4D_{1/2}$, respectively. The designation is augmented with two quantities: a prefix that distinguishes terms of the same type and multiplicity; and a superscript 'o' denoting that the configuration contains an odd number of p, f, h, etc. electrons. This notation is used in Appendix B to describe laser transitions.

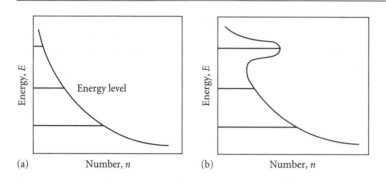

Figure 3.2 Distribution of energy in a species with (a) thermodynamic equilibrium and (b) a population inversion

Distribution of Energy

The normal distribution of energy in a population of species is given by the Maxwell–Boltzmann equation, illustrated in Fig. 3.2a. The ratio of the numbers, N_1 and N_2, populating two energy levels, E_1 and E_2, respectively, is

$$\frac{N_2}{N_1} = \exp - (E_2 - E_1)/kT \tag{3.1}$$

where k is Boltzmann's constant $(1.381 \times 10^{-23}\,\text{J K}^{-1})$ and T is the absolute temperature. At a given temperature, the number of species occupying higher energy levels decreases monotonically. As the temperature is increased, the *number* of species occupying higher energy levels increases, but the *form* of the distribution remains the same; the population of a lower level exceeds that of a higher level to give a condition of *thermodynamic equilibrium*. There is no driving force for energy to be released from the system, only for it to be redistributed internally.

Population Inversion

Normal populations occur naturally. However, a distribution can be disturbed artificially, such that the number of species occupying a higher energy level exceeds that of a lower level, Fig. 3.2b. This may be achieved by exciting or 'pumping' the population by using an external energy source. A *population inversion* is thus created – a prerequisite for laser light generation. A driving force now exists for energy to be released from the system. In the case of a laser, this energy is released in the form of light. Note the contrast between creating a population inversion and the effect of merely raising the temperature of the system; in the latter case the Maxwell–Boltzmann distribution is maintained, and so a driving force for the release of energy is not created.

EXCITATION

A population inversion may be achieved by using a variety of energy sources to excite the species. Electrical, optical and chemical means are the most common in industrial lasers. Gaseous species absorb radiation over discrete ranges of wavelength (lines), and so electrical excitation, which produces energy over a relatively broad range, is common in gas lasers. Solids are not easily excited electrically, but optical pumping can be highly efficient in solid state lasers. Chemical methods are generally more difficult to control, but are effective means of excitation in chemical lasers.

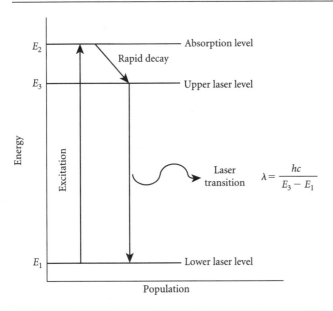

Figure 3.3   Energy transitions in a three-level laser system

In order to create a population inversion efficiently, it is necessary for the species to possess a large group of upper *absorption* energy levels, so that energy can be absorbed over an appreciable frequency range. These rapidly and efficiently feed more stable lower energy levels, which are termed the *upper laser levels*, or states. Below these lie the *lower laser states*. Laser light generation involves transitions from the upper to the lower laser states. In order to maintain a population inversion, the lifetime of the lower laser state must be shorter than that of the upper state. In addition, the rate of population of the upper state must be greater than that of the lower state.

Energy Level Transitions

Einstein proposed that light consists of bundles of wave energy, termed photons. It was originally thought that photons and species may interact only by *absorption* of a photon (with a corresponding increase in energy), or *spontaneous emission* of a photon originally in a higher energy state (leading to a reduction in energy). (Energy can also be reduced without the emission of a photon; a process known as non-radiative decay.) However, Einstein concluded that there must exist a third mechanism of interaction – *induced* or *stimulated* emission – in which an excited species could be stimulated to emit a photon by interaction with another photon. This is the basis of light amplification by stimulated emission of radiation, from which the acronym *laser* is formed.

The simplest form of laser is based on transitions between two energy levels, E_2 and E_1, which represent the ground and excited states. The ammonia maser and diode laser are examples of two-level systems. However, it is difficult to obtain useful light amplification in this type of system because as species in the upper laser level emit radiation their number approaches that of the species in the ground state, and absorption falls towards zero. For this reason, industrial lasers are often based on three and four energy level systems.

In a three-level laser system, illustrated in Fig. 3.3, excitation is achieved by pumping to the E_2 absorption level, or levels. If an energy level E_3 exists, which lies slightly below E_2, rapid non-radiative decay can occur to the level E_3 with little loss of energy. E_3 becomes the upper laser level. Laser emission

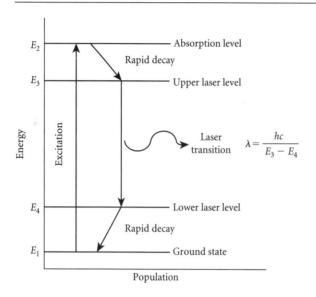

Figure 3.4   Energy transitions in a four-level laser system

then takes place between the levels E_3 and E_1. A number of conditions must be fulfilled for this type of laser to operate. First, the energy required for excitation must be relatively high, because more than half of the entire population of the species must be raised out of the ground state (which may be the lower laser level). Second, the transition $E_3 \rightarrow E_1$ must be very probable. Third, the species must be able to remain in the E_3 state longer than the E_1 state in order to build up and maintain a population inversion. If the lower laser level in a three-level system is the ground state, a population inversion is more difficult to achieve, and output is limited to pulsed operation.

A population inversion can be generated more easily if the laser transition terminates in a state that is not the ground state. This is the case in a four-level system, illustrated in Fig. 3.4. Species are excited to the level E_2, followed by rapid non-radiative decay to a lower level, E_3. The laser transition occurs between levels E_3 and a second intermediate level, E_4. Rapid relaxation to the ground state, E_1, is then desirable for efficient operation. The potential for four-level operation is much higher than three-level operation because the threshold pump energy is considerably lower, since it is not necessary to invert the entire population. Since the laser transition is to an intermediate level which is normally unpopulated, a four-level laser can operate in continuous wave mode.

LIGHT AMPLIFICATION

We have seen how light may be generated by stimulated emission of photons. However, a laser works on the principle of light *amplification* by stimulated emission. Amplification can only occur if emission takes place in a suitable device – the *optical cavity*. Amplification is achieved when stimulated emission increases the number of photons circulating in the optical cavity, illustrated schematically in Fig. 3.5.

The amplification achieved is the *gain* of the system. If the circulating power in a laser is restored to its original value after a round trip in the optical cavity, then the round trip gain is equal to the round trip loss; this is known as the *threshold gain*. If the loss is greater than the gain then the laser will not produce light. Positive gain is the second requirement for laser light generation – the first being a population inversion.

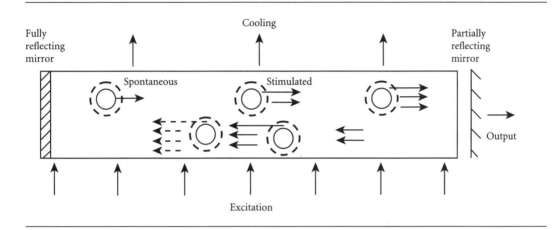

Figure 3.5 Schematic illustration of amplification by stimulated emission of radiation

There are two types of gain; saturated and unsaturated. Unsaturated gain, often referred to as small-signal gain, occurs with small input values. In contrast, with high excitation the number of photons passing through the optical cavity significantly depletes the population inversion, and the gain is reduced, or saturated. The gain is independent of the excitation power. It is the circulating power, and hence the output power, which increases with increasing excitation power.

OUTPUT

In a light bulb, electrons in the atoms and molecules of the filament are pumped to higher levels by electrical excitation. Electrons fall randomly to lower levels independently of one another, emitting light with a random collection of wavelengths (colours). Since many electrons are constantly falling to different levels, a range of wavelengths is produced, and the net result is the production of white light. The light that is produced is emitted in random directions.

In contrast, in the process of *stimulated* emission, a photon collides with another excited species, causing it to release its photon prematurely. Photons travel in the same direction until the next collision, thereby building a stream of increasing density, Fig. 3.5. Photons have the same phase (temporal and spatial properties), frequency and polarization. Laser light is therefore *coherent* and *monochromatic*, and has *low divergence* and *high brightness*.

Laser light can take the form of a continuous wave (CW), a pulse, or a train of pulses. The length of the pulse can vary from a tenth of a second to a few femtoseconds (10^{-15} s). Pulses may be produced at a rate of between one and several thousand per second. The average power may vary between milliwatt and kilowatt levels, with peak power attaining the order of gigawatts. Some lasers can be tuned to produce a range of wavelengths.

Efficiency

A number of efficiency values can be defined when considering laser light generation. The *fluorescent quantum efficiency*, η_f, is the ratio of the number of species participating in the laser transition to the number of species raised from the ground state. (Species in the excited state may decay to states other than the upper lasing state, and atoms in the upper lasing state may decay to states other than the lower lasing state.) The *thermodynamic efficiency*, η_t, is the ratio of the amplification energy to the energy required for excitation. Since laser photons have less energy than the excitation source, the thermodynamic efficiency lies below 1. An overall *wall plug efficiency*, η_w, can also be defined:

$$\eta_w = \eta_t \cdot \eta_f. \tag{3.2}$$

Typical wall plug efficiencies for material processing lasers can be found in Table B.1 (Appendix B).

CONSTRUCTION AND OPERATION OF COMMERCIAL LASERS

So far we have considered the theoretical aspect of laser light generation. Now we examine means of constructing and operating a practical laser.

A laser requires four basic components to operate: an *active medium* in which light can be amplified by stimulated emission of radiation; a means to *excite* the medium – the excitation or 'pumping' source – to maintain the population inversion; a means to provide optical *feedback* – the optical cavity; and an *output* device to enable usable amounts of beam energy to exit the laser.

Additionally, a laser requires power and control systems, means of cooling the active medium and an interface for operation.

Active Media

Industrial lasers are normally classified by the active medium, which may be a gas, an insulating solid, a semiconductor or a liquid.

Gases

Gases possess a number of properties that account for their popularity in industrial lasers: they can be excited directly with an electric current; they are homogeneous; they allow flexibility in the design of the resonator, which can be scaled easily; they can be manipulated aerodynamically, facilitating stability; propagation of the beam is unimpeded; and they are relatively inexpensive. Laser emission from gases is well defined, and occurs in three discrete parts of the electromagnetic spectrum: ultraviolet, visible and infrared.

The noble gases neon, argon, krypton and xenon, and mixtures of helium and neon, are the active media in *neutral atom* gas lasers. Neutral atoms produce light in the range between mid-ultraviolet and near infrared. The energy levels for infrared emission lie close to the limit of atomic ionization. As a result, the atom is excited to a high energy level, which in turn means that the photon emitted has a relatively small amount of energy. The fluorescent quantum efficiency is therefore low in atomic gases. Not every excited species is able to produce laser light; the ratio of those that can to those that cannot varies as the cube of the wavelength, thus favouring the infrared end of the electromagnetic spectrum. However, both the emitted energy and the maximum working temperature vary approximately with the reciprocal of the wavelength. Neutral atom gas lasers therefore incorporate relatively weak discharges and have moderate gain and power output. They can be used for fine scale, low power precision material processing.

The excitation energy of an *ion* is larger than a neutral atom; ionized gas lasers therefore produce short wavelength light, in the range between mid-ultraviolet and visible. High current densities are required for excitation since energy is used to ionize the atom and then to excite it. These lasers consequently have high plasma temperatures, and require substantial cooling to operate. Ions of noble gases, notably argon and krypton are used in commercial designs, producing ultraviolet light that is suitable for fine-scale material processing.

Molecules produce relatively long wavelength light in the range between visible and far infrared. The relevant transitions are those between vibrational and rotational energy levels in the molecule. Two types are possible: transitions between vibrational states of the same electronic level, as in carbon dioxide; and transitions between vibrational states of different electronic states, as in nitrogen. The vibrational levels of the electronic ground state are close to the ground state of the molecule. The photon energy is therefore a significant fraction of the excitation energy, resulting in a relatively high value of quantum efficiency; almost all of the electrons present in the discharge participate in the excitation process. Diatomic molecules are less suitable for continuous laser emission because of the unfavourable lifetime of such molecules excited to vibrational levels of the electronic ground state. Two molecules are particularly good emitters: carbon monoxide (around $5\,\mu m$); and carbon dioxide (around $10\,\mu m$). Moderate current densities are involved, but the ability to use large volumes allows multikilowatt output power to be obtained, hence the development of high power lasers for material processing based on these media.

Transitions can also take place between electronic states in a *metal vapour*. There are three basic types of metal vapour laser: metal ion; recombination and neutral atom. The helium–cadmium and helium–selenium lasers are well-developed examples of the metal ion type. Recombination lasers are still under development (strontium is being investigated). Of the neutral atom types, copper and gold vapour lasers are the most popular, although lasers based on lead, manganese and barium have been examined. Emission occurs via relatively high energy transitions between excited states and low lying ground states, which result in the production of visible or ultraviolet light.

The term *excimer* is derived from excited dimer, which refers to a diatomic molecule formed by a chemical reaction after one or both of its constituents have been excited. The term has come to be used to describe the association of two different atoms as well: a rare gas (neon, argon, krypton or xenon) and a halogen (fluorine, chlorine, bromine or iodine) – which strictly should be termed an *exciplex* (excited complex). The rare gas and the halogen are excited to form positive and negative ions, respectively, which are attracted and combine to form an excimer. An inert gas, normally helium or neon, regulates energy transfer. Since the molecule has dissociated, the ground state is empty, and a population inversion is readily formed. As the excimer loses its energy and returns to the ground state, it emits a photon of ultraviolet light, and the molecule dissociates into free atoms that are available to take part in the excitation process again. Transitions between both electronic and vibrational energy states result in emission. Since the lifetime of the excited state is on the order of nanoseconds, only pulsed output can be obtained.

Liquids

The active medium in most liquid lasers is a fluorescent organic dye dissolved in a solvent that flows through the laser. These dyes are large complex molecules, which have a large number of vibrational and rotational energy levels that blend together into energy bands. Emission has been obtained from about 50 dyes, providing a wide selection of lasing wavelengths. By combining several dyes, output wavelengths covering the visible spectrum may be produced. Their useable lifetimes range from hours to months, depending on the dye and the means of excitation. Liquid active media have advantages over gases and solids: they can be prepared more easily (solids require a high degree of optical homogeneity); and they contain a higher density of atoms than gases.

When the molecule drops from one broad electronic state to another, the wavelength of light emitted depends on the start and end points. The emission bandwidth can therefore be very broad – up to 100 mm in some dyes. The laser bandwidth can be selected by limiting the bandwidth of feedback provided by the resonator using prisms, gratings, birefringent filters, and other devices, thus enabling the output to be tuned.

Insulating Solids

The active media in solid state lasers (not including semiconductor lasers) comprise a host material doped with ions of transition or rare earth elements. (Colour centre, or F-centre lasers are a small group of laser-pumped sources with crystalline active media – including potassium chloride with lithium or sodium – that contain defects, which cause intense absorption, but are rarely used for material processing.) The term 'solid state' indicates that the active medium in the laser is a solid, rather than a gas or a liquid – it should not be confused with the terminology used in electronics. In comparison with gas lasers, solid state lasers require no mechanical devices for media circulation, complex heat exchangers, or vacuum and gas-supply systems. However, the thermal conductivity of the host determines the amount of heat generated, which limits the working range through thermal lensing. A beam of relatively high divergence is produced from solid active media because inhomogeneities in the active medium cannot be smoothed out, as in circulating gas lasers.

Suitable hosts are crystalline materials and glasses that are stable, hard and optically isotropic, and which possess sufficient tensile strength to be used in a variety of shapes. Materials are required to have a high thermal conductivity and low thermal expansion coefficient, for thermal stability, and must be able to accept dopant ions in substitutional sites. Yttrium aluminium garnet ($Y_3Al_5O_{12}$), referred to as YAG, sapphire (Al_2O_3), calcium fluoride (CaF_2), and silicate and phosphate glasses meet these criteria. YAG has a particularly good combination of low thermal expansion and high thermal conductivity, and is the host in the Nd:YAG laser – a popular solid state laser for material processing, available with output up to the multikilowatt level. Glasses can be doped to higher concentrations than YAG, with good uniformity, and can be produced in larger sizes with a greater variety of geometries. Glasses are particularly suitable for pulsed lasers.

The dopant contains an interior unfilled shell of electrons, which leads to a narrow emission bandwidth; this is favourable for laser operation since it leads to high gain and reduces the requirements on the population inversion necessary for operation.

The host material determines the characteristics of the available energy levels, and therefore the exact wavelength of light generated. When ions are embedded in a solid, they can absorb radiation over a much wider band of wavelengths. Laser transitions occur between low lying energy levels, the nature of which are determined by the forces acting on and between the electrons in partially filled electronic shells. There are three principal interactions to consider: coulomb forces acting between the electrons; crystal field interaction; and a coupling between electron spin and orbital angular momentum, known as spin-orbit coupling. Coulomb forces are normally the largest; they split the single electron configuration into a number of levels. In transition metals, the crystal field interactions are the next largest since the partially filled electronic shells are not shielded. This splits the term energy levels into further levels. In the case of transition metals, the spin-orbit interaction is relatively small, and is not considered here. In rare earth ions, the spin-orbit interaction is stronger than the effects of the crystal field. The crystal field then acts to produce further splitting of the multiplets. The wavelength range of light produced by solid state lasers covers the visible and infrared, between about 300 and 3000 nm.

Solid active media enable relatively small lasers to be constructed, with no gas flow maintenance requirements. However, heating limits the power that can be generated and the beam quality is relatively poor at high power.

Semiconductors

In contrast to the single energy levels found in isolated atoms, electrons in semiconductors occupy broad *bands* of energy levels. Each band comprises a number of closely spaced levels, which originate from the superposition of all the energy levels of the atoms making up the solid. The equilibrium atomic separation results in a sequence of bands separated by energy gaps. The most important features for laser light generation are the uppermost occupied band, the first empty band, and the gap in between, which are termed the valence band, the conduction band, and the energy gap, respectively.

Electrons can be excited from the valence band to the conduction band. In this way electrical conduction can take place via the motion of electrons in both bands. The absence of an electron in a band can be considered as a 'hole' that has a positive charge. In a pure semiconductor material the number of electrons and holes are equal. The number can be changed by adjusting the temperature, or by doping the semiconductor with atoms whose valences differ from that of the host material. If tetravalent silicon is doped with pentavalent phosphorus, each phosphorus atom replaces one of the silicon atoms, and four of its five valence electrons are used to satisfy the bonding requirements of its four neighbours. The remaining electron is not used in bonding, and is only weakly bonded to the phosphorus atom, and so it is readily detached, and promoted to the conduction band. Such dopants are known as donors, and the material is known as *n-type*, where *n* denotes negative, referring to the electron density. Dopants that have a valence one less than the host – acceptors – can also be added to form *p-type* material (*p* denotes positive). Light is emitted when electrons drop from the conduction band and occupy, or recombine with, a hole in the valence band, to form a neutral atom in the crystal lattice. The energy of this transition determines the wavelength of the light generated. The high concentration of electronic states in the bands provides high gain.

A junction can be made by placing *n*- and *p*-type materials together. Since there are more electrons in the conduction band of the *n*-material than the *p*-material, electrons flow from the *n* to *p* conduction bands. Conversely, holes flow from the *p*- to the *n*-valence bands. The simplest junction comprises one *p*-doped and one *n*-doped layer of group III and group V compounds, such as gallium arsenide (GaAs). This is referred to as a *homojunction* laser, since it comprises layers of the same basic material.

For practical reasons, most semiconductor lasers are of the *heterojunction* type. Heterojunctions comprise several layers of different semiconductor materials, based mainly on gallium, aluminium and indium in compounds of arsenide, phosphide or antimonide. The optical cavity is limited to a narrower region around the *p–n* junction because of the difference in refractive index between the layers. Less current is therefore needed for excitation, and heat build-up is reduced. Most designs of diode laser are based on blocks of semiconductor that may be no more than 1 mm square and 100 μm in thickness. Techniques such as liquid phase epitaxy are used to grow the thin layers of semiconductor crystal used in commercial diode lasers. Semiconductors are combined in arrays to attain multikilowatt output for material processing.

Excitation

As mentioned above, a population inversion is achieved in the active media of industrial lasers by using electrical, optical and chemical means. (Electron beams were also used in the early days of laser development.)

Electrical Pumping

Direct current (DC) excitation is relatively compact, simple and cheap, and comprises a high voltage transformer and rectifier with a large smoothing capacitor. A glow discharge is created in a gaseous medium by electrons emitted at the cathode, which travel though the gas under the action of an electric field. Excitation is achieved through the collision of energetic electrons with gas atoms and molecules. Since the electrodes must be placed inside the resonator, impurities are formed from reactions with the gases, and the electrodes must be cleaned at regular intervals. The discharge can become unstable, and an arc may form, which allows thermal equilibrium to be achieved, preventing laser action.

Power generated by radio frequency (RF) excitation can be coupled capacitatively through dielectric materials, such as quartz glass, into a gas mixture. The electrodes can therefore be mounted on the outside of discharge tubes. There is no electrode wear or contamination, which results in steadier discharge, lower gas consumption, and longer maintenance intervals. The discharge is also more homogeneous since the potential difference is distributed across the entire electrode surface. Capacitors can be used to limit the discharge current, rather than the ballast resistors of DC designs, thereby reducing resistive losses. Higher pulse frequencies and pulse modulation are possible, increasing the flexibility of operation. Since the separation of the electrodes is equal to the diameter of the tube, only a relatively low potential difference is required. However, an RF power source is more expensive than a DC source, and the resonator must be screened against emitted interference radiation. The power supply conversion efficiency is lower, resulting in higher electrical consumption than DC-excited lasers of similar output.

Alternating current (AC) excitation in practical gas lasers refers to frequencies in the range up to several hundred kHz. (Frequencies in the MHz range fall in the category of RF excitation.) AC excitation results in a rapidly changing electric field that promotes the conditions for maintenance of a glow discharge, allowing high electrical power densities to be produced and compact resonators to be constructed. The electrodes are mounted outside the discharge tubes, and so AC excitation enjoys similar benefits to RF excitation. The high spatial homogeneity of the discharge results in good beam quality and the wide stability range allows greater design freedom.

Optical Pumping

The difficulties associated with exciting insulating solids electrically mean that optical methods are preferred. The most common excitation sources are flashlamps, arc lamps and semiconductor lasers.

Flashlamps are glass or quartz tubes filled with a gas. Xenon is used when the output of the laser is on the order of watts. Krypton is more appropriate when a low current density discharge is needed, such as in continuous wave operation. Flashlamps provide a source of high intensity light, but a large part of the emission is not absorbed by the active medium, and is wasted as heat. The pulse repetition rate of a flashlamp is generally below 200 Hz. Flashlamps normally last between 500 and 1000 hours in continuous use.

Arc lamps are favoured when continuous wave operation is required. The lamps are filled with xenon and krypton. A higher pumping energy is required for CW operation because of the lower photon flux in the laser. Lifetimes between 400 and 1500 hours are typical.

As the cost of semiconductor lasers decreases, the use of diodes as a means of optical excitation is increasing. Diodes emit light at a fixed wavelength, which can be chosen to match the absorption bands of the active medium, resulting in significantly higher pumping efficiency than flashlamps and arc lamps. Diodes may also be located in novel orientations with respect to the active medium to maximize pumping efficiency.

Chemical Pumping

The energy produced by a chemical reaction, normally in the gaseous or liquid states, is used as the excitation method in chemical lasers. The reaction is initiated and sustained by a plasma or flashlamp, often via a mechanism involving photodissociation. Energy is transferred efficiently to the active medium by resonance.

Optical Cavity

In order to sustain laser action in a practical device, it is necessary to enclose the excited medium in an *optical cavity*. This is a container bounded by two mirrors. Five principal cavity parameters can be varied to optimize the output from a simple two-mirror optical cavity: the separation between the mirrors; the radii of curvature of the mirrors; and the reflectivities of the mirrors.

The simplest type of optical cavity is the Fabry–Perot interferometer – a container bounded by two parallel plane mirrors. Light travelling along the axis is reflected back and forth, the principal condition being that the spacing corresponds to an integral number of wavelengths. However, plane mirrors require exact alignment, and the reflective coating is critical in maintaining laser action. Spherical mirrors with a large radius of curvature are less sensitive to alignment, while still being able to fill the cavity, and are therefore commonly used in practical lasers.

The number of mirrors in the optical cavity is minimized to reduce losses that arise from imperfect reflection, and instabilities caused by temperature fluctuations. Cavities can be folded to lengthen the photon path, which increases power output while maintaining a small footprint, but the need for additional mirrors limits the number of folds in practical designs. Mirrors are normally made from water-cooled copper plated with silicon or gold, and have a large radius of curvature, typically tens of metres. The curvature of the mirrors determines the wavefronts that will oscillate in the cavity, and hence the modes of the beam that are supported.

Photons are repeatedly reflected through the active medium, which has two effects: the probability of stimulated emission is increased through an increase in the residence time of the photons; and feedback allows the emitted wave to grow coherently. Photons that do not travel parallel to the optical axis of the laser are quickly lost from the system; as a result the beam has low divergence. Reflections that are out of phase are lost through destructive interference, which maintains the coherence of the beam. Photons that do travel parallel to the axis have their path length considerably extended by optical feedback provided by the mirrors, before leaving the laser. This not only serves to amplify photon generation, but also produces a collimated beam of light. Losses in the cavity arise from a number of sources: transmission through the output coupler (the useful output); scattering by optical inhomogeneities in the active medium; absorption and scattering by the mirrors; diffraction around the perimeter of the mirrors; and absorption in the active medium by energy levels not involved in the laser transition.

The term *resonator* is used here to denote the combination of the optical cavity and the excitation device, together with the structure that holds and maintains the integrity of the optics. The resonator also includes devices that are inserted into the optical path to provide features such as pulsing capability, polarization control and mode control.

Stability of the Optical Cavity

The geometrical arrangement of the mirrors leads to the possibility of a large number of potential cavities capable of sustaining laser action. Cavities are classified as stable or unstable. A simple differentiation is that the beam converges in a stable cavity, whereas it diverges in an unstable cavity.

The main advantage of a *stable* cavity is that a fundamental mode (beam intensity distribution) can be generated that has standard measurable characteristics. High power, high order modes with a central intensity peak that are useful for material processing can be generated in a stable cavity. Rays of light are focused in a stable cavity. The mode remains constant as the beam propagates, or is focused. However, since the light rays pass though a waist between the mirrors, a stable cavity has a relatively small effective mode volume, which limits the power that can be generated. Examples of stable cavities are illustrated in Fig. 3.6.

A *confocal* cavity uses two spherical mirrors of equal radius, with coincident foci. This relaxes alignment tolerances somewhat, and reduces diffraction losses. A *spherical-flat* configuration is popular in high power lasers because of the relative ease of alignment and its good mode-filling characteristics. The radius of curvature of the focusing mirror is several times the cavity length. The beam can be extracted from a stable cavity by a partially reflecting transmissive window.

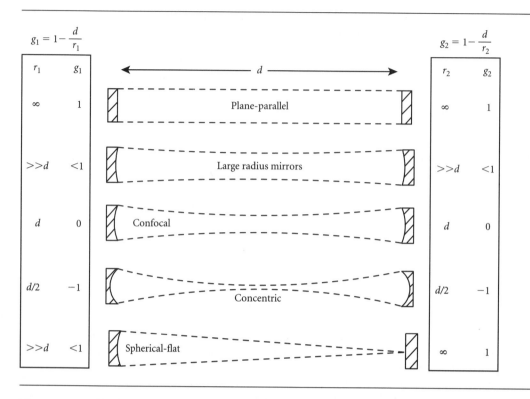

Figure 3.6   Stable optical cavities (r is the radius of curvature of a mirror and g is a geometrical factor used to characterize the stability of the cavity)

In an *unstable* optical cavity, radiation is not confined to a narrow beam, but is defocused as it is reflected between the cavity mirrors, filling the entire cross-section of at least one mirror. Rays of light diverge within the cavity. Examples of unstable cavities are illustrated in Fig. 3.7.

The most widely used unstable cavity design is the *positive branch confocal* cavity, which comprises a large concave mirror and a smaller convex mirror, around which the beam exits the cavity. The disadvantage with the *negative branch* design, which comprises two concave mirrors, is that the beam is brought to focus between the mirrors, reducing the active volume, and causing disruptive plasma generation. The diameter of the intensity distribution is determined by the internal apertures. The beam is normally extracted from an unstable cavity by using an annular scraper mirror. Unstable cavities are capable of producing a variety of beam intensity distributions, but the size of the resonator normally means that all but the lowest order distributions are eliminated. The intensity distribution in the output beam is annular when a scraper mirror is used. Unstable cavities have a number of advantages over stable designs: a single beam intensity distribution is possible even with a wide cavity; energy can be extracted from a large resonator volume in short resonators; and partially reflecting elements, which are expensive and sensitive to operating conditions, can be eliminated. The annular intensity distribution from an unstable resonator fills in at its centre as it propagates, or when it is focused. This can result in a near-Gaussian beam intensity distribution at the focus. Unstable cavity design is a compromise between power and beam quality.

As a general rule, a cavity is stable if the centre of curvature of one mirror, or the mirror itself, but not both, falls between the other mirror and its centre of curvature. In mathematical terms, a cavity is

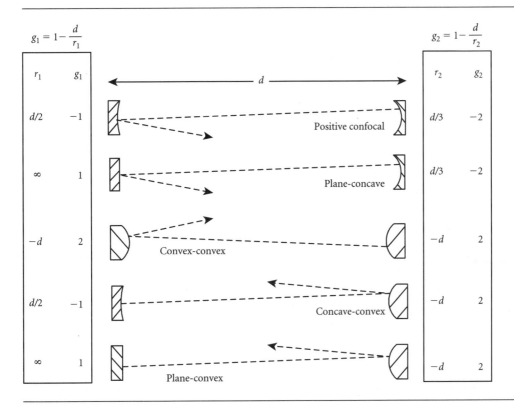

$$g_1 = 1 - \frac{d}{r_1}$$

$$g_2 = 1 - \frac{d}{r_2}$$

r_1	g_1		r_2	g_2
$d/2$	-1	Positive confocal	$d/3$	-2
∞	1	Plane-concave	$d/3$	-2
$-d$	2	Convex-convex	$-d$	2
$d/2$	-1	Concave-convex	$-d$	2
∞	1	Plane-convex	$-d$	2

Figure 3.7   Unstable optical cavities (r is the radius of curvature of a mirror and g is a geometrical factor used to characterize the stability of the cavity)

stable if the following condition is met:

$$0 \le \left[1 - \frac{d}{r_1} \right] \left[1 - \frac{d}{r_2} \right] \le 1 \tag{3.3}$$

where d is the distance between the mirrors and r_1 and r_2 are the radii of curvature of the mirrors. This can be written:

$$0 \le g_1 g_2 \le 1 \tag{3.4}$$

where $g_i = 1 - d/r_i$. The conditions required for stability, and various cavity configurations are shown in Fig. 3.8. Note that stable cavity designs that lie close to the stability boundary may become unstable with slight changes, such as a variation in the mirror curvature caused by thermal expansion.

An etalon – a piece of glass fabricated such that the two surfaces are parallel – may be inserted in the optical cavity to ensure that it operates in a single mode. The etalon effectively operates like an internal Fabry–Perot interferometer.

Resonator Support

The optical cavity and other components of the resonator are supported by a structure designed to minimize relative movements, both linear and angular, which would result in instabilities in power output, beam mode and pointing. The structure is made from a material with a very low coefficient of thermal expansion, such as Invar, in order to minimize changes in cavity dimensions during operation.

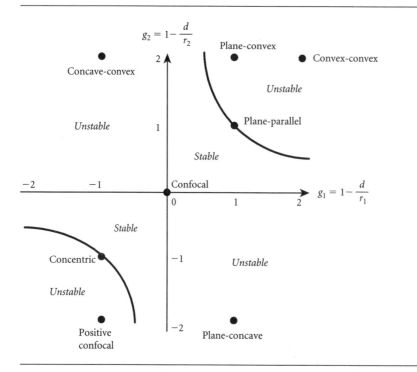

Figure 3.8   Conditions for stable and unstable cavities in terms of the spacing of the cavity optics, d, and the radii of curvature, r_1 and r_2

The resonator structure is also insulated from mechanical forces arising from dimensional changes and vibrations.

Output Devices

Light is extracted from the optical cavity through an output coupler. This is a 'window', which may be a partially transmitting solid, a fully transmitting gas, or a diffraction grating.

Transmissive solid output couplers are popular in relatively low power lasers since they can be made to be wavelength specific, transmitting a fraction of the cavity light within a limited range of frequency. The remainder is reflected back into the cavity. An antireflection coating, such as lead fluoride, is applied in order to achieve the desired reflectivity.

A scraper mirror is used to extract the beam from higher power unstable resonators. The beam is focused through an orifice across which an isolating curtain of high velocity, dry, compressed air flows – an aerodynamic window.

The broad emission spectrum of some lasers enables the output to be tuned if a diffraction grating is used for the output coupler.

OUTPUT

Earlier, we learned that laser light has four main characteristics that differentiate it from the light produced from, for example, an electrical light bulb: it is monochromatic; coherent, has low divergence; and has high brightness.

Monochromatic light effectively has a single wavelength – light is emitted in a well-defined segment of the optical spectrum. In practice, an industrial laser operates in a very narrow band of wavelength around a central peak. Emission is said to occur on several *lines* within a narrow band. The monochromatic nature of laser light is the basis of applications such as measurement, alignment and holography.

The beam from the laser normally converges to a waist as it leaves the resonator, where its diameter is a minimum, after which it diverges along the beam path. The tendency for the beam diameter to expand away from the waist is a measure of the beam divergence. Low divergence is the property that enables a laser beam to retain high brightness over a long distance, and is the basis of alignment systems. (A beam of red light emitted from a laser situated on earth may be only about 1 kilometre wide when it reaches the moon, situated at a distance of 390 000 km.)

Coherent radiation comprises waves travelling with the same wavelength, amplitude and wavefront. It is a measure of the degree to which light waves are in phase in both time and space. Laser radiation has high coherence. Spatial coherence is a measure of the difference in the spatial position of waves. Coherent laser light is up to 100 000 times higher in intensity than incoherent light of equivalent power, since the divergence, or dispersion, of energy is very low as the beam propagates from the laser. Because light propagates with a fixed velocity, a temporal coherence can be defined, which is a measure of the difference in time between waves emitted from a single source that produce stationary interference patterns. Coherence is the basis of applications in measurement and holography.

Thermal mechanisms of material processing take advantage of the high brightness (high power density) of a laser beam. *Athermal* (photonic) mechanisms are based on the short wavelength (high energy) of the beam, and the short duration of the pulses that can be produced. The beam characteristics influence the beam propagation and focusability, and therefore have an important effect on the suitability of the beam for material processing.

The characteristics of the emitted beam are determined by the cavity optics, the optical properties of the active medium, and apertures and devices placed within the resonator. The beam can also be manipulated using optical devices placed outside the resonator. A propagating light wave must satisfy the complex wave equation

$$\nabla^2 U - \frac{1}{c^2}\frac{\partial^2 u}{\partial t^2} = 0 \qquad (3.5)$$

where U is the complex amplitude of the wave, and takes the form

$$U(r,t) = a(r)\exp[i\varphi(r)]\exp[i2\pi vt]. \qquad (3.6)$$

The intensity, $I(r)$ is given by $I(r) = |U(r)|^2$.

Spatial Mode

Two spatial modes are commonly used to describe the beam: longitudinal and transverse. They are essentially independent of each other, since the transverse dimension in a resonator is normally considerably smaller than the longitudinal dimension.

Only light with a wavelength that satisfies the standing wave condition, $q\lambda = 2d$, will be amplified in the cavity, where q is a large integer referring to the number of nodes in the longitudinal standing wave, d is the cavity length (mirror separation), and λ is the wavelength. The longitudinal mode number is large in industrial lasers and is normally ignored when characterizing the beam since it has little influence on the essential beam characteristics and performance. The transverse electromagnetic mode (TEM) is of far greater significance.

The TEM describes the variation in beam intensity with position in a plane perpendicular to the direction of beam propagation. It characterizes the intensity maxima in the beam remote from its

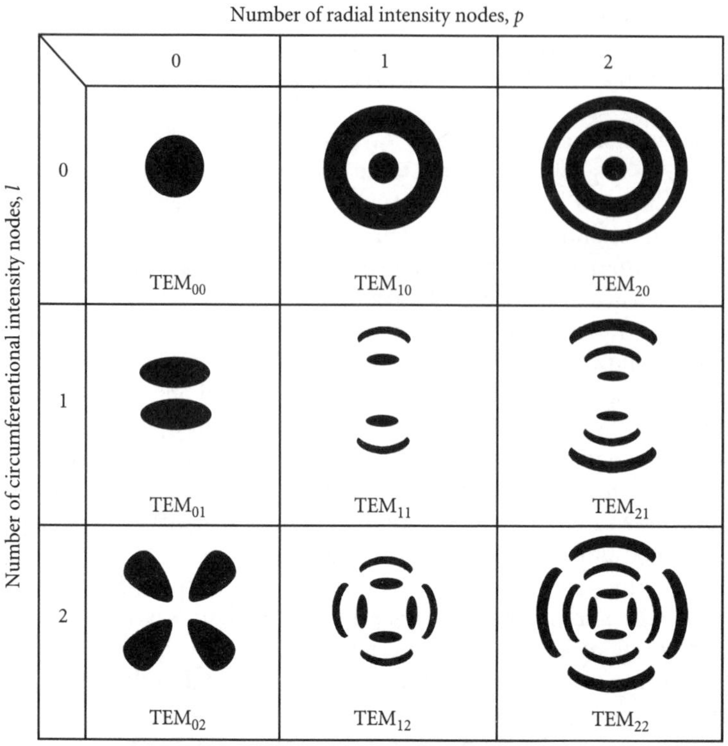

Number of radial intensity nodes, p

Figure 3.9 Transverse sections of transverse electromagnetic modes of circular symmetry

central axis. The TEM is determined by: the geometry of the cavity; the alignment and spacing of internal cavity optics; the gain distribution and propagation properties of the active medium; and the presence of apertures in the resonator. In gas lasers, gas flow and electrical discharge also influence the mode generated. The TEM is described by a set of subscripts that depend on the symmetry of the beam.

Cylindrical Symmetry

For a beam with cylindrical symmetry, the subscripts of the TEM are p, l and q. q denotes the number of nodes (field zeros) in the standing wave pattern along the longitudinal (z) axis, and is not normally quoted. p and l indicate the number of nodes along the radius of the transverse beam section, and around the circumference of the central power ring, respectively, as illustrated in Fig. 3.9.

Mathematically, a cylindrical mode is defined by a Gaussian distribution multiplied by Laguerre polynomials, to denote the Laguerre–Gaussian mode. The equation for the complex amplitude of the mode is:

$$U_{l,m}^{LG}(r, \phi, z) = C_{lm}^{LG}(1/w) \exp\left[-ik\frac{r^2}{2R}\right] \exp\left[\frac{-r^2}{w^2}\right] \exp[-i(l + m + 1)\psi]$$

$$\times \exp[-i(l - m)\phi](-1)^{\min(l,m)} \left[\frac{r\sqrt{2}}{w}\right]^{|l-m|} L_{min(l,m)}^{|l-m|}\left(\frac{2r^2}{w^2}\right)$$

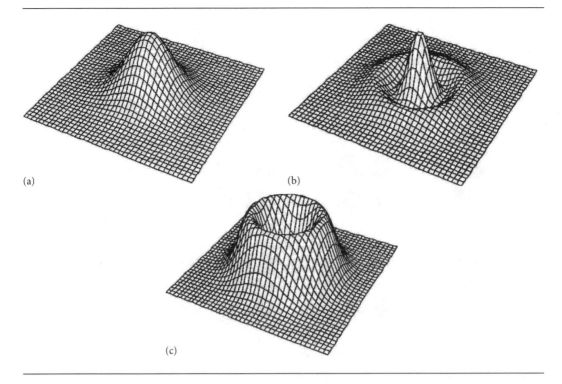

(a)

(b)

(c)

Figure 3.10   Intensity distributions in cylindrical (a) TEM$_{00}$, (b) TEM$_{10}$ and (c) annular (TEM$_{01*}$) beam modes

$$L_n(x) = \frac{e^x}{n!} \frac{d^n}{dx^n}(x^n e^{-x})$$

$$C_{lm}^{LG} = \left(\frac{2}{\pi \, l! m!}\right)^{1/2} min(l, m)! \tag{3.7}$$

The lowest order mode, TEM$_{00}$, refers to a beam with a Gaussian intensity distribution about a central peak, illustrated in Fig. 3.10a. The diameter of a TEM$_{00}$ beam of cylindrical symmetry can be defined by the points at which the intensity, I, has fallen to a given fraction of the peak intensity, I_0. The fractions $1/e$ or $1/e^2$ are often quoted for safety standards and manufacturing specifications, respectively, at which points the intensities have fallen to 36.8% and 13.5% of the peak, respectively.

The first order mode, TEM$_{10}$, refers to a central intensity distribution surrounded by an intensity annulus, Fig. 3.10b. Definitions of higher order beam diameters have been proposed, although there is currently no accepted standard. This is a major obstacle to comparing focused spot sizes between different beam modes. (The beam waist is not a good description of beam diameter since it is dependent only on the laser cavity, and is independent of the beam mode.)

A beam with an annular intensity distribution can be produced from an unstable optical cavity because of the nature of the scraper mirror used to extract the beam. This is often referred to as TEM$_{01*}$, illustrated in Fig. 3.10c. TEM$_{01*}$ is strictly not a true mode since the intensity distribution changes between the near and far fields. An asterisk denotes the superimposition of two degenerate modes, TEM$_{01}$ and TEM$_{10}$, one rotated 90° about its axis relative to the other, which combine to form a composite intensity distribution of circular symmetry. An annular beam is characterized by a magnification, M, defined by the ratio of the outer diameter to the inner diameter. The magnification determines the focusability of the beam from an unstable resonator, in much the same way that the

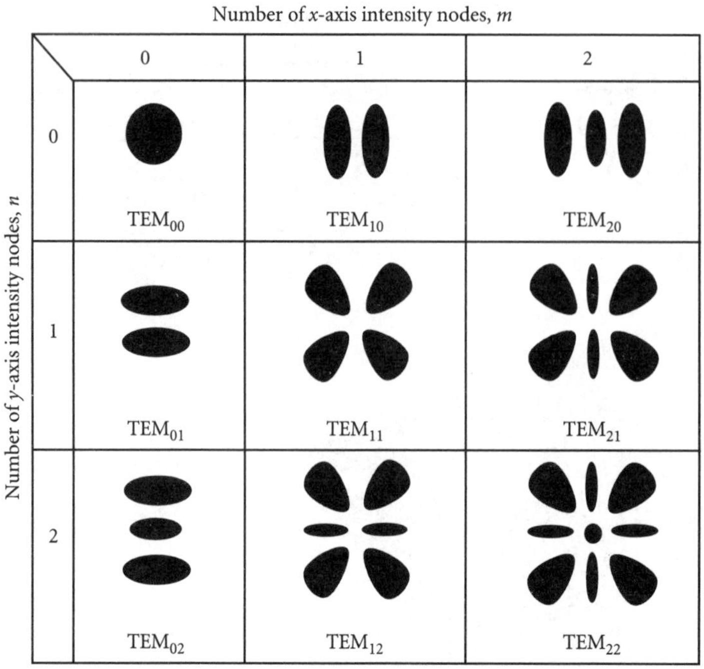

Number of x-axis intensity nodes, m

Figure 3.11   Transverse sections of beam modes of rectangular symmetry

indices defining the transverse electromagnetic mode characterize beam focusability from a stable resonator. Focusability increases with magnification, whereas maximum power is typically obtained with relatively low values of M lying between 1.6 and 1.7. Unstable cavity designs are therefore a compromise between beam power and beam focusability.

Rectangular Symmetry

Cavities containing mirrors of circular cross-section normally produce cylindrically symmetrical transverse modes. However, square mirrors, the presence of Brewster-angle windows, or mirror misalignments often cause optical cavities to oscillate with rectangular symmetry. End-pumped solid state lasers can produce rectangular modes because resonance can be sustained along off-axis ray paths within the cavity. In the case of rectangular symmetry, the subscripts m, n and q denote the number of nodes in the x, y and z directions, respectively, of a transverse section of the spatial intensity profile. The x direction is defined as the wider dimension, and the measurement is taken across the entire width/length of the pattern. Figure 3.11 shows a number of transverse beam modes of rectangular symmetry.

Mathematically, the rectangular mode is constructed by multiplying a Gaussian distribution by Hermitian polynomials, to define the Hermite–Gaussian mode. The equation for the complex amplitude of the mode is

$$U_{m,n}^{HG}(x, y, z) = C_{mn}^{HG}(1/w) \exp\left[-ik\frac{x^2 + y^2}{2R}\right]$$

$$\times \exp\left[-\frac{x^2 + y^2}{w^2}\right] \exp[-i(m + n + 1)\psi] H_m\left(\frac{x\sqrt{2}}{w}\right) H_n\left(\frac{y\sqrt{2}}{w}\right)$$

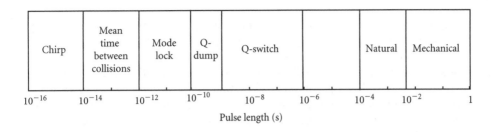

Figure 3.12 Time scales of pulsed laser output

$$H_m(u) = (-1)^m e^{u^2} \frac{d^m}{du^m}(e^{-u^2})$$

$$C_{mn}^{HG} = \left(\frac{2}{\pi n! m!}\right)^{1/2} 2^{-N/2}. \tag{3.8}$$

Again, the lowest order mode, TEM_{00}, refers to a beam with a Gaussian intensity distribution about a central peak. Higher order modes contain intensity peaks along the x axis, the y axis, or both.

Fresnel Number

The size of the aperture in relation to the cavity length determines the dominant mode produced by the laser. A narrow cavity favours low order mode operation because higher order modes are attenuated by the inner walls of the cavity. The Fresnel number, N_F, is a measure of the tendency for a stable laser cavity to operate low order modes:

$$N_F = \frac{a^2}{\lambda d} \tag{3.9}$$

where a is the radius of the smallest aperture in the system, λ is the wavelength, and d is the length of the cavity. An optical cavity with a low Fresnel number favours low order mode operation.

Temporal Mode

The temporal mode of the light emitted from a laser is determined by the number of energy levels in the active medium, their lifetimes, and the source of excitation. Three-level lasers such as ruby and dyes naturally produce pulses of light. Carbon dioxide and Nd:YAG lasers (four-level systems) are able to produce continuous wave output, but the temporal output mode may be changed through the use of various devices that can be inserted into the resonator, illustrated in Fig. 3.12 giving pulsed output down to the femtosecond (10^{-15} s) scale.

The simplest form of pulsed output is obtained by gating, or chopping the beam, which may be achieved through modulation of the excitation power (normal pulse), or by external attenuation of a CW beam. The pulse is characterized by its peak power, shape, length and repetition rate. From these quantities, the pulse period (the reciprocal of the repetition rate), the pulse energy (the area under the power–time plot), and the duty cycle (the ratio of the pulse width to the pulse period), can be obtained. Enhanced pulses, or super pulses, denote the superimposition of a high power spike pulse on the leading edge of a gated or normal pulse, or on a CW beam.

Q-switching

The Q-factor, or quality factor, of a laser cavity is the ratio of the energy stored to the energy lost per cycle. A high value of Q indicates that energy is easily stored within the cavity, whereas a low value means that the contained energy will emerge rapidly. Q-switching is a technique by which short pulses of high peak power can be created in a laser with a continuous excitation source. The optical cavity contains a shutter, which when closed prevents laser action. However, during this time excitation energy continues to be absorbed, and more species are raised to the upper laser energy level, which increases the population inversion. When the shutter is opened a large number of excited species become available for stimulated emission, producing a large burst of energy until the upper level is depleted. The Q-switching mechanism must operate rapidly in comparison with the build-up of laser oscillations. The Q-switch is placed in the resonator between the more reflective back mirror and the active medium in order to maximize the blocking of laser action. Q-switching is commonly used with lasers in which the lifetime of the upper lasing level is long enough to store a significant amount of energy during blocking, notably high power solid state lasers. The technique is used in marking applications and some drilling operations. Four types of Q-switch are commonly used in lasers: mechanical; acousto-optic; electro-optic; and dye.

Mechanical Q-switches, such as rotating mirrors and mechanical choppers, were the first to be developed. Optical losses within the cavity are high except for the brief interval when the mirrors are parallel, or the chopper does not block the beam path. Feedback can also be provided by total internal reflection from the inner surface of a prism. These were popular in the early days of lasers. However, because of the rapid mechanical action, their mechanical reliability is poor, they are limited in the length of pulse that can be produced, and they are difficult to synchronize with outside events. The pulse length produced is relatively long, on the order of milliseconds.

An acousto-optical switch is an alternative that is based on the change in the refractive index of a medium caused by the mechanical strains that are induced by an acoustic wave as it travels through the medium. In effect the acoustic wave sets up a diffraction grating that can be used to deflect a beam of light. The medium is a transparent block of material, such as fused quartz, to which is attached a piezoelectric transducer. It is used for pulses on the order of microseconds.

Electro-optical Q-switches are based on the behaviour of polarized light as it passes through certain electro-optic materials when they are subjected to electric fields. Light passes through the material unchanged when no voltage is applied. However, when a voltage is applied, the polarization of the light is rotated 90°. A second element is a polarizer, which passes light that is in its original form, but not in its rotated form. When the voltage is applied, energy is stored in the population inversion. When the voltage is switched off, the stored energy is emitted as a pulse. Practical examples are the Pockels cell, which is a crystalline wave plate made from a material such as potassium dihydrogen phosphate, and the Kerr cell, which uses a liquid medium, such as nitrobenzene, to provide the phase retardation. Electro-optical Q-switches are more expensive than acousto-optical devices, but they can be used to produce very short pulses, on the order of nanoseconds.

Passive Q-switching is based on the action of saturable-absorbers. These are materials (often dye solutions) whose absorption depends on the incident light intensity. At high intensities the dye bleaches and allows light to pass through, thus allowing the formation of a giant pulse after a significant amount of energy has been stored. They are simple and inexpensive, but suffer from pulse jitter, dye degradation and synchronization difficulties.

Cavity Dumping

A cavity-dumped laser contains a switch and two fully reflecting mirrors. When the switch is open, laser action is permitted, and energy is built up within the cavity. A voltage is then applied to the switch, causing the polarization of the light passing through it to be rotated. Light is then emitted as a cavity-dumped pulse. The output coupler is thus the switch and not one of the mirrors. The pulse length is proportional to the round-trip time for photons, i.e. the length of the cavity. Pulses of nanosecond duration can typically be produced.

Cavity dumping is used to produce high energy pulses in lasers in which the lifetime of the upper lasing level is too short to enable Q-switching to be used, e.g. in argon and krypton ion lasers. It can also be used to produce very short or very high frequency pulses from lasers in which the desired pulse characteristics cannot be obtained through Q-switching alone.

Mode Locking

The number of axial modes in the output of a laser beam increases with the strength of excitation. By exciting the medium to just above a threshold level, light of a single wavelength is produced. However, if the excitation energy is increased significantly above the threshold level, emission at several wavelengths can then be produced. Mode locking refers to the use of a modulating optical element to lock particular oscillation modes into phase, thus producing a train of pulses. The optical element is modulated at a frequency that matches the time taken for photons to travel the length of the optical cavity and back. The modulator is opened once per round trip, letting the pulse through. The resonance of these locked modes results in very short pulses of high intensity. Individual pulses can have different pulse lengths.

Mode locking is achieved using a fast optical gate. Electro-optic, acousto-optic and dye switches can be used. In the case of the last type, the dye absorbs the radiation in the cavity, except when all the modes are in step, both in phase and in spatial location. When this occurs, more energy is supplied to the dye than it can absorb and dissipate in the time of passage of the pulse through the dye. All the dye molecules are simultaneously in an excited state, making it saturated and allowing a part of the pulse to pass through. The length of the cell containing the dye is chosen to achieve the necessary condition for saturation. Once such a mode-locked pulse occurs, part of it is fed back into the laser and with sufficient gain it saturates the dye each time it passes through. Thus, once started, the pulse remains in the laser as long as pumping continues.

The duration of the mode-locked pulses depends on several factors, including the bandwidth of light generated and the effectiveness of the modulator. Nd:YAG lasers, with relatively narrow bandwidths, produce mode-locked pulses of duration 30–60 ps, whereas high bandwidth dye lasers can produce pulses of duration 0.1 ps. In comparison with Q-switching and cavity dumping, mode locking produces the shortest pulse durations. Mode locking can be combined with other pulsing techniques, or used on its own.

Chirping

Chirping is the rapid changing – in contrast to long-term drifting – of the frequency of an electromagnetic wave, most often observed in pulsed operation of a source. It is a pulse compression technique that uses frequency modulation during the pulse. It is used in femtosecond-scale pulsed lasers, such as the Ti:sapphire source.

Frequency Multiplication

The fundamental wavelength of light produced by a laser, λ, is related to the energy of the photons, E, through the formula $E = hc/\lambda$, where h is Planck's constant (6.626×10^{-34} J s^{-1}) and c is the velocity of light (2.998×10^8 m s^{-1}). The wavelength can be converted into a frequency, v, by using the formula $v = c/\lambda$.

Some crystalline materials and liquids interact with light in a manner that results in the generation of a new frequency that is a multiple of the fundamental. Thus light of one wavelength can be transformed into light of another frequency. Frequency multiplication occurs in materials that exhibit a non-linear response to an electric field. An analogy is an atom in a crystal that is bound in a potential well that acts like a spring. In a linear crystal, as light interacts with the electrons, they are displaced by an amount proportional to the energy of the light. In a non-linear crystal this proportionality does not exist; when the electrons are displaced, the restoring force is no longer proportional to the driving energy. Oscillations then occur at frequencies other than that of the incident light, producing harmonics. An electromagnetic component can be produced that oscillates at twice the rate of the original wave, often with a polarization orthogonal to the fundamental. In order to obtain output of useful power – up to 50% of the incident intensity – a direction in the crystal is found at which the velocity of the fundamental beam matches that of the harmonic, i.e. the phases are matched. Thus green light of wavelength 532 nm can be produced from 1064 nm infrared light produced by an Nd:YAG laser.

Light with third and higher order harmonics can also be generated, normally in multiple-step processes. The efficiency decreases with increasing order. The main benefit of harmonic generation for material processing is the improved absorptivity of most metals at shorter wavelengths.

Raman Effect

When a beam of monochromatic light passes through a transparent substance the beam is scattered. The scattered light is not monochromatic, but has a range of wavelength that is shifted relative to that of the incident light. In terms of quantum theory, as a stream of photons collides with a particular molecule the photons will be deflected without change in energy if collisions are perfectly elastic. However, if energy is exchanged between the photon and the molecule, the collision is inelastic. The molecule can gain or lose discrete amounts of energy in accordance with quantal laws; the energy change must coincide with a transition between two molecular energy levels. The effect, known as Raman shifting, is used to change the frequency of laser light, enabling the output to be tuned.

Propagation

The propagation axis denotes the centreline of the beam of radiation. It is determined by the orientation of the resonator optics. The propagation of a Gaussian beam may be expressed in terms of the increase in beam radius, r_B, with distance from the laser, or the distance beyond the focal point of a focusing optic. The hyperbolic envelope created is known as the beam caustic. The radius of a beam can be written as a function of the radius of the focused beam, r_f, and the distance from the focal point, z:

$$r_B = r_f \left[1 + \left(\frac{\lambda z}{\pi r_f^2} \right)^2 \right]^{1/2} \tag{3.10}$$

where λ is the wavelength of the beam.

The beam radius can only be defined uniquely for a fundamental Gaussian beam. However, a modification to account for higher beam modes of order TEM_{pl} can be written

$$r_B = r_f (2p + l + 1)^{1/2} \left[1 + \left(\frac{\lambda z}{\pi r_f^2} \right)^2 \right]^{1/2} \tag{3.11}$$

where r_B defines the radius of a circle containing a defined amount of the beam. It can be considered comparable with the $1/e$ definition of a TEM_{00} beam.

Waist

The beam waist refers to the minimum diameter of the beam. The location of the waist, z_w, in relation to the output coupler, is given by:

$$z_w = \frac{dg_1(1 - g_2)}{g_1 + g_2 - 2g_1g_2} \tag{3.12}$$

where g_1 and g_2 are defined by the characteristics of the optical cavity, as shown in Fig. 3.8, and d is the length of the optical cavity.

The waist normally lies inside the optical cavity. If the optical cavity contains a flat mirror the waist is located at the mirror.

Focused Spot Size

The diameter of a focused beam is directly proportional to its wavelength and inversely proportional to the numerical aperture of the objective lens. The numerical aperture is a value that depends on the diameter of the focusing optic, its radius of curvature and the material from which it is made. If we wish to minimize the focused spot diameter, we select a beam of short wavelength and an objective lens with a large numerical aperture. More details of the properties of a focused beam can be found in Chapter 4.

Rayleigh Length

The Rayleigh length or range, z_R, is the distance along the path of propagation from the beam waist to the plane in which the beam diameter exceeds the beam waist diameter by a factor of $\sqrt{2}$. It characterizes the near field or collimated region of the beam, and is defined for a TEM_{00} mode beam as

$$z_R = \frac{4\pi r_f^2}{\lambda} \tag{3.13}$$

where λ is the wavelength and r_f is the radius of the focused beam. Beyond the Rayleigh length the beam will expand at a constant rate or angle – the far field beam divergence. A Gaussian beam has the largest Rayleigh range, and the smallest far field divergence.

The Rayleigh range for higher order mode beams can be expressed in terms of the beam's quality factor, K (which is defined below):

$$z_R = \frac{4\lambda}{\pi} \frac{f^2}{K} \tag{3.14}$$

where f is the focal number of the optic, given by $f = F/d_B$. (F is the focal length and d_B the diameter of the optic.) It is a useful scale unit for measuring propagation distance beyond an optic.

Radius of Curvature

One of the characteristics of laser light is its coherence. If a surface is constructed containing all the points of common phase in a Gaussian beam, that surface would be a sphere with a particular radius of curvature, R. As the beam propagates, the radius of curvature changes; it is infinite at the beam waist, decreasing sharply after the waist to a minimum at the Rayleigh length, after which it increases again.

At large distances it is equal to the distance from the waist. The variation of R with distance from the waist, z, is given by:

$$R = z \left[1 + \left(\frac{\pi r_f^2}{\lambda z} \right)^2 \right] \tag{3.15}$$

where λ is the wavelength and r_f is the beam radius at the waist.

If the mirror curvature exactly matches the radius of curvature of a Gaussian beam, the energy in the wave, which travels perpendicular to the wavefront, will be reflected back on itself, and the resonator will be stable.

Fields

The terms *far field* and *near field* are frequently used when describing laser beams. The intensity distribution across the transverse beam cross-section at the exit of the laser is known as the near field mode. The beam propagates from the laser according to the laws of optics, but diffraction effects tend to modify the intensity distribution. Eventually a point is reached at which the beam has spread to such a degree that its area is considerably larger than that predicted by optical calculations, and diffraction effects totally dominate the intensity distribution. This is the far field distribution, and generally occurs at a distance of around five Rayleigh lengths from the beam waist.

Divergence

Divergence is a measure of the tendency for the beam to spread as it propagates from the laser. Since the beam emitted from many commercial gas lasers is symmetrical, divergence is normally measured in plane angles (radians), rather than solid angles (steradians). The divergence, θ, of a Gaussian beam of wavelength λ, after it has passed through the beam waist of diameter d_B, is given by:

$$\theta = \frac{2}{\pi} \frac{\lambda}{d_B}. \tag{3.16}$$

The larger the beam waist diameter, the smaller the divergence.

If the distance between the laser and the workpiece is large, the beam divergence should be small, preferably less than 1.0 mrad (half angle). Short wavelength lasers are therefore better for small divergence applications. A TEM$_{00}$ beam mode has the lowest beam divergence. Low divergence results in a smaller focused spot and a greater depth of focus.

Any system that moves optics along the beam path must take divergence into account, since the size of the beam at the focusing optic varies. Divergence is typically 1 mrad for a TEM$_{00}$ beam and 20 mrad for a multimode beam. A value of 2–3 mrad is common for industrial CO_2 lasers. Beam divergence has implications for the size of the optics that must be used; the beam can grow significantly over several metres in a large workstation. For example, the 35 mm diameter TEM$_{20}$ beam emitting from a CO_2 laser can grow to around 100 mm over a path length of 40 m.

Quality

The minimum size to which a laser beam can be focused is the diffraction limit, which refers to the minimum diameter of a (Gaussian) TEM$_{00}$ beam, given by λ/π, where λ is the beam wavelength. The quality of a beam is a measure of its focusability (spot size and focal length), and can be measured in various ways.

The K factor expresses beam focusability in terms of that of a TEM_{00} beam:

$$K = \frac{\lambda}{\pi} \frac{4}{d_B \theta} \qquad (3.17)$$

where d_B is the diameter of the incident beam and θ is the full beam divergence angle. $K = 1$ for a TEM_{00} beam, and is less than 1 for higher beam modes. Industrial gas laser beams typically have K values in the range 0.7–0.2. The closer the K factor is to 1 the better the quality of the laser beam. This notation is particularly popular in Germany.

An analogous beam quality system uses the M^2 notation, where

$$M^2 = \frac{\pi}{\lambda} \frac{d_B \theta}{4}. \qquad (3.18)$$

$M^2 = 1/K$ for a TEM_{00} beam ($K = 1$). A beam with an M^2 value of 1.2 can be thought of as being 1.2 times diffraction limited, and would produce a focused waist diameter 20% larger than the TEM_{00} mode. M can be calculated approximately for higher modes of circular symmetry, TEM_{pl}, using the formula

$$M = \sqrt{2p + l + 1}. \qquad (3.19)$$

For a pure TEM_{20} beam, $M^2 = 5$. The lower the M^2 value, the higher the beam quality. This notation is particularly popular in the United States. (M^2 is used rather than M because it represents the ratio of the divergence angle to that of a TEM_{00} beam.) The M^2 factor is sometimes referred to as the Q-factor (distinct from the cavity quality factor). (The M^2 factor should not be confused with the magnification of an annular beam, for which the symbol M is normally used.)

The beam parameter product, BPP, is normally quoted when discussing the quality of a laser beam produced from a solid active medium, or delivered from a fibre optic. The beam parameter product is proportional to the beam diameter (the fibre diameter of a fibre optic) and the beam divergence angle, and is defined as

$$BPP = \frac{d_B \theta}{4} = M^2 \frac{\lambda}{\pi}. \qquad (3.20)$$

BPP is measured in units of mm · mrad.

The benefits of a high beam quality for material processing are three-fold. A small focal diameter gives high process efficiency, low energy input, and narrow cut kerfs and welded seams. A slim processing head can be constructed, which provides high processing flexibility. A large working distance enables remote processing to be carried out in several locations sequentially, and a large depth of focus improves tolerance to the position of the focal plane during processing.

Bandwidth

The emission line used for laser operation has a finite spectral width. A certain amount of line broadening, or bandwidth, is associated with any form of electromagnetic radiation. The distribution of frequencies about the line defines the emission line shape. The bandwidth defines the degree of monochromaticity of the beam. It can be measured in terms of wavelength, frequency, wave numbers or coherence length. The wave number refers to the number of wavelengths that fit into 1 cm. The coherence length is the distance over which the laser remains sufficiently coherent to produce interference fringes; it is inversely proportional to the bandwidth expressed in frequency or wavelength, and is equal to the reciprocal of bandwidth in wave numbers. In terms of wavelength, there are a number of discrete lines either side of the central peak. The bandwidth measurement is made at an intensity of half that of the peak, and is defined as the full width half maximum (FWHM) measurement.

In a homogeneously broadened laser each individual atom has a bandwidth equal to the total laser bandwidth. If a particular photon can interact with one of the atoms, it can interact with all of them. Homogeneous broadening is mainly caused by collisions of gas molecules with each other, other species in the mixture, or the walls confining the gas. These cause perturbations in the energy of the photons emitted. As the pressure in the laser increases, the bandwidth of the laser increases because of the increased number of collisions in a given time. In solid state lasers, thermal line broadening is caused by interactions between the laser species and vibrations in the crystal lattice. Lattice vibrations are quantized, and referred to as phonons. Bandwidth is normally relatively easy to reduce in a homogeneously broadened laser because all the atoms can still contribute to stimulated emissions in the narrower bandwidth.

Heterogeneous broadening refers to a condition by which different atoms contribute to the gain at different frequencies. This may be a result of inhomogeneities in the active medium, such as defects, which cause the environment to vary from point to point in a solid. The Doppler effect is a source of heterogeneous broadening, and is significant in most gas lasers. Since the individual atoms are moving in random directions, at random speeds, their total emission covers a range of frequencies, in the same way that sound from a moving object changes frequency depending on the relative motion with the observer. The hotter the gas, the broader the bandwidth. Gaussian lines are generally broadened heterogeneously. Those atoms that contribute to gain outside the reduced bandwidth cannot be stimulated to emit in the narrowed bandwidth, and therefore the total power is reduced.

The bandwidth may be made narrower by cooling the active medium to reduce thermal broadening in a solid state laser, or to reduce Doppler broadening in a gas laser. The feedback of the resonator may also be modified to control the laser bandwidth. This may be achieved by using mirrors with a narrow bandwidth of reflection. A prism inside the cavity may also be used to direct only light at the centre of its bandwidth towards the mirrors. One of the mirrors may be replaced with a grating, which reflects different wavelengths at different angles. When aligned correctly, only light with a wavelength at the centre of the population inversion is reflected.

Coherence

A light source emits a sequence of light quanta, each of a certain length – the coherence length. This is the distance that the light will travel before its coherence changes. The coherence length, l_{coh}, depends on the wavelength, λ, and the bandwidth, $\Delta\lambda$:

$$l_{coh} = \frac{\lambda^2}{\Delta\lambda}. \qquad (3.21)$$

Two waves can only interfere when light quanta from the same emission process interact. Therefore, natural light sources which have a very large bandwidth, and consequently a very short coherence length, do not exhibit interference phenomena.

A coherence time, t_{coh}, can then be defined:

$$t_{coh} = \frac{l_{coh}}{c} \qquad (3.22)$$

where c is the velocity of light. Coherence time can be understood as the temporal difference between two light waves, originating from the same source, that produce stationary interference patterns. This aspect of coherence considers only point sources and interference between temporal shifted waves, and therefore is referred to as temporal coherence.

Brightness

Brightness, B, is a measure of the intensity of light at a particular location. It is defined as the emitted power, q, per unit area, A, per unit solid angle, Ω:

$$B = \frac{q}{A\Omega}. \qquad (3.23)$$

Brightness depends on the intensity of the source and the extent to which the light diverges after leaving the source. Since the laser can produce very high levels of power in very narrowly collimated beams, it is a source of high brightness energy.

Intensity

The intensity, I, obtained by focusing a beam of light is directly proportional to the brightness, B:

$$I = B\frac{\pi w_0^2}{F^2} \qquad (3.24)$$

where F is the optic focal length and w_0 is the beam waist diameter.

The beam intensity can be related to temperature through the Stefan–Boltzmann law:

$$I = \sigma T^4 \qquad (3.25)$$

where σ is the Stefan–Boltzmann constant (5.67×10^{-8} J m^{-2} s^{-1} K^{-4}), and T is the absolute temperature of the emitting surface. Thus it can be seen that the intensity of the sun, approximately 10^4 W cm^{-2} corresponds to a surface temperature of approximately 6500 K, whereas an intensity of 10^6 W cm^{-2}, used in keyhole welding, produces a temperature of approximately 20 500 K – sufficient to vaporize any known metal.

Note that the intensity of a beam can be increased by focusing, but the brightness cannot.

Polarization

Light is composed of electric and magnetic waves oscillating in orthogonal planes. The polarization of light characterizes the relationship between the plane of oscillation of the electric field and the direction of propagation. Only the electric field is normally considered since it is the most important when considering interactions with materials. The polarization of a laser beam affects the amount of light absorbed in many material processing applications.

The plane of incidence is the plane that contains the incident beam and the normal to the surface. If the electric vector of the light lies in the plane of incidence, the light is said to be *p*-polarized. If it is normal to the plane of incidence, it is *s*-polarized (from the German word *senkrecht*, meaning vertical or perpendicular). If it is at any other angle, it may be resolved into components of *s*- and *p*-polarization. Light sources with *s*- and *p*-polarization interact differently with a material surface as the angle of incidence, φ, increases. When $\varphi = 0$, *s* and *p* cannot be differentiated, and so the reflectivity is the same for all polarizations. As φ increases, the reflectivity for *s*-polarized light increases smoothly, until it becomes unity at 90°, or grazing incidence. In contrast, the reflectivity of *p*-polarized light decreases monotonically until it becomes zero. The angle at which complete absorption occurs is called the Brewster angle. Beyond this angle, reflectivity increases sharply until it too reaches unity at grazing incidence. The Brewster angle, φ, is related to the index of refraction, n, by

$$\varphi = \tan^{-1}(n). \qquad (3.26)$$

Many different states of polarization are possible; two are illustrated in Fig. 3.13.

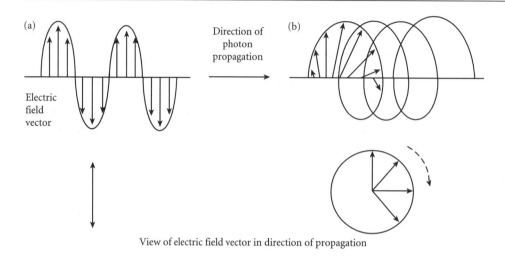

Figure 3.13   (a) Linear (plane) polarization and (b) circular polarization shown in terms of the direction of the electric field vector

In the state of linear, or plane, polarization, Fig. 3.13a, the electric field oscillates in one plane only. If the plane is vertical, the light is said to be vertically polarized. Similarly, in horizontally polarized light the electric vector oscillates in a horizontal plane. The s and p components of polarization are in phase. Plane polarization is often found in commercial lasers, and originates from the reflections within the optical cavity.

Circular polarization, Fig. 3.13b, describes light in which the electric field vector has a constant amplitude and rotates with a constant angular velocity around the axis of propagation. It may rotate clockwise or anticlockwise. The s and p components are 90° out of phase. Light of linear polarization may be converted to circular polarization through the use of a quarter wave plate (Chapter 4).

Elliptical polarization describes light in which the electric field vector rotates with a constant angular velocity around the axis of propagation, and the absolute value of the field vector also varies regularly. The locus of the projection of the electric field vector describes an ellipse.

If the electric field oscillates in random directions, the light is randomly polarized. Such radiation can be thought of as comprising two orthogonal linearly polarized waves of fixed directions whose amplitudes vary randomly over time and with respect to each other. Normal sunlight is randomly polarized.

Beam polarization affects the amount of energy absorbed by the material, and hence the efficiency and quality of laser processing. This is described for cutting in Chapter 14 and welding in Chapter 16. The polarization state of the beam must be established in order to optimize the processing parameters.

LASERS FOR MATERIAL PROCESSING

Lasers for material processing may be classified by: active medium (gas, liquid or solid); output power (mW, W or kW); wavelength (infrared, visible and ultraviolet); operating mode (CW, pulsed, or both); and application (micromachining, macroprocessing etc.) – to name a few. Since the state of the active medium determines the principal characteristics of the laser beam for material processing, it is used here as a primary means of classification: gases (atoms, molecules, ions and excimers); liquids (principally organic dyes); and solids (insulators and semiconductors). This categorization is shown in Fig. 3.14.

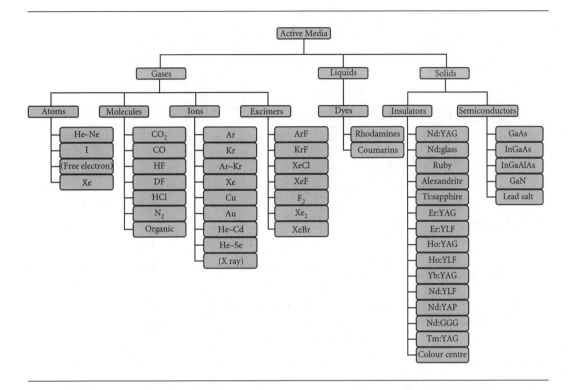

Figure 3.14 Lasers for material processing categorized by the type of active medium (lasers shown in parenthesis are allocated to the most representative group)

Within each state of active medium, the lasers are presented here in a format that facilitates understanding of the mechanism of light generation (electronic transitions in atoms, vibrational transitions in molecules etc.). The final ordering represents their popularity for industrial material processing (ArF, KrF, XeCl …). Figure 3.14 can be used as a guide to locate a particular laser in this chapter: the lasers are presented in the order shown. Further details of each laser can be found in Appendix B.

Figure 3.15 shows an alternative means of presenting a selection of material processing lasers, as a chart with axes of wavelength and average power. Operating regions of different lasers can thus be distinguished, and power levels appropriate for material processing selected. Graphical presentation facilitates understanding of the relationships between variables, and is a central feature of the book. We return to the use of charts and diagrams in laser material processing in Chapter 6.

ATOMS

Light generated in gaseous active media of atoms is considered first since atomic transitions are well defined and relatively straightforward. Transitions take place between *electronic* energy levels separated by a gap large enough to produce photons of high energy, corresponding to wavelengths in the ultraviolet and visible regions of the electromagnetic spectrum.

Helium–Neon

The active medium in the helium–neon (He–Ne) laser is neon, which is typically present in quantities less than 15% – the balance being helium. The first step involves exciting helium in an electrical discharge. Helium atoms then transfer energy to neon atoms by resonance, promoting them to higher

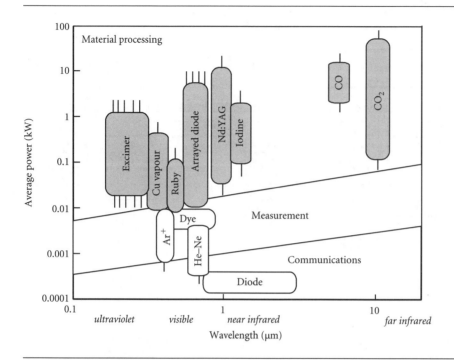

Figure 3.15 A selection of commercial lasers characterized by wavelength and average power, shown on a background of applications (lines indicate the principal output wavelengths, and those used principally in industrial material processing are shaded)

energy levels. The laser operates on a four-level principle (absorption, upper laser, lower laser, and ground states), illustrated in Fig. 3.4, and so continuous wave light can be produced with high efficiency.

The optical cavity is small in diameter. A large ratio of surface area to volume is desirable in order to maintain the population inversion by allowing Ne atoms to lose energy through collisions with surfaces, enabling them to participate in excitation again. Excitation is produced in a longitudinal DC electrical discharge. The gain is relatively low, and therefore the transmission of the output coupler is normally low, around 1–2%. Only very small losses can be tolerated in the cavity, and so high quality mirrors are needed.

Output close to infrared was obtained in the first He–Ne laser, but it was the discovery of visible red radiation of high beam quality that resulted in myriad applications, including barcode readers at supermarket checkouts, pointers, surveying equipment, scientific research, holography and light shows. Modern He–Ne lasers for illumination purposes produce output of a few milliwatts in a variety of colours: green (543 nm); yellow (594 nm); orange (612 nm) and infrared (1523 nm). Early material processing applications included laser printers and medical procedures. However, because of their low cost and compactness diode lasers have replaced them in many such applications. But the high beam quality is difficult to obtain with diode laser output – this is the key to using He–Ne lasers in modern methods of material processing on the microscopic scale.

Iodine

Atomic iodine is the active medium in the chemical oxygen iodine laser (COIL) – a member of the chemical laser family. Molecular oxygen is first excited by an exothermic chemical reaction between gaseous chlorine and an aqueous solution of hydrogen peroxide and potassium hydroxide. Molecular iodine (I_2) is added to form a gas mixture comprising about 1% I_2, which is converted to the atomic form (I) through dissociation by energy transfer from the excited oxygen. Energy is then transferred by resonance between metastable excited oxygen and atomic iodine, which is pumped to the excited state. Near infrared light of wavelength 1315 nm is generated through subsequent electronic transitions in iodine. The lower energy transition level is the ground state, where molecular iodine forms, allowing the population inversion to be maintained. The temperature must be maintained around 150°C to sustain laser action.

A high beam quality is produced because of the gaseous nature of the active medium. The output wavelength lies in the range corresponding to minimum power loss in silica; it is therefore suitable for fibreoptic beam delivery. Hydrogen peroxide and iodine are consumed, necessitating replenishment during operation. The resonator can be scaled relatively easily to achieve high power levels.

Multikilowatt iodine lasers have the potential to compete with CO_2 and Nd:YAG sources in comparable materials processing applications, such as cutting, welding, machining and surface treatment providing that their operating costs can be reduced. The near infrared wavelength of the beam provides advantages over far infrared CO_2 laser radiation: a smaller focused spot can be produced, and energy is coupled more efficiently with metals. Fibreoptic transmission of very high power levels provides advantages over Nd:YAG laser light. As part of the US Air Force Airborne Laser project, units capable of generating pulsed megawatt power levels have been installed in a fleet of Boeing 747 aircraft; one is shown in Fig. 2.5. The beam is focused through a lens mounted in a turret in the nose to destroy missiles moments after launch – laser material processing in space.

Free Electron

A beam of electrons generated in an accelerator can be passed through an array of magnets, causing the electrons to be bent back and forth (wiggled). The electrons emit radiation, based on the conservation of momentum. The frequency of radiation can be changed by varying the electron energy, the magnetic field, or the spacing of the magnets. Heat is carried away in the electron beam itself.

Light in the wavelength range 500–8000 nm can be produced. The wavelength and spectral width of the light is dependent on the number and spacing of the magnets. If the electron beam is pulsed to emit a spatial series of bunches, with a separation corresponding to the emission wavelength, the output becomes coherent. The emission bandwidth is not dependent on the optical bandwidth of a material such as a gas, liquid or solid, and so light can be emitted over a much wider wavelength range than is possible with a conventional laser.

Free electron lasers operating in the visible to mid-infrared regions of the electromagnetic spectrum have been constructed. The variable temporal structure and broadband tuneability of laser make it potentially suitable for a variety of surgical applications. When used to vaporize cells, less damage is caused to surrounding tissue than with conventional lasers.

Xenon

Atomic xenon (Xe) is mixed with argon or helium to form the active medium of the Xe laser. It emits radiation in the far infrared on a number of distinct lines. Beam quality is high because of the gaseous nature of the active medium. Pulses with energy on the order of tens of joules, of microsecond duration, can be produced. They are used for material processing by thermal mechanisms, and applications are being developed that are similar to those for chemical lasers (discussed below).

MOLECULES

Energy transitions in molecules are those between vibrational and rotational energy levels. Transitions between vibrational states of the same electronic level are possible, as in carbon dioxide. Also possible are transitions between vibrational states of different electronic levels, as in nitrogen.

Carbon Dioxide

Gaseous carbon dioxide (CO_2) is present in amounts between 1 and 9% as the active medium of commercial CO_2 lasers. The remaining volume comprises helium (60–85%), nitrogen (13–35%) and small amounts of other gases; the exact composition depends on the design of the optical cavity, the gas flow rate and the output coupler used. High gas purity is necessary, typically 99.995% for helium and nitrogen, and 99.990% for carbon dioxide.

Nitrogen increases the efficiency of excitation by facilitating the absorption of energy, which is subsequently transferred to the CO_2 molecule. Carbon dioxide *may* be excited directly in an electric discharge, but molecules are excited to states in addition to the upper laser level, and so the efficiency of the process is low. A more efficient means is by indirect excitation via excited N_2 molecules. Nitrogen is a diatomic molecule and has only one mode of vibration, which can be induced easily by collision with high energy electrons in the discharge. The vibrational levels of N_2 lie close to the upper laser level of CO_2, and the lifetime of N_2 in the excited state is long; the probability that energy is transferred from N_2 to CO_2 by resonance is therefore high. This two-step process is more rapid than direct excitation, and results in a four-fold increase in laser power.

Helium is added to expedite cooling, which is necessary if the gas mixture is to continue stimulated emission. Excited CO_2 molecules lose energy in the form of heat by colliding with helium atoms. Sufficient energy is lost such that the CO_2 molecules return to the ground state, becoming available for excitation again. The high thermal conductivity of He (around six times that of CO_2 and N_2) enables energy in the gases to be conveyed away from the discharge region. A more stable and uniform discharge is thus produced, allowing a higher working pressure to be used, which also aids in the generation of a high power beam.

Pollutants are generated during operation. Hydrogen ions – generated from the decomposition of water vapour – destabilize the discharge and degrade the operating efficiency. Hydrocarbons decompose forming carbon deposits on mirrors, reducing the gain of the laser. Nitrogen oxides, formed from reactions between dissociated nitrogen and oxygen (itself produced by the dissociation of CO_2 into CO and O_2), are harmful for the operation of the laser. Water vapour is added in small quantities to reduce CO_2 dissociation. This increases the lifetime of the gas. The addition of xenon in small amounts increases output power and efficiency, mainly because of its effect on the electron energy distribution, which is favourable for the vibrational excitation of CO_2 and nitrogen.

The CO_2 laser operates on a four-level basis, illustrated in Fig. 3.4. Photons are generated by transitions between modes of vibration in the linear triatomic CO_2 molecule. The molecule has three distinct vibrational modes: bending, symmetric stretching, and asymmetric stretching, illustrated in Fig. 3.16, which are associated with frequencies of 2.0, 4.2 and 7.0×10^{13} Hz, respectively. In the asymmetric stretching mode the two oxygen atoms move in the same direction, while the carbon atom moves in the opposite direction. In the symmetric stretching mode the two oxygen atoms move in opposite directions while the carbon atom is stationary. The bending mode comprises two degenerate vibrations. In all cases, the centre of mass of the molecule does not move. The vibrational state is denoted by three quantum numbers (v_1, v_2 and v_3) that represent the number of vibrational quanta (the level

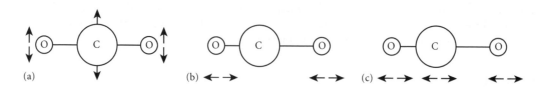

Figure 3.16 Modes of vibration of the carbon dioxide molecule: (a) bending; (b) symmetric stretching; (c) asymmetric stretching

of excitation) in the symmetric stretching, bending and asymmetric stretching modes, respectively. (A superscript to the bending mode number represents additional quanta for the rotational mode.)

The initial stage of light generation involves vibrational excitation of CO_2 molecules from the ground state (00^00) into the asymmetric stretching mode (00^01) by both inelastic collisions with low energy electrons, and resonant energy transfer from vibrationally excited N_2 molecules. The laser transition involves a change in the mode of vibration from the (00^01) state to the symmetric stretching state (10^00). This results in a loss of energy, emitted in the form of an infrared photon with a wavelength near 10.6 μm. (Energy may also be lost through a transition to the bending mode (02^00) with the emission of a photon of wavelength near 9.6 μm, but the probability of this transition is only around 5% of that of the 10.6 μm transition. In addition, the resonator is designed to favour 10.6 μm radiation.) The quantum efficiency of the 10.6 μm transition is relatively high (around 40%) hence the interest in this system for high power output.

For the process to continue, the CO_2 molecule must return to its ground state so that it can be excited again. This can occur by a number of mechanisms. Energy can be transferred by resonance to other CO_2 molecules, such as those in the (02^00), (01^10) or (00^00) states; energy is then redistributed without a total loss. In contrast, non-resonant collisions with the walls of the resonator, other CO_2 molecules, or foreign atoms result in the conversion of the energy of (01^10) molecules into heat, resulting in a loss in total energy. Helium is a particularly effective agent in this respect, which explains its high volume in the gas mixture.

Electrical methods are the most common means of exciting industrial CO_2 lasers. Both direct current (DC) and alternating current (AC) techniques are used. AC excitation may be high frequency (HF, 20–50 kHz), medium frequency, or radio frequency (RF, 2–100 MHz). In the last category, excitation frequencies of 13.56 and 27.12 MHz are popular in commercial designs.

Commercial CO_2 lasers are available in five basic configurations, which characterize the geometry of gas flow in the optical cavity: sealed; transversely excited atmospheric pressure; slow axial flow; fast axial flow; and transverse flow. Typical characteristics of these designs are given in Table 3.1.

Sealed

The optical cavity of a sealed CO_2 laser is made from a large bore glass tube or a square metal or dielectric tube about 2 mm in width. The latter is often referred to as a waveguide laser, since the

Table 3.1 Characteristics of commercial CO_2 laser designs

	Sealed	TEA	Slow axial flow	Fast axial flow	Transverse flow
Optical cavity design	Stable	Stable/ unstable	Stable	Stable/ unstable	Unstable
Gas, He–N_2–CO_2–N_2/ O_2–CO (vol. %)	72–16– 8–0–4	72–16– 8–0–4	72–19– 9–0–0	67–30– 3–0–0	60–25– 10–5–0
Gas flow rate (m s^{-1})	–	–	5–10	300	20
Gas pressure (mbar)	6–14	1000	6–14	70	50
TEM	TEM$_{00}$, multimode	TEM$_{00}$, multimode	TEM$_{00}$, multimode	TEM$_{00}$, multimode	Multimode
Gain (W cm^{-3})	20–30	0.5	0.5	5–10	4–6
Gain (W m^{-1})	50	100	100	1000	6000
Wall plug efficiency (%)	5–15	5–20	5–15	5–15	5–10
Cooling	Conduction	Conduction	Conduction	Convection	Convection
Ergonomics	Portable	Portable	Fixed	Fixed	Fixed

internal cavity surfaces are highly reflective and are an active element of the cavity. A totally reflecting focusing mirror and a partially transmitting output coupler bound the cavity, which is unstable and is permanently aligned.

RF excitation, applied transverse to the resonator axis, is preferred; the source is small, a larger volume can be excited producing more power, and contamination caused by electrode sputtering is avoided, which means that continuous gas replenishment is not needed. However, the discharge causes CO_2 to dissociate into CO and O_2, which reduces output power and corrodes internal parts. Hydrogen or water may be added to regenerate CO_2. A heated nickel cathode may be used to catalyse the recombination reaction. The gas mixture is cooled during low duty cycle applications by conduction through the cavity walls and natural convection via external fins. Forced air or liquid cooling is used in more demanding applications.

The narrow design of the optical cavity produces a high quality beam mode. The gain per metre length of discharge is relatively low because of the narrow cavity, but the gain per unit volume is high because the entire cavity section can be used to generate light. The beam is normally extracted as a square wave pulse of high frequency, high peak energy and low average power. Because of their construction, sealed lasers have a particularly stable output and mode. The power available from a sealed cavity is limited by two factors: the restricted volume of gas which can be excited in units of a practical size, and the rate at which heat can be removed by conduction.

Since the resonator contains no moving parts, and gas flow is not necessary, no external gas connections are required. The head can therefore be transported easily, and mounted on a robot to provide a high degree of processing flexibility. All of the power generated in the laser may then be used for processing since complex optical trains are not required for beam delivery. The laser can operate for many thousands of hours before the gas mixture needs to be changed. Sealed CO_2 lasers are relatively cheap, and are capable of marking, fusing, scribing and engraving the surface of a wide range of materials, as well as through-thickness cutting, trimming, welding and perforating thin sheet materials. Such lasers are used widely in desktop manufacturing systems, and are being used increasingly in surgical procedures.

A pseudo-sealed CO_2 laser design is also available, capable of producing multikilowatt output power. The unstable optical cavity is RF excited via two parallel copper electrodes of large surface area, which produce a relatively high power density. The electrodes are water cooled and their spacing is close; they are therefore able to dissipate heat generated in the gas (referred to as diffusion cooling). A conventional gas circulation system is therefore not required for cooling. The gas consumption is only around $2\,L\,hr^{-1}$ in comparison with about $85\,L\,hr^{-1}$ for fast axial flow designs of the same output, which means that only a small cylinder mounted near the head is required, reducing maintenance requirements. (Such lasers can be used for 12 months continuously.) The unit generates a high quality beam mode ($M^2 < 1.43$) and occupies 15% of the volume of a fast axial flow laser of equal power. A schematic illustration of a CO_2 slab design is shown in Fig. 3.17a, and a production laser is shown in Fig. 3.17b.

Transversely Excited Atmospheric Pressure

The gas mixture used in a transversely excited atmospheric pressure (TEA) carbon dioxide laser is given in Table 3.1. Carbon monoxide and hydrogen may be added to counteract the dissociation of carbon dioxide, and to produce a more uniform discharge and increase output power. A gas mixture that reduces the pulse decay time is used when high pulse rates are required. Gas pressures up to several atmospheres enable high power levels to be generated per unit volume of laser gas.

The gas mixture is excited by an electrical discharge applied transverse to the optical axis. Since the gas pressure is relatively high, large voltages are required for excitation. The electrodes, which are profiled to give a more uniform field, may be placed longitudinally along the optical axis to reduce the potential difference required.

(a)

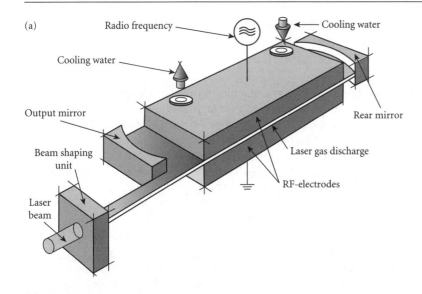

(b)

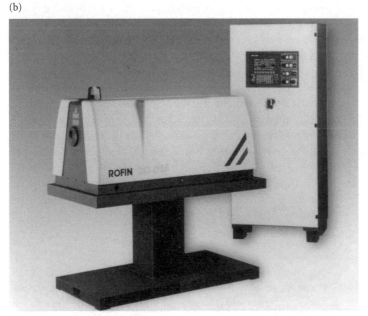

Figure 3.17   Diffusion-cooled CO_2 slab laser: (a) schematic; (b) Rofin DC 025. (Source: Friedrich Bachmann, Rofin, Hamburg, Germany)

Only pulsed output is possible from a TEA laser, since discharge instabilities are easily produced in the high pressure gas environment, which degrade the output power. The beam may be TEM_{00} or multimode, and is typically several square centimetres in cross-sectional area. By using very short discharge times, pulses of energy on the order of joules can be produced with a duration in the range tens of nanoseconds to microseconds. Repetition rates are limited to a few pulses per minute. Mode locking enables short (nanosecond) pulses of high peak power (1–50 MW) to be produced at a rate of 20–100 Hz.

TEA lasers have a small power supply and a lightweight laser head, and are used for marking – product coding of aluminium cans and plastic packages are popular applications.

Slow Axial Flow

The gas mixture used in slow axial flow (SAF) lasers contains a relatively high amount of helium to facilitate cooling, Table 3.1. The gas pressure is similar to that of a sealed unit. The optical cavity is constructed from glass tubes several centimetres in diameter. For a given temperature rise, the power generated is proportional to the tube length; long optical cavities are therefore necessary in high power units. The optical cavity normally consists of a totally reflecting spherical mirror, whose focal point is situated in the plane of a partially reflecting output coupler located at the opposite end of the cavity.

As the name suggests, gas flows relatively slowly in the SAF design, in an orientation parallel to the optical axis. Gas circulation allows contaminants generated in the discharge region (mainly carbon monoxide and oxides of nitrogen) to be removed. The low flow rate allows the laser gas to be heated quickly, which reduces start-up time, but the gas temperature must be kept below about 250°C to maintain gain in the resonator. Heat is conducted from the gas through the walls of the discharge tube, which may be cooled by air, oil or water; narrow tubes are therefore used to maximize cooling.

Excitation is normally achieved with a DC source of several tens of thousands of volts. DC is preferred since features such as sputter contamination and size, which require attention in sealed lasers, can be overcome by continuous gas circulation. Output power increases with increasing discharge current, up to a point at which the heating effect becomes significant. The optimum discharge current depends on the gas pressure and the tube diameter.

Since gas flow and light generation are coaxial, the beam propagates in the direction of the mean thermal gradients, and variations in gain are averaged out along the beam path, producing a stable beam mode. This construction, in combination with a stable mirror alignment and the limited tube diameter, enables a high quality TEM_{00} beam to be generated. Two kilowatt CW units are available – a value that corresponds to a resonator length of about 3 metres, above which the cost of floor space makes the unit uneconomical to operate. (If the cavity is folded, the mechanical complexity is increased and it becomes difficult to obtain a good beam quality.) Relatively high energy pulses can be generated.

Because of the simplicity of the design, running and maintenance costs are low. Units with an output power in the range 150–750 W account for most sales. Slow axial flow CO_2 laser designs are ideal sources for fine cutting, scribing, precision drilling, and pulsed welding applications.

Fast Axial Flow

The output power of the axial flow laser can be increased by increasing the diameter of the optical cavity and raising the gas flow rate – features of the fast axial flow (FAF) design. Turbine blowers are used to circulate the gas mixture at high speed. (Roots blowers were used in early FAF designs; these were the source of many problems that initially harmed the reputation of this type of laser.) The gas composition is maintained by continuously adding a small amount of make-up gas into the mixture; this represents the gas consumption of the laser. Gas is cooled by passage through a heat exchanger containing deionized water (to avoid voltage imbalances).

The optical cavity of the FAF laser has many features in common with the SAF design. The maximum tube diameter is limited by mechanical and thermal distortions, as well as the generation of high order beam modes. In order to minimize the footprint, the resonator tubes are folded in various geometries with the use of mirrors; vertical zig-zag arrangements, superimposed vertical squares, an inclined triangle and an octagon have all been used in commercial designs. Each section has its own excitation and gas ports in order to maintain a high flow rate. The radius of curvature of the back mirror is large (several metres). This enables power output to be increased, but leads to the production of modes other than the preferred TEM_{00}. Apertures can be used to limit the mode to TEM_{00} with an attendant loss in power. Small adjustments in the curvature of the back mirror can be made to change the beam mode in some designs. A hemispherical cavity configuration produces a superior mode, but again at the expense of lower power. A solid coated zinc selenide output coupler (highly transparent to far infrared radiation) is normally used for a power output less than about 10 kW; higher values require an aerodynamic window.

Both DC and RF excitation are used in commercial FAF designs. The anode of a DC system is a cylinder located inside the tube in a coaxial geometry. The cathode is situated downstream at the end of the discharge section. The discharge is stabilized using orifices that generate shockwaves, which remove high density electron clusters. Rapid expansion of gas downstream of the orifice also has a cooling effect, which enables a higher power output to be achieved. In RF-excited units the electrodes are mounted outside the discharge tubes.

A stable low order beam mode ($0.4 < K < 0.75$) can be generated because of the gain-smoothing effect of turbulent gas flow. The output, which may be continuous or pulsed, is stable to within $\pm 2\%$ over periods of several hours. Pulsing is achieved mechanically in DC-excited designs. Pulse and super pulse operation are possible by electrical means with RF excitation, enabling a peak power 2–10 times that of the continuous value to be obtained. The folding geometry of the cavity determines the beam polarization; if the angle between the beam and the mirrors is close to $45°$, then the perpendicular component is preferentially reflected, producing a state of linear polarization. A preferred state of polarization gives the beam specific material processing properties in different directions of travel.

FAF lasers are used in a wide range of material processing applications, including welding, cutting and surface treatment. Pulsed output is useful for initiating cuts, drilling, and perforating sheets. Commercial FAF units are limited to about 20 kW because high gas flow rates require complex blower technology, and high gas pressures lead to difficulties in maintaining a glow discharge. A schematic illustration of an FAF laser is shown in Fig. 3.18a, and a production laser shown in Fig. 3.18b.

Transverse Flow

A basic requirement for generating high power is the ability to excite a large volume of gas in a given time. This can be achieved by circulating the gas *across* a cavity of large cross-sectional area. Such designs are referred to as transverse flow (TF). This geometry has a number of advantages over the FAF design. Gas flow rates are typically one tenth those of fast axial flow designs, which reduces the requirements on the blowers, and reduces flow rate losses that lead to increased temperature, loss of population inversion and reduced beam power. Pressure differentials are lower, and since gas resides in the discharge volume for a shorter time, more power per unit length of cavity can be generated, enabling considerably shorter cavities to be constructed.

A positive branch unstable cavity, constructed using one concave and one convex mirror (Fig. 3.7), is normally used in systems that produce output above 6 kW. The high cavity gain permits laser transitions, despite the relatively high loss in the cavity. The beam is extracted (normally through an aerodynamic window) using an annular curved scraper mirror. The exclusive use of reflective optics is another factor that allows high values of power to be generated and extracted.

The axis of excitation is normally oriented perpendicular to both the gas flow and the optical axis. This geometry provides unimpeded gas flow and a short discharge path (typically around 5 cm), which enables a relatively low working voltage to be used – between 10 and 20% of fast axial flow designs. Both DC and RF excitation are suitable for TF designs. Direct current excitation uses a sophisticated segmented electrode to maintain a uniform discharge over the large resonator area, to reduce the possibility of arcing. In most designs the cathode is water cooled. Radio frequency excitation requires only two electrodes, which can be placed outside the discharge region. The discharge may be stabilized using a turbulence generator or by using an electron beam. The latter technique, in which a wide beam of high energy electrons assists the production of a uniform volume discharge, is expensive.

(a)

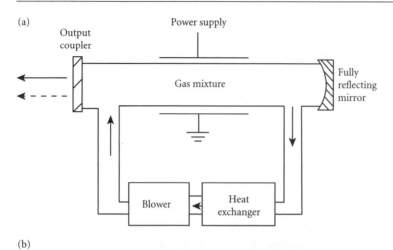

(b)

Figure 3.18   Fast axial flow CO_2 laser: (a) schematic; (b) Trumpf TLF 15000. (Source: Sven Ederer, Trumpf, Ditzingen, Germany)

Preionization is necessary if the main discharge is to fill the laser volume. Cooling is via a heat exchanger containing deionized water.

The distribution of power in the beam depends on the geometry of the optical cavity and the method by which the beam is extracted. The output from an unstable cavity has a characteristic annular intensity distribution, generated by the annular scraper mirror. A stable cavity produces a multimode beam comprising a mixture of low order transverse modes because of its relatively high Fresnel number. Transverse discharge inhomogeneities can result in an asymmetric beam that is larger than one comprising low order stable modes, and which has a higher divergence (2–3 mrad). Relatively high M^2 values (>5) are common. High power stability is within ±2% and ±5% over minutes and hours, respectively.

In comparison with other laser designs, transverse flow lasers can be made relatively easily in modules, enabling designs to be scaled to high power outputs. The capital cost per kW is lower, and the compact design results in a smaller footprint. The use of a metal container and all-reflecting optics

(a)

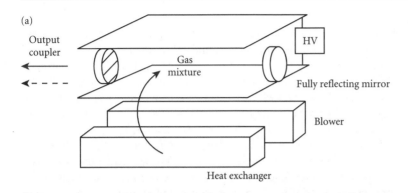

(b)

Figure 3.19 Transverse flow CO_2 laser: (a) schematic; (b) UTIL (now Prima Industrie) 24 kW. (Source: Robert Murray, Prima Industrie, East Hartford, CT, USA)

allows the optical cavity to be rugged in design. Gas usage is lower than a fast axial flow laser, and lower operating voltages can be used than in DC designs. However, pulsing, other than by mechanical means, is difficult, and the beam mode is of lower quality than axial flow designs. Such high power lasers are commonly used for material processing operations such as thick section welding and large area surface treatment. A schematic illustration of a TF laser is shown in Fig. 3.19a and a production laser shown in Fig. 3.19b.

Gas Dynamic

Although the gas dynamic carbon dioxide laser is not a commercial design, it is of interest for material processing since continuous power levels to 100 kW have been produced. The population inversion is created by thermodynamic means, rather than electrical. A fuel is mixed with an oxidizer to yield combustion products suitable for laser transitions; combustion of CO and CH_4 with O_2 and N_2 yields CO_2, N_2 and H_2O. At high temperatures, most of the energy is stored as vibrational excitation in nitrogen molecules. As the gas expands through a supersonic nozzle, this energy is rapidly transferred to CO_2 molecules through collisions, exciting them to the (00^01) level to create a population inversion with the (10^00) level. Certain high energy states are momentarily frozen, creating a non-equilibrium condition that promotes the population inversion.

Carbon Monoxide

The active medium in the carbon monoxide (CO) laser is CO gas in a mixture of helium and nitrogen, with small amounts of xenon. The gas ratio by volume lies around 1:20:1 ($CO:He:N_2$), maintained at an operating pressure of about 100 mbar in CW units and higher values in TEA designs. Nitrogen and helium play similar roles to those in the carbon dioxide laser: efficient excitation and cooling, respectively. Xenon is added to change the average electron energy, thereby increasing the fraction of electrical power transferred to molecular vibrational levels, which increases the power generation efficiency by about 30%. A small amount of oxygen also enhances performance.

The diatomic CO molecule has only a single vibrational mode, in contrast to the three vibrational modes of the carbon dioxide molecule. The quantum efficiency is close to 100% because the lower laser level of a given transition can serve as the upper laser level of a subsequent transition, which means that light can be produced from several pairs of levels to support a population inversion. Emission therefore occurs on a range of discrete infrared lines, with wavelengths between 5.2 and 6.5 μm.

The active medium must be cooled to cryogenic temperatures below 77 K to maintain a population inversion. This may be achieved by conductive or aerodynamic cooling. Low power lasers (below 100 W) are cooled by gases and circulating fluids. Higher power designs are usually cooled in a liquid nitrogen or Freon® heat exchanger. (Freon is a registered trademark belonging to E.I. du Pont de Nemours & Company (DuPont).) Cooling can also be achieved by adiabatic expansion in a supersonic nozzle. After power has been extracted from the resonator, the laser gas is recompressed by a diffuser and a pumping system.

Commercial units are often based on modified CO_2 resonator designs – the changes that are required include: substitution of appropriate optics; insertion of the correct gas mix; and the addition of cooling equipment. The cavity can take the form of a tube or waveguide, which can be sealed or transport flowing gas. A solid output window with a low absorption coating is located at one end of the cavity. The resonator design is scaleable to provide high power levels. Low power designs are based on large area slab waveguides: a large bore refractory ceramic tube 1 m in length and 25 mm in diameter can produce about 20 W, which can be increased to 300 W by scaling the dimensions three-fold. Excitation is electrical, achieved by DC and RF means. A wall plug efficiency close to 25% is possible (around twice that of an equivalent CO_2 laser) because of the high quantum efficiency, providing that the means of cooling is energy efficient. The beam quality (M^2) typically lies between 2 and 2.5.

In comparison with the carbon dioxide laser, the shorter wavelength of CO laser light possesses a number of advantages for material processing. Short wavelength light is absorbed more readily by metals and ceramics and so a lower power level can be used for processes that involve heating and melting. The beam can be focused to a spot of smaller diameter, giving a higher power density that facilitates keyhole formation in penetration welding, and piercing to initiate cutting. The plasma generated when welding metals is white (in contrast to the blue plasma generated through metal ionization during CO_2 laser welding), which means that a larger fraction of the beam is transmitted through the plasma to the workpiece. Transmissive optical materials are available with a high damage threshold, and there is greater potential for fibre optic beam delivery.

The development of CO lasers has been hindered by two factors: the need for cooling to operate efficiently; and instabilities in the gas mixture. CO dissociates into C and O as a result of the electrical discharge, necessitating the use of a flowing gas mixture to maintain composition during long-term operation.

Hydrogen Fluoride

The hydrogen fluoride (HF) laser is a chemical laser that combines heated hydrogen (produced in a combustion chamber similar to a rocket engine) with fluorine gas (produced by thermal decomposition of compounds such as sulphur hexafluoride) to form excited HF molecules. Far infrared emission occurs on lines between 2600 and 2900 nm in wavelength, the exact line depending on the chemical composition of the reacting gases.

Hydrogen fluoride lasers are under consideration for space-based missile defence systems, although the emitted wavelengths transmit poorly through the lower regions of the atmosphere since they are absorbed by water. (Propagation through upper regions and in space is superior.) Interest lies in the fact that the reactants used can be stored for long periods of time, and waste products and heat can be exhausted into space. Pulsed power levels on the megawatt scale have been generated. Systems are being designed that can be operated and maintained from the ground.

Deuterium Fluoride

The deuterium fluoride (DF) laser uses a chemical reaction between atomic fluorine and deuterium to produce the active medium. It is chemically the same as the hydrogen fluoride laser, however, the increased mass of heavy deuterium shifts the output lines to between 3500 and 4000 nm, which is superior for transmission through the lower atmosphere.

A megawatt-class CW chemical laser, known as MIRACL (Mid-Infrared Advanced Chemical Laser), produces spectra distributed among ten lasing lines between 3600 and 4200 nm. Beam shaping optics have been used to form a 14 cm square beam that can disable a satellite the size of a refrigerator located 260 miles above the earth's surface. Such military applications have been a significant driving force in the development of chemical lasers.

Hydrogen Chloride

Light can also be produced from excited hydrogen chloride (HCl). This was the first chemical laser to be demonstrated. HCl can be generated in a mixture of chlorine and hydrogen through pulsed photodissociation of chlorine using a flashlamp, or by the reaction of nobelium (No) with ClO_2. Output is on lines between 3600 and 4000 nm in wavelength.

Nitrogen

Nitrogen gas, at a pressure of between 0.03 and 1 bar, is the active medium in the nitrogen laser. The gas is normally circulated, but low repetition rate pulsed designs can be sealed.

Nitrogen lasers generate light through transitions between both electronic and vibrational energy levels. The resonator gain is high, which means that only one mirror needs to be used; feedback is not required for laser action. The lifetime of the upper level is short, while that of the lower level is long – CW operation is therefore not possible, but pulsed operation is. Pulse widths are short, because as soon as the laser transitions begin, the population of the terminal state increases rapidly, and after a few nanoseconds the population inversion is reduced to a level at which laser action cannot be sustained (self-termination).

The optical cavity is similar to that of a TEA CO_2 laser. The active medium is excited using a fast high voltage discharge between electrodes placed transversely to the optical axis. The gas medium is cooled by passing through a heat exchanger, which may use circulating air.

Pulsed ultraviolet light of wavelength 337.1 nm is generated, in the form of a rectangular beam with dimensions between 2×3 and 6×30 mm. The pulse duration is on the order of nanoseconds and is produced at frequencies to 1 kHz, with a pulse energy limited to tens of millijoules, and an average power of several hundred milliwatts.

Nitrogen lasers were initially used as excitation sources for pulsed dye lasers, but because of their short wavelength they can also be used in non-linear spectroscopy, non-destructive testing and Raman scattering. Material processing on a microscopic scale can be performed. Nitrogen lasers show potential for laser-assisted chemical vapour deposition because of the focusability of the beam, enabling sharp, even deposits to be produced.

Organic

Organic molecular gas lasers emit on lines of wavelength between about $28\,\mu m$ and $2\,mm$. At the short wavelength end of the range, light is generated from vibrational–rotational transitions. Purely rotational transitions are involved in longer wavelength transitions. The active media possess a permanent dipole moment. Alcohols and other carbon-based compounds such as CH_3OH, $C_2H_2F_2$ and CD_3OD are commonly used. Optical pumping is preferred, since it permits precise selection of the initial excited state. This is normally achieved using an external CO_2 or N_2O laser, which is tuned to emit on a single line around $10\,\mu m$ in wavelength. They are primarily used in research (to diagnose fusion plasmas), astronomy and in studies of semiconductor materials.

IONS

Light is generated in ionized gases through electronic transitions. Since the excitation energy of an ion is larger than a neutral atom, ionized gas lasers produce light of shorter wavelength, in the range between mid-ultraviolet and visible.

The gas is present in a mixture with buffer gases such as neon or helium or both (depending on the emission line or lines required), at a pressure below atmospheric. The cavity design is similar for all types of gas ion laser. The discharge tube is normally made from a ceramic such as beryllia in lower power lasers (typically a few millimetres in diameter), but may be made from graphite or tungsten discs surrounded by quartz in higher power designs. Excitation is via an electrical arc discharge operated at a relatively low voltage and high current. The current required to create a population inversion is high – larger than that used in atomic gas lasers – because atoms must first be ionized before further collisions with electrons can excite them to the higher energy levels required for laser transitions. The overall efficiency is therefore relatively low, less than 0.1%. The mirrors bounding the cavity may be spherical or a combination of spherical and concave. A solid window is attached to one end of the tube, mounted at the Brewster angle to minimize reflection.

The average beam power obtained is on levels between milliwatts and watts. Beam quality is high, and has a narrow bandwidth. Output is CW, but pulses can be produced using mode locking. The cost of a unit is on the order of thousands of dollars.

Argon

Argon ion lasers excited by an electric discharge emit on several lines in the range yellow–ultraviolet. Emission of green and blue wavelengths is the most prominent. Power levels on the order of tens of watts can be obtained.

Argon ion lasers are used in light shows, anemometry, fluorescence excitation Raman spectroscopy, forensic medicine, research, printing (exposing plates for printing presses) and holography. They are also effective optical pumping sources for other lasers. The most common material processing applications include etching of semiconductors to form optical waveguides, and microscopic medical procedures involving ophthalmology and minimally invasive surgery.

Krypton

The krypton ion laser emits on a range of lines in the wavelength range from ultraviolet to infrared (390–780 nm). The main line is red, on which several watts can be obtained. The output is less than the argon ion laser because the gain in the resonator is lower. The uses of krypton ion laser output are similar to those of the argon ion laser.

Argon–Krypton

Lasers using mixtures of argon and krypton emit strongly in the red and blue regions, with a beam power on the order of watts. Applications are similar to those of the individual lasers.

Xenon

The xenon ion (Xe^{3+}) laser produces pulsed green and ultraviolet light. It is used in minimally invasive surgery since green light is absorbed well by tissue containing red blood cells.

Copper Vapour

Copper, gold, and elements in the same and adjacent columns of the periodic table can be used to produce active media in the form of ionized metal vapours. Metal vapour lasers emit visible light as a result of transitions between low-lying energy levels. In the copper vapour type, the active medium is formed at the anode of a plasma tube containing elemental copper and an inert buffer gas, by passing an electric current using a high voltage switch. Transitions between the upper energy states result in the emission of yellow and green light. The lower laser level is the ground state, and so light can only be generated for a short time before the population inversion is destroyed. The laser is therefore operated in pulsed mode. It takes about 25 μs to deactivate the terminal level (after which energy transitions can recommence), which limits the maximum pulse duration and repetition rate.

The plasma tube is typically 10–80 mm in diameter, bounded by a converging mirror and a plane output coupler. Both stable and unstable cavity designs are used. The cavity is easily scaled in power, while retaining a good beam mode, because of the high gain. Metal vapour is formed in a neon atmosphere at a pressure between 30 and 70 mbar. The operating temperature is high (between 1300 and 1600°C), which is a factor that has limited the development of commercial units.

Output is in the form of short pulses, tens of nanoseconds in duration, with a high repetition rate, up to 20 kHz, a relatively high average power (over 100 W), a peak power of several hundred kilowatts, and a pulse energy of up to 20 mJ. High temperature operation favours the production of yellow light. Frequency-doubling provides two ultraviolet wavelengths: 255.2 and 289.1 nm. (Applications then overlap those traditionally associated with excimer lasers – KrF lasers operate at 248 nm. However, the pulse repetition rates obtainable from the copper vapour laser are around a thousand times higher, resulting in significantly greater material removal rates in drilling applications, for example.)

Short wavelength light can be focused to a small, high intensity spot, ideal for micromachining of ceramics, reflective metals (e.g. aluminium, copper and brass), and polymers. Cuts with a typical kerf width of 2 μm can be made. Holes with an aspect ratio of 50 can be drilled.

Gold Vapour

The gold vapour laser operates on similar principles to the copper vapour laser. Red light is produced in pulses with an average power up to 10 W. The laser initially found similar applications to the copper vapour type, but it is relatively expensive since red light can now be produced from other sources more cheaply and conveniently.

Helium–Cadmium

The helium–cadmium (He–Cd) laser is one of a family of sources that use a metal with a low vaporization temperature as the active medium. Cadmium is heated with a filament to produce a vapour that is mixed with helium. Helium is ionized and excited in a pulsed discharge. Energy is then transferred from excited He to a neutral Cd atom by collision, causing the Cd atom to be ionized and excited further.

Light is produced by electronic transitions between the excited ionic state and the ionic ground state. The strongest output is on lines in the blue and ultraviolet ranges. Weaker red and green light can also be produced. Atoms remain in the excited state for only a short time, returning to the ground state where they accumulate, quickly removing the population inversion needed for laser action.

The He–Cd laser operates in pulsed mode, with an average power on the order of milliwatts. (The power output is often quoted in terms of an average CW.) The active medium has a high gain compared with similar lasers. TEM_{00} and multimode beams can be produced.

The helium–cadmium laser is used in Raman spectroscopy, to fabricate holographic gratings, and in photochemical material processing techniques such as stereolithography.

Helium–Selenium

The helium–selenium (He–Se) laser operates on similar principles to the He–Cd laser. Transitions between excited states of selenium produce radiation in the visible spectrum, between blue and red. Average power output on the order of tens of milliwatts can be obtained, but the laser is not used extensively in material processing because alternative lasers that are cheaper can produce light in this range of wavelength.

X-ray

X-rays occupy the wavelength region of the electromagnetic spectrum between roughly 0.01 and 10 nm. They are generated in the X-ray laser by excitation of a polished palladium or titanium target with a high energy light pulse around one nanosecond in length. Electrons are stripped from titanium and palladium atoms, producing plasma that contains excited ions with similar stable electron configurations to neon and nickel, respectively. Decay of high numbers of ions produces powerful 'soft' X-rays with a wavelength slightly longer than that used in medical imaging.

Coherent radiation of short wavelength and short pulse duration is suitable for applications in biology, chemistry and materials science. The X-ray laser was developed during the *Strategic Defense Initiative* as a means of destroying incoming missiles. Civilian alternatives lie in the manufacture of nanometre scale structures required in the fields of quantum electronics, and the construction of nanometre-scale robots (nanides).

EXCIMERS

The active medium in an excimer laser consists of a mixture of gases: a rare gas (1–9%); a halogen (0.05–0.3%); and an inert buffer gas (90–99%). The total gas pressure is around six times atmospheric. The gas mixture is circulated rapidly (up to $50\,\mathrm{m\,s^{-1}}$) and augmented to maintain the desired composition.

It is cooled in a heat exchanger and filtered, since changes in temperature and gas composition create difficulties in maintaining a stable beam mode. The laser gas slowly degrades – halogen is depleted and impurities including HF, CF_4, SiF_4 and CO_2 form – resulting in a gradual reduction of gain.

A plane-parallel optical cavity is used, made of materials such as aluminium, fluorocarbons and ultra pure ceramics that are resistant to corrosive halogens. A thin layer of aluminium nitride may be chemically deposited on an extruded aluminium chamber to enhance protection. The two windows bounding the cavity are typically made from polished magnesium fluoride; one has a highly reflective aluminium or dielectric coating on its rear surface, the other acts as the output coupler. The aperture has a cross-sectional area between 1 and 3 cm^2. Gain in the cavity is high, and so adequate feedback can be achieved with a resonator length of only about 100 cm using an output coupler of reflectivity 5–10%. The mirrors may be placed outside the optical cavity in order to avoid corrosion problems. Low beam divergence and good focusability can be achieved though the use of an unstable resonator.

The active medium is excited by a high voltage discharge applied orthogonal to the axis of the optical cavity and the gas circulation, in a similar manner to that in a TEA CO_2 laser. The electrodes are normally rectangular with a spacing of about 25 mm. Large cross-section discharges require a uniform concentrated electron density for optimum ionization; this can be achieved by preionization using X-ray, ultraviolet and electron beam sources. The discharge is operated at high peak currents with short current rise times, which places considerable loads on the switches and high voltage capacitors – thyratrons are used as high voltage switches. Excitation typically involves a 50–100 ns duration pulse with a voltage of 35–50 kV.

In commercial sources, light is generated in the form of ultraviolet pulses, with a wavelength in the range 150–350 nm, via transitions between electronic and vibrational energy levels. Since the spectral gain bandwidth is broad ($>100 \, cm^{-1}$), many transverse and longitudinal modes can oscillate simultaneously in a stable plane-parallel resonator. The beam quality is relatively low ($M^2 \approx 100$) but can be improved by an order of magnitude through the use of unstable resonator optics, at the expense of a reduction in power. A rectangular beam cross-section is common, ranging between 2×4 and 25×40 mm, with an intensity distribution that is approximately Gaussian in the shorter dimension but flatter in the longer dimension. The beam divergence is typically below 4 mrad. Beam homogenizers (arrays of lenses) are used to split the beam into segments which are superimposed in the aperture plane. The number of materials that transmit efficiently at ultraviolet wavelengths is limited; fused silica, fused quartz and crystalline halides are suitable for wavelengths above about 375 nm, while sapphire and crystalline fluorides of lithium, magnesium, calcium and barium are more appropriate for shorter wavelengths.

The precise nature of the output depends on the active medium. Light pulses have the following characteristics: energy in the range millijoules to joules; average power up to several hundred watts; repetition rates between about 20 and 1000 Hz; and pulse duration from a few to a hundred nanoseconds, giving peak power values up to 50 MW. Operational limits are determined by the rate of high speed switching (the thyratron) and the resonator length. The maximum gas flow velocity determines the maximum pulse frequency obtainable. Long pulses can be obtained by superimposing the output from several lasers. The conventional technique for reducing pulse length – Q-switching – cannot be used because of the lack of a suitable saturable absorber. Mode locking cannot be used either, since it relies on maintaining gain in the cavity for a substantial number of round trips – this cannot occur in excimer lasers because of the small number of passes in the high gain medium. Properties of the most common commercial excimer lasers – those based on argon fluoride, krypton fluoride, xenon chloride, and increasingly fluorine – are summarized in Table B.1 (Appendix B). Brief descriptions of the individual lasers follow here.

Short wavelength ultraviolet light has three main advantages for material processing: high absorption by many engineering materials; high spatial resolution; and high photon energy (similar to that of chemical bonds). Ultraviolet light can therefore be focused to a small spot size and located with high accuracy to process a wide range of materials, including metals, ceramics and polymers. A short wavelength also provides opportunities for photochemical material processing, in addition to the thermal processing mechanisms characteristic of infrared lasers. The short pulse width and high peak power reduce the heat affected zone in materials. Since the transparency of the plasma is proportional to $1/\lambda^2$, where λ is the wavelength, plasma shielding is less of a problem than with infrared laser light.

Excimer lasers have a wall plug efficiency of 1–2.5%. The running cost of an excimer laser is high in comparison with solid state and CO_2 lasers. The capital cost is also relatively high – from \$30 000 to \$200 000 (around \$1000 per watt). Excimer lasers initially replaced nitrogen lasers for pumping tuneable dye lasers for spectroscopic studies, and later competed with Nd:YAG lasers for pumping higher

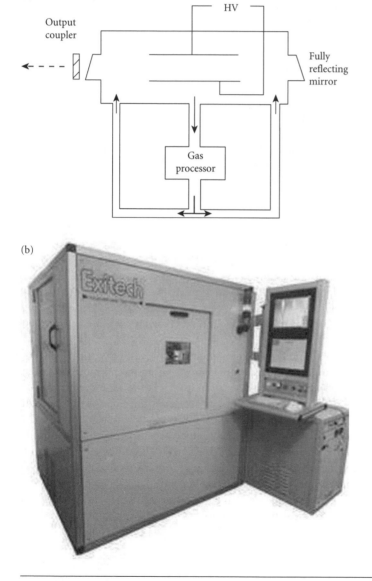

Figure 3.20 Excimer laser: (a) schematic; (b) Exitech M5000 micromachining tool. (Source: Adrian Baughan, Exitech Ltd, Oxford, UK)

power lasers. They are now used extensively in corrective eye surgery, microlithography, micromachining, marking, annealing, doping, vapour deposition, and other surface modification techniques with a wide range of metals, ceramics and polymers. A schematic illustration of an excimer laser is shown in Fig. 3.20a, and a production laser micromachining tool shown in Fig. 3.20b.

Argon Fluoride

The short wavelength of the argon fluoride excimer laser has resulted in it becoming the source of choice for corneal surgery including photorefractive keratectomy (PRK) and laser *in-situ* keratomileusis (Fig. 1.4, Chapter 1), as well as lithography.

Krypton Fluoride

The krypton fluoride laser has the highest intrinsic efficiency of excimer lasers, and found many applications shortly after its invention. Today its main application is in fine scale lithography.

Xenon Chloride

Xenon chloride is the optimum medium for discharge excitation, enabling high power pulses (1 kW average) of relatively long duration (150 ns) to be produced. The wall plug efficiency of 4.5% is the highest of the excimer laser family. Fused silica, fused quartz and several crystalline halides are used in beam delivery optics. Drilling and other machining operations are typical high power applications, while stereolithography is a major application of low average power units.

Xenon Fluoride

The xenon fluoride laser was investigated for early defence-related applications because the relatively long wavelength (350 nm) is transmitted well in the atmosphere.

Fluorine

The fluorine laser produces the shortest wavelength of commercial excimer lasers, and provides relatively high output power. It is being developed for future microlithography applications, particularly with polymers and glasses, photoresist and mask development, photochemistry and spectroscopy, and for testing of optics, coatings and metrology equipment.

Xenon

Xenon was the active medium in the first excimer laser to be demonstrated, but mixtures of rare gases and halides have proved to be more appropriate for industrial units.

Xenon Bromide

Xenon bromide formed the active medium of the first rare gas–halide excimer laser to be demonstrated. It is relatively inefficient as a laser, but is a good emitter of fluorescence, and so is a popular choice for lamps.

LIQUIDS

The active medium in most liquid lasers is a complex organic dye. Dye lasers operate on a fundamental four-level system, involving transitions between many split energy states created by vibration of the molecule.

Dye lasers are optically excited, using a flashlamp or another laser of shorter wavelength than the desired output. Xenon flashlamps arranged in a linear or coaxial geometry give a wall plug efficiency of up to 1%. Dye laser excitation, using nitrogen, excimer, copper vapour or frequency-multiplied Nd:YAG sources, produce light with 5–25% pumping efficiency. Argon or krypton ion lasers give around 10–20% conversion of light to pump energy. Dye lasers are high gain, requiring minimal oscillation to build up the beam, but must be cooled by a circulating medium.

Dye lasers can be tuned to emit at a variety of wavelengths in the range 300–1000 nm, by changing the angle of the grating used as the output coupler. They operate principally in pulsed mode, although models pumped using a CW ion laser can operate in CW mode. The pulse length depends principally on the type of excitation source. Flashlamp pumped dyes produce pulses of length 20–4000 ns, energy 0.05–50 J, with a repetition rate of 0.03–50 Hz, giving a peak power of several hundred kW and an average power in the range 0.25–50 W. Laser-pumped dyes typically produce pulses of length 3–50 ns, up to 10 kHz, with an average power 0.05–15 W, and peak pulse powers on the order of megawatts. The pulse characteristics are determined by the dye used, and can be modified by mode locking or cavity dumping. Frequency doubling and Raman shifting extend the range of operation even further.

Colour variability is important when treating materials whose absorptivity depends on wavelength. For example, blood does not absorb red light significantly, and so a different wavelength must be used for surgical procedures in blood-rich tissue.

Rhodamine

The chemical formula of Rhodamine B is $C_{28}H_{31}ClN_2O_3$. The energy-level structure of an organic dye molecule is correspondingly complex. The flashlamp-pumped Rhodamine 6G laser has a tuning range between 570 and 660 nm. It is used in cosmetic procedures to remove leg and spider veins, port wine stains and scars. Typical beam characteristics are: wavelength 585, 590, 595, 600 nm; pulse length 1.5 μs; and energy density 3–25 J cm^{-2}.

Coumarin

Coumarin has a chemical formula $C_9H_6O_2$, and is the dye used as the active medium in lasers intended for ablative processes. A flashlamp-pumped Coumarin dye laser produces a wavelength of 504 nm (green). During laser lithotripsy this type of beam has a large effect on a kidney stone and a small effect on the ureteral wall. When the stone absorbs the laser light, a small amount of heat is generated, which creates a cavitation bubble. The expansion and contraction of this bubble creates acoustic waves, which pass into the stone, resulting in fragmentation.

SOLIDS

Nd:YAG

The active medium in the Nd:YAG laser is a solid host material of yttrium aluminium garnet (YAG) doped with neodymium ions (Nd^{3+}). The host is a synthetic crystal with a garnet-like structure, and the chemical formula $Y_3Al_5O_{12}$. Neodymium ions take the place of yttrium ions in the garnet lattice – they are roughly the same size. Ions are present at concentrations around 1% by weight, which corresponds to a concentration of about 10^{20} ions cm^{-3}. The optimum concentration for continuous wave operation is around 0.8%, whereas 1.2% is more suitable for pulsed operation. The neodymium ion contains a partially filled $4f$ subshell, which provides the electrons for the laser transitions. The $4f$ subshell is shielded by filled $5s$ and $5p$ subshells. The active medium is in the form of a rod, or one of the novel geometries described below. The main advantage of YAG compared with other host materials is its good thermal stability.

Light is generated through transitions between energy levels of the neodymium ion. The laser is based on four-level operation, illustrated in Fig. 3.4: a ground level; absorption bands; an upper laser level; and a lower laser level. Excitation occurs by absorption of visible and infrared light at wavelengths of 730 and 800 nm, respectively. Ions are raised from the ground level, $^4I_{9/2}$, to the $^4S_{3/2}$ and $^4F_{7/2}$ (730 nm) and $^4F_{5/2}$ and $^3H_{9/2}$ (800 nm) absorption bands, with a quantum efficiency of up to 50%. Once in the higher energy level they undergo non-radiative relaxation (with the generation of heat) to a metastable upper laser level, $^4F_{3/2}$. The lower laser level is $^4I_{11/2}$, which is normally unpopulated at normal temperatures. It is therefore relatively easy to obtain a population inversion, resulting in a relatively low threshold for laser action. An excited ion that drops to the lower laser level emits a photon of wavelength 1064 nm. The lower laser level is depopulated by thermal transitions to the host, causing further heating of the YAG rod.

The optical cavity of lamp-pumped designs normally takes the form of an Nd:YAG rod. Rods are typically 8–10 mm in diameter and up to 200 mm in length. They are expensive, since they must be machined from boules that can take up to eight weeks to grow. Such a rod is capable of producing about 750 W of power, and so multiple rods are used in higher power units. A long rod produces a beam of low divergence, whereas a short rod possesses good mechanical stability and can be packaged easily. Rods of large diameter have a high energy conversion efficiency; rods of small diameter have low divergence. Compromises in the dimensions of the rod must therefore be made in commercial lasers. The optical cavity must be designed to compensate for thermal lensing of the rod caused by uneven heating, which limits scaling of power. A fully reflecting mirror is located at one end of the rod, and a partially reflecting output coupler at the other. The gain in a solid rod is normally considerably higher than a gas laser, and so cavity mirrors with lower reflectivity can be used. Dielectric mirrors feature in high power lasers, with gold-coated metallic mirrors being used in lower power lasers. An optical cavity based on rods may be arranged in stable or unstable configurations. A low order mode beam with restricted power is generated in a stable cavity. An unstable cavity can be constructed by using similar mirror designs to those in CO_2 lasers; larger amounts of power may then be generated at the expense of a reduction in beam quality and efficiency. However, the focal plane moves with changes in output power, which must be compensated for by using adaptive optics.

Excitation is produced by flashlamps, arc lamps or semiconductor lasers. Only lamp pumping is considered in this section – diode laser pumping is described later. Linear flashlamps may be arranged in various geometries in hollow reflective resonators. Lamps may be placed next to the rod in a closed-coupling geometry. Alternatively, the lamp and the laser rod may be placed at the two foci of an ellipse in order to maximize excitation. High power oscillators may be surrounded by arrays of flashlamps. For long pulse lengths (greater than 1 ms), the power supply current is stabilized to the required value, and the pulse length is determined by the time between switch on and switch off. Shorter pulse lengths use capacitor discharging techniques.

High power cavity designs are often based on the oscillator–amplifier principle. The oscillator is a conventional laser, but the amplifier is a rod without the feedback elements, which is pumped by a separate lamp. The amplifier sections do not generate light, but store energy when excited. As the beam from the oscillator passes through the amplifier section, much of the energy is extracted in the excited state. The arrangement of several rods in one resonator is the most advantageous design for multikilowatt units, since the power can be scaled to high levels without losing beam quality.

Around 50% of the electrical power consumed is dissipated as heat inside the rod. Convective air cooling is used in low power units, whereas in higher power designs deionized water flows through an annular transparent cooling jacket between the rod and the lamp. The removal of waste heat becomes a major concern when the continuous power of an Nd:YAG laser exceeds about 2 kW. Cooling induces a parabolic temperature gradient within the rod, which then acts as a thermal lens. (The refractive index of the rod depends on temperature and internal stress.)

The beam quality lies in the range 20–100 (M^2) for stable resonator modes from rod lasers. Single transverse mode operation can be obtained by inserting apertures that limit the power. The value of M^2 increases with an increase in power because heating introduces changes in the refractive index of the rod, and the effect of imperfections in the rod on optical behaviour increases. The transverse beam mode is often complex, and is difficult to describe in mathematical terms using the TEM notation. A more common measure of quality in such lasers is the beam parameter product (the product of the beam waist diameter and half the divergence angle), measured in mm · mrad.

Crystals of lithium iodate ($LiIO_3$) and lithium triborate (LiB_3O_5) can be inserted into the optical path to multiply the frequency of Nd:YAG laser light to generate harmonics. (The crystal only interacts with light polarized in a certain direction.) Thus the output wavelength can be halved to produce green light (532 nm), and divided by three to give ultraviolet light (355 nm).

Three temporal operating modes are possible: continuous wave; repetitive pulsing; and Q-switched pulsing. Multikilowatt power levels are available in CW operation. However, Nd:YAG lasers have traditionally been manufactured to take advantage of the ability of the YAG crystal to produce very high peak powers in very short duration pulses. The pulse length in a multiple element lamp-pumped laser is fixed by the length of the flashlamp pulse, which is typically on the order of milliseconds or microseconds. The pulse energy from such lasers is about 150 J, with pulse lengths up to 10 ms, and pulse repetition rates up to 50 Hz. The corresponding characteristics of single element units are: pulse energy up to 100 J; 0.5 ms pulse length; repetition rates up to 500 Hz; and a peak power up to 30 kW. As the repetition rate increases, the peak power available decreases, since excess heat must be removed. Q-switching enables 30 kW to be produced in a pulse of duration 100 ns.

Since Nd:YAG lasers can be operated in both continuous and pulsed mode, they possess flexibility for a wide range of material processing applications. The power available from CW units provides competition with CO_2 lasers in a variety of welding applications. Pulses of short duration and high peak power are particularly suitable for drilling applications. Frequency-doubled green light finds uses in material processing, particularly for machining colour-sensitive materials such as nitrides and polyimides. Ultraviolet frequency-tripled light competes with excimer output, and is finding applications in micromachining, marking polymers and glass, as well as in rapid manufacturing systems. The availability of fibreoptic beam delivery extends the range of application to complex geometry processing. The main disadvantages of the Nd:YAG laser, compared with the CO_2 laser, are: limited output power; low wall plug efficiency; and poorer beam quality. A schematic illustration of a lamp-pumped solid state laser is shown in Fig. 3.21a, and a production laser shown in Fig. 3.21b.

Nd:glass

Two families of glass are suitable hosts for neodymium ions: silicates, from which light of wavelength 1061–1062 nm can be generated; and phosphates that operate at 1054 nm. Glasses, which are relatively cheap, can be produced in larger sizes than YAG, and with a greater selection of geometries. Longer rods of high purity, optically uniform glass enable higher average power levels to be achieved. Glasses can also be doped to higher concentrations than YAG, with good uniformity, which allows more energy to be stored, so that short high power pulses can be produced. However, the thermal conductivity of glass is lower than that of YAG, and so adequate cooling is required to avoid distortion caused by thermal lensing. Phosphate-based glasses exhibit less optical distortion, but are less resistant to thermal fracture.

In comparison with the crystalline nature of YAG, glass is amorphous, which means that the line width of the neodymium ion transition is significantly broader; hundreds of axial modes operate

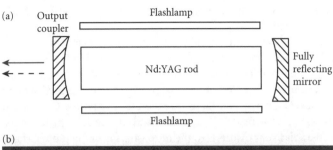

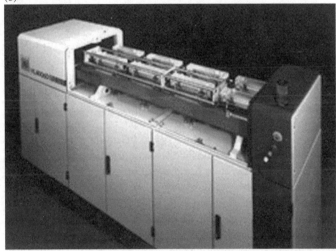

Figure 3.21 Lamp pumped Nd:YAG laser: (a) schematic; (b) HAAS 4 kW 4006 D. (Source: Sven Ederer, Trumpf, Ditzingen, Germany)

simultaneously. This leads to a higher lasing threshold, even though more energy can be stored and released in higher energy pulses.

Flashlamps are used to excite Nd:glass lasers. Optical distortion caused by thermal lensing limits the output available from the rod; the laser can only be operated in pulsed mode. Nd:glass laser performance is normally limited to low duty cycle pulse repetition rates, around one per second.

Three operating regimes can be defined for Nd:glass lasers: normal pulsed mode (pulse length 1–10 ms); Q-switched operation (pulse length on the order of tens of nanoseconds); and picosecond pulse length. To overcome the repetition rate limitation, manufacturers have resorted to methods such as extracting the energy from both ends of the rod. In Q-switched operation, the output energy of an Nd:glass laser is comparable with that of a ruby laser (described below), with similar pulse durations.

Nd:glass lasers produce spiked pulsed output, which is ideal for metal drilling applications. They are also used for spot welding and drilling of deep holes. Frequency-doubled output (532 nm) is also used in scientific research.

Ruby

The active medium in the ruby laser is a single-crystal host of sapphire (Al_2O_3), doped with small amounts (0.01–0.5%) of chromium ions (Cr^{3+}). The ion contains three electrons in the partially filled subshell, which gives ruby its characteristic pink colour, and provides the electrons for laser transition.

The low concentration ensures that the chromium ions are well separated, reducing the likelihood of interaction, which would lead to line broadening. Ruby has good thermal properties, and is unlikely to suffer fracture, particularly when water cooled.

The ruby laser operates on a three-level basis, illustrated in Fig. 3.3: a ground level (the lower laser level); absorption bands; and an upper laser level. Since the terminal state of laser action is the ground level, which is normally fully populated, a high excitation power is needed to produce a population inversion relative to the ground level – over 50% of the Cr^{3+} ions must be raised to the excited state to achieve laser action. The blue and green wavelengths of a flashlamp are used for excitation. Chromium ions are pumped from the $^4A_2$ ground level into the broad $^4F_2$ and $^4F_1$ bands of the absorption levels. Ions then relax very rapidly through non-radiative transitions to the more sharply defined upper laser levels, by transferring energy to the crystal lattice with the evolution of heat. The upper laser level has a relatively long lifetime, around 3 ms, which enables high amounts of energy to be stored, giving pulses of high peak power. This state then decays to the lower laser level over about 5 ms by emitting red photons of wavelength 694.3 nm.

The optical cavity is constructed from a ruby rod, typically between 3 and 25 mm in diameter, with a length up to 20 cm. (Longer rods are difficult to grow, and the internal parts of larger diameter rods are difficult to excite optically.) The cavity is bounded by one totally reflecting mirror and one partially reflecting mirror, which are normally flat, or slightly concave to limit the effects of thermal lensing.

In free-running operation, the ruby laser produces pulses of energy up to about 100 J, with millisecond pulse duration, giving a peak power of about 75 MW, with a repetition rate of one pulse per second. The multimode nature of the output leads to spikes of power, corresponding to emission bursts during excitation, which are superimposed on the pulse envelope. In the oscillator–amplifier configuration, pulse energies greater than 100 J can be obtained with multiple transverse modes. In Q-switched operation, pulses with several joules of energy are possible, with a length on the order of tens of nanoseconds, giving a peak power of about 100 MW and a repetition rate of around 1 Hz. The output can be mode locked using a dye because of the multimode operation, to give pulse trains a few hundred nanoseconds in length containing 20–30 pulses. Individual pulses can be 3 or 4 ps long, with individual pulse energies of approximately 1 mJ in a TEM_{00} beam. Output from a ruby laser is plane polarized if the crystal is cut such that the *c* axis lies perpendicular to the laser axis, but can be randomly polarized if the *c* axis lies parallel to the laser axis. The wall plug efficiency is relatively low, between 0.1 and 1%.

A high pulse energy and spiked output endows the ruby laser with good spot welding and drilling properties. It is not surprising that one of the first industrial applications was piercing of holes in diamonds for wire-drawing dies. However, a compromise between pulse energy and repetition rate is necessary – only one pulse can be generated every second, and so the average power available is limited. Consequently other types of laser, such as the pulsed Nd:YAG, are now favoured in many of the original material processing applications.

Modern ruby lasers are available as stand-alone devices, or packaged systems for specific applications. The beam can be delivered to the workpiece using mirrors, through an articulated arm mirror system, or through a fibre optic. Depending on the application, an end effector is supplied with an adjustable handpiece. This is a particularly important piece of equipment in medical applications – it determines the efficacy of treatment. Packaged systems also include software for control and monitoring.

One of the fastest growing application areas for the ruby laser is cosmetic surgery. Free running devices are used to remove unwanted hair – now a multimillion dollar business in the United States alone. Melanin in the hair absorbs red light, and converts it into heat, which is conducted into the hair follicle, destroying it. The millisecond pulse length matches the thermal relaxation time of the hair. The nanosecond pulse length of Q-switched devices is used in the removal of tattoos and skin blemishes since certain tattoo dyes and melanin strongly absorb red light, in contrast to the surrounding skin. Other applications take advantage of the visible, coherent nature of ruby laser light, and include interferometry, non-destructive testing, holography and plasma measurement. The price of a ruby laser depends on the complexity of the system into which it is built: a sophisticated unit for cosmetic surgery with articulated arm beam delivery can cost about $70 000, whereas a simpler laboratory device designed for spectroscopy typically costs around $10 000. The market for used machines is lively.

Alexandrite

Alexandrite is closely related to ruby, since the active ion is chromium. The host is $BeAl_2O_4$. Alexandrite laser output lies around 800 nm, is tuneable, and can be Q-switched. Excitation is normally by arc lamps or flashlamps. Power levels close to 100 W are available. The wavelength is shorter than that of the Nd:YAG laser, giving improved absorption properties with metals. Light of wavelength 755 nm from a flashlamp-pumped long pulse (3 ms) alexandrite laser is an effective means of removing hair follicles and skin pigmentations.

Ti:sapphire

The active medium in the Ti:sapphire laser is a host of sapphire (Al_2O_3), doped with small amounts (less than 0.0015% by weight) of titanium ions (Ti^{3+}). The sapphire host is robust; it has a high thermal conductivity, and is mechanically rigid and chemically inert. The laser transition takes place between the $E_{3/2}$ excited state and the $^2T_2$ ground state. The resonator can be configured in several amplifier stages to achieve high power levels.

The optical cavity comprises a Ti:sapphire rod, typically about 20 mm in length, optical elements to produce a short pulse length, and two focusing mirrors. Energy is absorbed over wavelengths between 450 and 600 nm, and so a wide range of pump wavelengths is possible. However, the short lifetime in the upper laser level (3.2 μs) leads to a high pump threshold, making flashlamp pumping difficult. A frequency-doubled Nd:YAG laser or a continuous wave argon laser (c. 500 nm wavelength) is therefore used for excitation.

Ti:sapphire output has a broad bandwidth, which allows it to be tuned between 680 and 1100 nm. Mode locking and chirped pulse amplification (CPA) are used to compress pulse lengths to the femtosecond (10^{-15} s) level.

Commercial machines for material processing are available with a wavelength around 800 nm. The output of the Ti:sapphire laser is characteristic: a short pulse length (on the order of femtoseconds); a high pulse energy (on the order mJ); a high repetition rate (a few kHz); and a high beam quality ($M^2 = 1.5$), which is close to TEM_{00}. The Ti:sapphire laser lies at the heart of photoablative micromachining systems. Electronic components are trimmed and layers selectively ablated with an accuracy on the order of 1 μm. Applications are found in medicine, electronics, and optoelectronics. The laser can be tuned to different wavelengths to treat different types of pigmented lesions on the skin. It is also used in photorefractive keratectomy (PRK) and lithotripsy, where the beam is guided to the kidney via an optical fibre. The output is also suitable for selectively removing coatings or deposits from buildings and sculptures.

Diode-pumped Solid State

Diode-pumped solid state (DPSS) lasers take advantage of the ability of diode lasers to optically excite active media in the form of insulating solids in a variety of geometries. In addition to YAG and glass, host materials used in DPSS lasers include yttrium lithium fluoride ($YLiF_4$, known as YLF) and perovskite ($YAlO_3$, YAP). Neodymium and other lanthanides such as holmium (Ho), erbium (Er) and thulium (Tm) are used as dopants.

DPSS lasers are efficient, reliable, long-lasting sources able to produce a multikilowatt beam with a quality that is superior to conventional lamp-pumped units. When combined with non-linear frequency conversion, these lasers can produce output that spans the spectrum from the ultraviolet to the mid-infrared. Operation can also range from subpicosecond pulses to continuous. They are small, which facilitates incorporation into moving laser processing systems. The wall plug efficiency can be up to three times that of lamp-pumped lasers, with maintenance intervals about 20 times longer. These factors provide DPSS lasers with a competitive advantage in applications such as material processing, medicine, metrology and remote sensing.

The diode laser pumps can be placed in a variety of orientations, which enables active media to be made in novel geometrical shapes, such as slabs, discs, fibres and tubes, discussed below. High pumping intensities are possible because the thermal gradients induced can be aligned with the direction of beam propagation. Efficient cooling can be achieved by placing heat sinks on appropriate faces of the active medium.

The active medium in a *slab* laser is a rectangular-shaped crystal, which is excited and cooled through its longitudinal faces. The beam is internally reflected at the slab walls, taking a zig-zag path through the active medium. In comparison with other geometries, a relatively large volume of active medium can be excited. Improved cooling reduces thermal loading, enabling greater power and a higher quality rectangular beam to be extracted. (M^2 values between 2 and 3 may be obtained from slab lasers.) The optical quality of the crystal is less critical than the rod design because beam irregularities are smoothed in the extended optical path. However, crystals in the form of slabs are more expensive than rods. Both flashlamp-pumped and diode-pumped slab designs have been constructed.

The geometry of the active medium in a *disc* laser is similar to that of a coin, with an aspect ratio (diameter:thickness) around 20. The aspect ratio is determined by the requirement for sufficient light amplification along the disc axis and adequate cooling through the face(s). One face of the disc is coated to create the optical cavity. High intensity excitation is possible because the thermal gradients induced are aligned with the axis of the disc. (The fracture limit of a disc scales inversely with its thickness.) Efficient cooling is achieved by placing a heat sink on one of the disc faces. The beam quality is high because the principal thermal gradient lies along one dimension. Output power can be scaled to multikilowatt levels without degrading beam quality – a notable advantage of this cavity design, and the reason for the interest in its use in production line welding and cutting operations.

The cavity of *tube* lasers is made by boring a cylinder into a rod and placing the flashlamp inside or outside the tube. The absorption of pumping energy is high and thermal lensing is low, such that a high power beam of high quality can be produced.

Active media can also be made in the form of *fibres*. The fibre is bounded by an end mirror and an output coupler. Fibre lasers may be pumped at their ends or along their length. The intensity of end pumping can be increased by cladding the fibre in a material of different refractive index: energy incident on the larger clad cross-section is internally reflected in the cladding (in a similar manner to a fibre optic), effectively pumping the fibre along its surface. By arranging fibres in modules, output power can be scaled while beam quality, which is determined by the dimensions and numerical aperture of the cavity, is maintained. Compact lasers can thus be produced without the need for a chiller, or output power can be scaled to multikilowatt levels with suitable cooling. The beam is of high quality (close to diffraction-limited) and has low divergence, because the ratio of the fibre diameter to its length is small.

Nd:YAG

Conventional Nd:YAG rods can be pumped using diodes (in addition to the lamps described earlier), which may be located at the ends of the rod or along its length. The former is more common in low power machines, in which a high quality beam mode can be generated. The gallium aluminium arsenide (GaAlAs) diode laser emits light of wavelength 807 nm, which corresponds with an absorption band of neodymium ions. The excitation efficiency is therefore high (30–40%). In comparison with lamp pumping, the low thermal load on the rod results in an improved beam quality, higher pulse rates, superior pulse repeatability and longer lifetimes.

The higher beam quality of diode-pumped Nd:YAG lasers provides a number of advantages for material processing: the smaller focused diameter gives higher power density; resonators and optics can be more compact; and larger working distances can be used. Diode-pumped Nd:YAG lasers are available with multikilowatt power levels for a variety of material processing applications. Frequency-doubled output challenges the conventional argon ion gas laser in the important blue–green portion of the spectrum for reprographics.

Er:YAG

The Erbium:YAG laser is ideal for cosmetic procedures on delicate skin, such as the hands and neck, and fine lines and wrinkles on the face and around eyes. This laser is also used to prepare dental cavities. Output from the Er:YAG laser can be frequency quadrupled to give pulsed blue–indigo light.

Er:YLF

Erbium can be doped to levels around 8% in a YLF host. Efficient diode pumping is achieved with a wavelength of 797 nm. The laser transition in the Er:YLF laser takes place between the $^4I_{11/2}$ (upper level) and $^4I_{13/2}$ (lower level), which results in the emission of a photon of wavelength 2800 nm. This wavelength lies close to the absorption peak of water molecules, and so the laser finds many applications in medicine.

Ho:YAG and Ho:YLF

Pulsed light from the Ho:YAG and Ho:YLF lasers is effective in lithotripsy as a means of removing urinary calculi (e.g. gallstones) by photochemical decomposition.

Yb:YAG

In comparison with the Nd:YAG crystal, the maximum doping level of Yb in YAG is higher (25% versus 1.5%), the absorption bandwidth is larger (reducing thermal loading), and the upper level lifetime is longer (enabling more energy to be stored). Yb:YAG has a maximum absorption efficiency near 940 nm, and so it can be pumped efficiently by InGaAs diodes, which are more robust than the AlGaAs diodes used to pump neodymium lasers. (When flashlamp pumping was the only means of excitation, the Nd:YAG laser had a competitive advantage.) Output is generated in wide emission bands, suitable for ultrashort pulse operation. Frequency doubling results in an output wavelength of 515 nm, providing the potential to replace the larger volume Ar ion laser, which emits a wavelength of 514 nm.

Suprakilowatt output of high beam quality can be obtained from a Yb:YAG disc several millimetres in diameter with a thickness less than 1 mm because of efficient excitation and cooling. Multikilowatt output is available from modules of Yb-doped fibres. The cost of such kilowatt-class lasers lies around that of comparable lamp-pumped YAG lasers, but less floor space is required, and maintenance intervals are longer. They are particularly suitable for cutting and welding.

Nd:YLF

YLF is the most common alternative to YAG as a host for neodymium doping; it has a lower thermal conductivity and is not as hard, but exhibits less thermal lensing and can operate continuously at room temperature.

Nd:YAP

The crystal anisotropy of YAP results in a small tuning range of wavelength. Nd:YAP lasers are used in dental procedures.

Nd:GGG

A laser made with gadolinium gallium garnet ($Gd_3Ga_5O_{12}$, GGG) doped with neodymium produces light of wavelength 1061 nm. GGG crystals can be grown more easily than YAG crystals, and so the possibilities of producing sources of high average power are greater. Such units are of interest in inertial confinement fusion (Chapter 17) and military laser systems.

Tm:YAG

Thulium-doped YAG solid state lasers that operate at 2 mm wavelength have many applications in medical, remote sensing and military technologies.

Colour Centre

Colour centres (or F-centres) are formed when molecules or ions are bound to neighbouring vacancies. Lithium fluoride doped with titanium and magnesium has been used to host molecular fluorine and F_3^+, which contain two electrons bound to two and three neighbouring anion vacancies, respectively. The colour centre is produced by irradiation with a femtosecond pulse. Visible light in the red–green range can be produced. Such lasers are suitable for the construction of miniature optical devices.

SEMICONDUCTORS

Before describing the workings of a semiconductor laser, it is worthwhile considering the terminology used today. The terms *semiconductor laser* and *diode laser* are often (correctly) used interchangeably. However, the term *laser diode* sometimes appears (incorrectly) in the same context. A laser diode refers to the combination of the active medium, photodiode chip used to control the power, and housing, which are combined with electronics and optics. Note also that a *light emitting diode* (LED) can be thought of as a laser diode without an optical cavity for feedback. The term *semiconductor laser* is used here when describing the physics of operation, and *diode laser* used when referring to commercially available units.

A semiconductor laser is an edge emitting device with a Fabry–Perot optical cavity, illustrated in Fig. 3.22a. The front and rear facets of the cavity are normally coated to act as mirrors, and the sides are roughened to reduce reflection back into the laser. (The beam may also be extracted from the top surface by creating an appropriate cavity.) Excitation is by electrical means, which results in direct injection of electrons into the active medium. Small (100 µm long) emitters are arranged in a bar about 1 cm in length. Many tens of watts of power can be extracted from a single bar. Laser output therefore comprises beams from a large number of individual sources, which creates a high beam divergence (because of diffraction effects) and a relatively poor beam quality in comparison with solid state laser output. The raw beam is suitable for surface treatment, but must be manipulated for penetration processing. A variety of cooling geometries have been designed, including backplane cooling of many laser diode bars by a single heat sink, or the use of individual heat sinks attached to each diode bar. Output may be delivered directly to the workpiece, or via a fibre optic.

(a)

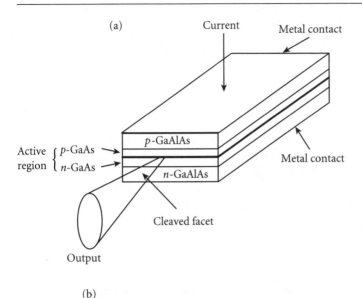

Current

Metal contact

p-GaAlAs

Active region { p-GaAs → n-GaAs →

n-GaAlAs

Metal contact

Cleaved facet

Output

(b)

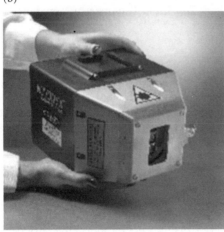

Figure 3.22 Diode laser: (a) schematic illustration of a Fabry–Perot double heterojunction semiconductor structure; (b) Nuvonyx ISL-4000L 4 kW InGaAlAs diode laser head. (Source: Tom Pallett, Nuvonyx Inc., Bridgeton, MO, USA)

Diode lasers originally found use in low power communications devices. In 1991, commercial diode lasers cost around $2000 per watt. By 1999 the price had fallen to $100 per watt, because of growing markets and investment by laser manufacturers. Diode lasers are consequently replacing other light sources for many medical, graphical and illumination applications.

Diode laser heads are now packaged with beam manipulation optics to give multikilowatt output for material processing – one is shown in Fig. 3.22b. The relatively short wavelength, scale of power output, and rapid control over power modulation provide benefits for processing materials. Performance is better than that obtained from an Nd:YAG laser of comparable power, provided that there are no particular absorption problems at the wavelength used. The laser head is sufficiently compact to be mounted directly on an articulated robot. Key benefits for users include compactness, high efficiency, high reliability, and low maintenance. Because diode lasers have high electrical-to-optical power conversion efficiency (up to 50%), they can deliver light for myriad applications based on heating or illumination at a fraction of the power consumption, cost and bulk of competing laser and non-laser sources. The nature of the application determines the most suitable wavelength and power, and hence the type of diode laser used.

GaAs

The principle of light generation in semiconductors was first demonstrated in gallium arsenide (GaAs). Early diode lasers took the form of a *p–n* junction comprising an *n*-type GaAs host into which atoms such as zinc were diffused to create a heavily doped *p* region. Low power, divergent, multimode, infrared light is produced from the homojunction GaAs laser. Only pulsed output can be obtained because CW operation results in overheating and damage, without active cooling. Heterojunction lasers, which comprise layers of different semiconductors, are therefore now used for material processing.

InGaAs

By doping with indium (In), a heterojunction laser capable of higher power output than the GaAs device can be constructed, shown schematically in Fig. 3.22a. The preferred wavelength for material processing is 940 nm.

InGaAlAs

Very stable output in CW or pulsed mode can be achieved from InGaAlAs diode lasers. CW output in the range 750–850 nm can be produced from a single unit. Multikilowatt power levels can be extracted from diode laser arrays. Output of wavelength 808 nm is favoured for material processing. The beam is normally elliptical, with quality values (M^2) of 1.02 and 1.3–100 parallel and perpendicular to the junction, respectively, depending on the injection current.

GaN

The gallium arsenide (GaN) laser was developed in 1995 for use in optical memory devices. Output lies in the blue–violet.

Lead Salt

The infrared output from the lead-salt laser can be tuned by adjusting the laser's temperature or the excitation current. These lasers are normally cryogenically cooled, but recent developments have removed the cooling requirements, which will lead to new applications in industry, research and process control. Such lasers are used for trace measurement of pollutants in the atmosphere and the analysis of reaction kinetics.

SUMMARY AND CONCLUSIONS

Laser light has unique properties: coherence (spatial and temporal); monochromaticity; low divergence; and high brightness. Lasers can be categorized according to the nature of the active medium (gas, solid or liquid). Gas media comprise molecules, atoms, ions and excimers. The principal liquid media are organic dyes. Solid media include insulating crystals and semiconductors. Laser light interacts with materials through thermal, photochemical, photoelectric and photophysical modes.

Designs for gas lasers are becoming more compact to reduce the floor space required, and more efficient in their use of gases, which reduces running costs, while output power is continually rising. The popularity of the carbon dioxide laser can be attributed to a number of favourable properties. Pulsed or continuous wave emission is produced in a high quality beam at suprakilowatt power levels. Far infrared light is transmitted readily in air, and is absorbed by a wide range of engineering materials. Designs, which are relatively simple and robust, are scaled easily to high power levels. The capital cost is relatively low (around $100 per watt). The wall plug efficiency is high (up to almost 20%), and consumable costs are low, leading to a relatively low cost of ownership.

Growth in the use of diode-pumped solid state lasers (DPSS) is expected to be vigorous because of the design opportunities that this form of excitation affords. Novel geometries of active media can be produced, with the design optimized for particular properties, such as output power, or beam quality. Multikilowatt DPSS lasers compete with gas and lamp-pumped solid state devices; the benefits that they offer (compactness, high power efficiency and high beam quality) will mean that they are likely to replace many of the traditional large-scale material processing sources.

The luminous performance of visible light emitting diodes (LEDs) has increased by a factor of ten every decade since the 1960s (Craford's law). The cost of diode lasers fell by a factor of ten during the last decade of the twentieth century. Diode lasers, now available in multikilowatt designs, are suitable for direct material processing. Their application can be expected to grow at a similar rate to LEDs (which are now rapidly replacing incandescent sources of light).

Ultrashort (femtosecond scale) pulsed sources, and high energy ultraviolet excimer lasers have improved the accuracy of athermal micromachining by several orders of magnitude. Lasers that once were thought to have insufficient power for material processing are increasingly finding uses in small-scale machining such as lithography. As photonics takes over from electronics, the use of such laser-based microfabrication techniques will increase.

Efforts continue to extend the wavelength range of lasers. The output from free electron sources, dyes, and certain solids can be tuned to given wavelengths. High energy short wavelength output is available from X-ray lasers.

When sources become sufficiently compact, and are packaged into dedicated turnkey systems, their field of application grows rapidly. For example, as soon as laser-based surgical and cosmetic procedures could be performed by practitioners using commercial turnkey systems, exponential growth was experienced worldwide. Such 'packaging' of lasers in material processing systems is the subject of the next chapter.

FURTHER READING

LASERS: PRINCIPLES, CONSTRUCTION AND APPLICATION

Arecchi, F.T. and Schulz-Dubois, E.O. eds (1972). *Laser Handbook*. Amsterdam: North-Holland.

Baur, D.R. (1997). *Lasers: Theory and Practice*. Tunbridge Wells: Elektor Electronics.

Bellis, J. ed. (1980). *Lasers – Operation, Equipment, Application and Design*. New York: McGraw-Hill.

Bertolotti, M. (1983). *Masers and Lasers*. Bristol: Adam Hilger.

Bloom, A.L. (1968). *Gas Lasers*. New York: John Wiley & Sons.

Brown, R. (1968). *Lasers, Tools of Modern Technology*. London: Aldus Books.

Carruth, J.A.S. and McKenzie, A.L. (1986). *Medical Lasers: Science and Clinical Practice*. Bristol: Adam Hilger.

Duley, W.W. (1983). *Laser Processing and Analysis of Materials*. New York: Plenum Press.

Hall, D.R. and Jackson, P.E. eds (1989). *The Physics and Technology of Laser Resonators*. Bristol: Adam Hilger.

Hecht, J. (1992). *The Laser Guidebook*. 2nd ed. Blue Ridge Summit: TAB Books.

Hitz, C.B. (1985). *Understanding Laser Technology*. Tulsa: PennWell Publishing.

Hitz, C.B., Ewing, J.J. and Hecht, J. eds (2001). *Introduction to Laser Technology*. 3rd ed. New York: John Wiley & Sons.

Kogelnik, H. and Li, T. (1966). Laser beams and resonators. *Proceedings of IEEE*, **54**, (10), 1312–1329.

Kuhn, K.J. (1998). *Laser Engineering*. New York: Prentice Hall.

Milonni, P.W. and Eberly, J.H. (1988). *Lasers*. New York: John Wiley & Sons.

Niku-Lari, A. and Mordike, B.L. eds (1989). *High Power Lasers*. Oxford: Pergamon Press.

Oakley, P.J. (1999). *Introduction to Engineering Lasers*. Dordrecht: Kluwer.

O'Shea, D.C., Callen, W.R. and Rhodes, W.T. (1977). *An Introduction to Lasers and Their Applications*. Reading: Addison-Wesley.

Siegman, A.E. (1986). *Lasers*. Herndon: University Science Books.

Siegman, A.E. (2000). Laser beams and resonators: the 1960s. *IEEE Journal of Special Topics in Quantum Electronics*, **6**, (6), 1380–1388.

Siegman, A.E. (2000). Laser beams and resonators: beyond the 1960s. *IEEE Journal of Special Topics in Quantum Electronics*, **6**, (6), 1389–1399.

Silvfast, W.T. (1996). *Laser Fundamentals*. Cambridge: Cambridge University Press.

Svelto, O. (1998). *Principles of Lasers*. 4th ed. New York: Plenum Press.

Young, M. (1984). *Optics and Lasers*. 2nd ed. Berlin: Springer.

Vaughan, J.M. (1989). *The Fabry Perot Interferometer: History, Theory, Practice, and Applications*. Bristol: Adam Hilger.

Weber, M.J. ed. (2003). *Handbook of Lasers*. Boca Raton: CRC Press.

Wilson, J. and Hawkes, J.F.B. (1987). *Lasers: Principles and Applications*. New York: Prentice Hall.

CARBON DIOXIDE LASERS

Adams, M.J. (1993). CO_2 gas lasers: engineering and operation. In: Crafer, R.C. and Oakley, P.J. eds *Laser Processing in Manufacturing*. London: Chapman & Hall. pp. 115–139.

Crafer, R.C., Greening, D. and Brown, M. (1996). *Design, Construction and Operation of Industrial CO_2 Lasers*. Dordrecht: Kluwer.

Duley, W.W. (1976). *CO_2 Lasers: Effects and Applications*. New York: Academic Press.

Evans, J.D. ed. (1990). *Selected Papers on CO_2 Lasers*. SPIE Milestone Series, Volume MS22. Bellingham: SPIE.

Witteman, W.J. (1987). *The CO_2 Laser*. Berlin: Springer.

CARBON MONOXIDE LASERS

Bhaumik, M.L., Lacina, W.B. and Mann, M.M. (1972). Characteristics of a CO laser. *IEEE Journal of Quantum Electronics*, **QE-8**, (2), 150–160.

DYE LASERS

Duarte, F.J. ed. (1991). *High Power Dye Lasers*. Berlin: Springer.

EXCIMER LASERS

Ewing, J.J. (2003). Excimer lasers at 30 years. *Optics and Photonics News*, **14**, (5), 26–31.

Gower, M.C. (1993). Excimer lasers: principles of operation and equipment. In: Crafer, R.C. and Oakley, P.J. eds *Laser Processing in Manufacturing*. London: Chapman & Hall. pp. 163–187.

Stein, H.A., Cheskes, A.T. and Stein, R.M. (1997). *The Excimer: Fundamentals and Clinical Use*. 2nd ed. New York: Slack Inc.

FREE ELECTRON LASERS

Saldin, E.L., Schneidmiller, E. and Yurkov, M. (1999). *The Physics of Free Electron Lasers*. Berlin: Springer.

METAL VAPOUR LASERS

Ivanov, I.G., Latush, E.L. and Sem, M.F. (1996). *Metal Vapour Ion Lasers*. Chichester: John Wiley & Sons.

Little, C.E. (1999). *Metal Vapour Lasers*. Chichester: John Wiley & Sons.

NEODYMIUM LASERS

Nonhof, C.J. (1988). *Material Processing with Nd-Lasers*. Ayr: Electrochemical Publications.

IODINE LASERS

Brederlow, G., Fill, E. and Witte, K.-J. (1983). *The High-power Iodine Laser*. Berlin: Springer.

SEMICONDUCTOR LASERS

Agrawal, G.P. and Dutta, N.K. (1993). *Semiconductor Lasers*. New York: Van Nostrand Reinhold.

Casey, H.C. and Panish, M.B. (1978). *Heterostructure Lasers*. Orlando: Academic Press.

Gurevich, S.A. ed. (1998). *High Speed Diode Lasers*. River Edge: World Scientific.

Holonyak, N. Jr (2000). From transistors to light emitters. *IEEE Journal of Special Topics in Quantum Electronics*, **6**, (6), 1190–1200.

Kressel, H. and Butler, J.K. (1977). *Semiconductor Lasers and Heterojunction LEDs*. New York: Academic Press.

Thompson, G.H.B. (1980). *Physics of Semiconductor Laser Devices*. Chichester: John Wiley & Sons.

Vasil'ev, P. (1995). *Ultrafast Diode Lasers: Fundamentals and Applications*. Norwood: Artech House.

SOLID STATE LASERS

Iffländer, R. (2000). *Solid-state Lasers for Materials Processing: Fundamental Relations and Technical Realizations*. Berlin: Springer.

Koechner, W. (1999). *Solid-State Laser Engineering*. 5th ed. Berlin: Springer.

Scheps, R. (2001). *Introduction to Laser Diode-pumped Solid State Lasers*. Bellingham: SPIE.

ULTRAFAST LASERS

Diels, J.-C. and Rudolph, W. (1996). *Ultrashort Laser Pulse Phenomena: Fundamentals, Techniques and Applications on a Femtosecond Time Scale*. San Diego: Academic Press.

ULTRAVIOLET LASERS

Duley, W.W. (1996). *UV Lasers: Effects and Applications in Materials Science*. Cambridge: Cambridge University Press.

Elliott, D.J. (1995). *Ultraviolet Laser Technology and Applications*. San Diego: Academic Press.

Laude, L.D. ed. (1994). *Proceedings of the NATO ASI Excimer Lasers*. Norwell: Kluwer Academic Publishers.

CHAPTER 4

SYSTEMS FOR MATERIAL PROCESSING

INTRODUCTION AND SYNOPSIS

The beam emitted from a laser is rarely suitable for material processing in its raw form – it is seldom the desired size, and often the intensity distribution is not appropriate for the process. It needs to be manipulated into a suitable processing tool. Transmissive, reflective and diffractive optics have been designed to be incorporated into beam delivery systems of varying sophistication. Similarly, systems for handling workpieces are available in a wide range of configurations. To match the most appropriate beam delivery and work handling systems to create a laser processing centre, many requirements must be taken into consideration – flexibility, accuracy, productivity and the nature of the components being some of the most important. Auxiliary equipment may then be added to the centre for tasks such as computer-aided design and manufacturing, beam diagnostics, process monitoring, and automatic adjustment of processing parameters. On-line adaptive control systems then become possible. The current goal is to use centres that require as little human intervention as possible.

This chapter considers the requirements of systems for laser material processing. Optical components, beam delivery techniques, methods of workpiece manipulation, and process monitoring devices are examined. The factors involved in the selection, construction and acceptance of industrial material processing systems are then discussed, together with safety requirements.

OPTICS

The raw beam emitting from a high power CO_2 laser, for example, may be between 15 and 70 mm in diameter, possessing a range of possible modes. Normally the beam from a laser must be modified into a suitable energy source for material processing. Up to a power level of about 5 kW, the beam can be manipulated using transmissive optics, whereas with a higher power level more robust water-cooled reflective devices are required.

OPTICAL TERMINOLOGY

The characteristics of optical devices are referred to using specific terminology related to the properties of the laser beam. Further information about the characterization of the beam can be found in Chapter 3.

Focal Length

The strength of a lens is measured by its focal length. The focal length is the distance from centre of lens to the focal point. The effective focal length is the distance from plane of clear aperture (front of lens) to the focal plane. The working distance is the distance from the back face of the lens to the focal plane. The distance between the point on the back surface of the lens to the focal point, measured along the optical axis, is known as the back focal length. The effective focal length is the distance that the designer uses to calculate the curvature of the lens. It is the distance from the principal plane in which an incoming ray is bent towards the focal point.

Optics with a large focal length have a greater depth of field, which gives a higher tolerance for irregularities in surface flatness. The optic can be placed a greater distance from the workpiece, reducing the chance of damage, and providing more room for other equipment. However, a large focal length produces a larger focused spot size with a lower power density. An optic with a short focal length produces a smaller focused spot, but extreme precision is needed to realize the full focusing potential, and the proximity to the workpiece may also make access more difficult.

Focal Number

The focal number, or f-number, of a focusing optic characterizes its focusing ability. The f-number defines the convergence angle of the beam:

$$f = \frac{F}{d_B} \tag{4.1}$$

where F is the focal length of the optic and d_B is the diameter of the beam or aperture, whichever is smaller. Typical focusing optics for multikilowatt laser systems lie in the range $f/6$ to $f/10$. A beam with a high f-number can be angled closer to a vertical member.

Beam Diameter at Focus

The minimum theoretical diameter, d_f, to which a laser beam of original diameter, d_B, and mode TEM_{00} can be focused is

$$d_f = \frac{4\lambda F}{\pi d_B} = \frac{4\lambda f}{\pi}. \tag{4.2}$$

The effect of the beam mode on the minimum spot diameter can be expressed in terms of the beam quality factor, K (Chapter 3):

$$d_f = \frac{4\lambda}{\pi} \frac{f}{K}. \tag{4.3}$$

Equations (4.2 and 4.3) illustrate the benefit of using an optic with a short f-number, a laser source of short wavelength, and a low order beam mode, in obtaining a small focused spot diameter.

The diffraction-limited spot size at focus, d_f, can be calculated from diffraction theory to give:

$$d_f = 2.44 \frac{\lambda F}{d_B} (2M + 1)^{1/2} \tag{4.4}$$

where λ is wavelength, F is the focal length of the optic, d_B is the diameter of the incident beam, and M is the number of oscillating modes (not to be confused with the magnification M of an annular beam.)

Depth of Focus

The depth of focus is also known as the depth of field. The depth of focus, z_f, is a measure of the change in the waist of the beam either side of the focal plane. It is often defined as the distance along the axis over which the focused spot size increases by 5%, or the distance over which the intensity exceeds half the intensity at focus. For a TEM_{00} beam it is defined as

$$z_f = \frac{8\lambda}{\pi}\left(\frac{F}{d_B}\right)^2 = \frac{8\lambda f^2}{\pi} = 2fd_f. \tag{4.5}$$

For a higher order beam mode of quality K, the depth of focus defined by separation of the points at which the beam waist is $\sqrt{2}d_f$ is given by

$$z_f = \frac{4\lambda}{\pi K}\left(\frac{F}{d_B}\right)^2. \tag{4.6}$$

The depth of focus is proportional to the square of the spot size, i.e. a smaller spot size leads to a shorter depth of focus. A compromise is often sought between these two features in practical applications – a small spot size to give a high power density, but a large depth of focus for through-thickness processing.

As a rough guide, the depth of focus of a lens is approximately 2% of its focal length.

TRANSMISSIVE OPTICS

Transmissive optics are normally easy to set up and enable relatively simple laser heads to be constructed. The improved focusing properties can result in smaller focused spots than a reflective optic of similar focal length. However, with a power level above about 5 kW, thermal lensing occurs. This originates from a combination of a change in shape through thermal expansion, and a change in the refractive index with increasing temperature. The range of power is also limited because lenses can only be cooled around the edges (although a limited amount of cooling can be produced by blowing a gas across the face). Lenses are also particularly sensitive to contamination. Particles of dirt can be burned onto the surface, producing localized heat sources and diffracted light. The lens is normally sandwiched between two rings of indium wire to form an airtight seal. Lenses should be used with care in sensitive applications, since thermal lensing affects the position of the focal plane, which could be critical when a small focused spot is required.

Materials

Conventional glass lenses cannot be used with far infrared CO_2 radiation because glass is not transparent to this range of wavelength. There are three general classes of candidate material: semiconductors; alkali halides; and non-oxide glasses. Semiconductors comprise a group II and a group IV element, such as zinc selenide, cadmium telluride, cadmium sulphide, gallium arsenide and zinc sulphide. The properties of such materials are given in Table 4.1. Zinc selenide (ZnSe) has become the material of choice with far infrared radiation for a number of reasons. It also transmits to some extent in the visible region, and so can be used in conjunction with red He–Ne laser light (although the different refractive indices for infrared and visible light produce different focusing characteristics). It remains essentially transparent to far infrared light up to several hundred degrees centigrade. It also possesses a high thermal conductivity, which reduces thermal loading. Alkali halides such as potassium chloride and sodium chloride transmit over a broad spectrum, and are five to ten times cheaper than ZnSe, but are hygroscopic and therefore have poor service properties under conditions of high humidity. Also, they cannot be coated easily, and therefore do not have the light transmittance and focusing ability of ZnSe.

Table 4.1 Properties of transmissive materials used with 10.6 μm wavelength light

	Zinc selenide	Germanium	Gallium arsenide
Transmission range (μm)	0.5–18	2–14	1–12
Reflective losses at surfaces (%)	30	53	45
Refractive index	2.40	4.00	3.24
Refractive index thermal coefficient (K^{-1})	64	277	149
Absorption coefficient (cm^{-1})	<0.0015	<0.030	<0.010
Thermal conductivity ($W\,cm^{-1}\,K^{-1}$)	0.18	0.59	0.48
Thermal expansion coefficient ($K^{-1} \times 10^{-6}$)	7.57	6.1	5.7
Hardness (Knoop 50 g)	130	692	750
Rupture modulus ($MN\,m^{-2}$)	55.2	93.1	138

Borosilicate crown glass (BK7) is used in optics for near infrared radiation, such as that from the Nd:YAG laser. It has excellent optical and thermal properties. Pyrex, synthetic fused silica, or a special glass ceramic called Zerodur can also be used. Glass optics are less expensive and higher in optical quality than optics for CO_2 laser radiation. Inexpensive glass cover slides can be used to protect optics from ejected material from the processing zone.

Quartz or fused silica is preferred for excimer laser optics because of their superior short wave transmittance. Crystalline MgF_2 or CaF_2 can also be considered.

Coatings

The simplest form of anti-reflection (AR) coating is a vacuum-deposited single thin film, with a thickness equal to one quarter of the wavelength of the light. The refractive index is equal to $\sqrt{n_s}$, where n_s is the refractive index of the substrate material. This results in destructive interference between light reflected from the substrate and from the coating surface, with the effect that no light is reflected. The coating can be made thicker, provided that it is always an odd multiple of a quarter wavelength, although the absorptivity increases.

Enhanced reflection (ER) coatings are also made by vacuum deposition. In their simplest form they comprise alternating quarter-wavelength layers of materials of differing refractive index, such that light reflected from alternate layers interferes constructively. As many as 20 pairs of layers may be added, producing a reflectance of over 98.8%.

Absorption of beam energy causes heating and expansion, which has two main effects: the refractive index increases with temperature, shortening the focal length; and differential expansion of the lens occurs (greater at the centre than the edge), both of which result in thermal lensing. Absorption occurs mainly at the lens surface, as a result of contaminants. Water cooling at the edges is the main method of reducing temperature rises, but contributes to the thermal gradient in the optic. Gas can be blown across the surface, although it is difficult to cool the lens uniformly. The optimum lens thickness is a balance between an increased capability to conduct heat with thickness, and an increasing absorption.

Beam Splitters

A beam splitter allows a selected fraction of the incident energy to be reflected with the remainder being directed to another location. In most cases beam splitters are angle, wavelength and polarization sensitive.

LENS PARAMETERS

Positive lenses bend parallel light rays so that they converge; negative lenses cause parallel rays to diverge. The desired spot size is the parameter with most influence in determining the type of lens required. The most important geometrical parameters of a lens are: edge thickness; effective focal length; back focal length; and working distance. The majority of material processing is carried out using the positive meniscus or the plano-convex lens.

Plano-convex Lens

One surface of a plano-convex lens is flat, and the other is the convex surface of a sphere. They are the most economical focusing elements available, since they can be made easily and require little raw material in their manufacture. They are used primarily when spot size and spherical aberration (discussed later) are not critical. Low f-number optics are limited in performance because their short focal length dictates that they have a steeper curve (tighter radius) on their back face, which gives a substantial change in lens cross-section over the diameter of the lens. This results in uneven thermal expansion across the lens, as well as increased spherical aberration, both of which alter focal length and focused spot size. They are therefore a better choice when a focal length above 20 cm is required, although are available with effective focal lengths in the range 2.5–25 cm. The edge thickness is typically 3–4 mm, but lenses up to 7.5 mm in thickness may be used in high pressure gas applications. The convex side must face the incoming beam in order to minimize spherical aberration. An example is shown in Fig. 4.1a.

Positive Meniscus Lens

A positive meniscus lens has two spherical surfaces. They are designed to reduce spherical aberration and thereby produce a smaller spot size for a given focal length than a plano-convex lens. They are recommended for operations where an f-number less than 3 is required, since their more uniform thickness reduces spherical aberration effects. They are typically available with effective focal lengths in the range 1–10 inches (25.4–254 mm). However, they are more expensive to manufacture than plano-convex lenses. An example is shown in Fig. 4.1b.

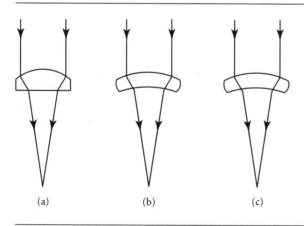

(a)   (b)   (c)

Figure 4.1   Focusing lenses (a) plano-convex, (b) positive meniscus, (c) aspheric

Aspheric Lens

At least one surface of an aspheric lens has a radius of curvature that varies with position on the surface to produce minimum spherical aberration and spot size. The lens is manufactured to suit the laser beam diameter and beam characteristics. Rays are brought to focus at a clean, well-defined focal point, since the surface is not spherical. A unique feature of this type of lens is that as the beam diameter at the lens increases, the focused spot diameter decreases monotonically. With plano-convex and meniscus lenses the focused spot diameter passes through a minimum value, after which it increases because of spherical aberration. Aspheric lenses are available with focal lengths below 127 mm (5 in). This design is particularly useful for cutting where a concentrated heat source and narrow kerf are desired. Total power requirements can be reduced, and the reduced energy input results in less distortion. The cost of an aspheric lens is roughly 2–3 times that of positive meniscus and plano-convex lenses. Varying levels of asphericity can be produced, referred to as well corrected, overcorrected, doubly overcorrected, and high order correction. An example is shown in Fig. 4.1c.

Polarizers

The polarization of light is determined by the relationship between the phases of the two polarized components (Chapter 3). By changing this relationship, the polarization can be changed. A birefringent device can be used to retard the phase of one component. The birefringent material has two refractive indices, called the ordinary and extraordinary indices. Light in one polarization sees the ordinary index, while the orthogonally polarized light sees the extraordinary index. Since the speed of light in a medium depends on the refractive index, the two components travel through the medium at different speeds, emerging out of phase. If the length of the material is selected such that a one quarter phase shift is introduced, then linearly polarized light can be converted into circularly polarized light. Such a device is known as a quarter-wave plate.

The fraction of light reflected by a surface depends on the angle of incidence and the polarization of the light (Chapter 3). For the horizontal component of polarization, the reflectivity increases gradually with angle of incidence. For the vertical component, reflectivity decreases gradually, reaching a minimum at which the reflection is zero, after which it increases. The angle of zero reflection is the Brewster angle. At the Brewster angle, the angle of reflection is perpendicular to the angle of refraction, and therefore the electric vector in the plane of incidence cannot be reflected since there is no component perpendicular to itself. The refractive index, n, is equal to the tangent of the Brewster angle. A Brewster window can be placed inside a laser to introduce a loss of approximately 30% for one component of polarization, but no loss for the other component. This preferential reflection is normally sufficient to restrict polarization to one particular direction.

Beam Collimators

In order to ensure consistent performance, a constant power density must be maintained at the work-piece, irrespective of the location of the final focusing optic relative to the laser (the beam path). Beam expanders and condensers control the focused beam diameter by expanding or condensing the laser beam, thus changing its divergence. Most use multiple optical components which may be positioned relative to one another in order to produce the desired effect. The spacing between the optical components is referred to as the collimation adjustment. The diameter of any optic should be at least 1.5 times the beam diameter, as measured at the $1/e^2$ contour. Two types of expander are used, referred to as Galilean and Keplerian designs, as illustrated in Fig. 4.2.

A typical Galilean design of beam collimator is shown in Fig. 4.2a, in which negative and positive lenses are combined to reduce or enlarge the beam, arranged such that the focal point lies in front of the negative lens. It is a compact construction that can be water cooled.

In the Keplerian design of a transmissive beam collimator, two positive focusing lenses are used, with a focal point that lies between them, Fig. 4.2b. The beam is focused before expanding and entering the second lens. A disadvantage of the Keplerian design is that the beam is focused within the instrument, which may cause air breakdown or thermal distortion with high power beams. This design can also be used to extract a TEM_{00} mode from higher order modes.

In many applications the beam collimator is used to reduce the beam divergence and to ensure that the expanded beam fills the clear aperture of the final optic. It can be used to increase the working distance of a lens by increasing the beam diameter and the focal length of the lens. The magnification of the beam expander can be expressed as the diameter of the output beam divided by the diameter of the input beam.

Axicon Lens

Axicon lenses are a family of optics which image an on-axis point source to a range of points along the axis. An axicon used in laser processing normally has one conical surface, and focuses the beam to an annular shape. It can be used to cut a hole of a given size, or for transformation hardening of an internal circular wall.

Beam Integrator

A transmissive beam integrator is constructed using a faceted lens. The facets split the beam into components (typically $6 \times 6\,mm$), which are recombined in the focal plane to produce a heating pattern 6 mm square of uniform intensity distribution. This can be used in conjunction with meniscus or plano-convex lenses to create a variety of spot sizes. A cylindrical lens allows a rectangular heating pattern to be produced. Smaller facets give a smaller heating pattern, but are more costly to produce, and diffraction effects have a negative effect on the intensity distribution in the plane of integration. Larger facets mean that the beam intersects fewer facets, giving a poorer intensity distribution in the integrated spot. The design is simple to use, and cheaper than the equivalent constructed using mirrors

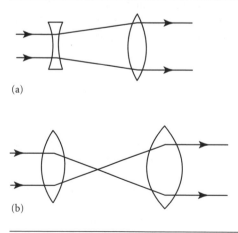

(a)

(b)

Figure 4.2   Transmissive beam collimators (a) Galilean, (b) Keplerian

(described below). Thermal aberrations do not affect performance significantly. The integrator is aligned along an optical axis of the system and is as simple to use as a lens. Such heating patterns are employed in transformation hardening and cladding.

Fibreoptics

Fibreoptic beam delivery is used with Nd:YAG lasers since the near infrared beam is transmitted efficiently by silica glass. The small beam delivery optic can easily be mounted on a robot, allowing flexible three-dimensional processing. With improvements in the repeatable accuracy of robots, robot controlled fibreoptic delivered beams are now common. This also enables fixturing to be optimized without the constraints of beam accessibility. There is also little constraint on relative movement of the work and the laser.

Fibreoptic beam delivery is currently unique to Nd:YAG lasers in a production environment, and around 80% of Nd:YAG lasers use this method of beam transmission in material processing applications. Many mirror-based beam delivery systems require elaborate beam alignment procedures, are susceptible to contamination and power absorption, and require complex motion systems when large multi-axis work envelope stations are used. Once the laser beam is focused into the fibre, it is the fibre diameter that determines the output beam diameter. The cone angle of the output beam matches that of the input beam. Therefore the beam is focused to a size that most closely matches the fibre diameter using as long a focal length lens as possible to give as small a focus cone angle as possible. The acceptance angle of the fibre is about 12°. The sine of the acceptance angle is referred to as the numerical aperture (NA). The acceptance cone refers to the acceptance angle rotated about the fibre axis, and lies around 24°. The beam must be focused to a diameter less than that of the fibre. The beam must then be recollimated on exiting the fibre.

There are two main types of fibre suitable for Nd:YAG laser beam transmission, illustrated in Fig. 4.3. A step index fibre comprises a low refractive index core, usually made from silica glass, clad with a high refractive index sleeve to evenly redistribute bending stresses, surrounded with jackets for

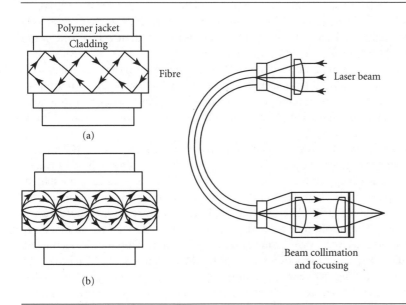

Figure 4.3   Fibreoptic beam delivery: (a) step index fibre; (b) graded index fibre

protection and continuity detection. Beam transmission occurs through total internal reflection at the boundary between the core and the cladding. The minimum core diameter is typically 0.4 mm. The exterior diameter is around 5 mm. A top-hat intensity distribution is produced. In a graded index fibre, there is a radial decrease in the refractive index from the centre, following the shape of a parabola, such that light is guided without a discontinuous core/cladding interface. This is obtained by doping the core by varying amounts. The beam is periodically focused down the fibre. The fibre is thus able to retain some of the input beam mode structure. Graded index fibres are more expensive, but are able to produce a more intense beam, giving, for example, greater penetration in welding. The step index fibre is normally used in material processing for economic reasons, since the output beam properties are sufficient for most processes. There are also single-mode and elliptical fibres.

Core diameters up to 1 mm are available. Smaller diameters are difficult to use because of the high power density on the fibre end and the difficulty of coupling the laser beam into the core. Above 1 mm, flexibility is impeded and bending losses are introduced. Since the minimum focused spot size, and subsequently the maximum power density, is proportional to the core diameter, efforts are being made to reduce the core diameter while retaining the power transmission capability. Most fibre loss occurs at the in- and outcoupling optics, and is typically around 10%. The minimum radius of curvature is proportional to the core diameter. In general, it lies between 100 and 300 times the core diameter. The beam delivery head is normally oriented at a small angle to the normal to the workpiece in order to avoid reflected light being transmitted back down the fibre.

The focused spot size, d_f, is given by

$$d_f = d_{fibre} \frac{F_l}{F_r} \qquad (4.7)$$

where d_{fibre} is the fibre diameter, and F_l and F_r are the focal lengths of the launch and recollimation lenses, respectively. The focused spot size may therefore be reduced by reducing the focal length of the focusing lens. The same effect may be produced by increasing the focal length of the recollimating lens.

The fibre is connected to the laser using a suitable connector. Light is emitted from the fibreoptic with a uniform power distribution, irrespective of input mode, in a cone with an angle only slightly greater than the input angle. Light will leak if the acceptance angle is exceeded, or if it is bent in too tight a radius; this is prevented by using a coaxial limiting structure.

Within the fibre itself, the attenuation of a multikilowatt Nd:YAG beam is on the order of 3–4 dB per km of fibre length. The losses associated with each insertion and extraction typically amount to 3–5% of the transmitted power. Therefore, a total power loss of 7–12% can be expected in a 20 m cable. The limiting factor in determining the length of a single fibre is the size of the original boule from which the fibre is drawn.

The laser may be remote from the work; some installations have fibre delivery of over 200 m. This is a good safety feature, allowing the laser to be placed outside a clean room, or alternatively outside a hostile environment such as in nuclear applications. One laser may be rapidly switched, or time shared, between a number of working locations. This allows different types of processing to be carried out at different locations, and improves 'up time'. Alternatively, the laser beam may be shared between a number of fibres simultaneously. The output may be used, for example, to make welds simultaneously in different locations, or to produce several weld beads to reduce distortion. Fibre networks may be reconfigured with ease to suit changes in production layout or product requirements. The beam size remains constant, irrespective of changes in operating parameters. The only slight disadvantage is that the brightness at the workpiece may be reduced in comparison with a mirror-delivered beam. This is because the focused spot size is determined by the fibre diameter, and the focal length of the focusing lens. The smallest useable spot diameter is typically half the fibre diameter.

The absorption of silica fibre increases with a decrease in wavelength; excimer laser light at wavelengths above 250 nm can be transmitted modest distances through quartz fibres. Articulated arms containing reflective optics are used when fibreoptic beam transmission is not possible.

Line Focus

If a line source of energy is required, cylindrical lenses with one or two surfaces shaped like the sides of a cylinder are used. Light is focused in only one dimension, transforming a circular beam into a line.

Aberrations

The behaviour of lenses deviates from the theoretical behaviour with monochoromatic light for a number of reasons. These affect the focusing properties and performance of the lens in high precision work.

Chromatic aberration arises because of the variation in refractive index with wavelength, such that different wavelengths are brought to focus at different positions. Multi-element lenses assembled from different glasses combat this (achromats). Chromatic aberration is insignificant in single wavelength laser light.

As the focal length of a lens decreases, the curvature of the lens surface increases. Rays incident on the outer portion of the lens (marginal rays) undergo more refraction than those close to the optical axis of the lens (paraxial rays). Marginal rays are thus focused in a plane closer to the lens than paraxial rays. This spreading of the focused beam is spherical aberration. It is most common in low f-number lenses, and sets a minimum practical value. Doublet lenses can be used to reduce spherical aberration.

Astigmatism occurs when a beam is incident on a lens at an angle to the optical axis, even when the beam is centred on the lens. The vertical and horizontal components of the beam focus at different locations. The best compromise focus lies between the foci. A larger focused spot is produced than by an on-axis beam.

REFLECTIVE OPTICS

The main advantage of reflective optics is that they do not suffer significantly from distortion since they can be uniformly water cooled from the back. They are more durable in an industrial environment, operating with higher powers, and can easily be cleaned. However, reflective optics are more difficult to set up, and require a precision laser head to ensure correct alignment with the incoming beam. Diffraction effects can result from diamond turning marks.

Materials

Internal and external CO_2 laser optics are normally made of copper (coated or uncoated), molybdenum, or silicon. Oxygen-free high conductivity (OFHC) copper is commonly used for high power reflective optics because of its high reflectivity when polished (99%) and high thermal conductivity. The surface must be polished to a smoothness less then $\lambda/20$, where λ is the wavelength of the incident light. Copper is easily damaged because of its softness. It is used for multikilowatt beams, but is not generally used for low powers because of difficulties in polishing, which lead to added cost. Metal optics are normally water cooled to maintain a constant mirror temperature in order to prevent distortion. A helical groove can be used to lead water under a thin copper membrane from the centre of the mirror to the edge. This geometry results in a minimum temperature gradient at the mirror surface. Molybdenum is more rugged than copper, but has a lower reflectivity. Silicon is also suitable,

Table 4.2 Reflectivities of materials to 10.6 μm wavelength light with normal incidence

Surface	Reflectivity (%)
Copper – uncoated	98.6
Copper – gold coated	98.9
Copper – silver coated	99.1–99.5
Molybdenum	97.0
Silicon – gold coated	98.9
Silicon – silver coated	99.1–99.7

because its very low coefficient of thermal expansion maintains stability. It is relatively cheap, but is limited to powers below 3 kW. Coated aluminium mirrors have the advantage of low mass in moving applications, although the thermal conductivity is lower. The reflectivities of such materials are given in Table 4.2.

Coatings

Copper mirrors are frequently coated with gold or silver. Gold coatings are normally applied to prevent pitting corrosion, marginally improve reflectivity, and give an abrasive resistance that lies between a molybdenum and a dielectric coating. Silver is over 99% reflective, depending on enhancements. However, it tarnishes easily and so is mainly used for internal optics and with coatings. A sputtered molybdenum coating may be applied to increase durability, but molybdenum only reflects around 97% of CO_2 laser light, resulting in significant losses after multiple reflections, as well as potential mirror distortion.

Multilayer dielectric coatings can improve reflectance to greater than 99.8%. These consist of alternating layers of a quarter-wave film of higher and lower refractive index than the substrate. These layers cause the reflected light to combine constructively and to be reflected efficiently. They also improve wear resistance. They are used mainly with Nd:YAG lasers.

Surfaces should be kept extremely clean. Fingerprints and dust, for example, lead to localized absorption of the beam, resulting in thermal distortion. Dry nitrogen or air is recommended to remove dust. Surfaces may also be cleaned by covering the surface with a solvent such as alcohol or acetone, and carefully dragging a piece of lens tissue across the surface.

Beam Turning Mirrors

The direction of the beam is changed by using plane turning mirrors. A large complex workstation may contain up to ten turning mirrors. About 1% of the incident light is absorbed by each mirror and so one objective of system design is to minimize the number of turning mirrors by considering the handling requirements of the workpiece. The path between the mirrors is enclosed in flight tubes. The internal atmosphere must be controlled carefully in order to avoid beam distortions, particularly in dirty environments.

Beam Splitting Mirrors

As mentioned earlier, a beam splitter allows the beam to be divided into several components. These may be used for simultaneous processing at several different workstations. Alternatively, the power

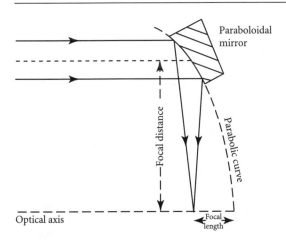

Optical axis

Figure 4.4   Paraboloidal mirror

distribution in a heating pattern can be tailored to a particular component. When hardening the teeth of gear wheels, for example, the root area requires up to ten times more power than the flanks. Thus by splitting the beam into four components, and recombining two in the root and one on each of the flanks, and adjusting the interaction times, the desired hardened profile can be obtained.

Spherical Mirrors

A spherical mirror gives best results with paraxial rays, i.e. rays parallel and close to the optical axis. The angle of incidence should be within 10° of the optical axis, otherwise the focused spot will be too large and have a pear-shaped footprint. This is normally achieved by positioning a flat fold mirror below and to one side of it. This makes the head more bulky and less convenient to use. Spherical mirrors are therefore not normally used where precision focusing is required. But they are used in laser resonators, since the wavefronts are spherical.

Paraboloidal Mirrors

The surface of a paraboloidal mirror is formed from a paraboloid of revolution. It is often referred to as a parabolic mirror. The mirror is used to simultaneously deflect the beam through 90°, as illustrated in Fig. 4.4. In a Z-fold optic, the incoming beam is reflected and expanded using a convex mirror, and then directed onto a paraboloidal mirror, which focuses the beam. The main advantage of a paraboloidal mirror is that a smaller, circular spot is produced, with a higher power density. These mirrors can be integrated into the beam handling system by taking the place of the last turning mirror, thus increasing flexibility. The number of turning mirrors can then be kept to a minimum, reducing total beam losses. The main drawback is that they are more expensive, since the parabola section must be machined by diamond turning a larger piece of material. Precise alignment of the laser beam is critical if a diffraction-limited beam free of spherical aberration is desired. A 50% reduction in peak irradiance can occur with a misalignment of 4 mrad. Paraboloidal mirrors are mainly used in welding applications because of their robustness. Lenses are more suitable for cutting, since they are less sensitive to misalignment.

Deformable Mirrors

A deformable focusing mirror contains PC-controlled piezocrystals, which allow the surface topography to be changed within a range of about $10\,\mu m$, allowing a wide range of beam shapes to be produced electronically, without changing the intensity profile. Thus the shape of the heating pattern can be changed without changing the working distance of the mirror. Alternatively, identical beam shapes can be created at different positions along the beam path.

Beam Collimators

Mirrors may also be used in beam collimators to change the diameter of the beam, as illustrated in Fig. 4.5. Thermal lensing is eliminated, which is a significant advantage with high power beams. However, a disadvantage is that the spherical mirrors are often tilted at a small angle to the optical axis. This creates an offset in the transmitted beam, making installation of this type of beam expander more difficult. The Galilean design performs beam size changes in the shortest optical distance. Reflective collimators are used to maintain uniform beam properties through a laser workstation by maintaining a constancy of beam divergence. This allows a constant focal position and focused beam size to be maintained over a large working volume, or for focus characteristics to be changed dynamically during processing.

The divergence angle, θ_2, of the beam after exiting a beam expander is:

$$\theta_2 = \frac{f_1}{f_2}\theta_1 \tag{4.8}$$

where f_1 and f_2 are the focal lengths of the entrance and exit optics, respectively, and θ_1 is the divergence angle of the beam on entry.

Line Focus Mirrors

A reflective line focus optic comprises a deflecting mirror, a line focusing mirror to produce a one-dimensional integration line, and a transforming mirror that projects the line. By varying the working distance, the length of the line can be changed. By sliding the transforming mirror towards the focus of the line focusing mirror, the focus of the line changes, which also produces a variation in length. Typical focus geometries vary between 8 mm × focal length and 35 × 10 mm. A wide, short beam gives high power density, which is ideal for rapid surface treatment. Longer beams increase interaction time and are suitable for homogenization treatments.

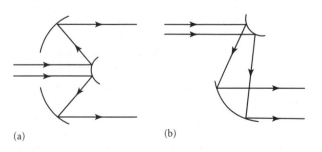

(a) (b)

Figure 4.5 Reflective beam collimators (a) Galilean, (b) Keplerian

Axicon Mirrors

An axicon mirror is designed to produce a particular axial distribution of light by reflection. A variation of the axicon mirror, the waxicon, is used to convert the annular distribution of an unstable resonator into a near Gaussian distribution.

Toric Mirrors

A toric mirror is used for treating internal and external surfaces of pipes. It requires an annular incoming beam, which it reflects radially.

Optical Kaleidoscopes

An optical kaleidoscope produces a heating pattern with a relatively uniform intensity distribution, suitable for surface treatment. It is constructed as a rectangular cylinder, normally surrounded by a water-cooling jacket, illustrated in Fig. 4.6. The beam enters the top and undergoes multiple reflection inside the cylinder, emerging as a heating pattern with the same dimensions as the exit aperture of the cylinder. Condensing and imaging lenses may be placed at the entry and exit, respectively, in order to accept a range of incident beams and to vary the size of the heating pattern. Kaleidoscopes can also be used in conjunction with a fibre optic for Nd:YAG beam delivery. They are relatively cheap and produce a uniform beam intensity distribution. However, multiple reflections result in high power

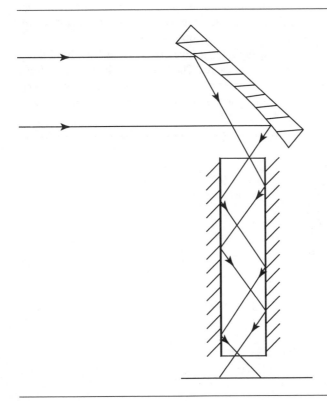

Figure 4.6   Optical kaleidoscope

losses (typically 20–30%), and efficient cooling is required. The unit must be positioned close to the workpiece for a sharply defined heating pattern (working range typically 0–10 mm), making it difficult to treat complex geometry components. The system can be used for both CO_2 and Nd:YAG processing, mainly for transformation hardening.

Integrating Mirrors

A beam integrating mirror comprises a concave reflecting surface composed of about 30 facets which reflect the part of the beam striking each facet to a common focal plane. Each partial beam has the same size and shape, determined by the facet dimensions, and passes through the focal plane in the same spot. The resulting superposition gives a heating pattern of uniform density, as illustrated in Fig. 4.7. The mirror can handle a wide variety of incoming beam modes, and alleviates problems that would be caused by changes in the output mode of the beam. A well-defined heating pattern can only be produced with a fixed dimension and the working distance is short, which limits flexibility. The dimensions can only be changed through the use of auxiliary focusing elements, which add substantially to the cost.

Polarizing Mirrors

A quarter-wave plate comprises one to four mirrors that have thin film coatings that act as a 90° phase shifter. If the *s*- and *p*-components are 90° out of phase, then circular polarization results. The plate converts linearly polarized light into circularly polarized light. The device must be precisely aligned to the polarization direction of the incoming beam to function correctly. It can be used for uniform cutting around circles.

A zero phase shift mirror has thin film coatings that produce a path length difference of an integral number of wavelengths for both components of polarization. Thus there is no phase shift on reflection, and therefore uniform reflection of both polarization components occurs.

A back reflection preventer is used when processing highly reflective materials, such as aluminium, in order to prevent light being reflected back into the laser cavity. The linearly polarized beam from the laser is directed onto a polarizing mirror oriented to reflect the beam completely. The light is then circularly polarized using a 90° phase retarder and impinges on the workpiece. Back-reflected, circularly

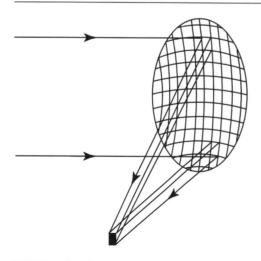

Figure 4.7   Integrating mirror

polarized light then experiences another 90° phase shift on going through the circular polarizer, and is thus converted into linearly polarized light, but in a direction 90° to the input beam. It is therefore totally absorbed by the polarizer.

Orientation devices are available that use axis motion data to establish the direction of beam travel, and through motorized mirrors maintain the plane of polarization in the plane of processing, in order to maximize processing speed.

Beam Rastering

Beam scanning systems normally contain two mirrors that oscillate about two orthogonal axes, to produce a two-dimensional Lissajous heating pattern, as illustrated in Fig. 4.8. A single oscillating mirror can be used to produce a line heat source. Mirrors with a large f-number (100–150) are used, to provide a large working distance. The frequency of oscillation typically lies in the range 100–400 Hz. Such a system may be combined with control of beam power to produce heating patterns with a range of spatial and temporal intensity distributions. The pattern geometry may be varied during processing. These features provide flexibility to treat a wide range of different component geometries. The maximum power that can be handled is currently limited to about 5 kW because of limits placed on the size and mass of moving mirrors. There is a brief dwell time at the extremities of the heating pattern, which may result in non-uniform processing, but which can be removed through the use of a mask.

In the rotating polygon rastering design, the beam is focused onto one or two orthogonal rotating polygon mirrors to produce a uniform heating pattern.

The beam shape and intensity can be changed through the use of various beam rotation devices. The beam may be split using a prism mirror, each component being directed towards a parabolic focusing mirror. The beams are focused in a plane above the workpiece surface, and rotated to produce two heating patterns. The diameter of rotation can be adjusted to vary the size of the heating pattern. This system can be used to produce a wider bead in laser welding, and to increase penetration depth in laser cutting, for example.

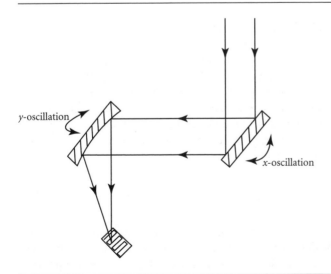

Figure 4.8   Beam rastering mirrors

Waveguides

Since most non-metals strongly absorb far infrared radiation, it is difficult to find a suitable fibreoptic material for the CO_2 laser beam. There is no doubt that the development of a suitable fibre would have a dramatic effect on the application of CO_2 lasers. Instead, waveguides are used.

Hollow waveguides can be divided into those composed of an inner metallic or metallic/dielectric layer, and those in which the innermost dielectric layer has a refractive index below one at a wavelength of $10.6\,\mu$m. A flexible nickel tube of internal diameter 2 mm with a germanium coating has been developed to transmit up to 1 kW of CO_2 power with low transmission loss. However, the low order mode of the beam cannot be preserved, producing a poorer quality beam in comparison with mirror delivery. A gold plated steel guide bent to a curvature of 20 cm has been demonstrated to transmit 91% of an 800 W incident beam. Hollow waveguides constructed using germanium/silver/nickel structures have been constructed for transmitting power levels above 1 kW. Hollow sapphire fibres have transmitted power levels above 1 kW. However, the material is stiffer than metallic structures, restricting the minimum radius of curvature. Two types of infrared fibre have been developed for very low power CO_2 laser surgery. One has a solid core and is coated in silver halide. The other has a hollow core. The latter is the more robust and shows promise for higher power beam delivery.

DIFFRACTIVE OPTICS

Diffractive optics are made by diamond turning reliefs, and by creating patterns using photosensitive films. Typical multikilowatt laser systems produce TEM_{00} or TEM_{01*} mode beams, which may not be the ideal distribution for a given application. Diffractive phase elements, sometimes referred to as kinoforms, change the intensity profile of a laser mode into a more application-specific shape, such as a circular, rectangular, or semi-annular shaped top hat. This can be achieved either by placing the optic near the work surface or within the laser cavity itself.

Customized diffractive optics for specific transformations can be made using CAD packages that include Fourier transform algorithms. Symmetric or asymmetric patterns can be made with equal ease, unlike conventional optics.

LASER PROCESSING CENTRES

A laser processing centre comprises a number of subsystems, or modules, which typically include: a CAD facility; a laser system; motion control; input/output; beam diagnostics and machine vision. The laser system is the beam source and the beam delivery optics. The laser processing centre comprises the above and workpiece manipulation. Laser processing centres are available in a range of sizes and complexities. Historically, they were derived from conventional machine tools – the cutting head was simply replaced with a head that delivered the laser beam. As demands for higher accuracy and greater flexibility grew, so did the sophistication of the systems.

COMPUTER-AIDED DESIGN

A computer-aided design (CAD) facility for off-line programming is an essential component of all types of centre – the centre is too expensive to be down while drawings are being made. CAD systems allow processing to be simulated and optimum beam paths to be constructed. The design of components can also be optimized. Potential collisions can be avoided. The required features are drawn in the CAD system and then translated into motion control commands for execution. An example of a CAD

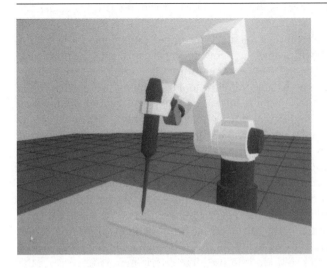

Figure 4.9   CAD simulation of laser cutting of a plate using a robot-mounted beam delivery head

simulation of laser cutting an ellipse in a plate using a robot-mounted beam delivery head is shown in Fig. 4.9.

CONSTRUCTION

Before we look at the different types of laser processing centre that are available, it is instructive to consider their construction; many features are common to different systems and affect performance markedly. By understanding their features and characteristics, the selection of an appropriate laser processing centre becomes easier.

The mechanical frame of the laser system should have a high bending and torsion stiffness to minimize deviations of the beam path. Moving parts should have a low mass to maximize acceleration and deceleration. The majority of drive systems use electric motors and mechanical transmission elements, which introduce additional moments of inertia, elasticity, and non-linearities such as backlash, all of which limit acceleration and accuracy. A ballscrew drive mechanism comprises a screw with a nut containing a ball-bearing tube. The screw rotates, moving the nut along its length to a desired position. It is used in many smaller systems. In large systems it can be difficult to maintain accuracy. The mechanism is normally protected using bellows or a telescopic cover. In rack and pinion drives, racks can be fixed together to extend their length, but the gears must not have any play, for accuracy. Belt drives incorporate toothed belts, which help to reduce backlash.

Of all the drives available, linear direct drive motors are the most suitable. They are constructed from only two parts: the coil assembly (slider) and the magnet assembly (stator). They provide simplicity, small size, low inertia loads and enable a stiffer system to be constructed. Benefits include accuracy, repeatability, smoothness, higher throughput, low maintenance and reduced energy consumption with no friction and no wear. The problem of backlash, common to gear and belt-driven systems, is eliminated, as is the problem of wear in ballscrews. Speeds of up to $170 \, \text{m min}^{-1}$ are possible, although the most distinctive feature of linear motors is the acceleration that can be achieved (up to $30 \, \text{m s}^{-2}$). The accuracy and repeatability are typically $\pm 0.02 \, \text{mm}$.

The system is isolated from external influences such as vibrations and temperature fluctuations. Short beam paths should be used where possible to minimize beam expansion and maximize stability. Optical components should be as simple as possible, and have accurate and unchanging orientation relationships to maintain alignment. The system is isolated from its surroundings using screening, doors and windows.

A high positioning accuracy is necessary because of the small size of the beam in operations such as cutting and welding, and the precision with which small components in particular must be processed. The beam delivery head can be considered as the tool centrepoint used in conventional machining systems. Movement in the different axes should be able to follow tool centrepoint movements, while maintaining the orientation of the beam to the workpiece surface, and a particular stand-off distance. Rapid movement in all the axes is required (higher than conventional machining), with a wide range of speeds and high values for acceleration and deceleration. Servomotors should have a low inherent moment of inertia.

A beam collimator is essential for workstations with a large envelope of movement. It can be used to adjust and maintain a particular beam size throughout a workstation. The beam size may be fixed, positioning the beam waist at the mid-point of the beam path, or motorized to maintain a desired beam size at the focusing optic. In the latter case, the necessary adjustments may be made either by detecting the beam size at the focusing optic, or by calculating it from the known position of the focusing optic. Such systems can then also compensate for the effects of changes in the properties of optics such as thermal lensing. An alternative solution is to use piezoelectric motors to decrease the radius of curvature of the turning mirrors when long path lengths are involved.

Workstations are normally controlled by a computer numerical control (CNC) system with an operator interface. CNC is a self-contained NC system for a single machine tool. Direct numerical control (DNC) systems serve several machine tools through a central computer. CNC has become more widely used for manufacturing systems mainly because of its flexibility and the lower investment required. The preference is becoming even stronger because of the availability and declining costs of computers. The open architecture of the PC is ideal for the implementation of overall control of systems operation. The control system is normally programmed in sequential command blocks. The block cycle times should be as short as possible to avoid delays between commands, and should be able to manage synchronous command facilities. It should be possible to input information from sensors for on-line adaptive control. Modern industrial lasers normally contain some type of embedded processor that enables the laser and process parameters such as power, repetition rate and gas flow rate to be controlled directly. Most also allow for remote control via a communications protocol.

TYPES OF LASER PROCESSING CENTRE

The various combinations of laser processing centres are listed in Table 4.3. They are classified according to the geometry of processing (two dimensional or three dimensional), movement of the workpiece, and movement of the laser beam.

The most suitable laser processing centre for a particular application depends on a number of factors: the size of the part to be processed; the complexity of the part; the nature of beam transmission; and the precision required.

Gantries

In a gantry system, the beam is delivered through a head mounted on a column that can be moved in a Cartesian coordinate system over several metres. An articulated arm located at the lower end of the column normally provides two rotational axes, giving a total of 5 degrees of freedom. Such systems have high stiffness and are suitable for processing large heavy sections, which are only moved during loading and unloading, but which require a large operating envelope. A high degree of accuracy is possible. They are normally used where the workpiece is highly accessible from the upper side. Typical applications are trimming of automobile body panels, and accurate profiling in the aerospace industries.

Table 4.3   Combinations of laser beam and workpiece handling systems for two- and three-dimensional laser material processing

Processing geometry	Workpiece	Laser beam	Example centre
Two dimensional	Moving	Moving	x table, y optic
		Stationary	x–y table
	Stationary	Moving	Flying optic
		Stationary	–
Three dimensional	Moving	Moving	Robot-mounted workpiece and beam
		Stationary	Robot-mounted workpiece
	Stationary	Moving	Robot-mounted beam
		Stationary	–

A flat bed cutter is an example of a flying optic gantry system designed to operate primarily in two dimensions. The optics are able to move rapidly (about 40 m min^{-1}) in both the x and y axes. Adjustment in the z direction is available to set the height of the beam delivery head, which can be controlled automatically using various methods described below. A beam collimation system may be needed to ensure a constant beam size over the table area. The axis drives are typically high torque, low maintenance, totally enclosed, brushless, AC servo motors used with precision ground recirculating ballscrews. Systems with dual laser heads provide increased productivity. Flat bed cutters have traditionally incorporated a CO_2 laser, but Nd:YAG systems are also available.

Multi-axis Beam Direction

An alternative to a gantry system for medium-sized components is beam delivery through arms fixed to a base that moves along Cartesian axes, with a head that is able to rotate in two planes. The benefits of this system are high mechanical rigidity, high dynamic accuracy, and high volumetric accuracy which enables rapid contouring in three dimensions. The working envelope is typically $1 \times 0.75 \times 0.75$ m, which can be traversed at speeds up to about 2 m min^{-1}. Such systems are designed for processing components the size of white goods, engine combustion liners and automotive prototypes. An example is shown in Fig. 4.10.

Articulated Robots

Industrial robots with six or more axes may be combined with a laser in various ways: light parts of complex geometry can be manoeuvred under a fixed beam; the beam can be delivered to the workpiece through the links and joints of an articulated arm using internal optics; the robot may guide a fibreoptic beam delivery head; or the robot may guide a compact mounted laser source. An example of an articulated robot guiding a beam delivery head is shown in Fig. 4.11.

Robotic manipulation is ideal for steering the beam delivery head of a fibre optic from an Nd:YAG laser. It provides a high degree of flexibility for processing three-dimensional components. The use of robots in laser processing applications is currently limited by the power that can be transmitted by internal optics (around 5 kW), the weight of the part (around 15 kg), and the repeatable accuracy of the robot (approx. 0.15 mm). A light, compact laser head can also be mounted on the end of an articulating robot, connected to a supply cabinet through an umbilical cord. When the robot is used to direct an external articulated beam guide, the mirrors may be located inside the structure of the robot such that the beam path is completely enclosed, and is relatively short.

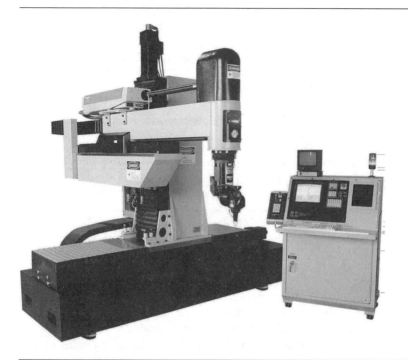

Figure 4.10 Laserdyne Prima 790 Beam Director. (Source: Iain Ferguson, Laserdyne Prima, Maple Grove, MN, USA)

Figure 4.11 Motoman articulated robot with beam delivery head. (Source: Damien Harris, CSIRO, Adelaide, Australia)

Moving Laser

The resonator and optics of a low inertia, rugged laser can be mounted on a mobile portal or cantilever-type machine, and moved in the x and y axes to achieve the required processing geometry. The construction is relatively simple, and the beam path is constant, allowing the designer to minimize distortions introduced by beam transmission. Such a system is suitable for use with large workpieces, but heavy and expensive machinery for laser movement might be needed.

Moving Table

A moving table comprises an x–y stage with CN-controlled movement in both axes. This is appropriate for small components, e.g. those with a footprint less than 1×1 m. It is conceptually the simplest means of moving the workpiece. Stages vary in size up to about 1.5×3 m. Ballscrew drives on precision ground ways or rods provide an accuracy of typically 1–100 μm with traverse rates of about 5 m min^{-1}. Linear motor drives provide accelerations up to $5\,g$; these are limited by the mass of the table. There are no restrictions on the size of the laser source. The workpiece may be fixtured to the table using a jig for welding components, or a vacuum for material cutting.

Moving Workpiece

In a simple moving workpiece centre, the table and the laser are fixed, and the workpiece and ancillary equipment such as fume extraction are mounted on a moving carriage. In its simplest form, the beam is directed to the workpiece using a single stationary mirror, enabling the beam path and length to be fixed. This type of centre is suitable for small, low inertia workpieces, or for large and immobile lasers, and is commonly used for one and two axes movement over about 1×1 m. By providing one long axis, the centre is suitable for welding, for example, pipes and profiles. The simplicity of construction and constancy of beam path allow the designer to minimize the number of mirrors and to eliminate problems caused by variable beam path lengths. The movement of heavy workpieces can lead to tolerance difficulties.

Hybrid Centres

Hybrid centres can be constructed such that the workpiece moves in the x axis and the optics move in the y axis. They typically combine a gantry-mounted y axis with potentially unlimited x axis motion. The beam delivery path is shorter than a full moving optic system, and the positioning accuracy is higher than full workpiece movement systems. Hybrid systems limit the dimensions of motion, which reduces floor space demands. Important factors are the weight carrying capacity of the x table and the low inertia of the beam delivery optics.

PROCESS MONITORING AND CONTROL

Laser material processing is particularly suited to integrated on-line monitoring and control. It is highly automated, and beam–material interactions produce characteristic emissions that can be used to monitor conditions in the processing zone with a variety of techniques. Highly automated industries, such as the automotive sector, are large potential markets for automated process monitoring and control systems. If such systems are only used for diagnostics, the system is said to be open loop. If they are used to control process variables during processing, a closed-loop system is created. Before considering techniques relevant to different processes, we consider means of measuring the properties of the beam itself.

BEAM DIAGNOSTICS

Beam diagnostic devices enable reference data for the beam to be established in both its raw and manipulated forms. The beam diameter and mode are particularly important, since they are often the only data available when comparing beams from different sources. These properties have a significant effect on processing characteristics, particularly welding and drilling in which the beam is focused. The precise effect of optical devices on the beam properties can be established, e.g. the location of the focal plane of a focusing lens, the focused spot diameter, or the intensity distribution produced from a beam integrating optic. Temporal fluctuations in the beam mode and power can be detected. These may be desirable, e.g. when drilling with optimized pulse shaping, or unintentional, e.g. degradation of source power during processing. Such factors become important when processes are running for extended periods of time.

Power

The power of the emitted laser beam is normally established off-line using calorimeters. Common designs are the 'hockey puck' (a calibrated absorbing disc placed under the beam) and the cone pyrometer, which acts as a black body absorber. The accuracy of measurement of such designs should be borne in mind – variations of $\pm 15\%$ in the multikilowatt range are not unusual. The implication of this should be appreciated – a 10% error in the measurement of a 6 kW beam can result in a loss of weld penetration, for example. This is particularly important when processing data are to be transferred between different locations. Standard calibration sources are available and should be used wherever possible.

Intensity Profile

The simplest method of obtaining a rough impression of the symmetry of a raw or partially focused beam is to char materials. Wood and firebrick are suitable for multikilowatt beams, whereas thermally sensitive paper is more appropriate for low intensity measurements. By vaporizing a polymethyl methacrylate (PMMA) block, information on the intensity profile within the beam can be obtained. (Care must be taken that the vapour produced does not ignite by using suitable fume extraction and a gas purge.) A split block can also be used in order to obtain an immediate impression of the beam profile. The extent of vaporization is dependent on the incident power and exposure time, and so the technique is best used to make comparative measurements, with a set power and exposure time. Expanded polystyrene is a more suitable material for beam powers below 100 W and when short interaction times are involved.

An aperture can be used to profile a cylindrically symmetrical beam with a Gaussian intensity distribution. The aperture is opened to measure the total power of the beam, and then reduced until the measured power is 86% of the total. This represents the diameter at which the intensity has fallen to $1/e^2$ of its peak value.

On-line power measurement includes the leaky back mirror, which extracts a small amount of the cavity beam in order to monitor on-line the extracted power. Silicon mirrors allow some of the incident light to pass through, which can be collected by detectors. Sampling at various points can be used to establish the beam mode. In conjunction with a motorized output mirror, on-line beam mode control may be achieved.

In the scanning rod, or rotating wire instrument, a molybdenum needle with a highly reflecting surface sweeps across a transverse section of the beam at high speed (around $25 \, \mathrm{m \, s^{-1}}$). A very small fraction of the beam (0.8%) is reflected into two single element detectors, arranged such that they record the intensity distribution in the x and y axes. The total beam power, beam diameter, beam wander and a three-dimensional image of the beam intensity distribution can be constructed. The instrument can be placed immediately after the laser or just before the workpiece. The main advantages of this design are that only a small portion of the beam is reflected, and the measurement is obtained quickly. It is therefore suitable for on-line use, for example as a sensor in an adaptive control system to regulate laser power.

An alternative to the scanning rod device uses a rotating hollow needle, which contains a small hole at one end, through which a sample of the beam is taken. The beam is reflected along the needle axis to a detector located on the axis of rotation. By scanning the hole across the beam, an intensity profile is built up. Measurements can be made on the raw and focused beams, and at positions either side of the focal plane. The image is recorded, and can be displayed in various forms such as a 3D profile and sections containing given percentages of the intensity (e.g. 86% at the $1/e^2$ contour). The main advantage of the technique is that a complete beam intensity measurement is obtained. It is particularly useful for establishing the intensity profile of a focused beam from an unstable resonator, which is difficult to calculate and represent mathematically. The principal drawback is that the time required for a complete scan is rather long, and it is difficult to use in an on-line application.

A rotating drum device containing a pattern of pinholes can be mounted in front of the detector. A portion of the beam is directed towards a focusing optic and then to a sensor. The main advantage of this device is that it is able to measure down to 1 micrometre, for focused beam analysis, although it only produces a single line profile with minimal numerical data, and can only be used reliably with CW lasers. It is used for stable beams since it is a sequential scanning technique.

Charge coupled device (CCD) cameras can instantaneously measure a two-dimensional pinhole scan, and show an averaged structure over a given time. They can be used with pulsed beams. Standard CCD arrays are sensitive to light in the wavelength range 190–1100 nm. Excimer and Nd:YAG laser beams can therefore be analysed.

Height Sensing

High quality laser welding and cutting require that the beam focal plane be located accurately at a particular position in the material, i.e. that a specified gap between the optic and workpiece be maintained. Workpiece tolerances, thermal warping, and the accuracy of beam delivery mean that tolerances are often exceeded. The two main types of height sensing devices available are based on contact and non-contact techniques.

Mechanical contact techniques use ball-bearings, sliding feet, or a capacitive ring. They are insensitive to plasma build-up in the beam–material interaction zone, and are relatively inexpensive.

The most common type of on-line non-contact sensor measures the capacitance of an electrical field between a nozzle tip and the workpiece, which is at ground potential. The nozzle is electrically isolated from the rest of the laser beam delivery. Movement in the z axis compensates for changes in height. Capacitive height sensing is affected by plasma formation, and is therefore difficult to use with welding. Scanning laser techniques and triangulation can also be used for height control.

Seam Tracking

Imperfections can be introduced in welded joints from various sources including: workpiece tolerances; positioning errors caused by clamping; deformation during welding; interpolation errors between control system points; and positioning errors of the beam delivery system. The diameter of the focused beam is small and so the high speed and accuracy of laser welding dictate that motion control be implemented automatically. Some form of seam following system is therefore required in applications where the cumulative positioning error is large in comparison.

Normally a CNC unit controls motion. Precise workpiece manufacture, stronger clamping, and more accurate beam delivery may not be justified economically in some cases, and may be technically impossible in others. In such cases a seam tracking system is an appropriate choice. Various types

Figure 4.12   Seam tracking using a scanning visible laser beam during laser welding of a hollow box beam

of equipment are available to locate the position of a joint for welding. All comprise a sensor unit, an interface, a motion control unit, and the software necessary to interpret and act on the signals received.

The joint can be characterized by a 1 mW HeNe beam which is scanned at up to 20 Hz. Reflected information is detected by a CCD array and interpreted by triangulation, in order to position the laser head in the lateral and vertical directions. The information can also be used to calculate the width and cross-sectional area of the gap for filler wire feed control if necessary. The sensor is positioned in front of the beam such that real time adjustments can be made. Tracking with an overall accuracy of ±0.2 mm is possible. This can be used for laser welding in all but the most precise applications. An example of this type of system in use is shown in Fig. 4.12.

Alternatively, line illumination can be generated from a visible laser source, which when projected onto a seam produces a pattern of structured light. This is detected by a camera and is used to build up a picture of the object that is interpreted in a similar way to scanning laser systems.

PROCESS MONITORING

Encoders, tachometers, accelerometers, interferometers and the laser Doppler anemometer can be used to establish the position, speed and acceleration of a workstation table, but the laser process itself emits a wide range of signals that can be monitored and interpreted to characterize the state of processing. Practical systems for process monitoring and control are described for the individual processes in the chapters that follow. Here we consider the principles that underlie such systems.

Pyrometry

Measurement of surface temperature provides valuable information in a number of processes. Thermocouples may be attached to the surface, or a pyrometer can be directed at points of interest.

A common form of pyrometric analysis uses emissions at two wavelengths, e.g. 1300 and 1700 nm, and provides a temperature conversion within an operating range, e.g. 500–1600°C. Pyrometry is useful in controlling surface temperature during transformation hardening, and can be incorporated into on-line adaptive control systems that adjust process variables such as laser power and traverse rate to maintain a required surface temperature. Pyrometry is also a means of controlling the temperature in the molten region during surface melting and cladding. Infrared detection of the upper side of a weld is used to identify spatter formation, and measurements from the underside of a weld can be used for penetration monitoring. Laser drilling can be monitored by observing the temperature in the interaction zone: uncontrolled melt expulsion can thus be reduced, and the maintenance of a fully penetrating hole ensured.

Ultraviolet Radiation

In processes that generate a plasma, there is a relationship between the mechanism of processing and the frequency and intensity of the plasma. In keyhole welding, emission in the wavelength range 200–400 nm originates from a small region just above the keyhole, which remains approximately constant when a good weld quality is maintained, and can be used as a quality monitor. Signals emitted from the plasma contain information about the weld bead size, penetration, pool collapse, humping and vapour generation, as well as the effects of variations in process variables on the weld bead depth and the condition of the surface. The state of the plasma is an indication of the energy transfer into the workpiece; a blue plasma indicates a keyhole penetration mechanism in the case of CO_2 laser welding of steels, and a loss of signal indicates hole formation in thin sheet welding. Plasma flashes indicate plasma shielding, which may be arise from insufficient gas flow. In commercial devices, ultraviolet emissions are analysed to monitor weld pool characteristics, and infrared emissions used to detect weld spatter.

Plasma inspection is also used as the basis for on-line quality control systems for laser cutting of ceramics. When the beam interacts with the ceramic, the ultraviolet signal rises, and subsequently falls during the quiescent period as the plasma quenches. If the quiescent period is too short to allow the plasma to quench, the incidence of cracking increases. The ultraviolet signal can therefore be used to select the pulse frequency in order to control the length of the quiescent period.

Acoustic Emission

Acoustic emission originates from changes in thermal gradients, martensitic transformation, void generation, crack formation and propagation, and rapid pressure fluctuations within a keyhole. Martensitic transformation produces a continuous acoustic signal, of intensity dependent on the transformed volume. Other sources more commonly produce transient burst signals. By attaching a piezoelectric sensor to a plate mounted above the weld, pressure changes produced by intensive vaporization can be detected rapidly. When weld penetration changes from partial to full, the frequency of the sound generated changes, which can be used as a measure of penetration. Discontinuities in a material can be detected by using ultrasonic waves generated by a piezoelectric transducer. The technique can also be used to detect defects in welds, such as porosity and cracking. Acoustic sensing can be used to monitor the erosion front condition during drilling, as well as the hole depth.

Ultrasonic Analysis

Discontinuities in a material can be detected by using ultrasonic waves generated by a piezoelectric transducer. The geometry of a molten region may be characterized by analysing the time of flight between two transducers placed adjacent to and on opposite sides of the pool. The technique can be used to detect defects in welds, such as porosity and cracking.

COMMISSIONING AND ACCEPTANCE

COMMISSIONING

Before considering the factors that must be evaluated in the successful implementation of laser material processing, it is instructive to look at the history of laser processing and to analyse why some processes have not taken off as quickly as was expected.

History

In the early days, few manufacturers of laser systems worked together with makers of work handling equipment. Incompatibilities meant that neither the laser nor work handling were used in the manner for which they were designed. A poor knowledge of the process and its mechanisms often led to the use of inappropriate processing parameters that resulted in the failure of a process – in many cases the result of a simple error such as the choice of a processing gas. Laser processing was not presented to the man-ufacturing industry in the same way as other materials processing techniques, i.e. as a turnkey system that incorporated flexible geometry processing, support for CAD, a database of processing information, and means of automatic process and quality control. The necessary skills in many disciplines could therefore be difficult to find in one person, or even a team. Manufacturers were often more familiar with cheaper conventional processes, and some lacked the foresight to appreciate the implications for new designs that laser processing offered. A lack of generic laser processing information accounted for the reluctance of designers to apply this new technology. It was often difficult to quantify some of the benefits arising from laser processing, such as product quality, rather than just productivity gains. Productivity increases may even have been considered to be potential causes of other problems, such as reliability and bottlenecks in existing production lines. Fortunately, these initial drawbacks have been addressed through the availability of improved laser centres, dedicated training programmes, and a greater appreciation of the implications of laser processing. But choosing the correct laser centre is still a challenge.

Design of the Centre

One of the first factors to take into account when considering laser-based manufacturing is the nature of the product. Laser processing centres are often designed simply to replicate an existing procedure with existing components. Manufacturers therefore often wish to avoid change of materials and procedures. In many cases this proves profitable: in the case of laser cutting of flat sheets, the cost of the centre is normally recouped in less than one year by the increase in productivity and the improvement in quality. However, to maximize the benefits of a new installation, opportunities for new product design and processing routes should be examined. Design encompasses the choice of materials, possibilities for the redesign of part geometries, and improvements to the aesthetics of the finished product. By considering the manufacturing operation in reverse, i.e. by starting with an optimized product and working back through the stages of production, laser-based operations may provide novel means of manufacturing. In recent years, the largest growth has been experienced in novel manufacturing techniques that could not be carried out by any other means.

A centre may be dedicated to a particular task, as are most conventional machine tools, or used for a number of different processes. Dedicated centres can be optimized to perform a single task repeatedly by integrating the laser with standard beam and work handling equipment. If the slowest part of a manufacturing operation is parts handling, indexed tables can be used for different stages of the process: the processing operation itself; loading the part; and unloading the part. At the opposite extreme, a single laser may be used in a number of different operations. For example, a robot can be programmed to select a cutting head and perform a programmed cutting operation, and then pick a head for welding, which is controlled by a different programme that includes the relevant commands. The flexibility of the laser for performing a variety of tasks is an aspect that has not been exploited to the full.

The geometry of the component and the complexity of the process determine whether processing needs to be carried out in two dimensions, or three. Flat sheet cutting is an example in which two-dimensional flying optic beam delivery is the most appropriate solution – the sheets can be fixed such that they are flat enough for cuts to be made by moving the beam rapidly over large areas in only two axes. Symmetry in components is an important factor in selecting the beam delivery and work handling: rotational symmetry, for example, enables a part to be processed in a simple turning unit under a stationary beam. If the component is large and heavy, a flexible beam handling system that can move around a stationary workpiece is more suitable. The optimum centre for processing a complex light component in three dimensions may comprise three-dimensional movement of both the beam and the workpiece. Such considerations enable certain processing centre geometries to be eliminated from a list of candidates in the initial stages.

The working envelope plays an important role in the selection of the beam delivery system: movements of one or two metres can be handled by fixed robots, whereas larger traverses require the use of gantries. The process tolerances required (accuracy and repeatability) also dictate specifications for the beam delivery system (and sometimes the work handling equipment). The cost of a gantry required for a large working envelope increases substantially if small tolerances are to be maintained over the entire range of traverses. In general, a short beam path minimizes misalignments and the need for beam correcting devices. Environmental effects on the beam are minimized by enclosing the beam in nitrogen-purged flight tubes. A turning mirror should be at least twice the diameter of the beam incident on it.

The combination of processing depth and productivity requirements are the main factors in determining the laser power required. Rules of thumb are available for processes such as welding and cutting, and databases have been created for other treatments. Processing is normally carried out at about 90% of the laser output power. The requirement for traverse rate should be chosen to provide a comfortable margin to compensate for expected variations in the depth of treatment, and unexpected variations of about 10% in the laser output power.

The most suitable type of laser depends on the power required and any wavelength-specific features of the component. The environment in which it is to operate also plays an important role in the choice – a rugged design is required if the equipment is subject to vibrations. If floor space is limited, the options for mounting the laser should be considered: vertically; horizontally; or overhead. If power levels above about 5 kW are required, the choice is currently restricted to a CO_2 source, or the output from a combination of Nd:YAG sources. If precision processing is on the scale of micrometres, only the short wavelength light from an excimer laser (or frequency multiplied output from a solid state laser) is suitable. The type of laser dictates the electrical and cooling requirements. Service contracts are frequently not paid the attention they deserve. Service frequency and costs, and the turnaround time after service, should be considered; a breakdown that can only be fixed by a qualified engineer may result in lost production for several days.

When processes compete for a particular application, the selection is likely to be made on the basis of cost. However, selection could also be made using other less tangible factors such as environmental concern, labour unions, and the willingness to employ new technology. The introduction of laser processing has wide-reaching implications. There must be commitment to the technology at management and shop floor levels. Staff requirements may be altered, and existing staff will need to be trained in the

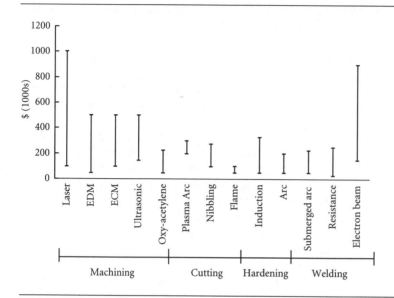

Figure 4.13   Capital costs of various processes

new technology. Unions may be opposed to such changes, particularly if increased automation leads to a reduction in the number of staff required. Safety regulations for the operation of equipment (discussed below) must be considered. However, change presents opportunities. A new company image may be created, based on the high technology image of laser processing. Delivery flexibility can be increased, as can the possibilities for just-in-time manufacturing.

Economics

A rule of thumb is often used to estimate the cost of rectifying mistakes at various stages of manufacture: if it costs $1 at concept, it costs $10 at design, $1000 at production, and $10 000 at product recall. Clearly the opportunities to evaluate new concepts when a new manufacturing technology is considered should be explored in depth. Although laser processing is often regarded to require a large investment – ranges of capital costs for various processes are shown in Fig. 4.13 – the potential for reducing the total cost by eliminating mistakes during the latter stages of production is significant.

The cost of laser processing can be broken down into fixed and variable costs. A summary of these components is given in Table 4.4.

The fixed costs include: labour (laser operators and engineers); capital investment (total or individual equipment costs); maintenance (a percentage of investment or fixed annual cost); depreciation (normally linear over a fixed term, e.g. 5 years); and floor space.

The variable costs are the running costs associated with the application. The laser costs include laser gases, cavity mirrors, electrical consumption, chiller electrical consumption and the output window (which can break). Costs associated with the beam delivery include directing optics, focusing optics and beam delivery gas. Workstation costs comprise electricity and pressurized air. Process costs vary considerably. Process gases are required for shielding and plasma suppression, assisting, and composition change in welding, cutting and alloying, respectively. Filler material is used in welding. Other costs include pretreatment techniques such as prebending and preheating, and post-treatment heating and mechanical working.

Table 4.4   Components of running costs for typical industrial laser centres

	Capital (5 year depreciation)	Maintenance (fixed agreement)	Consumables (unit costs)
Laser	X	X	Electricity Laser gases He, N_2, CO_2 (CO_2 lasers) Flashlamps (Nd:YAG lasers) Cavity mirrors
Cooling	X	–	Electricity
Beam delivery	X	–	Mirrors, lenses
Work handling	X	X	Electricity
Safety	X	–	–
Floor space	(X)	–	$\$ \, m^{-2} \, year^{-1}$
Process	X	–	Gases Filler materials Coating materials
Labour	–	–	$\$ \, h^{-1}$

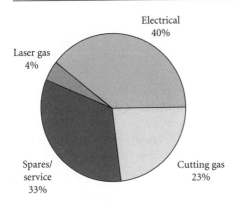

Figure 4.14   Cost breakdown for cutting steel using a 600 W CO_2 laser

A breakdown of the costs of oxygen-assisted cutting of structural steel using a 600 W laser is shown in Fig. 4.14. This can be used as a guide to estimating the running costs of a laser cutting operation. (The basis of the breakdown is 16 hours' use per day, 230 days per year, and a 50% duty cycle.) A comparison of the costs of tungsten inert gas and laser welding with both 6 and 9 kW lasers is given in Table 4.5. It can be seen from Table 4.5 that the higher capital cost of laser welding is offset by the substantially lower costs of labour and consumables, and the increase in productivity. Such cost breakdowns can be used to compare laser processing with conventional methods. Further details of costing techniques for individual processes are given in the chapters that follow. The examples provided here are intended to highlight the general factors that should be considered at the stage of commissioning a laser processing centre.

Table 4.5   Cost comparison for TIG and laser welding (6 kW and 9 kW) of pipeline

Cost factor	TIG	Laser	
		6 kW	9 kW
Welding speed (m min^{-1})	0.0635	0.635	1.397
Production rate (joints h^{-1})	1.0	10	11
Labour cost ($ joint^{-1})	47.50	5.40	5.10
Consumable cost ($ joint^{-1})	7.20	1.00	1.40
Capital cost ($ joint^{-1})	0.60	24	28
Total cost ($ joint^{-1})	55.30	30.40	34.50

ACCEPTANCE

Much expense and time can be saved by designing a thorough acceptance testing schedule, to be completed both at the place of manufacture and after installation of the centre.

Static testing should include measurements of the accuracy and repeatability of the beam at locations within the working envelope, performed using laser interferometry. Dynamic measurement includes the tolerance during movement, out-of-roundness, backlash, overshoot and deviations in programmed values.

SAFETY

The descriptions and advice given in this section should be treated as general in nature. Mandatory requirements for laser processing centres, such as controls, personal protective equipment and procedures, are described in standards that are listed in Appendix F. It is essential that operators know the requirements for the safe use of lasers, and that a laser safety officer ensures that those requirements are fulfilled.

HAZARDS

The basic hazards to humans associated with laser processing concern damage to the eyes, skin and respiratory system.

Eyes

The cornea is the main focusing element of the eye; the lens is used for fine focusing. The space between the cornea and lens is filled with a watery liquid known as the aqueous humour, and the interior of the eyeball is filled with the jelly-like vitreous humour. The retina is the screen upon which the cornea and lens project an image. The cornea, aqueous humour, lens, and vitreous humour are the ocular media. The eye can detect and focus light in the range of wavelength 350–750 nm, corresponding to the spectrum of colours in the rainbow. The eye also has an ocular focus region that includes wavelengths from the end of the red to 1400 nm.

Light from an Nd:YAG laser (1064 nm) falls in the ocular focus region, and can therefore be transmitted through the cornea and lens, and focused on the retina. This makes it particularly dangerous. Light from a CO_2 laser falls in the far infrared (10 600 nm), and is absorbed by the cornea. Mid-ultraviolet light (180–315 nm) shows similar absorption characteristics. Near ultraviolet radiation (315–390 nm) passes through the cornea and aqueous humour, but is absorbed by the lens. Continuous wave lasers can cause damage by thermal processes that overheat the tissue. Pulsed lasers cause damage from acoustic and vibrational shocks which rupture blood vessels in the retina.

It is normal procedure to make a retinal examination when an individual begins and terminates work in a laser facility.

Skin

Skin burns are caused by radiation from high power infrared lasers. At all wavelengths, exposure to the skin may result in erythema, skin cancer, skin aging, dry skin effects, and photosensitive reactions.

Respiratory System

Respiratory system damage can be caused by the evolution of hazardous materials. Small particles 0.02–0.2 μm in diameter are soluble in wet linings of the respiratory system, and thus pose a health threat when absorbed into the bloodstream. Poisonous vapours are produced when burn prints are made in plexiglass and when polymers are cut. Fine metal fumes are generated during welding and cladding. Ozone production is associated with plasmas. Adequate ventilation and fume extraction are therefore required.

Electrical

The majority of serious accidents in laser facilities are in fact associated with the electrical system of the laser rather than the beam itself. Power supplies must be screened to prevent electrocution. Voltages greater than 15 kV may generate X-rays. Interlocks and provisions for discharging capacitor banks before entry must be included.

Chemical

Burns can also be caused by chemicals and cryogenic coolants. Gases, fumes and dyes can constitute a fire hazard; the use of flammable materials should be avoided, and flame resistant enclosures used. Bottles of compressed gas can act as missiles if they are knocked over – they should always be securely fastened to a retainer.

RISK MANAGEMENT

Risk assessment is a management tool that allows managers to check that safety and health policies are effective, and provides a record that clearly shows the justification for arrangements that have been put into place. Risk assessment should form the basis of both the specification of safety equipment and the content of safety procedures.

A risk assessment must be carried out for any procedures that fall outside standard operating procedures (SOPs). Risk assessments are documented and regularly reassessed. They help to prioritize actions in safety management planning. They also ensure that the most important potential problems are recognized so that these can be addressed first, and that other significant problems are not overlooked. The precautions taken to minimize risk from these potential problems are noted. Risk is assessed as the product of the likelihood of an event that will cause harm and the severity of harm caused. Once the necessary precautions have been taken and noted, the risk assessment is kept for future reference.

LASER CLASSIFICATION

Continuous wave lasers are classified according to their wavelength and average power. Additional requirements are placed on repetitively pulsed lasers: the exposure time inherent in the design; the total energy per pulse; the pulse duration; the pulse repetition frequency; and the emergent beam radiant exposure.

Lasers are divided into four classes. The higher the category, the more powerful and dangerous the laser. Virtually all the lasers used for material processing fall into the two highest categories. The accessible emission limit (AEL) is the maximum accessible emission level permitted within a particular class, and is used in the definition of the class. The maximum permissible exposure (MPE) is the maximum level of laser radiation to which a human can be exposed without adverse biological effects to the eye or skin. A pulsed laser is defined as one that delivers energy in single or multiple pulses which are less than or equal to 0.25 s in duration.

Class 1

Class 1 lasers or laser systems are considered incapable of producing damaging levels of radiation during operation and maintenance, either because the output of the laser is very low or because of installed safeguards. Most commercial laser systems for material processing are sold as Class 1 products although they contain Class 4 lasers. However, the emitted radiation is enclosed, and is not accessible. The benefit of a Class 1 laser system is that it can be installed anywhere and no eye protection is required for workers. A Class 1M laser presents no risk for skin damage, but could result in damage to the eyes if viewed with optical instruments such as binoculars.

Class 2

Low power lasers or systems that produce accessible radiation fall into Classes 2 or 2M. Class 2 lasers emit visible radiation (350–750 nm) of low power (below 1 mW). They cannot damage a person accidentally, but could damage the retina if viewed directly for more than 0.25 s, which is the time of the normal blink reflex. For CW lasers the power must not exceed 1 mW. Class 2 is the only class that applies exclusively to visible lasers operating in CW and pulsed modes. Infrared lasers cannot therefore be classified as Class 2. The emission from Class 2M lasers and systems is not intended for viewing, and as with Class 1M lasers, damage to eyes could result if viewed with optical instruments.

Class 3

Class 3 lasers are subdivided into Class 3R and Class 3B. Class 3R lasers may emit visible or invisible radiation or both, but the unfocused beam must not present hazardous levels of radiation. However, Class 3R lasers do present a hazard if the beam is focused. For wavelengths outside the visible range, these lasers cannot exceed five times the Class 1 AEL. In the visible range, they may not exceed five times the Class 2 AEL (5 mW). Class 3B lasers mark the border between medium and high power devices. The beam from a visible Class 3B laser may be safe to view on a diffuse surface for less than 10 seconds, at a distance greater than 130 mm. However, eye protection should be worn since the blink reflex is not sufficient to prevent damage. The power of a visible Class 3B laser must not exceed 0.5 W. Exposure to 0.5 W for periods longer than 0.25 s is considered dangerous.

Class 4

Eye protection must be worn with Class 4 lasers operating above 0.5 W, or which produce radiant energy greater than 0.125 J within an exposure time of less than 0.25 s. Fire and skin hazards are also associated with Class 4 lasers. They may be incorporated into a Class 1 centre with suitable engineering.

PROTECTIVE MEASURES

Screening

When designing a laser system, the manufacturer must determine the level of radiation that may be accessible during normal operation. This determines the class of the laser, and also the requirements for controls, indicators and warnings.

All laser products require a protective housing that prevents access to radiation in excess of the limits of Class 1, in areas that do not require access under normal operating conditions. Enclosures complete with the appropriate interlocking, illuminated signs and labelling are used to turn Class 4 lasers into Class 1 installations. Screens, curtains, roller blinds and window blocks are designed as passive guards to enclose an area where Class 3B or Class 4 lasers are in use, to protect against accidental exposure to the laser beam. Window blocks are acrylic filters that enable the processing zone to be viewed while stopping laser radiation from coming out. The type of filter depends on the frequency of the laser radiation. A beam attenuator, e.g. a shutter, and an emission indicator (visible or audible), are required on all lasers above Class 1. Operating controls must be located such that the user is not exposed to radiation. Safety interlocks are normally installed to prevent access to radiation above the Class 1 level. A remote interlock connector must be installed on all Class 3B and 4 laser systems, such that an emergency stop switch or similar device may be installed. Such systems must also have a key control, which cannot be removed during operation.

Protective Eyewear

Absorptive eyewear is the most commonly used type of eye protection. The laser beam is blocked by absorption of the relevant frequency. Glass absorption filters can easily be made specific to a particular wavelength or power, which reduces colour distortion. However, plastic filters that incorporate an organic dye for absorption are lighter, cheaper, and more comfortable to wear.

SUMMARY AND CONCLUSIONS

The raw beam emitted from a laser is delivered to the workpiece via either mirror-based paths or a fibreoptic cable. The beam can then be manipulated into various shapes and intensity distributions through the use of refractive, reflective and diffractive optics for specific processing applications. The laser source, beam delivery optics, and means of manipulating the workpiece comprise the laser processing centre. A wide range of centres are available commercially – the choice depends on a number of factors: the size of the part to be processed; the complexity of the part; the nature of beam transmission; and the precision required. The safety requirements associated with laser processing centres are described in standards, and it is the duty of a laser safety officer to ensure that safety procedures are followed.

As processing becomes more automated, process monitoring and control equipment are being integrated into the physical structure and control system of laser centres. Health and safety legislation is also playing an important role in achieving the goal of many manufacturers – fully automated turnkey processing centres that can be operated in the same way as conventional manufacturing equipment.

FURTHER READING

OPTICS

Bass, M. ed. (1994). *Handbook of Optics*. New York: McGraw-Hill.

Das, P.K. (1991). *Lasers and Optical Engineering*. New York: Springer-Verlag.

Dicket, F.M. and Holswade, S.C. eds (2000). *Laser Beam Shaping: Theory and Techniques*. New York: Marcel Dekker.

Hecht, J. (2001). *Understanding Fibre Optics*. 4th ed. New York: Prentice Hall.

SYSTEMS

Henderson, A. (1997). *Guide to Laser Safety*. Boston: Kluwer.

Sliney, D.H. ed. (1993). *Laser Safety Guide*. 9th ed. Orlando: Laser Institute of America.

VDI Technologiezentrum. (1992). *3D-Processing with High Power CO_2 Lasers*. Düsseldorf: VDI Verlag. (In German.)

ENGINEERING MATERIALS

INTRODUCTION AND SYNOPSIS

Some materials have been used in engineering for millennia: ages in evolution are named after metals and alloys; stone, flint and pottery (natural ceramics) depict the culture of civilizations; and the use of wood and bone (natural composites), and skins and fibres (natural polymers) have been essential to both civilian and military success. Many continue to be used in their natural states, with very little processing. However, methods of alloying as well as thermal and mechanical processing provide means of designing materials. Designers seek lighter materials that can be produced more economically, and have lower environmental impact. Today's engineering materials can still be grouped into four broad classes: metals and alloys; ceramics and glasses; polymers; and composites. However, there are important differences: modern engineering glasses and ceramics are tough; polymers can be made with properties that mimic those of metals; and composites can be designed with novel materials and morphologies. As architects have done for centuries, material engineers are now drawing inspiration from structures found in nature to make tomorrow's materials – foams, cellular formations and sponges are notable examples.

In this chapter we consider engineering materials on three structural levels, all of which play a role in the interaction between materials and laser beams. The atomic structure, on the order of nanometres (10^{-9} m), describes the arrangements of individual atoms or molecules and the nature of bonding. Atomic packing and bonding influences material properties, and electronic configurations play a central role in determining the interaction between the photons of the laser beam and the material. The microstructure, on the order of micrometres (10^{-6} m), is governed by arrangements of atoms and molecules in discrete phases. Microstructural changes are induced through the thermal cycles generated during laser processing, which can be illustrated using various phase transformation diagrams. The macrostructure, on the order of millimetres (10^{-3} m), is used as the basis for design calculations to determine engineering performance. Fundamental mechanical and thermal properties are influenced by structure on all three levels. The chapter also introduces the most common industrial materials. Their composition, structure and properties are described, to reveal the philosophy that underlies their design. Strengthening mechanisms, introduced by alloying or thermomechanical treatments, are described – these are affected by the thermal cycles induced during laser processing. These materials make regular appearances in the chapters that follow. Those properties that control the thermal response of materials are displayed on charts, which indicate behaviour during laser processing and play an important role in the selection of materials for laser processing.

ATOMIC STRUCTURE

The atomic structure of engineering materials describes the bonding between atoms, the packing of the atoms, and the distribution of electrons.

METALS AND ALLOYS

Bonding

The atomic structure of metals and alloys is determined by the metallic bond. Electrons are freely shared by atomic nuclei (positive ions). On average, each nucleus is surrounded by sufficient electrons to maintain a full outer shell. Electrostatic attraction between the positive nuclei and the negative electron 'cloud' constitutes the metallic bond. Repulsion between species (electrons or nuclei) of the same charge maintains the equilibrium that creates the atomic structure. Electrons can travel freely in the cloud, and so the metallic bond is non-directional. An approximate value for the metallic bond energy is given in Table 5.1.

Atomic Arrangement

In a pure metal, atoms are the same size, and the metallic bond is non-directional. Atoms are ordered on a three-dimensional lattice in an optimum arrangement to minimize energy. To simplify the description of the atomic arrangement, atoms can be idealized as hard spheres. In close-packed lattices, the spheres order by touching as many adjacent atoms as possible. Two such close packed structures are face-centred cubic (f.c.c.) and close-packed hexagonal (c.p.h.) arrangements of atoms. Each atom is surrounded by 12 nearest neighbours, and is said to have a coordination number of 12. The structure that forms is the one that possesses the lowest energy at a given temperature and pressure. Atoms in the unit cell of the f.c.c. lattice are located at the corners and face centres of a cube. Alternatively, energy may be minimized by packing with a more open structure, such as the body-centred cubic (b.c.c.) arrangement, in which each atom is surrounded by eight nearest neighbours. Atoms in the unit cell of the b.c.c. lattice are located at the corners and centre of the cube. Some directionality of bonding can be identified in more open structures. The lattice may also be distorted in one or more directions, forming tetragonal structures based on f.c.c. and b.c.c. arrangements. Crystallographic planes of atoms can be identified within such lattices, which reflect the stacking sequence.

Table 5.1   Bond energies

Nature	Type	Energy $(J \times 10^{-19})$
Metallic	—	5
Ionic	—	3
Covalent	C—C	5.76
	C—O	5.92
	C—H	6.88
	O—H	7.68
	C=C	10.24
Van der Waals	–	0.03

Interaction of Metals and Alloys with a Laser Beam

The electron cloud in metals and alloys provides a large number of energy levels to which electrons may be promoted on collision with the photons of a laser beam. The energy of near infrared, visible and ultraviolet radiation is similar to that of electron transitions. Transition metals, which are bound by inner electrons, have many unfilled orbitals that can participate in energy absorption. Copper absorbs in the blue portion of the visible spectrum, and so appears red when illuminated with white light. Alkali metals and alloys such as aluminium and magnesium that are bound by outer electrons have fewer energy levels, and absorb poorly in the visible region, appearing white. At a given wavelength of incident radiation, absorption data spread over a range, which in the visible region is determined by the number of unfilled orbitals. As the wavelength of radiation increases, and the radiation energy decreases correspondingly, both the magnitude and width of the absorption range decrease as fewer electron transitions are available to interact with photons. Processing of metals and alloys is therefore more efficient with short wavelength laser light.

CERAMICS AND GLASSES

Ceramics are crystalline, inorganic non-metals. Glasses are amorphous (non-crystalline) solids with short range order, but no long range order.

Ceramics include some of the oldest of constructional materials – granite, stone and porcelain. The name is derived from Cerami, a district of ancient Athens where pottery was manufactured. Today's ceramics may be categorized according to the nature of the principal bonding mechanism (ionic or covalent); or divided into three broad groups according to their application (natural, domestic and engineering). We consider the categories of bonding first to illustrate the differences between ionic, covalent and metallic bonding.

Bonding

Ionic ceramics typically comprise compounds of a metal and a non-metal. Alumina (Al_2O_3), zirconia (ZrO_2) and sodium chloride (NaCl) are examples of ionic ceramics. Atoms that are able to gain electrons to form negatively charged anions are electronegative. Those that dissociate easily into positively charged cations and electrons are electropositive. An ionic bond is formed by the electrostatic attraction between electronegative and electropositive atoms. An average ionic bond energy is given in Table 5.1. By donating and accepting electrons, stable full outer shells of electrons are formed in the participating atoms. Ionic bonding is essentially non-directional.

Covalent ceramics are pure elements (diamond or silicon), or compounds of two non-metals, e.g. silicon and oxygen, which form silica (SiO_2). A covalent bond forms when two or more atoms share electrons in order to achieve stable, filled electron shells. Covalent bonds are strongly directional because the shared pair of electrons occupies a volume that resembles an oriented dumbbell. Covalent bonds are formed between atoms that have similar values of electronegativity.

Atomic Arrangement

The lattice structure of a ceramic depends on two factors: the number of atoms of each element required to maintain electrical neutrality; and the relative sizes of the ions (the ionic radius). If the ratio of the cation radius to the anion radius lies in the range 0.155–0.225, a coordination number of three is the most probable, with the anions arranged at the corners of a triangle around the cation. For ratios in the range 0.225–0.414, a coordination number of four is favoured, and a tetrahedron of anions is formed around the cation. Similarly, ratios in the range 0.414–0.732, and those greater than 0.732 favour coordination numbers of six and eight, respectively. Thus the atomic structure of alumina comprises crystals of hexagonal structure with close-packed layers of oxygen ions (O^{2-}) with the aluminium ions (Al^{3+}) occupying interstices such that each is surrounded by six oxygen ions.

Glasses are based on silica in the form of tetrahedra of $(SiO_4)^-$. These units may be linked by sharing corners to produce three-dimensional frameworks and chains, or two-dimensional rings and layers. The different ways of linking units result in random networks, with short range order, but no long range order – an amorphous structure.

Interaction of Ceramics and Glasses with a Laser Beam

The electrons in ceramics and glasses are bound. The energy of photons is principally absorbed by resonance of bound electrons through coupling to high frequency optical phonons. Phonons can be thought of as lattice vibrations, which take discrete values in the same way as electrons. Crystalline solids have strong phonon absorption bands in the infrared region of the spectrum, and CO_2 laser radiation is absorbed well. Absorption is generally weak over intermediate wavelengths, but increases rapidly in the ultraviolet region because electronic energy transitions then become available. Impurities in ceramics and glasses can cause sharp changes in absorptivity. Engineering ceramics, such as silicon nitride and silicon carbide, absorb well in the range from the visible to the infrared.

POLYMERS

Bonding

A simple polymer, such as polyethylene, comprises linear chains of the monomer ethylene (C_2H_4). More complex polymers comprise networks of cross-linked large monomers. The atoms in the chain and networks are covalently bonded. Examples of covalent bond energies are given in Table 5.1. A limited degree of structure can be formed through secondary weaker bonds such as Van der Waals bonds (Table 5.1), which arise from random temporal and spatial variations of charge in atoms and molecules.

Atomic Arrangement

The degree of polymerization (or density) describes the number of monomers in the chains, and is typically between 10^3 and 10^5. On heating, the chains can move relative to each other, enabling the polymer to soften and be formed. Such polymers are thermoplastics – the largest class of engineering polymer. By replacing one H atom of the ethylene monomer by a side group (or radical), different types of polymer can be formed. For example, if the radical is Cl, polyvinyl chloride is obtained; if it is CH_3, polypropylene is formed; and if C_6H_5 is substituted, polystyrene forms. The configuration of the polymer describes the way in which the atoms are arranged, and cannot be changed without breaking bonds. The conformation describes the arrangement of atoms about the chain: isotactic (all side groups on one side of the chain); syndiotactic (regularly alternating side groups); or atactic (random alternation of side groups).

More complex polymers, such as epoxies, are made up of networks of large monomers in which each chain is cross-linked at various locations to other chains. The cross-linking occurs because the chains react with one another, or small linking molecules, removing radicals to provide sites for links to form. These are thermosets, which if heated soften, but do not melt (instead they decompose).

Linear polymers, such as high density polyethylene, can form crystals because the molecules have no side groups or branches. Highly crystalline polymers, such as high density polyethylene, is around 80%

crystalline; the crystalline regions are separated by amorphous (unordered) glassy regions. A crystalline polymer has a relatively well-defined melting temperature at which the volume changes rapidly.

Amorphous polymers contain large side groups, atacticity, branching and cross-linking, all of which hinder crystallization. Molecules are continually rearranged, increasing the volume of the polymer. The extra volume created by this motion is the free volume, which gives viscous flow. The side units are able to rotate around the single bonds of the backbone, causing an increase in volume and flexibility. As the temperature decreases, the free volume decreases until a point is reached when no free volume remains. This is the glass transition temperature, T_g, below which the polymer becomes a glass in which molecules are bound together into an amorphous solid.

Elastomers are lightly cross-linked polymers, which can be extended considerably and reversibly. The 'backbone' of the polymer chain may be carbon, as in the case of natural rubber and butyl rubbers, or oxygen (polypropylene oxide), silicon (fluorosilicone) or sulphur (polysulphide). Thermoplastic elastomers comprise alternating hard and soft blocks, and can be repeatedly softened by heating, unlike conventional elastomers.

Interaction of Polymers with a Laser Beam

Polymers absorb radiation through resonant vibration of molecular bonds. Absorptivity of far infrared radiation (e.g. a CO_2 laser beam) is high in most organic materials. Energy is absorbed at the surface, and transmitted through the polymer by classical conduction. Radiation in the range near ultraviolet to near infrared (e.g. Nd:YAG and diode laser wavelengths) is transmitted, unless the polymer contains an absorbing pigment or filler. The photon energy of ultraviolet radiation produced by the shorter wavelength excimer lasers (Table B.1, Appendix B) is higher than, or similar to, the covalent bond energy of many organic materials given in Table 5.1; chemical bonds can therefore be broken, without the generation of heat, providing a means of athermal processing (Chapter 7). This interaction is the basis of a large number of laser micromachining techniques.

COMPOSITES

Atomic Structure and Bonding

No class of engineering material exemplifies better the way that nature can influence the design of structures for given applications than composites. Wood is made up of strong fibrous chains of cellulose embedded in a matrix of softer lignin. Teeth and bone are composed of hard inorganic crystals of hydroxyapatite or osteones in a tough matrix of organic collagen. Today's man-made composites adhere to the same design principles. A strong, stiff reinforcement is embedded in a tough, softer matrix. The reinforcement takes the form of regular arrangements of fibres, or random dispersions of whiskers and particles. The matrix material and the morphology of the reinforcing phase provide a convenient means of classifying composites: a metal, polymer or ceramic matrix, with reinforcing constituents of fibres, whiskers or particles. The bonding corresponds to that of the component materials.

Interaction of Composites with a Laser Beam

Composites are able to absorb laser radiation through various mechanisms, depending on the class of material used for the matrix and the reinforcement. The mechanisms correspond to those of the component materials. Differential absorption of energy between the matrix and the reinforcement is often the source of instabilities, and so there are currently few examples of the use of composites in laser processing. However, the potential is enormous if advantage can be taken of this variation in behaviour.

MICROSTRUCTURE

The properties of industrial materials can often be improved if they are processed thermally, mechanically, or both. They may also be alloyed with one or more elements. The atomic structure is transformed by such treatments – atoms are rearranged and new lattices are created, and host atoms mix with alloy atoms in various ways. Distinct *phases* are formed. These phases are typically observed on the micro-scale (10^{-6} m), and influence the microstructure and properties of materials.

Phase transformations are induced in engineering materials by changes in composition, temperature or pressure, or combinations of all three. Most laser-based processing is carried out at constant pressure, and so we are primarily interested in the structural changes that result from thermal cycles and modifications to composition. Phase transformations occur because of a *thermodynamic* driving force that causes one phase to be more stable than others under a given set of conditions. Under *equilibrium* conditions, the phase transformation proceeds to completion. However, the *rate* of transformation in thermally activated processes is governed by a *kinetic* mechanism, such as diffusion. The equilibrium microstructure may then not have time to form, and a metastable phase forms instead. We consider equilibrium transformations first.

EQUILIBRIUM TRANSFORMATIONS

Phase diagrams (also known as constitution or equilibrium diagrams) are maps that show the equilibrium microstructure of an alloy or ceramic over ranges of composition and temperature. They have many uses: they illustrate solidification sequences; they display the range of temperature and composition over which phases are stable; they indicate the temperature at which a phase transformation occurs with a given composition; and they provide a basis for inferring the behaviour of materials under non-equilibrium conditions, such as those present during laser processing. Phase diagrams can take many forms, depending on the mutual solubility of the components.

Complete Solid Solubility

A binary alloy contains two components. Copper and nickel, for example, are mutually soluble in the solid state over the entire composition range, and form a continuous *solid solution*. *Substitutional* solid solutions form when the solvent atoms replace atoms of the host lattice, e.g. zinc atoms in copper (brasses). This can occur in systems with similar atomic sizes (typically a difference of less than 14%), similar electrochemical behaviour, and those in which an increase in electron concentration can be accommodated. For these reasons, copper and transition metals form extensive solid solutions. *Interstitial* solid solutions form when the solvent atoms are small enough to occupy holes between the (often close-packed) atoms of the host lattice, e.g. carbon atoms in the iron lattice (steels).

Consider the phase diagram of the copper–nickel system, shown in Fig. 5.1. Pure copper (denoted by the left-hand ordinate) is solid until heated to 1083°C, at which temperature it melts. Similarly, pure nickel melts at 1453°C (right-hand ordinate). In contrast, alloys of copper and nickel melt and freeze over a temperature range. Consider a liquid alloy containing 75 wt% Cu and 25 wt% Ni (used to make coins). On cooling, the first solid material forms at a temperature of 1220°C. Solidification is complete when the temperature has fallen to 1150°C. This temperature interval is the solidification range for this alloy. An alloy containing 65 wt% Ni and 35 wt% Cu (used as the basis for aeroengine turbine blades) solidifies over a smaller temperature range, between 1410 and 1365°C. The locus of the temperatures of initial solidification over the entire composition range is the liquidus. Similarly, the locus of the temperatures at which solidification is complete is the solidus. At any intermediate temperature in the solidification range, the compositions of the liquid and solid, respectively, adjacent to the solidification

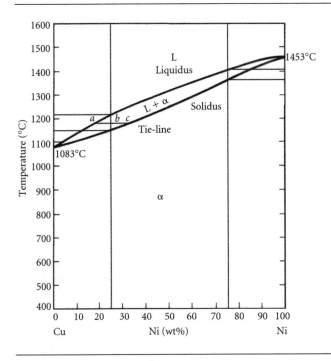

Figure 5.1   Copper–nickel phase diagram showing mutual solid solubility

front are found by constructing a horizontal tie line joining the liquidus and solidus, as illustrated in Fig. 5.1.

The first solid to form is richer in Ni than the alloy composition, and the final liquid to solidify is richer in Cu. At any temperature during solidification, the composition of the liquid is that given by the intersection of the tie line and the liquidus, and the composition of the solid is that given by the intersection of the tie line and the solidus. The weight fractions of solid and liquid, W_s and W_l, respectively, are found by applying the *lever rule* to the tie line:

$$W_l = \frac{c - b}{c - a} \quad \text{and} \quad W_s = \frac{b - a}{c - a}.$$

Solidification occurs by the formation of 'seed' crystals, or nuclei, in the melt. The lower the temperature at which solidification occurs, the more nuclei form. The solid nuclei grow in the melt along preferred crystallographic planes in the form of spiked dendrites with a characteristic atomic arrangement. When neighbouring crystals meet they form a grain boundary. The distance between the grain boundaries is the grain size.

Figure 5.1 is an equilibrium diagram. On solidification, the solidified alloy is able to maintain a homogeneous distribution of alloying elements through solid state diffusion. If solidification occurs too rapidly to allow homogenization by solid state diffusion, the individual layers that solidify retain their original composition, and cored dendrites form, with a centre rich in Ni and outer regions rich in Cu. Rapid cooling also affects the morphology of the solidified dendrites: the dendrite arm spacing reduces as the cooling rate increases. Similarly, on heating at a high rate, the composition of the solid at temperatures between the solidus and liquidus is not able to change fast enough by solid state diffusion to match that predicted by the phase diagram.

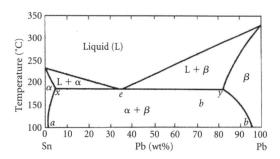

Figure 5.2   Lead–tin phase diagram showing a eutectic reaction

Partial Solid Solubility

Mutual solid solubility is often observed only within limited ranges of alloy components. Outside these ranges a *eutectic* or *peritectic* reaction occurs.

Eutectic Systems

The lead–tin system, from which soft solders are made, is an example of limited mutual solid solubility with a eutectic reaction. Alloys with a lead content up to *a* in Fig. 5.2 solidify as the solid solution α. Similarly, alloys with a tin content up to *b* solidify as the solid solution β. The mechanism of solidification is the same as that described for complete solid solubility.

The composition at which liquid can exist at the lowest temperature in the system is the *eutectic* composition, *e*, illustrated in Fig. 5.2; the temperature is the eutectic temperature. An alloy of composition *e* solidifies at the constant eutectic temperature as a eutectic of the two solid solutions α and β, with compositions *x* and *y*, and volume fractions *ey:ex*, respectively. Alloys in the range *x* to *e* solidify as dendrites of α surrounded by the eutectic of α and β. The α dendrites are referred to as primary α; the remainder as eutectic α. Both have the composition *x*. α dendrites continue to grow until the liquid composition reaches the eutectic composition. Alloys in the range *e* to *y* solidify correspondingly with primary β and eutectic.

At high cooling rates, coring can occur in the same manner as described for complete solid solubility. Excess solute can combine with the solvent to form compounds of low melting temperature. These are normally present at grain boundaries, and are locations of weakness within the microstructure. The morphology of the phases formed is also dependent on the cooling rate: the higher the cooling rate, the finer the microstructure.

Peritectic Systems

A peritectic reaction involves the formation of a solid phase from a liquid and a second solid phase. Figure 5.3 illustrates a peritectic system.

The line joining *w*, *z* and *u* is the liquidus, and that joining *w*, *x*, *y* and *u* is the solidus. T_p is the peritectic temperature. Alloys containing 0–10 wt% B solidify as the solid solution α. Those containing 65–100 wt% B solidify as the solid solution β. Rapid cooling causes coring, and can introduce some β into α alloys that approach *x* wt% B.

Consider an alloy of composition (1) in Fig. 5.3. Solidification starts at the temperature T_1 with the growth of dendrites of α of composition *w*. When the temperature has fallen to T_p, the dendrites

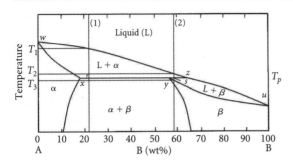

Figure 5.3   A system of A and B containing a peritectic reaction

have a composition x and the liquid a composition z with weights in the proportion α:liquid $= rz:rx$. The peritectic reaction then occurs causing all the remaining liquid to form β of composition y. The final structure consists of dendrites of α with a partial or complete network of β. The ratio $\alpha:\beta = ry:rx$. The effect of rapid cooling is to produce coring in the α and a similar variation in composition in the β, since apart from at the solid–peritectic interface the peritectic reaction depends on solid state diffusion. The volume fraction of β can be increased by rapid cooling.

Consider an alloy of composition (2) in Fig. 5.3. The first stage of solidification is the same as in alloy (1): solidification begins at T_2 and when the peritectic temperature is reached the alloy consists of dendrites of α of composition x and liquid of composition y in the ratio α : liquid $= sz:sx$. At the peritectic temperature all the α reacts with some of the liquid to form β of composition y, with the ratio β:liquid $= sz:sy$. As the temperature falls from T_p to T_3, the remaining liquid solidifies directly as β on that already formed. The composition of the β formed by the peritectic reaction, and that subsequently between T_p and T_3, changes along the line yu by absorption and diffusion of B from the liquid. Under equilibrium conditions the structure at T_3 is homogeneous β of composition s. Rapid cooling results in heterogeneous grains and can cause some α to be retained in the centre of the grains.

The iron–carbon phase diagram shown in Fig. 5.4 illustrates the peritectic reaction of delta ferrite and liquid to form austenite.

Eutectoid Systems

A eutectoid reaction describes the transformation of a solid solution into two new solid phases, normally with a fine lamellar morphology. The iron–carbon phase diagram shown in Fig. 5.4 contains a eutectoid reaction at 723°C when austenite of carbon content 0.8 wt% transforms to a lamellar mixture of ferrite (0.02 wt% C) and cementite (6.67 wt% C), known as pearlite.

Peritectoid Systems

The peritectoid reaction describes the formation of a single solid phase from the reaction of two other solid phases.

PHASE DIAGRAMS FOR ENGINEERING MATERIALS

Phase diagrams for many systems of engineering materials can be described by combining diagrams for solid solutions, eutectic and peritectic reactions over specific composition ranges. They may be used to estimate the course of phase transformations during thermal cycles, from which modifications

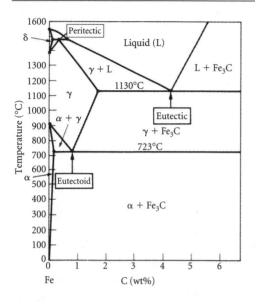

Figure 5.4   The iron–carbon phase diagram extending to the composition of cementite, illustrating peritectic, eutectic and eutectoid phase transformations

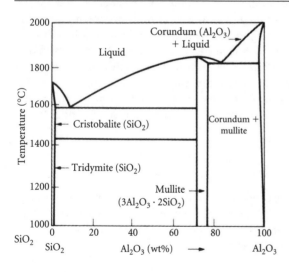

Figure 5.5   Phase diagram for the ceramic system silica–alumina

induced under non-equilibrium conditions may be inferred. A phase diagram for the ceramic system SiO_2 and Al_2O_3 is shown in Fig. 5.5.

NON-EQUILIBRIUM TRANSFORMATIONS

We have considered phase transformations occurring under equilibrium conditions until now. The driving force for transformation is *thermodynamic*. Sufficient time is assumed to be available for

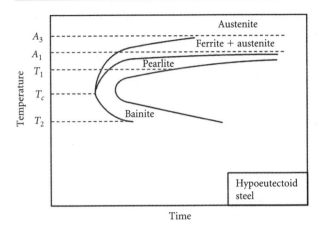

Figure 5.6   Time–temperature–transformation diagram for a hypoeutectoid steel

thermally activated mechanisms such as diffusion to occur. However, only small amounts of energy are absorbed during laser processing, and so heating and cooling rates are high, and transformations can take place under conditions far from equilibrium. The *kinetics* of transformation must then be considered. Phase diagrams can also be constructed for such non-equilibrium conditions.

Isothermal Transformation

The progress of a transformation at a given temperature may be presented in the form of a temperature–time–transformation (TTT) diagram, illustrated in Fig. 5.6 for a hypoeutectoid steel. We consider reactions in steels first because they have been studied in the greatest detail, and transformation diagrams have been produced for a wide range of compositions. However, the principles described apply equally to all materials in which the mechanisms of transformation are analogous.

Let the equilibrium transformation temperature at which a second constituent, pearlite, forms from austenite and becomes stable be A_1. If transformation occurs at a temperature T_1, with a small undercooling (the difference between T_1 and A_1), there is a relatively small thermodynamic driving force, but transformation can occur rapidly because the high temperature enables diffusion to occur rapidly. With a large undercooling at a temperature T_2, for example, a high thermodynamic driving force exists, but the kinetics of the transformation are slow. In both cases transformation takes a relatively long time. At an intermediate temperature, T_c, the optimum combination of thermodynamics and kinetics results in a minimum time for transformation, producing the 'nose' of a characteristic C-curve.

Transformation on Heating

The solid state formation of austenite on heating can be represented in a time–temperature–austenitization (TTA) diagram, shown in Fig. 5.7 for a hypoeutectoid steel.

The effect of rapid heating, as is the case during laser surface hardening (Chapter 9), is to raise the temperatures of transformation. For example, the A_{c1} temperature, at which pearlite transforms to austenite, is superheated by several tens of degrees. The transformation involves the diffusion of carbon from the cementite lamellae of pearlite, which is a kinetic mechanism that requires a certain time to be

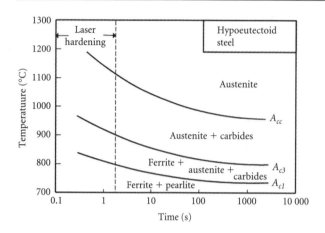

Figure 5.7 Time–temperature–austenitization diagram for a hypoeutectoid steel

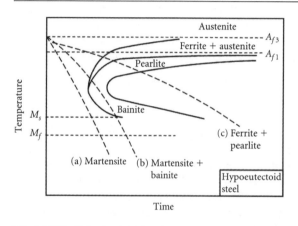

Figure 5.8 Continuous cooling transformation diagram for a hypoeutectoid steel

completed. If the time available is restricted, a higher temperature is required to increase the rate of diffusion. Similarly, carbides present in the microstructure may require even higher degrees of superheating, particularly if the mechanism of dissolution is controlled by the slow diffusion of a metal species.

Transformation on Cooling

The solid state decomposition of austenite on cooling can be represented in a continuous cooling transformation (CCT) diagram, shown in Fig. 5.8 for a hypoeutectoid steel. This is similar in form to the TTT diagram, but is constructed for conditions of cooling, rather than isothermal transformation. Cooling curves have been superimposed in Fig. 5.8 to illustrate the effects of cooling rate on the microstructural constituents formed from austenite decomposition.

Figure 5.8 illustrates that with a very low cooling rate, sufficient time is available for the phases predicted by the equilibrium diagram to form. As the cooling rate increases, metastable phases such as martensite are able to form through athermal reactions. During laser hardening (Chapter 9) we wish to induce a cooling rate similar to that denoted (a) in Fig. 5.8 to produce a hard martensitic surface. However, during laser welding (Chapter 16), a microstructure produced by a cooling rate denoted (c) in Fig. 5.8 might be more appropriate to avoid the likelihood of cracking associated with hard phases.

SOLIDIFICATION

Solidification Morphology

The shape of the solid–liquid interface controls the development of microstructural features during solidification. The nature and stability of the solid–liquid interface depend on the thermal and compositional conditions that exist adjacent to the interface. The most important variables in determining the morphology of solidification are the growth rate, the temperature gradient in the liquid, the undercooling, and the alloy composition. Depending on these conditions, the interface may grow by planar, cellular, or dendritic mechanisms.

Alloys rarely solidify with a planar growth front because temperature gradients in the liquid and growth rates are difficult to control with the accuracy required.

If the temperature gradient ahead of an initially planar interface is reduced, the first stage in the breakdown of the interface is the formation of a cellular structure. As the first protrusion forms, solute is rejected laterally, lowering the equilibrium solidification temperature, causing recesses to form, which in turn initiate the formation of other protrusions. The protrusions develop into long arms, or cells that grow in the direction of heat flow. At sufficiently low temperature gradients, secondary and tertiary arms develop, forming a dendritic morphology.

Constitutional supercooling describes the condition in which the composition of the liquid ahead of the solidifying interface is such that liquid exists below its equilibrium solidification temperature because of the rejection of solute from the solid. In such a case, a planar solidification front is again no longer stable – protrusions can grow into liquid regions with a liquidus temperature above that of the tip of the protrusion. The criterion for such plane front instability is

$$\frac{G_l}{R} < \frac{\Delta T_0}{D_l}$$

where G_l is the temperature gradient in the liquid, R is the solidification front growth rate, ΔT_0 is the equilibrium solidification temperature range, and D_l is the solute diffusion coefficient in the liquid.

The temperature gradient and growth rate are important in the combined forms GR (cooling rate) and G/R since they influence the scale of the solidification substructure and solidification morphology, respectively.

Solidification Phases

The room temperature microstructure that can be expected with solidification and cooling rates associated with arc fusion welding of stainless steels can be predicted by using various diagrams. The Schaeffler diagram uses an abscissa showing the concentration of ferrite-stabilizing elements in terms of a chromium equivalent, and an ordinate that denotes the concentration of austenite-stabilizing elements in terms of a nickel equivalent, Fig. 5.9. Room temperature microstructures containing 3–8 vol.% ferrite have the minimum susceptibility to solidification cracking. The diagram is constructed for arc fusion welding processes – the cooling rates associated with solidification following low energy laser processing are orders of magnitude higher, and so such diagrams should be used with caution when predicting solidification microstructures or selecting steel compositions. This is discussed further in Chapter 16.

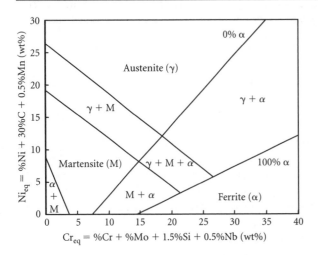

Figure 5.9 The Schaeffler diagram showing room temperature microstructure of stainless steels after arc fusion welding (Schaeffler, 1949)

INDUSTRIAL MATERIALS

Laser processing has been applied to a wide range of engineering materials. In the chapters that follow, laser-processed materials are categorized into classes (metals and alloys, ceramics and glasses, polymers and composites), which are subdivided into types (carbon–manganese steels, alloy steels etc.), and sometimes further differentiated by grades (e.g. 2000 and 6000 series aluminium alloys). If we are to understand the effects of laser processing on industrial materials, it is important to appreciate the philosophy behind the design of the materials: an observation from treatment of one material may enable a prediction of the response of an analogous material from a different class to be made, for example. This section on industrial materials can be used as a reference for the composition, microstructure and properties of the laser-processed grades described in the chapters that follow.

METALS AND ALLOYS

Metals and alloys account for over half the engineering materials in use today. The most common groups of engineering metals and alloys are carbon–manganese steels, alloy steels, cast irons, stainless steels, tool steels, aluminium alloys, magnesium alloys, titanium alloys, nickel alloys and copper alloys. They are popular because they possess attractive combinations of properties: a favourable ratio of strength to density; good formability and weldability; and relatively low cost. The various standards and designations used to identify individual alloys are listed in Appendix C.

Carbon–Manganese Steels

The popularity of steels as engineering materials can be attributed to a number of factors. They are relatively easy to manufacture using conventional techniques such as casting, forging, stamping and machining. Their cost is relatively low. A wide range of mechanical properties may be produced by thermal and mechanical treatments, enabling optimum combinations of strength, ductility and toughness to be achieved. Today, steels account for over 80% by weight of metallic engineering alloys.

Carbon–manganese steels, with microstructures consisting of ferrite and pearlite, are often referred to as *structural* steels. They may be classified by their carbon content as *hypoeutectoid* and *hypereutectoid*, depending on whether the carbon content is less than or greater than 0.8 wt%, respectively. Hypoeutectoid steels are further classified here according to carbon content: *low carbon* grades contain less than 0.1 wt%; *medium carbon* contain 0.1–0.25 wt%; and *high carbon* 0.25–0.8 wt%. They are produced in sections up to several centimetres in thickness, and used extensively in the construction and general engineering industries. Steels with grain sizes on the order 7–10 μm and ultimate tensile strengths above 700 N mm^{-2} are now common. Hypereutectoid steels are generally used in highly alloyed conditions in applications where hardness is the principal requirement, e.g. cutting tools.

Steels are alloyed for one or more of the following reasons:

- to produce solid solution hardening (C, N, Mn, B);
- for deoxidization during manufacturing (Si, Al);
- to modify the morphology of inclusions (Ce, Ti, Ca, Zr);
- to form carbides and nitrides to pin grain boundaries (Ti, Nb, V, W, Mo, Cr);
- to increase the amount of pearlite (C, Mn);
- to stabilize ferrite (Cr, Si, Al, W, Mo, Ti, V);
- to stabilize austenite (C, Mn, Ni, N);
- to refine grain size (Mn);
- to improve toughness (Ni);
- to increase hardenability (C, Mn, Cr, Mo, Ni, B);
- to increase hardness (C);
- to improve corrosion resistance (Cr, Mo);
- to improve oxidation resistance (Cr, Cu);
- to hinder grain growth (Nb, V, Ti, Cr, Al);
- to improve machinability (Mn); and
- to reduce segregation and the occurrence of solidification cracking (Mn).

Carbon is the most important alloying addition in structural steels. It is the simplest and cheapest means of increasing strength, which it does by increasing the volume fraction of pearlite formed. However, both impact toughness and weldability decrease as the carbon content is raised. This can present problems during laser processing, when cooling rates can be an order of magnitude greater than conventional arc processes. Carbon also strengthens the matrix by solid solution hardening since its atomic volume is sufficiently small to allow it to occupy the octahedral interstitial sites in both the ferrite and austenite lattices. Solubility is greater in austenite than ferrite because of the larger size of the interstices. Carbon increases hardenability and lowers both the martensite start (M_s) and martensite finish (M_f) temperatures. This encourages the formation of hard twinned martensite on quenching from austenite. The hardness of martensite is primarily determined by its carbon content. In addition to cementite (Fe_3C), carbon combines with other alloying additions to form carbides.

Nitrogen also strengthens through solid solution hardening, but its presence is generally undesirable because it increases the tendency for strain ageing, and hence brittleness. It combines with other elements to form nitrides, which can pin grain boundaries to maintain a small austenite grain size during thermal processing.

Manganese is present in structural steels, partitioned in ferrite and cementite. It has a mild deoxidizing effect. It combines with sulphur to form manganese sulphide inclusions, thus reducing the degree of sulphur segregation to grain boundaries, which may promote solidification cracking. It also increases the volume fraction of pearlite formed. In amounts between 0.8 and 1.6 wt% it increases strength without impairing toughness, partly because it lowers the austenite transformation temperature, resulting in grain refinement. Manganese also increases hardenability.

Silicon is used as a deoxidizing agent during the steelmaking process, to produce killed steel. If insufficient silicon is present to combine with all the dissolved oxygen, bubbles of carbon monoxide form during solidification, which result in undesirable macrosegregation on solidification. Such unkilled steels are known as rimming steels. Silicon produces a significant strengthening effect in ferrite.

Microalloyed structural steels contain small additions of titanium, vanadium, aluminium and niobium (less than 0.15 wt%), which form carbides, nitrides or carbonitrides around 200 nm in diameter. These precipitates are stable in austenite, and restrict grain growth during thermal processing. The rapid thermal cycles induced by laser processing have a beneficial effect on restricting precipitate dissolution and coarsening. A small ferrite grain size is produced on cooling, which increases both yield strength and toughness.

Phosphorus and sulphur are impurity elements that are present in small amounts in most steels. Phosphorus exists in solid solution in ferrite, increasing strength, but it has a particularly noticeable effect on decreasing toughness. Sulphur has a very low solubility in iron, and normally exists in the form of manganese sulphide. Impurity limits designed for arc welding steels might not be applicable to laser welding because of the higher cooling rates involved, and care should be taken when applying guidelines.

Low carbon steels contain less than 0.1 wt% C. They are used in the normalized, hot rolled, or cold rolled conditions. Structural strength is not of prime importance; applications include automotive bodies, stampings, and cladding for buildings. The microstructure comprises ferrite and pearlite; the volume fraction of pearlite increases with carbon content. The requirement for improved corrosion resistance in sheet steel applications in the automotive industry has led to the widespread introduction of coated steels. Zinc coatings are produced by electrogalvanization, galvannealing, and hot dip galvanizing. Polymer coatings can also be deposited on a zinc-coated sheet base. Generally, the thickness of electrolytic zinc and iron–zinc alloy coatings is less than 10 μm, but thicker coatings (>20 μm) produced by hot dipping are used in regions requiring added protection. Uncoated low carbon sheet steels can be laser cut and welded easily, but precautions are necessary when coatings are present within lap joints.

Medium carbon steels respond well to heat treatment, but as the carbon content rises the fracture toughness falls rapidly. They are therefore normally used in the hardened and tempered condition. Plain carbon steels can only be heat treated in small sections with rapid quenching. Deeper hardening requires the use of more highly alloyed steels. Medium carbon steels are used in crankshafts, axles, wheels, connecting rods, crane hooks, die blocks, turbine rotors and gears. They can be laser welded, normally with the aid of preheating, and also laser hardened.

High carbon steels are brittle, and are therefore normally alloyed with elements such as chromium and vanadium to improve toughness. These are known as tool steels, and are considered later. Hardening is normally the only laser process used with such steels.

Alloy Steels

Alloy steels contain deliberate alloying additions, in addition to carbon and manganese. Chromium, copper, molybdenum, nickel, silicon, niobium, titanium, tungsten and vanadium are added to impart specific properties. Alloy steels are classified here according to the total level of deliberate alloying additions: low alloy steels contain less than 2 wt%; medium alloy steels contain 2–10 wt%; and high alloy steels contain more than 10 wt%.

The carbon content in alloy steels is typically less than 1 wt%. The size of metallic atoms such as manganese, nickel and chromium is similar to iron, and so they are able to occupy substitutional sites in the iron lattice. Chromium is added in amounts typically between 1 and 2.5 wt% in order to increase hardenability, thus increasing the thickness of material that can be hardened. The most noticeable effect of nickel in alloy steels is to promote toughness at low temperatures. For example, the addition of 9 wt% Ni provides toughness down to $-200°C$. Molybdenum is added to increase hardenability, but it is also a strong carbide-forming element. The stable carbide must be dissolved, and molybdenum must be available in solid solution in austenite for it to be effective in increasing hardenability. The sulphide inclusion content is usually modified to produce fine dispersions of round inclusions rather than deleterious stringers, by the addition of Ce Ti, Ca or Zr. Steels alloyed with Cr are used for power generation and high temperature applications, in which resistance to creep is important. The most common steels are the Cr–Mo steels containing 0.5–2.25 wt% Cr and 0.5–1.0 wt% Mo. Vanadium may also be added (approximately 0.25 wt%), and higher Cr contents (5 and 9 wt%) are used in some applications. High strength low alloy (HSLA) steels are produced by a combination of low alloy additions and thermomechanical treatment during production to produce a range of enhanced mechanical properties (e.g. yield strength in the range 450–850 N mm^{-2}). Alloy steels respond well to laser hardening.

Stainless Steels

There are four basic types of corrosion resistant or 'stainless' steel, which are characterized by the dominant room temperature microstructure: austenitic; ferritic; duplex; and martensitic. In practice, austenitic stainless steels contain a small amount of ferrite. A fifth category – stainless steels that respond to precipitation hardening – can also be defined. Stainless steels are often considered to be expensive, but normally prove economically sound when all the costs – including lifetime, maintenance and repair – are taken into consideration.

Chromium forms an oxide film in many environments, which protects the steel from corrosion. In normal aqueous media, 12 wt% Cr provides protection. Under aggressive conditions, the chromium content might need to be well in excess of 12 wt%. Chromium stabilizes ferrite. Nickel is added to stabilize austenite, providing toughness, particularly at low temperatures. Molybdenum in amounts of about 2.5 wt% increases the corrosion resistance of austenitic stainless steels in chloride-containing environments such as sea water. When heated into the temperature range 420–870°C, carbides can be precipitated, primarily along grain boundaries. This is known as sensitization, or weld decay, and results in chromium-depleted regions, with reduced corrosion resistance. Niobium and titanium are used as stabilizers in some grades to overcome this problem. The rapid thermal cycle of laser welding is beneficial in reducing weld decay.

Austenitic stainless steels contain 18–25 wt% Cr, 8–20 wt% Ni, and up to 0.2 wt% C. The most common contain 18 wt% Cr and 9–10 wt% Ni. The solidified microstructure comprises austenite or a dual phase austenite–ferrite mixture. A ferrite content between 3 and 8 vol.% endows the steel with good resistance to solidification cracking. However, ferrite may be preferentially attacked in some corrosive environments, and may transform to the brittle sigma phase if exposed to temperatures in the range 540–930°C. Ferrite is also detrimental to fracture toughness in cryogenic environments. Austenitic stainless steels can only be hardened by mechanical working. They possess high corrosion resistance, particularly to pitting caused by chlorides. They can be cleaned easily, are easy to fabricate, have a good appearance, are non-magnetic, and have good toughness and ductility at low temperatures. Of the stainless steels in use today 80–90% are of the austenitic type. They are used when corrosion resistance and toughness are the primary requirements, e.g. in car wheel covers, wagons, fasteners, chemical and food processing equipment, heat exchangers, oven liners, aircraft exhaust manifolds and pressure vessels. They are suitable candidates for laser cutting and welding.

Ferritic stainless steels contain 13–25 wt% Cr, and up to about 0.2 wt% C. Ferrite is formed on cooling from the liquid. Ferritic stainless steels are used when moderate corrosion resistance is required, although their resistance to stress corrosion is particularly high. They are magnetic and are not generally considered to be heat treatable. They are used in decorative trims, light-gauge sheet and acid tanks, and many laser-cut and laser-welded applications can now be found, e.g. the components of car exhaust systems. Ferritic steels containing 40 wt% Cr have high resistance to chloride stress-corrosion cracking, and a lower cost in comparison with duplex stainless alloys, because of the lower nickel content.

Duplex stainless steels appeared in early 1930s. They typically contain 22 wt% Cr, 5 wt% Ni, 3 wt% Mo and 0.17 wt% N, and solidify as ferrite, with austenite islands forming at grain boundaries during solid state transformation. A microstructure containing equal amounts ferrite and austenite exhibits a good balance of toughness, and resistance to local and stress-corrosion cracking. The addition of nitrogen has a number of beneficial effects: it stabilizes the duplex microstructure at high temperature; it increases the stability of austenite with respect to intermetallic phase formation (e.g. sigma) in the range 800–1000°C; and it imparts corrosion resistance in chloride-containing neutral or oxidizing acid media. Even higher alloy additions are used in super duplex grades – typically 25 wt% Cr, 7 wt% Ni, 3 wt% Mo, 0.17 wt% N and 1 wt% Cu; these have particular resistance to pitting corrosion. Duplex stainless steels can cost ten times more than structural steels, and are used where high corrosion resistance is required, such as in chemical and food processing equipment, and paper and pulp machinery. There is much interest in the properties of laser-welded duplex steels because the rapid thermal cycles help to maintain the desirable 50–50 mixture of ferrite and austenite, and microstructural control can be exercised through the use of appropriate shielding gases.

Martensitic stainless steels contain 11.5–18 wt% Cr and up to 1.2 wt% C. Additional alloying with about 5 wt% Ni and up to 2 wt% Mo enhances corrosion resistance. Since austenite is formed on solidification, the steels can be hardened by heat treatment, although they are often delivered in the annealed condition for ease of machining, with a microstructure of chromium carbides in a ferrite matrix. In some oil and gas applications, 13 wt% Cr weldable martensitic stainless steels have the potential to replace duplex stainless steels with a material cost saving of about 50%. Popular uses include steam turbine blades, cutlery, surgical instruments, valves, screws, springs, machinery, blades, bolts, bearings, nozzles, and races. Fine parent microstructures, in which carbides are uniformly distributed, are good candidates for laser hardening.

Precipitation hardening stainless steels may be austenitic, duplex or martensitic, depending on the alloying additions. Copper, molybdenum, aluminium, titanium and niobium are added to form precipitates. The alloys are solution treated, quenched and aged at temperatures between 400 and 500°C. The precipitation hardening agents are copper, Ni_3Ti, Ni_3Al and $NiAl$. Martensitic steels normally contain 4–7 wt% Ni to maintain the M_s temperature above room temperature. Reasonable levels of ductility and fracture toughness are obtained, making the alloys suitable for use in valve parts, shafts, landing gear parts, aircraft and nuclear components.

Tool Steels

Some tool steels are actually cast irons. A useful distinction is that a steel solidifies as austenite, whereas a cast iron solidifies as a eutectic. Tool steels may be classified according to the method by which they are hardened (water, oil or air), their properties (e.g. shock resisting), or their composition (plain carbon, alloy etc.). A mixture of the above is often used in practice.

Plain carbon tool steels contain up to 1.2 wt% C; they derive their hardness through quenching in cold water. They are hard, but brittle. Alloy tool steels contain stable hard carbides, formed by alloying with manganese, chromium, molybdenum, tungsten and vanadium. Manganese tool steels typically contain 0.7–1.0 wt% C and 1–2 wt% Mn, and are hardened by oil quenching and tempering. Toughness can be increased by replacing some of the manganese with chromium. The addition of vanadium refines the grain size, improving toughness and shock resistance. Chromium, tungsten and molybdenum form carbides that are stable at high temperatures, and are added to hot working tool steels, such as high speed steels (named after the high machining speeds that they tolerate). Punches, shear blades, blanking tools, pneumatic tools, shear blades and mandrels are made from tool steels, which may be selectively laser hardened.

Cast Irons

The term *cast iron* covers a range of ferrous alloys containing 2–4.5 wt% C. Cast iron can be produced in many different forms, depending on the pig iron used, the casting conditions, the alloying additions, and post-casting heat treatment. The carbon may be present combined with iron as cementite, or as graphite in the form of flakes or spheroids. Graphite is the thermodynamically stable form of carbon, but the slow kinetics of its formation during cooling through the eutectic range and solidification mean that cementite is produced in practical treatments. A process for spheroidizing graphite was developed in 1948, leading the way for a range of spheroidal graphite irons, also known as ductile or nodular irons. Rapid cooling and the presence of carbide-forming elements such as chromium favour cementite formation. The terminology used with cast irons varies among countries and industries, and is explained in Appendix C. Although microstructural terminology is the most exact, the names used here are those most commonly cited in relation to laser processing: *nodular iron* – spheroidal graphite in a matrix of pearlite, ferrite, bainite or martensite; *grey iron* – flake graphite in a matrix of ferrite or pearlite; *white iron* – a microstructure containing cementite; *malleable iron* – a white cast iron that has been heat treated to improve ductility; and *alloy iron* – graphite-containing or graphite-free alloyed iron. On the basis of strength/cost, cast iron is second only to structural steel among metals and alloys. It has the advantage of good casting properties, and in the grey form, good machinability, which allow it to be used in components of complex shape.

The higher carbon content of cast irons leads to a reduction in melting temperature in comparison with steels, which must be considered when selecting laser processing parameters. A low melting temperature improves the castability of the material. Alloying additions are made to control the solidification behaviour and the as-cast properties of cast irons. Phosphorus may be present up to 1.5 wt% in order to form an intermediate phase, Fe_3P, which solidifies at about $950°C$, thus lowering the solidification temperature significantly, improving fluidity and reducing shrinkage. Silicon promotes the formation of graphite, and increases the eutectic temperature, whereas chromium promotes cementite formation. Cerium and magnesium promote spheroidization of graphite, and are used in the production of nodular irons. Aluminium prevents oxidization at elevated temperatures. Nickel promotes graphite formation and reduces the sensitivity of the microstructure to cooling rate. Calcium silicide refines the microstructure of flake graphite cast iron by creating nucleation sites for the flakes. Such irons have a refined microstructure (known as Meehanite), and are used to produce high quality castings. The majority of products are used in the as-cast form, with little post-cast working, apart from machining. Cast irons that have fine parent microstructures, in which carbon can be dissolved on heating to form martensite on quenching, are candidates for laser hardening.

The microstructures of nodular irons comprise spheroidal graphite nodules in a matrix of ferrite, pearlite, bainite, retained austenite, or a mixture of these. They combine cheapness with ease of casting and are used in machine parts that require good ductility, such as crankshafts, turbine casings, gears, brake drums, machine components, moulds, heavy duty piping, and highly stressed piston rings. Ferritic matrix nodular irons possess relatively high impact resistance and machinability, but reduced tensile properties. Pearlitic nodular iron is relatively hard, shows moderate ductility, and has a high tensile strength. In ferritic–pearlitic nodular cast iron, the matrix comprises pearlite with rims of ferrite making up about 50% of the volume. Bainitic nodular iron is produced by austenitizing nodular iron at $900°C$, followed by quenching to a temperature between 300 and $450°C$. It possesses high tensile strength combined with good ductility and is used in crankshafts, railway wagons and gears. Austempered ductile iron (ADI) is produced by austenitizing at $815–920°C$ and tempering at

230–400°C to produce a microstructure of bainite and retained austenite. ADI is almost twice as strong as normal grades of nodular iron, and is a well-established gear material, also finding uses in crankshafts and connecting rods.

Grey irons contain flakes of graphite in a matrix of ferrite, pearlite or austenite. The graphite flakes act as stress raisers, which cause localized plastic flow at low stresses, and initiate fracture in the matrix at higher stresses. Grey iron therefore exhibits poor elastic behaviour and fails in tension without significant plastic deformation. However, the presence of graphite flakes gives grey iron excellent machinability, damping characteristics and self-lubricating properties. High strength grades contain less carbon and more silicon, which promotes the formation of graphite as fine flake clusters. Ferritic grey iron contains more silicon than pearlitic types, and is produced either by using a lower cooling rate, or by annealing pearlitic varieties. Austenitic grey irons are used when tensile strength is not the primary requirement, but corrosion resistance is desirable. Applications for grey cast iron are engine cylinder heads, gearboxes, gears and machine tool slide ways, which utilize thin and complex cast sections for water-cooling passages and the damping characteristics for quietness. General engineering castings benefit from the ease of casting, the comparatively simple pattern equipment and the shorter lead times to produce castings.

Malleable irons typically contain 2.5 wt% C and 0.6–1.0 wt% Si, which promote the decomposition of cementite during heat treatment, without producing graphite flakes on casting. The microstructure thus consists of irregularly shaped nodules of graphite called 'temper carbon' in a matrix of ferrite or pearlite, or both. The presence of graphite in a more compact or sphere-like form gives malleable iron a favourable combination of ductility and strength. Malleable irons are often referred to by names that describe the appearance of the fracture surface; for example, blackheart and whiteheart. Blackheart malleable iron, often referred to as ferritic malleable iron, is produced by heat treatment of white iron (850°C for up to a week) in a neutral environment. It has a microstructure of rosette-shaped graphite nodules in a matrix of ferrite. It combines the casting and machining properties of grey iron, enabling intricate shapes to be produced, with mechanical properties similar to structural steel. It is used in vehicles, agriculture and general engineering. Whiteheart malleable iron is made by heating white iron in an oxidizing atmosphere, producing a decarburized surface of cementite. The material has a higher tensile strength than blackheart, and lower ductility. It is used in the production of railway wheels.

Alloy irons include nodular graphite irons, grey and white irons, and are used for machine tool beds, cams, piston rings, pistons, and cylinders. Meehanite describes a family of easily machinable alloy irons. Ni-hard alloy irons contain nickel and chromium, for example 8 wt% Cr and 6 wt% Ni. Both have a microstructure of iron and chromium carbides in a matrix of martensite and bainite. Such irons can be used as cast, but heat treatment improves the hardness and resistance to surface cracking and spalling. High chromium cast irons typically have compositions of 15 wt% Cr and 3 wt% Mo, or 23–28 wt% Cr, and possess a good combination of abrasion resistance and toughness. In some cases they may be used as cast, but are normally air hardened to develop the optimum properties.

White irons contain low levels of graphite-forming elements, such as silicon, in order to promote the formation of cementite. The melt is cooled rapidly, so that carbon is precipitated as cementite rather than graphite. It is therefore extremely hard and abrasion resistant, but very brittle. White iron is rarely used in its pure form, but is alloyed to improve ductility and toughness. These irons are limited in application because of the lack of impact resistance and the difficulty in maintaining the structure in thicker sections. They are mainly adopted as the starting material for malleable cast irons. However, they are used for wear-resistant parts such as grinding parts, crushing equipment and brake shoes.

Aluminium Alloys

The British scientist Sir Humphrey Davy established the existence of the element aluminium in 1807. He called it 'aluminum' – the spelling that is still used in the United States. The Danish physicist

H.C. Oersted first produced aluminium in the laboratory in 1825. Various chemical production processes were developed in Europe, until 1886 when Hall in the United States, and Heroult in France independently perfected an electrolytic method for producing aluminium from aluminium oxide (alumina). In 1888 the German Karl Bayer developed an economical method of producing alumina from bauxite ore, which led to the availability of cheap aluminium for the first time. The widespread, and increasing, use of aluminium and its alloys in engineering applications today can be attributed to a combination of low density, high modulus, formability, weldability and good corrosion properties.

Aluminium alloys are categorized as wrought (heat treatable and non-heat treatable), and cast. Wrought alloys account for about 85% of aluminium use, and are made into finished and semi-finished products by mechanical forming processes, such as rolling, forging and extrusion.

Alloying elements have many effects in aluminium alloys. They:

- improve fluidity and casting characteristics (Si);
- form strengthening precipitates (Si, Cu, Mg, Zn, Li);
- cause solid solution strengthening (Cu, Mn, Mg, Fe);
- improve ductility and castability (Mn);
- improve work hardening characteristics (Mg);
- reduce solidification cracking (Fe); and
- reduce density and increase modulus (Li).

The elements listed above form the basis of *alloy series*, which are designated in Appendix C, together with heat treatment procedures that generate tempers.

The primary interest in laser processing of aluminium alloys has been in their welding characteristics. Aluminium alloys are more difficult to laser weld than ferrous alloys because of their physical and thermal properties. This is discussed further in Chapter 16.

The impurity content of the high purity 1000 series aluminium alloys is below 0.3 wt%. Such high purity material is used for electrical conductors. Commercial purity aluminium typically contains 99.3–99.7 wt% Al. It can be strengthened by strain hardening, has good corrosion properties and takes decorative finishes. 1100 is used in sheet metal work, packaging, and cooking utensils.

The primary alloying addition in wrought 2000 series alloys is copper, which is present in binary alloys up to 6.3 wt%. Copper provides strengthening by solid solution hardening, and by precipitation hardening through the formation of metastable semicoherent plates of θ' (Al_2Cu). The first precipitation-hardened aluminium alloy contained copper, manganese, magnesium and silicon, and was discovered accidentally in 1906 by the German metallurgist Alfred Wilm. Precipitation hardening can be intensified through the addition of up to 1.6 wt% Mg, accompanied by a reduction in copper content to about 4.4 wt%. Strengthening in Al–Cu–Mg alloys originates from the formation of metastable incoherent S′ laths (Al_2CuMg) after a temper such as T3 (Appendix C). Aluminium–copper–lithium alloys have been developed in recent years in an attempt to reduce weight and provide alternatives to traditional aerospace alloys and new composite materials. Relatively low Cu/Li ratios result in the formation of the equilibrium T1 phase (Al_2CuLi), as well as the metastable δ' (Al_3Li) and θ' precipitates. Such alloys possess high strength, high stiffness and low density, but their use has been hindered by anisotropic mechanical properties. 2000 series alloys are used in aircraft lower wing and fuselage sections.

Wrought 3000 series alloys contain manganese in amounts up to about 1.2 wt%. Heat treatment has little effect on their mechanical properties – strength is increased primarily through solid solution hardening. A limited amount of strengthening may be obtained by the addition of magnesium (up to about 1 wt%) as a result of solution treatment, quenching and cold working, and subsequent annealing, to produce a fine dispersion of (Mn, Fe)Al and α (Al–Fe–Mn–Si), which inhibits recrystallization. 3003 finds uses in chemical plant, tubes, and heat exchangers. 3004 is mainly used in beverage can bodies.

The principal alloying addition in wrought 4000 series alloys is silicon, which lowers the melting temperature in amounts up to 12.6 wt%. The main use of such alloys is as filler wires, notably 4043 and 4047, which contain 5 and 12 wt% Si, respectively. Such compositions solidify over relatively small temperature ranges, preventing solidification cracking.

General purpose and structural 5000 series wrought alloys contain 1–5.5 wt% Mg. Magnesium is a solid solution strengthening element, and endows alloys with high work hardening characteristics. In annealed material, the yield strength is approximately proportional to magnesium concentration. In order to increase strength further, Mn (0.1–1.0 wt%) or Cr (0.1–0.25 wt%), or both, are added. 5000 series alloys are frequently used in the O temper condition. Strength may also be increased by work hardening. 5083 finds many marine applications, and 5182 is used for the ends of beverage cans.

Magnesium and silicon are the principal alloying additions in wrought 6000 series alloys, present in amounts up to 1.3 wt%. Higher strength can be achieved through increased silicon addition. A ratio by weight of Mg:Si of 2:1 is desirable to optimize tensile properties. Precipitation hardening is the principal strengthening mechanism, achieved through the formation of metastable Mg_2Si, notably the needle-shaped β'' precipitate. Two-stage hardening treatments have been developed to improve mechanical properties; one novel possibility is to use the paint baking cycle of automobile alloys as a secondary hardening mechanism. Other alloying additions include manganese and chromium, which enhance strength and control grain size. Copper may also be added to increase strength, but in amounts below 0.5 wt% to maintain corrosion properties. 6000 series alloys for structural applications possess medium strength, and have good corrosion resistance. 6005 is used in automobile body sheet and 6013 is a damage tolerant alloy used in strengthening airframe stringers, whereas 6061 and 6082 have good extrusion properties and are used in architectural sections.

Wrought 7000 series alloys contain zinc (4–8 wt%) and magnesium (1–3 wt%). Both elements are highly soluble in solid aluminium. The addition of magnesium produces a marked increase in precipitation hardening characteristics. Copper additions (1–2 wt%) increase strength by solid solution hardening, and form the basis of high strength aircraft alloys. The addition of chromium, typically to 0.3 wt%, improves stress-corrosion cracking resistance. Peak hardness is normally achieved through the use of duplex heat treatments to intensify precipitation hardening. For example, 7075 in the T651 condition contains Guinier–Preston zones (*c.* 7.5 nm) and semicoherent monoclinic η' precipitates of $MgZn_2$ (*c.* 15 nm). The 7000 series alloys are among the strongest aluminium alloys available, and are used predominantly in aerospace applications. 7020 alloys find applications in car bodies, rockets, and rail cars.

The most common wrought 8000 series alloys are those containing lithium, which have been developed since the 1950s for the aerospace sector as replacements for 2000 and 7000 series alloys, and as competitors to carbon fibre composite materials and other alloys. Each weight percent addition of lithium reduces the density by around 3% and increases the Young's modulus by about 6%. Thermal conductivity is also reduced. Other alloying additions include copper, magnesium and zirconium, which can produce strengthening precipitates of S (Al_2CuMg), T1 (Al_2CuLi) and T2 (Al_6Li_3Cu). Casting of specialist 8000 series alloys produces strengthening dispersoids of, for example, $Al_{12}(Fe,V)_3Si$ in a fine-grained aluminium matrix. However, high cost, reductions in ductility and toughness, and a high degree of anisotropy, in comparison with conventional aluminium alloys, have hindered commercial application of aluminium–lithium alloys. 8090 is used in airframes and the space shuttle fuel tank.

The 300 series of casting alloys have been developed, containing silicon with copper or magnesium to impart good fluidity and filling ability. When solidified they possess high strength, ductility and corrosion resistance, and can be precipitation hardened.

The 400 series of casting alloys, in which silicon is the principal alloying addition, are the most important casting alloys. Silicon increases the fluidity of the melt, reduces the melting temperature, has a low density, decreases the contraction associated with solidification, and is a cheap raw material. Excess silicon precipitates as plates, which increase wear resistance. The strength of cast products can be increased by the addition of up to 0.35 wt% Mg, which makes the alloy heat treatable. 443 is used in sand and permanent mould castings. 413 is used in wear-resistant die-castings such as pistons, transmission casings, and connecting rods.

Magnesium Alloys

Magnesium and its alloys have the lowest density of the common engineering alloys, possess high strength, rigidity and dimensional stability, and are easily machined. Their thermal conductivity is high. They are easily formed, with about 90% of magnesium alloys being cast using various techniques. (Molten magnesium must be kept under a gas such as SF_6 to avoid explosion.) World production of magnesium is relatively small, but the potential market for cast magnesium components is large.

Magnesium alloys may be also classified as casting and wrought types, although some can be used in both cast and wrought form.

The functions of the principal alloying additions are as follows:

- solid solution and precipitation hardening (Al, Zn, Mn, As); and
- grain refinement (Zr).

Magnesium–aluminium–manganese alloys derive their strength from solid solution and precipitation hardening. They possess good weldability, and are used in sheet metal fabrication. Cast AM60A is used in archery bows, baseball bats, and car wheels.

Magnesium–aluminium–zinc alloys can be strengthened through solid solution hardening; zinc is the second most potent solid solution hardening element in magnesium alloys. The addition of zinc also confers precipitation hardening characteristics on these alloys. Magnesium alloys containing rare earth elements (e.g. zirconium and thorium), zinc and silver are used in both cast and wrought forms. Zirconium strengthens through grain refinement. Cast AZ81A is used for car wheels, transmission housings, plastics moulds; AZ91B for cylinder head covers, carburettors, fans and housings; and AZ91C – the most common Mg–Al–Zn alloy – is used for hand trolleys. The alloy QE22A, containing arsenic and rare earth elements, maintains a high yield strength at high temperature, and is used in aircraft gearboxes. The alloy ZE41A, which contains zinc and rare earth elements, is replacing the AZ91 series of alloys in suspensions, chassis, bearings and gearboxes. Wrought AZ31B is forgeable, drawable and weldable, and is used in monocoque sheet and extrusions, and racking.

Titanium Alloys

The density of titanium is about one-half that of steel, and its Young's modulus lies between those of aluminium and iron. It has a relatively low coefficient of thermal expansion, is non-magnetic, does not exhibit a ductile–brittle transition, and has good biocompatibility. Alloys possess good fatigue resistance, but have a high coefficient of friction both against themselves and against other metals, which means that they suffer from low wear resistance. Titanium easily forms a protective oxide coating in a water environment, which provides good corrosion resistance. However, at temperatures above 500°C oxygen and nitrogen are absorbed rapidly, leading to potential embrittlement problems. The main drawback of titanium is its cost – about five times that of steels and aluminium alloys. The aerospace and defence industries are the largest users of titanium alloys.

Crystal structure is used as a basis for classifying titanium alloys. Titanium can exist in the hexagonal close packed alpha (α) phase, or the body-centred cubic beta (β) phase. Alloys are therefore classified α, β or $\alpha + \beta$ (alpha–beta). In pure titanium the α phase is stable up to 883°C, above which β becomes stable. Commercially pure titanium contains 99.0–99.5 wt% Ti, the main impurities being iron, carbon, oxygen, nitrogen and hydrogen. Pure titanium for industrial use is α phase, and is classified according to the oxygen and iron content.

The principal alloying addition in α titanium alloys is aluminium. Tin, oxygen and nitrogen also stabilize α. Near α titanium alloys comprise small amounts of β phase dispersed in an α matrix, achieved through the addition of 1–2 wt% of β-stabilizing elements. Rapid cooling from the α phase field produces martensitic α, Widmanstätten α, or both. Both structures exhibit higher strength and lower ductility than fine-grained α. At elevated temperatures some alloys precipitate the gamma phase, which increases high temperature strength. Ti–5Al–2.5Sn is used for aircraft compressor blades and ducting. Ti–8Al–1Mo–1V finds applications in airframes and jet engine components.

The two-phase $\alpha + \beta$ alloys are normally alloyed with more than two elements including aluminium, copper, molybdenum, silicon, vanadium and zirconium. Strength may be increased through the formation of titanium-rich intermetallic phases on ageing, as well as a degree of solid solution hardening. However, in practice it is not possible to cold form α and $\alpha + \beta$ alloys. The most widely used $\alpha + \beta$ alloy is Ti–6Al–4V, which contains 10–50 vol.% β. It has relatively high strength up to 300°C, good hot workability, and good weldability if the volume fraction of the β phase is less than about 0.2. Alpha–beta alloys are used in high temperature aerospace applications such as aircraft turbine and compressor blades where aluminium alloys would overage and lose strength. The alloy Ti–6Al–4V is also used in surgical prostheses and sports goods. The alloy Ti–6Al–6V–2Sn finds uses in structural aircraft parts and rocket motor cases, and Ti–7Al–4Mo is used to make missile forgings and aircraft jet engine components.

Beta alloys contain sufficient β-stabilizing elements to possess a microstructure containing only β. Beta titanium alloys have excellent cold formability, and can be hardened by heat treatment. Highly alloyed β titanium alloys are not as common as $\alpha + \beta$ and α alloys. The β alloys Ti–11.5Mo–6Zr–4.5Sn and Ti–13V–11Cr–3Al are used for high strength fasteners and aircraft and sheet components.

Nickel Alloys

Nickel has a relatively high melting temperature of 1453°C. It has a face-centred cubic lattice, which confers ductility. It is ferromagnetic at room temperature. Nickel forms alloys readily, both as a solute and a solvent. An adherent oxide film forms in oxidizing environments, which resists corrosion by alkalis. Thermal expansion coefficients of the nickel–chromium-base alloys lie mid-way between those of ferritic and austenitic steels, and their conductivity is about the same as that of austenitic stainless steel. Nickel alloys have high tensile strength, which can be maintained to relatively high temperatures. These materials were traditionally used in wrought form, but improvements in investment casting have increased the number of types available. They have good machining properties, and can be joined by welding, brazing and soldering. Two thirds of the nickel produced in the world is used in stainless steels.

The metallurgy of nickel alloys is complex. Nickel is used industrially in a relatively pure form, but is also alloyed with copper, chromium, iron, molybdenum, aluminium, titanium, niobium and tungsten. Alloys are added to impart high temperature strength through three basic mechanisms:

- solid solution strengthening (Cr, W, Re);
- precipitation strengthening through the formation of stable intermetallic compounds (Al, Ti, Ta, Nb); and
- precipitation strengthening and high temperature oxidation resistance (Ta, Th).

Solid solution strengthened alloys include those containing copper and chromium. Nickel and copper are mutually soluble in the solid and liquid states, and have high strength and excellent corrosion resistance in a range of media including sea water, hydrofluoric acid, sulphuric acid, and alkalies. They are used for marine engineering, chemical and hydrocarbon processing equipment, valves, pumps, shafts, fittings, fasteners, and heat exchangers. Nickel–chromium alloys may also be alloyed with limited amounts of iron, molybdenum and niobium, and additional strength may be obtained from cold working. Wrought alloys are widely used in the petrochemical industry. Alloys containing chromium and various hard carbides have been developed for coating applications. Nickel–molybdenum alloys have outstanding resistance to hydrochloric and sulphuric acids in the as-welded condition, as well as excellent thermal stability.

Precipitation strengthened characteristics result from the addition of aluminium, beryllium, silicon and titanium to nickel–chromium–iron alloys. Intermetallic compounds such as NiAl and Ni_3Al provide high strength coupled with good resistance to creep, fatigue and corrosion at elevated temperatures, as a result of the ordering that reduces atomic mobility at elevated temperatures. These wrought alloys are used for forged and fabricated gas turbine and aerospace components, which require high strength at temperature.

Copper Alloys

Copper has the highest electrical and thermal conductivity of all commercially available metals. It has good formability and machinability as a result of its relatively low hardness. Pure copper is ductile and weak, but can be strengthened by alloying, mechanical working, and in a small number of cases precipitation hardening. Corrosion resistance, especially in marine environments is good, and its attractive appearance leads to uses in sheathing, plumbing, tubing, roofing and cladding. High electrical conductivity indicates high reflectivity to infrared radiation, even in the liquid state.

Alloys of copper are categorized according to composition: copper alloys, with no deliberate alloying additions; brasses, containing zinc; bronzes, containing tin (and sometimes phosphorus); aluminium bronzes, containing aluminium and iron; silicon bronzes, containing silicon and zinc; beryllium bronzes, containing beryllium with small amounts of cobalt and nickel; and cupro-nickels, containing nickel.

High purity copper is produced by electrolytic refining. Oxygen-free copper is an important grade that can be used in atmospheres containing hydrogen, since water, which would crack the copper, cannot be produced.

Brasses contain up to about 43 wt% Zn, and are available in both casting and wrought forms. If the zinc content lies below 35 wt% the alloy will solidify with a single phase alpha microstructure, which has high ductility and can be cold worked. A higher zinc concentration produces a two-phase alpha–beta or duplex microstructure that must be hot worked.

Bronzes containing up to 8 wt% Sn solidify with a single-phase alpha microstructure that is suitable for cold working. The brittle delta phase is formed in alloys containing higher amounts of tin, although such alloys have good casting properties. Gunmetal describes a group of brasses that contain zinc, or lead, to improve machining properties. Alloys with a large solidification interval can exhibit segregation problems, which may be alleviated by homogenizing tempering near 650°C.

Aluminium bronzes with up to 9 wt% Al solidify as single phase alpha, which can be cold worked. Alpha–beta duplex alloys, containing 9–10 wt% Al, are mainly used for castings. Their strength is equivalent to that of medium carbon steel, and their good corrosion properties are exploited in marine applications. They also have low magnetic permeability, low coefficient of friction, resistance to softening at elevated temperatures, and are non-sparking. Silicon bronzes generally contain 1–4 wt% Si, solidifying as single-phase alpha that can be cast or worked. Beryllium bronzes fall into two groups, containing around 0.4 and 1.8 wt% Be, together with small additions of cobalt.

Copper and nickel are mutually soluble in the solid state over the entire composition range, and form a continuous solid solution (Fig. 5.1). Cupro-nickels therefore have a single-phase alpha microstructure over all ranges of composition. When zinc is added, nickel silver is produced, which confers a silver colour.

The most important properties of engineering alloys with respect to laser processing can be found in Table D.1 (Appendix D).

CERAMICS AND GLASSES

Natural Ceramics

Natural ceramics are minerals formed through geological processes. Limestone and granite are well-known examples. Limestone is a sedimentary rock containing calcium carbonate ($CaCO_3$). It is used in roadbeds and gravel roads, building and landscape construction, and the manufacture of cement. Granite is a hard natural igneous rock formation comprising aluminium silicates in minerals such as potassium feldspar (($K,Na)AlSi_3O_8$), plagioclase feldspar ($CaAl_2Si_2O_8$ to $NaAlSi_3O_8$), biotite mica ($(K(Fe,Mg)_3\ AlSi_3O_{10}\ (OH)_2$), and quartz ($SiO_2$). Because granite is hard and considerably cheaper than marble, it is often used to make decorative pieces and kitchen tabletops.

Domestic Ceramics

Domestic ceramics are made from clays, which are compounds of alumina (Al_2O_3), silica (SiO_2) and water. They are shaped, dried and fired at high temperatures in order to remove the water. On cooling the structure comprises crystalline silicates in a glassy silica matrix. They are hard but brittle, because they contain pores and microcracks.

Alumina is the main component of the principal ore of aluminium (bauxite), and the main component of the gems ruby and sapphire. It is hard, electrically insulating, abrasion resistant, and finds uses in porcelain figures and utensils, crucibles, insulators and medical implants. It is also used as the host in a number of solid state lasers. Alumina is also considered as an engineering ceramic.

Cement commonly refers to powdered materials that develop strong adhesive qualities when combined with water. Constructional cement comprises compounds of lime (CaO), alumina and silica. The raw materials used in Portland cement – the most important form – are calcium oxide (44%), silica (14.5%), alumina (3.5%), ferric oxide (3%), and magnesium oxide (1.6%). After mixing with water the cement hydrates and hardens.

Engineering Ceramics

Modern high performance engineering ceramics started to become commercially available in the 1980s, principally to replace metals and alloys in high temperature environments. Their low density has been exploited in a variety of novel turbine applications. Many engineering ceramics have been designed to compete with metals. They are made from compounds such as oxides, nitrides and carbides, and are produced in the crystalline state. They are light, tough, strong, and resist corrosion, oxidation and wear at high temperatures. They are therefore difficult to work, and must be formed using novel techniques. Engineering ceramics are fine ceramics that are capable of withstanding loads.

Alumina (Al_2O_3) can be made with a variety of compositions and microstructures for specific uses. Applications include electrical insulators, spark plugs, crucible linings and hip joints.

Beryllia (BeO) has a unique combination of excellent electrical insulating properties and high thermal conductivity. It is also corrosion resistant. However, the high toxicity of beryllia and the high cost of the raw material have limited its application. It is used as an electronic substrate because of its high thermal conductivity and good electrical resistivity, which provide an effective heat sink. It is also used to make optical cavities for lasers.

Boron carbide (B_4C) is the third hardest material, behind diamond and cubic boron nitride. It has a low density and good resistance to chemicals. Bulk components are formed using hot isostatic pressing. Coatings can be formed by vapour deposition. Applications include nozzles, armour plating and ceramic tooling dies.

Boron nitride (BN) is a man-made ceramic with refractory qualities. Its physical and chemical properties are similar to carbon. Graphite-like BN is soft and is a good lubricant. The cubic (diamond) form is hard and abrasive. Boron nitride has high strength and low electrical conductivity. It is used to make insulators, crucibles, welding tips, nozzles and sputtering targets.

Magnesia (MgO) has high temperature strength, low electrical conductivity, high thermal conductivity, and good corrosion resistance. It is used in heating elements, thermocouple tubes and crucible linings. It is transparent to infrared radiation, which means that it can be used to make windows for such lasers.

Silicon carbide (SiC) is stiff, abrasion resistant, and semiconducting. Stiffness makes it suitable as a reinforcing fibre in composite materials. Abrasion resistance leads to uses such as grinding paste, cutting tools and engine parts. Its semiconducting properties are exploited in wafers.

Silicon nitride (Si_3N_4) is a covalently bonded crystalline ceramic. It possesses good creep strength, oxidation resistance, low density and a low coefficient of expansion. It is used as a reinforcing fibre in composites, and in aeroengine components such as discs and turbine blades.

Because of the similarity in size between the tetrahedral sites of aluminium oxide and silicon nitride, the two may be alloyed to make sialon (silicon aluminium oxynitride). Sialon combines the strength, hardness, fracture toughness and low thermal expansion of silicon nitride with the corrosion resistance, good high temperature strength and oxidation resistance of alumina. It finds uses in immersion heater and burner tubes, feed tubes in aluminium die-casting, and fixtures for welding and brazing.

Tungsten carbide (WC) is used in cutting tools – a typical composition being 97 wt% WC and 3 wt% Co.

Zirconia (ZrO_2) is tough, corrosion resistant and insulating, with a low coefficient of friction. It is used for cutting tools, medical implants and engine components.

The properties of ceramics relevant to laser processing are given in Table D.3 (Appendix D).

Glasses

Soda lime glasses comprise about 70 wt% SiO_2, 10 wt% CaO and 15 wt% Na_2O, and are used in windows, bottles and jars. They also contain metal oxides, which modify the networks by replacing some of the covalent bonds between the tetrahedra with lower energy non-directional ionic bonds. The viscosity of molten soda lime glass is thus reduced, enabling it to be worked at high temperature. Such glasses have a relatively high value of thermal expansion, which causes large stresses to be built upon rapid heating or cooling.

Borosilicate glasses contain about 80 wt% SiO_2, 5 wt% Na_2O and 15 wt% B_2O_3, and are used in optics, cooking equipment and chemical glassware. They have low ductility, low thermal expansivity, low thermal conductivity, and so have poor resistance to thermal shock. The tensile properties are affected markedly by microscopic defects and scratches. Borosilicate glasses were developed to withstand thermal shocks through the addition of boric oxide (B_2O_3) and small amounts of alumina (Al_2O_3). They are used in transmissive optical components for near infrared Nd:YAG laser light, but they absorb far infrared CO_2 laser light.

Metallic glasses are produced by quenching liquid metals so rapidly that crystals do not have time to form, and an amorphous structure is retained in the solid state. They have interesting properties, which can include good corrosion and wear resistance. They have been produced by melting of very thin surface films by using a rapidly pulsed laser beam.

The properties of glasses relevant to laser processing are given in Table D.2 (Appendix D).

POLYMERS

Industrial polymers broadly fall into one of three categories: thermoplastics; thermosets; and elastomers. The polymers described below are the most popular used in laser processing. There are many more – the reader is referred to the texts listed at the end of the chapter for further examples. The properties of polymers relevant to laser processing are given in Table D.4 (Appendix D).

Thermoplastics

Polyethylene is the most widely used polymer. Grocery bags, shampoo bottles, children's toys, and bullet-proof vests are some of the most popular applications. Low density polyethylene (LDPE) contains branches of polyethylene on the main structure, whereas linear (high density) polyethylene (HDPE) contains no branches. HDPE is stronger than LDPE, but the latter is cheaper and easier to make.

Polypropylene (PP) has a low density and high melting point, which make it well suited for use in packing, handles for pots and pans, kitchenware (and even banknotes used in Australia). It is resistant to attack by many chemical solvents and bacterial growth, making it suitable for medical equipment such as disposable syringes.

Polyvinylchloride (PVC) is a hard polymer that is made softer and more flexible by the addition of phthalates (plasticizers). In its hard form it is used in construction because it is cheap and easy to assemble, and is replacing traditional building materials. It is the material used to make gramophone records. In its soft form it is used for clothing, upholstery and sheeting.

Polystyrene is used in numerous products, normally in the form of an expanded foam, which is a mixture of 5% polystyrene and 95% air. This is the material of which cups, takeaway food containers and building insulation are made.

Acrylonitrile butadiene styrene (ABS) is tough, hard and rigid, and has good chemical resistance and dimensional stability. It is used to make light moulded products such as pipes, golf club heads (because of its good shock absorbance), computer housings (electroplated on the inside), automotive interior and exterior trim enclosures, and toys including Lego bricks.

Acetal is a high performance engineering polymer. It has high strength, modulus, and resistance to impact and fatigue, and finds uses as a weight-saving replacement for metal parts.

Nylon is a polyamide. Nylon fibres are used in fishing lines, carpet fibres and rope, and to make many synthetic fabrics. Solid nylon is used as an engineering material in machine parts, such as gears and bearings.

Polyetheretherketone (PEEK) is a tough, strong, and stiff polymer that can be loaded for a long time without suffering any damage. PEEK has very high application temperatures, and high chemical resistance. It is expensive, and is mainly used together with glass or carbon fibres, where high mechanical and thermal properties are required. When not reinforced, PEEK is used as insulation material for cables and wires, because of its excellent fire and high temperature resistance.

Polycarbonates (PC) are strong, stiff, hard, tough, transparent engineering thermoplastics that can maintain rigidity up to 140°C and toughness down to −20°C. The material is amorphous, and possesses excellent mechanical properties and high dimensional stability. It is used to make compact discs, riot shields, vandal-proof glazing, baby feeding bottles, electrical components, safety helmets and headlamp lenses.

Polyurethane is a unique material that possesses the elasticity of rubber combined with the toughness and durability of metal. It has replaced metals in sleeve bearings, wear plates, sprockets, and rollers, with benefits such as weight reduction, noise abatement and wear improvement. Polyurethane is best known as the polymer used to make foams.

Acrylics such as polymethylmethacrylate (PMMA) 'perspex' have about half the density of glass, are impact resistant, and unaffected by sun or salt spray. Its clarity and stability make it very suitable for the manufacture of measuring burettes and in sheet form it may be cemented to produce tanks and trays. When heated to decomposition it emits acrid smoke and irritating fumes.

Thermosets

Phenolics have good mechanical strength and dimensional stability, good machinability, low density, and possess good heat, wear and corrosion resistance. They are used in terminal boards, switches, bearings, gears, washers and transformers. Phenolics may be used in their liquid form in laminating veneers, fabrics, and paper.

Melamine formaldehyde (MF) thermoset products are popular members of the amino resin family, which includes urea and thiourea. They are strong, lightweight and tough, and can be moulded into domestic table and kitchenware or used in laminated counter tops.

Epoxies are rigid, clear, very tough, chemical resistant polymers with good adhesion properties, low curing temperatures and low shrinkage. They are used as adhesives and coatings, and for encapsulation of electrical components. They are finding increasing use in aerospace applications requiring bonding of composite materials.

Elastomers

Elastomers are lightly cross-linked polymers, which can be extended considerably and reversibly. Natural rubber and butyl rubbers have a 'backbone' polymer chain of carbon. In others, oxygen (polypropylene oxide), silicon (fluorosilicone) or sulphur (polysulphide) form the backbone. The glass transition temperature of elastomers is well below room temperature, removing the effect of the light cross-linking.

COMPOSITES

The mechanical and thermal properties of composites relevant to laser processing are given in Table D.5 (Appendix D).

Metal Matrix Composites

Metal matrix composites (MMCs) may be classified by the morphology of matrix reinforcement, which may be dispersion strengthened, particle reinforced or fibre reinforced.

The majority of metal matrix composites (MMCs) are based on aluminium, reinforced with particulate silicon carbide. The stiffness of aluminium may thus be increased by more than 50% in particulate composites, and doubled in fibre-reinforced systems, with a reduction in density. Titanium, nickel, cobalt, copper and magnesium matrices are also being actively investigated. Such materials are produced by powder metallurgy, mechanical alloying, liquid metal pressure forming, stir casting, squeeze casting, spray deposition and reactive processing. Their high specific strength and stiffness make them particularly attractive materials for aerospace structures. Much work is being done in the field of SiC fibre-reinforced titanium alloys, as replacements for superalloys in disc and blade applications. Nickel and cobalt-based MMCs were originally developed for rocket combustion chambers and nozzles. Fibre-reinforced copper alloys are being developed for heat exchanger applications in hypersonic aircraft engines.

The performance of automotive components depends on a combination of stiffness, good elevated temperature fatigue properties and wear resistance performance can be enhanced through the use of aluminium and magnesium MMCs. The most widely used types have a matrix of aluminium reinforced with particulate silicon carbide (SiC), which is relatively cheap and has a low density.

Full commercial exploitation has been hampered by a number of factors. Reliable fabrication processes have been difficult to develop, mechanical properties can be degraded through interfacial reactions, and secondary processing methods, e.g. joining, are only now being developed. A gap exists in the understanding of properties between the materials designer and the applications engineer. MMCs offer improved performance, but at an increased cost, which must be justified.

Polymer Matrix Composites

Polymer matrix composites (PMCs) are characterized by the morphology of the reinforcement: filler strengthened or fibre reinforced. They comprise a thermoset matrix, which is reinforced with fibres or whiskers of a material of very high tensile strength. The first fibre reinforced polymers for aerospace application were developed for the Spitfire, when shortages of aluminium were feared, and were demonstrated successfully, but never used in the construction of the aircraft.

Polyacrylonitrile fibre-reinforced composites find many aerospace applications, and also make up more than 75% by weight of the chassis of a Formula 1 car. PMCs have had a large impact in the design and manufacture of sports equipment, e.g. fishing rods, golf clubs and tennis rackets. A typical aerospace application involves a thermally cured epoxy thermoset matrix that is reinforced with a high volume fraction of carbon, boron, glass or aromatic polyamide (aramid or Kevlar) fibres. Each type of fibre endows the composite with a particular combination of properties and cost.

Ceramic Matrix Composites

Conventional monolithic ceramics have attractive high temperature properties, but many are brittle. A reinforcing phase can be introduced to act as a site to arrest crack propagation, and enhance toughness, producing a ceramic matrix composite (CMC). Particles, whiskers and fibres have all been proposed as reinforcing materials, with silicon carbide being a popular choice. Such materials show potential in a number of aeroengine applications.

Concrete, a particulate composite of aggregate and a cement binder, made its appearance in constructions in the early twentieth century. Modular building technologies using concrete were introduced in the 1950s.

PROPERTIES OF MATERIALS

The principal mechanical and thermal properties of engineering materials are described below, and representative values are listed in Appendix D. Material properties play a dominant role in determining the interaction between the laser beam and engineering materials, dictating the processing mechanisms. Many material properties change with temperature; although this would at first sight appear to present problems in modelling, it is an interesting phenomenon that could be exploited in novel laser treatments.

ABSORPTION COEFFICIENT

The absorption coefficient describes the attenuation of an incident beam with depth in a material. For a beam of incident energy E_0, the transmitted energy, E, at a depth z, is given by the Beer–Lambert law:

$$E = E_0 \exp(-az)$$

where a is the absorption coefficient (around 5×10^5 cm^{-1} for metals in the visible spectrum). Absorption therefore takes place in a very shallow region, with a depth only a fraction of the wavelength of the incident radiation. The absorption length is defined as the distance over which the intensity is reduced by a factor $1/e$.

ABSORPTIVITY

Laser light impinging on the surface of a material may be absorbed, reflected, transmitted, or re-radiated. On the macroscopic scale, absorptivity is a measure of the fraction of incident radiation absorbed. As the absorption coefficient indicates, radiation is absorbed by electrons in the upper 10^{-6} to 10^{-5} cm of the surface – the electromagnetic skin depth. The absorption mechanism is known as the inverse *Bremsstrahlung* (braking radiation) effect. Energy is subsequently transferred into the material by a mechanism that depends on the energy of the photons. The photon energy of material processing lasers that emit with a wavelength that lies above the ultraviolet region of the electromagnetic spectrum is relatively low, Table B.1 (Appendix B), which means that classical thermal conduction through collision with lattice defects and other electrons is the dominant heat transfer mechanism. As the interaction time approaches that of the mean free time of electrons (10^{-13} s in a conductor), classical thermal conduction laws are no longer valid and athermal processing mechanisms, associated with rapid picosecond (10^{-12} s) and femtosecond (10^{-15} s) pulses, apply.

The variation of absorptivity with wavelength for various materials is shown schematically in Fig. 5.10.

The absorption of laser light by a surface varies with the angle of incidence. For vertical incidence (angle $= 0°$), the beam component oriented parallel to the incidence plane (R_p) and the normal component (R_s) are absorbed equally. With increasing angle of incidence the absorption of R_p increases, while that of R_s decreases. At the so-called Brewster angle, the absorption of R_p reaches a maximum, while it drops to zero for R_s (the Brewster effect).

As the temperature of a material changes, the absorptivity can increase or decrease, depending on its optical properties and modifications to the surface, e.g. oxidation reactions or phase transformations. For many metals and alloys in the solid phase, the absorptivity in the infrared range is well approximated by an increase relationship with temperature. The absorption of CO_2 laser radiation by a polished steel

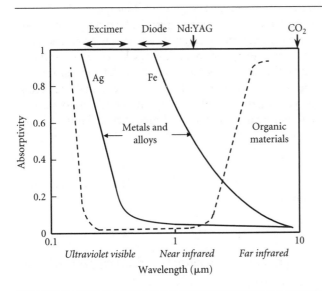

Figure 5.10   Schematic variation of absorptivity with wavelength for metals and alloys and organic materials

surface is around 4% at room temperature, increasing to more than 30% at the melting temperature, reaching around 90% at the vaporization temperature.

Absorptivity also varies with surface roughness. A rough surface presents a greater surface area to the laser beam, and causes light to be reflected several times, thereby increasing the total absorptivity.

SPECIFIC HEAT CAPACITY

The specific heat capacity of a material is a measure of the energy required to raise its temperature by one Celsius degree at constant pressure. It is expressed in units of $J\,kg^{-1}\,K^{-1}$, or as a volumetric quantity as $J\,m^{-3}\,K^{-1}$. The term 'heat capacity' is normally used when describing molar quantities, and has units $J\,mol^{-1}\,K^{-1}$. The volumetric specific heat capacity for homogeneous materials at room temperature is about $3 \times 10^{-6}\,J\,m^{-3}\,K^{-1}$. The heat capacity of metals and alloys increases with temperature until it reaches a limiting value of $25\,J\,mol^{-1}\,K^{-1}$. For ceramics and glasses the heat capacity increases with temperature to about $1000°C$, above which it remains approximately constant. In polymers it increases steadily until the glass transition temperature is reached.

THERMAL CONDUCTIVITY

Thermal conductivity is the rate at which heat flows through a material. It is important in *steady state* thermal processes. Thermal conductivity is directly proportional to the amount of energy present (the volumetric heat capacity), the number and velocity of energy carriers (electrons and phonons), and the amount of energy dissipation (the amount of scattering or the attenuation distance of lattice waves, i.e. the mean free path). Metals and alloys have high values of thermal conductivity because there are many carrier electrons, which can move easily, and have a large mean free path. As the temperature rises, the amount of energy dissipated increases by collisions, and thermal conductivity decreases. The main carriers in ceramics and glasses are phonons, which can be thought of as lattice vibrations that occur on discrete energy levels or quanta. Electrons are restrained in ionic and covalent bonds, and so cannot participate in thermal conduction at low temperatures. Strong periodic bonds transfer lattice waves efficiently, thermal conductivity is highest in materials that have an orderly structure comprising single elements, or elements of similar atomic weight. Differences in atomic size result in greater lattice scattering. In materials with two-dimensional layered structures, such as graphite, thermal conductivity in the direction of bonding is high, but van der Waals forces acting in the perpendicular direction attenuate vibrations, lowering conductivity. The mean free path in ordered ceramics is inversely proportional to temperature, and so thermal conductivity decreases as temperature increases. Glasses are amorphous, and so have a relatively short mean free path that does not change significantly with temperature. The increase in heat capacity with temperature is the mechanism responsible for the increase in thermal conductivity with temperature in glasses. Most polymers have low values of thermal conductivity because electrons are bound in covalent bonds, molecular sizes are large, and the degree of crystallinity is small.

DENSITY

Close packing of atoms results in high density and a high melting temperature. This accounts for high values in metals and alloys, and low values in polymers.

THERMAL DIFFUSIVITY

Thermal diffusivity, a, is the ratio of the energy transmitted by conduction to the energy stored in unit volume of material:

$$a = \frac{\lambda}{\rho c}$$

where λ is the thermal conductivity and ρc is the volumetric heat capacity.

It is often referred to as the diffusion coefficient for heat. Thermal diffusivity determines how rapidly a material will accept and conduct thermal energy, and is important in characterizing transient thermal processing. This may result from pulsed laser treatment, or a moving heat source. It determines the thermal penetration in a material, being particularly important in laser heating processes. A useful rule of thumb relates the depth of thermal penetration, z, to the heating time, t, and the thermal diffusivity, a: $z = \sqrt{(4at)}$.

The diffusivity of alloys is generally lower than that of the pure metal in the alloy; stainless steel is particularly low in comparison with plain carbon steels.

COEFFICIENT OF THERMAL EXPANSION

The relationship between the change in length, Δl, and a change in temperature, ΔT, is expressed as the coefficient of thermal expansion, α:

$$\alpha = \frac{\Delta l}{l_o \Delta T}$$

where l_o is the length at room temperature. The coefficient of thermal expansion is controlled by atomic and molecular vibration; as the temperature increases, the amplitude of vibration increases.

In close-packed structures, such as those found in metals and alloys, the increase in atomic vibration is accumulated in neighbouring atoms, producing relatively high expansion of the lattice. Ceramics with predominantly ionic bonding also form close-packed structures, and so also exhibit high values of thermal expansion. Covalently bonded structures contain spaces in which vibrations can be accommodated, reducing the thermal expansion coefficient. The thermal expansion characteristics of glasses are controlled by composition, structure and thermal history. Thermal shock resistance can be imparted by heat treatment of glasses. Anisotropic structures exhibit different values for expansivity along different axes. Polymers and elastomers stretch extensively on heating before failing.

TRANSFORMATION TEMPERATURES

The principles of laser material processing depend on the mechanisms of processing, i.e. heating, melting and vaporization, which occur in the solid, liquid and vapour states.

Empirical formulae have been derived for the most important phase transformation temperatures in steels. A selection can be found in Table E5.1 (Appendix E). Close packing of atoms results a high melting temperature. Alkali metals such as sodium and potassium are bound by outer low energy electrons, resulting in weak bonding. These metals therefore have low strength and low melting temperatures. Transition metals such as chromium, iron and tungsten are bound by inner electrons, and exhibit high values of strength and melting temperature. The covalent nature of the bonding in silica glasses endows them with high strength and stability, and a high softening temperature.

A softening, or 'melting' temperature, T_m, can be identified in polymers that contain a high degree of crystallinity. This is accompanied by an increase in energy – the latent heat of melting. In amorphous polymers, as the temperature is raised there is a continuous increase in heat capacity until the glass transition temperature, T_g, is reached, at which point the rate of energy absorption increases, without a discontinuity. No latent heat of melting can be associated with amorphous polymers. The melting temperature is generally about $1.5 \times T_g$.

The vaporization temperature for metals and alloys is approximately twice the melting temperature.

LATENT HEAT OF MELTING

The latent heat of melting for crystalline materials corresponds approximately to the energy change from free vibration in the solid state to free rotation in the liquid state, without a change in temperature. The ratio L_m/T_m (the entropy of fusion) is approximately constant, with a value similar to the gas constant ($8.314\,\text{J}\,\text{mol}^{-1}\,\text{K}^{-1}$) for most metals (Richard's rule). Materials with high bond strengths have high values for both latent heat of fusion and melting temperature. The relationship holds for materials in which secondary bonding forces, such as van der Waals bonding, are not significant.

LATENT HEAT OF VAPORIZATION

The latent heat of vaporization represents the energy required to convert a liquid to a gas at the vaporization temperature, without a change in temperature. The ratio L_v/T_v (the entropy of vaporization) is approximately constant, with a value about one order of magnitude greater than that of the entropy of fusion, about $83.14\,\text{J}\,\text{mol}^{-1}\,\text{K}^{-1}$ for most metals (Trouton's rule). (If we assume that the average coordination number in close-packed structures is 10, and that the liquid state can be formed by the breaking of one bond, then it is reasonable to assume that about ten times as many bonds must be broken to form a gas.) Again, materials with high bond strengths have high values for both latent heat of vaporization and vaporization temperature. Trouton's rule fails for very small, light molecules (He and H) or for substances which have strong hydrogen bonding in the liquid state (e.g. water).

MATERIAL PROPERTY CHARTS

The mechanical and thermal properties of common engineering materials are given in Appendix D. Ashby (1989) describes means of charting and relating the basic mechanical and thermal properties of engineering materials by using Material Property Charts. These provide valuable information on the behaviour of materials during laser processing, enabling suitable materials for various processes to be identified. Charts are presented here that display the thermal properties of the different classes of engineering material and their subsets.

THERMAL CONDUCTIVITY VS HEAT CAPACITY

Figure 5.11 illustrates the relationships between thermal conductivity, λ, and the volume heat capacity, ρc. Metals and alloys, which possess high values of both λ and ρc, appear in the top right-hand portion of the chart. Copper, silver and aluminium possess the highest values of λ, and so are candidates for processing using a high power density beam, which is able to melt material before energy is transported from the interaction zone by conduction. However, absorptivity to infrared radiation is inversely proportional to λ (hence the use of these materials in carbon dioxide laser optics), and so they are difficult to treat with low power density infrared lasers. Alloys of titanium and nickel, as well as stainless steels, have relatively low values of λ in comparison with other metals and alloys because they contain foreign atoms in solid solution. These are therefore candidate materials for stable processing with a wide range of laser techniques. Materials with high values of ρc absorb and store large quantities of energy, and are more susceptible to deformation during thermal processing.

Polymers and organic composites such as wood lie in the lower left-hand portion of Fig. 5.11. These materials are high absorbers of laser light in the range infrared to ultraviolet. They show strong potential for laser processing.

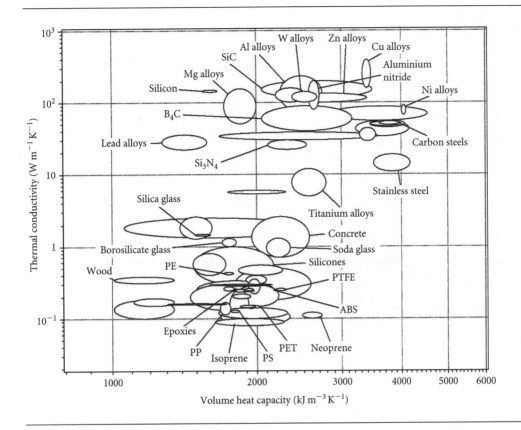

Figure 5.11 Thermal conductivity plotted against volume heat capacity for engineering materials. (Source: CES 4.1 2003 Granta Design, grantadesign.com)

Glasses are found in the central region of Fig. 5.11. These transmit light in the near infrared to ultraviolet range, and so are difficult to process with such laser wavelengths (hence their use in optics for Nd:YAG and excimer lasers). However, far infrared laser light *is* absorbed, which means that glasses have potential for material processing with CO_2 lasers. Ceramic properties fall in similar regions.

THERMAL CONDUCTIVITY VS THERMAL DIFFUSIVITY

Thermal conductivity governs steady state heat flow, whereas thermal diffusivity controls transient heat flow. The information shown in Fig. 5.11 is replotted in this framework, Fig. 5.12. Most classes of engineering material lie along the diagonal corresponding to a volume heat capacity on the order of 3×10^6 J m^{-3} K^{-1}. This value is constant since the volume associated with each heated atom is approximately constant in most homogeneous materials. Deviations occur in inhomogeneous materials such as composites and foams.

The most notable feature of Fig. 5.12 is that the properties of engineering ceramics such as silicon nitride and boron carbide are similar to those of metals and alloys, and so such materials are expected to respond similarly to steady state and transient laser heating.

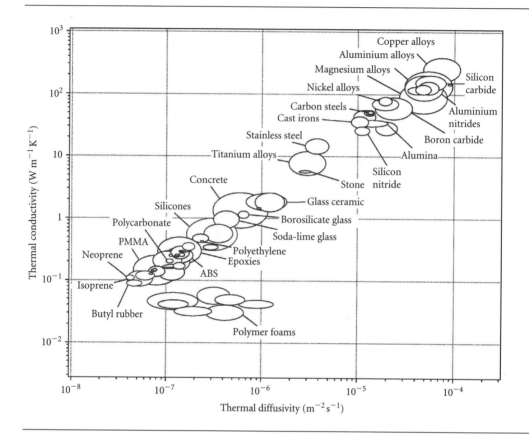

Figure 5.12 Thermal conductivity plotted against thermal diffusivity for engineering materials. (Source: CES 4.1 2003 Granta Design, grantadesign.com)

MECHANICAL STRAIN VS THERMAL STRAIN

In Fig. 5.13, mechanical strain is characterized by strength divided by Young's modulus, and thermal strain is represented by the product of linear expansion coefficient and melting temperature. Materials that appear in the top right-hand portion of the figure tolerate large amounts of thermal strain before failing, which is likely to occur before the onset of melting. In contrast, materials appearing in the lower left-hand region fail mechanically at low levels of thermal strain. Concrete, for example, fails under low thermal loading – a property that is exploited in the process of laser scabbling, described later. Metals and alloys, which appear in the central region of Fig. 5.13, can be deformed permanently by heating before melting occurs, and are popular choices in the process of laser forming.

Figure 5.13 shows that materials with similar properties to concrete, which have low values for both mechanical strain and thermal strain, will accommodate little thermal expansion before failing. Polymers, in contrast, have high values of both of these property groups, and therefore will be able to expand extensively on heating before failing.

SUMMARY AND CONCLUSIONS

Interactions between laser beams and materials can be considered on three levels: the arrangement of atoms; clusters of phases and constituents in the microstructure; and the properties that result from the

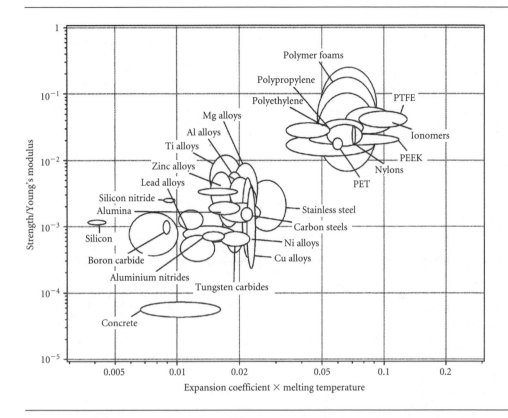

Figure 5.13 Mechanical strain (strength/Young's modulus) plotted against thermal strain (expansion coefficient × melting temperature) for engineering materials. (Source: CES 4.1 2003 Granta Design, grantadesign.com)

macrostructure. Atomic structure and bonding influence the mechanisms of absorption of the laser beam photons by the material. Microstructural changes induced through the thermal cycles created during laser processing can be represented in various forms of phase transformation diagram. Equilibrium diagrams illustrate the phases that are stable under equilibrium conditions, when the driving force for transformation is thermodynamic. Non-equilibrium transformation diagrams are available for rapid heating, solidification and cooling associated with laser processing. Both types of diagram provide valuable information on the stability of phases during thermal treatments, and can be used to predict the outcome of laser processing. Industrial materials are seen to fall into four broad classes: metals and alloys; ceramics and glasses; polymers; and composites. The distinctions lie in the composition and atomic structure, which endow the material classes with characteristic sets of properties. The properties of the most popular industrial materials are described as an aid to understanding their behaviour during laser processing, and also as a means of identifying materials for novel laser-based treatments.

By plotting various mechanical and thermal properties of engineering materials on charts, we are able to distinguish the material classes since they occupy characteristic regions. The charts may be used for material selection with given criteria, and to predict the behaviour of materials during processing. They provide an understanding of the mechanisms of laser processing, and are referred to in the chapters that follow.

FURTHER READING

ENGINEERING MATERIALS

Ashby, M.F. (1999). *Materials Selection in Mechanical Design*. 2nd ed. Oxford: Butterworth-Heinemann.

Ashby, M.F. and Jones, D.R.H. (1996). *Engineering Materials 1*. 2nd ed. Oxford: Butterworth-Heinemann.

Ashby, M.F. and Jones, D.R.H. (1998). *Engineering Materials 2*. 2nd ed. Oxford: Butterworth-Heinemann.

Bolton, W. (1989). *Engineering Materials Pocket Book*. Oxford: Newnes.

Charles, J.A., Crane, F.A.A. and Furness, J. (1997). *Selection and Use of Engineering Materials*. Oxford: Butterworth-Heinemann.

Cottrell, A.H. (1964). *Mechanical Properties of Matter*. New York: Wiley.

Gordon, J.E. (1988). *The New Science of Strong Materials*. London: Penguin Books.

Leaver, K.D., Anderson, J.C., Rawlings, R.D. and Alexander, J.M. (2003). *Materials Science for Engineers*. 5th ed. Boca Raton: CRC Press.

Martin, J. (2002). *Materials for Engineering*. 2nd ed. Leeds: Maney Publishing.

Shackelford, J.F. and Alexander, W. (2000). *The CRC Materials Science and Engineering Handbook*. 3rd ed. Boca Raton: CRC Press.

Wyatt, O.H. and Dew-Hughes, D. (1974). *Metals, Ceramics and Polymers*. Cambridge: Cambridge University Press.

METALS AND ALLOYS

Angus, H.T. (1976). *Cast Iron: Physical and Engineering Properties*. 2nd ed. London: Butterworths.

Brandes, E.A. and Brook, G.B. eds (1997). *Smithells Metals Reference Book*. 7th ed. Oxford: Butterworth-Heinemann.

Busk, R. (1987). *Magnesium Products Design*. New York: Marcel Dekker.

Cottrell, A. (1975). *An Introduction to Metallurgy*. 2nd ed. London: Edward Arnold.

Easterling, K.E. and Porter, D.A. (1992). *Phase Transformations in Metals and Alloys*. 2nd ed. London: Van Nostrand Reinhold.

Honeycombe, R.W.K. and Bhadeshia, H.K.D.H. (1995). *Steels: Microstructure and Properties*. 2nd ed. London: Edward Arnold.

Llewellyn, D.T. (1992). *Steels: Metallurgy and Applications*. 2nd ed. Oxford: Butterworth-Heinemann.

Pickering, F.B. (1978). *Physical Metallurgy and the Design of Steels*. London: Applied Science Publishers.

Polmear, I.J. (1995). *Light Alloys: Metallurgy of the Light Metals*. 3rd ed. London: Edward Arnold.

Reed-Hill, R.E. and Abbaschian, R. (1992). *Physical Metallurgy Principles*. 3rd ed. Boston: PWS-Kent Publishing.

Robb, C. (1987). *Metals Databook*. London: The Institute of Metals.

Street, A. and Alexander, W. (1998). *Metals in the Service of Man*. London: Penguin Books.

Thelning, K.-E. (1975). *Steel and its Heat Treatment*. 2nd ed. London: Butterworths.

CERAMICS AND GLASSES

Engineered Materials Handbook, Volume 4, Ceramics and Glasses (1991). Metals Park: ASM International.

Green, D.J., Clarke, D.R., Suresh, S. and Ward, I.M. (1998). *Introduction to the Mechanical Properties of Ceramics*. Cambridge: Cambridge University Press.

Rawson, H. (1991). *Glasses and their Applications*. London: The Institute of Metals.

Richerson, D.W. (1992). *Modern Ceramic Engineering*. 2nd ed. New York: Marcel Dekker.

POLYMERS

Clegg, D.W. and Collyer, A.A. (1993). *The Structure and Properties of Polymeric Materials*. London: The Institute of Materials.

Engineered Materials Handbook, Volume 2, Engineering Plastics (1988). Metals Park: ASM International.

Mark, J.E. ed. (1999). *Polymer Data Handbook*. New York: Oxford University Press.

COMPOSITES

Daniel, I.M. and Ishai, O. (1994). *Engineering Mechanics of Composite Materials*. New York: Oxford University Press.

Engineered Materials Handbook: Volume 1 – Composites (1987). Metals Park: ASM International.

Harris, B. (1999). *Engineering Composite Materials*. 2nd ed. London: The Institute of Materials.

Hull, D. and Clyne, T.W. (1996). *An Introduction to Composite Materials*. 2nd ed. Cambridge: Cambridge University Press.

LASER–MATERIAL INTERACTIONS

Allmen, M.v. and Blatter, A. (1995). *Laser-beam Interactions with Materials*. Berlin: Springer.

Bertolotti, M. ed. (1983). *Physical Processes in Laser–Materials Interactions.* New York: Plenum Press.

Duley, W.W. (1976). *CO$_2$ Lasers: Effects and Applications.* New York: Academic Press.

Ready, J.F. (1971). *Effects of High-power Laser Radiation.* New York: Academic Press.

BIBLIOGRAPHY

Ashby, M.F. (1989). Materials selection in conceptual design. *Materials Science and Technology*, **5**, (6), 517–525.

Schaeffler, A.L. (1949). Constitution diagram for stainless steel weld metal. *Metal Progress*, **56**, (11), 680–680B.

CHAPTER 6

LASER PROCESSING DIAGRAMS

INTRODUCTION AND SYNOPSIS

It is said that a picture is worth a thousand words. Nothing demonstrates this better than a well-organized scientific diagram. Diagrams provide clarity in a concise and convenient form. In Chapter 3, we saw how two characteristic variables of industrial laser beams – wavelength and average power – could be used to display the operating conditions of various commercial lasers. Areas of application could also be shown in that framework, enabling those lasers suitable for material processing – or other uses – to be identified, as illustrated in Fig. 3.15. Such a diagram enables a type of laser and an appropriate power level for material processing to be selected. In Chapter 5, various properties of engineering materials were plotted in the form of charts, which differentiate the behaviour of different material classes to thermal treatment, enabling materials to be ranked quickly for given applications. We now apply this graphical approach to the *interaction* of laser beams and engineering materials by identifying the principal process variables and processing mechanisms, and determining the level of sophistication required to achieve the desired result – the prediction of processing parameters.

The construction of various charts and diagrams for laser material processing is described in this chapter. A distinction is made between a *chart* and a *diagram*. Charts, which are empirical, display experimental data on a set of axes, from which information can be derived. Diagrams, in contrast, have an underlying physical basis, normally in the form of a mathematical model, which enables constitutive relationships between process variables to be displayed. (But empirical data may be included in such diagrams to verify model predictions.)

Two further methods are used to construct graphical representations of laser–material interaction. First, empirical observation is combined with mathematical modelling; a model suggests the form of constitutive relationships, but empirical methods are used to establish the precise functional relationships. This leads to the construction of model-based empirically-informed diagrams. Second, calibration techniques are used to eliminate the effects of poorly known mechanisms and constants on the results of processing; variations are predicted about a known condition (an experimental data point that is obtained from a known mechanism of processing). The level of accuracy required determines the most suitable approach – empirical or purely model-based methods are normally sufficient for presenting overviews that display trends, whereas a calibrated method involving both modelling and empirical data might be needed for predicting practical processing parameters. This chapter concentrates on the first two approaches to illustrate the construction and use of charts and diagrams in the context of laser material processing in general. In later chapters, which deal with individual processes, the latter two methods are used to generate diagrams that are used to predict processing parameters.

EMPIRICAL METHODS

A large number of variables influence the interaction between a laser beam and a material – over 140 can be identified for welding alone. At the finest level of detail, when drawing up a procedure specification, all must be fixed within certain tolerances so that processing can be performed repeatedly to a particular requirement. It might appear that any attempt to characterize a process is likely to be arduous, and potentially intractable. However, generalizations may be made to simplify the approach to achieve useful results. The first task is to identify the *principal process variables*. These have the greatest effects on the principal processing *mechanisms*.

PROCESS VARIABLES AND PROCESSING MECHANISMS

Consider a stationary laser beam of circular cross-section that heats (without melting) the surface of a large block of material, as illustrated in Fig. 6.1a. Following an initial transient when the beam is switched on, the heat flow becomes *steady state*; energy absorbed by the surface is balanced by that conducted into the block, and the temperature field around the heat source becomes constant. The principal process variables are the beam power, q, the beam radius, r_B, and the material properties. Often in laser processing we wish to establish the conditions required to raise the surface temperature by a given amount. An empirical relationship between the two variables could be established by systematically varying q and r_B until the required condition is obtained.

However, the principal process variables can conveniently be grouped to form the power density, E:

$$E = \frac{q}{\pi r_B^2}. \tag{6.1}$$

The power density is easily calculated, and has a physical meaning that can be understood intuitively. The power density can be increased four-fold by quadrupling the power, or by halving the beam radius. By identifying this variable group, a smaller subset of experiments can be used to establish that it is the power density that determines the peak surface temperature attained. This in turn determines the principal mechanism of thermal interaction – *heating*, *melting* or *vaporization* – as illustrated in Figs 6.1a, b and c, respectively. For example, power densities on the order of 10, 100 and 10 000 W mm^{-2} are required to heat, melt and vaporize structural steel, respectively, using a CO_2 laser beam.

Consider now the case of a laser beam traversing the surface at a rate v, as illustrated in Fig. 6.2. At any point in the material, *transient* heat flow is experienced, i.e. the temperature field changes with time.

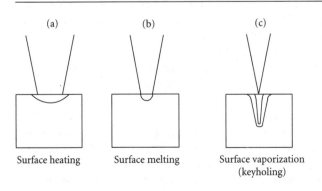

Figure 6.1   The effect of power density on the interaction mechanism between a laser beam and a large block of material under steady state conditions

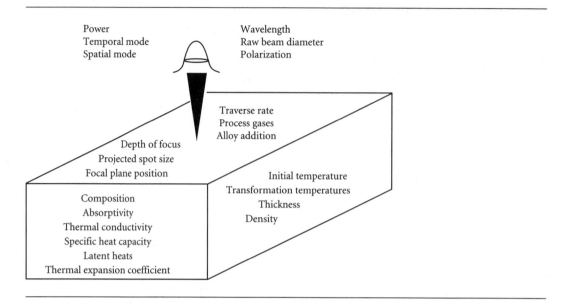

Power
Temporal mode
Spatial mode

Wavelength
Raw beam diameter
Polarization

Traverse rate
Process gases
Alloy addition

Depth of focus
Projected spot size
Focal plane position

Initial temperature
Transformation temperatures
Thickness
Density

Composition
Absorptivity
Thermal conductivity
Specific heat capacity
Latent heats
Thermal expansion coefficient

Figure 6.2   Laser beam traversing a material showing the principal process variables of laser material processing

The power density is then no longer unique in determining the processing mechanism – the heating time must also be considered. The heating time can be approximated by the beam interaction time, τ:

$$\tau = \frac{2r_B}{v}. \tag{6.2}$$

E and τ clearly influence the processing mechanism, but we do not yet know the exact form of the relationship. This can be established by plotting experimental data on a chart.

An Empirical Process Chart

Applied (incident) beam power density (W mm^{-2}) and beam interaction time (s) are used as the ordinate and abscissa, respectively, of the chart shown in Fig. 6.3. The product of these two quantities, the energy density (J mm^{-2}), appears as a series of diagonal contours. Experimental data (Ion *et al.*, 1992) for five common methods of laser processing of metals and alloys are plotted in this framework. Details of the individual processes are provided in the chapters that follow; they are not needed at this stage. Logarithmic scales are used to cover practical ranges of parameters.

Experimental data for different processes cluster in discrete regions of Fig. 6.3. (The overlap between some clusters is discussed below.) The clustering of data indicates that the variable groups power density and interaction time discriminate between different processes.

The relative positions of the data clusters with respect to the axes appear logical. Hardening, which is a solid state process involving only heating, is carried out with relatively low values of power density (to avoid surface melting), but high values of interaction time (to allow diffusion and microstructural homogenization to occur). In contrast, keyhole welding requires a high power density to form a narrow, deeply penetrating vapour cavity, but the interaction time is short (i.e. the welding speed is high) to maximize productivity.

Data cover a range of interaction time from 10^{-4} s, which approaches the minimum for noticeable heat transfer by thermal diffusion, to 10 s, beyond which the rate of laser processing normally becomes

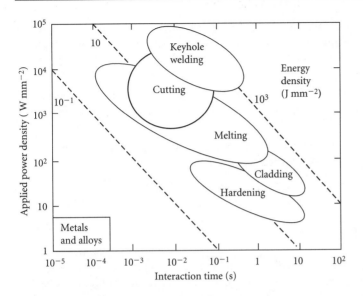

Figure 6.3 Experimental data for five methods of thermal laser processing of metals and alloys (regions bounded by solid boundaries) and contours of energy density (broken lines)

uneconomically low. The limits of data on the power density axis approximate those for surface heating ($10\,W\,mm^{-2}$) and surface vaporization ($10^4\,W\,mm^{-2}$), for a CO_2 laser beam illuminating a thick steel block under steady state conditions (Fig. 6.1).

Data for heating (hardening) are bounded by energy density contours of 1 and 100 $J\,mm^{-2}$. Most of the data for melting processes (cladding, melting and some cutting techniques) lie between contours of 10 and 1000 $J\,mm^{-2}$. Data for vaporization processes (keyhole welding and other cutting techniques) fall between the 100 and 10 000 $J\,mm^{-2}$ contours. On average, it can be seen that an order of magnitude increase in energy density is required to change the processing mechanism from heating to melting, and from melting to vaporization. This is observed over wide ranges of power density and interaction time.

The overlap of some clusters of process data in Fig. 6.3 can be explained as follows. The data are gathered from a large number of sources. Process parameters are not always reported in a standard manner. For example, the radius of a beam with a Gaussian power distribution can be defined as the point at which the power has fallen to $1/e$ or $1/e^2$ of its peak value – a difference of 2.7. This introduces a possible error of 7.4 in the calculated power density. In addition, processing is rarely carried out under optimum conditions, and so data might not be truly representative. Most importantly, the data represent a range of metals and alloys that have different thermal properties. A set of processing conditions that cause titanium to melt might only result in heating of copper, which has a thermal conductivity more than ten times greater. Although such data points appear in the same position on the chart, the principal process mechanisms are different. We shall return to this point later.

Figure 6.3 serves many useful functions. It allows data for common methods of laser processing to be presented in a framework that has a physical meaning. It distinguishes processing parameters for different processes, and enables them to be selected easily. It indicates that the energy density of the laser beam plays a role in determining the processing mechanism. Often, particularly when purchasing a laser, it is important to gain an overall impression of its performance in a number of tasks – this can be established quickly by using Fig. 6.3.

The data in Fig. 6.3 also raise a number of important questions:

■ The data span ranges of process parameters that induce processing mechanisms resulting from thermal interaction. Could processing mechanisms involving non-thermal interaction be induced by using process parameters that lie outside these ranges? (We shall return to this point in the next chapter.)

■ Energy density contours bound clusters of processing data, but they do not appear to be aligned with the clusters. Could a more discriminating set of contours be constructed to represent the data more accurately?

■ Process parameters may only be *derived* from the empirical data presented in the chart. Could the chart be modified to enable process parameters to be *predicted*?

Answers to these questions can be found through the use of *process modelling*.

MODEL-BASED METHODS

If a picture is worth a thousand words, then it could be argued that a model provides the clarity of a thousand pictures. A model is an idealization of an object, process, phenomenon or organization. One of the most effective models is the map of the London Underground transport system, now widely regarded as a seminal piece of design. In the first diagram, published in 1908 (Garland, 1994), colours were used to represent different lines (even though the trains were not painted in those colours). Colours evoke associations: the green District line passes through leafy suburbs; and the red Central line is the arterial route across the city. When the map was designed, routes were drawn to follow their true course. As more lines were built and stations added, this approach became unwieldy. Routes were then approximated by using straight lines and simple geometrical shapes, which allowed people to visualize geographical relationships (smoothing out actual bends and detours). In today's diagram, stations are presented in the correct sequence, but the distances are not represented accurately, to maintain the clarity of the map. However, journey times in the central region can still be estimated by allowing about two and a half minutes between stations. The map demonstrates the connectivity between different lines (by locating stations some distance from their geographical position). Despite these idealizations, the diagram is effective; it is visually distinctive, and works as a convenient means of planning a journey and navigating the system.

PROCESS MODELLING

Models of laser processing relate the process variables – the beam characteristics, the material properties and the processing parameters – to the characteristics of the product. Effective models combine justifiable idealizations with essential relationships that are obtained from the underlying principles of physics and chemistry. Models therefore have the predictive power that empirical approaches lack.

Models have many uses. They provide a better understanding of the process and the interactions between the process variables. The process can be simulated, thus reducing (but not eliminating) the need for expensive testing during the optimization and certification phases of procedure development. Modelling is a valuable tool in the design process, particularly in the selection of material and processing parameters and the scheduling of production sequences. Models provide a means for a process to be optimized, by identifying a condition that maximizes an aspect of the process. Results obtained from models provide valuable input to on-line adaptive process control systems, which play an increasing role in intelligent laser processing.

Solutions to a model can be obtained by using analytical or numerical methods. Analytical models that use justifiable assumptions yield explicit relationships that enable the effects of variations in process variables on the product characteristics to be visualized and established quickly. Numerical techniques require fewer assumptions, can produce more exact results if input data are known accurately, but require more sophisticated methods of solution, and often do not allow explicit relationships to be written. The detail needed in the model, and the method by which the formulation is solved, should correspond to the complexity of the problem, the reliability of input data, the accuracy required in the results, and the computing power and time available.

The approach adopted here is predominantly analytical. The methods described can be solved by using a desktop personal computer, and results obtained within one minute. An accuracy of ±5% is sought, which reflects the typical accuracy of data available for modelling.

A MODEL-BASED PROCESS DIAGRAM

Consider again the empirical process chart shown in Fig. 6.3. The variable groups (power density and interaction time) distinguish conditions for different processes. However, the precise role of the derived group (energy density) in determining the processing mechanism is not clear.

Mathematical modelling of heat flow provides a means of establishing functional relationships. Section 2 of Appendix E contains analytical formulae that describe the temperature field in a material below a surface heat source. Equation (E2.2) describes the surface temperature rise, $T - T_0$, in a semi-infinite body in terms of the absorbed power density, Aq/A_B, the time, t (represented by the interaction time, τ), the material thermal conductivity, λ, and thermal diffusivity, a. For a power density, E, and a time, τ, equation (E2.2) (Appendix E) can be written in logarithmic form:

$$\log E = \log(T_p - T_0) + \log\left(\frac{\pi^{\frac{1}{2}}\lambda}{2a^{\frac{1}{2}}}\right) - \frac{1}{2}\log \tau. \tag{6.3}$$

By substituting material properties for structural steels, which can be found in Appendix D ($\lambda = 30\,\mathrm{J\,s^{-1}\,m^{-1}\,K^{-1}}$, $a = 9.2 \times 10^{-6}\,\mathrm{m^2\,s^{-1}}$, $T_{Ac1} = 996\,\mathrm{K}$, $T_m = 1800\,\mathrm{K}$, $T_v = 3100\,\mathrm{K}$), a series of contours representing the conditions required to obtain a given peak surface temperature in structural steel can be constructed. The most important contours for structural steels are the temperature of solid state formation of austenite (996 K), surface melting (1800 K), and surface vaporization (3100 K). These are plotted in Fig. 6.4 using equation (6.3).

A fundamental difference between Figs 6.3 and 6.4 is that a physical model has been used to construct a relevant set of peak surface temperature contours (instead of plotting an arbitrary quantity – the energy density). There is now justification for comparing the model results with experimental data.

However, before we make a comparison, it is important to note that the physical model describes heat flow *within* the material. Not all of the incident energy is *absorbed* by the material – some is reflected, some is transmitted, and some may be reradiated. Therefore, in order to compare experimental data with model predictions, the *incident* power density of different processes must be converted to the *absorbed* power density. This is achieved by multiplying the incident power by a fraction A, the absorptivity. A varies with the wavelength of laser light, the material and other processing conditions, which were discussed in Chapter 5. But, for this purpose, an average value can be used to characterize each process: 0.5; 0.3; 0.5; 0.5 and 0.8 for hardening, melting, cladding, cutting and keyhole welding, respectively (Ion *et al.*, 1992).

Once the experimental data for hardening, melting, cladding and keyhole welding are corrected to represent the absorbed power density, the clusters align with the theoretical contours of peak surface temperature, as shown in Fig. 6.4. Although the data clusters appear to be shifted slightly above the contours (this observation is explained below), the process data fall in the correct sequence of heating, melting and vaporization. We can therefore infer that these processes are governed by a principal processing mechanism, which is controlled by the peak surface temperature. Data for cutting, however,

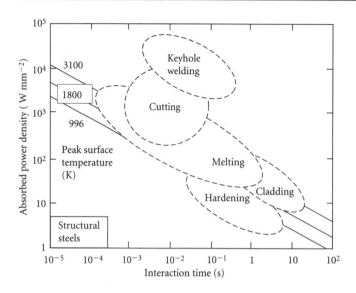

Figure 6.4   Theoretical contours of surface peak temperature for solid state transformation of pearlite to austenite (996 K), melting (1800 K), and vaporization (3100 K) for structural steels (solid lines), together with experimental data for processing of metals and alloys (regions bounded by broken lines)

appear not to align with the contours. This suggests that cutting can be performed by using more than one principal mechanism – melting, vaporization, or both – which is the case in practice.

Figure 6.4 illustrates the physical basis of the mechanisms of processing, and the functional relationship between the axes. It has *predictive* power. Optimum processing parameters can then be estimated for any process provided that the fundamental processing mechanism is known. However, the theoretical peak temperature contours are constructed for a single group of materials (structural steels) and so the diagram is strictly only valid for that group. We now consider the use of dimensional analysis to identify dimensionless groups of process variables, which may be used to broaden the scope of modelling to take a wider range of engineering materials into account.

DIMENSIONLESS GROUPS OF PROCESS VARIABLES

Many physical problems containing a large number of parameters may be simplified by identifying *dimensionless groups* of process variables. By considering the temperature field and the equations in sections 2 and 3 of Appendix E, the following groups describing the dimensionless beam power, q^*, and the dimensionless beam traverse rate, v^*, can be defined:

$$q^* = \frac{Aq}{r_B \lambda (T_m - T_0)} \quad \text{(dimensionless beam power)} \tag{6.4}$$

$$v^* = \frac{v r_B}{a} \quad \text{(dimensionless traverse rate)} \tag{6.5}$$

where A is the material absorptivity, q is the incident beam power, r_B is the beam radius, λ is the thermal conductivity, T_m is the melting temperature, T_0 is the initial temperature, v is the traverse rate, and a is the thermal diffusivity.

Equation (E2.19) in Appendix E gives the thermal cycle, $T(t)$, induced at a depth, z, in a semi-infinite body, below the centre of a moving, continuous wave, circular Gaussian laser beam. Equation (E2.19) contains two reference parameters. The first, t_0, is a characteristic heat transfer time, related to the beam width, and defined by $t_0 = r_B^2/(4a)$ where r_B is the radius at which the beam intensity has fallen to $1/e$ of the peak value. The second, z_0, is a characteristic length; its function is to limit the surface temperature to a finite value because energy is input over a finite time – the true surface is taken to be a distance z_0 below the 'model' surface. (Note how the spread of the beam is characterized by a heat transfer time, and the time for energy to be absorbed is represented by a heat transfer distance.)

Equation (E2.19) can be expressed in terms of the dimensionless variable groups defined above, and the following additional variable groups:

$$T^* = \frac{T - T_0}{T_m - T_0} \quad \text{(dimensionless temperature rise)} \tag{6.6}$$

$$t^* = \frac{t}{t_0} \quad \text{(dimensionless time)} \tag{6.7}$$

$$z^* = \frac{z}{r_B} \quad \text{(dimensionless depth)} \tag{6.8}$$

to give:

$$T^* = \frac{(2/\pi)\,(q^*/v^*)}{[t^*(t^*+1)]^{\frac{1}{2}}} \exp - \left[\frac{(z^* + z_0^*)^2}{t^*} \right]. \tag{6.9}$$

The means for determining z_0^* (which is a function of v^* only) is as follows. The normalized time, t_p^*, taken to attain the peak temperature is found by differentiating equation (6.9) with respect to time, and solving the resulting quadratic equation:

$$t_p^* = \frac{1}{4} \left\{ 2(z^* + z_0^*)^2 - 1 + [4(z^* + z_0^*)^4 + 12(z^* + z_0^*)^2 + 1]^{\frac{1}{2}} \right\}. \tag{6.10}$$

Substitution of t_p^* into equation (6.9) yields the normalized peak temperature, T_p^*. The parameter z_0^* is found by equating T_p^* to a standard solution for the peak temperature produced by a stationary circular Gaussian beam acting for a time equal to the beam interaction time, equation (E2.8) (Appendix E), which in dimensionless form is

$$(T_p^*)_{z^*=0} = \left(\frac{1}{\pi} \right)^{\frac{3}{2}} q^* \tan^{-1} \left(\frac{8}{v^*} \right)^{\frac{1}{2}}. \tag{6.11}$$

Equation (6.9) is set equal to equation (6.11) with $t^* = t_p^*$ and $z^* = 0$ to find z_0^*.

We may now calculate the normalized peak surface temperature rise, T_p^*, in terms of the dimensionless beam power, q^*, and the dimensionless traverse rate, v^*.

A DIMENSIONLESS MODEL-BASED PROCESS DIAGRAM

The dimensionless variable groups v^* and q^* may be used as the abscissa and ordinate, respectively, of a dimensionless model-based diagram. Regions corresponding to different process mechanisms can now be plotted as contours on the diagram. Surface melting corresponds to a normalized surface temperature, T_p^*, of 1. The onset of surface vaporization corresponds approximately to a normalized surface temperature of 2. By solving equation (6.9) in the manner described above, contours corresponding to the onset of melting and vaporization at the surface can be constructed, as shown in Fig. 6.5.

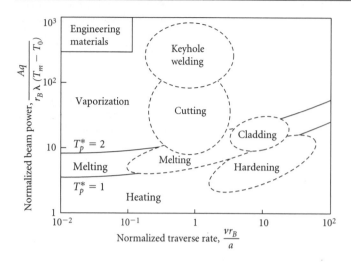

Figure 6.5 Normalized model-based laser processing diagram showing contours for the onset of surface melting and vaporization constructed from a surface heat flow model (solid lines); experimental data for metals and alloys, ceramics and glasses, and polymers are shown as regions bounded by broken lines

Experimental processing parameters for metals and alloys, ceramics and glasses, and polymers are also plotted in Fig. 6.5. Data are converted into dimensionless form by using material properties given in the tables of Appendix D. The data cluster in regions, as in the previous diagrams. However, by incorporating the thermal properties of the materials (thermal conductivity and thermal diffusivity) into dimensionless variables, the data span on each axis has been halved compared with Figs 6.3 and 6.4. This indicates that these variable groups improve the differentiation between processes. The experimental data for hardening, melting and cladding – processes that are almost exclusively carried out on metals and alloys, whose properties are well known and behave predictably at the process temperatures – show the best agreement with the model predictions of thermal mechanisms.

Figure 6.5 provides a comprehensive overview of the surface processing mechanisms – heating, melting or vaporization – that are induced by the thermal interaction of the laser beam and engineering materials. A model-based approach works well for a process for which the underlying process mechanism is relatively simple and well understood, and processing is carried out under well-defined conditions. For now it is all we shall consider – the foundations for laser processing diagrams have been laid. We return to model-based diagrams in the following chapters to describe individual processes in more detail, using additional techniques to determine thermal cycles at depth, the effects of melting, and vaporization on temperature fields, and the temperature fields induced around through-thickness heat sources.

SUMMARY AND CONCLUSIONS

Charts and diagrams can be constructed for a range of engineering materials that describe the conditions required to achieve the principal thermal mechanisms of laser processing – heating, melting or vaporization.

The simplest are charts based on empirical data. The framework (axes) of an empirical chart can be established by considering the principal process variables. By plotting experimental data in such a framework, processes may be compared, and groups of processing parameters that influence the processing mechanisms can be elucidated. Empirical charts provide insight into the principles of a process, and are often the first stage in understanding the underlying physical principles. But processing parameters can only be derived from empirical charts.

The basis of a model-based diagram is a physical model that describes the functional relationships between the principal process variables (axes). The model endows the diagram with predictive capability. Analytical models of heat flow have been used here. Such modelling reveals fundamental dimensionless groups of process variables, which simplify analysis while maintaining the level of accuracy required. The resulting overview diagram has generality and a wide scope for application.

The diagrams presented in this book are only as accurate as the material properties and models from which they are constructed. However, by careful choice of thermal properties (Appendix D), and the use of calibration techniques (described later) they provide a rapid means of establishing an initial set of processing parameters from which to begin development of a procedure.

The charts and diagrams are also useful pedagogical tools that illustrate graphically the effects of changes in process parameters on properties. Their use in specific techniques of laser material processing is demonstrated in the chapters that follow. These processes are presented in order of the approximate surface temperature rise. However, before considering such thermally based processes, we first use the framework developed to consider a family of processes in which the temperature rise is negligible, and structural change occurs through *athermal* processing mechanisms.

FURTHER READING

MATHEMATICAL MODELLING

Bass, M. (1983). Laser heating of solids. In: Bertolotti, M. ed. *Physical Processes in Laser–Materials Interactions*. New York: Plenum Press. pp. 77–115.

Carslaw, H.S. and Jaeger, J.C. (1959). *Conduction of Heat in Solids*. 2nd ed. Oxford: Clarendon Press.

Cerjak, H. and Bhadeshia, H.K.D.H. eds. *Mathematical Modelling of Weld Phenomena*. London: The Institute of Materials. Published biennially since 1992.

Christensen, N., Davis, V. de L. and Gjermundsen, K. (1965). Distribution of temperatures in arc welding. *British Welding Journal*, **12**, 54–75.

Dowden, J.M. (2001). *The Mathematics of Thermal Modeling: An Introduction to the Theory of Laser Material Processing*. Boca Raton: Chapman & Hall/CRC.

Grong, Ø. (1997). *Metallurgical Modelling of Welding*. London: The Institute of Materials.

Ion, J.C. (1992). Modelling of laser material processing. In: Belforte, D. and Levitt, M. eds. *The Industrial Laser Handbook*. New York: Springer. pp. 39–47.

Mazumder, J. (1987). An overview of melt dynamics in laser processing. In: Kreutz, E.W., Quenzer, A. and Schuöcker, D. eds *Proceedings of the Conference High Power Lasers*. Vol. 801. Bellingham: SPIE. pp. 228–241.

BIBLIOGRAPHY

Garland, K. (1994). *Mr Beck's Underground Map*. Harrow Weald: Capital Transport.

Ion, J.C., Shercliff, H.R. and Ashby, M.F. (1992). Diagrams for laser materials processing. *Acta Metallurgica et Materialia*, **40**, (7), 1539–1551.

CHAPTER 7

ATHERMAL PROCESSING

INTRODUCTION AND SYNOPSIS

In Chapter 6, a simple analysis based on classical heat flow during laser processing enabled the *thermal* effects of the principal process variables on the treatment of materials to be predicted in terms of three principal processing mechanisms: heating, melting and vaporization. Most of the processes described in this book are based on these three mechanisms. However, in Chapters 3 and 5 we noted that the energy of photons from short wavelength lasers, such as excimer sources, is on the same order of magnitude as the energy of chemical bonds in molecules. An interaction can occur between the beam and the material by resonant transfer of energy with no change in temperature – an *athermal* mechanism. Similarly, the mean free time between collisions of species in materials is 10^{-12}–10^{-14} s, and so a beam with a pulse duration on the order of femtoseconds (10^{-15} s) is able to induce mechanisms of interaction that do not follow the laws of classical thermal conduction. Such interactions form the basis of families of athermal processes. Their importance cannot be underestimated – they are actively being researched and are key enabling technologies for future applications, particularly in micromanufacturing. Lasers are tools with which concepts associated with nanoscience are being turned into products based on nanotechnology.

This chapter is intended as an introduction to processes that have been developed based on athermal photochemical, photoelectric and photophysical mechanisms of laser beam–material interaction. These are processes in which a temperature rise in the material is considered undesirable (or is practically negligible), and which does not contribute to the principal mechanism of processing. The chapter is not typical of the process chapters in the book (which describe individual methods of laser treatment) – it covers diverse processes whose parameters exceed the ranges typical of thermal interactions by orders of magnitude; from extremely low power densities associated with laser printing to ultrashort interaction times used in optical lithography. The link between the processes is the lack of a temperature rise, rather than a specific thermal processing mechanism. The principles of athermal processing mechanisms are explained, which form the basis of a number of industrial applications that highlight the advantages over conventional thermal laser processes (in cases where a comparison with a conventional process is possible). Predictive laser processing diagrams are difficult to construct for athermal processes because mechanistic modelling must be performed on the atomic scale, and is beyond the scope of this book. However, the framework used to construct the chart in Fig. 6.4 (Chapter 6) provides a convenient means of presenting athermal processing parameters, enabling them to be compared with those of thermal processing mechanisms. The chapter provides an overview of the diversity of available athermal processes, and allows them to be placed in context with thermal processes.

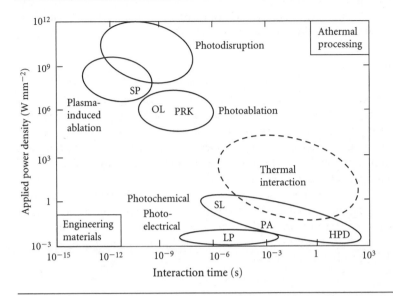

Figure 7.1   Athermal mechanisms of laser material processing: photophysical, photochemical and photo-electrical mechanisms are regions bounded by solid lines; LP: laser printing; HPD: haematoporphyrin derivative; PA: photochemical annealing; SL: stereolithography; OL: optical lithography; PRK: photorefractive keratectomy; SP: shock processing (the region of thermal processing is shown with a broken boundary for comparison)

PRINCIPLES

The three principal mechanisms of athermal processing are based on photoelectrical, photochemical and photophysical interactions between the laser beam and engineering materials. The operating regions of athermal processing methods are illustrated in Fig. 7.1; the operating region of thermal interaction is shown for comparison.

PHOTOELECTRIC EFFECTS

The photoelectric effect refers to the emission of electrons from the surface of a material in response to incident light. Energy contained within the incident light is absorbed by electrons within the material, giving the electrons sufficient energy to be emitted. In the laser printer, described below, laser light is used to modify the static electric field of a drum to create an image of the page to be printed.

PHOTOCHEMICAL EFFECTS

Photochemistry involves interactions between the photons of laser light and the chemical bonds that hold engineering materials together. Bonds may be made or broken by such interactions.

Making Chemical Bonds

Certain liquid resins are low molecular weight monomers that can chain react to form long polymers through interaction with ultraviolet light. In the case of composite resins used as restorative materials in dentistry ('white' fillings), light of wavelength 488 nm produced by the argon ion laser is optimal for the activation of the initiator in the resin (camphoroquinone). The polymers produced are solid, and can be formed on a scale on the order of micrometres because of the size of the laser beam. This mechanism is the basis of stereolithography, which is described below.

Breaking Chemical Bonds

The ultraviolet wavelength of excimer lasers is absorbed well by most materials and the beam photon energy is on the same order of magnitude as chemical bond energies (Table 5.1, Chapter 5), enabling energy to be transferred efficiently by resonance. Excimer lasers are therefore an efficient means of breaking chemical bonds – the basis of micromachining. Short wavelength light is diffracted to a smaller extent than longer wavelengths, and so a greater degree of control can be exercised over the application of the beam, enabling small features to be machined.

Short duration (femtosecond) pulses from solid state lasers produce such high intensity that strong non-linear absorption of photons can occur in materials that would otherwise be transparent to such wavelengths. The time scale of interaction is shorter than the mean free time between collisions in materials, and so interactions occur that do not follow the laws of classical heat conduction.

Both mechanisms of breaking bonds are used in micromachining processes described later.

Biostimulation

Biostimulation refers to a group of biological processes that do not result principally from a rise in temperature, but from an interaction between the photons from a low power laser (mW cm^{-2}) and an organic material. In one of the earliest trials of photodynamic therapy, haematoporphyrin derivative was administered, and malignant lesions exposed to red light, which responded to treatment. The tumors treated included carcinomas of the breast, colon, prostate, malignant melanoma. The precise mechanisms have yet to be elucidated, but it is believed that a photochemical mechanism via direct absorption of respiratory chain molecules is involved.

PHOTOPHYSICAL EFFECTS

Photoablation

Photoablation is the decomposition of material by direct breaking of molecular bonds when exposed to high energy ultraviolet laser irradiation. Excimer sources are particularly suitable since they produce a power density in the range 10^5–10^8 W mm^{-2} and a pulse duration between 1 and 100 ns. Very clean precise ablation is therefore obtained with a lack of damage to adjacent material. An audible report can be heard. Since most non-metallic materials absorb ultraviolet light in a very thin surface layer typically 0.5–1 μm deep, the main applications lie in refractive eye surgery (Fig. 1.4) and micromachining of man-made polymers and ceramics. Material is ejected by the rapid build-up of pressure at the ablation interface, without damage to underlying material.

Plasma-induced Ablation

Plasma-induced ablation results in material removal by the formation of an ionizing plasma (a mixture of loosely bound ions and electrons). The plasma expands away from the material in a very highly ionized state, taking most of the energy away with it. Consequently, very little energy is left behind to create an undesirable melt that can solidify. There is no appreciable melt phase; the material undergoes a transition directly into the vapour phase. The duration of the pulse of light is so short that there is insufficient time for heat to propagate into the surrounding material. It is a clean process, which can be heard. Pulsed Nd:YAG, Nd:YLF and Ti:sapphire lasers are sources of choice for plasma-induced ablation. The power density lies between 10^9 and 10^{11} W mm^{-2} with pulse durations between 100 fs and 500 ps. Medical applications include refractive corneal surgery and well-defined removal of caries from tooth tissue without evidence of thermal or mechanical damage.

Photodisruption

Photodisruption is the fragmentation and cutting of materials by mechanical forces. Plasma sparking is observed together with the generation of shockwave cavitation and jet formation. Pulsed Nd:YAG, Nd:YLF and Ti:sapphire are the most common lasers used, operating with power densities in the range 10^9–10^{14} W mm^{-2} and pulse durations between 100 fs and 200 ns. Photodisruption is used in applications such as micromachining.

INDUSTRIAL APPLICATION

A selection of industrial applications of athermal laser processing can be found in Table 7.1. The biomedical and microelectronics industries are the traditional driving sectors for developments in such procedures.

Table 7.1   Industrial application of laser-induced structural change by athermal photoelectric, photochemical and photophysical mechanisms: PRK photoreactive keratectomy; LASIK laser *in-situ* keratomileusis

Industry sector	Mechanism	Material	Laser	Application	Reference
Various	Photochemical curing	Photopolymer	He–Cd	Rapid prototyping	(Jacobs, 1995)
Biomedical	Biostimulation	Tumours	Diode	Necrosis	(Dougherty et al., 1978)
	Photochemical curing	Composites	Ar ion	Dental fillings	(Fleming and Maillet, 1999)
	Photoablation	Si_3N_4	Solid state	Microsieves	(van Rijn et al., 1999)
	Photoablation	Corneal tissue	ArF excimer	PRK	(Trokel et al., 1983)
	Photoablation	Corneal tissue	ArF excimer	LASIK	(Slade, 2002)
Domestic goods	Annealing	Nylon 6	ArF excimer	Carpet fibres	(Lizotte and O'Keeffe, 1998)
	Microdrilling	Polyimide	KrF excimer	Inkjet nozzles	(Booth, 2003)
Electronics	Micromachining	Silicon	KrF excimer	MEMS devices	(Gower, 2001)
	Photolithography	Silicon	KrF excimer	DRAM chips	(Levinson and Arnold, 1997)
	Micromachining	Diamond	Excimer-pumped dye	Thermal spreaders	(Shirk et al., 1998)
Power generation	Shock processing	12% Cr steel	Q-switched Nd:glass	Steam turbine blades	(Peyre et al., 1998)

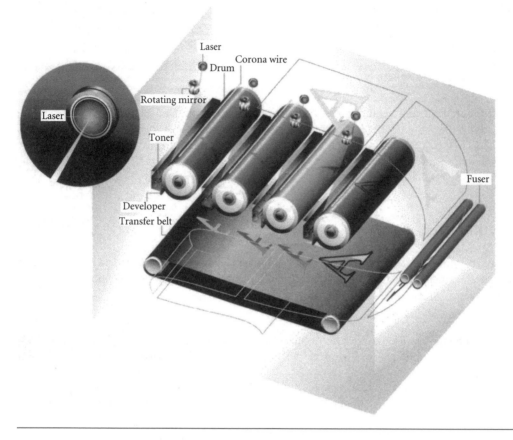

Figure 7.2   Schematic illustration of a single pass colour laser printer. (Source: Michelle Wista, Fuji Xerox Printers, Frenchs Forest, NSW, Australia)

LASER PRINTING

Single pass colour laser printing (LP) is based on a photoelectric effect induced by the laser beam. The printer contains four cylindrical drums, each coated with a material that is sensitive to light, as illustrated in Fig. 7.2. The drums first rotate past four corona wires, which creates a negative charge on their surfaces. Separate milliwatt diode laser beams are scanned across each drum by a rapidly rotating mirror, depicting the information in the image. Where the beam contacts the drum, the electrostatic charge changes from negative to positive and a latent image is created. Each drum is paired with a developer, containing cyan, magenta, yellow or black toner. The toner is negatively charged, and is attracted to the positive latent image. The latent image is then transferred to a belt. The rotation of the drums is timed such that each drum contacts the transfer belt to align the four images. Paper is then fed across the belt, and the image is transferred. Finally the paper passes through a fuser where rollers heat and meld the toner into the paper. Permanent images with a resolution of over 1000 dots per inch (dpi) can be generated in seconds.

STEREOLITHOGRAPHY

Product quality, time-to-market, and cost are critical factors in highly competitive manufacturing industries. The most expensive and time-consuming phase of the development process is often the transfer of a new product design from the 'drawing board' to the testbed. Stereolithography Apparatus (SLA®), based on a photochemical mechanism of making chemical bonds, was invented by Charles Hull in 1986 to provide a means of producing highly accurate, functional models of components in a few hours instead of days. (SLA is a registered trademark of 3D Systems, Valencia, CA, USA.) The starting point is a three-dimensional computer model created using a computer-aided design (CAD) package. The model is then sliced into hundreds or thousands of cross-sections, which are used to guide the light from an ultraviolet laser across the surface of a vat of a liquid photopolymer.

Photosensitive liquid resins, such as acrylics or epoxies are used as the starting material. These are low molecular weight monomers that can chain react to form long polymers. Continuous wave ultraviolet output from helium–cadmium or argon ion lasers initiates polymerization, and cures the solid formed. Typical ranges of operating parameters are shown in Fig. 7.1. Laser light has a considerably narrower spectral bandwidth than conventional light sources, which improves curing efficiency, and enables the laser beam to be focused to the diameter required for high precision curing. The light cures selected regions of the surface forming a solid layer. The solid is supported on a perforated platform that is gradually lowered into the vat. The device then moves up approximately 0.1–0.5 mm to the next layer. The model is submerged in the liquid resin and smoothed out, covering the hardened layer. Successive slices are polymerized until the object is formed. The object can be stiff, to represent an alloy, ceramic or thermoset polymer, or flexible to mimic the properties of an elastomer depending on the resin used. The process is one member of a family of laser-based rapid prototyping and manufacturing techniques. SLA models allow designers, sales people and customers to examine parts for aesthetics, ergonomics and manufacturability. Tolerances on parts can be established for satisfactory assembly. Materials can be chosen with similar properties to the final product to assist in testing. Prototype tools can also be manufactured. SLA is offered as a job shop service, often with a turnaround time of one day, and is the most widely used application of rapid prototyping in the medical field.

OPTICAL LITHOGRAPHY

Moore's law originates from an interview given in 1965 by Dr Gordon Moore (then head of research at Fairchild Semiconductor Corporation) to *Electronics* magazine, in which he predicted where the fledgling semiconductor industry would be in 10 years' time. Moore observed that the number of microcomponents that could be placed in an integrated circuit (with the lowest manufacturing cost) was doubling every year, and that the trend was likely to continue. In 1975 Moore, a founder of Intel, predicted that the doubling would continue every two years; in fact it currently occurs every 18 months. The technology that has enabled Moore's law to hold today is optical lithography, which, together with innovative circuit design, has allowed the size of features to be reduced.

Optical lithography enables microelectronic circuit patterns to be etched on the surface of a silicon wafer. The excimer laser is the source of choice – it has allowed designers to develop lenses that have higher resolution by taking advantage of an ultraviolet wavelength and a narrow spectral bandwidth in comparison with the mercury lamps used previously. Resolution is improved as a direct result of the smaller size of a focused spot of shorter wavelength, and also as a consequence the use of high numerical aperture optics. Features below 15 nm in size are expected to be produced with the use of ArF excimer and extreme ultraviolet radiation sources.

There are two types of exposure tools currently used for the mass production of microchips: the stepper and the scanner. The stepper exposes an area using a stationary lens, wafer and mask. The stepper repeats the procedure until the entire wafer surface is exposed. The scanner traverses a narrow slit of light over the mask and the wafer, which move in opposite directions.

MICROMACHINING

There are many advantages to micromachining with lasers producing pulses on the scale of femtoseconds: high accuracy and detail can be achieved; shapes can be programmed rapidly for prototype production; all types of engineering material can be treated; and serial or batch production is possible.

A number of conditions must be met for lasers to produce high intensity ultrashort pulses for photophysical processing: the laser medium must have a broad spectral bandwidth; the laser should be able to be mode locked; pulse spreading caused by gain in the active medium and components should be eliminated; and a method is needed to amplify the laser energy without damaging the optical components. Two practical systems are of interest: the Ti:sapphire laser (735–1053 nm) and the excimer-dye laser (220–300 nm and 380–760 nm). The use of chirped pulse amplification and pulse compression techniques enables a pulse energy of up to 125 mJ to be produced from the Ti:sapphire source, and 15 mJ from the excimer-dye configuration, corresponding to a focused power density on the order of $10^{14}\,\mathrm{W\,mm^{-2}}$.

Printed Circuit Boards

Figure 7.3 shows a flexible multilayer printed circuit board in which the top (polymer) layer has been ablated in specific locations using femtosecond laser pulses to expose a copper layer underneath to form the circuit board.

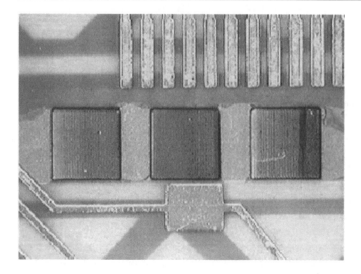

Figure 7.3 Flexible multilayer printed circuit board. (Source: Philippe Bado, Clark-MXR, Inc., Dexter, MI, USA)

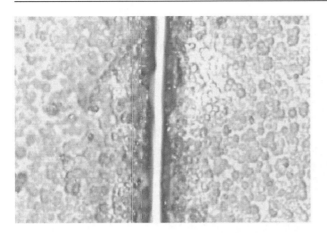

Figure 7.4   Entrance to a slot machined in a piezoceramic of thickness 115 μm, using 400 pulses of energy density 17 J cm^{-2} and duration 25 ns made using an ArF excimer laser. (Source: Jan Brune, Lambda Physik AG, Microlas Application Laboratory, Göttingen, Germany)

Slotting

Since ultraviolet excimer laser light is absorbed well by most materials, and the pulse duration can be selected with high accuracy, materials that would be difficult to machine using conventional techniques can be laser processed. Figure 7.4 shows a thin piezoceramic that has been slotted by excimer laser ablation. The material is too brittle to be slotted using mechanical means, and the thermal effects of other laser slotting techniques introduce unacceptable defects into the material.

Figure 7.5 shows channels machined in a sheet of the nickel–iron alloy Invar of thickness 1 mm. The channel on the left was machined using pulses of length 200 fs with 0.5 mJ of energy per pulse, delivered from a Ti:sapphire laser. The channel on the right was machined using nanosecond pulses. The advantages of femtosecond machining are clear: cleanness; straighter edges; absence of a recast layer; and a higher machining efficiency. These advantages originate from the rapid creation of vapour and plasma phases, without the formation of a liquid phase.

Nanostructuring

Ultraviolet laser light is a particularly suitable tool for machining engineering materials by ablation on the scale of nanometres, enabling Micro Electro-Mechanical Systems (MEMS) and Micro-Optical Electro-Mechanical Systems (MOEMS) to be fabricated. A microlens can be made directly by selective ablation of polymethylmethacrylate (PMMA), for example. Dielectric masks are used to define the geometry of structures illuminated by laser light – they are themselves fabricated by excimer laser ablation. Surface features on the order of 100 nm can be produced on a variety of engineering materials using subpicosecond pulses of ultraviolet light.

Corrective Eye Surgery

An output of 193 nm from the argon fluoride laser is the source of choice for ophthalmic procedures involving photorefractive keratectomy because of the high controllability of the pulse and the ablation accuracy associated with the short wavelength. The parameter ranges associated with the procedure are shown in Fig. 7.1.

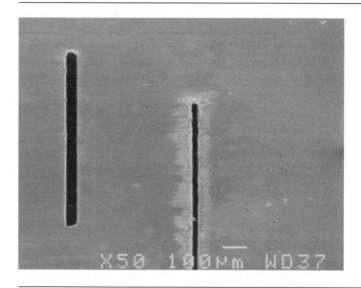

Figure 7.5 Femtosecond (left) and nanosecond (right) slotting of Invar of thickness 1 mm. (Source: Philippe Bado, Clark-MXR, Inc., Dexter, MI, USA)

Photorefractive keratectomy

Photorefractive keratectomy (PRK) involves the removal of tissue (usually less than 5%) from the surface of the central area of the cornea. This decreases the curvature of the cornea, which reduces the focal length, thus reducing short sight (myopia). The laser treatment itself is very brief and is carried out under local anaesthetic. Each laser pulse removes a layer of tissue around one quarter of a micrometre in thickness. The whole treatment process takes less than 15 minutes. PRK can also be used for treating long sight (hyperopia) by removing a circle of peripheral corneal tissue, thus increasing the curvature of the cornea.

PRK should not be confused with radial keratotomy (RK), which is a technique that was pioneered in the former Soviet Union in the 1970s for correcting short sight by making a series of radial cuts around the outer edge of the cornea with a hand-held blade.

Laser in-situ keratomileusis (LASIK)

LASIK is a development of PRK, and was described in Chapter 1. A keratome (a device containing a high-speed rotating disc) is used to cut a layer of superficial corneal tissue (a natural polymer) about 0.15 mm in thickness, which is folded aside, Fig. 1.4 (Chapter 1). Corneal tissue is exposed to pulses of ultraviolet ArF laser light, as in PRK. Movement of the beam over the eye is computer controlled for precision profiling. Myopia and hyperopia are treated using the principles described above for PRK.

SHOCK PROCESSING

The effects of laser shock processing (LSP) originate from plastic deformation generated by shockwaves that result from the rapid expansion of a surface plasma. The plasma is traditionally generated at the material surface by very short, intense pulses of laser light, 3–30 ns in length, with a power density on the order of 10^8 W mm^{-2}, Fig. 7.1. A thin surface layer of material is rapidly vaporized, and ejected, generating shockwaves in a region within a few micrometres of the surface, with a pressure up to several tens of GPa. These waves cause plastic deformation, which lead to work hardening, an increase in hardness in hardenable materials, and the creation of a compressive residual stress state. The shock pressure can be magnified by a factor of 10–100 by confining the plasma expansion by coating the material with a dielectric medium. The hardening mechanism is thus a mechanical effect rather than a thermal effect. Improvements have been attributed to the dense, uniform and stable dislocation and precipitate structures produced, as well as the residual compressive stress state.

A Q-switched Nd:glass laser can also deliver the high power density and short interaction time required for shock hardening. Effective surface hardening has been demonstrated to a depth of several millimetres in structural steels, stainless steels, aluminium alloys, titanium alloys and nickel-based superalloys. Alloys that cannot be strengthened by heat treatment can be hardened using LSP. The depth of treatment can be closely controlled. Soft materials such as aluminium and powdered materials can be hardened with little or no degradation of machined surfaces.

Shot peening is the equivalent conventional hardening method. It is much cheaper, but it exhibits several disadvantages such as superficial particle inclusion, accessibility problems and surface roughness modification. Although LSP was developed in the early 1970s, its application was hindered by the lack of suitable laser sources. New solid state lasers capable of generating high power density pulses of short duration have led to a resurgence of interest in the process, particularly for improving the fatigue strength of aircraft engines, cleaning of moulds, and medical prostheses.

SUMMARY AND CONCLUSIONS

The principles of laser-based processes that are based on athermal photochemical, photoelectric and photophysical mechanisms of laser beam–material interaction have been described, and illustrated with a number of industrial applications. Their range is diverse: from ultraviolet light curing of photosensitive polymers to femtosecond-scale pulsed machining of metals. Yet a link between these processes can be found – material processing in the absence of a temperature rise; this is the principle underlying the novel opportunities that such processes offer.

The biomedical and microelectronics industries are driving much of the growth in athermal laser processing. The demand for laser eye surgery continues to grow, and new, more sensitive lasers are being developed to expand the range of procedures that can be offered. Demands for increased miniaturization drive the need for laser systems capable of processing on the nanoscale. The laser will play an important role in developing nanoscience into applications based on nanotechnology. This chapter has been an introduction to the variety of processes that have been developed, which are 'pushing the envelope' of laser material processing.

FURTHER READING

OPTICAL LITHOGRAPHY

Yen, A. ed. (2003). *Proceedings of the Conference Optical Microlithography XVI*. Vol. 5040 Bellingham: SPIE.

PHOTOPHYSICAL PROCESSING

Boulnois, J.-L. (1986). Photophysical processes in recent medical laser developments: a review. *Lasers in Medical Science*, **1**, 47–66.

Shirk, M.D. and Molian, P.A. (1998). A review of ultrashort pulsed laser ablation of materials. *Journal of Laser Applications*, **10**, (1), 18–28.

Zhao, J., Huettner, B. and Menschig, A. (2001). Microablation with ultrashort laser pulses. *Optics and Laser Technology*, **33**, 487–491.

SHOCK PROCESSING

Montross, C.S., Wei, T., Ye, L., Clark, G. and Mai, Y.-W. (2002). Laser shock processing and its effects on

microstructure and properties of metal alloys: a review. *International Journal of Fatigue*, **24**, 1021–1036.

STEREOLITHOGRAPHY

Jacobs, P. ed. (1995). *Stereolithography and Other RP&M Technologies*. Dearborn: Society of Manufacturing Engineers.

Miserendino, L.J. and Pick, R.M. eds (1995). *Lasers in Dentistry*. Chicago: Quintessence Press.

EXERCISES

1. Athermal laser processing involves the interaction of bonds and photons. Describe the nature of the four principal types of bonding found in metals and alloys, ceramics and glasses, and polymers. Which would you expect to dominate in each class of material?

2. What are the main mechanisms of absorption of laser beam photons for different classes of engineering material?

3. By considering the properties of lasers given in Appendix B, which would be most suitable for stereolithography using liquid photopolymers?

4. What are the characteristics of lasers required for athermal photoablation?

5. By considering the atomic structure of and energy levels of different classes of engineering material, suggest why the mechanism of ablation of metals by ultrashort (10^{-15} s) laser beam pulses is different from that of semiconductors and insulators.

BIBLIOGRAPHY

Booth, H.J. (2003). Recent applications of pulsed lasers in advanced materials processing. In: *Proceedings of the European Materials Research Society Symposium on Photonic Processing of Surfaces, Thin Films and Devices*, 10–13 June 2003. Strasbourg: E-MRS.

Dougherty, T.J., Kaufman, J.E., Goldfarb, A., Weishaupt, K.R., Boyle, D. and Mittleman, A. (1978). Photoradiation therapy for the treatment of malignant tumors. *Cancer Research*, **38**, (8), 2628–2635.

Fleming, M.G. and Maillet, W.A. (1999). Photopolymerization of composite resin using the argon laser. *Journal of the Canadian Dental Association*, **65**, (8), 447–450.

Gower, M.C. (2001). Laser micromachining for manufacturing MEMS devices. In: Helvajian, H., Janson, S.W. and Laermer, F. eds. *Proceedings of the Conference MEMS Components and Applications for Industry, Automobiles, Aerospace, and Communication*. Vol. 4559. Bellingham: SPIE. pp. 53–59.

Jacobs, P.F. (1995). Stereolithography accuracy, QuickCast™ & rapid tooling. In: Mazumder, J., Matsunawa, A. and Magnusson, C. eds. *Proceedings of the Laser Materials Processing Conference (ICALEO '95)*. Orlando: Laser Institute of America. p. 194.

Levinson, H.L. and Arnold, W.H. (1997). Optical lithography. In: Rai-Choudhury, P. ed. *Handbook of Microlithography, Micromachining, and Microfabrication, Volume 1: Microlithography*, Bellingham: SPIE Optical Engineering Press. pp. 11–138.

Lizotte, T. and O'Keeffe, T. (1998). Excimer surface treatment behind stain-resistant carpet. *Industrial Laser Review*, **13**, (7), 9–10.

Peyre, P., Scherpereel, X., Berthe, L. and Fabbro, R. (1998). Current trends in laser shock processing. *Surface Engineering*, **14**, (5), 377–380.

Shirk, M.D., Molian, P.A. and Malshe, A.P. (1998). Ultrashort pulsed laser ablation of diamond. *Journal of Laser Applications*, **10**, (2), 64–70.

Slade, S.G. (2002). *The Complete Book of Laser Eye Surgery*. New York: Bantam. p. 188.

Trokel, S.L., Srinivasan, R. and Braren, B.A. (1983). Excimer laser surgery of the cornea. *American Journal of Ophthalmology*, **96**, 710–715.

van Rijn, C.J.M., Nijdam, W., Kuiper, S., Veldhuis, G.J., van Wolferen, H. and Elwenspoek, M. (1999). Microsieves made with laser interference lithography for micro-filtration applications. *Journal of Micromechanics and Microengineering*, **9**, 170–172.

CHAPTER 8

STRUCTURAL CHANGE

INTRODUCTION AND SYNOPSIS

In the previous chapter, methods of laser processing based on athermal mechanisms were described. Structural change may also be induced in engineering materials through various *thermal* mechanisms associated with classical heat transfer. Changes in phase or state occur because of a *thermodynamic* driving force that causes one form to be more stable than others under a given set of conditions. Thermodynamic mechanisms are activated by a change in temperature. The boundary of the heat affected zone (HAZ) in a structural steel weld, for example, can be defined by the A_{c1} isotherm, at which temperature pearlite transforms to austenite on heating. Within the transformed region, the rate (or extent) of transformation depends on the *kinetics* of the transformation mechanism. The amount of grain growth at a location in the HAZ depends on the amount of diffusion that occurs during the thermal cycle experienced at that location. Such fundamental changes underlie a large number of laser-based techniques of material processing. An understanding of thermodynamic and kinetic changes involved in laser processing provides an insight into the process mechanism, from which a process model can often be developed.

This chapter describes the thermodynamics and kinetics of thermally induced structural change, and illustrates their application in a number of laser-based processes. (Some processes have been developed to such a degree that they are the subjects of individual chapters, which follow.) Analytical models of heat flow are developed to describe the thermal cycles induced by laser heating. These are combined with models of changes in phase and state to construct diagrams that can predict the extent of structural change. Industrial applications reveal that laser-induced structural change underlies novel procedures ranging from microsurgery to vapour deposition of diamond (which is an excellent heat sink for microelectronic components).

PRINCIPLES

The mechanisms of structural change considered here are those based on thermodynamic and kinetic change. They underlie a diverse assortment of laser-based processes, many of which present novel solutions to existing manufacturing problems, as well as providing opportunities for innovation.

THERMODYNAMIC CHANGE

The thermodynamics of phase transformations were described in terms of phase diagrams in Chapter 5. Since most laser processing is carried out under conditions of constant pressure, it is the effect of variations in temperature on structural change that is considered here.

A method of modelling the *surface* temperature rise induced in a material by a discrete surface energy source was presented in Chapter 6. That model is extended here to describe thermodynamic change *at depth* in a material, and to characterize melting and vaporization.

During a solid state structural change resulting from a thermodynamic transformation, the laser beam is required to heat material at a given depth to a given peak temperature, without the onset of surface melting. Following the dimensionless variable group approach introduced in Chapter 6, the peak temperature T_p is normalized using the melting temperature T_m to give the normalized temperature rise T_p^*:

$$T_p^* = \frac{(T_p - T_0)}{(T_m - T_0)} \tag{8.1}$$

where T_0 is the initial temperature of the material. Similarly, the transformation depth l is normalized using the beam radius r_B to give the normalized depth, l^*:

$$l^* = \frac{l}{r_B}. \tag{8.2}$$

We also describe the processing variables in terms of the dimensionless beam power q^* and the dimensionless beam traverse rate v^*, defined in Chapter 6:

$$q^* = \frac{Aq}{r_B \lambda (T_m - T_0)} \tag{8.3}$$

$$v^* = \frac{v r_B}{a} \tag{8.4}$$

where A is the material absorptivity, q is the incident beam power, r_B is the beam radius, λ is the thermal conductivity, v is the traverse rate, and a is the thermal diffusivity.

Let us assume that a peak temperature corresponding to 60% of the melting temperature is required to induce a desired thermodynamic structural change in a material. The theoretical limits of the process are defined by the need to heat the surface of the material to the transformation temperature ($T_p^* = 0.6$), but not above the melting temperature ($T_p^* = 1$). Equations (6.9–6.11) (Chapter 6) may be used to find the values of q^* and v^* that result in normalized surface temperatures of 0.6 and 1. (The *surface* condition is represented by $z^* = 0$ in equations (6.9) and (6.10).) Equations (6.9–6.11) may also be used to predict the *depth* at which a normalized peak temperature of 0.6 is attained; we solve for the normalized depth l^* at which $T_p^* = 0.6$ for a given q^* and v^*. Thus an explicit relationship may be obtained between q^* and v^* for a given value of l^*. This model is used to construct the process overview diagram shown in Fig. 8.1.

The solid boundaries in Fig. 8.1 represent the conditions required for the surface to attain the melting temperature ($T_p^* = 1$) and the transformation temperature ($T_p^* = 0.6$), which denote the limits of processing and define the operating window for the process. Within the operating window, contours representing values of l^* are shown as broken lines.

Practical processing parameters for a given material (laser beam power, q, traverse rate, v, and transformation depth, l) may be extracted from Fig. 8.1 by substituting an appropriate value for the beam size, r_B, and values for the material properties A (absorptivity), λ (thermal conductivity), T_m (material melting temperature), T_0 (initial material temperature) and a (thermal diffusivity), which can be found in Tables D.1–D.5 (Appendix D). The diagram may be modified to describe structural change activated when a different peak temperature is required by substituting an appropriate value for T_p^*.

The model can be adapted to processes in which surface melting is induced, by incorporating a characteristic molten geometry and the latent heat of melting, from which the net energy input to the base material may be obtained. This is described in Chapter 11, where it is applied to various processes

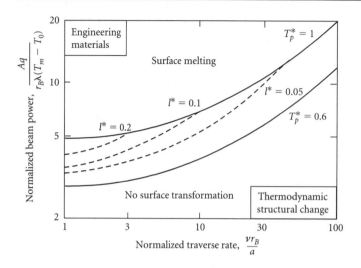

Figure 8.1 Model-based diagram of thermodynamic structural change caused by a temperature rise corresponding to 60% of the melting temperature, induced by a surface energy source. The operating window is bounded by solid contours that represent the onset of surface transformation ($T_p^* = 0.6$) and the onset of surface melting ($T_p^* = 1$). Within the operating window, broken contours represent normalized depths of transformation, l^* (l/r_B), at which $T_p^* = 0.6$

based on surface melting to prescribed depths. An analogous approach may be used to model surface vaporization – in this case both molten and vaporized geometries are characterized, and the latent heats of melting and vaporization incorporated into the model. Models may also be developed for heating patterns other than discrete surface energy sources; the temperature fields induced during keyhole welding, for example, may be modelled in terms of a through-thickness line source of energy, described in Chapter 16. For now, we shall consider the model described above for heating from a discrete surface energy source, to demonstrate its usefulness in solid state processing, and to illustrate is application in practical processing.

Calibration

Before discussing kinetic change, it is worth considering one of the most powerful features of a model-based laser processing diagram – its ability to illustrate the relationships between the variable groups, which enable the effect of *changes* in the values of the groups on the result of processing to be quantified. However, to extract *absolute* values of practical processing parameters from a normalized overview such as Fig. 8.1, the material properties must be known accurately. This is seldom the case – few material properties are known with accuracy, and may vary in an unknown manner with changes in temperature. Such poorly known process 'constants' can be established using a *data point*. This is an experimental observation of a processing result obtained under known conditions. In the case of structural change, this could be an experimental measurement of the depth of a known isotherm produced using a set of known process parameters. In this way the depth contours shown in Fig. 8.1 may be calibrated to a data point, thus obtaining an *effective* value for an unknown constant in the process model (often a 'lumped' term that contains the absorptivity of the material). Providing that a reasonable value for the unknown constant is obtained from the calibration – a value below unity for the effective absorptivity of a heating process, for example – the calibration enables the effects of *variations* in the process variables on the result of processing to be established about the known condition. Examples of the use of data points to construct practical processing diagrams are given below and in the chapters that follow.

KINETIC CHANGE

Thermodynamics describe the *probability* of structural change. Kinetics quantify the *extent* of structural change. The kinetics of a process determine the number of thermally activated species participating in the change, expressed in terms of the temperature T, the activation energy for change Q, and structural constants. Diffusion, grain growth and precipitate coarsening are examples of thermally activated transformations occurring during a temperature–time cycle. The amount of transformation may be quantified in terms of the *kinetic effect* of the thermal cycle.

The Kinetic Effect

The kinetic effect I of a thermal cycle $T(t)$ is related to the number of diffusive jumps that occur during the cycle, and can be written

$$I = \int_0^\infty \exp - \frac{Q}{RT(t)} dt \qquad (8.5)$$

where Q is the activation energy for diffusion (J mol^{-1}), and R is the gas constant (8.314 J mol^{-1} K^{-1}). The kinetic effect can be calculated by evaluating equation (8.5) using numerical methods.

However, the majority of diffusive jumps take place at temperatures close to the peak of the thermal cycle, which may be approximated as a parabolic profile (Ion *et al.*, 1984). Equation (8.5) may then be written explicitly in terms of the peak temperature, T_p, the time taken to attain the peak temperature, t_p (the time constant), and a quantity α (a function of T_p, Q and R):

$$I = \alpha t_p \exp - \left(\frac{Q}{RT_p}\right). \qquad (8.6)$$

Expressions for α and t_p are given in equations (E4.1) and (E2.14), respectively (Appendix E), for a thermal cycle generated by a point source of surface energy. By substituting for α and t_p in equation (8.6) we obtain

$$I = \left(\frac{2\pi RT_p}{Q}\right)^{1/2} \frac{1}{2\pi\lambda e} \frac{Aq}{v} \frac{1}{T_p - T_0} \exp - \left(\frac{Q}{RT_p}\right) \qquad (8.7)$$

where λ is the thermal conductivity, e is the base of natural logarithms (2.718), and v is the traverse rate.

The corresponding relationship for a line source of through-thickness energy is obtained by the same method, using equations (E4.1) and (E3.6), respectively, for α and t_p to give

$$I = 2\left(\frac{\pi RT_p}{Q}\right)^{1/2} \frac{1}{4\pi\lambda\rho ce} \left(\frac{Aq}{vd}\right)^2 \frac{1}{(T_p - T_0)^2} \exp - \left(\frac{Q}{RT_p}\right) \qquad (8.8)$$

where d is the plate thickness. The thermal cycle, parabolic approximation and kinetic effect are illustrated in Fig. 8.2.

An objective of laser-induced kinetic change is often to achieve transformation as rapidly as possible, or to the greatest possible extent, which means that the peak temperature approaches the melting

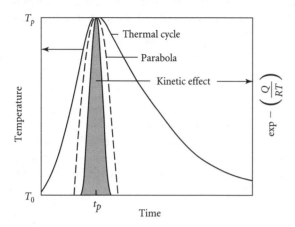

Figure 8.2   Thermal cycle, parabolic approximation, and kinetic effect of a thermal cycle

temperature T_m, i.e. $T_p \approx T_m$. Equations (8.7) and (8.8) then become

$$I = \left(\frac{2\pi RT_m}{Q}\right)^{1/2} \frac{1}{2\pi \lambda e} \frac{Aq}{v} \frac{1}{T_m - T_0} \exp -\frac{Q}{RT_m} \tag{8.9}$$

and

$$I = 2\left(\frac{\pi RT_m}{Q}\right)^{1/2} \frac{1}{4\pi \lambda \rho c e} \left(\frac{Aq}{vd}\right)^2 \frac{1}{(T_m - T_0)^2} \exp -\frac{Q}{RT_m}, \tag{8.10}$$

respectively.

The value of $Q/(RT_m)$ is characteristic of a class of engineering material, for substitutional self-diffusion. It takes values of 15–20 for pure engineering metals, 22–27 for ceramics, with diamond having a value between 31 and 36 (Brown and Ashby, 1980). These are useful representative values that simplify analysis in many applications involving thermally activated processes.

Equations (8.9) and (8.10) suggest the form of constitutive relationships between the principal processing variables, but a calibration procedure is required to establish the precise functional relationship.

Calibration

Assume, for the sake of illustration, that a known amount of kinetic change I^* occurs during a thermal cycle of peak temperature T_p^* and time constant t_p^* in a material in which the activation energy for the change is Q. Equation (8.6) then becomes

$$I^* = \alpha^* t_p^* \exp -\frac{Q}{RT_p^*}. \tag{8.11}$$

Consider now a different thermal cycle of peak temperature T_p and time constant t_p induced in the same material. The amount of kinetic change is given by equation (8.6). By dividing equation (8.6) by equation (8.11) we obtain

$$\frac{I}{I^*} = \frac{\alpha t_p}{\alpha^* t_p^*} \exp -\frac{Q}{R}\left(\frac{1}{T_p} - \frac{1}{T_p^*}\right). \tag{8.12}$$

The amount of kinetic change may then be calculated for any thermal cycle, relative to that of the known condition (I^*, t_p^* and T_p^*) using equation (8.12). The use of this calibration is described below for kinetic processes that underlie diffusion, grain growth and precipitate coarsening.

To simplify the analysis further, we assume that the peak temperature of the thermal cycle approaches the melting temperature for both the data point and the condition sought, i.e. processing is carried out under conditions that lead to the maximum amount of structural change, during which $T_p \approx T_m$. By substituting equation (8.9) into equation (8.12) for the known and desired conditions we obtain

$$\frac{I}{I^*} = \frac{q}{q^*}\frac{v^*}{v}$$

(8.13)

for a surface point energy source. Similarly, equations (8.10) and (8.12) yield

$$\frac{I}{I^*} = \left(\frac{qv^*d^*}{q^*vd}\right)^2$$

(8.14)

for a through-thickness line energy source. (Note that in this case q^* and v^* represent calibration values of power and traverse rate, respectively, and should not be confused with the dimensionless groups q^* and v^* used previously.)

Equations (8.13) and (8.14) show that for a given material, an explicit relationship between q and v exists for a given amount of kinetic change and a particular calibration. These relationships are now used to analyse thermally induced structural change, and to construct corresponding laser processing diagrams.

PROCESSES

Two thermally induced processes are considered here: diffusion and grain growth. The principles are applicable to all forms of structural change in which the extent of the transformation can be expressed in terms of thermally activated diffusive mechanisms. Processes such as sintering, precipitate coarsening and dissolution, thermally induced phase transformations, and vapour deposition may thus be modelled in terms of the principal processing variables. Equations describing such processes can be found in Appendix E.

DIFFUSION

Laser induced diffusion is the basis of processes such as surface carburizing of steels and doping of semiconductors. We would like to be able to predict the depth at which the concentration of the diffusing species has attained a given value. A well-known expression for the characteristic diffusion depth l (the depth at which the concentration has reached half of the surface concentration) produced after a time t at a fixed temperature is

$$l \approx \sqrt{Dt}$$

(8.15)

where D is the temperature-dependent diffusion coefficient, written $D = k\exp - Q/(RT)$, where k is a kinetic constant, Q is the activation energy (the activation energy for volume diffusion of the rate-controlling species), R is the gas constant and T is the absolute temperature. Equation (8.15) may then be written

$$l^2 = kt\exp - \frac{Q}{RT}.$$

(8.16)

The characteristic diffusion depth produced by a thermal cycle of peak temperature T_p and time constant t_p may be written in terms of the kinetic effect described above:

$$l^2 = k\alpha t_p \exp{-\frac{Q}{RT_p}}. \tag{8.17}$$

The unknown constant k is eliminated by calibrating equation (8.17) to an experimental data point – an observation that a characteristic diffusion depth l^* is produced after a thermal cycle characterized by a peak temperature T_p^* and time constant t_p^* – to give

$$\frac{l^2}{l^{*2}} = \frac{\alpha t_p}{\alpha^* t_p^*} \exp{-\frac{Q}{R}\left(\frac{1}{T_p} - \frac{1}{T_p^*}\right)}. \tag{8.18}$$

Equation (8.18) may then be used to calculate the characteristic diffusion depth produced by any given thermal cycle. The right-hand side of equation (8.18) is I/I^*, which for a surface point energy source is given by equation (8.13). Equation (8.18) may then be written

$$\frac{l^2}{l^{*2}} = \frac{qv^*}{q^*v}. \tag{8.19}$$

Canova (1986) observed that carbon from a surface coating of graphite diffused to a characteristic depth of 0.17 mm in Armco iron when heated by a square CO_2 laser beam of side 12 mm, with an incident power of 1.098 kW, traversing at a rate 0.036 m min^{-1}. The peak temperature attained at the surface during carburizing was close to the melting temperature. This data point is used in the construction of Fig. 8.3.

Figure 8.3 shows contours of constant diffusion depth l (broken lines) calculated using equation (8.19), which are calibrated to the data point indicated by a star. The conditions required for the onset of surface melting are described by equations (6.9)–(6.11) (Chapter 6) – the normalized coordinates obtained from these equations are converted to the practical processing variables q and v by substituting the following values for the properties of iron (Table D.1, Appendix D):

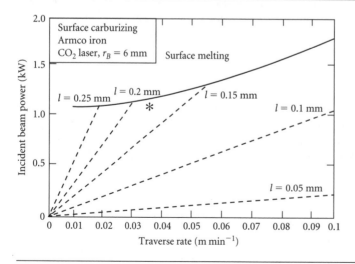

Figure 8.3   Parameter selection diagram for laser carburizing of Armco iron, showing model-based contours of characteristic diffusion depth (broken lines) and the boundary denoting the onset of surface melting (solid line). The contours are calibrated to an experimental data point (*)

$\lambda = 32.5\,\mathrm{W\,m^{-1}\,K^{-1}}$; $a = 7.5 \times 10^{-6}\,\mathrm{m^2\,s^{-1}}$; $T_m = 1810\,\mathrm{K}$; and $T_0 = 298\,\mathrm{K}$. The diagram is constructed for a beam half-width of 6 mm, corresponding to that used in establishing the data point.

Figure 8.3 may be used to expedite the development of a laser carburizing procedure by predicting initial sets of processing parameters rapidly. Existing data may be extrapolated to conditions that lie outside current operating windows, providing an indication of processing limits. Contours of diffusion depth may be constructed for materials other than iron by using a data point for the given material to calibrate the depth contours.

GRAIN GROWTH

One advantage of laser-based material processing over conventional methods is that the energy input is relatively low, and so thermally activated processes such as grain growth may be minimized to retain many of the properties of the base material. Grain growth is assumed to be diffusion controlled, driven by surface energy, and requires no nucleation. The rate of grain growth, dg/dt, at a constant temperature T is given by (Avrami, 1939; Johnson and Mehl, 1939):

$$\frac{dg}{dt} = \frac{k}{g}\exp-\frac{Q}{RT} \tag{8.20}$$

where g is the grain size, k is a kinetic constant, Q is the activation energy for grain growth, and R is the gas constant. The grain size after a time t is then

$$g^2 - g_0^2 = kt\exp-\frac{Q}{RT} \tag{8.21}$$

where g_0 is the initial grain size. By writing equation (8.21) in terms of the kinetic effect of a thermal cycle of peak temperature T_p and time constant t_p, and calibrating the equation to an experimental data point (an observation that a grain size g^* is produced after a thermal cycle characterized by a peak temperature T_p^* and time constant t_p^*), the unknown constant k can be eliminated, and we obtain

$$\frac{g^2 - g_0^2}{g^{*2} - g_0^2} = \frac{\alpha t_p}{\alpha^* t_p^*}\exp-\frac{Q}{R}\left(\frac{1}{T_p} - \frac{1}{T_p^*}\right). \tag{8.22}$$

Equation (8.22) may then be used to calculate the grain growth produced by any given thermal cycle.

Consider grain growth adjacent to the fusion line in the HAZ of a fully penetrating laser weld. The right-hand side of equation (8.22) is I/I^*, which for a through-thickness line energy source is given by equation (8.14), allowing equation (8.22) to be written

$$\frac{g^2 - g_0^2}{g^{*2} - g_0^{*2}} = \left(\frac{qv^*d^*}{q^*vd}\right)^2, \tag{8.23}$$

where d is plate thickness. Ion (1985) measured a prior austenite grain diameter of 41 μm adjacent to the fusion line of a CO_2 laser weld in a niobium microalloyed steel of thickness 1 mm made using an incident power of 1 kW and a traverse rate of 1.2 m min^{-1}. This data point is used in the construction of Fig. 8.4.

Figure 8.4 shows contours of constant grain diameter g calculated using equation (8.23), which are calibrated to the data point indicated by a star. The diagram may be used to predict the maximum amount of grain growth expected during welding, enabling a suitable set of processing parameters to be obtained when grain growth is a criterion used to accept a weld. The figure may also be used to extrapolate existing data to conditions that lie outside current operating windows. Contours of grain

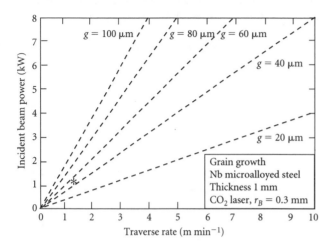

Figure 8.4   Parameter selection diagram for laser-induced grain growth in the heat affected zone adjacent to the fusion line of a niobium microalloyed steel. Model-based contours of grain diameter are shown as broken lines. The contours are calibrated to an experimental data point (*)

size may be constructed for other materials by using an appropriate data point to calibrate grain growth in a similar manner to that described above for diffusion. The effect of plate thickness on grain growth may also be taken into account by substituting the relevant value into equation (8.23).

SINTERING

The thermodynamic driving force for sintering is the large surface energy of fine powders. The kinetics of diffusion determine the rate at which sintering and subsequent densification occur. When powders are compacted and heated to a temperature around two thirds of the melting temperature, particles bond together forming necks, which grow to reduce the surface area, causing the powder to densify. A small amount of porosity remains in the final product.

The rate of sintering, $d\rho/dt$, is given by

$$\frac{d\rho}{dt} = \frac{k}{p^n} \exp -\frac{Q}{RT} \tag{8.24}$$

where ρ is the density, p is the particle size, k is a kinetic constant, n is a constant (about 3 for sintering), Q is the activation energy for sintering (normally the activation energy for grain boundary diffusion), R is the gas constant, and T is the absolute temperature. Equation (8.24) may be used as a basis for modelling the process, in a similar manner to the way that equation (8.20) is used to model grain growth. Sintering diagrams may then be constructed using the modelling and calibration techniques outlined above.

PARTICLE COARSENING

The standard equation relating the change in particle radius $(p - p_0)$ to the coarsening time t at a constant temperature T is (Lifshitz and Slyozov, 1961; Wagner, 1961):

$$p^3 - p_0^3 = \frac{k}{t} \exp -\frac{Q}{RT}. \tag{8.25}$$

Equation (8.25) may be written for a thermal cycle:

$$p^3 - p_0^3 = \frac{k\alpha t_p}{T_p} \exp - \frac{Q}{RT_p}. \tag{8.26}$$

Proceeding as before, the unknown kinetic constant k may be eliminated by calibrating to a known condition:

$$\frac{p^3 - p_0^3}{p^{*3} - p_0^3} = \frac{\alpha t_p}{\alpha^* t_p^*} \frac{T_p^*}{T_p} \exp - \frac{Q}{R} \left(\frac{1}{T_p} - \frac{1}{T_p^*} \right), \tag{8.27}$$

and equation (8.27) used to model the process and construct diagrams of particle coarsening.

Heat treatable aluminium alloys derive much of their strength from a fine dispersion of second phase precipitates that hinder dislocation movement. Equation (8.27) may be used to estimate the degree of softening that occurs during laser processing. Similarly, the optical properties of doped fibres are dependent on the size of embedded nanoparticles; the method may be used to estimate the amount of particle coarsening during laser annealing.

PROPERTIES OF STRUCTURALLY-CHANGED MATERIALS

The principles of thermodynamic and structural change described above underlie many practical techniques, some of which are described below. Structural change induced by laser heating is the basis of solid state processing. One method – surface hardening – is so well developed that it is the subject of a complete chapter (Chapter 9). Laser-induced melting forms the basis of a family of surfacing techniques that includes remelting, alloying and cladding (Chapters 11 and 12). Vaporization is a fundamental mechanism of keyhole welding, cutting and machining, which are described in Chapters 14, 16 and 17. In those chapters, the model of Chapter 6 is extended to describe the temperature rise at depth in a material in which surface melting occurs, as well as the temperature fields induced by a *through-thickness* source of energy. Here we consider a selection of processes that are based on structural change induced by thermodynamic and kinetic effects of heating, melting and vaporization, and the materials with which they are used. Many are novel, and still under development, but have the potential to revolutionize manufacturing, particularly on the scales of micrometres and nanometres.

MICROSTRUCTURAL CHANGE

Annealing is a kinetic process in which heating is used to change the microstructure through diffusion. The examples given in Table 8.1 illustrate the importance of laser annealing of materials used in the microelectronics industries.

Semiconductors are created by implanting atoms whose valences differ from that of the host material. Ion implantation is a common technique. However, the implantation process introduces defects, which have a significant effect on the performance of the semiconductor. Laser annealing is a method of repairing ion-implantation damage, which results in significantly increased electrical activity in the semiconductor. A high crystal quality is required for efficient operation – epitaxial recrystallization may be induced by laser annealing.

In another application of laser annealing, the relationship between optical properties and nanoparticles radius has been studied in glasses containing dispersed colloid gold nanoparticles. The peak plasmon wavelength shifts continuously to longer wavelengths with increasing average particle radius. The desired particle size may be obtained by laser heating. An estimate of the processing parameters may be obtained using the model for particle coarsening described above, together with appropriate material properties.

Table 8.1   Examples of laser-induced annealing

Material	Laser	Effect	Reference
Amorphous Ge–N alloys	XeF	Optical diffraction	(Mulato *et al.*, 2002)
CdS_xSe_{1-x} particles in a glass matrix	Nd:YAG ($\lambda/2$)	Particle coarsening	(Fukumi *et al.*, 1994)
Ductile iron	Nd:YAG	Novel alloy formation	(Wang and Bergmann, 1995)
Fe:LiNbO$_3$	XeCl	Increase in the optical absorbance of optoelectronic devices	(Sorescu *et al.*, 1995)
Ge films	Nd:YAG (Q-switched)	Conductivity changes	(Srivastava *et al.*, 1986)
InP–Zn$^+$		Recrystallization	(Vitali *et al.*, 1996)
Ion-implanted amorphous Si	Nd:YAG (Q-switched)	Pattern writing	(Celler *et al.*, 1978)
Nb$_3$Ga and Nb$_3$Al	CO$_2$	Superconductivity	(Kumakura *et al.*, 1986)
p-InSb thin films	Ruby (Q-switched)	Increased Hall mobility	(Srivastava *et al.*, 1983)
Polycrystalline Si films	–	Epitaxial growth	(Voutsas *et al.*, 2003)
Si–As, Sb	Ar$^+$	Recrystallization	(Williams *et al.*, 1978)
Si–B, P, As, and Sb	Ruby (Q-switched)	Recrystallization	(White *et al.*, 1978)

LASER-ASSISTED VAPOUR DEPOSITION

Laser-assisted vapour deposition (LAVD) combines thermodynamic mechanisms of laser-induced vaporization with kinetic mechanisms of subsequent solid deposition and film growth. Vaporization may be achieved using chemical or physical techniques. Chemical methods involve photolytic processes based on photochemical mechanisms of breaking chemical bonds. Excimer lasers are therefore sources of choice for such processes. Physical mechanisms involve pyrolytic processes based on photothermal interactions, for which higher power Nd:YAG and CO$_2$ lasers are often more suitable. Reactions may occur homogeneously within a gas or liquid phase, or heterogeneously at gas–solid or liquid–solid interfaces or surfaces. Processing is carried out in a high vacuum chamber containing the component materials.

Both techniques produce a low energy input to the substrate on which the solid film is grown. Heat treatment of the substrate is therefore restricted or virtually eliminated. Temperature-sensitive materials can be processed, such as GaAs and InP semiconductors, and polymer films. In addition to film deposition, LAVD can be used for patterning, writing and physicochemical modification of surfaces, through the activation or enhancement of chemical reactions. The advantages in comparison with conventional methods include: high deposition rates; accurate deposits; very thin, fine-grained deposits; deposition of difficult materials such as domestic ceramics and diamond; selected areas of the source material can be evaporated; and contamination is eliminated.

Silicon carbide is an attractive choice for heat-resistant coatings and semiconductors because of mechanical strength, thermal conductivity and resistance to irradiation. It can be applied by sputtering,

Table 8.2   Examples of laser-assisted vapour deposition: N/S not specified

Reactants	Deposit	Laser	Reference
N/S	Al–Mg–B–Ti	N/S	(Tian et al., 2002)
$Cr(CO)_6$	$(Cr_{1-x}Ti_x)_2O_3$	ArF	(Jacobsohn et al., 1991)
$Fe(CO)_5$ and NH_3	Ultrafine γ'-Fe_4N and γ-Fe powders	N/S	(Zhao et al., 1995)
Gold–cobalt electrolysis	Epoxy–Cu substrate	N/S	(Roos et al., 1986)
N_2, H_2 and $TiCl_4$	TiN	CO_2	(Silvestre et al., 1994; Silvestre and Conde, 1998)
O_2 gas environment	Oriented Bi–Ca–Sr–Cu–O thin films	Excimer	(Fork et al., 1988)
	Ta_2O_5	KrF	(Mukaida et al., 1993; Watanabe et al., 1993)
	Ta_2O_5	Nd:YAG ($\lambda/2$)	(Zhang and Boyd, 2000)
Si_2H_6 and C_2H_2	SiC	ArF	(Silvestre et al., 1994)

ion plating and chemical vapour deposition (CVD). Laser-assisted CVD has the advantage over these processes that lower amounts of defects and impurities are formed. Table 8.2 provides an overview of the materials and lasers used in LAVD.

As with laser annealing, the microelectronics industries have much to gain from the use of LAVD. Patterns can be written in a single-step process, with lateral dimensions down to the submicron range. Such patterns are conventionally produced by large-area processing techniques such as traditional CVD with the aid of mechanical masking – a process that requires several production steps.

NANOSCALE MATERIAL PRODUCTION

Materials may be produced on the scale of nanometres by laser-induced thermodynamic change. As a raw material in the form of nanoparticles such materials provide the engineer with novel building blocks. As a finished product – a thin film deposited by laser-assisted vapour deposition, for example – they underlie innovative methods of fabrication.

Nanoparticles exhibit surface and volume effects that are absent in particles on the scale of micrometres. They may be sintered rapidly at relatively low temperatures, and formed superplastically. Nanoparticles are used in catalysts, aerosols, paints, cosmetics, sunscreens, electronic inks and structural ceramics, and a wealth of procedures based on their unique biomedical, magnetic and electronic properties. Nanoparticles with high purity, a narrow particle size distribution, and an equiaxed microstructure may be produced by laser ablation using the high energy photons of the ArF excimer laser. Alternatively, they can be precipitated from solution through the interaction of a CO_2 laser beam with liquid solutions. Various processes and materials are listed in Table 8.3.

Carbon nanotubes are tubular molecules of carbon with properties that make them suitable for small-scale electronic and mechanical applications. The nanotube has a structure similar to that of a fullerene, with a diameter on the order of a few nanometres and a length that can be many orders of magnitude larger. The tubes may be closed at either end by caps containing pentagonal rings of carbon atoms. The tube may have a single wall, or comprise many walls. Carbon nanotubes can be produced by laser ablation of graphite in the inert gas atmosphere of a furnace. The ultraviolet

Table 8.3 Examples of laser-based nanoscale material production: N/S not specified

Material	Laser	Product	Reference
Ag	CO_2	Nanoparticles	(Subramanian *et al.*, 1998)
Al_2O_3–ZrO_2	Nd:YAG	Nanocomposite powders	(Yang and Riehemann, 2001)
B–BN nanoballoons	ArF	Biocompatible components	(Komatsu *et al.*, 2001)
C nanoropes	–	Superconductors	(Kociak *et al.*, 2001)
C nanotubes	ArF	Transistors	(Thess *et al.*, 1996; Suda *et al.*, 2003)
Cubic BN	CO_2	Hard coatings	(Molian and Waschek, 1993)
Diamond	CO_2	Potential solid lubricant	(Molian *et al.*, 1995)
SiC, sapphire	ArF	Composite production	(Huang *et al.*, 1995)

photons generated by an ArF excimer laser have sufficient energy to remove carbon atoms, which then recombine in the form of nanotubes. The addition of a metal catalyst enables uniform single-wall tubes with a high aspect ratio to be produced with a relatively high yield. Nanotubes naturally align themselves into 'ropes' that are held together by van der Waals forces. Nanoballoons of amorphous boron coated with crystalline boron nitride have also been produced, which have potential biomedical applications.

The structure of a nanotube affects its electrical conductivity: it can act as a conductor or semiconductor. Its thermal conductivity is high along the tube axis, but low in a transverse plane – selective heat sinks may therefore be constructed based on these properties. Their strength along the tube axis is also extremely high, enabling high tensile strength fibres to be produced.

INDUSTRIAL APPLICATION

The examples listed in Table 8.4 illustrate the diversity of applications for which laser-induced structural change has been used. This overview serves as an introduction to a number of industrial processes described below.

ANNEALING OF CARPET FIBRE

Neuman Micro Technologies Inc. (Concord, NH, USA) uses ArF excimer lasers for photochemical annealing of the surface of Nylon 6, which makes the fibres less porous to stains. The breaking and reforming of surface chemical bonds that occurs means that there are very few free electrons that can bond with foreign molecules, and so liquids form beads on the surface. The laser typically produces 600–800 mJ per pulse, with a repetition rate of 200 Hz. Each pulse irradiates 20 mm of fibre, which means that feed rates of up to $4\,\mathrm{m\,s^{-1}}$ are possible. An additional benefit is the ability to eliminate traditional solvents and chemicals used in surface treatment.

PHOTODYNAMIC THERAPY

The energy of a photon may be absorbed by a molecule, which is excited to a higher state, facilitating its interaction with other structures. The most common medical use of this mechanism is the excitation of haematoporphyrin derivatives (HpD), which are selectively retained by tumour tissue (Chapter 7).

Table 8.4 Industrial application of laser-induced structural change; SLS: selective laser sintering; HIP: hot isostatic pressing; PT: phase transformation; LAVD: laser-assisted vapour deposition

Industry sector	Process	Material	Laser	Application	Reference
Aerospace	SLS	Cermets	CO_2	Turbine sealings	(Das et al., 1998a)
	SLS	High temperature materials	CO_2	Aerospace components	(Zong et al., 1992)
	SLS	Ti–6Al–4V	Nd:YAG, CO_2	Aerospace components	(Ramos, 2001)
	SLS/HIP	Ti–6Al–4V	CO_2	AIM-9 Sidewinder missile housing	(Das et al., 1998b)
Automotive	Annealing	Steel	CO_2	Tyre reinforcement wire	(Minamida et al., 1990)
	SLS	Polymers	CO_2	Ford windscreen wiper cover	(Ogando, 1994)
Biomedical	Annealing	Bone	Diode	Photocoagulation of osteoid osteoma	(Gangi et al., 1998)
Domestic goods	Annealing	Nylon	ArF	Stain-resistant carpet	(Lizotte and O'Keeffe, 1998)
Electronics	Annealing	CDA-510 phosphor bronze	Nd:YAG	Contact springs	(Tice, 1977)
	LAVD	C	CO_2	Rapid prototyping	(Duty et al., 1999)
	LAVD	Diamond	CO_2	Semiconductor heat sinks	(Molian, 1994)
	LAVD	III–V, II–VI compounds	CO_2, ruby	Semiconductor films	(Ban and Kramer, 1970)
	LAVD	PbO	Nd:glass	Semiconductor gas sensors	(Baleva and Tuncheva, 1994)
	LAVD	ZnS	CO_2	Electroluminescent displays	(Nosaka et al., 1994)
	PT	Dye	Diode	CD writing	(Bouwhuis et al., 1985)
	PT	Metallic alloy	Diode	CD rewriting	(Bouwhuis et al., 1985)
Machinery	LAVD	TiN on AISI M2	CO_2	Hardened tools	(Silvestre et al., 1994)
Power generation	LAVD	C nanohorns	–	Fuel cells	(Williams, 2001)

The porphyrin molecule absorbs red photons (about 630 nm in wavelength) emitted from a dye or diode laser. Energy is then transferred to molecular oxygen, which in turn reacts with the tumour tissue, causing it to be destroyed. Photodynamic therapy is just one of a large number of medical procedures based on structural change induced in the natural polymers and composites in the human body.

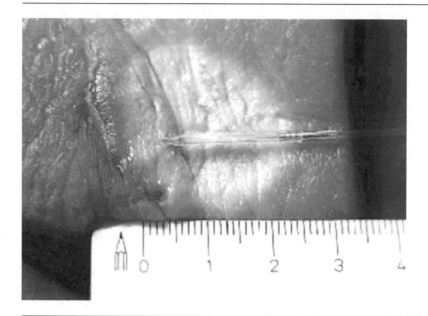

Figure 8.5 Coagulated region (light area) produced in damaged muscle tissue by laser-induced interstitial thermotherapy. (Source Markolf Niemz, University of Heidelberg, Germany)

LASER-INDUCED INTERSTITIAL THERMOTHERAPY

Localized tissue coagulation, initiated by laser heating, forms the basis of a tissue treatment known as laser-induced interstitial thermotherapy (LITT). LITT is used in the treatment of tumours of the retina, brain, prostate, liver and uterus, and is one of a family of techniques involving minimally invasive surgery, without bleeding. Near infrared light from an Nd:YAG or diode laser (which is readily absorbed by tissue) is delivered to the tumour via a fibre optic inserted through a catheter. The assembly is known as the applicator. The end of the fibre is etched in order to diffuse light to produce even illumination. Diseased tissue is heated to about 60°C, at which temperature it decays (a process known as necrosis). Simultaneously, blood cells coagulate, preventing severe haemorrhaging. Treatment is normally carried out using several watts of continuous wave power, for a period of several minutes, resulting in the treatment of tissue diameters of up to 40 mm. Figure 8.5 shows coagulation in muscle tissue achieved using CW Nd:YAG laser light. The tissue was treated for 5 minutes at 5.5 W. Tissue that appears white is completely coagulated. Larger volumes may be treated by using several fibres, whose arrangement determines the geometry of the treatment zone.

SELECTIVE LASER SINTERING

Selective Laser Sintering (SLS®) is a member of the family of rapid prototyping processes. (SLS is a registered trademark of 3D Systems, Valencia, CA, USA.) The laser beam induces sintering in powder-based materials by heating to create sections of a solid model layer by layer, in a similar fashion to stereolithography (Chapter 7). The system comprises a heating laser (e.g. a 50 W CO_2 unit for polymer powders, or a 1 kW unit for metals), a part chamber, and a control system. The part chamber consists of a platform, a powder cartridge, and levelling roller. A thin layer of material is first spread across the platform. The laser beam forms a two-dimensional cross-section of the part by sintering the material. A wetting agent can be used – the production of low melting temperature phases is an important mechanism in the sintering process. The platform then descends one layer in thickness and the levelling roller spreads material from the powder cartridge across the platform, where the next cross-section is sintered to the previous layer. Once the model is complete, it is finished by removing any loose material and smoothing the visible surfaces.

The benefits of SLS to a manufacturer are similar to those of stereolithography, namely a significant reduction in the time and cost of producing prototypes or production batches, with such a high degree of flexibility that the techniques are sometimes referred to as freeform modelling. However, a greater variety of materials can be used with SLS than with stereolithography: metals and alloys, and thermoplastic materials such as nylon, glass-filled nylon, and polystyrene are all suitable. Complex geometry components can be produced rapidly in sequence or in parallel, depending on the laser beam configuration. Functionally graded parts may be designed, in which properties of discrete regions are tailored to the application. Product flexibility is high and minimal tooling is required. Near net shaping is possible. The cost of a system is about \$300 000, with powders costing around \$100 kg^{-1} for polymers and \$60 kg^{-1} for metals and alloys.

LASER-INDUCED VITRIFICATION

Ceramic and glass tableware is decorated using colours dispersed in organic liquids or with patterns printed onto an organic film. In both cases, the decorated ware must be heated in a kiln to burn away the organic material and fix the colour to the item. A plate or cup can be given several decoration firings during its manufacture if the pattern is complex. Decoration firing is often carried out in batches of several thousand items and takes approximately 8 hours. Laser processing could be an alternative to kiln firing since lasers are a discrete, efficient, clean and flexible heat source – decoration firing operations could be eliminated. Multi-coloured bands could be heated to remove the organic binder and fused to unglazed plates. Coloured glaze layers could also be applied to pottery surfaces. Potential cost savings include reduced energy (about 90%), reduced process time and decreased labour. The simplified production route would improve production flow by replacing a batch process. Greater responsiveness and agility in the manufacturing operation are long-term goals of ceramic and glass manufacturers; the innovation would provide greater opportunities for flexibility in product design.

FUEL CELLS

Electronic applications of carbon nanotubes include: fine electron guns used in miniature cathode ray tubes for field emission displays; semiconductor devices comprising tubes of different diameter; and computer storage devices. The high aspect ratio of nanotubes, and their ability to hold other materials, provides novel means of producing fuel cells. Fuel cells create electrical energy by electrochemically reacting hydrogen and oxygen to produce water vapour and heat, without combustion. Some fuel cells require pure hydrogen as a fuel source; others can run on methane (natural gas) or other hydrocarbons. Such fuel cells are being evaluated by government authorities and private enterprises, with the aim of developing a 'hydrogen economy' to reduce our reliance on fossil fuels.

OPTICAL DISCS

A recordable compact disc (CD-R) comprises many layers: a strong flexible polycarbonate substrate; a coating of the recording medium (a green–blue cyanine dye); a reflective layer; a protective coating;

and uppermost a surface layer that can be labelled. An unrecorded disc reflects light since the dye is transparent and only the reflective layer is visible through the substrate. Data are stored in a continuous spiral track of flat reflective lands (representing a binary one) and non-reflective 'pits' (zero) with a total length of about 5 km. The pits are created by irradiation with a train of modulated optical pulses from a milliwatt-class infrared diode writing laser. The length of the pit represents the information coding. Pulses of light pass through the substrate and heat the dye, causing it to become opaque, which creates a permanent non-reflective pit. Data are retrieved by detecting the sequence of reflected pulses as the disc passes across a lower power infrared diode reading laser of wavelength 780 nm.

In a rewritable CD (CD-RW), the non-reflecting pits are created by structural transformations in a metallic alloy of silver, antimony, tellurium and indium. When the alloy is heated by the writing laser to above 600°C it melts, forming an opaque amorphous phase (the pit). Below about 200°C it crystallizes and is translucent. However, if solidification occurs quickly, the amorphous phase is retained at room temperature, and the pit remains, signifying a binary zero. The pit can be removed by an erase laser that is not powerful enough to cause melting, but which can heat the compound to its crystallization temperature. The erase laser thus restores the compound to its translucent crystalline state, effectively erasing the encoded zero. This clears the disc so that new data can be written.

Increased demands for higher amounts of data storage led to the development of the digital versatile disc (DVD). Data are again stored in the form of amorphous pits (the minimum pit length being 400 nm) and crystalline lands in a track that has a pitch (spacing) of 780 nm. The DVD is read by an infrared diode laser of wavelength 650 nm.

Blue–violet gallium nitride diode lasers of even shorter wavelength (405 nm) are used in the Blu-ray Disc[TM], shown in Fig. 2.7. (Blu-ray Disc is a trademark of Sony Corporation, Japan.) The shorter wavelength allows the length of the pits to be reduced to 150 nm since the beam can be focused to a smaller spot. An increase in the numerical aperture of the focusing lens from 0.6 for the DVD to 0.85 for the Blu-ray Disc also contributes to a reduction in track pitch to 320 nm.

SUMMARY AND CONCLUSIONS

Thermodynamic and kinetic mechanisms of structural change are the basis of many applications of lasers, particularly when microfabrication is involved. Providing that the underlying mechanisms of processing are understood, relatively simple thermodynamic and kinetic process models may be constructed. When calibrated to a limited amount of experimental data, modelling can be used to predict operating conditions that lie outside current regimes. The models may also be used to construct laser processing diagrams for structural change, which aid in the selection of processing parameters, and are a useful tool for understanding the effects of changes in process variables on the results of processing. The mechanisms of the more novel processes are not yet understood as well as the processes described in the chapters that follow, but progress in such areas will enable the power of process modelling to be applied in understanding and developing new procedures.

Perhaps the most notable conclusion to be drawn from this chapter is that the use of lasers has enabled the fundamentals of nanoscience to be developed into practical applications of nanotechnology in a surprising number of industrial processes. As laser technology progresses, the scope for its use in applications ranging from biomedical procedures to fuel cells will increase. But of even greater importance, opportunities for novel methods of material processing will become apparent. Few involved in the application of infrared lasers for transformation hardening of steel components in the 1960s would have imagined that the same principle of phase transformation would be used a few decades later to create a rewritable compact disc capable of storing the contents of an encyclopedia.

FURTHER READING

LASER-ASSISTED VAPOUR DEPOSITION

Allen, S.D. (1981). Laser chemical vapor deposition: a technique for selective area deposition. *Journal of Applied Physics*, **52**, (11), 6501–6505.

Andrews, D.L. (1997). *Lasers in Chemistry*. 3rd ed. Berlin: Springer.

Bäuerle, D. (1986). *Chemical Processing with Lasers*. Berlin: Springer.

Chrisey, D.B. and Hubler, G.K. eds (1994). *Pulsed Laser Deposition of Thin Films*. New York: John Wiley & Sons.

Conde, O. and Silvestre, A.J. (2004). Laser-assisted deposition of thin films from photoexcited vapour phases. *Applied Physics A*, **79**, 489–497.

Evans, D.K. ed. (1989). *Laser Applications in Physical Chemistry*. New York: Marcel Dekker.

Jackson, T.J. and Palmer, S.B. (1994). Oxide superconductor and magnetic metal thin film deposition by pulsed laser ablation: a review. *Journal of Physics D: Applied Physics*, **27**, (8), 1581–1594.

Mazumder, J. and Kar, A. (1995). *Theory and Application of Laser Chemical Vapor Deposition*. New York: Plenum Press.

Skouby, D.C. and Jensen, K.F. (1988). Modeling of pyrolytic laser-assisted chemical vapor deposition: mass transfer and kinetic effects influencing the shape of the deposit. *Journal of Applied Physics*, **63**, (1), 198–206.

MEDICAL PROCEDURES

Alster, T.S. (2000). *Manual of Cutaneous Laser Techniques*. Philadelphia: Lippincott Williams and Wilkins.

Berlien, H.-P. and Müller, G.J. eds (2002). *Applied Laser Medicine*. New York: Springer.

Berns, M.W. and Nelson, J.S. (1988). Laser applications in biomedicine. *Journal of Laser Applications*, **1**, (Fall), 34–39.

Carniol, P.J. ed. (2001). *Facial Rejuvenation: From Chemical Peels to Laser Resurfacing*. New York: John Wiley & Sons.

Carniol, P.J. ed. (2000). *Laser Skin Rejuvenation*. Philadelphia: Lippincott Williams and Wilkins.

Catone, G.A. and Alling, C.C. (1997). *Laser Applications in Oral and Maxillofacial Surgery*. Philadelphia: Saunders.

Fitzpatrick, R.E. and Goldman, M.P. (2000). *Cosmetic Laser Surgery*. Philadelphia: Mosby.

Gerber, B.E., Knight, M. and Siebert, W.E. eds (2001). *Lasers in the Musculoskeletal System*. Berlin: Springer.

Goldberg, D.J. (2000). *Laser Hair Removal*. London: Martin Dunitz.

Karu, T. ed. (1998). *The Science of Low Power Laser Therapy*. London: Martin Dunitz.

Lanigan, S.W. (2000). *Lasers in Dermatology*. London: Springer.

Miller, P.D., Eardley, I. and Kaplan, S.A. eds (2001). *Benign Prostatic Hyperplasia: Laser and Heat Therapies*. London: Martin Dunitz.

Niemz, M.H. (2002). *Laser–Tissue Interactions: Fundamentals and Applications*. 2nd ed. Berlin: Springer.

Ohshiro, T. (1996). *The Role of the Laser in Dermatology*. New York: John Wiley & Sons.

Puliafito, C.A. (1995). *Laser Surgery and Medicine: Principles and Practice*. New York: Wiley-Liss.

NANOFABRICATION

Feynman, R. (1961). There's plenty of room at the bottom. In: Gilbert, H.D. ed. *Miniaturization*. New York: Reinhold. pp. 282–296.

Köhler, M. and Fritzsche, W. (2004). *Nanotechnology: An Introduction to Nanostructuring Techniques*. New York: John Wiley & Sons.

OPTICAL DISC TECHNOLOGY

Bouwhuis, G., Braat, J., Huijser, A., Pasman, J., van Rosmalen, G. and Schouhamer-Immink, K. (1985). *Principles of Optical Disc Systems*. Bristol: Adam Hilger.

RAPID PROTOTYPING

Jacobs, P.F. (1995). *Stereolithography and Other RP&M Technologies: From Rapid Prototyping to Rapid Tooling*. Dearborn: Society of Manufacturing Engineers.

Kai, C.C. and Leong, K.F. (1998). *Rapid Prototyping: Principles and Applications in Manufacturing*. New York: John Wiley & Sons.

Proceedings of the European Conferences on Rapid Prototyping, held annually since 1992.

Proceedings of the International Conferences on Rapid Prototyping, held annually since 1989.

SELECTIVE LASER SINTERING

Matsumoto, M., Shiomi, M., Osakada, K. and Abe, F. (2002). Finite element analysis of single layer forming on metallic powder bed in rapid prototyping by selective laser processing. *International Journal of Machine Tools and Manufacture*, **42**, 61–67.

EXERCISES

1. It is observed that carbon from a surface coating of graphite diffuses to a characteristic depth of 0.17 mm when heated by a laser beam of incident power 1098 W, traversing at a rate 0.036 m min^{-1}. Calculate the characteristic depth of carburizing produced in the same material with the same laser beam operating with a power of 528 W and traversing at a rate of 0.05 m min^{-1}.

2. A prior austenite grain diameter of 41 μm is measured adjacent to the fusion line of a through-thickness laser weld made in a steel of thickness 1 mm using an incident power of 1 kW and a traverse rate of 1.2 m min^{-1}. Calculate the minimum traverse rate of a similar laser beam of power 6 kW in a plate of the same material, but of thickness 6 mm, if a grain diameter of 60 μm is not to be exceeded adjacent to the fusion line. Assume that the initial grain size during grain growth is zero.

3. Which lasers are most suitable for chemical and physical vapour deposition and why?

4. Some thermodynamic processes of structural change are based on melting and vaporization. It is observed that the values of the quantities L_m/T_m and L_v/T_v are approximately constant for metals and alloys, the latter being about ten times that of the former, where L and T are the latent heat and transition temperature, respectively, with m and v referring to melting and vaporization, respectively. Why would you expect the quantities to be constant? Suggest an origin for the difference of a factor of 10. Would you expect these relationships to hold for non-metals, and why (not)?

5. Data are stored on compact discs by laser writing patterns of 'pits' and 'lands' along a continuous spiral path. What feature(s) of the laser would you consider modifying to enable greater amounts of data to be stored?

BIBLIOGRAPHY

Avrami, M. (1939). Kinetics of phase change I. *Journal of Chemical Physics*, 7, 1103–1112.

Baleva, M. and Tuncheva, V. (1994). Laser-assisted deposition of PbO–films. *Journal of Materials Science Letters*, 13, 3–5.

Ban, V.S. and Kramer, D.A. (1970). Thin films of semiconductors and dielectrics produced by laser evaporation. *Journal of Materials Science*, 5, 978–982.

Bouwhuis, G., Braat, J., Huijser, A., Pasman, J., van Rosmalen, G. and Schouhamer-Immink, K. (1985). *Principles of Optical Disc Systems*. Bristol: Adam Hilger.

Brown, A.M. and Ashby, M.F. (1980). Correlations for diffusion constants. *Acta Metallurgica*, 28, 1085-1011.

Canova, P. and Ramous, E. (1986). Carburization of iron surface induced by laser heating. *Journal of Materials Science*, 21, 2143–2146.

Celler, G.K., Poate, J.M. and Kimerling, L.C. (1978). Spatially controlled crystal regrowth of ion-implanted silicon by laser irradiation. *Applied Physics Letters*, 32, (8), 464–466.

Das, S., Fuesting, T., Brown, L., Harlan, N., Lee, G., Beaman, J.J., Bourell, D.L., Barlow, J.W. and Sargent, K. (1998a). Direct SLS processing of

cermet composite sealing components. *Materials and Manufacturing Processes*, 13, (2), 241–261.

Das, S., Wohlert, M., Beaman, J.J. and Bourell, D.L. (1998b) Producing metal parts with selective laser sintering/hot isostatic pressing. *Journal of Materials*, 50, (12), 17–20.

Duty, C., Jean, D. and Lackey, W.J. (1999). Laser CVD rapid prototyping. In: *Proceedings of the American Ceramic Society Meeting*, January 1999, Cocoa Beach, FL, USA. pp. 347–354.

Fork, D.K., Boyce, J.B., Ponce, F.A., Johnson, R.I., Anderson, G.B., Connell, G.A.N., Eom, C.B. and Geballe, T.H. (1988). Preparation of oriented Bi–Ca–Sr–Cu–O thin films using pulsed laser deposition. *Applied Physics Letters*, 53, (4), 337–339.

Fukumi, K., Sakaguchi, T., Mori, H. and Sakata, T. (1994). Effect of laser induced crystallization of semiconductor particle CdSx7Se1-x in glass matrix on reflectivity of phase-conjugated signal. *Journal of Materials Science Letters*, 13, 1727-1728.

Gangi, A., Dietmann, J.L., Clavert, J.M., Dodelin, A., Mortazavi, R., Durckel, J. and Roy, C. (1998). Laser photocoagulation for osteoid osteoma treatment. *Revue de Chirugie Orthopédique*, 84, 676–684.

Huang, C.M., Xu, Y., Xiong, F., Zangvil, A. and Kriven, W.M. (1995). Laser ablated coatings on ceramic fibers for ceramic matrix composites. *Materials Science and Engineering*, **A191**, 249–256.

Ion, J.C., Easterling, K.E. and Ashby, M.F. (1984). A second report on diagrams of microstructure and hardness for heat-affected zones in welds. *Acta Metallurgica*, **32**, (11), 1949–1962.

Ion, J.C. (1985). *Modelling the microstructural changes in steels due to fusion welding*. PhD thesis, Luleå University of Technology, Sweden. ISSN 0348-8373.

Jacobsohn, E., Zahavi, J., Rosen, A. and Nadiv, S. (1991). Laser-induced deposition of thin chromium oxide films. *Journal of Materials Science*, **26**, 1861–1866.

Johnson, W.A., and Mehl, R.F. (1939). Reaction kinetics in processes of nucleation and growth. *Transactions of the American Institute of Mining and Metallurgical Engineering, Iron Steel Division*, **135**, 416–458.

Kociak, M., Kasumov, A.Yu., Guéron, S., Reulet, B., Khodos, I.I., Gorbatov, Yu.B., Volkov, V.T., Vaccarini, L. and Bouchiat, H. (2001). Superconductivity in ropes of single-walled carbon nanotubes. *Physical Review Letters*, **86**, (11), 2416–2419.

Komatsu, S., Shimizu, Y., Moriyoshi, Y., Okada, K. and Mitomo, M. (2001). Nanoparticles and nanoballoons of amorphous boron coated with crystalline boron nitride. *Applied Physics Letters*, **79**, (2), 188–190.

Kumakura, H., Togano, K., Tachikawa, K., Yamada, Y., Murase, S., Nakamura, E. and Sasaki, M. (1986). Synthesis of Nb_3Ga and Nb_3Al superconducting composites by laser beam irradiation. *Applied Physics Letters*, **48**, (9), 601–603.

Lifshitz, I.M. and Slyozov, V.V. (1961). The kinetics of precipitation from supersaturated solid solutions. *Journal of the Physics and Chemistry of Solids*, **19**, 35–50.

Lizotte, T. and O'Keeffe, T. (1998). Excimer surface treatment behind stain-resistant carpet. *Industrial Laser Review*, **13**, (7), 9–10.

Minamida, K., Kido, M., Ishibashi, A., Mogami, S. and Sasaki, S. (1990). Surface annealing of steel wires for automotive tires by CO_2 laser with cone shaped focusing mirror. In: Ream, S.L., Dausinger, F. and Fujioka, T. eds *Proceedings of the Laser Materials Processing Conference (ICALEO '90)*. Orlando: Laser Institute of America/Bellingham: SPIE. pp. 460–468.

Molian, P.A. and Waschek, A. (1993). Laser physico-chemical vapour deposition of cubic boron nitride thin films. *Journal of Materials Science*, **28**, 1733–1737.

Molian, P.A. (1994). CO_2 laser deposition of diamond thin films on electronic materials. *Journal of Materials Science*, **29**, 5646–5656.

Molian, P.A., Janvrin, B. and Molian, A.M. (1995). Synthesis of fluorinated diamond films using a laser technique. *Journal of Materials Science*, **30**, 4751–4760.

Mukaida, M., Osato, K., Watanabe, A., Imai, Y., Kameyama, T. and Fukuda, K. (1993). Densification of Ta_2O_5 film prepared by KrF excimer laser CVD. *Thin Solid Films*, **232**, 180–184.

Mulato, M., Zanatta, A.R., Toet, D. and Chambouleyron, I.E. (2002). Optical diffraction gratings produced by laser interference structuring of amorphous germanium–nitrogen alloys. *Applied Physics Letters*, **81**, (15), 2731–2733.

Nosaka, Y., Tanaka, K., Fujii, N. and Igarashi, R. (1994). Preparation of ZnS thin films from solution mist by laser irradiation. *Journal of Materials Science*, **29**, 376–379.

Ogando, J. (1994). Rapid prototyping moves toward rapid tooling. *Plastics Technology*, (January), 40–44.

Ramos, J. (2001). Laser-based freeform fabrication. *Industrial Laser Solutions*, **16**, (8), 9–10.

Roos, J.R., Celis, J.P. and van Vooren, W. (1986). Combined use of laser irradiation and electroplating. In: Draper, C.W. and Mazzoldi, P. eds *Proceedings of the NATO ASI Laser Surface Treatment of Metals*, 2–13 September 1985, San Miniato, Italy. Dordrecht: Martinus Nijhoff. pp. 577–590.

Silvestre, A.J., Conde, O., Vilar, R. and Jeandin, M. (1994). Structure and morphology of titanium nitride films deposited by laser-induced CVD. *Journal of Materials Science*, **29**, 404–411.

Silvestre, A.J. and Conde, O. (1998). TiN films deposited by laser CVD: a growth kinetics study. *Surface and Coatings Technology*, **101**, 153–159.

Sorescu, M., Knobbe, E.T., Martin, J.J., Barrie, J.D. and Barb, D. (1995). Excimer-laser and electron-beam irradiation effects in iron-doped lithium–niobate. *Journal of Materials Science*, **30**, 5944–5952.

Srivastava, G.P., Tripathi, K.N. and Sehgal, N.K. (1983). Effect of laser irradiation on the electrical properties of polycrystalline p-InSb thin films. *Journal of Applied Physics*, **54**, (10), 6055–6057.

Srivastava, G.P., Tripathi, K.N. and Sehgal, N.K. (1986). Effect of laser irradiation on the electrical properties of amorphous germanium films. *Journal of Materials Science*, **21**, 2972–2976.

Subramanian, R., Denney, P.E., Singh, J. and Otooni, M. (1998). A novel technique for synthesis of silver nanoparticles by laser–liquid interaction. *Journal of Materials Science*, **33**, 3471–3477.

Suda, Y., Utaka, K., Bratescu, M.A., Sakai, Y. and Suzuki, K. (2003). Carbon nanotube formation by ArF excimer laser ablation. In: Fotakis, C., Koinuma, H., Lowndes, D.H. and Stuke, M. eds *Proceedings of the 7th International Conference on Laser Ablation (COLA'03)*, 5–10 October 2003, Hersonissos, Crete, Greece.

Thess, A., Lee, R., Nikolaev, P., Dai, H., Petit, P., Robert, J., Xu, C., Lee, Y.H., Kim, S.G., Rinzler, A.G., Colbert, D.T., Scuseria, G.E., Tománek, D.,

Fischer, J.E. and Smalley, R.E. (1996). Crystalline ropes of metallic carbon nanotubes. *Science*, **273**, 483–487.

Tian, Y., Womack, M., Molian, P., Lo, C.C.H., Anderegg, J.W. and Russell, A.M. (2002). Microstructure and nanomechanical properties of Al–Mg–B–Ti films synthesized by pulsed laser deposition. *Thin Solid Films*, **418**, 129–135.

Tice, E.S. (1977). Laser annealing of copper alloy 510. *IEEE Journal of Quantum Electronics*, **QE-13**, (9), 813–814.

Vitali, G., Rossi, M., Pizzuto, C., Zollo, G. and Kalitzova, M. (1996). Low power pulsed laser annealing of Zn^+-implanted InP: first endeavours. *Materials Science and Engineering*, **B38**, 72–76.

Voutsas, A.T., Limanov, A. and Im, J.S. (2003). Effect of process parameters on the structural characteristics of laterally grown, laser-annealed polycrystalline silicon films. *Journal of Applied Physics*, **94**, (12), 7445–7452.

Wagner, C. (1961). Theory of precipitate ageing during re-solution (Ostwald ripening). *Zeitschrift der Elektrochemie*, **65**, 581–591. (In German.)

Wang, H.M. and Bergmann, H.W. (1995). Annealing laser-melted ductile iron by pulsed Nd:YAG laser radiation. *Materials Science and Engineering*, **A196**, 171–176.

Watanabe, A., Mukaida, M., Imai, Y., Osato, K., Kameyama, T. and Fukuda, K. (1993). Morphology and structure of tantalum oxide deposit prepared by KrF excimer laser CVD. *Journal of Materials Science*, **28**, 5363–5368.

White, C.W., Christie, W.H., Appleton, B.R., Wilson, S.R., Pronko, P.P. and Magee, C.W. (1978). Redistribution of dopants in ion-implanted silicon by pulsed-laser annealing. *Applied Physics Letters*, **33**, (7), 662–664.

Williams, J.S., Brown, W.L., Leamy, H.J., Poate, J.M., Rodgers, J.W., Rousseau, D., Rozgonyi, G.A., Shelnutt, J.A. and Sheng, T.T. (1978). Solid-phase epitaxy of implanted silicon by cw Ar ion laser irradiation. *Applied Physics Letters*, **33**, (6), 542–544.

Williams, M. (2001). *NEC develops fuel cell for handelds*. [Internet]. Available from: <http://edition.cnn.com/2001/TECH/ptech/09/03/nec.pda.fuel.cell.idg/> [Accessed 3 March 2004]

Yang, X.-C. and Riehemann, W. (2001). Characterization of Al_2O_3–ZrO_2 nanocomposite powders prepared by laser ablation. *Scripta Materialia*, **45**, 435–440.

Zhang, J.-Y. and Boyd, I.W. (2000). Pulsed laser deposition of tantalum pentoxide film. *Applied Physics A: Materials Science & Processing*, **70**, (6), 657–661.

Zhao, X.Q., Zheng, F., Liang, Y., Hu, Z.Q., Xu, Y.B. and Zhang, G.B. (1995). Growth of γ'-Fe_4N and γ-Fe ultrafine powders synthesized by laser-induced pyrolysis of mixtures of $Fe(CO)_5$ and NH_3. *Materials Letters*, **23**, (4-5-6), 305–308.

Zong, G., Wu, Y., Tran, N., Lee, I., Bourell, D.L., Beaman, J.J. and Marcus, H.L. (1992). Direct selective laser sintering of high temperature materials. In: Marcus, H.L., Beaman, J.J., Barlow, J.W., Bourell, D.L. and Crawford, R.H. eds *Proceedings of the Conference Solid Freeform Fabrication*, The University of Texas at Austin. pp. 72–85.

CHAPTER 9

SURFACE HARDENING

INTRODUCTION AND SYNOPSIS

Laser transformation surface hardening, or *laser hardening*, is an autogenous method of producing wear-resistant patterns on discrete surface regions of components. A shaped laser beam is scanned across the component to heat, *but not melt*, the surface, as illustrated in Fig. 1.7. A temperature rise of about 1200°C is ideal. The underlying bulk material acts as an efficient heat sink, causing rapid cooling. The hardness, strength, wear, fatigue and lubrication properties of the surface can be improved, while desirable bulk properties, such as toughness and ductility, remain unaffected. Ferrous alloys are particularly suitable for laser hardening because quenching produces hard martensite. (The process may be used for other materials in which a hardening phase change is induced by a thermal cycle.) Until the mid-1990s only CO_2 lasers were able to deliver the power density required for economic rates of hardening large areas. Then multikilowatt Nd:YAG and (a few years later) diode laser sources became available, which produce beams that are absorbed more readily by metal surfaces, resulting in improvements in process efficiency, and which led to an increase in the scope of application. The principal objective of laser hardening is to produce a surface with a required hardness to a prescribed depth with the highest coverage rate possible.

About 50 joules of energy are required to harden one cubic millimetre of material. Most laser hardening is carried out with a beam power density on the order of 10 W mm^{-2} and a beam interaction time of about 1 second. These values typically translate into a beam power of about 1 kW, a beam area of about 100 mm^2, and a traverse rate of about 10 mm s^{-1}. The thermal cycles induced are an order of magnitude shorter than those of conventional hardening processes, which enables both hardenability and hardness to be increased. Hardness values up to about 1000 HV can be achieved to a depth of about 1.5 mm in alloys of high hardenability, before surface melting occurs. A common hardness requirement of wear-resistant surfaces is a value of 55 HRC (610 HV) – this can be achieved by laser hardening a steel containing about 0.25 wt% carbon.

In this chapter, we consider the principles of laser hardening and compare them with those of conventional hardening processes. Features of the process are identified that enable the characteristics of suitable applications for laser hardening to be determined, in preference to competing hardening methods. The properties of laser-hardened materials are presented so that comparisons between materials and processes can be established. The aim of process modelling is to predict hardening depth and surface hardness – this is achieved by analysing temperature fields and phase transformations, and using analytical modelling to construct laser hardening diagrams. Selected industrial applications reveal the philosophy behind the choice of the process, and indicate its potential for developing novel solutions to manufacturing problems involving components subject to wear.

PRINCIPLES

The principles of laser hardening are similar to those of conventional through-hardening, although the time scales involved are typically an order of magnitude shorter. The laser beam is shaped into an appropriate pattern for the required hardened region and scanned over the component. Hardenable ferrous alloys are the most suitable for laser hardening. The temperature at a required depth is raised to fully austenitize the microstructure, without melting the surface. During the peak of the thermal cycle the microstructure is homogenized by solid state diffusion. On cooling, austenite transforms to martensite with a uniform carbon distribution and hardness. An additional benefit is the development of a compressive residual stress state at the surface because of the 4% volume increase associated with the transformation of austenite to martensite.

The laser beam induces rapid thermal cycles in surface regions of the substrate. An equilibrium iron–carbon phase diagram *may* be used to estimate critical transformation temperatures on heating. However, during laser heating phase transformations take place under conditions far from equilibrium. Heating rates on the order $1000 \, \mathrm{K \, s^{-1}}$ are common, and steep temperature gradients are present, which cause phase transformation temperatures to increase above equilibrium values, discussed in Chapter 5. Such superheating should be considered in accurate modelling work, but equilibrium values can be used for the practical treatments described here.

Diffusion-controlled changes during the peak of the thermal cycle are dependent on the *kinetic effect* of the thermal cycle, which was in Chapter 8, and which is significantly smaller in laser hardening than conventional hardening. Homogenization of austenite is controlled by the diffusion of carbon from dissolved pearlite colonies in hypoeutectoid steels, or the diffusion of metallic elements from dissolving carbides in alloyed steels. Inhomogeneous austenite, undissolved carbides (particularly those containing metallic elements), and untransformed ferrite are present in regions of the substrate that experience a small kinetic effect. The extent of austenite grain growth is also reduced as a result of a small kinetic effect. Localized peaks in concentration can be created, which may lead to local phase transformations, or incipient melting.

Phase transformations on cooling may be established from an appropriate continuous cooling transformation (CCT) diagram, as described in Chapter 5. Austenite can transform into ferrite, pearlite, bainite or martensite, depending on the severity of the quench. The composition determines a number of factors in the process: the hardenability (the presence of austenite-stabilizing elements increases the likelihood of the formation of martensite on cooling) the hardness of martensite (a high carbon content increases the hardness); and the presence of retained austenite (excessive amounts of austenite-stabilizing elements increase the amount of retained austenite at room temperature). Since grain growth is limited during laser hardening, a CCT diagram that has been constructed for a small prior austenite grain size should be used to predict the microstructure.

PROCESS SELECTION

Laser hardening competes with many established methods of surface treatment. Engineers and managers are often more familiar with a large number of competing techniques, such as induction hardening, flame hardening and gas carburizing, for which technical and economic data are readily available. Methods are therefore needed to assist in making informed decisions concerning process selection.

CHARACTERISTICS OF LASER HARDENING

The most noticeable characteristic of laser hardening is its low energy input. Only discrete surface regions are heated – the bulk of the component remains cold – enabling hardening to occur through self-quenching; no external quenching medium is required. This means that treatments may be carried out in sensitive environments, without the need to dispose of waste products. Discrete hardened regions up to about 1.5 mm in depth can be produced before the surface melts. The hardened region is constrained by untransformed material, which creates a compressive residual stress state at the surface; the resistance to fatigue crack initiation is therefore improved, endowing good wear properties. Low energy input results in low distortion and minimal surface disruption – this is particularly advantageous when thin sections are hardened. Post-treatment machining operations, which may in some cases account for about 30% of production costs, can be reduced or eliminated. In many cases, treatments may therefore be carried out on finished components. The alloying additions in many conventionally hardenable steels were chosen to achieve deep hardening through high energy input treatment. The thermal cycles induced during laser hardening are typically an order of magnitude shorter, which means that it is often possible to reduce the level of expensive alloying additions to a steel, while maintaining hardenability. In addition, highly alloyed steels, which may not be suitable for other hardening processes such as gas carburizing, may be laser hardened. The rapid thermal cycle also limits grain growth, giving a tougher, harder surface.

The shape and location of the heating pattern can be controlled accurately. Discrete regions of surfaces can therefore be treated, which reduces the total energy input, thus reducing distortion and the need for post-treatment finishing operations. Novel hardened patterns can also be produced because of the flexibility of laser beam processing; they may be designed for specific applications, e.g. to reduce fatigue crack initiation under a particular loading geometry, or to retain softer regions within a hardened zone, where lubricants can accumulate. The accurate control of power delivery during processing avoids the build-up of heat at external corners and edges, which would otherwise melt. The ability to guide the beam accurately eliminates the need for masking operations. A variety of component geometries can be treated because only software changes to the treatment programme are needed to produce a wide range of hardened zone shapes.

A beam of light can be guided and manipulated quickly and in a flexible manner. Beam delivery is amenable to automation, which means that laser hardening can be incorporated into an existing production line relatively easily. The beam can be split to serve a number of workstations simultaneously or in sequence, or switched between applications. The beam-on time can therefore be maximized, leading to an increase in productivity and improved processing economics. The beam can also be directed into inaccessible locations, by using either mirrors or fibreoptic delivery. New component designs may then be considered, since the hardening process can be designed around the functionality of the component, rather than being limited by the component geometry. The working depth of laser hardening is large, and processing can be carried out in air – there is no limit on the size of components that can be treated. The beam can be manipulated around a large component, or alternatively a small component can be moved relative to a stationary beam. The most economical combination of laser and workstation can then be designed (Guidelines can be found in Chapter 4). Tooling is simple, and the lack of physical contact between the tool and workpiece eliminates tool wear. Downtime for maintenance is thus minimized.

The capital cost of a laser system is orders of magnitude greater than most conventional techniques. An economic application therefore requires a combination of favourable factors such as high productivity, lower product cost, improved quality, greater production flexibility, high added value, or novel application. The information in this chapter will help in assessing the suitability of laser processing for a particular task.

The characteristics described above can be quantified. Others factors are qualitative. The high technology image of laser processing may be exploited as a powerful marketing tool, for example. The value of such features may not be immediately obvious, but should not be underestimated.

COMPETING METHODS OF SURFACE HARDENING

The energy required for heating during hardening processes may be produced by a variety of sources. The characteristics of competing methods are defined by the power density that can be produced, and the total energy input. In making comparisons, it is important to consider the hardening process in its entirety: the cost of the energy source; the cost of the component materials; the precision required; the area to be hardened; the cost of post-hardening finishing operations; and the environmental cost of disposing of quenchants. Novel laser hardening techniques, which enable new materials and designs to be used, often provide both technical and economic benefits that might not appear obvious if the process is approached from the viewpoint of conventional methods of surface hardening.

Induction Hardening

Induction hardening is based on the generation of magnetic fields by an alternating current. When a metal component is placed inside the coil, eddy currents are induced in surface regions of the object, which in turn generate heat because of the resistance of the metal. The depth of heating depends on the strength of the magnetic field, the magnetic properties of the material, and the proximity of the coils to the object. Cooling may be carried out in air or by quenching in oil or water if a deep hardened zone is required. Hardened regions up to about 5 mm can be produced without surface melting, with relatively high coverage rates. A comparison of the hardened regions produced by the two processes is shown in Fig. 9.1; the shallower, darker regions were produced by laser hardening. Induction hardening is easily incorporated into a production environment, has a relatively low cost, and is particularly suitable for rotationally symmetric components such as gear wheels and shafts, especially when high volumes of identical components are to be treated. Codes of practice have been written that specify the required hardness and geometry of the treated region, enabling the process to be automated by using preset values.

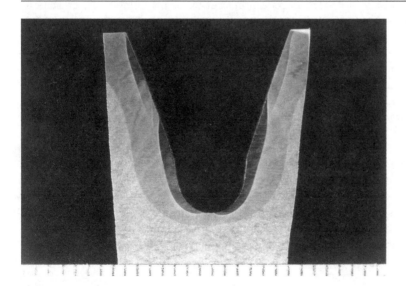

Figure 9.1   Comparison of laser hardening (darker regions) and induction hardening (lighter regions). Scale indicates mm

Laser hardening becomes competitive when the variety of component geometries is large, or when components are of complex geometry, since the beam pattern can be changed easily and locations that are difficult to access can be hardened. Shallow cases with low distortion can be produced more easily and with greater control. Steels that may have insufficient hardenability for the thermal cycles induced during induction hardening may often be laser hardened because of the more rapid cooling rate produced. Codes of practice, which address the higher hardness that can be obtained, as well as the greater control of a laser-hardened geometry, are needed if laser hardening is to become more competitive in applications currently dominated by induction hardening.

Flame Hardening

Flame hardening is a well-known, flexible process with a very low capital cost. The flame from a welding torch, often oxyacetylene, is used to austenitize the component surface, which is then quenched in water or oil. The process can be carried out in manual, semi-automatic or fully automatic modes. In addition to the steel grades recommended for induction hardening, flame hardening can also be used for partial hardening of alloy and tool steels. Gear wheels, lathe guides, plates, blanking tools, moulds and camshafts have traditionally been hardened using this method. The energy input is relatively high, which can lead to the need for an external quenchant to generate the cooling rate required for hardening, and distortion of the large heated volume is often significant. It can also be difficult to control the geometry of the hardened zone because of the imprecise nature of the application of energy, which may lead to undesirable surface melting. As with induction hardening, the range of steels that may be hardened using a flame is limited.

Laser hardening is significantly more energy efficient than flame hardening, and competes when high accuracy is required and the cost of post-treatment machining must be minimized. Many of the advantageous characteristics of laser hardening, which make it competitive with induction hardening also apply to flame hardening.

Arc Hardening

A welding arc or plasma arc can be used as the energy source for surface hardening. Again, the capital cost is very low, and the technology is simple and flexible. Surfaces can be hardened to depth without surface melting because of the relatively low power density of the energy source.

With regard to the controllability of the process, the choice of materials, and the amount of post-hardening finishing work required, the advantages of laser hardening are similar to those of flame hardening. The greater control of an arc, in comparison with a flame, may offset the higher cost of the former. Similarly, the higher cost of a laser is offset by the increase in precision and productivity that can be gained over both processes.

Arc Lamp Hardening

The light from an arc lamp can be concentrated into a heating pattern suitable for hardening. Large heating patterns can be produced, making the source suitable for large, plane areas. However, complicated surface geometries are difficult to treat because of the sophistication of the optics required. An arc image intensifier, producing infrared light, could prove to be a competitive technique, both economically and technically, to carbon dioxide laser hardening.

However, the cost of laser equipment continues to decrease, and new high power laser sources are becoming available, which can be integrated into automatic work and beam handling equipment. This provides flexibility that arc lamp hardening lacks.

Electron Beam Hardening

By defocusing or scanning an electron beam, a heat source with similar characteristics to a laser beam can be created, and similar characteristics of hardening can be achieved. The beam can be manipulated simply, and no absorptive coating is needed. Since processing is carried out in a vacuum, oxidation during processing is eliminated, which enables a uniform hardened depth to be achieved with repeatability.

However, the size of the vacuum chamber limits the part size that can be treated, and the vacuum pumping cycle reduces the production rate. X-rays are produced, and so protective shielding is needed. Laser hardening competes when large parts need to be hardened quickly.

Gas Carburizing and Nitriding

The heat treatments described above are autogenous, i.e. no additional material is used – the hardness is generated from the composition of the base material. Hardening treatments based on exposing components to gaseous media containing hardening elements can be applied to low carbon, low alloy steels to produce a relatively thick, wear-resistant surface on a softer, tougher core. Carburizing, often referred to as case hardening, is normally carried out on low carbon steels in solid media, salt baths, or gases at temperatures between 825 and 925°C, i.e. when the steel is in the austenite phase. Carbon is transported in the gaseous phase, usually in the form of carbon monoxide, and diffuses into the surface, after which the component is quenched in single- or multiple-step treatments. The principles of nitriding are similar, involving the diffusion of nitrogen at temperatures between 500 and 590°C, i.e. in the ferrite phase region of the steel, when no phase transformation occurs on cooling. The four main methods of nitriding are based on gas, salt bath, plasma and powder media. In the carbonitriding process, both nitrogen and carbon are diffused into the surface at austenitic temperatures. Boriding is the name given to the process of producing boride particles in surface regions by means of boron diffusion in gaseous, liquid or solid media. The process is carried out between temperatures of 800 and 1000°C for several hours, after which the parts are allowed to cool. Very high hardness values may be attained, typically between 1600 and 2000 HV, but the hardened depth is small, typically less than 0.1 mm. Such element diffusion processes are most suitable for parts in which the entire surface requires hardening, and which are designed such that distortion caused by heating is not a problem. The equipment cost is relatively low.

Processes that require the whole component to be heated are normally carried out in batches, involving the transportation and stocking of materials, which can create difficulties in an environment of mass production. In addition, if certain areas are not to be hardened, masking is required, which may be time consuming and expensive. Carburizing is also becoming expensive because of environmental considerations of handling and disposing of waste compounds. A comparison between gas carburizing and laser hardening for hardening of gears has demonstrated the technical (Howes, 1993) and economic (Anon., 1985; Bhattacharyya and Seaman, 1985) advantages of the latter, particularly with large components on which small hardened regions are required.

ECONOMIC ANALYSIS

A simple economic analysis can be used to find the maximum area of a component that can be hardened using a laser in preference to a batch process such as gas carburizing. Consider the outer curved surface of a solid cylindrical shaft of length l and radius r. Assume that the cost of gas carburization is proportional to the weight of the shaft, and that the cost of laser hardening is proportional to the area to be hardened. Let the cost of carburizing be $C_c \cdot W$, where C_c is the cost of carburizing per unit weight, and W is the weight of the component. Similarly, let the cost of laser hardening be $C_l \cdot f \cdot A$, where C_l

is the cost of laser hardening per unit area, A is the total curved surface area of the shaft, and f is the fraction of surface area to be hardened. Equating the costs gives $C_c \cdot W = C_l \cdot f \cdot A$. For a shaft of radius r, length l, and density ρ, for which $W = \pi r^2 l \rho$ and $A = 2\pi rl$, we obtain $C_c \cdot \pi r^2 l \rho = C_l \cdot f \cdot 2\pi rl$. The area fraction of the shaft for which the cost of the two treatments is equal is then

$$f = \rho r \frac{C_c}{C_l}. \tag{9.1}$$

By assuming that laser hardening costs \$2.62 per 100 cm^2, and that gas carburizing costs \$1.10 per kg (Bhattacharyya and Seaman, 1985), equation (9.1) indicates that if less than one third of the surface of a steel shaft (density 7960 kg m^{-3}, radius 10 mm) requires treatment, then laser hardening is less expensive than gas carburizing. A similar analysis may be applied to other geometrical shapes.

PRACTICE

Successful application of laser hardening requires an understanding of the principles of the process and an appreciation of the effects of practical process variables on the properties of the processed material. The principal process variables of laser hardening are shown in Fig. 6.2 (Chapter 6). Here we consider the process variables in three groups: material properties, beam characteristics, and processing parameters.

MATERIAL PROPERTIES

Composition

A common requirement for surface-hardened components is a minimum hardness value of 55 HRC (610 HV) present to a specified depth. The data of Krauss (1978) indicate that hardened carbon–manganese steels with a carbon content above about 0.25% meet this criterion. (The presence of alloying additions that promote hardenability may enable this carbon content limit to be reduced.) When selecting a material, it is also important to consider how composition affects thermal properties. As the thermal conductivity of the alloy increases, the peak surface temperature decreases, and the depth of thermal penetration in a given time increases. Process variables for hardening of cast iron, for example, must therefore be adjusted to take into account its relatively high thermal conductivity and low melting temperature in comparison with structural steels (Table D.1, Appendix D) to maintain hardening to a given depth without the onset of surface melting. Methods for achieving this are described later in the chapter.

Geometry

The geometry of the component influences the distribution of heat flow. As a rule of thumb, the component thickness should be at least ten times the depth of the desired hardened region, to enable self-quenching to occur without significant bulk heating effects; the energy source can then be treated as a true surface source, and the temperature fields modelled accordingly. Edges are potential sources of heat build-up – the beam should be kept a prescribed distance from an edge in order to prevent overheating or melting. This may be achieved by removing the absorptive coating adjacent to the edge (discussed below), or by using a beam with an intensity profile that tails off towards its edge to reduce edge heating. External corners have a large ratio of heated surface area to volume, and so are prone to overheating. The opposite is true of internal corners.

Absorptivity

A polished metal surface reflects between 90 and 98% of the incident light of a perpendicular CO_2 laser beam. The absorptivity of the surface must therefore be increased if far infrared light is to be used as the energy source, to improve process efficiency and reduce reflection. This can be achieved in a number of ways: by roughening the surface; by using linearly polarized light; by directing the beam at the Brewster angle (about 70° from the normal); by preheating the substrate; or by applying an absorptive surface coating. An aerosol spray coating of colloidal graphite is the most popular means of increasing absorptivity because it can be applied and removed easily with high repeatability. A uniform layer 10–20 μm in thickness is optimum, enabling the absorptivity to be increased to over 80%. If the coating is too thin, it will burn off quickly, exposing the substrate, thereby dramatically reducing the absorptivity. If the coating is too thick, vaporization of volatile elements in the spray can affect processing through the formation of plumes. A shroud of inert gas helps to stabilize the surface condition, maintaining the absorptivity at a constant value.

BEAM CHARACTERISTICS

Wavelength

For many years, only CO_2 laser beams were able to deliver the combination of power density and interaction time necessary for transformation hardening. Multikilowatt Nd:YAG and diode lasers, with shorter wavelengths, now provide significant advantages; as the beam wavelength decreases, the absorptivity of a metal surface increases (Chapter 5), and so an absorptive coating might not be necessary. This simplifies the operation considerably – the cost of applying and removing absorptive coatings can be the factor that makes CO_2 laser hardening uneconomical in comparison with other methods of surface hardening.

Power

The power level used normally lies in the range 1–3 kW. A high power level enables high traverse rates to be used, with correspondingly high coverage rates. However, the practical window of traverse rate is then reduced because the risk of both overheating, leading to surface melting, or insufficient peak temperature with no hardening, increases. The robustness of the process is thus reduced. For these reasons an incident power of about 1 kW is normally recommended. Materials of high hardenability may be processed with a lower power density and a higher interaction time, in order to achieve a homogeneous case with significant depth. Conversely, materials with low hardenability are processed with higher power density and lower interaction times in order to generate the rapid cooling rates required for martensite formation, at the expense of a shallower case.

Power Density

A heating pattern may be created by simply defocusing the beam. This is a satisfactory solution for many engineering applications. The sectional depth profile of the hardened region can be approximated as the mirror image of the beam intensity distribution, with a reduced amplitude and some rounding of the edges that result from lateral heat flow. Thus a beam with a Gaussian power distribution typically produces a hemispherical hardened section, and a uniform power distribution produces a rectangular hardened section, with rounded edges.

By using beam-shaping optics (Chapter 4) the geometry of the hardened sections can be varied, and it is possible to harden with higher coverage rates. The beam aspect ratio, i.e. the ratio of width to length, may be varied to produce the desired thermal cycle and section geometry – a high ratio gives a rapid thermal cycle and a wide hardened track, and vice versa. If a uniform depth profile of constant width is needed, a kaleidoscope is the cheapest solution. If a high power density is needed, an integrating mirror absorbs less power, but is more expensive. Diffractive optics can also be used to transform the raw beam into a suitable heating pattern, although these are also relatively expensive. Complex or variable geometries may require the use of a beam rastering system, which allows the shape and power distribution to be varied, but again is expensive, and is normally used with relatively low power levels. A toric mirror and an annular beam can be used for hardening both the inner and outer surfaces of rotationally symmetric components, such as shafts and cylinders. A single beam may be split into several components for producing spot hardened patterns. Lenses have been used in such a manner to defocus and manipulate Nd:YAG laser light from a fibre.

A uniform distribution of power in the beam profile maximizes case depth, the uniformity of depth, and coverage rate. In contrast, ring and Gaussian profiles minimize distortion, although the latter can lead to centreline melting. The optimum distribution of power within the heating pattern consists of a leading edge with a high power density, with the power tailing off towards the trailing edge. This induces a thermal cycle that provides sufficient time for microstructural homogenization, but has a sufficiently high cooling rate for martensite formation. By using such a heating pattern, the energy required to harden unit volume of material is minimized. However, the optics required to produce such a heating pattern are expensive, and so the method is normally reserved for large volume applications.

Beam Interaction Time

A beam interaction time in the range 0.01–3 s is typical of transformation hardening. The length of the beam in the travel direction is fixed by the power density and track width requirements. Hence the required traverse rate can be found by dividing the length by the interaction time. Traverse rate is the variable that is normally changed when fine tuning the process, to achieve the required hardened depth and degree of homogenization.

PROCESSING PARAMETERS

Workstation

If the component is large, and cannot be moved easily, the beam is traversed over the component by using a suitable optic. If the component is small, and can be manipulated under a stationary beam, the sophistication and cost of the optics can be reduced considerably. A hybrid system, in which both the component and the beam are moved, provides the highest degree of flexibility, but with the highest cost. Suitable systems for different components are described in Chapter 4.

When large areas are to be hardened, the normal procedure is to place individual tracks adjacent to one another in sequence. In such cases, lateral heat flow from one track results in tempering of the previous track. The hardened surface thus contains bands of softer tempered material. It is normally desirable to avoid such structures, although in certain applications the softer zones may be useful for retaining lubricant. Higher power lasers enable larger heating patterns to be produced, increasing coverage rates thus avoiding the need for multiple passes and tempering.

Process Gases

Process gas serves two functions in transformation hardening. It shields the interaction zone, thus preventing oxidation, which can increase absorptivity in an undesirable, uncontrollable manner, and which may result in overheating or melting. The process gas also protects the optics from smoke and any other contaminants produced during processing. Argon and nitrogen are common choices since their relatively high density blankets the interaction zone effectively. Gas flow rates of around $20\,L\,min^{-1}$ are normally used, depending on the area to be covered, delivered either coaxial with the beam or from an external nozzle.

Adaptive Control

Most applications of transformation hardening are carried out using a fixed set of processing parameters. However, there may be occasions when changes need to be made to processing variables during the treatment. Such changes may be intentional – the section size of the component may change significantly – or unintentional – the properties of the surface coating may change as a result of processing. Systems for controlling the temperature at the surface of a component have been constructed to adaptively control laser hardening. The surface temperature can be measured using a pyrometer, and used to calculate the transformed depth by means of a mathematical model. The beam power is adjusted during processing until the desired surface temperature is attained. The increase in resistivity of a ferritic steel following transformation to martensite can also be used to provide information about the depth of a hardened track. Adaptive control is essential in automated hardening of complex geometries where heat build-up along edges or in corners could lead to melting. Such on-line control allows product quality to be maximized, and the influence of external factors to be minimized.

PROPERTIES OF SURFACE-HARDENED MATERIALS

Data for laser-hardened steels are given in Table 9.1. The types of steel are described in Chapter 5 and the designations used for individual materials and their properties are explained in Appendix C. The data are provided to show the expected hardness that can be achieved in different steels, and the trends that can be expected. Data from the references may be used to establish the precise process variables and techniques used to treat different steels and components.

CARBON–MANGANESE STEELS

Hypoeutectoid

Consider the microstructure of a hypoeutectoid steel containing 0.35% carbon. It consists of pearlite colonies surrounded by proeutectoid ferrite. On heating, pearlite transforms to austenite by dissolution of the cementite lamellae, followed by growth of the austenite transformation front into regions of high carbon concentration, at a rate controlled by carbon diffusion between the lamellae. Ferrite transforms by nucleation and growth of austenite at internal ferrite grain boundaries, at a rate controlled by carbon diffusion over greater distances associated with the size of the ferrite colonies.

The phase diagram in Fig. 9.2 shows that under equilibrium conditions, pearlite begins to transform to austenite at 723°C (A_{c1}), and that transformation of ferrite is complete at about 800°C (A_{c3}). The high heating rate experienced during laser heating (on the order of $1000\,K\,s^{-1}$) results in superheating of A_{c1} and A_{c3}, typically by about 30 and 100°C, respectively. The increase in A_{c3} is greater than A_{c1} because the diffusion distance for transformation of ferrite is larger than that of pearlite. In practice

Table 9.1 Surface hardness of laser-hardened steels: N/S not specified; *quenched in liquid nitrogen

Type	Grade	C (wt%)	Hardness (HV)	Reference
C–Mn	0.1% C	0.10	390	(Bach *et al.*, 1990)
	JIS S15C	0.18	420	(Hino *et al.*, 1997)
	0.2% C	0.20	450	(Bach *et al.*, 1990)
	JIS S25C	0.26	495	(Hino *et al.*, 1997)
	0.3% C	0.30	525	(Bach *et al.*, 1990)
	JIS S35C	0.36	580	(Hino *et al.*, 1997)
	En8	0.36	680	(Steen and Courtney, 1979)
	AISI 1035	0.39	600	(Peng and Ericsson, 1998)
	JIS S45C	0.44	625	(Hino *et al.*, 1997)
	AISI 1045	0.45	700	(Skrzypek, 1996)
	AISI 1045	0.45	656	(Ion and Anisdahl, 1997)
	AISI 1045	0.45	750	(López *et al.*, 1995)
	C 45	0.46	700	(Bach *et al.*, 1990)
	JIS S55C	0.54	665	(Hino *et al.*, 1997)
	C 60	0.62	875	(Bach *et al.*, 1990)
	N/S	0.95	1000	(Hill *et al.*, 1974)
	N/S	1.00	880	(Li *et al.*, 1986)
Low alloy	49 MnVS 3	0.47	750	(Ion *et al.*, 1992)
Medium alloy	AISI 4130	0.26	530	(Peng and Ericsson, 1998)
	AISI 4130	0.28	880	(Doong *et al.*, 1989)
	X 38 CrMoV 5 1	0.38	750	(Bach *et al.*, 1990)
	AISI 4340	0.40	690	(Gregson, 1988)
	AISI 5135	0.40	910	(Selvan *et al.*, 1999)
	42 CrMo 4	0.40	664	(Ion and Anisdahl, 1997)
	42 CrMo 4	0.43	800	(Bach *et al.*, 1990)
	42 CrMo 4	0.45	700	(Merrien *et al.*, 1992)
	AISI 4140	0.45	750	(Peng and Ericsson, 1998)
	50 NiCr 13	0.51	900	(Bach *et al.*, 1990)
	56 NiCrMoV 7	0.54	950	(Bach *et al.*, 1990)
	50 NiCr 13	0.55	844	(Ion and Anisdahl, 1997)
	90 MnCrV 8	0.90	1000*	(Bach *et al.*, 1990)
	100 Cr 6	0.95	1000*	(Bach *et al.*, 1990)
	X 100 CrMoV 5 1	1.00	750	(Ion and Anisdahl, 1997)
High alloy	X 155 CrVMo 12 1	1.55	850*	(Bach *et al.*, 1990)
	X 210 Cr 12	2.09	950*	(Bach *et al.*, 1990)
	X 210 Cr W 12	2.10	900*	(Bach *et al.*, 1990)
Martensitic stainless	AISI 410	0.12	560	(Com-Nougué and Kerrand, 1985)
	AISI 410	0.12	550	(Camoletto *et al.*, 1991)
	AISI 440	0.61	585	(Ion *et al.*, 1991)
	AISI 410	0.23	560	(Roth and Cantello, 1985)
Tool	AISI A6	0.22	760	(Miller and Wineman, 1977)
	X 30 CrMoN 15 1	0.3	700	(Heitkemper *et al.*, 2003)
	AISI O1	0.45	940	(Ion and Anisdahl, 1997)
	AISI O1	0.45	870	(Miller and Wineman, 1977)
	JIS SK5	0.80	850	(Kawasumi, 1978)
	JIS SK5	0.83	900	(Kikuchi *et al.*, 1981)
	Assab DF-2	N/S	N/S	(Yang *et al.*, 1990)

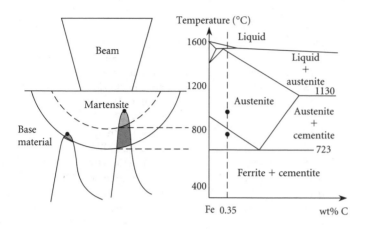

Figure 9.2   Thermal cycles at two positions in a hypoeutectoid steel during laser hardening and their relationship to the iron–carbon phase diagram

this means that a higher temperature must be achieved than predicted by the equilibrium diagram for complete homogenization to be achieved, and so the process variables may be varied accordingly.

On cooling, the microstructure of the transformed region comprises martensite near the surface, proeutectoid ferrite and martensite (containing small undissolved cementite plates) close to the transformation boundary, and an intermediate region of ferrite and carbides in a martensitic matrix, Fig. 9.3. The A_{c1} isotherm is identified by the boundary between transformed and untransformed pearlite.

If the beam traverse rate is too high, the surface peak temperature attained is too low for complete austenitization, and the kinetic effect too low for extensive homogenization.

The hardness of the surface depends primarily on the carbon content of the material. The hardness of martensite increases linearly with carbon content, from about 300 HV at 0.05 wt% C to about 750 HV at 0.5 wt% C (Krauss, 1978). A higher carbon content results in the retention of austenite at room temperature (Troiano and Greninger, 1946), which reduces the hardness. These carbon composition limits are affected by the presence of other alloying additions – both are reduced by the addition of manganese, chromium, nickel and molybdenum (Andrews, 1965).

Hypereutectoid Steels

Laser heating of spheroidized hypereutectoid steels, containing globular cementite in a ferrite matrix, causes austenite to nucleate at the cementite–ferrite interface in the form of a shell around the cementite. The shell expands by cementite dissolution and redistribution of carbon. Carbon-rich austenite subsequently transforms to martensite, although austenite regions of particularly high carbon concentration may be retained as concentric rings at room temperature.

ALLOY STEELS

Hypoeutectoid low alloy steels respond well to laser hardening, Table 9.1. Alloying additions raise the maximum hardness that can be obtained, and increase the depth of the uniformly hardened region, in comparison with a carbon–manganese steel of similar carbon content. Alloying additions influence both the thermodynamics and kinetics of phase transformations. They increase the incubation time for pearlite formation, thus stabilizing austenite, which in turn increases the hardenability and the depth

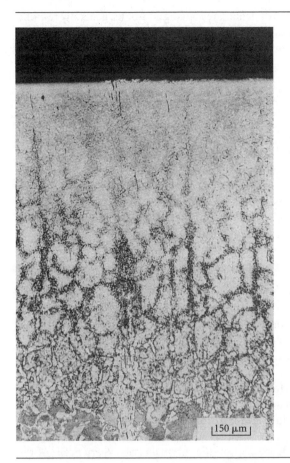

Figure 9.3   Microstructure of a laser-hardened hypoeutectoid steel showing homogeneous martensite (surface), mixtures of martensite, ferrite and carbides (middle), and ferrite and martensite (lower)

of hardening possible. However, they stabilize carbides, by reducing the diffusion rate of carbon in austenite, which can result in a reduction in the carbon available for martensite hardening. (Nickel in particular reduces the rate of carbon diffusion, increasing hardenability.) Carbides pin austenite grain boundaries, limiting austenite grain growth, providing more nucleation sites for pearlite, thus reducing hardenability. Alloying additions affect the phase transformation temperatures, notably A_{c1} and A_{c3}, and the start and finish temperatures for martensite formation, M_s and M_f, respectively. A lower M_s temperature results in a deeper case. The gradient of the hardness profile is increased by any factor that increases hardenability. Retained austenite may be attributed to the segregation of austenite-stabilizing elements.

The effect of alloying additions on thermal conductivity should also be considered – an increase in thermal conductivity increases the case depth but reduces the surface temperature. A low austenitizing temperature results in a small amount of retained austenite, which is beneficial, but also limits the degree of homogenization, and so a compromise peak temperature is required.

Medium alloy steels, with alloying additions between 2 and 10%, may be hardened by laser heating, although retained austenite is frequently observed in the as-hardened microstructure. A steel such as 42 CrMo 4 is popular, in which hardness levels around 700 HV may be obtained. Quenching in liquid nitrogen is one means of transforming any retained austenite to martensite, enabling hardness values up to 1000 HV to be achieved.

Highly alloyed steels, containing more than 10% of deliberate alloying additions, generally respond poorly to transformation hardening because of the amount of austenite retained in the room temperature microstructure. Hardness can be increased by quenching in liquid nitrogen, although the effect is lower than with medium alloy steels.

TOOL STEELS

Transformation hardening of oil-hardening tool steels produces microstructures containing lath martensite, with carbides and retained austenite. A more homogeneous hardened zone is produced in materials with finer quenched and tempered initial microstructures, compared with those in the as-received spheroidized condition. As their name suggests, air hardening tool steels are designed to be hardened without the use of external quenchants. Although such steels can be hardened readily, the hardness of the transformed region decreases. This is a reflection of the effect of the alloying additions on the microstructure produced. Hardness data can be found in Table 9.1.

STAINLESS STEELS

Only martensitic grades of stainless steel contain sufficient carbon to be laser hardened, Table 9.1. The degree of hardening is strongly dependent on the initial condition of the steel, as well as its chromium content. Laser heating of martensitic stainless steels in their annealed delivery condition (a microstructure of ferrite and carbides) causes a significant amount of carbide dissolution. Sufficient carbon is released to form martensite on cooling, although austenite may be retained in regions of particularly high carbon concentration. Despite a lower carbon content, similar levels of hardness can be developed in AISI 410 with an initial microstructure of tempered martensite, because the carbon distribution is finer and homogenized more readily during treatment, resulting in the formation of martensite with little retained austenite.

Corrosion properties, however, are strongly dependent on the thermal cycle experienced. At the centre of a hardened track, the kinetic effect of the thermal cycle may be sufficient to enable enough chromium to diffuse from the dissolved carbides and go into solid solution to provide a limited amount of corrosion protection. At the edges of the tracks, however, insufficient bound chromium is released, and pitting corrosion around undissolved carbides is observed (Ion *et al.*, 1991).

Figure 9.4 illustrates the relationship between the maximum surface hardness and the carbon content for a range of stainless and other steels. A linear relationship up to a carbon content of about 0.5 wt% is observed. The scatter is primarily caused by the effects of alloying elements on the hardenability of the steels. Above this carbon content, austenite is retained in the microstructure, resulting in a plateau in the maximum hardness. In addition, carbon may be bound in alloy precipitates in alloyed steels. In high chromium steels delta ferrite remains in the matrix. Figure 9.4 is a useful means of estimating the carbon content required to achieve a given surface hardness.

CAST IRONS

The different types of cast iron were described in Chapter 5. The amount of pearlite present determines their response to laser hardening – the greater the amount, the higher the hardenability. Variations in composition and thermal conductivity also affect the response of cast irons to hardening: a high thermal conductivity means that the surface temperature remains relatively low, and hardening can be achieved more easily with a lower risk of surface melting. In contrast, a high carbon content lowers the melting temperature, which limits the depth of hardening that can be achieved before surface melting occurs. Hardness data for laser-hardened cast irons can be found in Table 9.2.

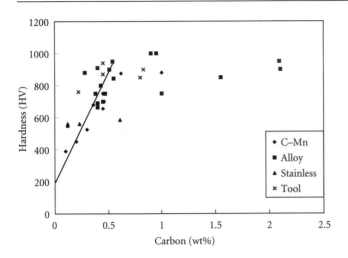

Figure 9.4 Hardness as a function of nominal carbon content for various laser-hardened steels

Table 9.2 Surface hardness of laser-hardened cast irons: N/S not specified

Type	Grade	Hardness (HV)	Reference
Pearlitic grey	ASTM 40	700	(Yessik and Scherer, 1981)
	N/S	715	(Miller and Wineman, 1977)
	N/S	750	(Amende, 1984)
	N/S	800	(Fouquet et al., 1990)
	N/S	900	(Hwang et al., 2002)
	Grade 17	720	(Trafford et al., 1983)
Austenitic grey	DIN GGL 25	530	(Wiesner and Eckstein, 1987)
Pearlitic nodular	DIN GGG 60	950	(Amende, 1984)
	DIN GGG 40	650	(Amende, 1984)
	N/S	950	(Papaphilippou and Jeandin, 1996)
	N/S	680	(Sandven, 1981)
	N/S	620	(Asaka et al., 1987)
Ferritic–pearlitic nodular	ASTM 80-55-06	960	(Mathur and Molian, 1985; Molian and Mathur, 1986)
	JIS FCD 45A	900	(Fukuda et al., 1985)

Grey Irons

Laser heating can be used to harden most types of grey iron. Graphite flakes can be dissolved during typical thermal cycles because of their relatively high ratio of surface area to volume. The thermal conductivity of grey iron, with its interconnecting graphite flakes, is higher than that of nodular iron, which means that a relatively deep hardened case is produced. Heat is transported from the surface to the bulk rapidly, limiting the surface temperature, reducing the likelihood of surface melting. This widens the range of useable processing parameters, providing a greater tolerance to changes during processing. In all cases, the extent of transformation increases with an increase in beam interaction time and an increase in surface peak temperature.

Pearlitic grey iron is relatively easy to harden because the pearlite matrix readily transforms to homogeneous austenite on heating. On cooling, the surface microstructure comprises coarse martensite, retained austenite and undissolved graphite flakes with rims of ledeburite where incipient melting occurred. The amount of retained austenite decreases at greater depths, while the amount of incompletely transformed pearlite increases. Cementite lamellae, martensite, carbides and ferrite are observed at the interface between the transformed zone and the base material. Typical values lie in the range 700–900 HV, Table 9.2. At the surface of Meehanite, dissolution of graphite is not complete. Coarse martensite is observed with retained austenite in regions of high carbon content. Rims of ledeburite are observed at the edge of the flakes where carbon has diffused, causing localized melting.

Ferritic grey iron is not readily hardened by laser heating because the carbon is dispersed in discrete graphite flakes, and the diffusion time for transformation to austenite on heating is limited. Rims of martensite and retained austenite form around the flakes, but no matrix hardening occurs. Austenitic grey iron has limited hardenability, since it is designed such that austenite is stable at room temperature.

Nodular Irons

Graphite is present in nodular iron in the form of discrete spheroids, in a matrix of pearlite, ferrite or a mixture of the two. In contrast to the continuous conducting lamellae found in grey iron, thermal conduction occurs mainly through the matrix in nodular irons. The thermal conductivity of nodular iron is therefore lower than that of grey iron, and so a steeper temperature gradient is created. In addition, since nodules dissolve more slowly than lamellae (because of the smaller ratio of surface area to volume) relatively long treatment times are required to cause sufficient redistribution of carbon to produce transformation hardening. The result is that the surface is easily overheated and melted. The tolerance on processing parameters is therefore narrower than that for hardening grey iron.

The fine matrix of pearlitic nodular iron transforms readily to austenite on heating. Carbon then diffuses from the graphite nodules into the surrounding austenite, creating a carbon concentration gradient. In regions adjacent to the nodule, where the carbon concentration is sufficiently high to lower the melting temperature below the local peak temperature, incipient melting at the rims of the nodules occurs. This is most noticeable on the upper surface of a nodule where the peak temperature is higher. On solidification, such regions transform into ledeburite. Farther from the nodule surface, where the carbon concentration is sufficient to depress the M_f temperature below room temperature, a rim of retained austenite is observed on cooling. In regions of lower carbon content, even farther from the surface, martensite is formed on cooling. Regions of very low carbon content transform to ferrite on cooling. Thus rims of ledeburite, retained austenite, martensite, or a mixture of these are observed around the graphite nodules, as illustrated in Fig. 9.5. The extent of transformation depends on the time available for carbon diffusion, and the temperature at which this occurs, and so increases with an increase in beam interaction time and surface temperature.

Ferritic nodular iron is not readily hardened by laser heating because the carbon content of the matrix is low, and carbon is dispersed in relatively stable nodules. Regions containing very low concentrations of carbon transform to ferrite, rather than martensite, on cooling. Very slow heat treatment, which would not be considered economically viable by using a laser, enables hardness values in the range 400–500 HV to be achieved.

The transformed microstructure of a nodular iron with a matrix of ferrite and pearlite comprises partially dissolved graphite nodules with a thin rim of ledeburite (where incipient melting occurred), surrounded by a rim of martensite.

A bainite matrix, produced by heat treating a pearlitic nodular iron, is a particularly suitable microstructure for laser hardening. Bainite is readily austenitized on heating, and transforms to homogeneous martensite on cooling.

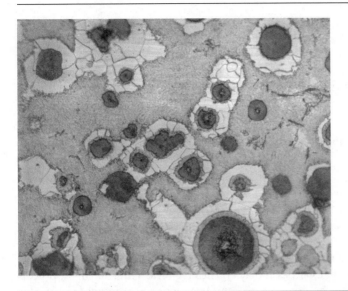

Figure 9.5   Microstructure of laser-hardened region of nodular iron showing graphite nodules (dark), a rim of ledeburite (grey), thicker rims of retained austenite (white) and a martensite matrix (Molian and Mathur, 1986)

Malleable Irons

Malleable irons may also be hardened by laser heating. However, the low amount of widely dispersed carbide limits the hardness that can be achieved without surface melting. Preferential melting occurs around the temper carbon, producing an irregular surface. Provided that processing parameters are chosen to allow sufficient time for diffusion of carbon from temper graphite, a martensitic structure can be produced by laser heating of ferritic malleable iron.

Alloy Irons

Cast irons that are alloyed with elements such as chrome and molybdenum to enhance hardenability respond particularly well to transformation hardening. The hardness profile is characterized by an extended flat plateau, which falls sharply to the base material hardness. The hardened zone is thus defined unambiguously, and compliance with specified hardness requirements can be established easily.

NON-FERROUS MATERIALS

Zirconium Alloys

Alloys of zirconium and niobium are used for pressurized tubing in the power generation industry. Laser heating of Zr–2.5Nb has been observed to produce a martensitic transformation, resulting in an increase in hardness from 275 to 350 HV in a surface layer to a depth of 10 μm (Amouzouvi *et al.*, 1995).

Titanium Alloys

Improvements in the fatigue life of Ti–6Al–4V are achieved by laser heating, based on the reduced fraction of α in the microstructure, and a reduction in grain size (Konstantino and Altus, 1999). The heat affected zone below the laser-melted region of the $(\alpha + \beta)$ titanium alloy Ti–10V–2Fe–3Al transforms to β on heating, which partially transforms to orthorhombic martensite (α'') on cooling (Abboud and West, 1992).

LASER HARDENING DIAGRAMS

The mechanism of laser hardening is based on the well-known principles of heat treatment of ferrous alloys. Thermal cycles are induced that cause solid state phase transformations, which result in the formation of hardening phases. Extensive collections of data, such as Continuous Cooling Diagrams, Jominy hardness curves, and experimental results can therefore be drawn upon when developing process models. The principal difference is that thermal cycles take place within time scales that are often an order of magnitude shorter than conventional heat treatment. The goal of modelling laser hardening is to predict the geometry and properties of the hardened region as a function of the process variables. Analytical models that describe heat flow, phase transformations and property development are used.

Analytical solutions for the temperature fields can be obtained by making a number of justifiable assumptions about the heat source and the material: the laser beam can be considered to be a well-defined surface heat source; the material can be considered to behave as a semi-infinite body, with material properties that do not vary with direction or temperature; and heat flow may be considered to be steady state, i.e. the isotherms surrounding the energy source appear stationary to an observer located at the source. Solutions for temperature fields around moving point, disc and rectangular energy sources yield equations for the thermal cycles in the material, and can be found in Table E2.1 (Appendix E). Expressions for peak temperature, heating and cooling rates, and thermal dwell times can then be obtained, which indicate that heating rates of the order $10^3 \, \mathrm{K \, s^{-1}}$ and thermal cycles of less than one second are common during laser hardening. Analytical solutions yield explicit relationships between process variables and temperature fields, and provide a reliable basis for models of solid state structural transformations. They are used in the treatment that follows.

The normalized process diagram shown in Fig. 6.5 (Chapter 6) indicates that an analytical model-based approach works well when used to predict process variable groups for laser hardening; hardening data fall below the surface melt boundary, although some data appear below the austenitization boundary, for reasons discussed earlier. The model is now extended to describe the temperature field at depth in the material.

Process Overview

The laser beam is required to heat material to the A_{c3} temperature to a given depth, l, to induce austenitization without melting the surface. Following the dimensionless variable group approach outlined in Chapter 6, the peak temperature is normalized with the melting temperature T_m, and the depth with the beam radius r_B:

$$T_p^* = \frac{(T_p - T_0)}{(T_m - T_0)} \quad \text{(dimensionless peak temperature rise)} \tag{9.2}$$

$$l^* = \frac{l}{r_B} \quad \text{(dimensionless depth).} \tag{9.3}$$

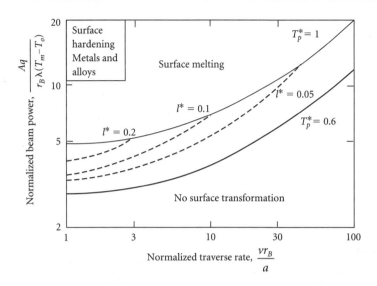

Figure 9.6 Model-based overview of laser surface hardening: the axes are dimensionless groups of processing variables; the operating window is defined by solid boundaries that correspond to normalized peak surface temperatures, T_p^* (T_p/T_m), of 0.6 (the onset of surface transformation) and 1 (the onset of surface melting); and contours for hardening to various normalized depths, l^*, (l/r_B) are shown as broken lines

We assume that some superheating of the A_{c3} temperature occurs, such that a value corresponding to 60% of the melting temperature is required for austenitization (807°C). The theoretical limits for the process are defined by the need to heat the surface to the transformation temperature ($T_p^* = 0.6$), but not above the melting temperature ($T_p^* = 1$). Equations (6.9)–(6.11) (Chapter 6) may be used to obtain the values of q^* and v^* that result in normalized surface temperatures of 0.6 and 1. The surface condition is represented by $z^* = 0$ in equations (6.9) and (6.10). Equations (6.9)–(6.11) may also be used to predict the depth at which a normalized peak temperature of 0.6 is attained; the equations are solved for the normalized depth, l^*, at which $T_p^* = 0.6$ for a given q^* and v^*.

Figure 9.6 shows a normalized processing diagram constructed using this model. Solid boundaries represent the conditions required for the surface to attain the melting temperature ($T_p^* = 1$) and the transformation temperature ($T_p^* = 0.6$), which denote the limits of processing, and define the operating window for the process. Within the operating window, contours representing various values of l^* are shown as broken lines.

Figure 9.6 has many uses. Groups of process parameters can be selected and optimized readily for a range of ferrous alloys. The suitability of novel materials for laser hardening, or other heating processes in which the surface does not reach the melting temperature, can be assessed rapidly. The diagram possesses predictive power in practical situation applications.

Process Parameter Selection

The dimensionless axes and contours of Fig. 9.6 can be transformed into parameters for a given material by substituting thermal properties and selecting a beam size to be used for processing. Incident beam power, traverse rate and hardened depth are the variables of greatest practical interest. For example, the high temperature material properties for a pearlitic grey iron are given in Table D.1 (Appendix D) and are as follows: $\lambda = 75 \, \text{W m}^{-1} \, \text{K}^{-1}$ and $a = 2.1 \times 10^{-5} \, \text{m}^2 \, \text{s}^{-1}$. The melting temperature, T_m, is

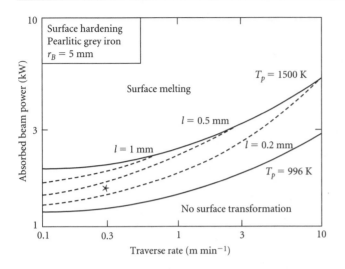

Figure 9.7 Laser hardening diagram for a pearlitic grey iron treated with a beam of width 5 mm: the theoretical operating window is bounded by solid lines, representing the onset of surface melting (1500 K), and surface transformation (996 K); contours of transformed depth ($T_p = 996$ K) are shown as broken lines; and the diagram is calibrated using a data point (an experimental measurement) indicated by a star

1500 K, and processing is carried out at room temperature ($T_0 = 298$ K). We assume that a beam width of 10 mm is used ($r_B = 5$ mm) to maximize the coverage rate. Thus Fig. 9.7 can be constructed by substitution of these values into the dimensionless variable groups of Fig. 9.6.

Recall that in Fig. 6.4 an adjustment was made in the value of power used, for experimental data to be plotted on a theoretical diagram, because only a fraction of the incident power (represented by the absorptivity) is absorbed by the material. The effective absorptivity can be established by using a calibration – a single experimental measurement in which a hardened depth of 0.5 mm (l) is measured after laser hardening with a beam of width 10 mm, a traverse rate of 0.3 m min^{-1}, and an *incident* power of 2.5 kW. When the depth contours of Fig. 9.7 are calibrated to this data point of hardened depth and traverse rate, a value of 1.7 kW can be read from the ordinate. The absorptivity is therefore 0.68 (1.7/2.5) – a value typical of CO_2 laser hardening using an absorptive graphite coating.

Figure 9.7 can be used to establish initial sets of processing parameters for laser hardening of specific materials with given heating patterns. Candidate materials and procedures can be sifted rapidly. The diagram can also be used to obtain an initial set of processing parameters, based on a specified component hardness and hardened depth. It is often the case that the width of the hardened track required is specified, and so the parameter selection diagram is a useful practical tool for estimating the remaining processing parameters. The diagram can be constructed rapidly on a personal computer for a variety of beam sizes and materials by simply changing the input variables.

Processed Material Properties

Figure 9.8 is constructed for a vanadium-microalloyed forging steel containing 0.47% carbon, in the same manner as Fig. 9.7.

Figure 9.8 is also calibrated to an experimentally measured data point. Figure 9.3 shows a section from an experimental trial obtained by CO_2 laser hardening of the steel 49 MnVS 3, by scanning a 1 kW beam of width 7 mm and length 5 mm, at a rate of 0.1 m min^{-1}. A hardened depth of 0.92 mm is

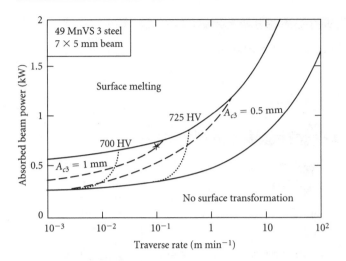

Figure 9.8 Laser hardening diagram for the steel 49 MnVS 3 treated with a beam of width 7 mm and length 5 mm: the theoretical operating window is bounded by solid lines; contours of the A_{c3} transformed depth are shown as broken lines; contours of hardness are shown as dotted lines; and the data point used to calibrate the diagram to a single experimental hardened depth is indicated by a star

observed in Fig. 9.3. When the depth contours of Fig. 9.8 are calibrated to this data point of hardened depth and traverse rate, a value of 0.6 kW is obtained for the absorbed power. The absorptivity in this case is 0.6 (0.6/1).

The hardness contour is obtained by using the analytical methods described in Section 5 of Appendix E (Table E5.1). The phase transformation temperatures are calculated using equations (E5.1)–(E5.6). The carbon equivalent is calculated using equation (E5.7), from which critical cooling times for phase transformations are calculated using equations (E5.8)–(E5.11). The absorbed beam power is fixed and equations (E5.12)–(E5.18) are solved by iterating the traverse rate until a required hardness (e.g. 700 HV) is attained. The procedure is repeated for different values of absorbed beam power to construct the 700 HV contour shown in Fig. 9.8. The value of hardness is then changed (e.g. to 725 HV) and the process repeated to generate the 725 HV contour. Figure 9.7 shows that if a homogeneous hardened depth of 0.5 mm is required with a hardness of 725 HV in this steel, an absorbed beam power of 0.7 kW should be used with a beam traverse rate of 0.3 m min^{-1}, with this beam geometry. Absorbed beam power can be converted to incident beam power by dividing by the absorptivity established from the experimental data point (0.6).

This technique may be used to generate families of diagrams for a given material processed with different beam widths. Alternatively, it can be used to generate families of diagrams for different materials processed with a given beam geometry.

SOFTWARE FOR PROCEDURE DEVELOPMENT

A PC-based system for procedure development in laser hardening (ProLaser) has been described (Ion and Anisdahl, 1997). It is based on the models used to generate the diagrams described above. The software prompts the user for known data. These include the material composition, and a given beam power and geometry, i.e. the variables that are normally kept constant during practical laser hardening. The requirements of the processed material are then entered, for example the depth of hardening and

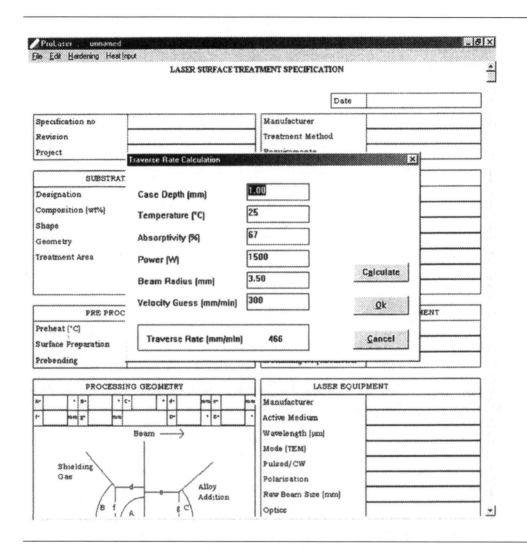

Figure 9.9 ProLaser software for procedure development in laser hardening (Ion and Anisdahl, 1997)

the maximum hardness. The analytical models are then used to establish the unknown parameter – i.e. the traverse rate – that should be used. The process may be repeated to rapidly assess the effects of changes in process variables (material composition, beam characteristics and process parameters) on the properties of the hardened region. Material compositions, procedures such as preheating, and beam geometries may be filtered quickly to establish an initial set of parameters for qualification testing. The system also includes a facility for archiving and retrieving procedures for future use, in a similar manner to welding procedures. The user interface is shown in Fig. 9.9.

INDUSTRIAL APPLICATION

The information in Table 9.3 shows that the automotive and machinery industries have played a dominant role in the industrial application of laser hardening. They have been able to demonstrate both economic and technical benefits in a large number of applications.

Table 9.3   A selection of industrial applications of laser hardening

Industry sector	Component	Material	Reference
Automotive	Axle bearing seat	AISI 1035 steel	(Anon., 1982)
	Blanking die	Tool steel	(Shibata, 1987)
	Camshaft lobes	Cast iron	(Obrzut, 1978; Seaman, 1986)
	Clutch plate rivet	Steel	(Stanford, 1980; Bellis, 1980)
	Conical shaft	Steel	(Sandven, 1981)
	Crankshaft fillet	Steel	(Shibata, 1987)
	Cylinder bore	Cast iron	(Amende, 1989)
	Cylinder liners	Alloy cast iron	(Anon., 1978; Seaman, 1986)
	Engine valve	Alloy steels	(McKeown *et al.*, 1990)
	Gear teeth	Steel	(Creal, 1980a; Sandven, 1981; Gregory, 1996)
	Hand brake ratchet	Low C steel	(Bellis, 1980)
	Motor shaft	Low C steels	(Bellis, 1980)
	Piston ring	Steel	(Creal, 1980b)
	Piston ring groove	Cast iron/steel	(Seaman and Gnanamuthu, 1975; Bransden *et al.*, 1986; Asaka *et al.*, 1987)
	Shaft	Steel	(Stanford, 1980)
	Spacer	Malleable iron	(Seaman and Gnanamuthu, 1975)
	Splined shaft	AISI 1050 steel	(Taylor, 1979)
	Spring component	Spring steel	(Stanford, 1980)
	Steering gear housing	Malleable iron	(Wick, 1976; Miller and Wineman, 1977)
	Valve guide	Grey iron	(Yessik and Schmatz, 1975)
	Valve seat	Steel	(Ready, 1997)
Domestic goods	Typewriter interposer	AISI 1065 steel	(Seaman, 1986)
	ATM ball-rail tracks	N/S	(Morley, 1995)
Machinery	Cutting edge	Steel	(Bellis, 1980; Taylor, 1979)
	Capstan	AISI 1045 steel	(Gregory, 1995)
	Gear teeth	AISI 1060/low alloy steels	(Gutu *et al.*, 1983)
	Mandrel	Martensitic stainless steel	(Gregory, 1996)
	Pipe internal surface	Steel	(Anon., 1982)
	Roll flute crowns	Steel	(Creal, 1980a)
	Tool bed	Cast iron	(Stanford, 1980; Ion *et al.*, 1999)
Power generation	Turbine blade edge	Martensitic stainless steel	(Camoletto *et al.*, 1991; Brenner and Reitzenstein, 1996)
Railway	Diesel engine cylinders	Cast iron	(Strong, 1983; Eberhardt, 1983)

In order to gain some insight into the decisions taken when laser hardening is considered for industrial application, two cases are examined in detail. The first is one of the initial applications to be publicized; which reveals much about the reasoning behind adopting a new manufacturing technology. The second is a modern application, and illustrates the direction in which the process is being developed for greater user-friendliness.

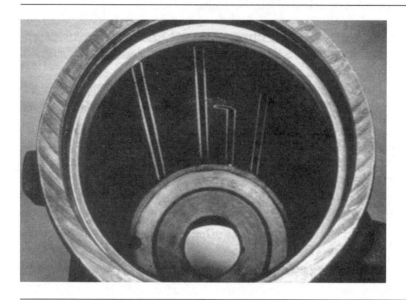

Figure 9.10 Wear-resistant tracks produced on the internal surface of a steering gear housing. (Source: ASM International)

AUTOMOBILE STEERING GEAR HOUSING

Around 1970, it was realized that automobiles of the future would feature air conditioning systems, pollution control devices, radial tyres and energy absorbing bumpers as standard features. This would lead to higher front end loads and increased wear in the power steering gear housings. The housing encases a rack-actuated piston (mounted on a recirculating ballscrew), which reciprocates in the bore. The housing is made from malleable cast iron, with a microstructure comprising graphite nodules in a ferrite matrix, with less than 20% pearlite. The material has good machinability, with a hardness of HV 120–175, good impact properties, and adequate tensile strength. The bore of the housing is 152 mm long, with an internal diameter of 89 mm, and weighs 6.3 kg (Wick, 1976; Miller and Wineman, 1977). A number of potential materials and methods of hardening were considered.

A higher casting hardness could have been obtained by using a cast iron with a higher carbon content, such as a pearlitic malleable or nodular type. This was rejected because of the increased demands placed on machining, which would have resulted in increased tooling costs. Both furnace hardening and induction hardening were rejected because of the excessive distortion the processes would introduce. Surface nitriding was rejected because it required an intermediate annealing operation prior to finish machining, leading to cost and energy debits.

Worn housings were examined to establish the locations of major abrasion. In 1971, tests began into laser hardening using a 500 W Photon Sources CO_2 laser. The TEM_{01*} mode laser beam was focused, directed down the axis of the housing, and turned through 90° using a polished copper turning mirror. A circular heating pattern, of diameter 2 mm, was thus produced on the internal housing surface. By traversing the heating pattern vertically at different rates around 25 mm s^{-1}, hardened tracks were produced, with widths varying between 1.5 and 2.5 mm, and a hardened depth between 0.25 and 0.35 mm. The surface temperature attained, between 628 and 1093°C, combined with the time of beam interaction causes sufficient dissolution of the graphite nodules to release carbon, enabling a surface hardness of 700 HV to be obtained after treatment. A pattern of five axial stripes was produced, spaced between 15 and 25 mm. Four can be seen in Fig. 9.10. The absorptivity of the surface was

Figure 9.11 Diode laser hardening of an alloy steel machine component: power 2 kW, traverse rate 1 m min^{-1}. (Source: Henrikki Pantsar, Lappeenranta University of Technology, Finland)

increased by applying a manganese phosphate coating by immersing the component in a series of baths. Heating caused the coating to change colour – a blue–black centre with a grey border gave a good indication of satisfactory heat treatment.

The benefits of the process stemmed mainly from the reduction in hardening costs and the higher quality of the hardened component. A reduction in wear of 80% was achieved, with almost no distortion, since only about 28 g of the 6.3 kg housing was affected by heat. Since softer castings were used, machinability was also improved. In September 1974, around 30 000 housings were being treated daily.

MACHINE TOOL COMPONENT

Figure 9.11 shows a machine tool component made from 42 CrMo 4 being hardened using the beam from a multikilowatt diode laser. No absorptive coating is necessary since the absorptivity of the diode laser beam to the metal surface, allowing for controlled surface oxidation, is around 60%. The laser is mounted on a robot, and positioned numerically. An adaptive control system measures the surface temperature, which is maintained at 1300°C through a feedback system linked to the beam power, thus ensuring a uniform hardening depth of 0.5 mm with a surface hardness of about 800 HV. Only the region in which wear resistance is required is hardened. Distortion is minimal because of the low energy input. The component is ready for assembly after hardening, with no need for machining.

SUMMARY AND CONCLUSIONS

Laser hardening was one of the first laser-based processes to be industrialized. However, because the technical and economic benefits of the process could not immediately be appreciated, and there were already a large number of familiar hardening processes available, its potential was not realized as quickly as many had anticipated. The principles of laser hardening are similar to those of conventional autogenous hardening processes, the main difference being that laser-induced thermal cycles are an order of magnitude shorter. Transformation temperatures are elevated on heating, and the time available for the microstructure to be homogenized (necessary for a uniformly hardened surface) is limited, but cooling occurs rapidly by self-quenching. Ferrous materials with a fine microstructure and a carbon content between about 0.25 and 0.5 wt% are particularly suitable for laser hardening. Well-defined hardened regions are produced in materials of high hardenability, notably those designed for hardening through austenitizing and quenching in air. Laser hardening is particularly suitable for high volume production of small, discrete hardened regions on large parts, or those of irregular section that would distort during conventional heat treatment. Parts with widely varying shapes and sizes may be treated with a single laser beam, through software control, and parts may be hardened in their final machined condition. Analytical equations describing temperature fields and phase transformations, combined with empirical hardness formulae, provide sufficient accuracy to be used for constructing laser hardening diagrams, which aid in developing preliminary processing procedures.

The main advantages of laser hardening originate from the highly controllable application of low amounts of optical energy, which reduce distortion, eliminate post-treatment finishing, and provide a high degree of processing flexibility. The process can be used as a simple substitute for an existing hardening technique, but the most profitable applications are those that add significant value to a product, or enable it to be made more quickly and cheaply through the use of new materials and component designs. The major drawbacks are the high capital cost of equipment, a continuing lack of familiarity in comparison with established hardening techniques, and the lack of quality assurance codes. These factors are being addressed through the application of more efficient laser sources, flexible optics, CAD and control programs, which can be integrated into turnkey systems, thus increasing the scope of potential application.

FURTHER READING

Amende, W. (1984). Transformation hardening of steel and cast iron with high power lasers. In: Koebner, H. ed. *Industrial Applications of Lasers*. New York: John Wiley & Sons Ltd. pp. 79–98.

Amende, W. (1985). *Hardening of Materials and Machine Components with a High Power Laser*. Düsseldorf: VDI Verlag. (In German.)

Gregson, V.G. (1983). Laser heat treatment. In: Bass, M. ed. *Laser Materials Processing*. Amsterdam: North Holland Publishing Co. pp. 201–233.

Ion, J.C. (2002). Laser transformation hardening. *Surface Engineering*, **18**, 14–31.

Molian, P.A. (1986). Engineering applications and analysis of hardened data for laser heat treated ferrous alloys. *Surface Engineering*, **2**, (1), 19–28.

EXERCISES

1. A stationary circular laser beam of uniform power distribution irradiates the surface of a carbon steel, a polymer and concrete, such that a steady state of heat flow is attained, but melting does not occur. Which material would experience the greatest limiting surface peak temperature rise? Assume that the absorptivity of each material is equal.

2. A finite pulse of energy from a stationary circular laser beam of uniform power distribution irradiates the surface of a carbon steel, polyethylene and concrete, such that transient heat flow occurs, without melting. Which material would experience the greatest depth of thermal penetration? Assume that the absorptivity of each material is equal.

3. The relationship between the hardness obtained in plain carbon and alloy steels and the nominal carbon content is approximately linear up to a carbon content of about 0.5 wt%, above

which it reaches a plateau. Suggest reasons for the limit in hardness that can be achieved by laser hardening.

4. How do changes in the principal processing variables of laser hardening affect: (a) the surface temperature, (b) the depth of hardening and (c) the hardness of laser-hardened low alloy steels? Consider the material properties, beam characteristics and processing parameters.

5. Rank the following materials in order of decreasing suitability for laser hardening: ferritic nodular cast iron, pearlitic grey cast iron, ferritic grey cast iron, medium alloy steel, and pressed automobile body steel. Give reasons for your selection.

6. The following components are to be laser hardened: (a) the linear bearing surfaces of width 15 mm on a machine tool slide of length 2 m; (b) the contacting surfaces of the teeth of a gear wheel of diameter 5 mm; and (c) the internal surface of a pipe in a nuclear reactor. Which types of laser and beam delivery system would be most suitable for each component? Justify your choices.

7. Use Fig. 9.6 to calculate the limits on beam traverse rate when hardening the surface of a medium carbon steel, without causing surface melting, by using a Gaussian beam of power 1 kW and diameter 5 mm. Assume the following material constants: absorptivity $= 0.75$, thermal conductivity $= 30 \, \text{J s}^{-1}\text{m}^{-1}\,\text{K}^{-1}$, melting temperature $= 1800 \, \text{K}$, initial temperature $= 298 \, \text{K}$, thermal diffusivity $= 9.2 \times 10^{-6} \, \text{m}^2 \, \text{s}^{-1}$.

8. A set of spheres with different radii are to be hardened. Calculate the largest sphere radius for which laser hardening is more economical than gas carburizing, given that the cost of laser hardening is \$2.62 per 100 cm^2 and the cost of gas carburizing is \$1.10 per kg. Assume a value of 7764 kg m^{-3} for the density of the spheres. What other factors need to be considered in the choice of hardening process?

BIBLIOGRAPHY

Abboud, J.H. and West, D.R.F. (1992). Laser surface melting of Ti–10V–2Fe–3Al alloy. *Journal of Materials Science Letters*, **11**, 1322–1326.

Amende, W. (1984). Transformation hardening of steel and cast iron with high-power lasers. In: Koebner, K. ed. *Industrial Applications of Lasers*. Chichester: John Wiley & Sons. pp. 79–97.

Amende, W. (1989). Surface treatment of machine tools and automobile components with a laser. *VDI-Z*, **131**, (6), 80–83. (In German.)

Amouzouvi, K.F., Clegg, L.J., Styles, R.C., Mannik, L., Ma, T.-C., Brown, S.K. and Gu, B.-W. (1995). Microstructural changes in laser hardened Zr–2.5Nb alloy. *Scripta Metallurgica et Materialia*, **32**, (2), 289–294.

Andrews, K.W. (1965). Empirical formulae for the calculation of some transformation temperatures. *Journal of the Iron and Steel Institute*, **203**, (7), 721–727.

Anon. (1978). Laser treating of cylinder liners for diesel engines to increase wear resistance. *Industrial Heating*, **45**, (9), 38–39.

Anon. (1982). Industrial lasers offer novel solutions to heat-treatment problems. *Metallurgia*, **49**, (3), 121.

Anon. (1985). Laser vs gas carburising: comparative cost estimates. *Heat Treating*, **27**, (3), 26–27.

Asaka, Y., Kobayashi, H. and Arita, S. (1987). Laser heat treatment of piston ring groove. In: Arata, Y. ed. *Proceedings of the Conference Laser Advanced Materials Processing (LAMP '87)*, 21–23 May 1987, Osaka, Japan. Osaka: High Temperature Society of Japan and Japan Laser Processing Society. pp. 555–560.

Bach, J., Damascheck, R., Geissler, E. and Bergmann, H.W. (1990). Laser transformation hardening of different steels. In: Bergmann, H.W. and Kupfer, R. eds *Proceedings of the 3rd European Conference on Laser Treatment of Materials (ECLAT '90)*. Coburg: Sprechsaal Publishing Group. pp. 265–282.

Bellis, J. (1980). *Lasers: Operation, Equipment, Application and Design*. New York: McGraw-Hill. pp. 124–135.

Bhattacharyya, S. and Seaman, F.D. (1985). Laser heat treatment for gear application. In: Mukherjee, K. and Mazumder, J. eds *Proceedings of the Symposium on Laser Processing of Materials*. Warrendale: The Metallurgical Society of AIME. pp. 211–223.

Bransden, A.S., Gazzard, S.T., Inwood, B.C. and Megaw, J.H.P.C. (1986). Laser hardening of ring grooves in medium speed diesel engine pistons. *Surface Engineering*, **2**, (2), 107–113.

Brenner, B. and Reitzenstein, W. (1996). Laser hardening of turbine blades. *Industrial Laser Review*, **11**, (4), 17–20.

Camoletto, A., Molino, G. and Talentino, S. (1991). Laser hardening of turbine blades. *Materials and Manufacturing Processes*, **6**, (1), 53–65.

Com-Nougué, J. and Kerrand, E. (1985). Laser transformation hardening of chromium steels: correlation between experimental results and heat flow modelling. In: Mazumder, J. ed. *Proceedings of the Materials Processing Symposium (ICALEO '84)*. Toledo: Laser Institute of America. pp. 112–119.

Creal, R. (1980a). Laser specialist discovers boom market in heat treating. *Heat Treating*, **12**, (12), 24–27.

Creal, R. (1980b). Research institute cuts risks in shopping for lasers. *Heat Treating*, **12**, (12), 20–22.

Doong, J.-L., Chen, T.-J. and Tan, Y.-H. (1989). Effect of laser surface hardening on fatigue crack growth rate in AISI-4130 steel. *Engineering Fracture Mechanics*, **33**, (3), 483–491.

Eberhardt, G. (1983). Survey of high power CO_2 industrial laser applications and latest developments. In: Kimmitt, M.F. ed. *Proceedings of the 1st International Conference on Lasers in Manufacturing*. Bedford: IFS Publications. pp. 13–19.

Fouquet, F., Nicot, C., Renaud, L. and Merlin, J. (1990). Laser surface hardening of a pearlitic grey cast iron. In: Mordike, B. and Vannes, A.B. eds *Proceedings of the 6th International Conference on High Power Laser Applications (Laser-6)*. Gournay-sur-Marne: IITT-International. pp. 73–78.

Fukuda, T., Kikuchi, M., Yamanishi, A. and Kiguchi, S. (1985). Laser hardening of spheroidal graphite cast iron minimizes distortion while increasing wear resistance. *Industrial Heating*, **52**, (3), 41–43.

Gregory, R.D. (1995). 'Toughening up' capstan surfaces. *Wire Technology International*, **23**, (6), 67–68.

Gregory, R.D. (1996). Job shop laser heat treating. *Industrial Laser Review*, **11**, (8), 13–15.

Gregson, V.G. (1988). Hardness and hardness vs depth for laser heat treated AISI 4340 steel. In: Ream, S.L. ed. *Proceedings of the 6th International Congress Applications of Lasers and Electro-optics (ICALEO '87)*. Bedford: IFS Publications. pp. 169–177.

Gutu, I., Mihăilescu, I.N., Comaniciu, N., Drăgănescu, V., Denghel, N. and Mehlmann, A. (1983). Heat treatment of gears in oil pumping units reductor. In: Fagan, W.F. ed. *Proceedings of the Conference Industrial Applications of Laser Technology*. Vol. 398. Bellingham: SPIE. pp. 393–397.

Heitkemper, M., Bohne, C., Pyzalla, A. and Fischer, A. (2003). Fatigue and fracture behaviour of a laser surface heat treated martensitic high-nitrogen tool steel. *International Journal of Fatigue*, **25**, 101–106.

Hill, J.W., Lee, M.J. and Spalding, I.J. (1974). Surface treatments by laser. *Optics and Laser Technology*, **6**, (6), 276–278.

Hino, M., Hiramatsu, M., Akiyama, K., Kawasaki, H., Tsujikawa, M. and Kawamoto, M. (1997). Surface hardening of carbon steel using high powered YAG laser. *Materials and Manufacturing Processes*, **12**, (1), 37–46.

Howes, M.A.H. (1993). Laser system for heat treating gears. In: Singh, J. and Copley, S.M. eds *Proceedings of the International Conference Beam Processing of Advanced Materials*. Metals Park: The Minerals, Metals and Materials Society. pp. 229–245.

Hwang, J.-H., Lee, Y.-S., Kim, D.-Y. and Youn, J.-G. (2002). Laser surface hardening of gray cast iron used for piston ring. *Journal of Materials Engineering and Performance*, **11**, (3), 294–300.

Ion, J.C. and Anisdahl, L.M. (1997). A PC-based system for procedure development in laser transformation hardening. *Journal of Materials Processing Technology*, **65**, 261–267.

Ion, J.C., Kauppila, J. and Metsola, P. (1999). Laser-based fabrication: opportunities for novel design and manufacturing. In: Kujanpää, V. and Ion, J.C. eds *Proceedings of the 7th Nordic Conference on Laser Material Processing (NOLAMP '7)*. Lappeenranta: Acta Universitatis Lappeenrantaensis. pp. 38–47.

Ion, J.C., Moisio, T.J.I., Paju, M. and Johansson, J. (1992). Laser transformation hardening of low alloy hypoeutectoid steel. *Materials Science and Technology*, **8**, (9), 799–803.

Ion, J.C., Moisio, T., Pedersen, T.F., Sørensen, B. and Hansson, C.M. (1991). Laser surface modification of a 13.5% Cr, 0.6% C steel. *Journal of Materials Science*, **26**, 43–48.

Kawasumi, H. (1978). Metal surface hardening CO_2 laser. *Technocrat*, **11**, (6), 11–20.

Kikuchi, M., Hisada, H., Kuroda, Y. and Moritsu, K. (1981). The influence of laser heat treatment technique on mechanical properties. In: *Proceedings of the 1st International Laser Conference (ICALEO)*, Anaheim, CA, USA. Toledo: Laser Institute of America.

Konstantino, E. and Altus, E. (1999). Fatigue life enhancement by laser surface treatment. *Surface Engineering*, **15**, (2), 126–128.

Krauss, G. (1978). Martensitic transformation, structure and properties in hardenable steels. In: Doane, J.V. and Kirkaldy, J.S. eds *Proceedings of the Symposium Hardenability Concepts with Application to Steel*. Warrendale: American Institute of Mechanical Engineers. pp. 229–248.

Li, W.-B., Easterling, K.E. and Ashby, M.F. (1986). Laser transformation hardening of steel – II. Hypereutectoid steels. *Acta Metallurgica*, **34**, (8), 1533–1543.

López, V., Fernández, B., Belló, J.M., Ruiz, J. and Zubiri, F. (1995). Influence of previous structure on laser surface hardening of AISI 1045 steel. *ISIJ International*, **35**, (11), 1394–1399.

Mathur, A.K. and Molian, P.A. (1985). Laser heat treatment of cast irons – optimization of process variables: part 1. *Transactions of the ASME*, **107**, (7), 200–207.

McKeown, N., Steen, W.M. and McCartney, D.G. (1990). Laser transformation hardening of engine valve steels. In: Ream, S.L., Dausinger, F. and Fujioka, T. eds *Proceedings of the Laser Materials Processing Conference (ICALEO '90)*. Orlando: Laser Institute of America. pp. 469–479.

Merrien, P., Lieurade, H.P., Theobalt, M., Baudry, G., Puig, T. and Leroy, F. (1992). Fatigue strength of laser beam surface treated structural steels. *Surface Engineering*, **8**, (1), 61–65.

Miller, J.E. and Wineman, J.A. (1977). Laser hardening at Saginaw steering gear. *Metal Progress*, **111**, (5), 38–43.

Molian, P.A. and Mathur, A.K. (1986). Laser heat treatment of cast irons – optimization of process variables part II. *Journal of Engineering Materials Technology*, **108**, 233–239.

Morley, J. (1995). CO_2 lasers go to the bank. *Photonics Spectra*, **29**, (7), 20–21.

Obrzut, J.J. (1978). Heat treaters gear up for the new demands. *Iron Age*, **221**, (28), 39–42.

Papaphilippou, C. and Jeandin, M. (1996). Spot laser hardening. *Journal of Materials Science Letters*, **15**, 1064–1066.

Peng, R.L. and Ericsson, T. (1998). Effect of laser hardening on bending fatigue of several steels. *Scandinavian Journal of Metallurgy*, **27**, 180–190.

Ready, J.F. (1997). *Industrial Applications of Lasers*. 2nd ed. San Diego: Academic Press. pp. 373–380.

Roth, M. and Cantello, M. (1985). Laser hardening of a 12%-Cr steel. In: Kimmitt, M.F. ed. *Proceedings of the 2nd International Conference on Lasers in Manufacturing*. Bedford: IFS Publications Ltd. pp. 119–128.

Sandven, O. (1981). Laser surface transformation hardening. In: *Metals Handbook*. 9th ed. Metals Park: American Society for Metals. pp. 507–517.

Seaman, F.D. (1986). Laser heat-treating. In: Belforte, D. and Levitt, M. eds *The Industrial Laser Annual Handbook*. Tulsa: PennWell Books. pp. 147–157.

Seaman, F.D. and Gnanamuthu, D.S. (1975). Using the industrial laser to surface harden and alloy. *Metal Progress*, **108**, (3), 67–74.

Selvan, J.S., Subramanian, K. and Nath, A.K. (1999). Effect of laser surface hardening on En 18 (AISI 5135) steel. *Journal of Materials Processing Technology*, **91**, 29–36.

Shibata, K. (1987). Laser applications at Nissan. In: *Applications of Laser Processing in Automobile Fabrication and Related Industries*, 3–4 December 1987. Abington: The Welding Institute.

Skrzypek, H. (1996). Effect of laser surface hardening on permeability of hydrogen through 1045 steel. *Surface Engineering*, **12**, (4), 335–339.

Stanford, K. (1980). Lasers in metal surface modification. *Metallurgia*, **47**, (3), 109–116.

Steen, W.M. and Courtney, C. (1979). Surface heat treatment of En 8 steel using a 2 kW continuous-wave CO_2 laser. *Metals Technology*, **6**, (12), 456–462.

Strong, E.J. (1983). How General Motors decided to heat-treat with lasers on the assembly line. *Laser Focus*, **19**, (11), 172–180.

Taylor, J. (1979). Heat treatment hots up with lasers. *Metalworking Production*, **123**, (9), 138–142.

Trafford, D.N.H., Bell, T., Megaw, J.H.P.C and Bransden, A.S. (1983). Laser treatment of grey iron. *Metals Technology*, **10**, 69–77.

Troiano, A.R. and Greninger, A.B. (1946). The martensite transformation. *Metal Progress*, **50**, (8), 303–307.

Wick, C. (1976). Laser hardening. *Manufacturing Engineering*, **76**, (6), 35–37.

Wiesner, P. and Eckstein, M. (1987). Laser hardening of steel and cast iron. *Welding International*, **1**, 986–987.

Yang, L.J., Jana, S. and Tam, S.C. (1990). Laser transformation hardening of tool steel specimens. *Journal of Materials Processing Technology*, **21**, 119–130.

Yessik, M. and Scherer, R.P. (1981). Practical guidelines for laser surface hardening. In: Metzbower, E. ed. *Source Book on Applications of the Laser in Metalworking*. Metals Park: American Society for Metals. pp. 219–226.

Yessik, M. and Schmatz, D.J. (1975). Laser processing at Ford. *Metal Progress*, **107**, (5), 61–66.

CHAPTER 10

DEFORMATION AND FRACTURE

INTRODUCTION AND SYNOPSIS

Deformation and fracture are normally undesirable features of thermal processing with a laser beam; distortion and cracking lead to expensive reworking, or render a component unusable, in an application such as welding. However, a group of processes has been developed in which deformation and fracture are controlled to such a degree that they provide a means of processing materials in a novel manner, which would be difficult or impossible using conventional manufacturing methods. Controlled deformation can be induced in homogeneous or heterogeneous materials by tracking the laser beam over discrete surface areas to create differential thermal expansion of regions at different temperatures; it is particularly suited to materials that have a high coefficient of thermal expansion, or composites in which the components have differing coefficients. Controlled fracture is achieved by inducing a temperature field that creates a stress state in which fracture is initiated; this may be at predetermined locations within a homogeneous crack-sensitive material such as a highly crystalline polymer, or at the interface between materials of differing thermal expansion. Nd:YAG, diode and CO_2 lasers are the most appropriate sources of infrared radiation for heating; the beam can be controlled with the accuracy required to achieve a precise degree of deformation or fracture at discrete locations.

The principles of controlled deformation and fracture are described first in this chapter. Experimental processing data indicate the most popular materials used to date, and highlight the advantages of laser-based processing over conventional methods. An analytical model of deformation is calibrated to experimental data and is used to construct a deformation diagram with predictive capabilities in terms of both processing parameters and materials with potential for treatment. Several industrial processes based on controlled deformation and fracture are described, including forming of sheets and plates, delicate cleaning techniques, internal three-dimensional 'sculptures', and large-scale surface material removal. With the sensitive process monitoring equipment now available, such processes are suited to automation and adaptive control, and have potential for application in many industrial sectors.

PRINCIPLES

DEFORMATION

Deformation can be induced in a controlled manner in sheets and plates by tracking the laser beam across one side of the material. Temperature gradients are developed through the material thickness, which induce stresses because of the differential expansion of adjacent 'layers' that are at different temperatures. Materials such as stainless steels and the light alloys of aluminium, magnesium and titanium have a high coefficient of thermal expansion (Table D.1, Appendix D) – such sheet materials therefore deform significantly when laser heated. The most important beam variables are the energy absorbed per unit length, the configuration of the heating source, and the treatment sequence.

Laser-induced deformation occurs by four principal mechanisms: the temperature gradient mechanism (TGM); the buckling mechanism (BM); the upsetting mechanism (UM); and the surface melting mechanism (SMM). The principal mechanism activated depends on the laser processing parameters, the workpiece geometry, and the material properties.

The TGM is based on the difference between rapid laser surface heating and slow conduction of heat into the bulk. It is the principal deformation mechanism when the beam diameter impinging on the material surface is similar to the material thickness. A steep thermal gradient is created through the material thickness, which results in differential thermal expansion. During the initial stages of heating, expansion of the top surface is greater than that of the bottom surface, which creates a bending moment that develops a small amount of plastic tensile strain at the heated surface. As heating continues, compressive strain is generated in regions in which thermally induced stress exceeds the temperature-dependent flow stress of the material. During cooling, material in upper regions contracts, generating a bending angle, and the sheet is deformed towards the beam. This mechanism is used to deform sheets up to several millimetres in thickness along straight lines. Below a threshold value, the angle of bending is proportional to the number of beam passes.

The BM is invoked when the laser beam diameter is larger than the material thickness (typically by a factor of 10). The temperature gradient through the material thickness is negligibly small. During heating, large compressive stresses are generated in the direction of beam travel, which induce strains that lead to buckling. The whole sheet can be buckled by traversing the beam across its entire length. The part can be bent towards, or away from, the beam. The nature of deformation depends in a complex manner on the processing parameters, but is repeatable. This mechanism can be used to form thin sheets along straight lines.

The UM is a variation of the BM in which the beam diameter is relatively small in comparison with the material thickness. The sheet is heated uniformly throughout its thickness. Thermal expansion is restricted by surrounding material, leading to uniform compression, which reduces the material length while increasing its thickness. The thickness can be increased further by repeating the treatment. Tubular sections can be bent using this mechanism by heating one face.

The SMM has a similar effect to the TGM. The section below the beam first expands, then melts, solidifies and contracts. Repeated tracking induces stresses and strains to create the shape required.

FRACTURE

Thermally induced fracture may also be initiated by a variety of mechanisms. By heating surface regions comprising a coating and substrate with different thermal expansion coefficients, the stress at the interface may cause the coating to separate from the substrate. If locations within a homogeneous bulk material that has some degree of transparency to the laser beam are heated, fracture may be initiated by the differential expansion of discrete regions in which different peak temperatures are attained. The data in Fig. 5.13 (Chapter 5) indicate that materials such as concrete tolerate little thermal expansion before failing mechanically, and so are candidate materials for controlled fracture.

A residual tensile stress is generated on a surface heated by a tracking laser beam, which may be sufficient to initiate fracture at the surface. Under an appropriate state of stress the crack is able to propagate through the material thickness. This mechanism of controlled fracture is the basis of effective through-thickness cracking that follows the path of the laser beam, and is used as a means of parting hard, brittle materials such as domestic ceramics. (Fracture may also be initiated by applying a bending stress after the beam has tracked – this is the scribing mechanism of laser cutting, which is described in Chapter 14.)

PROCESS SELECTION

In contrast to many of the laser-based processes described in the book, which have been integrated into the production line of fabricating industries, laser-induced deformation and fracture is relatively new. There *are* applications in which comparisons can be made with conventional processes, but the most successful laser-based processes are those that present a novel solution to problems for which no conventional method exists. The subject of process selection is therefore considered qualitatively.

DEFORMATION

The most popular conventional method of deforming sheet materials is press-forming. Laser-induced deformation possesses characteristics that provide advantages in niche applications. Expensive stamping and forming dies are not required. Springback is eliminated, and distortion outside the region of deformation is minimal, which means that components can be formed without further processing. Since the method involves no contact, tool wear is eliminated and deformation can be produced in poorly accessible locations with line of sight. Procedures can therefore be developed that could not be performed using conventional equipment. However, capital costs are high and productivity is relatively low for standard configurations. The laser competes in novel applications, flexible component processing, and where high accuracy is required, particularly involving forming on the submillimetre scale.

FRACTURE

Laser-induced fracture competes with thermal, mechanical and chemical methods of material parting and removal. Comparisons can be made when the geometry of processing (surface or bulk), and the properties of the materials involved, are specified.

Coatings are removed conventionally using mechanical or chemical methods. Jets of abrasive particles remove the coating and a small layer of substrate with high rates of coverage, but large amounts of mixed waste products are generated. Chemicals dissolve coatings, but often require precautionary measures during treatment and present environmental problems when disposed. Laser-based coating removal is highly accurate and localized. However, the material removal rate is relatively low. Laser-based techniques are therefore preferred for removing coatings from delicate substrates, such as sculptures and artwork, in which the processing rate is acceptable.

Surface layers of a homogeneous material are normally removed using mechanical means. A mechanical scabbler comprises arrays of pneumatically or hydraulically driven pins that act like small jackhammers to break up a surface. The material removal rate is relatively high, but the depth of material removal cannot be controlled accurately. In contrast, the depth of laser surface heating can be controlled with great accuracy, which enables the treatment depth to be managed with greater precision, particularly in materials with a thermally homogeneous microstructure.

PRACTICE

DEFORMATION

The information in Table 10.1 indicates that infrared lasers are the sources of choice for laser-induced deformation, but that there is interest in using new laser designs, such as the Nd:VA active medium and the near infrared Yb-doped fibre laser. The type of laser is selected based on a number of criteria: the required energy input (which depends on the absorptivity of the material to the beam wavelength and

Table 10.1   Sources of data for laser-induced deformation

Material	Laser	Reference
AA1015	Nd:YAG	(Merklein et al., 2001)
AA6082	Nd:YAG	(Merklein et al., 2001)
AISI 1008	Nd:YAG	(Hu et al., 2002)
AISI 1010	CO_2	(Li and Yao, 2001; Liu et al., 2004)
AISI 1012	CO_2	(Cheng and Yao, 2001)
AISI 301	Nd:VA	(Zhang and Xu, 2003)
AISI 301	Nd:YAG	(Zhang et al., 2002)
AISI 301	Yb fibre	(Zhang and Xu, 2003)
AISI 304	CO_2	(Lee and Jehnming, 2002)
AISI 304	Nd:YAG	(Chan et al., 2000; Hu et al., 2002)
Al_2O_3/TiC	Nd:VA	(Zhang and Xu, 2003)
Al_2O_3/TiC	Nd:YLF	(Chen et al., 1998)
Al_2O_3/TiC	Yb fibre	(Zhang and Xu, 2003)
AlMg3	Diode	(López et al., 1997)
C–Mn steel	Nd:YAG	(Yu et al., 2001)
CR4 mild steel	CO_2	(Magee and De Vin, 2002)
Glass reinforced aluminium laminate	CO_2	(Edwardson et al., 2004)
Si	Nd:VA	(Zhang and Xu, 2003)
Si	Yb fibre	(Zhang and Xu, 2003)
St 14	Diode	(López et al., 1997)
Stainless steel	Diode	(López et al., 1997)
Stainless steel	Nd:YLF	(Chen et al., 1998)
Ti–6Al–4V	CO_2	(Reeves et al., 2003)
Various	Nd:YAG	(Dearden and Edwardson, 2003)

the temporal mode of the beam); requirements for the location and accuracy of deformation (light of a short wavelength may be focused to a small spot size); and the surface condition desired (a CW beam produces more bending than the pulsed laser, at the expense of changes in surface composition and thermomechanical damage). Controlled deformation is most easily obtained in metals and alloys with high coefficients of thermal expansion, but has also been used to deform composites and a small number of ceramic-based materials in which deformation on the scale of nanometres is required (such as alumina-based composites used in the reading heads of computer hard drives).

Typical deformation procedures involve a power level of several hundred watts, a beam diameter of several millimetres and a traverse rate on the order of metres per minute. The parameters used in practice depend on the material, the type of laser used, the degree of deformation, and the accuracy required.

Fracture

Brittle ceramics and glasses, which are notoriously difficult to part with accuracy, have been considered as potential candidates for fracture induced by the heating effect of infrared laser beams, Table 10.2. Ultraviolet excimer lasers have also been used to induce fracture, on the nanoscale, although the

Table 10.2 Sources of data for laser-induced fracture: N/S not specified

Material	Laser	Reference
Glass–ceramic wafers	N/S	(Elperin and Ruding, 2004)
Al_2O_3	CO_2	(Tsai and Chen, 2003; Tsai and Liou, 2003)
Anthracine microcrystals	KrF excimer	(Hosokawa *et al.*, 2002)
$NaNO_3$	Nd:YAG	(Langford *et al.*, 1999)
Ceramics	CO_2	(Rice *et al.*, 1975)
Au, Cu, W on Si substrate	N/S	(Arronte *et al.*, 2002)
Al_2O_3 on SiO_2 substrate	KrF excimer	(Pleasants and Kane, 2003)

mechanism of fracture is not understood well – it may be thermal heating or athermal breaking of chemical bonds.

The energy input required to fracture ceramics depends on their thermal and mechanical properties, which can be found in Table D.2 (Appendix D), and which may be compared with the properties of metals and alloys (Table D.1, Appendix D). The effect of absorptivity on the incident power required is taken into account (ceramics absorb infrared light more efficiently than metals and alloys). Practical procedures can be found in the references listed in Table 10.2.

Laser-induced fracture is an efficient means of separating materials with differing thermal properties (notably thermal expansion coefficient), but can be difficult to control as a means of parting *homogeneous* materials with a required accuracy on the order of micrometres. A successful application of interfacial fracture on such a scale is the removal of metallic and ceramic particles from the surface of ceramics and metals, respectively, through differential expansion on heating – an efficient means of obtaining the cleanliness required in semiconductor materials.

LASER DEFORMATION DIAGRAMS

The various mechanisms of laser-induced deformation have been modelled using numerical and analytical techniques, and the literature contains a wealth of experimental data. Analytical models of the temperature field and deformation by the temperature gradient mechanism are used here as the basis of a laser deformation diagram.

We consider simple bending of a plate along one axis. Bending is induced by tracking a laser beam of power q at a velocity v over the entire length of the plate. The beam heats a rectangular surface element of the substrate with a width equal to the beam diameter, $2r_B$, and depth l. The element is assumed to be heated to a temperature close to the melting temperature, T_m, without the onset of surface melting, to maximize the degree of bending. An energy balance for the power required to heat the element from T_0 to T_m gives

$$Aq = 2r_B l v \rho c (T_m - T_0) \tag{10.1}$$

where A is the material absorptivity, ρ is the density ($kg\,m^{-3}$), and c is the specific heat capacity ($J\,kg^{-1}\,K^{-1}$).

We assume that the linear expansion of the surface, and the subsequent contraction during cooling, is directly proportional to the depth of heating and the thermal expansion coefficient. Further, we assume that the angle of bending is proportional to the linear surface contraction.

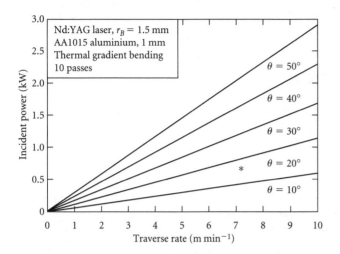

Figure 10.1   Parameter selection diagram for laser bending of aluminium, showing model-based contours of constant bend angle that are calibrated to an experimental data point (∗)

PARAMETER SELECTION

Equation (10.1) gives the form of the constitutive relationship between q and v for a given depth of heating. The precise functional relationship between q and v for a constant bend angle θ depends on the plate thickness and the bending geometry. It is established by calibrating equation (10.1) to an experimental measurement of the bending angle induced in a given material and plate thickness, with known processing parameters. Merklein (2001) observed that a bend angle of 16° is produced in a plate of AA1050 aluminium alloy of thickness 1 mm, by ten passes of an Nd:YAG laser beam of incident power 650 W and diameter 3 mm, traversing at a rate of $119\ \text{mm s}^{-1}$, and that the bend angle is directly proportional to the number of passes. This data point is plotted in Fig. 10.1 as a star. The calibration establishes an effective value for the absorptivity of the material to the laser beam (0.15). Contours of constant bend angle are then constructed using equation (10.1) and calibrated to this data point. (The depth of heating, l, acts as a dummy variable that is directly proportional to the angle of bending.) The boundary for the onset of surface melting, which limits the parameters that can be used for bending by the temperature gradient mechanism, was established for metals and alloys in Fig. 6.5 (Chapter 6), but it appears above the range of power shown in Fig. 10.1.

Figure 10.1 may be used to expedite the development of laser bending procedures for Nd:YAG laser heating of AA1015 aluminium plate using a beam diameter of 3 mm. By assuming that the degree of bending is directly proportional to the number of passes (which is reasonable until the bend angle exceeds about 50°), the required number of passes may be established for a desired bend angle and set of processing parameters. The contours shown in Fig. 10.1 may be extrapolated to predict bending parameters for conditions that lie outside current operating windows. The calibration data point can also be used to generate contours of bending angle for materials other than aluminium by substituting the appropriate material properties (Table D.1, Appendix D) into equation (10.1). It is then assumed that the bending mechanism and absorptivity of the material remain constant. This assumption may be removed by performing a new calibration for a different material.

Models for bending by mechanisms other than the thermal gradient mechanism may be constructed in a similar manner, and calibrated to an experimental data point.

The mechanism(s) of laser-induced fracture are not well understood at present, which means that it is difficult to construct an analytical model that could be calibrated, and used to develop diagrams for laser-induced fracture.

INDUSTRIAL APPLICATION

Industrial applications of laser-induced deformation and fracture are listed in Table 10.3. Examples are explained in more detail below.

FORMING

The objective of laser forming is to selectively heat areas of a workpiece such that deformation in one or many planes is induced in a controlled manner by one or more of the mechanisms described above. (Note that the term *laser forming* is also commonly used to describe a method of building up components by direct metal deposition techniques, which are described in Chapter 12.) Metal and alloy sheets respond particularly well to laser forming, which can be achieved with a variety of infrared lasers of relatively low power in applications ranging from household items to thick steel ship hull plates, which can also be straightened using the same process. Laser-induced deformation is readily applied in an automated production environment because the laser beam can be located accurately and automatic process monitoring and control systems ensure that deformation is controlled precisely. Laser forming is particularly suitable for producing small batches, and for shaping hard and brittle materials that would fracture when processed with conventional tooling. It can be considered as a method of rapid prototyping and manufacturing.

CONTROLLED FRACTURE

Figure 10.2 shows a three-dimensional image produced in an acrylic block by controlled laser micro-cracking. A digitized file is first created with the coordinates of the image. Frequency-doubled pulsed Nd:YAG laser beams of wavelength 532 nm and pulse energy 2.5 mJ are projected into the block at a

Table 10.3 Industrial applications of laser-induced deformation and fracture

Industry sector	Component	Material	Process	Reference
Domestic goods	Household items	Metals and alloys	Forming	(Silve *et al.*, 2000)
	Ornaments	PMMA Glass	Controlled fracture	(Laser Crystal Works, 2004)
Electronics	IC leadframe	AISI 304	Forming	(Chan *et al.*, 2000)
	Reed switches	Magnetic materials	Forming	(Verhoeven *et al.*, 2000)
	Semiconductors	Si wafers	Cleaning	(Apostol *et al.*, 2002)
Heritage	Statues	Limestone	Cleaning	(Cooper *et al.*, 1995)
	Statues	Stone	Cleaning	(Mazzinghi and Margheri, 2003)
Power generation	Reactor wall	Concrete	Scabbling	(Johnston *et al.*, 1998; Savina *et al.*, 1999, 2000)
Shipbuilding	Hull plate	C–Mn steels	Forming	(Reutzel, 2001)

rate of 600 Hz such that they focus on the image coordinates. Thermal expansion causes a microcrack to form. The image is built up from within as the beams move and the microcracks join together. The surface of the block remains clear and unmarked because the power density at the surface is insufficient to cause any thermal deformation. Similar three-dimensional images can be produced in glass. Irregular surfaces may be accommodated by applying a liquid with the same refractive index as the substrate, such that a plane outer surface is presented to the impinging beam.

CLEANING

Laser light was first considered for cleaning surfaces in the early 1970s. Since then, various methods of laser cleaning have been developed, based on mechanisms of controlled fracture of the interface between the coating and the substrate, selective thermal vaporization of the coating (Chapter 17), and athermal ablation using excimer lasers (Chapter 7). Figure 1.10 (Chapter 1) shows a Q-switched Nd:YAG laser system that has been developed for cleaning stone monuments. The beam is directed into regions that are difficult to access via an articulated arm and a series of mirrors. Limestone, sandstone, plaster, aluminium, terracotta, textiles, paper, stained glass, paints and ivory are just some of the materials that have been cleaned successfully using the Nd:YAG laser. The Nd:YAG laser is popular because many contaminants absorb the 1064 nm wavelength efficiently. Laser beam heating may also be used to vaporize moisture and contaminants trapped at the interface; as the heated gas expands, the coating may be ejected, leaving the substrate intact. A coating may also be detached by the ultrasonic vibrations induced by the laser beam.

Figure 10.2   Three-dimensional image made by controlled laser microcracking in acrylic

Coatings are removed conventionally by chemical reaction or mechanical abrasion. Many sculptured materials and coatings are porous to chemicals; salts can form that can result in blooming later. Paints are designed to be durable, and so aggressive softening agents and solvents must be used in chemical processes, which expose staff to health risks, as well as generating a large amount of toxic waste that must be disposed of in an environmentally friendly way. Mechanical techniques are time consuming and require a high degree of operator skill since the coating and substrate are only differentiated by hardness, which can easily lead to substrate damage.

Laser cleaning by controlled fracture provides a number of advantages. Monochromatic light is absorbed selectively by different materials. A wavelength can be chosen that is absorbed strongly by the coating, leading to deformation and interfacial fracture, but which is reflected by the substrate, thereby efficiently removing the coating without damaging the substrate. Light imposes no direct mechanical forces, and so can be used with fragile workpieces. The size of the laser beam can be varied to suit the detail to be cleaned, thereby providing flexibility and accuracy. The operator has immediate visual feedback of the degree of cleaning, enabling the beam parameters to be adjusted immediately if necessary. The only waste produced is the coating itself.

Scabbling

The principles of mechanical scabbling were described earlier. Laser scabbling is based on controlled fracture initiated by differential thermal expansion of surface regions of a material. Concrete is heated efficiently by infrared laser radiation, but tolerates only low values of thermal strain before fracturing. The laser beam is formed into a suitably large heating pattern and traversed across the concrete. Surface regions expand, particularly at the leading edge of the beam, and fail by flaking away from the substrate.

Concrete structures in nuclear power plant are about 30 cm in thickness, but only the upper 6 mm of the walls and floors is normally radioactive. Rather than demolishing decommissioned plant, which would produce tonnes of radioactive waste, the preferred approach is to separate the contaminated material. Laser scabbling is an ideal solution: an Nd:YAG laser beam may be delivered remotely to hazardous locations via a fibreoptic cable. Far infrared laser light can also be used to remove contaminated organic coatings such as epoxy-based paints with a significant reduction in the volume of waste created.

Decontamination by sand blasting requires large amounts of sand, and produces considerably more contaminated waste product than coating. Liquid coating removal suffers from similar drawbacks, and the use of chemicals also produces mixed hazardous waste.

Summary and Conclusions

Both laser-induced deformation and fracture can be achieved through a number of different mechanisms that form the basis of a number of novel applications: non-contact sheet forming in two and three-dimensions; cleaning through differential expansion of coatings and substrates; material removal through differential expansion of laser-heated surface layers; and internal three-dimensional 'sculpturing' of acrylics and glasses. Materials with a high value of the index defined by the product of thermal expansion coefficient and Young's modulus, such as metals and alloys, respond well to laser-induced deformation. Materials with low values of thermal and mechanical strain, such as concrete, respond well to controlled fracture. A calibrated analytical model of deformation can be used to construct laser deformation diagrams for various materials and processing conditions, which expedites the development of laser forming procedures.

The advantages of laser-based methods of deformation and fracture lie in the low, controllable energy input. When used in conjunction with sensitive process monitoring and control equipment, such processes are ideally suited to automation, with potential for application in many industrial sectors. Perhaps the most interesting characteristic of the processes described is the innovative manner in which they may be applied, which in turn stimulates novel ideas.

FURTHER READING

FORMING

Pridham, M.S. and Thomson, G.A. (1995). Laser forming. *Manufacturing Engineer*, **74**, (3), 137–139.

CLEANING

Cooper, M.I. ed. (1998). *Laser Cleaning in Conservation: (An Introduction)*. Oxford: Butterworth-Heinemann.

Jetter, J. (2002). Laser surface cleaning. *Industrial Laser Solutions*, **17**, (4), 15–17.

Luk'yanchuk, B. (2002). *Laser Cleaning*. River Edge: World Scientific.

Prandoni, S., Salvadeo, P., Castelli, P. and Galli, R. (2002). Industrial laser cleaning. *Industrial Laser Solutions*, **17**, (11), 21–24.

EXERCISES

1. Use the data in Tables D.1–D.5 (Appendix D) to identify classes of engineering material that are likely to be suitable for laser forming.

2. By considering the data in Fig. 5.13 (Chapter 5) and Tables D.1–D.5 (Appendix D), explain the susceptibility of different classes of engineering material to laser-induced fracture.

3. A bend angle of 16° is produced in a plate of aluminium by ten passes of an Nd:YAG laser beam of incident power 650 W and diameter 3 mm, traversing at a rate of 7.14 m min^{-1}. Calculate the incident laser power required to induce an angle of bending of 10° in the same material after ten passes of the same laser beam diameter traversing at a rate of 5 m min^{-1}. Assume that bending occurs through the thermal gradient mechanism. Mention any other assumptions that are made.

4. A bend angle of 20° is produced in a plate of aluminium by ten passes of an Nd:YAG laser beam of incident power 570 W and diameter 3 mm, traversing at a rate of 5 m min^{-1}. Estimate the angle of bending produced in a plate of alumina of similar thickness by 15 passes of a CO_2 laser beam of incident power 500 W and diameter 5 mm traversing at 10 m min^{-1}. Assume that bending occurs through the thermal gradient mechanism. Mention any other assumptions that are made.

BIBLIOGRAPHY

Apostol, I.G., Ulieru, D.G., Dabu, R.V., Ungureanu, C. and Rusen, L. (2002). Laser cleaning in the process of electronic device production. In: Dumitras, D.C., Dinescu, M. and Konov, V.I. eds *Proceedings of the International Conference on Advanced Laser Technologies ALT'01*. Vol. 4762. Bellingham: SPIE. pp. 235–238.

Arronte, M., Neves, P. and Vilar, R. (2002). Modeling of laser cleaning of metallic particulate contaminants from silicon surfaces. *Journal of Applied Physics*, **92**, (12), 6973–6982.

Chan, K.C., Yau, C.L. and Lee, W.B. (2000). Laser bending of thin stainless steel sheets. *Journal of Laser Applications*, **12**, (1), 34–40.

Chen, G., Xu, X., Poon, C.C. and Tam, A.C. (1998). Laser-assisted microscale deformation of stainless steels and ceramics. *Optical Engineering*, **37**, (10), 2837–2842.

Cheng, J. and Yao, Y.L. (2001). Microstructure integrated modelling of multiscan laser forming. *Journal of Manufacturing Science and Engineering*, **124**, 379–388.

Cooper, M.I., Emmony, D.C. and Larson, J. (1995). Characterisation of laser cleaning of limestone. *Optics and Laser Technology*, **27**, (1), 69–73.

Dearden, G. and Edwardson, S.P. (2003). Some recent developments in two- and three-dimensional laser forming for 'macro' and 'micro' applications.

Journal of Optics A: Pure and Applied Optics, **5**, (July), S8–15.

Edwardson, S.P., Dearden, G., Watkins, K.G. and Cantwell, W.J. (2004). Forming a new material. *Industrial Laser Solutions*, **19**, (3), 16–20.

Elperin, T. and Ruding, G. (2004). Controlled fracture of nonmetallic thin wafers using a laser thermal shock method. *Journal of Electronic Packaging*, **126**, (1), 142–147.

Hosokawa, Y., Mito, T., Tada, T., Asahi, T. and Masuhara, H. (2002). Laser-induced expansion and ablation mechanisms of organic materials. *RIKEN Review: Focused on 2nd International Symposium on Laser Precision Microfabrication (LPM2001)*, (43), 35–38.

Hu, Z., Kovacevic, R. and Labudovic, M. (2002). Experimental and numerical modelling of buckling instability of laser sheet forming. *International Journal of Machine Tools and Manufacture*, **42**, 1427–1439.

Johnston, E.P., Shannon, G., Steen, W.M., Jones, D.R. and Spencer, J.T. (1998). Evaluation of high-powered lasers for commercial laser concrete scabbling (large scale ablation) system. In: Beyer, E., Chen, X. and Miyamoto, I. eds *Proceedings of the Laser Materials Processing Conference (ICALEO '98)*. Orlando: Laser Institute of America, Orlando. pp. A210–A218.

Langford, S.C., Dickinson, J.T. and Alexander, M.L. (1999). The production of sub-micron sodium nitrate particles by laser ablation. *Applied Physics A: Materials Science and Processing*, **69**, (7), S647–S650.

Laser Crystal Works (2004). Available from: <http://www.lasercrystalworks.com/laser.htm> [Accessed 30 July 2004].

Lee, K.-C. and Jehnming, L. (2002). Transient deformation of thin metal sheets during pulsed laser forming. *Optics and Laser Technology*, **34**, 639–648.

Li, W. and Yao, Y.L. (2001). Laser bending of tubes: mechanism, analysis and prediction. *Journal of Manufacturing Science and Engineering*, **123**, 674–681.

Liu, C., Yao, Y.L. and Srinivasan, V. (2004). Optimal process planning of laser forming of doubly-curved shapes. *Journal of Manufacturing Science and Engineering*, **126**, (1), 1–9.

López, C.D., Dubslaff, J., Höfling, R. and Aswendt, P. (1997). In: Beckmann, L.H. ed. *Proceedings of the Conference Temporal Characterization of Plasma cw High-power CO$_2$ Laser-matter Interaction: Contribution to the Welding Process Control*. Vol. 3097. Bellingham: SPIE. pp. 692–697.

Magee, J. and De Vin, L.J. (2002). Process planning for laser-assisted forming. *Journal of Materials Processing Technology*, **120**, 322–326.

Mazzinghi, P. and Margheri, F. (2003). A short pulse, free running, Nd:YAG laser for the cleaning of stone cultural heritage. *Optics and Lasers in Engineering*, **39**, 191–202.

Merklein, M., Hennige, T. and Geiger, M. (2001). Laser forming of aluminium and aluminium alloys – microstructural investigation. *Journal of Materials Processing Technology*, **115**, 259–265.

Pleasants, S. and Kane, D.M. (2003). Laser cleaning of alumina particles on glass and silica substrates: experiment and quasistatic model. *Journal of Applied Physics*, **93**, (11), 8862–8866.

Reeves, M., Moore, A.J., Hand, D.P., Jones, J.D.C., Cho, J.R., Reed, R.C., Edwardson, S.P. Dearden, G., French, P. and Watkins, K.G. (2003). Dynamic distortion measurements during laser forming of Ti–6Al–4V and their comparison with a finite element model. *Proceedings of the Institute of Mechanical Engineering Part B Journal of Engineering Manufacture*, **217**, (12), 1685–1696.

Reutzel, E. (2001). Laser-assisted ship hull plate forming demo conducted. *iMAST Quarterly*, (3), 7.

Rice, R.W., Mecholsky, J.J. and Spann, J.R. (1975). Laser induced fracture of ceramics. *Journal of Defense Research IV-B*, (January).

Savina, M.R., Xu, Z., Wang, Y., Leong, K. and Pellin, M.J. (1999). Pulsed laser ablation of cement and concrete. *Journal of Laser Applications*, **11**, (6), 284–287.

Savina, M.R., Xu, Z., Wang, Y., Reed, C. and Pellin, M.J. (2000). Efficiency of concrete removal with a pulsed Nd:YAG laser. *Journal of Laser Applications*, **12**, (5), 200–204.

Silve, S., Podschies, B., Steen, W.M. and Watkins, K.G. (2000). Laser forming – a new vocabulary for objects. In: Christensen, P., Denney, P., Miyamoto, I. and Watkins, K. eds *Proceedings of the Laser Materials Processing Conference ICALEO '99*. Orlando: Laser Institute of America. pp. F87–96.

Tsai, C.-H. and Chen, C.-J. (2003). Formation of the breaking surface of alumina in laser cutting with a controlled fracture technique. *Proceedings of the Institute of Mechanical Engineering Part B Journal of Engineering Manufacture*, **217**, (4), 489–497.

Tsai, C.-H. and Liou, C.-S. (2003). Fracture mechanism of laser cutting with controlled fracture. *Journal of Manufacturing Science and Engineering*, **125**, (3), 519–528.

Verhoeven, E.C.M., de Bie, H.F.P. and Hoving, W. (2000). Laser adjustment of reed switches: micron accuracy in mass production. In: *Proceedings of the Microfabrication Conference ICALEO 2000*. Orlando: Laser Institute of America. pp. B21–B30.

Yu, G., Masubuchi, K., Maekawa, T. and Patrikalakis, N.M. (2001). FEM simulation of laser forming of metal plates. *Journal of Manufacturing Science and Engineering*, **123**, (3), 405–410.

Zhang, X.R., Chen, G. and Xu, X. (2002). Numerical simulation of pulsed laser bending. *Journal of Applied Mechanics*, **69**, 254–260.

Zhang, X.R. and Xu, X. (2003). High precision microscale bending by pulsed and CW lasers. *Journal of Manufacturing Science and Engineering*, **125**, 512–518.

CHAPTER 11

SURFACE MELTING

INTRODUCTION AND SYNOPSIS

Laser surface melting comprises a family of processes that includes alloying and particle injection, in which the material surface is melted, but not intentionally vaporized, by a scanning distributed laser beam. Large-scale processing geometries are similar to those used in hardening (Chapter 9). Figure 11.1 shows the lobe of a camshaft being remelted – in this example the camshaft is rotated under a laser beam that is formed into a transverse line heat source of high power density. Alternatively, the beam may be tracked across the surface of large components that are not easily moved. Surface melting, and subsequent rapid resolidification, is a means of producing a refined or metastable microstructure in localized areas on a component, which have improved service properties such as resistance to wear, corrosion and oxidation, particularly at high temperatures. Ferrous alloys – notably highly alloyed steels

Figure 11.1 CO_2 laser remelting of the surface of a cast iron cam lobe to improve durability. (Source: Tero Kallio, Lappeenranta University of Technology, Finland)

and cast irons – respond particularly well to such treatment. Porous, inhomogeneous ceramics and ceramic-based composites are suitable candidates for laser-based glazing. Kilowatt-class infrared CO_2, Nd:YAG and diode lasers are the sources of choice for applications on the scale of millimetres, where deep melting and a relatively low solidification rate are desirable, because the power density required may be generated over a relatively large beam area. Practical millimetre-scale surface melting is carried out over wide ranges of power density (10–10^4 W mm^{-2}) and beam interaction time (1–10^{-4} s), as illustrated in Fig. 6.3 (Chapter 6). Pulsed laser output from Q-switched ruby and Nd:YAG sources, as well as ultraviolet excimer radiation, with a duration and energy on the order of tens or hundreds of nanoseconds and joules, respectively, is the preferred option for applications in which melting to a depth on the submicrometre scale is required (when producing amorphous metallic glasses, for example).

Various elements and compounds may be introduced prior to or during melting in the form of solids, liquids and gases. If the addition dissolves in the melted substrate to form a homogeneous alloy, the process is referred to as *surface alloying*. If insoluble particles are added to the melt, a particulate-reinforced composite surface may be formed by a process commonly referred to as *particle injection*. Ceramic particles such as carbides of silicon and tungsten, which have a limited amount of solubility during processing, are popular choices because the favourable characteristics of both surface alloying and localized composite formation may be exploited.

The principles of all three processes are similar. The main procedural variations are associated with the techniques of adding alloy materials. The processes compete with conventional means of melting, alloying and composite formation, but process characteristics can be identified that provide laser melting with a competitive edge in specific applications. A large amount of processed material data has been generated in the decades that these techniques have been researched, but the number of industrial applications is surprisingly small – the reasons for this are investigated. Surface melting induced by various surface energy sources lends itself to analytical modelling, which is used to construct diagrams that illustrate laser melting in overview, from which practical parameter selection diagrams can be extracted. Ongoing developments in process monitoring and control equipment are aimed at improving the user-friendliness of the processes, which is likely to lead to greater industrial uptake, particularly in small-scale applications for which there are few conventional methods of accurate localized melting that also have the potential for simultaneous alloying.

PRINCIPLES

The type of laser is selected according to the application: an infrared wavelength for large-scale application, or a shorter wavelength for microprocessing. The laser beam is formed into a geometry that is appropriate for the area to be treated using optics described in Chapter 4. The form of beam power – continuous wave (CW) or pulsed – is selected depending on the application: a traversing CW beam is suitable for treatment areas on the order of square millimetres, whereas a stationary pulsed source is appropriate for discrete processing to micrometre-scale depths.

HEATING

The energy input to the substrate is low in comparison with a conventional means of surface melting. Heating rates associated with laser melting are therefore orders of magnitude higher than conventional methods; equation (E2.14) in Appendix E indicates that they are on the order of 10^5 K s^{-1} for a typical linear energy input of 0.1 J m^{-1} associated with large-scale processing, rising to 10^9 K s^{-1} for nanosecond duration excimer laser pulses of energy 10 J. Melting therefore occurs rapidly, with an associated increase in the absorptivity of the material to the laser beam.

FORMATION OF THE MELT POOL

Radial temperature gradients on the order of 10^2–10^4 K mm^{-1} develop between the centre of the forming melt pool and the cooler solid/melt interface. In most materials, the coefficient of surface tension increases with a decrease in temperature; surface tension gradients are therefore induced that drive fluid flow from the centre of the melt towards the edges (Marangoni flow). (Certain elements invert the relationship between surface tension and temperature – as welding engineers will be aware – leading to a reversal of the fluid flow geometry.) Marangoni flow is the dominant convection mechanism in a laser melted pool since there are no Lorenz forces (forces induced on moving charged particles in the presence of magnetic and electric fields).

A number of well-known dimensionless groups of processing variables may be used to indicate the relative importance of the process mechanisms operating during laser melting. (These should not be confused with the dimensionless groups of process variables used in the process models described, which form the basis of the laser processing diagrams.)

The Prandtl number, Pr, represents the ratio of the rate of thermal convection to the rate of thermal conduction and is defined as $Pr = \mu/a$ where μ is the kinematic viscosity and a is the thermal diffusivity. Ceramics and polymers have large Prandtl numbers – they can be ablated with little thermal effect on the substrate, for example. In contrast, metals have low Prandtl numbers (0.15 for steel and 0.015 for aluminium); viscous friction is low, which means that they flow readily but heat is transferred to greater depths in underlying material. Metals are therefore difficult to ablate by laser heating without leaving solidified residue from melt flow.

The Reynolds number, Re, is the ratio of inertial force to viscous force, and may be defined as $Re = lv\rho/\mu$ where l is a characteristic length, v is the velocity, and ρ is the density. When its value lies below a critical value, fluid flow is laminar rather than turbulent. The Reynolds number indicates that surface melting at high speed encourages turbulent flow in the melt pool, which increases entrainment of a surrounding atmosphere and may result in a rougher solidified surface.

The Péclet number, Pe, relates the forced convection to the thermal conductivity, and is defined as $Pe = vl/a$ (the product of the Reynolds number and the Prandtl number). It is useful in relating the depth of melting to the traverse rate; conditions leading to a high Péclet number result in a low melt depth (a high traverse rate and a low thermal diffusivity).

The Marangoni number, Ma, is the ratio of the rate of convection and the rate of conduction. Typical Marangoni numbers for metals during laser melting lie in the range 10^3–10^6. Molten steel has a higher Marangoni number than molten aluminium, and so the Marangoni effect will be higher in the former, which aids homogenization by convection, but results in a wider, shallower melt pool.

The second Froude number, Fr_2, represents the ratio of inertial force to gravitational force, and is defined as $Fr_2 = v^2/gl$ where g is gravitational force. It applies to wave and surface behaviour. Conditions leading to a high Froude number (a high traverse rate, for example) lead to increased fluid flow, which may result in surface ripples or humps in solidified material. (This is particularly important in high-speed welding, in which longitudinal undulations in the weld bead are observed with high welding speeds.)

Liquid state diffusion coefficients for metal solutes in molten metals lie in the range 10^{-6}–10^{-7} m^2 s^{-1}. Therefore the diffusion length, or mixing range, for very short melt times is considerably smaller than the melt depth; homogenization then occurs mainly by convection.

COOLING AND SOLIDIFICATION

Temperature gradients within the molten pool control cooling rates, and when combined with solute gradients influence solidification microstructures. With relatively low cooling and solidification rates, primary solidification products may be estimated from equilibrium phase diagrams and charts such as the Schaeffler diagram. However, cooling rates during laser surface treatment are significantly higher. Equation (E2.13) in Appendix E predicts cooling rates on the order of 10^3 K s^{-1} for conditions typical of multikilowatt surface melting, rising to 10^{11} K s^{-1} for very low energy input surface skin processing. The solidification velocity is also high (0.01–1 m s^{-1}), as are the temperature gradients (10^4–10^6 K m^{-1}).

Four different types of reaction can occur during rapid solidification associated with laser surface melting: amorphous glass formation; primary crystallization; massive or polymorphous crystallization; and eutectic crystallization. Conditions for the formation of amorphous structures have been calculated (Bergmann *et al.*, 1981); these are material dependent but indicate that a rate on the order of 10^6 K s^{-1} is required for the volume fraction of crystallized material to be reduced to less than 10^{-6} in alloys of iron and boron. Solidification by crystallization from the melt was described in terms of phase diagrams in Chapter 5 for various degrees of solid solubility.

Planar, cellular, dendritic or eutectic solidification fronts may form, depending on the temperature gradient G, the rate of solidification R, and solute concentration gradients, as described in Chapter 5. Briefly, the criterion for planar solidification front stability is

$$\frac{G_l}{R} > \frac{\Delta T_0}{D_l} \tag{11.1}$$

where G_l is the temperature gradient in the liquid, R is the solidification front growth rate, ΔT_0 is the equilibrium solidification temperature range, and D_l is the solute diffusion coefficient in the liquid. Conditions of plane front stability are associated with a high temperature gradient. When the solidification rate is above about 1 m s^{-1}, there is insufficient time for solute diffusion, and a condition of absolute stability arises. Dendrites form with temperature gradients and solidification rates typical of practical laser melting G_l/R in which a remelted depth on the order of a millimetre is common.

The temperature gradient and growth rate are important in the combined forms G_l/R (the cooling rate) and G_l/R since they influence the scale of the solidification substructure and solidification morphology, respectively.

Relationships have been developed to describe microstructural features such as dendrite arm and lamellae spacings in terms of the cooling rate (Jackson and Hunt, 1966). For example, the secondary dendrite arm spacing d is related to cooling rate dT/dt through

$$d = k \left(\frac{dT}{dt} \right)^{-m} \tag{11.2}$$

where k is a kinetic constant and m has a value around 0.3.

PROCESS SELECTION

One reason for the limited industrial uptake of laser surface melting is the variety of conventional coating techniques with which manufacturers are generally more familiar. Here we consider the characteristics of laser-based processes, and compare them with those of conventional techniques, to identify niche applications.

CHARACTERISTICS OF LASER SURFACE MELTING

The integrity of surfaces produced by laser melting is high. The melted region exhibits low porosity and few imperfections, and has a sound metallurgical bond with the substrate. The energy input is low, causing little distortion to a component, which reduces or eliminates the need for post-melt finishing operations. Melting promotes homogenization of the surface composition, which means that materials

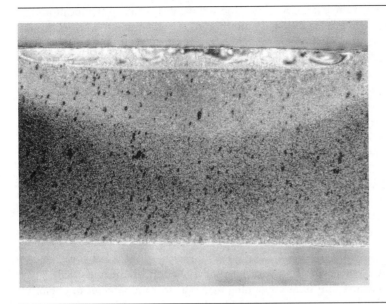

Figure 11.2   Transverse section through a C–Mn steel laser alloyed with chromium powder to impart corrosion resistance. The plate thickness is 6 mm

such as nodular cast irons may be hardened, and surface properties improved markedly over those of the parent material. The high cooling rate results in a fine solidified microstructure, which may contain non-equilibrium phases, novel precipitates and extended solid solubility. In cases of extremely high cooling rates, thin amorphous surface layers can be produced with a sound bond to the substrate – a unique feature of low energy input melting.

The composition and geometry of the remelted surface may be controlled accurately. Rapid autogenous melting results in limited loss of alloying elements. Material additions, whether in the form of a soluble alloy, insoluble particles, or a mixture can be managed precisely to obtain a prescribed final composition – an important factor if surface alloying for corrosion resistance is performed, since the composition determines the rate of reaction with the environment. An example of surface alloying for corrosion resistance is shown in Fig. 11.2. Laser surface alloying is flexible: a wide range of compositions and geometries can be produced in discrete locations, thus reducing the use of expensive strategic materials. The alloying addition can be concentrated at the surface, or uniformly distributed throughout the melt. Functionally graded surfaces may therefore be produced, with properties tailored to a particular application. By using ceramic particles such as carbides, nitrides and oxides of low solubility, a particle-reinforced surface composite can be produced *in situ*. Locations with restricted access can be treated, using either mirrors or fibreoptic beam delivery.

The main drawbacks of laser surface melting are the high equipment cost and the limited coverage rate – multiple overlapping tracks are needed for large areas. Applications are therefore characterized by high quality and high productivity, or they may constitute a novel manufacturing solution.

CONVENTIONAL METHODS OF SURFACE MELTING

Remelting

Tungsten inert gas (TIG) remelting is cheap and well known. However, it suffers from a number of disadvantages in comparison with laser remelting. TIG electrodes require adjustment prior to melting, and are consumed to a small extent during melting. It is more difficult to control the location and energy of a TIG arc to produce a prescribed melted pattern and treatment depth. Gas inclusions can lead to surface porosity. Environmental protection adds to operational costs. The relatively low quench rate may mean that a post-treatment nitriding stage is required in order to obtain a satisfactory hardness. The remelted surface might need to be made by patterned remelting, requiring a post-treatment grinding process to produce the final profile. Laser remelting competes when the remelt depth is less than about 1 mm, such that the distortion introduced is low (minimizing reworking), the surface is smooth, and the complete operation can be carried out in a single stage.

The negative characteristics of flame melting are similar to those of TIG remelting: low control, high energy input, high distortion and a low cooling rate. The process benefits from familiarity, ease of mobility and a low capital cost.

Lying between the two processes above is plasma remelting. Not surprisingly, the properties of the equipment and the characteristics of the processed material lie between flame and TIG remelting.

Alloying

Chemical and physical means of vapour deposition came into widespread use in the mid-1970s. They are particularly suitable for producing thin films containing materials that are difficult to obtain in a form suitable for conventional alloying. The deposition of hard, wear-resistant titanium nitride on titanium alloy components is popular, for example. Lasers are also used to assist the vapour deposition process, which is described in Chapter 8. Laser surface alloying competes when alloying to depths greater than about 1 mm is required, with a high coverage rate, and in situations where the enclosing chamber needed for vapour deposition would not be practical.

Ion beam techniques used for surface modification are generally divided into three classes: ion implantation; ion beam mixing; and ion beam assisted deposition. In the ion implantation process, charged ions, accelerated to an energy in the range 100–400 keV, are directed at the surface. The penetration depth depends on the acceleration energy, and is of the order of several hundred micrometres. The maximum solute concentration depends on the implantation dose, and is generally limited to 10–20%. Ion beam mixing occurs when a predeposited monolayer is bombarded to form the alloy. Relatively high concentrations can be produced, but the depth of alloying is limited. Ion beam assisted deposition consists of growing a vapour-deposited layer while implanting it with ions. Thicker coatings can be formed (> 1 mm) with good cohesion and adhesion to the surface. In all cases, the atomic scale of the process means that no defined interface is formed between the alloyed layer and the substrate. Laser surface alloying competes in applications similar to those in which it competes against vapour deposition.

ECONOMIC COMPARISON

The technique used in Chapter 9 for laser hardening may be applied to describe the economics of laser surface melting and to compare them with other methods of surface coating, by expressing the processing costs in terms of a cost per unit area for surface treatments, or a cost per unit weight for bulk treatments.

PRACTICE

MATERIAL PROPERTIES

The most suitable materials for surface melting are metals and alloys in which the desired properties cannot be obtained by heating alone, and inhomogeneous ceramics. Nodular cast irons with a matrix other than pearlite, for example, cannot be hardened to a significant degree by laser surface heating because insufficient time is available for solid state diffusion of carbon to occur through the matrix, as discussed in Chapter 9. The surface of domestic ceramics is often porous; it may be sealed by localized melting to form a barrier to contamination. The absorptivity of the material to the wavelength of the laser beam is important: surfaces that are highly reflective to the far infrared radiation of the CO_2 laser beam may need to be pretreated to enhance energy absorption, for example. The relationships between the absorptivity of different classes of engineering material and beam wavelength are described in Chapter 5.

If the objective is to produce an amorphous surface by remelting a metallic alloy, the alloy system should exhibit a deep eutectic (Chapter 5) to encourage glass formation at low temperatures.

Alloy materials may be introduced prior to processing (predeposition) or during melting (codeposition). The absorptivity of a predeposited alloy to the wavelength of the laser beam used should be high enough that absorption-enhancement techniques are not required. Its thermal conductivity should be high enough for energy to be transferred to the substrate for melting. Ideally, differences in melting temperature and latent heat of fusion between the alloy and the substrate should be minimized to obtain a homogeneous solidified surface. Large differences in the thermal expansion coefficient of the alloyed layer and substrate should be avoided – they can lead to cracking or the need for preheating.

Fewer restraints are imposed on the properties of materials used in the process of particle injection, since the principal interaction is between the laser beam and the substrate. The solubility of the particle in the substrate determines the properties of the reinforcement and the matrix composition in the ensuing surface composite. (Additional strength may be gained by the matrix through solid solution hardening resulting from partial dissolution of the particles.) Practical particle sizes typically lie in the range 40–150 μm.

BEAM CHARACTERISTICS

Multikilowatt infrared CO_2, Nd:YAG and diode lasers are used when melting to a depth in excess of about 0.5 mm is required. The power density for this scale of processing lies around 10^4 W mm^{-2} and the interaction time may be up to 1 second for deep melting. A reduction in the beam interaction time, produced by an increase in traverse rate or a decrease in beam size, enables finer microstructures to be produced. Shallow molten regions are readily generated with lower power infrared lasers, often operated in pulsed mode with an interaction time (pulse duration) of tenths or hundredths of a second. When shallow regions with fine microstructures are required over small areas, Q-switching is applied to Nd:YAG and ruby lasers, limiting the pulse duration to the range between microseconds and nanoseconds. The lower end of this range corresponds to the pulse duration of excimer lasers, which are absorbed well by all materials, and are used for high precision surface melting, often leading to the production of novel microstructures, at relatively high coverage rates. Beam shaping optics similar to those used in laser hardening are suitable for surface melting on the millimetre scale.

PROCESSING PARAMETERS

Inert gas shrouding is used to prevent oxidation and undesirable contamination of the melt pool during surface melting. Aluminium, for example, forms a high melting point oxide layer, which creates variations in absorptivity and surface tension gradients that can result in unstable processing and a poor molten pool geometry. Shielding gas can be delivered coaxially with the beam, or through a dish shroud, a side jet or a trailing jet. A flow of inert gas increases the cooling rate, leading to microstructural refinement, but it may also result in cracking of the solidified surface. Preheating may be used to reduce the cooling rate and the risk of cracking.

Deliberate alloy additions (in contrast to contamination) may be introduced in the form of deposited films or powders. Predeposited coatings required with infrared radiation are normally electroplated, plasma sprayed, or rolled onto the substrate. Thin films (less than 500 nm) used in alloying with shallow-melting Q-switched lasers are normally deposited using vacuum evaporation, sputtering, or ion implantation. Codeposition may take the form of a pneumatic powder feed or wire feed. Codeposition allows processing to be performed in a single step, in many orientations, and is therefore a more rapid and flexible means of alloying than predeposition. An alloying gas atmosphere may be created using a gas jet directed towards the melt pool, or by processing in a chamber. The method is a single step process. Alloying in the gas phase is relatively simple, but the range of gaseous alloy additions is limited.

PROPERTIES OF LASER-MELTED MATERIALS

METALS AND ALLOYS

The aim of remelting metals and alloys is to refine the microstructure to improve service properties such as wear, corrosion, oxidation and high temperature hardness.

Iron

Fundamental work has been performed using high purity iron, with the aim of understanding the mechanisms of remelting and alloying. Sources of data are given in Table 11.1.

Carbon–Manganese Steels

Improvements to wear, corrosion and high temperature oxidation properties are the main reasons for remelting and alloying C–Mn steels. Table 11.2 gives examples of the materials and lasers used.

Alloy Steels

Alloy steels are remelted and alloyed to improve wear and corrosion properties. Sources of data for laser melting of alloy steels can be found in Table 11.3.

Table 11.1   Sources of data for laser melting of iron

Material	Alloy addition	Laser	Reference
Fe	–	Ruby	(Lichtman and Ready, 1963)
Fe	B	Nd:YAG	(Pons *et al.*, 1986a,b)
Fe	Cr, Ni, Cr–Ni	CO_2	(Molian and Wood, 1983a,b,c)
Fe	Ni	CO_2	(Molian and Wood, 1982a,b)
Fe (Armco)	Air, N_2	XeCl	(Schaaf *et al.*, 1995)
Fe (Armco)	C	CO_2	(Canova and Ramous, 1986)
Fe (Armco)	Nitrides	CO_2	(Cordier-Robert *et al.*, 1998)

Table 11.2   Sources of data for laser melting of C–Mn steels: N/S not specified

Material	Alloy addition	Laser	Reference
AISI 1016	–	CO_2	(Chan *et al.*, 1987)
AISI 1018	–	CO_2	(Akkurt *et al.*, 1996)
AISI 1018	C	Nd:YAG	(Tayal and Mukherjee, 1994)
AISI 1018	Cr	CO_2	(Dahotre and Mukherjee, 1990)
AISI 1020	CO_2	Ni	(Chande and Mazumder, 1983a,b)
AISI 1040	–	CO_2	(Palombarini *et al.*, 1991; Rawers, 1984; Preece and Draper, 1981)
AISI 1045	–	Nd:YAG (Q-switched, $\lambda/2$)	(Altshulin *et al.*, 1990)
AISI 1045	–	CO_2	(Riabkina-Fishman and Zahavi, 1988)
AISI 1045	Cr, Ni	CO_2	(Riabkina-Fishman and Zahavi, 1990)
CK22	–	CO_2	(van Brussel *et al.*, 1991a)
C–Mn steel	AISI 316L	CO_2	(Ayers and Schaefer, 1979)
C–Mn steel	Carbides	N/S	(Gasser *et al.*, 1987)
C–Mn steel	Cr	N/S	(Gnanamuthu, 1979; Moore and McCafferty, 1981)
C–Mn steel	Cr	CO_2	(Dahotre and Mukherjee, 1990; Molian and Wood, 1982a; Molian, 1985; Molian and Wood, 1983a,c)
C–Mn steel	Cr–Ni	CO_2	(Chiba *et al.*, 1986)
C–Mn steel	Cr–Ni, Cr–Ni–P	CO_2	(Renaud *et al.*, 1991)
C–Mn steel	Ni	CO_2	(Molian and Wood, 1982b)
C–Mn steel	Ni	CO_2, Nd:YAG	(Pelletier *et al.*, 1991)
C–Mn steel	Ni, Cr	CO_2	(Molian and Wood, 1983b; Khanna *et al.*, 1995)
C–Mn steel	Ni, Cr, Al	CO_2	(de Damborenea and Vázquez, 1992)
C–Mn steel	Ni–P	CO_2	(Renaud *et al.*, 1990)
C–Mn steel	Ti	CO_2	(Ayers and Schaefer, 1979)
C–Mn steels	–	CO_2	(Hegge and De Hosson, 1987)
C–Mn steels	Cr	CO_2	(Molian and Wood, 1984)
C–Mn steels	TiC	CO_2	(Fasasi *et al.*, 1994)
C–Mn steels	TiC, WC	CO_2	(Ayers, 1981; Ayers *et al.*, 1981a,b)
En8	–	Diode	(Lawrence and Li, 2000a)
Fe–C, Fe–W, Fe–Cr	–	N/S	(Hegge *et al.*, 1990c)
Low C	–	CO_2, Nd:YAG	(Pelletier *et al.*, 1989)
Medium C	–	CO_2, Nd:YAG	(Pelletier *et al.*, 1989)
Steel	TiN	N/S	(Wiiala *et al.*, 1988)

Tool Steels

Laser melting of tool steels results in the dissolution of metal carbides and enrichment of the matrix. On solidification a microstructure comprising δ-ferrite, retained austenite, and an ultrafine precipitation of carbides such as cellular M_6C and plate-like MC is normally observed. A martensitic matrix can be formed in T1 and M2 steels. The abrasion resistance of tool steels can be improved to a level above that

Table 11.3   Sources of data for laser melting of alloy steels

Material	Alloy addition	Laser	Reference
16MnCrS5	Graphite	CO_2	(Grünenwald *et al.*, 1992)
16MnCrS5	WC	CO_2	(Grünenwald *et al.*, 1992)
16MnCrS5	WC/Co	CO_2	(Grünenwald *et al.*, 1992)
1C–1.5Cr, 0.38C–Ni–Cr–Mo	–	CO_2	(Carbucicchio and Palombarini, 1986)
2C–11Cr–0.62W	–	CO_2	(van Brussel *et al.*, 1991b)
9Cr–1Mo	Cr	CO_2	(Pujar *et al.*, 1993)
AISI 1770	Cr, Zr	CO_2	(Leech *et al.*, 1992)
AISI 4135	–	CO_2	(Wilde *et al.*, 1995)
AISI 4135	Pd, Pt, Au	CO_2	(Manohar and Wilde, 1995)
AISI 4340	–	CO_2	(Fastow *et al.*, 1990)
AISI 5210	–	CO_2	(Palombarini *et al.*, 1991)
AISI 9840	–	CO_2	(Palombarini *et al.*, 1991)
Fe–C, Fe–W, Fe–Cr	–	CO_2	(Hegge *et al.*, 1990c)
M-50	–	CO_2	(Breinan *et al.*, 1976; Greenwald *et al.*, 1979)

Table 11.4   Sources of data for laser melting of tool steels

Material	Alloy addition	Laser	Reference
1% C	–	CO_2	(Jiandong *et al.*, 1992)
AISI D2	SiC and Cr_3C_2	CO_2	(Gemelli *et al.*, 1998)
AISI M2	–	CO_2	(Molian and Rajasekhara, 1986; Strutt *et al.*, 1978; Strutt, 1980; Åhman, 1984)
AISI M2	CrB, VB_2	CO_2	(Rabitsch *et al.*, 1994)
AISI M2	WC	CO_2	(Riabkina-Fishman *et al.*, 2001)
AISI M42	–	CO_2	(Hlawka *et al.*, 1993)
AISI O1	–	CO_2	(Conners and Svenzon, 1983; Bande *et al.*, 1991)
AISI T1	–	CO_2	(Molian and Rajasekhara, 1986)
AISI T1	–	Nd:YAG	(Kusiński, 1988)

of conventionally treated material, as a result of the presence of retained austenite. Sources of data for laser melting of tool steels can be found in Table 11.4.

Stainless Steels

Surface melting of stainless steels is carried out to refine the microstructure and to desensitize regions of welds that have undergone weld decay through a weld thermal cycle. The main improvements resulting from surface alloying of stainless steels are an increase in hardness and subsequent improvement in wear and erosion properties. Sources of data can be found in Table 11.5.

AISI 304 austenitic stainless steel is known for its high corrosion resistance, but certain thermal cycles can cause chromium carbide ($Cr_{23}C_6$) precipitation at grain boundaries during slow cooling in

Table 11.5   Sources of data for laser melting of stainless steels

Material	Alloy addition	Laser	Reference
12% Cr steel	Co alloy	CO_2	(Com-Nougué and Kerrand, 1986)
AISI 303	–	CO_2	(Preece and Draper, 1981)
AISI 304	TiC	CO_2	(Ayers, 1984)
AISI 304	–	Nd:YAG	(Mudali *et al.*, 1995)
AISI 304	–	CO_2	(Anthony and Cline, 1977, 1978, 1979; Wade *et al.*, 1985; McCafferty and Moore, 1986; Nakao and Nishimoto, 1986; de Damborena *et al.*, 1989; Masumoto *et al.*, 1990a,b; Akgun and Inal, 1992a, 1995a,b; Lee *et al.*, 1995)
AISI 304	Mo		(McCafferty and Moore, 1986)
AISI 304	Mo	CO_2	(Tomie *et al.*, 1991)
AISI 310	B	CO_2	(Liu and Humphries, 1984)
AISI 316	–	Nd:YAG	(Mudali *et al.*, 1995)
AISI 316	–	CO_2	(Lamb *et al.*, 1985, 1986)
AISI 316L	Si_3N_4	CO_2	(Tsai *et al.*, 1994a,b)
AISI 420	–	CO_2	(Vilar *et al.*, 1990; Colaço and Vilar, 1996)
AISI 420	–	CO_2	(Lamb *et al.*, 1985, 1986)
AISI 430	–	CO_2	(Wade *et al.*, 1985)
AISI 430	Si_3N_4	CO_2	(Tsai *et al.*, 1994a,b)
AISI 440	–	CO_2	(Ion *et al.*, 1991)
AISI 440-C	–	CO_2	(Strutt *et al.*, 1978)
Austenitic	TiC	CO_2	(Ayers, 1984)
Duplex	–	CO_2	(Nakao *et al.*, 1988; Nakao, 1989)
Ferritic	Mo, Mo + B	CO_2	(Rieker *et al.*, 1989)
PH 15-5	–	CO_2	(Ozbaysal and Inal, 1994)

the temperature range 900–450°C, to produce Cr-depleted zones adjacent to the grain boundaries. If the Cr content falls below, 12% corrosion resistance is lost, and the region becomes sensitized. The effect is known as weld decay in the HAZ of welds, and can be reduced through alloying additions of carbon-stabilizing elements, e.g. Ti and Nb, or the use of a low carbon content steel. However, localized laser melting offers a method of treating only the sensitized regions of the weld, avoiding the need to solution treat the entire weldment. Laser surface melting of sensitized regions dissolves carbides, homogenizes the distribution of Cr through liquid state diffusion and convection, and removes sulphide inclusions. The rapid cooling rate prevents reprecipitation, and also results in a fine grain size. Improvements in intergranular corrosion resistance, and mechanical properties of sensitized type 304 stainless steel have thus been produced.

Surface melting of martensitic stainless steels dissolves carbides to produce a homogeneous highly alloyed melt zone. Cellular-dendritic solidification structures comprising δ-ferrite, austenite, martensite and carbides are normally observed. Both carbon and chromium depress the M_s and M_f temperatures, resulting in retained austenite at room temperature and a reduced hardness. Pitting corrosion resistance is enhanced as a result of microstructural homogenization and the residual stress state formed.

A study of a wide range of Fe–Cr–Ni ternary alloys subject to laser melting reveals solidification modes different to those found in conventionally solidified alloys (Nakao *et al.*, 1988). Rod-like eutectic structures form with compositions normally associated with primary ferrite solidification. Massive structures also form under certain conditions.

Cast Irons

Of all the cast irons, nodular iron with a ferrite or ferrite–pearlite matrix is the material most often treated by laser surface melting. Solid state heat treatment alone provides insufficient time for carbon homogenization from the dissolution of graphite nodules. Spheroidal and flake graphite cast irons can be hardened further by alloying with elements such as chromium, nickel, vanadium and cobalt to produce microstructures containing primary austenite dendrites and carbide-containing eutectics to produce improved corrosion and wear resistance. Procedures for laser melting of cast irons can be found in the references in Table 11.6.

On laser surface melting, the graphite nodules in nodular iron float towards the surface. The nodule surfaces melt, and carbon is released into the molten pool. The time available for dissolution determines the carbon concentration in the melt, which is homogenized by stirring. On solidification, austenite precipitates as dendrites. At the eutectic temperature, the remaining melt solidifies as ledeburite ($\gamma + Fe_3C$). On further cooling, austenite may transform to pearlite, bainite or martensite, depending on the cooling rate. There is insufficient time for graphite to nucleate. The cementite phase

Table 11.6 Sources of data for laser melting of cast irons

Material	Alloy addition	Laser	Reference
ASTM 40	–	CO_2	(Molian and Baldwin, 1986, 1988)
ASTM 60-40-18	–	CO_2	(Lopez *et al.*, 1994)
ASTM 80-55-06	–	CO_2	(Lopez *et al.*, 1994; Mathur and Molian, 1985; Molian and Mathur, 1985; Molian and Baldwin, 1986, 1988)
ASTM 100-70-03	–	CO_2	(Lopez *et al.*, 1994)
BS 400-12	–	CO_2	(Grum and Šturm, 1996, 2002)
Cr–Mo–Cu iron	–	CO_2	(Li *et al.*, 1987)
Ductile	–	CO_2	(Papaphilippou *et al.*, 1996; Gadag *et al.*, 1995)
Fe–2.06C–4.30Si (white), Fe–4.16C–0.97Si (grey)	–	CO_2	(Lima and Goldenstein, 2000)
GGG 60	–	CO_2	(Gasser *et al.*, 1987)
Grade 17 Meehanite GC	Cr, Ni, Co, Co–Cr	CO_2	(Tomlinson and Bransden, 1990)
Grey	–	CO_2	(Tomlinson *et al.*, 1987a; Chen *et al.*, 1986)
High Cr	–	CO_2	(Chen *et al.*, 1987)
Meehanite	–	CO_2	(Trafford *et al.*, 1983)
Ni-hard	–	CO_2	(Chen *et al.*, 1987)
Ni-resist	–	CO_2	(Chen *et al.*, 1987)
Nodular	–	CO_2	(Bergmann and Young, 1985)
Nodular grey	–	CO_2	(Ricciardi *et al.*, 1983)
Pearlitic cast iron	Ni	KrF	(Panagopoulos *et al.*, 1998)
Silal	–	CO_2	(Chen *et al.*, 1987, 1988)
Various	–	CO_2	(Leech, 1986)

imparts hardness and wear resistance. Incompletely dissolved graphite nodules may also be present in the solidified surface. The presence of graphite nodules at the surface imparts a beneficial effect on friction properties. An increased proportion of ledeburite as well as retained austenite is observed around the peripheries of such nodules.

Flake graphite in grey cast irons can normally be dissolved completely by laser melting, to produce a near eutectic composition, which can transform to fine ledeburite. The high carbon content stabilizes the austenite at room temperature. However, the laser-melted ledeburitic surface is unstable, and when subjected to tempering using a pulsed laser, for example, a layer of Fe-base alloy is formed, containing a very fine dispersion of graphite particles. Crack-free microstructures can be produced with a high hardness, which extend wear life.

At the surface of laser-melted alloy cast irons, a fine ledeburitic structure is observed where graphite flakes dissolve completely. This is white iron, which has a high hardness, but low ductility. Changes in hardness resulting from laser melting of alloy cast irons vary considerably depending on the major alloying additions and the processing parameters. Microstructures containing austenite, martensite, eutectic, graphite, cementite and alloy carbides in various proportions are produced.

Aluminium

The use of aluminium is growing, particularly in transport applications. However, the tribological properties of aluminium alloys are relatively poor. Laser surface melting of aluminium alloys has therefore been studied to improve wear and corrosion properties, Table 11.7. Aluminium alloys are further alloyed to produce intermetallic phases, such as Al_3Fe, $TiAl_3$, Al_3Ni and Al_3Ni_2, which enable the hardness to be increased, thus improving wear properties. Creep properties can be improved through the formation of a composite material with the alloyed layer strengthened by a fine dispersion of Al_3Ni. By alloying with silicon, angular primary silicon particles in an Al–Si eutectic matrix can be produced, possessing a hardness of 400 HV.

Magnesium Alloys

Nickel and silicon are the most effective alloying elements in magnesium alloys, since strengthening intermetallic compounds and eutectic microstructures are formed. Procedures can be found in the references of Table 11.8.

Titanium Alloys

Titanium can be hardened by alloying with elements that possess a number of critical properties. They should form a hard phase rich in titanium to minimize the alloy concentration. The hard phase should form through primary solidification since solid state diffusion is limited with the high cooling rates involved. The alloying element should have sufficient solubility in the α or β phase to allow solid solution hardening and to retain ductility. Alloying with carbon causes dendrites of TiC to form. By melting in a nitrogen atmosphere, surface alloys of TiN dendrites in an α-Ti matrix, with a golden colour and a hardness up to 2000 HV, can be produced. The strengthening intermetallic Ti_3Al is formed by adding aluminium to a molten titanium surface.

The addition of SiC to commercial purity titanium leads to the formation of TiC and Ti_5Si_3 in an α' matrix, giving a typical increase in hardness from 200 to 600 HV, with a corresponding improvement in the rate of erosion of about 25%. The injection of TiC particles into Ti–6Al–4V

Table 11.7   Sources of data for laser melting of aluminium alloys: N/S not specified

Material	Alloy addition	Laser	Reference
A1070	TiS_2, TiB_2, TiN, TiC, TiO_2	CO_2	(Matsuda and Nakata, 1990)
AA1050P–O	Si, Fe, Cu, Ni	CO_2	(Matsuyama and Shibata, 1994)
AA2014	Al–Ti–Ni	CO_2	(Liu *et al.*, 1995)
AA2024	–	CO_2	(Noordhuis and De Hosson, 1993)
AA2024, AA5052, AA6061	TiC, SiC, WC	CO_2	(Ayers, 1984)
AA3xxx	–		(Juarez-Islas, 1991)
AA4047	–	CO_2	(Hegge and De Hosson, 1990a; Noordhuis and De Hosson, 1992)
AA5000 series	Fe	CO_2	(Gjønnes and Olsen, 1994)
AA6061	Al–SiO_2	CO_2	(Zhou and DeHosson, 1994)
AA6061	Cr, Ni	CO_2	(Fu and Batchelor, 1998)
AA6061	SiC	–	(Giannetaki *et al.*, 1994)
AA7175	Cr	CO_2	(Almeida *et al.*, 1995a,b; Ferreira *et al.*, 1996)
AA7xxx	–	N/S	(Major *et al.*, 1993)
Al	–	CO_2	(Gureyev *et al.*, 1991)
Al	Cr	CO_2	(Jain *et al.*, 1981)
Al	Si	CO_2	(Ferraro *et al.*, 1986)
Al	SiC	CO_2	(Abbass *et al.*, 1988)
Al (99.5%)	Si, Fe, Cu, Ni	CO_2	(Shibata and Matsuyama, 1990)
Al (99.6%)	SiC	Nd:YAG	(Ocelík *et al.*, 2001)
Al (99.9%)	Al, Ti	CO_2	(Uenishi *et al.*, 1992)
Al (CP)	Ni	Nd:YAG	(Das *et al.*, 1992, 1994)
Al alloys	Si	CO_2	(Gnanamuthu, 1979)
Al alloys	SiC	CO_2	(Hu *et al.*, 1995)
Al–(0–40)Cu	–	CO_2	(de Mol van Otterloo and DeHosson, 1994)
Al–10.6Si	Si	CO_2	(Tomlinson and Bransden, 1994)
Al–11Si	Pb, Bi	N/S	(Svéda *et al.*, 2003)
Al–12Si	Si, Ni, Fe, Cu, Mn, Cr, Co, Mo, Ti	CO_2	(Tomlinson and Bransden, 1995)
Al–20Si	–	CO_2	(Hegge, and De Hosson, 1990a)
Al–26.5Cu	–	CO_2	(Gill *et al.*, 1992)
Al–32.7Cu	–	CO_2	(Zimmermann *et al.*, 1989)
Al–4.5Cu	–	CO_2	(Munitz, 1985)
Al–4Si	–	CO_2	(Hegge and De Hosson, 1990a,b)
Al–5Si–3Cu	Ni	CO_2	(Pelletier *et al.*, 1991)
Al–7.5Si	–	CO_2	(Hegge and De Hosson, 1990a,b)
Al–Si–Cu alloy	Ni	N/S	(Gaffet *et al.*, 1989)

leads to a reduction in the coefficient of friction from 0.4 to 0.16, as well as an increase in the matrix hardness. Treatment with TiC, SiC, and WC improves the resistance to abrasive wear; large carbide particles (over 100 μm in diameter) and high volume fractions (over 0.5) produce the greatest improvement.

Table 11.8   Sources of data for laser melting of magnesium alloys

Material	Alloy addition	Laser	Reference
AZ31B	–	KrF	(Koutsomichalis et al., 1994)
AZ31BH24	–	KrF	(Koutsomichalis et al., 1994)
AZ61	Al, Si, Ni, Cu	CO_2	(Galun et al., 1994, 1995)
AZ91	Al, Si, Ni, Cu	CO_2	(Galun et al., 1994, 1995)
Mg	Al, Si, Ni, Cu	CO_2	(Galun et al., 1994)
WE54	Al, Si, Ni, Cu	CO_2	(Galun et al., 1994, 1995)

Significant microstructural refinement is observed after laser remelting α titanium alloys, with the formation of α' martensite. Fine strengthening dispersoids of Y_2O_3 can be formed by melting $\alpha + \beta$Ti–10V–2Fe–3Al alloys.

Procedures for laser melting of titanium alloys can be found in the references given in Table 11.9.

Nickel Alloys

Nickel alloys are remelted to improve corrosion properties, and alloyed to form strengthening intermetallic compounds such as Ni_3Al. Data can be found in the references given in Table 11.10.

Copper Alloys

High temperature strength is a problem with copper, particularly noticeable in electrodes for resistance welding machines. The addition of chromium or aluminium results in solid solution strengthening and the formation of hardening precipitates such as Al_2Cu. The electrical properties of copper can also be improved by alloying with gold. The addition of chromium and nickel leads to an improvement in corrosion properties. Procedures can be found in the references given in Table 11.11.

Zirconium Alloys

Laser surface melting of zirconium alloys has been investigated for potential application in the nuclear industry. Sources of data can be found in Table 11.12.

Laser remelting produces a supersaturated α' martensite microstructure, with an associated increase in hardness. Melting also increases corrosion resistance in 10% $FeCl_3$, through the dissolution of second phase particles and the formation of a supersaturated solid solution, and reduces the propensity for pitting.

Other Non-ferrous Alloys

Other non-ferrous alloys, in addition to those above, can be laser melted to improve service properties through the formation of fine solidification microstructures. Sources of data are given in Table 11.13.

Amorphous Alloys

Metallic glasses have been produced in very thin surface layers of alloys in which a deep eutectic forms. Such amorphous alloys have particularly good service properties, notably corrosion resistance.

Table 11.9 Sources of data for laser melting of titanium alloys: N/S not specified

Material	Alloy addition	Laser	Reference
Ti	–	N/S	(Moore et al., 1979)
Ti	Al	CO_2	(Abboud and West, 1990, 1991a,b, 1992a,b, 1994)
Ti	N_2	N/S	(Akgun and Inal, 1992b)
Ti	Pd	N/S	(Draper et al., 1981c)
Ti	Si	CO_2	(Abboud and West, 1991c)
Ti	SiC	CO_2	(Abboud and West, 1991c; Abboud et al., 1993)
Ti (CP)	Al	Nd:YAG	(Li et al., 1995)
Ti (CP)	C	CO_2	(Walker et al., 1985)
Ti (CP)	N_2	CO_2	(Mridha and Baker, 1991; García and de Damborenea, 1998; Walker et al., 1985)
Ti (Grade 2)	N_2	CO_2	(Seiersten, 1988)
Ti alloys	C	CO_2	(Bharti et al., 1989)
Ti–SiC_p	N_2	CO_2	(Mridha and Baker, 1996)
Ti–13Nb–13Zr	N_2	N/S	(Geetha et al., 2004)
Ti–5.5Al–3.5Sn–3Zr–1Nb	N_2	CO_2	(Mridha and Baker, 1991)
Ti–5.5Al–3.5V	N_2	CO_2	(Seiersten, 1988)
Ti6Al4V	N_2	CO_2	(García and de Damborenea, 1998)
Ti–6Al–4V	–	CO_2	(Hu et al., 1995)
Ti–6Al–4V	–	KrF	(Jervis et al., 1993)
Ti–6Al–4V	C	CO_2	(Walker et al., 1985)
Ti–6Al–4V	N_2	CO_2	(de Damborenea et al., 1996; Robinson et al., 1995; Walker et al., 1985)
Ti–6Al–4V	SiC	CO_2	(Abboud and West, 1989)
Ti–6Al–4V	SiC	Nd:YAG	(Pei et al., 2002)
Ti–6Al–4V	TiN	CO_2	(Xin et al., 1998)
Ti–6Al–4V	Triballoy T-400	N/S	(Cooper and Slebodnick, 1989)
Ti–6Al–4V	WC	N/S	(Vreeling et al., 2002)
Ti–8Al–4Y	–	CO_2	(Konitzer et al., 1983)
Ti–Al–Mo alloy	–	N/S	(Hirose et al., 1995)

Materials and procedures can be found in the references given in Table 11.14. Cooling and heating rates and transformation curves have been modelled for the preparation of metallic glasses (Bergmann et al., 1981).

CERAMICS AND GLASSES

Porous ceramics are suitable for laser melting, which glazes the surface to improve wear properties and restricts contamination and the ingress of liquids. Mixed oxides are remelted to produce high quality optical coatings. Enamel powders may be melted to produce discrete surface patterns. Silicon wafers are alloyed with dopants to generate the *p–n* junctions on which semiconductors are based. Data can be found in the references given in Table 11.15.

Table 11.10 Sources of data for laser melting of nickel alloys: N/S not specified

Material	Alloy addition	Laser	Reference
Incoloy 800H	SiC	CO_2	(Zhu *et al.*, 1995)
Inconel 617	–	Nd:YAG	(Yilbas *et al.*, 2001)
Inconel X-750	Al_2O_3	CO_2	(Ayers and Tucker, 1980)
Ni	–	CO_2 (TEA)	(Dallaire and Cielo, 1982)
Ni	Ag	Nd:YAG (Q-switched)	(Draper *et al.*, 1981b)
Ni	Ag, Au, Pd, Sn, Ta	CO_2, Nd:YAG (Q-switched)	(Draper *et al.*, 1980a)
Ni	Al, Mo	CO_2	(Dey *et al.*, 1994)
Ni	Au	Nd:YAG	(Draper, 1986)
Ni	Au	N/S	(Draper *et al.*, 1980a,b; Draper *et al.*, 1981a,b)
Ni alloy	Al and SiC	N/S	(Jasim *et al.*, 1993)

Table 11.11 Sources of data for laser melting of copper alloys: N/S not specified

Material	Alloy addition	Laser	Reference
Bronze	–	N/S	(McCafferty and Moore, 1986)
Bronze (Al, CDA-614)	–	CO_2	(Draper *et al.*, 1980b)
Bronze (NiAl)	–	CO_2	(Oakley and Bailey, 1986)
Bronzes (Fe–Al)	–	CO_2	(Draper, 1981)
Cu	–	KrF	(Panagopoulos and Michaelides, 1992)
Cu	Al	Ruby	(Manna *et al.*, 1994)
Cu	Au	N/S	(Draper, 1986; Wang *et al.*, 1983)
Cu	Cr, Ni	N/S	(Draper *et al.*, 1982; Galantucci *et al.*, 1989)
Cu–3Ag–0.5Zr	–	N/S	(Singh *et al.*, 1997)
Cu–Ni	–	CO_2	(Draper and Sharma, 1981)

Table 11.12 Sources of data for laser melting of zirconium alloys

Material	Alloy addition	Laser	Reference
Zircaloy-2	–	CO_2	(Rawers *et al.*, 1991)
Zr, Zr-702, Zr-4	–	CO_2	(Reitz and Rawers, 1992)

COMPOSITES

Laser surface melting is an efficient means of producing surface composites through the injection of ceramic particles into the melt pool. The properties of existing composites may be improved by surface melting, principally through refinement of the matrix microstructure. Data can be found in the references given in Table 11.16.

Table 11.13 Sources of data for laser melting of various non-ferrous alloys

Material	Alloy addition	Laser	Reference
Ag, Al, Be, Cu, Pb	–	Ruby (Q-switched)	(Vogel and Backlund, 1965)
As alloys	–	CO_2	(Dahotre *et al.*, 1992)
Ca alloys	–	CO_2	(Dahotre *et al.*, 1992)
Pb–Sb alloys	–	CO_2	(Dahotre *et al.*, 1992)
Sn alloys	–	CO_2	(Dahotre *et al.*, 1992)
W, Mo	–	Nd:glass	(Kostrubiec and Walczak, 1991, 1994)
Zn–3.37 wt% Cu	–	CO_2	(Ma *et al.*, 2001)

Table 11.14 Sources of data for laser melting to produce amorphous surfaces: N/S not specified

Material	Alloy addition	Laser	Reference
Al	–	Nd:YAG (Q-switched)	(Bonora *et al.*, 1981)
$Al_{71}Cu_{29}$	–	He–Ne	(Yan and Gan, 1993)
$AlY_{31}Ni_{11}$ alloys	–	N/S	(Jost and Haddenhorst, 1991)
Fe–21% Cr–10% Ni	–	N/S	(Nakao, 1989)
Fe–4% C–10% Sn	–	CO_2	(Zhang *et al.*, 1988)
Nb–Ni alloys	–	N/S	(Vitta, 1991)
Nd–Fe–B	–	N/S	(Bingham *et al.*, 1993)
Pd–4.2Cu–5.1Si	–	CO_2	(Breinan *et al.*, 1976; Greenwald *et al.*, 1979)
$YBa_2Cu_3O_x$	–	CO_2	(Shih *et al.*, 1994)

Table 11.15 Sources of data for laser melting of ceramics and glasses

Material	Alloy addition	Laser	Reference
Al_2O_3	–	CO_2	(Lee and Zum Gahr, 1990)
Al_2O_3–MgO–SiO_2	–	CO_2	(Shieh *et al.*, 1995)
Cement (Portland)	–	Diode	(Lawrence and Li, 2000c)
Grout	–	Diode	(Lawrence *et al.*, 1998a,b)
Mixed oxides	–	CO_2	(Swarnalatha *et al.*, 1991)
PbO-free glass enamels	–	CO_2	(Hahn *et al.*, 1996)
Si wafers	As	Ar^+	(Nissim *et al.*, 1980)
Si wafers	B	Ruby (Q-switched)	(Narayan *et al.*, 1978)
Si wafers	P, B	Nd:YAG, Nd:YAG (Q-switched), CO_2	(Affolter *et al.*, 1978)
Ti, Al, SiC	–	CO_2	(Abboud and West, 1994)
WC–Co	–	CO_2	(Mueller *et al.*, 1989)
ZrO_2–20 wt% Y_2O_3/MCrAlY	–	CO_2	(Tsai and Tsai, 1998)
ZrO_2-base	–	CO_2	(Jasim *et al.*, 1991, 1992)

Table 11.16   Sources of data for laser melting of composites: N/S not specified

Matrix	Laser	Reference
Al_2O_3–ZrO_2–CeO_2	CO_2	(Chen *et al.*, 1995)
Cermets	N/S	(Kahrmann, 1988)
Concrete	Diode	(Lawrence and Li, 1999, 2000b)
Ni1/WC cermet	CO_2	(Tomlinson *et al.*, 1987b)
Ti, Al, SiC	CO_2	(Abboud, and West, 1994)
Zeolites	CO_2	(Sugimoto *et al.*, 1990)
ZrO_2-base	CO_2	(Jasim *et al.*, 1991, 1992)

LASER SURFACE MELTING DIAGRAMS

The complex processes of heat and fluid flow that govern convection in a moving molten pool can be simulated with numerical methods. Such information forms the basis of prediction of solidification microstructures and physical properties. Models can then be used to predict the effects of changes in process variables on the following properties of the melted region: the melt pool geometry; solidification rates; macrostructural features, such as the formation of surface ripples or humps; solidification microstructure; convectional mass transport; and solid state cooling rate; and solid state phase transformations. A complete process model would combine all these features, and link them with structure–property relationships of the surface. However, there are few such models; the most well known are probably those of Mazumder and co-workers (Chan *et al.*, 1987; Chande and Mazumder, 1983a,b).

Here, we extend the analytical methods for modelling temperature fields during laser heating described in Chapter 7 to the case of laser melting.

MELTING OVERVIEW

Above the melting temperature, additional latent heat of melting must be provided. By assuming that melting occurs uniformly to a depth l_m over the beam width $2r_B$, the power available to heat the material, $(Aq)_{net}$, is $(Aq)_{net} = Aq - 2r_B l_m v L_m$ where L_m is the volumetric latent heat of melting. In dimensionless terms this becomes

$$q_{net}^* = q^* - 2l_m^* v^* L_m^* \tag{11.3}$$

where $l_m^* = l_m/r_B$ and $L_m^* = L_m/(\rho c(T_m - T_0))$. For metals and alloys the entropy of fusion is constant, and the value of L_m^* lies around 0.4.

The peak surface temperature may then be found by using the net absorbed power in the temperature field equations given in Chapter 7, and eliminating the dimensionless melt depth l_m^* by noting that $T_p^* = 1$ when $z^* = l_m^*$. The upper boundary of the melting region occurs when the surface temperature equals the vaporization temperature, which corresponds approximately to a value of $T_p^* = 2$ (Chapter 5). These boundaries, together with contours for normalized melt depths, which are calculated in the same way as the hardening contours, are shown in Fig. 11.3.

Figure 11.3 is a comprehensive overview of the effects of variations in process variables on the geometry of surface melting in a range of metals and alloys. It is a useful tool for understanding and displaying the principles of the process. Practical parameter selection diagrams for individual materials may be extracted from Fig. 11.3.

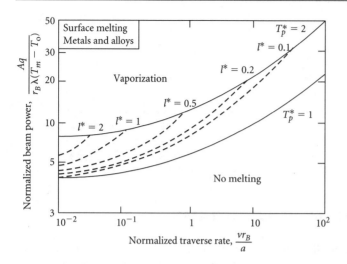

Figure 11.3 Model-based overview of laser surface melting of metals and alloys. The axes are dimensionless groups of processing variables. The operating window is defined by solid boundaries that correspond to normalised peak surface temperatures, T_p^* (T_p/T_m), of 1 (melting) and 2 (vaporization). Contours for melting to various normalized depths, l^*, (l/r_B) are shown as broken lines

Parameter Selection

The dimensionless axes of Fig. 11.3 may be converted to process variables for a particular material by substituting appropriate values for material properties, which can be found in Table D.1 (Appendix D), and by fixing the beam width. Figure 11.4 is thus constructed for the austenitic stainless steel AISI 304 for melting using a beam of width 12 mm ($r_B = 6$ mm), and the following material properties: $T_0 = 298$ K, $T_m = 1773$ K, $T_v = 3300$ K, $\lambda = 25.5$ W m^{-1} K^{-1}, $a = 7.2 \times 10^{-6}$ m^2 s^{-1}.

Calibration is an effective means of tailoring the model to a given set of processing conditions. The model can then be used to predict the effects of variations in process variables on the melt geometry in relation to a known condition (an experimental data point), as illustrated in Fig. 11.4. Model predictions can then be presented with greater precision. Figure 11.4 also serves as the starting point for developing practical processing diagrams for other processes, such as alloying.

Laser Surface Alloying Diagram

Figure 11.5 shows a calibrated model-based parameter selection diagram for laser surface alloying of structural steels using a beam of width 12 mm. The principal axes are dimensional process variables (kW and mm min^{-1}). The operating limit is defined by the solid boundary that corresponds to the peak surface temperature for melting. A contour for melting to a depth of 0.4 mm is shown as a broken line. The calibration point is shown with a star. The auxiliary chromium feed rate ordinate is calculated by applying a mass balance to the liquid. The auxiliary coverage rate abscissa is constructed geometrically.

Diagrams such as Figs 11.4 and 11.5 can be used to select parameters for the development of melting and alloying procedure specifications. They may be used to extrapolate data to predict the outcome of processing outside current operating windows.

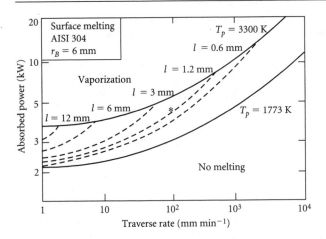

Figure 11.4 Calibrated model-based parameter selection diagram for laser surface melting of AISI 304 austenitic stainless steel using a beam of width 12 mm. The axes are practical process variables. The operating window is defined by solid boundaries that correspond to peak surface temperatures for melting (1773 K) and vaporization (3300 K). Contours for melting to various depths are shown as broken lines. The contours are calibrated to an experimentally measured depth of 0.15 mm for an incident beam power of 10 kW and a traverse rate of 100 mm min^{-1}, which is indicated by a star. The calibration reveals an effective value for absorptivity of 0.45, which may then be used to convert absorbed power into incident power

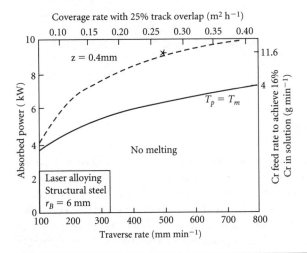

Figure 11.5 Calibrated model-based parameter selection diagram for laser surface alloying of structural steels using a beam of width 12 mm. The solid boundary is a model-based representation of the peak surface temperature for melting. A model-based contour for melting to a depth of 0.4 mm is also shown (broken line), which has been calibrated to the condition shown by a star. A mass balance is used to construct the right hand ordinate

INDUSTRIAL APPLICATION

The applications listed in Table 11.17 illustrate the relatively low industrial acceptance of laser surface remelting, alloying and particle injection, in comparison with other laser-based surface processes, such as hardening. However, a number of industrial applications lend themselves to analysis.

Table 11.17 Industrial applications of laser surface melting and alloying

Industry sector	Process	Application	Material	Improvement	Reference
Aerospace	Melting	Space shuttle main engine combustion chamber liner	NARloy-Z (Cu, 3Ag, 0.5Zr)	Crack resistance	(Singh *et al.*, 1994, 1995)
Automotive	Alloying	Valve seat	AISI 4815	Temper resistance	(Seaman and Gnanamuthu, 1975)
	Melting	Cams	Cast iron	Wear	(Olaineck and Lührs, 1996; Hamazaki, 1995)
	Melting	Piston ring	Alloy cast iron	Hardness	(Chauxuan, 1983)
Biomedical	Alloying	Implants	Ti	Wear	(Fink and Bergmann, 1990)
Construction	Roughening	Dimension stones	Granite	Aesthetics	(Pou *et al.*, 2002)
Machine tool	Boronizing	Gear cutting hob	AISI M1	Wear	(Han *et al.*, 1992)
	Melting	Pattern equipment	Grey iron	Wear	(Chithambaram *et al.*, 1989)
Material production	Melting	Dieless laser drawing	Ni-200	Finer diameter	(Li *et al.*, 2002)
Mining	Melting	Slurry pump paddle wheel	Grey iron	Erosion	(Ma *et al.*, 1997)

AUTOMOBILE ENGINE CAM LOBES

Modern engines use high lift camshafts to increase power and reduce fuel consumption. Sharper camshafts with a steep ramp and fall and a relatively high cam pitch are required. However, the sliding action of the cam lobe over the tappet (or rocker arm) in order to open and close the valves leads to very high stresses on the surface layers of the cam lobes. Camshafts are generally made from grey cast iron, which is a low cost, easily machinable material, but which has relatively poor wear properties. The surfaces of the cam lobes must therefore be hardened in order to engineer a product with an acceptable lifetime. Multikilowatt CO_2 lasers are used to remelt and harden cam lobes, as illustrated in Fig. 11.1.

Prior to laser treatment, the camshaft is preheated to around 450°C, normally by resistance heating. This enhances the absorptivity of the material to the laser beam, as well as preventing the formation of cracks in the resolidified microstructure. The laser beam, typically 6–9 kW, is formed into a line heat source of the same width as the cam, typically 10 mm. The cam is rotated beneath the beam, which is adjusted such that the line source follows the surface of the cam lobe. The rate of rotation is varied to maintain a constant melt depth over the profile, typically 1 mm, Fig. 11.6. A hard remelted ledeburitic surface layer is thus formed. It normally takes around a minute to treat the camshaft of a four cylinder, two valve engine (Olaineck and Lührs, 1996).

The traditional hardening method is TIG remelting, or in some cases the relatively expensive clear chill casting technique. In this particular application, the main advantages of laser remelting result from

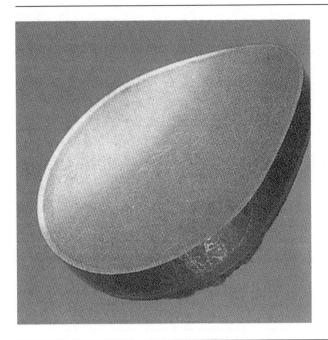

Figure 11.6   Section through a laser remelted cam lobe

the high quality of the treated component and the ease with which the process can be automated. The high hardness developed though the finer resolidified microstructure enables the conventional post-treatment nitriding stage to be eliminated. The smoothness and uniformity of the remelted surface, and reduced distortion of the component, allows grinding operations to be reduced by a factor of two. The whole process is CN controlled, which means that changes in the geometry and size of camshafts can be taken into account by software changes.

DESENSITIZATION OF WELD DECAY REGIONS IN STAINLESS STEELS

Weld decay in stainless steels results from the precipitation of carbides, notably along grain boundaries, when the material is subject to a thermal cycle, as is the case in welding. The cooling curve leading to precipitation during MIG welding is shown schematically in Fig. 11.7b.

Regions of weld decay may be homogenized through laser surface remelting, shown schematically in Fig. 11.7a. The cooling rate following melting is sufficiently high to avoid the reprecipitation of carbides (Fig. 11.7b), thus returning the corrosion resistance to that of the base material.

SPACE SHUTTLE MAIN ENGINE CHAMBER LINER

NASA (Huntsville, AL, USA) remelt the liner of the main engine main combustion chamber of the space shuttle (Fig. 11.8). The liner is made from a copper alloy containing 3 wt% Ag and 0.5 wt% Zr. Remelting results in a uniform and stable microstructure. Grain growth and grain boundary precipitation of intermetallic phases are observed in service in untreated inhomogeneous material, which cause a degradation in mechanical properties. The laser remelted surface is about 15% harder than the substrate, and possesses a higher thermal stability because of the extended solid solubility

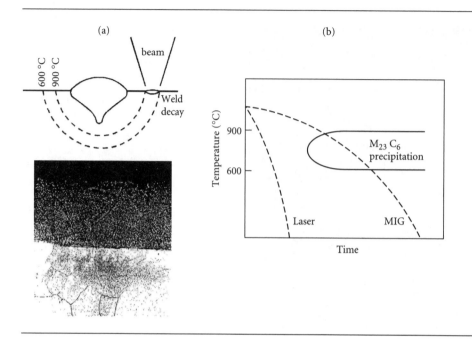

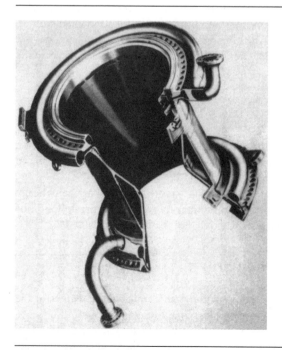

Figure 11.7 Desensitization of HAZ regions in AISI 304 that are susceptible to weld decay

Figure 11.8 Main combustion chamber liner for space shuttle main engine, which is laser remelted to improve mechanical properties. (Source: Biliyar Bhat, NASA Marshall Space Flight Center, Huntsville, AL, USA)

of zirconium in the copper matrix, and reduced second-phase precipitation. The overall benefit is an extension in the life of the liner.

SUMMARY AND CONCLUSIONS

The main benefits of surface melting, alloying and particle injection originate from the ability to produce fine remelted microstructures with good wear, corrosion and oxidation properties. The principles of all three processes are similar; procedural variations are associated with the techniques of adding alloy materials. Conventional means of melting, alloying and composite formation are often more familiar to production engineers, but process characteristics can be identified that provide laser melting with a competitive edge in particular applications. A large amount of processed material data has been generated, but the number of industrial applications is surprisingly small – laser alloying, for example, has yet to find a commercially viable application. Surface melting may be modelled analytically, and various laser melting diagrams can be constructed. Surface melting processes are currently not as widespread as transformation hardening since they are processes that rely on transformation to the liquid state, with the attendant change in surface absorptivity and the difficulty in maintaining a constant absorbed energy for consistent and reproducible processing.

However, processes based on laser surface melting provide unique means of precisely controlling the formation of microstructures. The surface chemistry can be tailored to specific applications. Once process automation becomes more user-friendly, the processes will have much to offer for industrial application, particularly in small-scale applications for which there are few conventional methods of accurate localized melting that also have the potential for simultaneous alloying.

FURTHER READING

SURFACE MODIFICATION

Mordike, B.L. (1996). Surface modification by lasers. In: Cahn, R.W., Haasen, P. and Kramer, E.J. eds *Materials Science and Technology. A Comprehensive Treatment. Vol. 15: Processing of Metals and Alloys*. Mississauga: John Wiley & Sons Canada Ltd. pp. 111–136.

Rubahn, H.-G. (1999). *Laser Applications in Surface Science and Technology*. Chichester: John Wiley & Sons.

Steen, W.M. (1983). Laser cladding, alloying and melting. In: Belforte, D. and Levitt, M. eds *The Industrial Laser Annual Handbook*. Tulsa: PennWell Books. pp. 158–174.

SURFACE MELTING

Kurz, W. and Trivedi, R. (1992). Microstructure and phase selection in laser treatment of materials. *Journal of Engineering Materials and Technology*, **114**, 450–458.

Mazumder, J. (1987). An overview of melt dynamics in laser processing. In: Kreutz, E.W., Quenzer, A. and Schuöcker, D. eds *Proceedings of the Conference High Power Lasers*. Vol. 801. Bellingham: SPIE. pp. 228–241.

Mazumder, J. and Kar, A. (1988). Nonequilibrium processing with lasers. *Opto Elektronik Magazin*, **4**, (3), 261–269.

GLAZING

Kear, B.H., Breinan, E.M. and Greenwald, L.E. (1979). Laser glazing – a new process for production and control of rapidly chilled metallurgical microstructures. *Metals Technology*, **5**, (4), 121–129.

ALLOYING

Draper, C.W. (1981). Laser surface alloying: a bibliography. *Applied Optics*, **20**, 3093–3096.

Draper, C.W. (1981). Laser surface alloying: the state of the art. In: Mukherjee, K. and Mazumder, J. eds *Proceedings of the Symposium Lasers in Metallurgy*, 22–26 February 1981, Chicago, IL, USA. Warrendale: The Metallurgical Society of AIME. pp. 67–92.

Draper, C.W (1982). Laser surface alloying: the state of the art. *Journal of Metals*, **34**, (6), 24–32.

Draper, C.W. and Poate, J.M. (1983). Laser surface alloying. In: Poate, J.M., Foti, G. and Jacobson, D.C. eds *Surface Modification and Alloying*. New York: Plenum Press. pp. 385–404.

Draper, C.W. and Ewing, C.A. (1984). Review. Laser surface alloying: a bibliography. *Journal of Materials Science*, **19**, 3815–3825.

Draper, C.W. and Poate, J.M. (1985). Laser surface alloying. *International Metal Reviews*, **30**, (2), 85–108.

EXERCISES

1. Describe the different modes of solidification possible during laser surface melting in terms of equilibrium and non-equilibrium phase transformations.

2. It is observed that a cast iron solidifies with a secondary dendrite arm spacing of $2\,\mu\text{m}$ after laser surface melting using an incident power of $2\,\text{kW}$ and a traverse rate of $1\,\text{m min}^{-1}$. Calculate the secondary dendrite arm spacing produced if the traverse rate is $0.7\,\text{m min}^{-1}$ and the incident power is $2.5\,\text{kW}$. Assume that the surface absorptivity remains constant.

3. By using Fig. 11.3, estimate the depth of melting expected when a nodular cast iron is treated at room temperature with a CO_2 laser beam of incident power $2\,\text{kW}$, beam diameter $5\,\text{mm}$, and traverse rate $0.1\,\text{m min}^{-1}$. Assume a value of 0.5 for the absorptivity of the material.

4. What are the main benefits to be gained by tailoring the surface of a component using laser surface alloying to meet its service requirements, in preference to making the component from bulk material of the required surface composition?

5. Show that during solidification of an equilibrium mixture of a binary alloy of A and B of average composition b, the weight fractions of liquid and solid, W_l and W_s, respectively, are given by

$$W_l = \frac{c - b}{c - a}$$

$$W_s = \frac{b - a}{c - a}$$

where a and c are the concentrations of B in the solid and liquid phases, respectively.

BIBLIOGRAPHY

Abbass, G., Steen, W.M. and West, D.R.F. (1988). Laser cladding with SiC particle injection. In: Sossenheimer, E.H. and Sepold, G. eds *Proceedings of the 2nd European Conference on Laser Treatment of Materials (ECLAT '88)*, 13–14 October 1988, Bad Nauheim, Germany. DVS Berichte 113. Düsseldorf: DVS-Verlag. pp. 76–78.

Abboud, J.H. and West, D.R.F. (1989). Ceramic-metal composites produced by laser surface treatment. *Materials Science and Technology*, **5**, (7), 725–728.

Abboud, J.H. and West, D.R.F. (1990). Laser surface alloying of titanium with aluminium. *Journal of Materials Science Letters*, **9**, (3), 308–310.

Abboud, J.H. and West, D.R.F. (1991a). Martensite formation in Ti–Al layers produced by laser surface alloying. *Materials Science and Technology*, **7**, (9), 827–834.

Abboud, J.H. and West, D.R.F. (1991b). Processing aspects of laser surface alloying of titanium with aluminium. *Materials Science and Technology*, **7**, (4), 353–356.

Abboud, J.H. and West, D.R.F. (1991c). Laser surface alloying of titanium with silicon. *Surface Engineering*, **7**, (2), 159–163.

Abboud, J.H. and West, D.R.F. (1992a). Laser surface melting of Ti–10V–2Fe–3Al alloy. *Journal of Materials Science Letters*, **11**, 1322–1326.

Abboud, J.H. and West, D.R.F. (1992b). Microstructures of titanium-aluminides produced by laser surface alloying. *Journal of Materials Science Letters*, **11**, 1675–1677.

Abboud, J.H., West, D.R.F. and Hibberd, R.H. (1993). Property assessment of laser surface treated titanium alloys. *Surface Engineering*, **9**, (3), 221–225.

Abboud, J.H. and West, D.R.F. (1994). Titanium aluminide composites produced by laser melting. *Materials Science and Technology*, **10**, (1), 60–68.

Affolter, K., Lüthy, W. and von Allmen, M. (1978). Properties of laser-assisted doping in silicon. *Applied Physics Letters*, **33**, (2), 185–187.

Åhman, L. (1984). Microstructure and its effect on toughness and wear resistance of laser surface melted and post heat treated high speed steel. *Metallurgical Transactions*, **15A**, 1829–1835.

Akgun, O.V. and Inal, O.T. (1992a). Desensitization of sensitized 304 stainless steel by laser surface melting. *Journal of Materials Science*, **27**, 2147–2153.

Akgun, O.V. and Inal, O.T. (1992b). Laser surface melting of Ti–6Al–4V alloy. *Journal of Materials Science*, **27**, 1404–1408.

Akgun, O.V. and Inal, O.T. (1995a). Laser surface melting and alloying of type 304L stainless steel. Part I. Microstructural characterization. *Journal of Materials Science*, **30**, 6097–6104.

Akgun, O.V. and Inal, O.T. (1995b). Laser surface melting and alloying of type 304L stainless steel. Part II. Corrosion and wear resistance properties. *Journal of Materials Science*, **30**, 6105–6112.

Akkurt, A.S., Akgün, O.V. and Yakupoğlu, N. (1996). The effect of post-heat treatment of laser surface melted AISI 1018 steel. *Journal of Materials Science*, **31**, 4907–4911.

Almeida, A., Anjos, M.A., Vilar, R., Li, R., Ferreira, M.G.S., Watkins, K.G. and Steen, W.M. (1995a). Laser alloying of aluminium alloys with chromium. *Surface and Coatings Technology*, **70**, 221–229.

Almeida, A., Que, Y.Y. and Vilar, R. (1995b). Microstructure of Al–4.3(at.%)Cr alloy produced by laser surface alloying. *Scripta Metallurgica et Materialia*, **33**, 863–870.

Altshulin, S., Zahavi, J., Rosen, A. and Nadiv, S. (1990). The interaction between a pulsed laser beam and a steel surface. *Journal of Materials Science*, **25**, (4), 2259–2263.

Anthony, T.R. and Cline, H.E. (1977). Surface rippling induced by surface-tension gradients during laser surface melting and alloying. *Journal of Applied Physics*, **48**, (9), 3888–3894.

Anthony, T.R. and Cline, H.E. (1978). Surface normalization of sensitised stainless steel by laser surface melting. *Journal of Applied Physics*, **49**, (3), 1248–1255.

Anthony, T.R. and Cline, H.E. (1979). Laser surface melting of stainless steel for corrosion protection. In: Ready, J.F. ed. *Proceedings of the Conference Laser Applications in Materials Processing*. Vol. 198. Bellingham: SPIE. pp. 82–91.

Ayers, J.D. and Schaefer, R.J. (1979). Consolidation of plasma-sprayed coatings by laser remelting. In: Ready, J.F. ed. *Proceedings of the Conference Laser Applications in Materials Processing*. Vol. 198. Bellingham: SPIE. pp. 57–64.

Ayers, J.D. and Tucker, T.R. (1980). Particulate–TiC-hardened steel surfaces by laser melt injection. *Thin Solid Films*, **73**, 201–207.

Ayers, J.D. (1981). Modification of metal surfaces by the laser melt/particle injection process. *Thin Solid Films*, **84**, 323.

Ayers, J.D., Tucker, T.R. and Schaefer, R.J. (1981a). In: Metzbower, E.A. ed. *Source Book on Applications of the Laser in Metalworking*, 18–20 April 1979, Washington DC, USA. Metals Park: American Society for Metals. p. 301.

Ayers, J.D., Schaefer, R.J. and Robey, W.P. (1981b). A laser processing technique for improving the wear resistance of metals. *Journal of Metals*, **7**, (8), 19–23.

Ayers, J.D. (1984). Wear behaviour of carbide injected titanium and aluminium alloys. *Wear*, **97**, 249–266.

Bande, H., L'Espérance, G., Islam, M.U. and Koul, A.K. (1991). Laser surface hardening of AISI O1 tool steel and its microstructure. *Materials Science and Technology*, **7**, (5), 452–457.

Bergmann, H.-W., Fritsch, H.U. and Hunger, G. (1981). Calculation of cooling and heating rates and transformation curves for the preparation of metallic glasses. *Journal of Materials Science*, **16**, 1935–1944.

Bergmann, H.W. and Young, M. (1985). Properties of laser melted SG iron. In: Kimmitt, M.F. ed. *Proceedings of the 2nd International Conference Lasers in Manufacturing (LIM-2)*, 26–28 March 1985, Birmingham, UK. Bedford: IFS (Publications) Ltd. pp. 109–118.

Bharti, A., Sivakumar, R. and Goel, D.B. (1989). A study in surface alloying of titanium alloys. *Journal of Laser Applications*, **1**, (2), 43–47.

Bingham, D., Jing, J., Campbell, S.J. and Cadogan, J.M. (1993). Laser glazing of Nd–Fe–B magnets. *Materials Science and Engineering*, **A160**, 107–111.

Bonora, P.L., Bassoli, M., Cerisola, G., De Anna, P.L., Battaglin, G., Della Mea, G. and Mazzoldi, P. (1981). On the corrosion behaviour of laser-irradiated aluminium surfaces. *Thin Solid Films*, **81**, 339–345.

Breinan, E.M., Kear, B.H. and Banas, C.M. (1976). Processing materials with lasers. *Physics Today*, **29**, (11), 44–50.

Canova, P. and Ramous, E. (1986). Carburization of iron surface induced by laser heating. *Journal of Materials Science*, **21**, 2143–2146.

Carbucicchio, M. and Palombarini, G. (1986). Surface structures produced in 1C–1.5Cr and 0.38C–Ni–Cr–Mo steels by high power CO_2 laser processing. *Journal of Materials Science*, **21**, 75–82.

Chan, C.L., Mazumder, J. and Chen, M.M. (1987). Three-dimensional axisymmetric model for convection in laser-melted pools. *Materials Science and Technology*, **3**, (4), 306–311.

Chande, T. and Mazumder, J. (1983a). Composition control in laser surface alloying. *Metallurgical Transactions*, **14B**, (6), 181–190.

Chande, T. and Mazumder, J. (1983b). Dimensionless parameters for process control in laser surface alloying. *Optical Engineering*, **22**, (3), 362–365.

Chauxuan, D. (1983). Laser heat treatment of piston rings. *Zeitschrift für Werkstofftechnik*, **14**, 81–85.

Chen, C.H., Altstetter, C.J. and Rigsbee, J.M. (1986). Laser processing of cast iron for enhanced erosion resistance. *Metallurgical Transactions*, **15A**, (4), 719–728.

Chen, Z.D., West, D.R.F. and Steen, W.M. (1987). Laser melting of alloy cast irons. In: Banas, C.M. and Whitney, G.L. eds *Proceedings of the 5th International Congress Applications of Lasers and Electro-optics (ICALEO '86)*. Bedford: IFS (Publications) Ltd/Berlin: Springer-Verlag. pp. 27–35.

Chen, Z.D., West, D.R.F., Steen, W.M. and Cantello, M. (1988). Laser surface hardening of alloy cast irons. In: *Heat Treatment '87: Proceedings of the International Conference Included in Materials '87*, 11–15 May 1987, London, UK. London: Institute of Metals. pp. 171–180.

Chen, Z.-H., Ho, N.-J. and Shen, P. (1995). Microstructures of laser-treated Al_2O_3–ZrO_2–$CeCO_2$ composites. *Materials Science and Engineering*, **A196**, 253–260.

Chiba, S., Sato, T., Kawashima, A., Asami, K. and Hashimoto, K. (1986). Some corrosion characteristics of stainless surface alloys laser processed on a mild steel. *Corrosion Science*, **26**, (4), 311–328.

Chithambaram, S., Madhusudana, K., Prabhakar, O. and Bharti, A. (1989). Laser surface remelt-hardening of gray iron for improving the wear life of pattern equipment. In: *Proceedings of the 93rd AFS Casting Congress*. Illinois: American Foundrymen's Society. pp. 115–126.

Colaço, R. and Vilar, R. (1996). Phase selection during solidification of AISI 420 and AISI 420C tool steels. *Surface Engineering*, **12**, (4), 319–325.

Com-Nougué, J. and Kerrand, E. (1986). CO_2 laser deposition of a cobalt-based alloy on 12% Cr steel. In: Quenzer, A. ed. *Proceedings of the 3rd International Conference Lasers in Manufacturing (LIM-3)*, 3–5 June 1986, Paris, France. IFS Publications: Bedford/Springer-Verlag: Berlin. pp. 191–195.

Conners, A. and Svenzon, M. (1983). Parameters for the laser processing of tool and carbon steels. *Thin Solid Films*, **107**, 297–304.

Cooper, K.C. and Slebodnick, P. (1989). Recent developments in laser melt/particle injection processing. *Journal of Laser Applications*, **1**, 21–29.

Cordier-Robert, C., Crampon, J. and Foct, J. (1998). Surface alloying of iron by laser melting:

microstructure and mechanical properties. *Surface Engineering*, **14**, (5), 381–385.

Dahotre, N.B. and Mukherjee, K. (1990). Development of microstructure in laser surface alloying of steel with chromium. *Journal of Materials Science*, **25**, 445–454.

Dahotre, N.B., McCay, M.H., McCay, T.D. and Kim, M.M. (1992). Laser cladding of Co-based hardfacing on Cu substrate. *Journal of Materials Science*, **27**, 6426–6436.

Dallaire, S. and Cielo, P. (1982). Pulsed laser treatment of plasma sprayed coatings. *Metallurgical Transactions*, **13B**, (9), 479–483.

Das, D.K., Paradkar, A.G. and Mishra, R.S. (1992). Microstructure and creep behaviour of laser surface alloyed aluminium. *Scripta Metallurgica et Materialia*, **26**, (8), 1211–1214.

Das, D.K., Prasad, K.S. and Paradkar, A.G. (1994). Evolution of microstructure in laser surface alloying of aluminium with nickel. *Materials Science and Engineering*, **A174**, 75–84.

de Damborena, J., Vazquez, A.J., Gonzalez, J.A. and West, D.R.F. (1989). Elimination of intergranular corrosion susceptibility of a sensitised 304 steel by subsequent laser surface melting. *Surface Engineering*, **5**, (3), 235–238.

de Damborenea, J.J. and Vázquez, A.J. (1992). Laser processing of a NiCrAlFe alloy: microstructural evolution. *Journal of Materials Science*, **27**, 1271–1274.

de Damborenea, J., Fernández, B., López, V. and Vázquez, A.J. (1996). Hard coatings by laser gas alloying of Ti6Al4V. *International Journal of Materials and Product Technology*, **11**, (3/4), 1996.

de Mol van Otterloo, J.L. and DeHosson, J.Th.M. (1994). Laser treatment of aluminium copper alloys: a mechanical enhancement. *Scripta Metallurgica et Materialia*, **30**, (4), 493–498.

Dey, G.K., Kulkarni, U.D., Batra, I.S. and Banerjee, S. (1994). Laser surface alloying of nickel by molybdenum and aluminium – microstructural studies. *Acta Metallurgica et Materialia*, **42**, (9), 2973–2981.

Draper, C.W., Preece, C.M., Jacobson, D.C., Buene, L. and Poate, J. (1980a). Laser alloying of deposited metal films on nickel. In: White, C.W. and Peercy, P.S. eds *Laser and Electron Beam Processing of Materials*. New York: Academic Press. pp. 721–727.

Draper, C.W., Woods, R.E. and Meyer, L.S. (1980b). Enhanced corrosion resistance of laser surface melted aluminium bronze D (CDA-614). *Corrosion-NACE*, **36**, (8), 405–408.

Draper, C.W. (1981). The use of laser surface melting to homogenize Fe–Al bronzes. *Journal of Materials Science*, **16**, 2774–2780.

Draper, C.W., Buene, L., Poate, J.M. and Jacobson, D.C. (1981a). Rutherford backscattering and marker diffusion to determine melt threshold in laser mirror damage studies. *Applied Optics*, **20**, 1730–1732.

Draper, C.W., Meyer, L.S., Buene, L., Jacobson, D.C. and Poate, J.M. (1981b). Laser surface alloying of gold films on nickel. *Applied Surface Science*, **7**, 276–280.

Draper, C.W., Meyer, L.S., Jacobson, D.C., Buene, L. and Poate, J.M. (1981c). Laser surface alloying for passivation of Ti–Pd. *Thin Solid Films*, **75**, 237–240.

Draper, C.W. and Sharma, S.P. (1981). The effect of laser surface melting on tin-modified Cu–Ni. *Thin Solid Films*, **84**, 333–340.

Draper, C.W., Jacobson, D.C., Gibson, J.M., Poate, J.M., Vandenberg, J.M. and Cullis, A.G. (1982). Laser and ion beam mixing Cr and Ni + Cr films on Cu. In: Appleton, B.R. and Celler, G.K. eds *Proceedings of the Conference Laser and Electron Beam Interactions with Solids*. Amsterdam: Elsevier North-Holland. pp. 413–418.

Draper, C.W. (1986). Laser surface alloying of gold. *Gold Bulletin*, **19**, (1), 8–14.

Fasasi, A.Y., Pons, M., Tassin, C., Galerie, A., Sainfort, G. and Polak, C. (1994). Laser surface melting of mild steel with submicronic titanium carbide powders. *Journal of Materials Science*, **29**, 5121–5126.

Fastow, M., Bamberger, M., Nir, N. and Landkof, M. (1990). Laser surface melting of AISI 4340 steel. *Materials Science and Technology*, **6**, (9), 900–904.

Ferraro, F., Nannetti, C.A., Campello, M. and Senin, A. (1986) Aluminium alloy surface hardening by laser treatment. In: Quenzer, A. ed. *Proceedings of the 3rd International Conference Lasers in Manufacturing (LIM-3)*, 3–5 June 1986, Paris, France. IFS Publications: Bedford/Springer-Verlag: Berlin. pp. 233–243.

Ferreira, M.G.S., Li, R. and Vila, R. (1996). Avoiding crevice corrosion by laser surface treatment. *Corrosion Science*, **38**, (12), 2091–2094.

Fink, U. and Bergmann, H.W. (1990). Laser surface treatment of implants. In: Bergmann, H.W. and Kupfer, R. eds *Proceedings of the 3rd European Conference on Laser Treatment of Materials (ECLAT '90)*, 17–19 September 1990, Erlangen, Germany. Coburg: Sprechsaal Publishing Group. pp. 451–460.

Fu, Y. and Batchelor, A.W. (1998). Laser alloying of aluminum alloy AA6061 with Ni and Cr. Part II. The effect of laser alloying on the fretting wear resistance. *Surface & Coatings Technology*, **102**, 119–126.

Gadag, S.P., Srinivasan, M.N. and Mordike, B.L. (1995). Effect of laser processing parameters on the structure of ductile iron. *Materials Science and Engineering*, **A196**, 145–151.

Gaffet, E., Pelletier, J.M. and Bonnet-Jobez, S. (1989). Laser surface alloying of Ni film on Al-based alloy. *Acta Metallurgica*, **37**, (12), 3205–3215.

Galantucci, L.M., Ruta, G. and Magnanelli, S. (1989). An experimental study on CO_2 500 W laser surface alloying of chromium on copper. In: Niku-Lari, A. and Mordike, B.L. eds *High Power Lasers*. Oxford: Pergamon Press. pp. 57–74.

Galun, R., Weisheit, A. and Mordike, B.L. (1994). Surface treatment of magnesium alloys with laser melting and laser alloying. In: *Proceedings of the 5th European Conference on Laser Treatment of Materials (ECLAT '94)*, 26–27 September 1994, Bremen-Vegesack, Germany. DVS Berichte 163. Düsseldorf: Deutscher Verband für Schweißtechnik. pp. 421–426.

Galun, R., Weisheit, A. and Mordike, B.L. (1995). Laser surface alloying of magnesium base alloys to improve the corrosion and wear resistance. In: Mazumder, J., Matsunawa, A. and Magnusson, C. eds *Proceedings of the Laser Materials Processing Conference (ICALEO '95)*. Orlando: Laser Institute of America. pp. 69–77.

García, I. and de Damborenea, J.J. (1998). Corrosion properties of TiN prepared by laser gas alloying of Ti and Ti6Al4V. *Corrosion Science*, **40**, (8), 1411–1419.

Gasser, A., Herziger, G., Kreutz, E.W. and Wissenbach, K. (1987). Surface melting of cast iron: laser parameters, processing conditions, surface morphology, metallurgy. In: Arata, Y. ed. *Proceedings of the Conference Laser Advanced Materials Processing (LAMP '87)*, 21–23 May 1987, Osaka, Japan. Osaka: High Temperature Society of Japan and Japan Laser Processing Society. pp. 451–458.

Geetha, M., Kamachi Mudali, U., Pandey, N.D., Asokamani, R. and Raj, B. (2004). Microstructural and corrosion evaluation of laser surface nitrided Ti–13Nb–13Zr alloy. *Surface Engineering*, **20**, (1), 68–74.

Gemelli, E., Gallerie, A. and Caillet, M. (1998). Improvement of resistance to oxidation by laser alloying on a tool steel. *Scripta Materialia*, **39**, (10), 1345–1352.

Giannetaki, P.E., Chryssoulakis, Y. and Ponthiaux, P. (1994). Laser surface treatment of aluminium alloy 6061 with silicon carbide powder injection: study of microstructure and wear behavior. *Plating and Surface Finishing*, **81**, (11), 52–56.

Gill, S.C., Zimmermann, M. and Kurz, W. (1992). Laser resolidification of the Al–Al_2Cu eutectic: the coupled zone. *Acta Metallurgica et Materialia*, **40**, (11), 2895–2906.

Gjønnes, L. and Olsen, A. (1994). Laser-modified aluminium surfaces with iron. *Journal of Materials Science*, **29**, 728–735.

Gnanamuthu, D.S. (1979). Laser surface treatment. In: Metzbower, E.A. ed. *Proceedings of the Conference Applications of Lasers in Materials Processing*, 18–20 April 1979, Washington DC,

USA. Metals Park: American Society for Metals. pp. 177–211.

Greenwald, L.E., Breinan, E.M. and Kear, B.H. (1979). Heat transfer properties and microstructures of laser surface melted alloys. In: Ferris, S.D., Leamy, H.J. and Poate, J.M. eds *Proceedings of the Conference Laser–Solid Interactions and Laser Processing 1978*. New York: AIP. p. 189.

Grum, J. and Šturm, R. (1996). Microstructure analysis of nodular iron 400-12 after laser surface melt hardening. *Materials Characterization*, **37**, (2–3), 81–88.

Grum, J. and Šturm, R. (2002). Comparison of measured and calculated thickness of martensite and ledeburite shells around graphite nodules in the hardened layer of nodular iron after laser surface remelting. *Applied Surface Science*, **187**, 116–123.

Grünenwald, B., Bischoff, E., Shen, J. and Dausinger, F. (1992). Laser surface alloying of case hardening steel with tungsten carbide and carbon. *Materials Science and Technology*, **8**, (7), 637–643.

Gureyev, D.M., Zolotarevsky, A.V. and Zaikin, A.E. (1991). Laser-arc hardening of aluminium alloys. *Journal of Materials Science*, **26**, 4678–4682.

Hahn, K., Buerhop, C. and Weißmann, R. (1996). Firing PbO-free glass enamels using the cw-CO_2 laser. *Glass Science and Technology*, **69**, (1) 1–6.

Hamazaki, M. (1995). Laser surface hardening of automotive cams. *Welding International*, **9**, (2), 158–163.

Han, Z.-F., Feng, Z.-G., Zhou, Z.-R. and Liou, J.-L. (1992). High speed laser boronising of high speed steel gear cutting hobs. *Surface Engineering*, **8**, (1), 66–67.

Hegge, H.J. and De Hosson, J.Th.M. (1987). The relationship between hardness and laser treatment of hypo-eutectoid steels. *Scripta Metallurgica et Materialia*, **21**, (12), 1737–1742.

Hegge, H.J. and De Hosson, J.Th.M. (1990a). Solidification structures during laser treatment. *Scripta Metallurgica et Materialia*, **24**, (3), 593–599.

Hegge, H.J. and De Hosson, J.Th.M. (1990b). Microstructure of laser treated Al alloys. *Acta Metallurgica et Materialia*, **38**, (12), 2471–2477.

Hegge, H.J., De Beurs, H., Noordhuis, J. and De Hosson, J.Th.M. (1990c). Tempering of steel during laser treatment. *Metallurgical Transactions*, **21A**, 987–995.

Hiraga, H., Inoue, T., Kamado, S. and Kojima, Y. (2001). Improving the wear resistance of a magnesium alloy by laser melt injection. *Materials Transactions*, **42**, 1322–1325.

Hirose, A., Arita, Y. and Kobayashi, K.F. (1995). Microstructure and crack sensitivity of laser-fusion zones of Ti–46mol% Al–2mol% Mo alloy. *Journal of Materials Science*, **30**, 970–979.

Hlawka, F., Marchione, T., Jacura, O. and Cornet, A. (1993). Characterisation of M42 type high speed steel after laser melting treatment. *Surface Engineering*, **9**, (4), 300–304.

Hu, C., Xin, H. and Baker, T.N. (1995). A semi-empirical model to predict the melt depth developed in overlapping laser tracks on a Ti–6Al–4V alloy. *Journal of Materials Science*, **30**, 5985–5990.

Ion, J.C., Moisio, T., Pedersen, T.F., Sørensen, B. and Hansson, C.M. (1991). Laser surface modification of a 13.5%Cr, 0.6%C steel. *Journal of Materials Science*, **26**, 43–48.

Jackson, K.A. and Hunt, J.D. (1966). Lamellar and rod eutectic growth. *Transactions of the Metallurgical Society of the American Institute of Mining Engineers*, **236**, 1129–1142.

Jain, A.K., Kulkarni, V.N. and Sood, D.K. (1981). Laser treatment of chromium films on aluminium at high power densities. *Thin Solid Films*, **86**, 1–9.

Jasim, K.M., Rawlings, R.D. and West, D.R.F. (1991). Pulsed laser processing of thermal barrier coatings. *Journal of Materials Science*, **26**, 909–916.

Jasim, K.M., Rawlings, R.D. and West, D.R.F. (1992). Operating regimes for laser surface engineering of ceramics. *Journal of Materials Science*, **27**, 1937–1946.

Jasim, K.M., Rawlings, R.D. and West, D.R.F. (1993). Metal-ceramic functionally gradient material produced by laser processing. *Journal of Materials Science*, **28**, 2820–2826.

Jervis, T.R., Zocco, T.G., Hubbard, K.M. and Nastasi, M. (1993). Excimer laser surface treatment of Ti–6Al–4V. *Metallurgical Transactions*, **24A**, 215–224.

Jiandong, H., Zhang, L., Yufeng, W. and Xiangzhang, B. (1992). Wear resistance of laser processed 1.0%C tool steel. *Materials Science and Technology*, **8**, (9), 796–798.

Jost, N. and Haddenhorst, H. (1991). Laserglazing of aluminium-based alloys. *Journal of Materials Science Letters*, **10**, 913–916.

Juarez-Islas, J.A. (1991). Analysis of the microstructure obtained by using unidirectional solidification, tungsten inert gas weld and laser surface melt traversing techniques in Al–Mn alloys. *Journal of Materials Science*, **26**, 5004–5012.

Kahrmann, W. (1988). Laser surface coating with cermets. In: Sossenheimer, E.H. and Sepold, G. eds *Proceedings of the 2nd European Conference on Laser Treatment of Materials (ECLAT '88)*, 13–14 October 1988, Bad Nauheim, Germany. DVS Berichte 113. Düsseldorf: DVS-Verlag. pp.119–120.

Khanna, A.S., Gasser, A., Wissenbach, K., Li, M., Desai, V.H. and Quadakkers, W.J. (1995). Oxidation and corrosion behaviour of mild steel laser alloyed with nickel and chromium. *Journal of Materials Science*, **30**, 4684–4691.

Konitzer, D.G., Muddle, B.C. and Fraser, H.L. (1983). A comparison of the microstructures of as-cast and laser surface melted Ti–8Al–4Y. *Metallurgical Transactions*, **14A**, 1979–1988.

Kostrubiec, F. and Walczak, M. (1991). Microhardness of the surface layer of tungsten and molybdenum after recrystallization with laser radiation. *Journal of Engineering Materials and Technology*, **113**, 130–134.

Kostrubiec, F. and Walczak, M. (1994). Thermal stresses in the tungsten and molybdenum surface layer following laser treatment. *Journal of Materials Science Letters*, **13**, 34–36.

Koutsomichalis, A., Saettas, L. and Badekas, H. (1994). Laser treatment of magnesium. *Journal of Materials Science*, **29**, 6543–6547.

Kusiński, J. (1988). Laser melting of T1-high speed tool steel. *Metallurgical Transactions*, **19A**, 377–382.

Lamb, M., Steen, W.M. and West, D.R.F. (1985). The pitting corrosion behaviour of laser surface melted 420 and 316 stainless steels. In: Mazumder, J. ed. *Proceedings of the Materials Processing Symposium (ICALEO '84)*, Toledo: Laser Institute of America. pp. 133–139.

Lamb, M., West, D.R.F. and Steen, W.M. (1986). Residual stresses in two laser surface melted stainless steels. *Materials Science and Technology*, **2**, (9), 974–980.

Lawrence, J., Li, L. and Spencer, J.T. (1998a). A two-stage ceramic tile grout sealing process using a high power diode laser – I. Grout development and material characteristics. *Optics and Laser Technology*, **30**, 205–214.

Lawrence, J., Li, L. and Spencer, J.T. (1998b). A two-stage ceramic tile grout sealing process using a high power diode laser – II. Mechanical, chemical and physical properties. *Optics and Laser Technology*, **30**, 215–223.

Lawrence, J. and Li, L. (1999). Surface glazing of concrete using a 2.5 kW high power diode laser and the effects of large beam geometry. *Optics and Laser Technology*, **31**, 583–591.

Lawrence, J. and Li, L. (2000a). Carbon steel wettability characteristics enhancement for improved enamelling using a 1.2 kW high power diode laser. *Optics and Lasers in Engineering*, **32**, 353–365.

Lawrence, J. and Li, L. (2000b). High power diode laser surface glazing of concrete. *Journal of Laser Applications*, **12**, (3), 116–125.

Lawrence, J. and Li, L. (2000c). The wear characteristics of a high power diode laser generated glaze on the ordinary Portland cement surface of concrete. *Wear*, **246**, (1–2), 91–97.

Lee, C.H., Kim, K.C. and Chang, R.W. (1995). Laser desensitisation of the heat affected zone (HAZ) in austenitic stainless steels. *Welding International*, **9**, (2), 121–127.

Lee, S.Z. and Zum Gahr, K.-H. (1990). In: Bergmann, H.W. and Kupfer, R. eds *Proceedings of the 3rd European Conference on Laser Treatment of Materials (ECLAT '90)*, 17–19 September 1990, Erlangen, Germany. Coburg: Sprechsaal Publishing Group. pp. 515–521.

Leech, P.W. (1986). Comparison of the sliding wear process of various cast irons in the laser-surface-melted and as-cast forms. *Wear*, **113**, 233–245.

Leech, P.W., Batchelor, A.W. and Stachowiak, G.W. (1992). Laser surface alloying of steel wire with chromium and zirconium. *Journal of Materials Science Letters*, **11**, 1121–1123.

Li, Z., Zou, J. and Xu, Z. (1987). Effect of laser processing on the cavitation erosion of Cr–Mo–Cu alloy cast iron. *Wear*, **119**, 13–27.

Li, M., Kar, A., Desai, V. and Khanna, A. (1995). High-temperature oxidation resistance improvement of titanium using laser surface alloying. *Journal of Materials Science*, **30**, 5093–5098.

Li, Y., Quick, N.R. and Kar, A. (2002). Dieless laser drawing of fine metal wires. *Journal of Materials Processing Technology*, **123**, 451–458.

Lichtman, D. and Ready, J.F. (1963). Laser beam induced electron emission. *Physical Review Letters*, **10**, 342–345.

Lima, M.S.F. and Goldenstein, H. (2000). Structure of laser remelted surface of cast irons. *Surface Engineering*, **16**, (2), 127–130.

Liu, C.A. and Humphries, M.J. (1984). Effects of process parameters on laser surface modification. In: Metzbower, E.A. ed. *Proceedings of the Materials Processing Symposium (ICALEO '83)*. Toledo: Laser Institute of America. pp. 108–117.

Liu, Z., Xie, C.S., Watkins, K.G., Steen, W.M., Vilar, R.M. and Ferreira, M.G.S. (1995). Microstructure and pitting behaviour of Al–Ti–Ni laser alloyed layers in 2014 substrates. In: Mazumder, J., Matsunawa, A. and Magnusson, C. eds *Proceedings of the Laser Materials Processing Conference (ICALEO '95)*. Orlando: Laser Institute of America. pp. 431–440.

Lopez, V., Bello, J.M., Ruiz, J. and Fernandez, B.J. (1994). Surface treatment of ductile irons. *Journal of Materials Science*, **29**, 4216–4224.

Ma, D., Li, Y., Wang, F.D. and Li, Z.Y. (2001). Laser resolidification of a Zn–3.37 wt.% Cu peritectic alloy. *Materials Science and Engineering*, **A318**, 235–243.

Major, B., Ciach, R., Handzel-Powierza, Z. and Radziejewska, J. (1993). Structure modification by laser remelting. In: Charissoux, C. ed. *Proceedings of the 5th International Conference on Welding and Melting by Electron and Laser Beams (CISFFEL-5)*, 14–18 June 1993, La Baule, France. Saclay: SDEM. pp. 409–416.

Manna, I., Abraham, S., Reddy, G., Bose, D.N., Ghosh, T.B. and Pabi, S.K. (1994). Laser surface alloying of aluminium on copper substrate. *Scripta Metallurgia et Materialia*, **31**, (6), 713–718.

Manohar, M. and Wilde, B.E. (1995). Laser surface alloying with noble metals to impede hydrogen ingress into AISI 4135 steel. *Corrosion Science*, **37**, (4), 607–619.

Masumoto, I., Shinoda, T. and Hirate, T. (1990a). Weld decay recovery by laser beam surfacing of austenitic stainless steel welded joints. *Transactions of the Japan Welding Society*, **21**, (1), 11–17.

Masumoto, I., Shinoda, T. and Hirate, T. (1990b). Laser beam surfacing to avoid weld decay. *Welding International*, **4**, (7), 515–520.

Mathur, A.K. and Molian, P.A. (1985). Laser heat treatment of cast irons – optimization of process variables: Part I. *Journal of Engineering Materials and Technology*, **107**, 200–207.

Matsuda, F. and Nakata, K. (1990). Laser surface alloying of aluminium. In: Russell, J.D. ed. *Proceedings of the 2nd International Conference on Power Beam Technology*, 23–26 September 1990, Stratford-upon-Avon, UK. Abington: The Welding Institute. pp. 297–303.

Matsuyama, H. and Shibata, K. (1994). Study of alloying by defocused laser beam. Surface alloying of aluminium by CO_2 laser irradiation (1). *Welding International*, **8**, (4), 274–278.

McCafferty, E. and Moore, P.G. (1986). Electrochemical behaviour of laser-processed metal surfaces. In: Draper, C.W. and Mazzoldi, P. eds *Proceedings of the NATO ASI Laser Surface Treatment of Metals*, 2–13 September 1985, San Miniato, Italy. Dordrecht: Martinus Nijhoff. pp. 263–295.

Molian, P.A. and Wood, W.E. (1982a). Formation of austenite in laser-processed Fe–0.2%C–20%Cr alloy. *Materials Science and Engineering*, **56**, 271–277.

Molian, P.A. and Wood, W.E. (1982b). Martensitic transformation of laser processed Fe–5%Ni alloy. *Scripta Metallurgica*, **16**, (11), 1301–1303.

Molian, P.A. and Wood, W.E. (1983a). Transformation behaviour of laser processed Fe–5%Cr, Fe–5%Ni and Fe–6%Cr–2%Ni alloys. *Journal of Materials Science*, **18**, 2555–2562.

Molian, P.A. and Wood, W.E. (1983b). Non-equilibrium phases in laser-processed Fe–0.2%wt.C–20%wt.Cr alloys. *Materials Science and Engineering*, **60**, 241–245.

Molian, P.A. and Wood, W.E. (1983c). Ferrite morphology in rapidly solidified ferritic Fe–Cr–C steels. *Scripta Metallurgica*, **17**, (4), 431–434.

Molian, P.A. and Wood, W.E. (1984). Rapid solidification of laser-processed chromium steels. *Materials Science and Engineering*, **62**, 271–277.

Molian, P.A. (1985). Structure and hardness of laser-processed Fe–0.2%C–5%Cr and Fe–0.2%C–10%Cr alloys. *Journal of Materials Science*, **20**, 2903–2912.

Molian, P.A. and Mathur, A.K. (1985). Laser heat treatment of cast irons – optimization of process variables, Part II. *Journal of Engineering Materials and Technology*, **108**, 233–239.

Molian, P.A. and Baldwin, M. (1986). Wear behavior of laser surface-hardened gray and ductile cast irons. Part 1 – sliding wear. *Journal of Tribology*, **108**, 326–333.

Molian, P.A. and Rajasekhara, H.S. (1986). Analysis of microstructures of laser surface-melted tool steels. *Journal of Materials Science Letters*, **5**, 1292–1294.

Molian, P.A. and Baldwin, M. (1988). Wear behavior of laser surface-hardened gray and ductile cast irons. Part 2 – erosive wear. *Journal of Tribology*, **110**, 462–466.

Moore, P., Kim, C. and Weinman, L.S. (1979). Topographical characteristics of laser surface melted materials. In: Metzbower, E.A. ed. *Proceedings of the Conference Applications of Lasers in Materials Processing*, 18–20 April 1979, Washington DC, USA. Metals Park: American Society for Metals. pp. 259–282.

Moore, P.G. and McCafferty, E. (1981). Passivation of Fe/Cr alloys prepared by laser-surface alloying. *Journal of the Electrochemical Society*, **128**, 1391–1393.

Mridha, S. and Baker, T.M. (1991). Characteristic features of laser-nitrided surfaces of two titanium alloys. *Materials Science and Engineering*, **A142**, (1), 115–124.

Mridha, S. and Baker, T.M. (1996). Metal matrix composite layers formed by laser processing of commercial purity $Ti–SiC_p$ in nitrogen environment. *Materials Science and Technology*, **12**, (7), 595–602.

Mudali, U.K., Dayal, R.K. and Goswami, G.L. (1995). Desensitisation of austenitic stainless steels using laser surface melting. *Surface Engineering*, **11**, (4), 331–335.

Mueller, H., Wetzig, K., Schultrich, B., Pimenov, S.M., Chapliev, N.I., Konov, V.I. and Prochorov, A.M. (1989). *In situ* laser irradiation of WC–Co hardmetals inside an SEM. *Journal of Materials Science*, **24**, 3328–3336.

Munitz, A. (1985). Microstructure of rapidly solidified laser molten Al–4.5 wt pct Cu surfaces. *Metallurgical Transactions*, **16B**, 149–161.

Nakao, Y. (1989). Surface treatment by laser quenching. *Welding International*, **3**, (7), 619–623.

Nakao, Y. and Nishimoto, K. (1986). Desensitization of stainless steels by laser surface heat-treatment. *Transactions of the Japan Welding Society*, **17**, (1), 84–92.

Nakao, Y., Nishimoto, K. and Zhang, W.-P. (1988). Effects of rapid solidification by laser surface melting on solidification modes and microstructures of stainless steels. *Transactions of the Japan Welding Society*, **19**, (2), 100–106.

Narayan, J., Young, R.T., Wood, R.F. and Christie, W.H. (1978). p–n junction formation in boron-deposited silicon by laser-induced diffusion. *Applied Physics Letters*, **33**, (4), 338–3410.

Nissim, Y.I., Lietoila, A., Gold, R.B. and Gibbons, J.F. (1980). Temperature distributions produced in semiconductors by a scanning elliptical or circular cw laser beam. *Journal of Applied Physics*, **51**, (1), 274–279.

Noordhuis, J. and De Hosson, J.Th.M. (1992). Surface modification by means of laser melting combined with shot peening: a novel approach. *Acta Metallurgica et Materialia*, **40**, (12), 3317–3324.

Noordhuis, J. and De Hosson, J.Th.M. (1993). Microstructure and mechanical properties of a laser treated Al alloy. *Acta Metallurgica et Materialia*, **41**, (7), 1989–1998.

Oakley, P.J. and Bailey, N. (1986). Laser surfacing of nickel aluminium bronze. In: Albright, C. ed. *Proceedings of the International Conference Applications of Lasers and Electro-optics (ICALEO '85)*. Bedford: IFS Publications/Berlin: Springer-Verlag. pp. 169–177.

Ocelík, V., Vreeling, J.A. and De Hosson, J.Th.M. (2001). EBSP study of reaction zone in SiC/Al metal matrix composite prepared by laser melt injection. *Journal of Materials Science*, **26**, 4845–4849.

Olaineck, C. and Lührs, D. (1996). Economic and technical features of laser camshaft remelting. *Heat Treatment of Metals*, **23**, (1), 17–19.

Ozbaysal, K. and Inal, O.T. (1994). Age-hardening kinetics and microstructure of PH 15-5 stainless steel after laser melting and solution treating. *Journal of Materials Science*, **29**, 1471–1480.

Palombarini, G., Sambogna, G. and Carbucicchio, M. (1991). The occurrence of austenitic layers in laser surface-melted iron alloys. *Journal of Materials Science*, **26**, 3396–3399.

Panagopoulos, C. and Michaelides, A. (1992). Laser surface treatment of copper. *Journal of Materials Science*, **27**, 1280–1284.

Panagopoulos, C.N., Markaki, A.E. and Agathocleous, P.E. (1998). Excimer laser treatment of nickel-coated cast iron. *Materials Science and Engineering*, **A241**, 226–232.

Papaphilippou, C., Vardavoulias, M. and Jeandin, M. (1996). Effects of CO_2 laser surface hardening in the unlubricated wear of ductile cast iron against alumina. *Journal of Tribology*, **118**, 748–752.

Pei, Y.T., Ocelik, V. and De Hosson, J.T.M. (2002). SiCp/Ti6Al4V functionally graded materials produced by laser melt injection. *Acta Materialia*, **50**, (8), 2035–2051.

Pelletier, J.M., Pergue, D., Fouquet, F. and Mazille, H. (1989). Laser surface melting of low and medium carbon steels: influence on mechanical and electrochemical properties. *Journal of Materials Science*, **24**, 4343–4349.

Pelletier, J.M., Renaud, L. and Fouquet, F. (1991). Solidification microstructures induced by laser surface alloying: influence on substrate. *Materials Science and Engineering*, **A134**, 1283–1287.

Pons, M., Galerie, A. and Caillet, M. (1986a). A comparison between ion implantation and laser alloying of pure iron for oxidation resistance improvement. Part 1 Boron alloying. *Journal of Materials Science*, **21**, 2697–2704.

Pons, M., Caillet, M. and Galerie, A. (1986b). A comparison between ion implantation and laser alloying of iron for oxidation resistance improvement. Part 2 Aluminium alloying. *Journal of Materials Science*, **21**, 4101–4106.

Pou, J., Trillo, C. and Pérez-Amor, M. (2002). Laser surface treatment of dimension stones. *Industrial Laser Solutions*, **17**, (9), 19–24.

Preece, C.M. and Draper, C.W. (1981). The effect of laser quenching the surfaces of steels on their cavitation erosion resistance. *Wear*, **67**, 321–328.

Pujar, M.G., Dayal, R.K., Khanna, A.S. and Kreutz, E.W. (1993). Effect of laser surface melting on the corrosion resistance of chromium-plated 9Cr–1Mo ferritic steel in an acidic medium. *Journal of Materials Science*, **28**, 3089–3096.

Rabitsch, K., Ebner, R. and Major, B. (1994). Boride laser alloying of M2 high speed steel. *Scripta Metallurgica et Materialia*, **30**, (2), 253–258.

Rawers, J.C. (1984). Impact testing of laser glazed Charpy samples. *International Journal of Fracture*, **25**, R61–R63.

Rawers, J., Reitz, W., Bullard, S. and Roub, E.K. (1991). Surface and corrosion study of laser-processed zirconium alloys. *Corrosion*, **47**, 769–777.

Reitz, W. and Rawers, J. (1992). Effect of laser surface melted zirconium alloys on microstructure and corrosion resistance. *Journal of Materials Science*, **27**, 2437–2443.

Renaud, L., Fouquet, F., Elhamdaoui, A., Millet, J.P., Mazille, H. and Crolet, J.L. (1990). Surface alloys obtained on mild steel by laser treatment of electroless nickel coatings. *Acta Metallurgica et Materialia*, **38**, (8), 1547–1553.

Renaud, L., Fouquet, F., Elhamdaoui, A., Millet, J.P., Mazille, H. and Crolet, J.L. (1991). Microstructural characterization and comparative electrochemical behaviour of Fe–Ni–Cr and Fe–Ni–Cr–P laser surface alloys. *Materials Science and Engineering*, **A134**, 1049–1053.

Riabkina-Fishman, M. and Zahavi, J. (1988). Structure and microhardness of laser-hardened 1045 steel. *Journal of Materials Science*, **23**, 1547–1552.

Riabkina-Fishman, M. and Zahavi, J. (1990). Laser surface treatments for promoting corrosion resistance of a carbon steel. *Materials and Manufacturing Processes*, **5**, (4), 641–660.

Riabkina-Fishman, M., Rabkin, E., Galun, R., Maiwald, T. and Mordike, B.L. (2001). Structure and composition of laser produced WC alloyed surface layers on M2 high-speed steel. *Journal of Materials Science Letters*, **20**, 1917–1920.

Ricciardi, G., Pasquini, F., Rudilosso, S. and Tiziani, A. (1983). Remelting surface hardening of cast iron by CO_2 laser. In: *Proceedings of the 1st International Conference Lasers in Manufacturing*, 1–3 November 1983, Brighton, UK. Bedford: IFS Publications/ Amsterdam: North-Holland. pp. 87–95.

Rieker, C., Morris, D.G. and Steffen, J. (1989). Formation of hard microcrystalline layers on stainless steel by laser alloying. *Materials Science and Technology*, **5**, 590–594.

Robinson, J.M., Anderson, S., Knutsen, R.D. and Reed, R.C. (1995). Cavitation erosion of laser melted and laser nitrided Ti–6Al–4V. *Materials Science and Technology*, **11**, (6), 611–618.

Schaaf, P., Emmel, A., Illgner, C., Lieb, K.P., Schubert, E. and Bergmann, H.W. (1995). Laser nitriding of iron by excimer laser irradiation in air and N_2 gas. *Materials Science and Engineering*, **A197**, L1–L4.

Seaman, F.D. and Gnanamuthu, D.S. (1975). Using the industrial laser to surface harden and alloy. *Metal Progress*, **108**, (8), 67–74.

Seiersten, M. (1988). Surface nitriding of titanium by laser beams. In: Sossenheimer, E.H. and Sepold, G. eds *Proceedings of the 2nd European Conference on Laser Treatment of Materials (ECLAT '88)*, 13–14 October 1988, Bad Nauheim, Germany. DVS Berichte 113. Düsseldorf: DVS-Verlag. pp. 66–69.

Shibata, K. and Matsuyama, H. (1990). Surface alloying of aluminium by CO_2 laser radiation. In: Bergmann, H.W. and Kupfer, R. eds *Proceedings of the 3rd European Conference on Laser Treatment of Materials (ECLAT '90)*, 17–19 September 1990, Erlangen, Germany. Coburg: Sprechsaal Publishing Group. pp. 681–688.

Shieh, Y.N., Rawlings, R.D. and West, D.R.F. (1995). Constitution of laser melted Al_2O_3–MgO–SiO_2 ceramics. *Materials Science and Technology*, **11**, (9), 863–869.

Shih, C.H.S., Molian, P.A., McCallum, R.W. and Balachandran, U. (1994). Laser surface refinement of $YBa_2Cu_3O_x$ superconductor. *Journal of Materials Science*, **29**, 1629–1635.

Singh, J., Jerman, J., Poorman, R. and Bhat, B.N. (1994). High energy beams enhance life of NASA Space Shuttle combustion chamber liner. *Advanced Materials and Processes*, **146**, (6), 60–61.

Singh, J., Kurvilla, A.K., Jerman, G. and Bhat, B. (1995). Life enhancement of NASA's Space Shuttle main engine–main combustion chamber liner by high energy beams. In: Mazumder, J., Matsunawa, A. and Magnusson, C. eds *Proceedings of the Laser Materials Processing Conference (ICALEO '95)*. Orlando: Laser Institute of America. pp. 221–230.

Singh, J., Jerman, G., Poorman, R., Bhat, B.N. and Kurvilla, A.K. (1997). Mechanical properties and microstructural stability of wrought, laser, and electron beam glazed NARloy–Z alloy at elevated temperatures. *Journal of Materials Science*, **32**, 3891–3903.

Strutt, P.R., Nowotny, H., Tuli, M. and Kear, B.H. (1978). Laser surface melting of high speed tool steels. *Materials Science and Engineering*, **36**, 217–222.

Strutt, P.R. (1980). A comparative study of electron beam and laser melting of M2 tool steel. *Materials Science and Engineering*, **44**, 239–250.

Sugimoto, K., Wignarajah, S., Nagase, K., Yasu, S., Kimura, K., Kasuya, M. and Kureha, S. (1990). Fundamental study on laser treatment of architectural materials. In: Ream, S.L., Dausinger, F. and Fujioka, T. eds *Proceedings of the Laser Materials Processing Conference (ICALEO '90)*. Orlando: Laser Institute of America/Bellingham: SPIE. pp. 302–312.

Svéda, M., Roósz, A., Sólyom, J., Kovács, Á. and Buza, G. (2003). Development of monotectic surface layers by CO_2 laser. *Materials Science Forum*, **414–415**, 147–152.

Swarnalatha, M., Stewart, A.F., Guenther, A.H. and Carniglia, C.K. (1991). Laser-fused refractory oxides for optical coatings. *Materials Science and Engineering*, **B10**, 241–246.

Tayal, M. and Mukherjee, K. (1994). Selective area carburising of low carbon steel using an Nd:YAG laser. *Materials Science and Engineering*, **A174**, 231–236.

Tomie, M., Abe, N., Noguchi, S., Kitahara, Y. and Sato, Y. (1991). Laser surface modification of stainless steel – alloying with molybdenum. *Transactions of the Japan Welding Research Institute*, **20**, (2), 189–193.

Tomlinson, W.J., Megaw, J.H.P.C., Bransden, A.S. and Girardi, M. (1987a). The effect of laser surface melting on the cavitation wear of grey cast iron in distilled and 3% salt waters. *Wear*, **116**, 249–260.

Tomlinson, W.J., Moule, R.T., Megaw, J.H.P.C. and Bransden, A.S. (1987b). Cavitation wear of untreated and laser-processed hardfaced coatings. *Wear*, **117**, 103–107.

Tomlinson, W.J. and Bransden, A.S. (1990). Laser surface alloying grey iron with Cr, Ni, Co, and Co–Cr coatings. *Surface Engineering*, **6**, (4), 281–286.

Tomlinson, W.J. and Bransden, A.S. (1994). Sliding wear of laser alloyed coatings on Al–12% Si. *Journal of Materials Science Letters*, **13**, 1086–1088.

Tomlinson, W.J. and Bransden, A.S. (1995). Laser surface alloying of Al–12Si. *Surface Engineering*, **11**, (4), 337–344.

Trafford, D.N.H., Bell, T., Megaw, J.H.P.C. and Bransden, A.S. (1983). Laser treatment of grey iron. *Metals Technology*, **10**, 69–77.

Tsai, H.L. and Tsai, P.C. (1998). Laser glazing of plasma-sprayed zirconia coatings. *Journal of Materials Engineering and Performance*, **7**, (2), 258–264.

Tsai, W.-T., Lai, T.-H. and Lee, J.-T. (1994a). Laser surface alloying of stainless steel with silicon nitride. *Materials Science and Engineering*, **A183**, 239–245.

Tsai, W.-T., Shieh, C.-H. and Lee, J.-T. (1994b). Surface modification of ferritic stainless steel by laser alloying. *Thin Solid Films*, **247**, 79–84.

Uenishi, K., Sugimoto, A. and Kobayashi, K.F. (1992). Formation of titanium aluminide on aluminum surface by CO_2 laser alloying. *Zeitschrift für Metallkunde*, **83**, (4), 241–245.

van Brussel, B.A., Noordhuis, J. and De Hosson, J.Th.M. (1991a). Reduction of the tensile stress state in laser treated materials. *Scripta Metallurgica et Materialia*, **25**, (7), 1719–1724.

van Brussel, B.A., Hegge, H.J., De Hosson, J.Th.M., Delhez, R., de Keijser, Th.H. and van der Pers, N.M. (1991b). Development of residual stress and surface cracks in laser treated low carbon steel. *Scripta Metallurgica et Materialia*, **25**, (4), 779–784.

Vilar, R., Colaço, R., Colin, D. and Conde, O. (1990). Laser surface treatment of a high-chromium martensitic steel. In: Bergmann, H.W. and Kupfer, R. eds *Proceedings of the 3rd European Conference on Laser Treatment of Materials (ECLAT '90)*, 17–19 September 1990, Erlangen, Germany. Coburg: Sprechsaal Publishing Group. pp. 377–387.

Vitta, S. (1991). The limits of glass formation by pulsed laser quenching in Nb–Ni alloys. *Scripta Metallurgica et Materialia*, **25**, (10), 2209–2214.

Vogel, K. and Backlund, P. (1965). Application of electron and optical microscopy in studying laser-irradiated metal surfaces. *Journal of Applied Physics*, **36**, (12), 3697–3701.

Vreeling, J.A., Ocelik, V. and De Hosson, J.T.M. (2002). Ti–6Al–4V strengthened by laser melt injection of WCp particles. *Acta Materialia*, **50**, (19), 4913–4924.

Wade, N., Koshihama, T. and Hosoi, Y. (1985). Improvement of oxidation resistance of stainless steels by laser surface melting. *Scripta Metallurgica*, **19**, (7), 859–864.

Walker, A., Folkes, J., Steen, W.M. and West, D.R.F. (1985). Laser surface alloying of titanium substrates with carbon and nitrogen. *Surface Engineering*, **1**, (1), 23–29.

Wang, Z.L., Westendorp, J.F.M. and Saris, F.W. (1983). Laser and ion-beam mixing of Cu–Au–Cu and Cu–W–Co thin films. *Nuclear Instruments and Methods*, **209/219**, 115–124.

Wiiala, U.K., Sulonen, M.S. and Korhonen, A.S. (1988). Laser hardening of TiN-coated steel. *Surface and Coatings Technology*, **36**, 773–780.

Wilde, B.E., Manohar, M. and Albright, C.E. (1995). The influence of laser surface melting on the resistance of AISI 4135 low alloy steel to hydrogen-induced brittle fracture. *Materials Science and Engineering*, **A198**, 43–49.

Xin, H., Watson, L.M. and Baker, T.N. (1998). Surface analytical studies of a laser nitrided Ti–6Al–4V alloy: a comparison of spinning and stationary laser beam modes. *Acta Materialia*, **46**, (6), 1949–1961.

Yan, H. and Gan, F. (1993). Microscopy study of laser-induced phase change in Al–Cu film. *Journal of Materials Science*, **28**, 5382–5386.

Yilbas, B.S., Khaled, M. and Gondal, M.A. (2001). Electrochemical response of laser surface melted Inconel 617 alloy. *Optics and Lasers in Engineering*, **36**, 269–276.

Zhang, J.G., Zhang, X.M., Lin, Y.T. and Jun, K. (1988). Laser glazing of an Fe–C–Sn alloy. *Journal of Materials Science*, **23**, 4357–4362.

Zhou, X.B. and DeHosson, J.Th.M. (1994). Metal–ceramic interfaces in laser coated aluminium alloys. *Acta Metallurgica et Materialia*, **42**, (4), 1155–1162.

Zhu, S.M., Wang, L., Li, G.B. and Tjong, S.C. (1995). Laser surface alloying of Incoloy 800H with silicon carbide: microstructural aspects. *Materials Science and Engineering*, **A201**, L5–L7.

Zimmermann, M., Carrard, M. and Kurtz, W. (1989). Rapid solidification of Al–Cu eutectic alloy by laser remelting. *Acta Metallurgica*, **37**, (12), 3305–3313.

CHAPTER 12

CLADDING

INTRODUCTION AND SYNOPSIS

Laser cladding is a melting process in which the laser beam is used to fuse an alloy addition onto a substrate. The alloy may be introduced into the beam–material interaction zone in various ways, either during or prior to processing. Figure 12.1 shows one technique, in which alloy powder is blown into the laser beam and is deposited as a molten coating on the rim of a valve. Very little of the substrate is melted and so a clad with the nominal alloy composition is created. Surface properties can then be tailored to a given application by selecting an alloy with good wear, erosion, oxidation or corrosion properties. The molten clad solidifies rapidly, forming a strong metallurgical bond with the substrate. Most substrates that tolerate laser melting are generally suitable for cladding: carbon–manganese and stainless steels, and alloys based on aluminium, titanium, magnesium, nickel and copper. Popular cladding alloys are based on cobalt, iron and nickel. They may also contain carbides of tungsten, titanium and silicon, or ceramics such as zirconia, which form a particle-reinforced metal

Figure 12.1 CO_2 laser beam cladding of a valve rim with cobalt-based hardfacing powder. (Source: Tero Kallio, Lappeenranta University of Technology, Finland)

matrix composite surface on solidification, endowing additional wear resistance. The type of laser used depends on the surface area to be covered, the thickness of clad required, and the complexity of the component. CO_2 lasers are ideal for large areas requiring clads several millimetres in thickness over regions with a regular geometry. A robot-mounted diode laser beam, or Nd:YAG laser light delivered via a fibreoptic cable, is more suitable for precision treatment of complex three-dimensional components requiring a coating less than one millimetre in thickness. The overall aim is to produce a clad with appropriate service properties, a strong bond to the substrate, with the maximum coverage rate, the minimum use of alloy addition, and minimal distortion.

The cladding data in Fig. 6.3 (Chapter 6) cluster around a power density of $100 \, W \, mm^{-2}$. This is around ten times that required for solid state heating, but slightly below that needed for other processes based on surface melting. The reason is that the alloy is introduced in the form of powder, wires, or foils, which are melted more easily than a solid flat surface because of the larger ratio of surface area to volume. In practice, a beam power around 2 kW is the minimum needed for economical cladding. Insufficient power results in limited melting of the alloy addition, whereas an excess of power results in excessive melting of the substrate and undesirable dilution of the clad. A beam interaction time on the order of 0.1 s provides sufficient time for mixing and homogenization of the clad, while enabling the surface to solidify and cool with a sufficiently high rate to form a fine microstructure with superior properties to the larger-grained clads produced by conventional hardfacing techniques. The geometry of the clad is established by the surface properties required. A small, discrete region can be clad in a single pass, in which case the beam width is fixed by the required clad width and the traverse rate may be up to $1 \, m \, min^{-1}$. If a large area is to be clad, the aim is to maximize the coverage rate by using the widest possible beam and the highest traverse rate that satisfy the requirements of power density and interaction time. Overlapping parallel tracks are often required in such cases.

Early trials of laser cladding were crude, but developments in equipment and lasers have led to the production of turnkey cladding systems, which find use in a wide range of industries. The principles of clad formation are similar to those of surface melting, described in the previous chapter. The main difference is the effect of alloy delivery on the complexity of processing. Analytical modelling can be used to predict the thickness of the deposited clad in terms of the principal process variables. Model calibration is particularly useful in characterizing laser cladding involving a large number of process variables. Reference data for clad surfaces illustrate the breadth of application, and enable procedures specific to material and laser combinations to be developed. Industrial applications illustrate the use of cladding for surfacing and repair, as well as its role in novel processes of rapid manufacturing.

PRINCIPLES

In contrast to surface alloying, described in the previous chapter, the aim of laser cladding is to melt a *thin* layer of substrate together with as much of the alloy addition as possible. Normally, sufficient time should be available for the components of the molten clad to homogenize before it solidifies. An exception occurs when the alloy feed contains a reinforcing phase, when the aim is to produce a particle-reinforced metal matrix composite on the surface. Vaporization is undesirable, since expensive alloying additions are lost.

The alloy addition may be introduced either before laser processing (predeposition) or during processing (codeposition). If the alloy is predeposited, the melt front moves rapidly through the addition until it reaches the interface with the substrate. A small part of the substrate melts and dilutes the alloy, as illustrated in Fig. 12.2a. The substrate then acts as an efficient thermal sink, causing rapid solidification. If the alloy is codeposited, a part of the laser beam melts the addition, and a part melts the substrate, as illustrated in Fig. 12.2b. In both cases, a strong metallurgical bond is formed between the alloy and the substrate. Under optimum conditions, dilution of the clad by the substrate is low, typically about 5%.

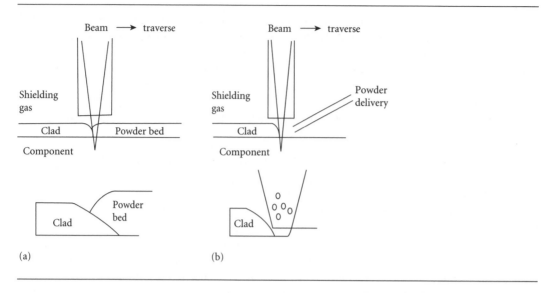

Figure 12.2 Illustrations of (a) predeposition and (b) codeposition of an alloy powder for laser cladding

During melting of the substrate and the alloy addition, steep temperature gradients form between the leading edge of the molten pool and its centre. Since surface tension is dependent on temperature, gradients in surface tension are created, which drive fluid flow (Marangoni flow). This, in combination with convective fluid flow, stirs the melt pool and homogenizes the composition of the clad.

In a predeposited alloy, after the melt has descended to the interface with the substrate it recedes and crosses the clad–substrate interface again only under conditions of high input energy and long interaction time. During blown powder delivery, the powder is melted by the beam and is deposited on the substrate where it solidifies almost instantaneously. This is one reason for the low dilution observed in such clads. The cooling rate is high, on the order of $10^4\,\mathrm{K\,s^{-1}}$. Fine dendritic morphologies are therefore observed in the solidified clad. The principal plane of solidification lies normal to surface, but the dendrite growth directions vary within the plane. The rapid cooling rate results in extended solid solubility in solidified clads – the composition of solute is higher than the solid solubility limit predicted by the equilibrium phase diagram.

The metallurgical bond between the clad and the substrate facilitates rapid transfer of energy into the substrate, which acts as an effective heat sink. A heat affected zone (HAZ) is therefore formed beneath the clad. The effects of solid state transformations in the substrate on its properties are important in determining the integrity of the clad–substrate interface – large differences in thermal expansion coefficient, for example, can cause cracking at the interface.

Thermal gradients in the clad and substrate induce residual stresses. Differential thermal contraction between the substrate and the clad determines the state of stress, which may be tensile or compressive. The magnitude of creep in the HAZ also influences the residual stress state.

PROCESS SELECTION

Laser cladding competes with a large number of conventional coating processes, particularly ones involving thermal spraying. By examining the characteristics of cladding, the attributes of components that favour the process on technical and economic grounds can be identified.

CHARACTERISTICS OF LASER CLADDING

The low energy input of laser cladding causes very little distortion to the component, which leads to a significant reduction in post-treatment machining operations. Close control of dilution from the substrate ensures the desired coating composition. The quality of the deposit is high – laser clads can be made with low porosity and few imperfections. High solidification and cooling rates result in fine solidified clad microstructures, which endow superior wear and corrosion properties to those of coarse microstructure. High cooling rates can result in the formation of beneficial metastable phases, and extended solid solubility, which also enhances properties. However, high cooling rates increase the sensitivity of the clad and HAZ to cracking, and precautions such as preheating might be necessary.

Discrete regions of surfaces can be treated because the beam can be delivered accurately, which reduces the total energy input, contributing to a reduction in distortion. The use of expensive alloying elements can be minimized since they are used only in surface regions, rather than throughout the bulk. The clad thickness can be controlled accurately, reducing the amount of post-treatment machining required. The maximum clad width is limited by the power density required for melting, and so overlapping passes are required for large area coverage. The process may be used both to build up coatings and to add material to specific locations, which may be applied readily in repair.

The laser beam can be manipulated easily, quickly and in a flexible manner. The process is therefore amenable to automation, beam sharing and beam switching. The beam can be directed into inaccessible locations, either using mirrors or fibreoptic beam delivery. A variety of component geometries can be treated by software changes to generate an optimum beam and traverse path. Automatic cladding machines have been developed from CNC equipment, which provide high-speed quantity production, consistent quality, and labour savings.

The principal drawbacks of laser cladding relate to cost and familiarity. The capital cost of the equipment is orders of magnitude higher than many conventional techniques, and the learning curve is knowledge intensive. Applications are required to be of sufficient volume, or add sufficient value, to be economically feasible. A large number of competing techniques are available, many of which are more familiar to engineers and managers responsible for production decisions. It can be difficult to obtain the technical and economic information needed to present a convincing case for changing to laser cladding.

COMPETING TECHNIQUES OF SURFACE COATING

A simplified comparison of the various techniques of surface coating is given in Table 12.1.

Table 12.1   Comparison of laser cladding and competing coating processes: SMA shielded metal arc; MIG metal inert gas; SA submerged arc; TIG tungsten inert gas; HVOF high velocity oxy-fuel

	Laser	SMA	MIG	SA	TIG	Plasma spraying	Flame	HVOF	Plasma arc
Thickness (mm)	0.2–2.0	1.6–10	1–6	2–10	0.5–3.0	0.1–0.2	0.8–2	0.3–1.5	1–5
Deposition rate ($kg\,h^{-1}$)	0.2–7	0.5–2.5	2.3–11	5–25	0.5–3.5	0.5–7	0.45–2.7	1–5	2.5–6.5
Distortion	Low	Medium	Medium	High	High	Low	High	Low	Medium
Precision	High	Low	Low	Low	Medium	Medium	Low	Low	Medium
Dilution (%)	1–5	15–25	15–20	10–50	10–20	5–30	1–10	Low	Medium
Integrity	High	High	High	High	Medium	Low		Medium	Medium	Medium

High velocity oxy-fuel spraying (HVOF) was developed in the early 1980s for spraying carbide coatings, but is now used for a range of metals and ceramics. The equipment comprises a combustion chamber, exit nozzle and input nozzles for oxygen, fuel gas, powder and carrier gas, and water cooling. Fuel gas and oxygen enter the gun, are premixed, and introduced into the combustion chamber. Powder to be deposited, typically around 40 μm in diameter, is fed either directly into the chamber, or downstream in the nozzle using an inert carrier gas such as argon. The powder is heated and accelerated along the nozzle by the rapidly expanding gases, reaching speeds up to 800 m s^{-1}. A layer about 10 μm in thickness is deposited on each pass, giving a final coating with low roughness. HVOF is an efficient method for coating a large area with a thickness up to about 1.5 mm. The cost of coating is relatively low in comparison with weld overlay. The high energy of the process results in coatings of relatively low porosity ($<1\%$), which have good adhesion. Laser cladding is competitive when small, discrete regions are to be clad: the lower energy input reduces distortion and post-treatment reworking and porosity can be eliminated.

Plasma spraying is used extensively to coat metal substrates for wear and corrosion protection. An ionized gas is produced by passing electrically generated plasma through a constricted anode. The alloy powder is injected into the hottest part of the plasma, heated to the molten state, and blown at high velocity onto the substrate. The technique is flexible, allowing a variety of coatings to be applied to a diverse range of components economically. The temperature of the substrate remains comparatively low, which reduces distortion, and permits the coating of low melting temperature metals such as zinc. Large areas may be coated quickly with a minimum of dilution from the substrate. However, the coatings may suffer from porosity in excess of 1%, chemical inhomogeneity, and weak bonding to the substrate, particularly in systems in which the coefficients of thermal expansion differ widely. Laser cladding competes when a high integrity, homogeneous coating free of porosity is required.

In plasma transferred arc (PTA) cladding, the coating powder is transported in an inert gas stream into the plasma of one arc, where it is melted. Molten powder is then transferred to the substrate by a main arc, and is laid down as a track. The width of the track may be increased by oscillating the arc at an angle to the travel direction. A high clad thickness and a large track width can be produced in a single pass. The clad is homogeneous, and a sound metallurgical bond is formed with the substrate. In comparison with laser cladding, the energy input is high, and so the HAZ is more extensive, distortion is greater, and dilution is higher. This may not be a concern with large area coverage, but it limits the suitability of the process for small components.

PRACTICE

The principal process variables of laser cladding are the beam characteristics (laser beam power, size and shape of the heating pattern and traverse rate), the material properties (those of the substrate and clad), and the processing parameters (the component geometry, and rate and geometry of alloy addition).

MATERIAL PROPERTIES

Substrate

Components made from carbon-manganese, alloy, stainless and tool steels are popular for laser cladding. Other suitable substrates include alloys based on aluminium, magnesium, cast iron, and nickel-base superalloys. The substrate must have sufficient thermal conductivity to act as an efficient heat sink to form a metallurgical bond. Elements that can combine with the alloy addition to form undesirable brittle compounds should be avoided. The dimensions of the component determine the number of passes required for coverage, and the geometry dictates the need for additional techniques to clad discontinuities such as corners and sharp edges. The surface is prepared prior to cladding by cleaning, degreasing and pickling, if contamination of the clad is undesirable.

Types of Alloy Addition

Suitable alloy additions can be characterized by a number of desirable properties. They absorb the wavelength of laser radiation efficiently to achieve melting, but do not vaporize to a significant degree. The composition is such that dilution with the substrate material creates an alloy that solidifies to form a strong metallurgical bond, without excessive formation of brittle phases at the interface. (Cladding of metals and alloys with most other metals and alloys is normally possible, but problems may arise when the clad and the substrate are from different classes of engineering material, e.g. a ceramic clad on a metal substrate.) Their thermal conductivity is sufficient to conduct heat into the substrate and their coefficient of thermal expansion is similar to that of the substrate to avoid cracking. Finally, they must be commercially available in the desired form (powder, wire etc.).

Most commercial hardfacing alloys can be classified into four different categories: cobalt-base; iron-base; nickel-base; and carbides of tungsten, titanium and silicon. Light metal alloy powders, such as aluminium, are also used for specific alloy substrates. The compositions of a number of popular cladding alloys are given in Table 12.2.

The most popular cobalt-base alloys are the Stellite family, which are familiar to industries involved in thermal spraying. The alloys are numbered according to the amounts of chromium, carbon, iron, silicon and nickel that they contain. Chromium forms carbides, which strengthen the cobalt matrix. Nickel is added to increase ductility. Other elements are added to improve various properties – tungsten and molybdenum have a large atomic radius, which distorts the lattice providing strength. They also form hard carbides readily. Stellites have high flow stresses and exhibit good resistance to oxidation, erosion, abrasion, and galling at temperatures up to about 90% of their melting temperature. They are primarily used to coat components in the chemical and power generation industries, and are available in powder, rod, electrode, chip, foil, mat and wire form.

Table 12.2 Alloys commonly used in laser cladding (Delcrome®, Deloro®, Stellite® and Tribaloy® are registered trademarks of Deloro Stellite Company, Inc.; Colmonoy® is a registered trademark of Wall Colmonoy Corporation)

	Cr	C	Si	Mo	Fe	Ni	Co	B	Mn	W
Cenium Z20	27	0.26	–	–	Balance	18.8	0.33	–	9.4	1.5
Delcrome 90	27	2.7	1.0	–	Balance	–	–	–	1.0	–
Colmonoy 4	10	0.4	2.4	–	2.8	Balance	–	2.1	–	–
Colmonoy 5	18	0.45	3.3	–	4.8	Balance	–	2.1	–	–
Deloro 60	15	0.9	4.5	–	4.5	Balance	1.0	3.5	–	–
Eutrolloy 16262	27.7	1.69	0.95	7.4	0.88	Balance	0.02		0.01	
Nicrobor 40	8.0	0.25	3.7	–	1.3	Balance	–	1.7	–	–
Tribaloy T700	15.5	–	3.4	32.5	–	Balance	–	–	–	–
Stellite 6	29	1.2	1.4	0.6	2.0	2.0	Balance	–	1.0	4.5
Stellite SF6	19	0.7	2.5	–	3.0	13	Balance	1.7	0.5	7.0
Stellite 158	26	0.7	1.2	–	–	–	Balance	0.7	–	–
Tribaloy 66	8.0	–	2.0	28	–	–	Balance	–	–	–

Nickel-base alloys commonly contain chromium, boron, carbon, silicon and aluminium. The formation of hard borides and silicon carbide improves hardness and wear resistance. Boron and silicon also improve the wetting characteristics of molten alloys. Aluminium combines with nickel to form hard intermetallic phases such as $NiAl_3$ and Ni_2Al_3. Nickel-base alloys are marketed under trade names such as Colmonoy and Triballoy. Colmonoys have specific wear-resistant characteristics with respect to impact, abrasion, erosion and galling, and contain chromium boride particles, which are second only to diamond on the hardness scale. They have good wear resistance and high hardness at elevated temperatures. Carbides, notably tungsten and chromium, can be added in limited amounts to provide additional wear resistance and to improve abrasion properties.

Iron-base alloys containing chromium, carbon, manganese and tungsten are used primarily for surfacing ferrous substrates. Carbides are formed, notably those of the type M_6C, which improve wear resistance. Other elements are present to improve oxidation and corrosion resistance. The use of laser cladding to produce corrosion-resistant surfaces on inexpensive carbon–manganese steel substrates is an attractive means of replacing expensive bulk stainless steel components for many industries.

Alloy Introduction

Predeposited alloy may be added in the form of a dry powder, wire, strip, foil, sheet, tape, mat or slurry. Powders are most widely used. The alloy addition is spread on the substrate, often with the aid of a binder. The laser beam heats and melts the addition, which is thermally isolated from the substrate, until the melt front reaches the interface with the substrate at which point a small part of the substrate melts and dilutes the alloy.

Predeposited powder can be applied easily to horizontal surfaces, but suffers from a number of limitations: the part complexity is restricted; cladding must be carried out in the downhand position; powder can be blown away; shielding is difficult; a uniform bed thickness is difficult to achieve; it is difficult to control process parameters to ensure good bonding; a high energy input is required; and *in situ* change of alloy composition is not possible. Some of these drawbacks may be overcome by using organic binders to locate the powder, but care must be taken as volatile compounds can generate porosity and other imperfections in the clad. The alloy may be predeposited by a thermal spraying technique, in which case the laser beam is used to melt the coating, which reduces porosity and results in the formation of a metallurgical bond with the substrate. However, this is a two-stage process, and so its commercial application is limited. Alloy paste may also be delivered ahead of the scanning laser beam. Mats can be located readily, and a uniform, controllable depth of clad achieved, but few alloys are available in such form. Chips and foils may be preplaced in the desired locations, and melt readily to form a good bond. The amount of material can be controlled easily, they are adaptable to contoured shapes, and porosity is generally avoided. However, hardfacing alloys have limited availability in chip and foil form, and the efficiency of energy usage is lower than with powder feeding. Predeposition was the most common means of introducing alloy in early investigations of cladding, but it suffers from one major drawback for industrial application: it is a two-stage, time-consuming process.

The most common means of introducing alloy into the laser beam *during* processing involve the use of powders and wires. Alloy powder is delivered into the beam using a gravity feed, or via a gas-assisted powder delivery system. The energy of the laser beam is partially absorbed by the powder, to a degree that depends on the powder composition, particle size and powder density. The laser beam energy that reaches the substrate creates a very shallow molten surface. A clad is built up by a combination of molten powder particles sticking to the substrate, and unmelted particles being absorbed in the molten surface. The metal powder spheroidizes on melting, creating a stirring action which homogenizes the coating and helps trapped gas bubbles to escape, thus reducing porosity.

The first codeposition cladding trials involved the introduction of alloy powder via a side nozzle. The simplest form of this powder feeding geometry is illustrated in Fig. 12.2a. The powder is injected into the melt pool at its leading, trailing or side edges. Shielding is provided by an inert gas delivered coaxially with the laser beam. Improved shielding may be obtained by delivering the powder via a feeder comprising two concentric nozzles; the inner containing the powder and carrier gas, the outer containing the shielding gas, discussed below. This has the added benefit of enabling the powder to be directed more accurately through the focusing effect of the gas on the powder stream. Multiple powder feed units may be used in order to create alloys of different composition *in situ*. Up to 90% of the powder feed may be melted using this technique – the remainder bypasses the beam – to produce an integral bond with the substrate.

The powder is typically 80–100 mesh, although a wide range of powder sizes are usable, e.g. 2 to 200 μm diameter. Many alloys are available in powder form for plasma spraying at relatively low cost, and powder supply systems designed for plasma spraying are often used. However, it is important to bear a number of important requirements in mind when selecting a unit specifically for laser cladding. It should be possible to maintain powder delivery rates to within $\pm$1% of a desired value, with a wide range of delivery rates. Typical rates are on the order 25 g min^{-1}, although for precision work lower flow rates may be required, which may result in an undesirable pulsed flow of powder with some plasma spraying equipment. The equipment should also have a facility to allow powders to be mixed in order to vary compositions. Water atomized powders are generally preferred to gas atomized powders because the former have a rougher surface that absorbs laser radiation more efficiently. In order to ensure the free-flowing characteristics of the powder, it is often desirable to bake the powders for several hours to remove moisture.

The powder flow rate is critical to the formation of an integral clad of the desired thickness. If the flow rate is too low, excessive melting of the substrate occurs, resulting in high dilution of the clad, and undesirable properties. If the flow rate is too high, the powder shields the substrate from the beam, and a metallurgical bond cannot then be formed – the powder forms molten globules on the surface. A theoretical optimum value can be calculated, but the efficiency of powder usage, which is the fraction of powder actually melted, is dependent on the processing parameters, and so experimental trials are required to establish a practical value.

There are a number of advantages to dynamic powder delivery over preplaced powder. Most importantly for commercial cladding, it is single-stage process. The alloy composition can be varied *in situ*, which enables surfaces of graded composition to be produced, providing a means of fabricating functionally engineered materials. The homogeneity is high, and abrasive particles such as carbides can be kept in the mixture. Very thin surface clads possessing a sound bond with the substrate may be produced using the side delivery technique. The main drawback of powder delivery using a directional nozzle is that it is difficult to use when cladding components of complex geometry – accessibility problems arise.

One solution for limited accessibility is to combine beam and powder delivery into a single unit. The unit typically contains three concentric nozzles. The beam emerges from the central nozzle. Powder is delivered with a carrier gas via an annular orifice concentric with the laser beam. The outer annulus delivers the shielding gas, which also acts to focus the powder stream to a specified position relative to the focal plane of the laser beam, as shown in Fig. 12.3. This type of powder delivery enables cladding to be carried out in any direction and in a variety of orientations, thus providing a means of treating complex three-dimensional geometries in various positions. However, the powder feed must be regulated carefully and aligned with the beam.

An alternative means of enabling cladding to be performed out-of-position is to introduce the alloy addition in the form of a wire feed (Alam and Ion, 2002). The laser beam is defocused onto the workpiece surface in order to melt a thin surface layer. Wire is fed into the molten region, and interacts with the beam to form the clad. Wire may be fed into the leading edge of the molten pool, in a direction parallel with, or perpendicular to, the direction of beam travel. This directionality means that the technique is most suitable for clads made in straight lines, but the robust nature of wire feeding provides a means of cladding in remote locations that might be difficult to access, such as the internal surfaces of

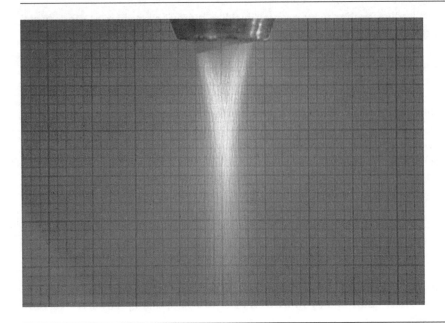

Figure 12.3 Coaxial alloy powder feed. (Source: Antti Salminen, Lappeenranta University of Technology, Finland)

pipes. A variety of hardfacing and stainless steel wires have been used successfully. Since all the wire is melted, deposition rates are high and accurately controllable. Productivity can be increased further by preheating the wire. However, some high performance alloys may not be available in wire form, and the process must be controlled carefully because of the constantly changing wire–beam interaction zone.

BEAM CHARACTERISTICS

Wavelength

CO_2 lasers have traditionally been used for cladding because for many years they were the only sources that could provide the power density required for melting. As multikilowatt Nd:YAG lasers with fibreoptic beam delivery became more widely available, the potential for lower cost, flexible coating of three-dimensional components was realized. The improved absorption of the shorter wavelength Nd:YAG laser beam by metallic alloy additions provides a means of increasing the process efficiency and increasing coverage rates. The absorptivity of powder additions to a multikilowatt diode laser beam is about twice that of a CO_2 laser beam, and the rectangular cross-section of the beam is an ideal shape for high coverage rates. The properties of clads produced by using diode lasers are similar to those produced by CO_2 and Nd:YAG lasers.

Power

For practical laser beam cladding a power density of about $100 \, \text{W mm}^{-2}$ and a beam interaction time of about 1 second are used. About 2 kW is the normal minimum laser beam power needed for cladding, since insufficient power will result in limited melting of the alloy addition, whereas too much power causes excessive melting of the substrate and dilution of the clad. Practical processing data cluster along the line corresponding to a process energy density of about $100 \, \text{J mm}^{-2}$ in Fig. 6.3 (Chapter 6), a value that represents the conditions required to obtain homogeneous melting of the alloy addition and good bonding with the substrate. A continuous wave beam is normally used to maximize the coverage rate. However, if low energy input is required, to minimize distortion of the component, then the laser beam may be pulsed.

Transient effects of power on the process stability are often noted at the beginning and end of the cladding run. At the start, the component is at room temperature, and insufficient beam power impinges to heat it significantly. The alloy addition might then not adhere, and delamination is common. It may therefore be necessary to start with a high power that preheats the component, followed by ramping down as cladding proceeds. Alternatively, powder can be introduced only when the beam has preheated the component to the required temperature.

Beam Delivery

Mirrors are normally used in laser cladding in preference to lenses to shape the beam geometry, in order to minimize the risk of damage, in laser cladding spatter from the interaction zone. A long focal length and hence a large stand-off is desirable caused by increasing the depth of focus, which is beneficial when cladding uneven surfaces. A beam with a uniform intensity distribution is normally used, to produce a clad of uniform thickness in a transverse section. This may be achieved through the use of oscillating and integrating mirrors, and other optical devices designed for creating uniform heating patterns, described in Chapter 4.

Heating Pattern

A wide heat source with a uniform distribution of energy is the most appropriate for producing wide clads of uniform thickness. The beam may be oscillated perpendicular to, or parallel with, the direction of traverse, with a frequency on the order of 1 kHz to produce a smoother clad surface. The surface may also be smoothed by remelting a small upper portion of the clad by using a second beam pass.

Traverse Rate

The traverse rate used in laser cladding is generally higher than that required for surface melting, since a particulate alloy addition can be melted more readily than a solid surface. The maximum traverse rate is determined by the rate at which powder can be melted, with sufficient power impinging on the substrate to cause limited surface melting. The rate of traverse of the beam affects the thickness of the clad produced. A decrease in clad thickness is associated with an increase in traverse rate, and vice versa.

PROCESSING PARAMETERS

Process Gas

A shielding gas must be used if the substrate or alloy is susceptible to oxidation. Gas is often introduced in a hood, but coaxial shrouds are also suitable. The intention is to shield without disturbing the powder flow. Argon is normally used, but nitrogen is an alternative if there is no danger of brittle nitride formation. The use of helium, particularly with high gas flow rates, causes instabilities in the beam–material interaction zone. The gas flow rate and nozzle geometry must be specified. The greatest problems in cladding arise from inadequate attention to the geometry of the powder delivery and the systems for delivery of the carrier gas and the shielding gas. A sketch included in the cladding procedure specification provides an unambiguous means of describing the setup.

Workpiece Manipulation

Small workpieces may be moved around a fixed head. Large workpieces are fixed. The principles of system design for various types of workpiece are outlined in Chapter 4.

Coverage of Large Areas

Several clad tracks are deposited sequentially when large areas need to be covered. In such cases it is important to establish the optimum amount of overlap between adjacent clad tracks. Insufficient overlap results in the formation of grooves in the clad surface, a lack of fusion at the 'toe' of the clad, and often an insufficiently thick clad. Excessive overlap reduces the coverage rate, and may lead to excessive preheating. The optimum amount of track overlap should be established through experimental trials. In general, an increment between clads of more than 3 mm leads to a significant groove on the clad surface under normal operating conditions.

Preheating

Preheating is used for two main reasons: to avoid cracking within the clad; and to deliberately increase the dilution of the clad by the substrate for composition control. Substrates may be preheated, and cooled after processing, in a furnace to reduce the cooling rate in order to avoid cracking. Preheating also enables the range of ferrous alloys to be extended, compared with those used for hardening. When high deposition rates are required, notably with coaxial powder feeding, insufficient laser beam power may impinge on the substrate to form a metallurgical bond, resulting in delamination of the clad. By preheating the substrate, the degree of bonding may be increased significantly. When preheating is used during cladding of large areas, a minimum interpass temperature must be specified in order to reduce the likelihood of cracking.

Post-clad Heat Treatment

If large areas are treated, particularly with a thick clad, the energy input is relatively high, which results in the formation of residual stresses in the component. A stress-relieving post-clad heat treatment may be necessary to avoid distortion. The temperature and duration of this treatment must be specified.

Post-clad Mechanical Treatment

Clads may be shot peened to induce a compressive state of residual stress, which improves the fatigue properties. A limited amount of machining might be required to meet dimensional tolerances, and to satisfy requirements for surface roughness.

Adaptive Control

Signals from the interaction zone may be detected by a photo diode. They relate to the following: the integrity of the clad bond; porosity; system faults; clad roughness; clad thickness; and substrate defects. Power may be controlled to ensure a constant clad dilution through the use of a laser beam analyser. Process models, expert analysis and monitoring systems form a basis for the development of parameter optimization (Li *et al.*, 1990a) and adaptive control (Li *et al.*, 1990b) systems for laser cladding.

PROPERTIES OF CLAD MATERIALS

A wide range of substrates and alloy additions have been investigated for laser cladding. Procedures have been developed for the most popular combinations, and the properties shown to be superior to those of coatings produced using conventional coating techniques. Popular cladding alloys were described earlier. Descriptions and designations of the various alloys used as substrates can be found in Chapter 5 and Appendix C, respectively.

QUALITY ASSURANCE

Standard specifications for cladding procedures have not yet been developed. Laser processing facilities therefore draw up their own Cladding Procedure Specification (CPS), which contains the relevant parameters and allowable ranges that are to be used for clads. A CPS may cover a specified range of substrate and alloy materials. Acceptance criteria, in the form of levels of tolerable imperfections, are specified by the end user. Once standardized, the CPS will refer to a Cladding Procedure Approval Record (CPAR), which contains details of the testing carried out during the approval process.

Imperfections

The most common imperfections found in laser clads are described below, together with practical means of avoiding them.

Lack of Fusion

Crevices caused by a lack of fusion are often observed at the toes of adjacent overlapping clads, particularly when the aspect ratio of the clad section (thickness/width) is high. A reduction in the powder feed rate and the rate of traverse of the workpiece enables molten clad material to flow into discontinuities to form an overlap region of high integrity.

Cracking

Cracking in the solidified clad layer is a common problem because popular alloy additions solidify to produce a coating of high hardness, often containing hard particles that act as stress raisers. Cracks originate from thermal stresses that develop because of the high thermal gradients present during cooling. These may be reduced by preheating the substrate. Differential thermal expansion of the solidified clad and the substrate can also cause cracking, as well as delamination of the clad from the substrate.

Distortion

Residual stresses are created in the substrate during cladding. These may result in distortion of the substrate in various planes, which increases the amount of post-process reworking required to meet dimensional tolerances. Distortion can be reduced through a reduction in the energy input of the process, by either lowering the power of a CW beam, or by pulsing the beam. In extreme cases, the substrate can be prebent to compensate for distortion arising during cladding.

Porosity

Porosity in the clad has many origins. Fine porosity is a result of degassing as the clad solidifies. Coarse porosity is dependent on the geometry of solidification of the clad – outer regions solidify first, enclosing pockets of gas in the centre. Agitation of the solidifying clad reduces such porosity. The interface between the clad and the substrate is also a source of porosity, caused by contamination of the substrate surface, which influences the wetting ability of the molten clad, and can result in vaporization of volatile organic compounds. The substrate should be cleaned thoroughly and degreased to eliminate such imperfections.

Dilution

A small amount of substrate must be melted in order to form a strong metallurgical bond. However, molten substrate mixes with the molten clad by convection, effectively diluting the composition of the clad. Dilution of about 5% is sought in laser cladding. Dilution can be controlled by ensuring efficient melting of the alloy addition through adjustments to the alloy delivery geometry and feed rate. In some cases, it may be desirable to incorporate a given amount of substrate into the clad to obtain specific properties, but process parameters must be controlled to such a degree that the reproducibility of the process is limited under such conditions.

COBALT ALLOYS

Stellites are the most common cobalt-base alloys used in laser cladding. They possess good mechanical, thermal and chemical properties, particularly at high temperatures. Most of the practice described in previous sections was developed by cladding Stellites onto steel substrates. Variations have been developed, mainly to increase the efficiency of alloy usage and to improve the crack resistance. They are too specific to be covered in a book of this nature, but can be found in the references listed in Table 12.3.

Stellites solidify with a dendritic microstructure containing carbides, notably the type M_7C_3, in various morphologies. The preferred plane of dendrite growth lies approximately perpendicular to the surface, although dendrites may grow in various orientations in this plane. The scale of the dendrites is determined by the solidification rate, which is related to the energy input of the process. As the process energy input is reduced, the solidification microstructure becomes finer, and a lower volume fraction of smaller interdendritic carbides forms. Improved service properties are associated with the finer, more uniform microstructure of the laser clad in comparison with conventional surfacing techniques. Hardness values on the order of 550 HV are typical of laser-clad Stellite 6. The hardness and wear resistance increases linearly with the volume fraction of carbides present in the clad.

The properties of the clad can be improved further by adding carbides, nitrides and borides to Stellite, which do not melt during cladding, forming a particulate-reinforced metal matrix composite at the surface. Tungsten carbide is a popular choice because of its low thermal expansion coefficient, high degree of wettability by cobalt, and high hardness.

NICKEL ALLOYS

Nickel-base alloys are suitable for applications in which high temperature strength and corrosion resistance are required in aggressive environments. They can be used in place of cobalt-base alloys in sensitive environments such as those associated with nuclear power generation. Nickel is also more readily available than cobalt, and is less expensive. Table 12.4 provides sources of information for laser cladding using such alloys.

The microstructure of nickel-base alloy clads is characterized by fine dendrites with interdendritic strengthening particles such as Ni_3Fe, Ni_3Si, B_4C, CrB, and Cr_3Si, depending on the alloy composition. The spacing of dendrite arms is significantly lower than that obtained using conventional coating techniques, reflecting the lower energy input of laser cladding.

Table 12.3 Sources of data for laser cladding using cobalt-base alloys: N/S not specified

Alloy addition	Substrate	Laser	Reference
Co-base	12% Cr steel	CO_2	(Com-Nougué et al., 1987)
Co-base	AISI 1010	CO_2	(Pilloz et al., 1992)
Co-base	AISI 304L	CO_2	(Pilloz et al., 1992)
Co-base	C–Mn steel	CO_2	(Monson and Steen, 1990)
Co–Cr–W–B–Si	C–Mn steel	CO_2	(Lugscheider et al., 1991)
Co–Ni–Cr–W–B–Si	C–Mn steel	CO_2	(Lugscheider et al., 1991)
Stellite 1	AISI 316	CO_2	(De Hosson and De Mol van Otterloo, 1997)
Stellite 6	CP Fe	CO_2	(Capp and Rigsbee, 1984)
Stellite 6	AISI 304	N/S	(Kaul et al., 2002)
Stellite 6	AISI 304L	CO_2	(Li and Mazumder, 1984; Yellup, 1995)
Stellite 6	AISI 316	CO_2	(De Hosson and De Mol van Otterloo, 1997)
Stellite 6	C–Mn steel	CO_2	(Matthews, 1983; Picasso et al., 1994; Li and Mazumder, 1984)
Stellite 6	C–Mn steel	Nd:YAG, CO_2	(Jansson et al., 2003)
Stellite 6	HY80	Nd:YAG	(Webber, 1992)
Stellite 12	Nimonic 75	CO_2	(Steen and Courtney, 1980)
Stellite 20	AISI 316	CO_2	(De Hosson and De Mol van Otterloo, 1997)
Stellite 21	AISI 316	CO_2	(De Hosson and De Mol van Otterloo, 1997)
Stellite 21	C–Mn steel	Diode	(Nowotny et al., 1998)
Stellite 156	C–Mn steel	CO_2	(Nurminen and Smith, 1983)
Stellite 158	C–Mn steel	CO_2	(Matthews, 1983)
Stellite F	AISI 304	CO_2	(Giordano et al., 1984)
Stellite SF6	AISI 304	CO_2	(Giordano et al., 1984)
Stellite SF20	AISI 316	CO_2	(De Hosson and De Mol van Otterloo, 1997)
Stellite SF6	C–Mn steel	CO_2	(Matthews, 1983)

IRON ALLOYS

Although not as widely investigated as cobalt- and nickel-base clad materials, clads have been made using iron-base alloys, primarily with the intention of improving the corrosion resistance of carbon–manganese steel substrates. Cladding procedures and clad properties can be found in the references listed in Table 12.5.

Fe–Cr–Mn–C clads on carbon-manganese steels show a high degree of grain refinement and an increase in solid solubility of alloying elements, which produces a fine distribution of complex carbides in a ferrite matrix. M_6C precipitates are formed, which have a higher hardness than the M_7C_3 precipitates associated with Stellites. Consequently the wear properties can be superior to those of Stellite clads.

Clads of Fe–Cr–W–C on mild steel also have fine microstructures, comprising primary dendrites of austenite with an interdendritic eutectic consisting of a network of M_7C_3 carbides rich in Cr in an austenitic phase. Hardness values around 900 HV are possible. Subsequent heat treatment may transform the microstructure into primary ferrite dendrites and a eutectic matrix of M_7C_3 networks distributed in a ferritic phase, which is slightly less hard.

Clads of Fe–Cr–C–W alloys contain predominantly diamond-shaped M_6C and $M_{23}C_6$ carbides, and have a hardness of about 900 HV. Preheating to 550–600°C can be used to eliminate cracks that typically form at the clad–substrate interface during cooling, although the hardness is reduced. A higher

Table 12.4 Sources of data for laser cladding using nickel-base alloys: N/S not specified

Alloy addition	Substrate	Laser	Reference
Inconel 718	Ni–Cr–Al–Hf	CO_2	(Singh *et al.*, 1987c)
Inconel 100	Ti–6Al–4V	CO_2	(Coquerelle *et al.*, 1986)
Nimonic 80A	Ti–6Al–4V	CO_2	(Coquerelle *et al.*, 1986)
Ni	C–Mn steel	CO_2	(Takeda *et al.*, 1985)
Ni–26% Al	Ni	CO_2	(Kaplan and Groboth, 2001)
Ni–Al	Nickel alloy	CO_2	(Abboud *et al.*, 1995)
Ni–Al bronze	AA333	CO_2	(Liu *et al.*, 1994a, b, c)
Ni-base	AISI 1010	CO_2	(Pilloz *et al.*, 1992)
Ni-base	AISI 304L	CO_2	(Pilloz *et al.*, 1992)
Ni-base	C–Mn steel	CO_2	(Monson and Steen, 1990)
Ni–Cr	Fe37	Nd:YAG	(Tuominen *et al.*, 2002)
NiCrAl	Stainless steel	CO_2	(Walther and Pera, 1978; de Damborenea and Vázquez, 1993)
Ni–Cr–Al–Hf	AISI 304	CO_2	(Li and Mazumder, 1984)
Ni–Cr–Al	AISI 316L	CO_2	(de Damborenea and Vázquez, 1992)
Ni–Cr–Al–Hf	C–Mn steel	CO_2	(Li and Mazumder, 1984; Singh and Mazumder, 1987b)
Ni–Cr–Al–Mo–Fe	Brass	Nd:YAG	(Tam *et al.*, 2002)
Ni–Cr–Al–Si–Y	Inconel 738	CO_2	(Marsden *et al.*, 1990)
NiCrBSi	16MnCrS5	CO_2	(Grünenwald *et al.*, 1992)
Ni–Cr–B–Si	C–Mn steel	Diode	(Conde *et al.*, 2002)
Ni–Cr–Co–Fe	AISI 304L	CO_2	(Pelletier *et al.*, 1993)
Ni–Cr–Fe–Ti	C–Mn steel	CO_2	(Tosto *et al.*, 1994)
Ni–Cr–Mo–Co	2Cr1Mo	CO_2	(Ono *et al.*, 1987)
Ni–Cr–Nb–B	C–Mn steel	CO_2	(Oberländer and Lugscheider, 1992)
Ni–Cr–Ta–B	C–Mn steel	CO_2	(Oberländer and Lugscheider, 1992)
Ni–Fe–Cr–Al–Hf	AISI 1016	CO_2	(Singh and Mazumder, 1987b, 1988)
Ni–26% Hf	Ni	CO_2	(Kaplan and Groboth, 2001)
Colmonoy 5	C–Mn steel	CO_2	(Takeda *et al.*, 1985)
Colmonoy 6	Austenitic	N/S	(Kaul *et al.*, 2003)
Delcrome 90	C–Mn steel	CO_2	(Matthews, 1983)
Deloro 60	AISI 1043	CO_2	(Wetzig *et al.*, 1998)
Deloro 60	AISI O2	CO_2	(Wetzig *et al.*, 1998)
Deloro 60	C–Mn steel	CO_2	(Matthews, 1983)
Eutrolloy 16262	AISI 304L	CO_2	(Yellup, 1995)
Hastelloy C	C–Mn steel	Nd:YAG	(Brandt *et al.*, 1995)
Nicrobor 40	AISI 1043	CO_2	(Wetzig *et al.*, 1998)
Nicrobor 40	AISI O2	CO_2	(Wetzig *et al.*, 1998)
Tribaloy 66	Copper	CO_2	(Dehm and Bamberger, 2002)
Tribaloy T-700	C–Mn steel	CO_2	(Matthews, 1983)
Tribaloy T-800	ASTM 387	CO_2	(Gnanamuthu, 1979)

Table 12.5   Sources of data for laser cladding using iron-base alloys

Alloy addition	Substrate	Laser	Reference
Fe, Cr	SS400	CO_2	(Kagawa and Ohta, 1995)
Fe–Cr_2O_3	AISI 304	CO_2	(Zhou and De Hosson, 1991)
Fe–Cr_2O_3	SAF 2205	CO_2	(Zhou and De Hosson, 1991)
Fe–Cr–C–W	AISI 1016	CO_2	(Choi *et al.*, 2000)
Fe–Cr–Mn–C	AISI 1016	CO_2	(Singh and Mazumder, 1986a, b, 1987a)
Fe–Cr–Mn–C	C–Mn steel	CO_2	(Eiholzer *et al.*, 1985)
Fe–Cr–Ni	C–Mn steel	CO_2	(Takeda *et al.*, 1985)
Fe–Cr–W–C	AISI 1016	CO_2	(Choi and Mazumder, 1994)
Fe–Cr–W–C	C–Mn steel	CO_2	(Nagarathnam and Komvopoulos, 1993)
AISI 316	C–Mn steel	CO_2	(Powell *et al.*, 1988)
AISI 316	En3	CO_2	(Powell *et al.*, 1988)
AISI 316L	C–Mn steel	CO_2	(Weerasinghe *et al.*, 1987)
Cenium Z20	AISI 304L	CO_2	(Yellup, 1995)
Hadfield steel	C–Mn steel	CO_2	(Pelletier *et al.*, 1995)

Table 12.6   Sources of data for laser cladding using chromium-base alloys

Alloy addition	Substrate	Laser	Reference
Cr–Ni–Mo	Cast iron	CO_2	(Belmondo and Castagna, 1979)
Cr, Ni	AISI 347	CO_2	(Kosuge *et al.*, 1987; Ono *et al.*, 1987)

power density results in larger carbide precipitates which gives high hardness, but relatively low wear resistance.

CHROMIUM ALLOYS

A limited number of investigations have been made using chromium-base alloys to create clads with improved wear characteristics on stainless steels and cast irons. Procedures and properties can be found in the references listed in Table 12.6.

ALUMINIUM ALLOYS

As discussed earlier, the use of alloys of aluminium in all forms of transportation is growing. However, the relatively poor wear properties of aluminium alloys limit their application. Laser cladding using other aluminium alloys containing silicon and aluminium oxide particles improves wear characteristics. Cast aluminium can also be clad with aluminium bronze powders, to improve wear resistance. The procedures are material specific because of the highly reflective nature of aluminium. Details can be found in the references listed in Table 12.7.

MAGNESIUM ALLOYS

Magnesium alloys are also being used in an increasing number of automobile applications, notably cylinder heads, gear assemblies, crank cases and chassis members. The driving forces for the use of magnesium alloys, particularly in transport applications are similar to those of aluminium. The limitations of magnesium alloys are also similar to those of aluminium. Surface treatment may provide the increase in hardness needed to achieve the necessary service properties. Cladding with Mg–Zr powder mixtures can be used to produce Mg–2 wt% Zr and Mg–5 wt% Zr surface alloys, which also exhibit improved corrosion properties. Procedures are material specific, and can be found in the references listed in Table 12.8.

COPPER ALLOYS

The motivation for cladding copper onto substrates is that the metal possesses low friction and good erosion resistance. It is often used in guide blocks and shoes, but is a relatively expensive material. Copper can be clad onto substrates through the addition of deoxidizers such as silicon and boron, which enables the copper to wet the substrate surface more effectively, as described in the references in Table 12.9.

BRONZES

Ni–Al bronzes clad onto substrates of high thermal conductivity solidify as martensitic structures with high densities of dislocations and stacking faults, which results in an improvement in wear properties.

Table 12.7 Sources of data for laser cladding using aluminium-base alloys

Alloy addition	Substrate	Laser	Reference
99.5Al–0.5Cu	ZM51–SiC$_p$	Nd:YAG	(Yue *et al.*, 2001)
Al–SiO$_2$	6061	CO$_2$	(De Hosson, 1995)
CP Al	Rapid prototyping	CO$_2$	(Koch and Mazumder, 1994)
Hypereutectic Al–Si	AlSi10	Nd:YAG, CO$_2$	(Arnold and Volz, 1999)
Mn–Al bronze	AA333	CO$_2$	(Kelly *et al.*, 1998)

Table 12.8 Sources of data for laser cladding using magnesium-base alloys

Alloy addition	Substrate	Laser	Reference
Mg–2 wt% Zr	Magnesium alloys	CO$_2$	(Subramanian *et al.*, 1991)
Mg–5 wt% Zr	Magnesium alloys	CO$_2$	(Subramanian *et al.*, 1991)

Table 12.9 Sources of data for laser cladding using copper-base alloys

Alloy addition	Substrate	Laser	Reference
Cu	Cast iron	CO$_2$	(Yang *et al.*, 1993)
Cu	Al alloy	CO$_2$	(Miyamoto *et al.*, 1997)

NIOBIUM ALLOYS

Alloys of niobium and aluminium are potential choices for applications requiring high temperature oxidation resistance and a low ratio of strength to weight. However, problems arise because the ordered Nb_xAl_y alloys are brittle at room temperature. Rapid solidification of niobium-base alloys by laser cladding increases ductility, and the niobium aluminide $NbAl_3$ oxidizes more slowly. Further details of these investigations can be found in the references given in Table 12.10.

CERAMIC METAL COMPOSITES

Ceramic metal composites produced on various substrates have two potential benefits: the surface is hardened; and the incorporation of hard particles into a ductile metallic matrix improves resistance to abrasive and erosion wear. Mixtures of WC and NiCrBSi powders clad onto substrates produce layers with a hardness in excess of 550 HV, with abrasive sliding wear properties comparable with sintered WC hard metals. Yttria partially stabilized zirconia clad onto mild steel acts as a thermal barrier in gas turbine engines with a hardness of 1500 HV (in contrast to the value of 800 HV obtained by plasma spraying). Ceramic coatings of oxides such as Cr_2O_3 endow steels with excellent corrosion resistance and wear properties. Molybdenum disilicide ($MoSi_2$) can also be clad onto steel substrates to produce a hardness in excess of 1000 HV. Further information can be found in the references listed in Table 12.11.

FUNCTIONALLY GRADED SURFACES

In situations where the clad and the substrate have very different properties, the interface region may be a potential area of weakness. In order to minimize the transition, clads with a gradated composition and structure may be produced. Functionally graded clads of nickel aluminide and iron aluminide have been produced by successive cladding of different compositions. A graded microstructure reduces the

Table 12.10 Sources of data for laser cladding using niobium-base alloys: N/S not specified

Alloy addition	Substrate	Laser	Reference
Nb–Al	N/S	CO_2	(Agrawal *et al.*, 1993)
$NbAl_3$	Nb	CO_2	(Haasch *et al.*, 1992; Sircar *et al.*, 1992)

Table 12.11 Sources of data for laser cladding using ceramics: N/S not specified

Alloy addition	Substrate	Laser	Reference
$MoSi_2$	C–Mn steel	Nd:YAG	(Ignat *et al.*, 2001)
Ti–6Al–4V	TiC–Ti MMC	Nd:YAG	(Man *et al.*, 2001)
TiC–Al	AA 2024, AA 6061	Nd:YAG	(Kadolkar and Dahotre, 2003)
TiC–Ni	AISI 1045	CO_2	(Lei *et al.*, 1995)
TiN, TiC	Aluminium clad	CO_2	(Ferraro *et al.*, 1986)
WC/Co	16MnCrS5	CO_2	(Grünenwald *et al.*, 1992)
Yttria–zirconia	C–Mn steel	CO_2	(Jasim *et al.*, 1990)
ZrO_2–Y_2O_3 ceramic/nickel	4Cr13	CO_2	(Pei *et al.*, 1995)
TiB–Ti	N/S	CO_2	(Kooi *et al.*, 2003)
SiO_2	AlCu4SiMg	CO_2	(Li and Steen, 1993)

likelihood of the interface acting as defect initiation site. Uses of functionally graded ceramic–metallic coatings include coatings for gas turbine components (Wolfe and Singh, 1998), and interfacial layers to enable joints to be made between metals and ceramics.

LASER CLADDING DIAGRAMS

Laser cladding is a relatively difficult process to model in detail, particularly when dynamic powder delivery is used. The interactions among the laser beam, the alloy addition, and the molten region are complex. A small change in the powder delivery geometry, for example, results in a large variation in the nature of the clad produced. The objective of the modelling described here is to provide a means of determining the effects of changes in process variables on the geometry of the clad. The model is used to construct two forms of laser cladding diagram: a process overview that illustrates trends among process variables for a class of engineering material; and a practical diagram for selecting processing parameters, which is specific to a given type of laser, beam size, clad and substrate, and which is calibrated using a single experimental measurement.

PROCESS OVERVIEW

The cladding geometry is modelled as follows. A transverse section of the clad is idealized as a rectangle of height l (the clad thickness) and width equal to the beam diameter, $2r_B$. The area of the clad section is then $2r_B l$. The clad is formed by traversing the beam at a rate v. Sufficient clad material is assumed to be present in the form of dynamically fed powder or wire, or a preplaced bed, mat or foil such that all the available energy is used to create the clad. An energy balance between the absorbed power, Aq, and the power used to heat and melt the clad material gives

$$Aq = 2r_B l v [\rho c (T_m - T_0) + L_m] \tag{12.1}$$

where ρ is the density (kg m^{-3}), c is the specific heat capacity (J kg^{-1} K^{-1}), L_m is the volumetric latent heat of melting (J m^{-3}), and T_m and T_0 are the melting and initial temperatures (K) of the clad material, respectively. In terms of the dimensionless variables defined in Chapter 6, equation (12.1) can be written

$$q^* = 2l^* v^* \left(1 + L_m^*\right) \tag{12.2}$$

where $q^* = Aq/[r_B \lambda (T_m - T_0)]$, $l^* = l/r_B$, $v^* = v r_B/a$, $L_m^* = L_m/[\rho c(T_m - T_0)]$ and $a = \lambda/(\rho c)$ where λ is the thermal conductivity of the clad material (J s^{-1} m^{-1} K^{-1}). The value of L_m^* lies around 0.4 for metals and alloys (Richard's rule). Equation (12.2) gives an explicit relationship between q^* and v^* for given values of l^* and L_m^*, and is the basis of the cladding overview diagram presented in Fig. 12.4.

The limits on process variables for cladding are determined by the need to melt the clad, but to avoid significant vaporization. Such conditions were described in Chapter 11 for surface melting, and are applied here to construct the contours of melting and vaporization shown in Fig. 12.4. Equation (12.2) is used to construct the contours of l^*.

Figure 12.4 provides a comprehensive overview of the effects of variations in process variables on the geometry of the clad in a range of metals and alloys. It is a useful tool for understanding and displaying the principles of laser cladding.

However, many assumptions were made in constructing Fig. 12.4. For example, heat conduction into the substrate is neglected, an idealized clad geometry is used, the absorptivity of the clad material to the laser beam is assumed to be 100%, and the effects of variations in process gas on processing conditions are not considered. In practical cladding, these assumptions and idealizations may be unrealistic. But they may be taken into account by calibrating the model using an experimental data point.

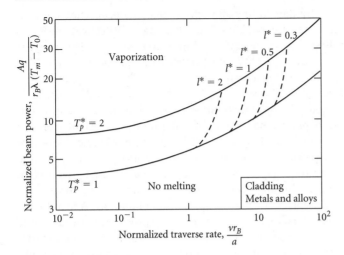

Figure 12.4 Normalized model-based overview of laser cladding of metals and alloys, showing boundaries defining the operating window for melting (solid lines) and contours of constant normalized clad thickness (broken lines)

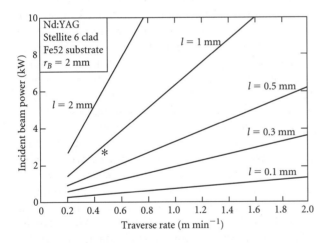

Figure 12.5 Parameter selection diagram for Nd:YAG laser cladding of Fe52 with Stellite 6 using a beam radius of 2 mm, showing model-based contours of constant clad thickness calibrated using an experimental data point (∗)

PARAMETER SELECTION

Calibration is an effective means of tailoring the model to a given set of laser cladding conditions. It enables the model to be used to predict the effects of variations in process variables on the clad geometry in relation to a known condition (the experimental data point). Model predictions can then be presented with greater precision, on a practical diagram constructed using linear axes of principal processing variables. The contours of clad thickness shown in Fig. 12.5 are constructed using equation (12.1).

Figure 12.5 is extracted from Fig. 12.4 by substituting appropriate material properties and constant processing parameters. It is constructed for Nd:YAG laser cladding of Fe52 with Stellite 6 using a beam radius of 2 mm, and calibrated to an experimental measurement of a clad thickness of 1 mm obtained using an incident beam power of 3 kW, and a traverse rate of 0.45 m min^{-1}. This data point is indicated in Fig. 12.5 by a star. The following values have been used as material properties for Stellite 6: $\rho = 8640$ kg m^{-3}; $c = 442$ J kg^{-1}K^{-1}; $T_m = 1613$ K; $T_0 = 298$ K; and $L_m = 232$ kJ kg^{-1}.

Figure 12.5 can be used to select parameters for the development of a cladding procedure specification. It may also be used to extrapolate data to predict the outcome of cladding outside current operating windows, e.g. when using an Nd:YAG laser capable of delivering 10 kW to the workpiece. A diagram valid for a different laser beam size may be produced by substituting the appropriate beam radius into equation (12.1). A non-rectangular-clad section may be described by using a representative section geometry, expressed in terms of the beam radius and clad thickness. Diagrams can be constructed for different clad materials by substituting appropriate material properties (Table D.1, Appendix D). Specific cladding conditions, such as a particular alloy feeding geometry and process gas, are taken into account through the data point calibration. The model only breaks down when the cladding mechanism changes from that valid for the data point; this can occur when the process gas is changed, for example.

INDUSTRIAL APPLICATION

The list of applications given in Table 12.12 illustrates the importance of the machine tool industry in the industrialization of laser cladding. CO_2 and Nd:YAG lasers have traditionally been used, in CW and pulsed modes, but there is increasing interest in the use of flexible multikilowatt diode sources. The cobalt-base Stellite alloys are popular clad materials, and are used in applications in which the working temperature is limited to about 500°C. Nickel-base alloys endow cheaper substrates with the surface properties of more expensive superalloys. Cladding was originally developed as an industrial surfacing technique, but has evolved into a cost-effective means of *repairing* expensive components as well. The process has been refined to the stage where it can now be used to manufacture near net-shaped components, particularly for aerospace applications that use novel combinations of materials.

Until recently, applications of laser cladding were limited to small areas, complex shapes, and very high value/high cost items. Today there is interest in laser cladding of larger components. With reductions in the cost of laser power, higher reliability, and increased output power, deposition of thicker claddings over larger areas has become cost effective. These changes have enabled lower value items to benefit from laser cladding, either during their original fabrication or as a repair technique. The power generation industry has begun to use laser cladding of boiler tubes and water walls in steam generators. This can be justified in terms of the environmental improvement resulting from the reduced use of chromium. Manufacturers of heavy equipment are using laser cladding to repair and refurbish high cost drive train and suspension components.

Today, manufacturers of machine tools appreciate the potential of CNC systems for cladding, and have developed turnkey systems for hardfacing components. These comprise an integrated vision system, the laser, and metal powder delivery with multiple axes of motion, all of which can be controlled automatically. The aerospace industry drives many current novel applications.

AEROENGINE TURBINE BLADES

The outer shroud of a jet engine is made up of a large number of turbine blades. The notch between adjacent blades experiences rubbing contact during operation. Laser cladding is now used extensively in both hardfacing to reduce local wear, and to rebuild worn components. The blade is cast from a nickel-base superalloy of high thermal resistance. A cobalt-base alloy is used as the hardfacing material. The entire flat notch surface and radius zone is covered, but mixing of the alloy and substrate is

Table 12.12 Industrial applications of laser cladding and metal deposition

Industry sector	Component	Clad	Reference
Automotive	Valve seat	Copper	(Belforte, 1994)
	Valve seat	Ni–Cr–B–Si and Co–Cr–W	(Aihua *et al.*, 1991)
	Valve rim	Stellite	(Küpper, *et al.*, 1990)
	Injection moulding tool	P20 tool steel	(Morgan, 2001)
Aerospace	Turbine blade	Stellite	(Macintyre, 1983a, b, c; Duhamel *et al.*, 1986; Merchant, 2002)
	Turbine blade	Nickel alloy	(Ritter *et al.*, 1992)
	Various	Titanium alloys	(Arcella and Froes, 2000)
	Compressor blade	Ni–Cr	(König *et al.*, 1992)
Machinery	Plough blades	Stellite 6	(Bruck, 1987)
	Deformation tool	Stellite 6	(van de Haar and Molian, 1987)
	Pulp comminution blades	Stellite	(Ion *et al.*, 1999)
	Tooling modification	P20 tool steel	(Morgan, 2001)
	Cutting tools	Iron-base	(Semionov, 1989)
	Pump bushing	Alloy–WC	(Blake and Eboo, 1985)
	Extruder screw	Al bronze	(Wolf and Volz, 1995)
	Deep drawing tool	Stellite SF6	(Haferkamp *et al.*, 1994)
Petrochemical	Drill rods	Stellites, Colmonoys	(Eboo and Lindemanis, 1985)
	Valves	Stellite 6	(Koshy, 1985)
Power generation	Gas turbine parts	Inconel 718	(Mehta *et al.*, 1985)
	Turbine blades and vanes	Cobalt-base	(Regis *et al.*, 1990)
	Turbine blades	Stellite	(Alam and Ion, 2002)
Shipbuilding	Catapult rails	Stellite	(Irving, 1991)

minimized to maintain the hardness of the deposit and to prevent a crack-sensitive alloyed zone being formed.

In one method (Duhamel *et al.*, 1986), a chip of hardfacing alloy is melted onto the notch region. Fixturing is simple and the deposit volume can be controlled accurately. The chip is placed on the notch and irradiated under an argon shield with a 1.5 mm diameter CO_2 beam for around 10 s with 4.5 kW. The chip melts and assumes a roughly spherical shape prior to attachment to the notch. Melting is controlled using a copper chill bar placed adjacent to the blade. The blade is machined to the final configuration. At one stage, Pratt & Whitney were hardfacing up to 800 blades of their JT8 and JT9 series engines each day, with an acceptance rate better than 99%.

Rolls-Royce introduced the blown powder method of laser cladding successfully in their RB 211 series engines. Previously the process was performed by highly skilled operators using TIG hardfacing. This process suffers from two main drawbacks – a significant amount of reworking is necessary to achieve the tolerances, and it is difficult to guarantee the correct amount of dilution. Laser cladding was cheaper, because of the greatly reduced processing time, and proved to be a reliable and reproducible process, giving less HAZ cracking and a high yield. The consumption of expensive cobalt-base hardfacing material was reduced by more than 50%.

Laser cladding is used by a number of organizations, including the Aircraft Engine Group of General Electric (Cincinnati, OH, USA) and Honeywell International Aerospace Services (Norcross,

Figure 12.6   Cladding of a steam turbine blade to improve resistance to erosion by water droplets. (Source: Juha Kauppila, Lappeenranta University of Technology, Finland)

GA, USA) to repair worn and damaged components of aeroengines. Clad material can be deposited using the blown powder technique at a rate at least twice that of conventional TIG or microplasma methods. In addition, the accuracy of the process reduces the time required for subsequent finishing operations. Systems incorporate a vision system for on-line beam adjustment.

An example of laser cladding a steam turbine blade from power generation equipment with Stellite, to improve resistance to erosion by water droplets is shown in Fig. 12.6.

ROTOR BLADES FOR PULP SCREENING MACHINERY

Ahlström Machinery (Karhula, Finland) manufactures a wide range of equipment for the paper and pulp making industry. One of their best selling products, the MODUScreen™F, fine screens pulp by removing unwanted impurities. Such impurities include knots, bark, shives, extractives and fibre bundles, as well as foreign matter such as sand, stones, ash, metal, plastics, rust and slime. The equipment can process up to 2000 tonnes of pulp each day.

Pulp is fed tangentially into the upper section of the screen, as illustrated in Fig. 12.7. A rotor drives a drum inside the screen cylinder, keeping it open. Blades are attached to the outside of the rotor, which increase turbulence in the mass, thus improving the efficiency of screening. Between six and 50 blades are used with one rotor, depending on its diameter, which can be between 0.4 and 1.5 m. The rotor is connected to a motor of power 15–500 kW, depending on the application and the size of the screen. The blades are approximately 120 mm in length, 60 mm in width, and 20 mm in thickness, and vary slightly in shape and sectional profile. Agitation is most effective when a small gap of 3 mm is present between the edge of the blades and the screen cylinder. In order to maintain the fine screening capacity and running characteristics of the equipment, the gap must be maintained with a tolerance of ±0.2 mm. Sharp corners are therefore machined on the blades, which are made from acid-resistant AISI 316 austenitic stainless steel.

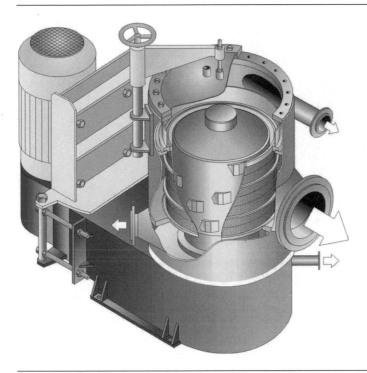

Figure 12.7   Ahlstrom MODUScreen™F fine screen pulp machinery, showing rotor with blades in centre cutaway section. (Source: Kalevi Jaaranen, Ahlstrom, Karhula, Finland; MODUScreen is a trademark of Ahlstrom Corporation, Finland)

During operation, foreign matter wears the edges of the blades. Blade wear, which is checked twice a year during regular servicing, reduces the fine screening capacity. Blades must typically be replaced after about two years when pulp of normal cleanliness is screened. About 100 screens are delivered each year, and on average 20–30 rotors must be replaced every year. By increasing the lifetime of the blades, the maintenance costs can be reduced.

Blade wear can be reduced by coating the blades with a hard, wear-resistant alloy. Manual metal arc welding had been used previously to hardface the blade surface with Stellite 6. This resulted in an increase in blade lifetime of between 10 and 20%, in comparison with untreated blades. However, in order to meet the dimensional tolerances, it was necessary to machine the blades after hardfacing. This is an expensive and time-consuming operation. An alternative, more accurate, cladding technique was therefore sought.

A hard, wear-resistant surface was produced on the blades using laser cladding. Grooves with the profile of an arc were first machined in the blade surface to a maximum depth of 0.5 mm. (This enabled the dimensional tolerances to be met after cladding without further machining.) The laser beam was formed into a heating pattern of width 7 mm and length 5 mm, with a uniform power distribution, using a molybdenum-coated integrating mirror. The beam power at the workpiece was 5.5 kW. The heating pattern was traversed across the blade at a rate of 0.8 m min^{-1}. Stellite 6 was delivered from a nozzle using argon, which carried the powder and shielded the molten clad. A powder delivery rate of 27 g min^{-1} was used. A clad track of width 7 mm was produced. The variation in clad thickness was in the range ±0.1 mm, which meant that no post-cladding machining was necessary. Blades were clad in batches of 12, by aligning the components under the traversing beam. The total processing time for each component was approximately 30 seconds. Examples of clad blades are shown in Fig. 12.8.

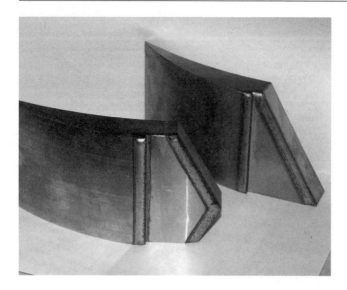

Figure 12.8 Stellite-clad rotor blades from a pulping machine. (Source: Timo Kankala, Lappeenranta University of Technology, Finland)

Figure 12.9 Die made by a direct metal deposition process based on laser cladding. (Source: David Belforte, Industrial Laser Solutions, Sturbridge, MA, USA)

The blades are installed in several machines, which have now been operating throughout the world for over 10 years. In comparison with untreated blades, the lifetime was increased by 10–20%. In comparison with MMA hardfacing, the machining time of 2 hours for a typical sized drum was eliminated. The laser beam can be manipulated to give a wide range of heating patterns, and can be used to clad blades of variable geometry, which provides the designer with freedom to design blade profiles that may enable the screening capacity of the drum to be increased further.

DIRECT METAL DEPOSITION

Various techniques involving the direct deposition of metals and alloys for fabricating near net-shaped components have been developed, as discussed in Chapter 1. The method is particularly suitable for making prototype parts that can be remodelled by making changes in the software, or small batches of products. Figure 12.9 shows a small die made using a direct metal deposition process.

SUMMARY AND CONCLUSIONS

Laser cladding has become the process of choice in many applications involving coating, repair and reworking. It is a particularly suitable process for the production of relatively small, discrete surface clads with moderate thickness (0.5–2.5 mm) and track width (2–15 mm). The process principles are similar to those of other surface melting processes, enabling the microstructure and properties of the clad surface to be established from data for rapid solidification of the alloy used. Practical processing has been simplified by the development of coaxial alloy delivery systems, which permit cladding of complex three-dimensional components in various orientations. Analytical modelling aids in understanding the process principles, and empirical data can be used to calibrate parameter selection diagrams to particular cladding geometries. Novel manufacturing methods, notably rapid manufacturing by various means of direct metal deposition, have been developed based on the process.

In comparison with conventional hardfacing processes, the energy input of laser cladding is low, which results in finer microstructures with superior properties, and minimal component distortion. In addition, in-service performance is good because of the smooth clad surface, the lack of defects in the clad, and the metallurgical bond to the substrate. The use of expensive, strategic alloying elements can also be reduced substantially. As with all laser-based processes, capital cost and unfamiliarity with the process are the main obstacles to industrial application. However, as more sophisticated turnkey systems become available, in which computer-generated models are integrated with automated beam and alloy delivery, the scope of application is likely to widen considerably.

FURTHER READING

Steen, W.M. (1978). Surface coating using a laser. In: Anderson, J.C. ed. *Proceedings of the Conference Advances in Surface Coating Technology*, 13–15 February 1978, London, UK. Abington: The Welding Institute. pp. 175–187.

Steen, W.M. (1983). Laser cladding, alloying and melting. In: Belforte D. and Levitt M. eds

The Industrial Laser Annual Handbook. PennWell Books. pp. 158–174.

Steen, W.M. (1986). Laser surface cladding. In: Draper, C.W. and Mazzoldi, P. eds *Proceedings of the NATO ASI Laser Surface Treatment of Metals.* Dordrecht: Martinus Nijhoff. pp. 369–387.

EXERCISES

1. Compare the advantages and disadvantages of laser cladding performed using predeposition and codeposition techniques.

2. A clad of thickness 0.5 mm is made at a rate of 0.6 m min^{-1} using an incident beam power of 2 kW and a beam diameter of 2 mm. Use an energy balance to estimate the fraction of incident energy absorbed to form the clad. Assume that the clad is rectangular in section with a width equal to the beam diameter. Use the following values for the clad material properties: $\rho = 8640$ kg m^{-3}; $c = 442$ J kg^{-1} K^{-1}; $T_m = 1613$ K; $T_0 = 298$ K; and $L_m = 232$ kJ kg^{-1}. What happens to the remainder of the incident energy?

3. Use the information provided in Fig. 12.5 to calculate the powder flow rate and the incident beam power required to produce a clad of thickness 1 mm and width 4 mm with an incident beam power of 3 kW. Assume that the powder density is 8640 kg m^{-3} and that 50% of the powder used in cladding is melted by the laser beam.

4. What are the benefits of using direct metal deposition by laser fusion for rapid prototyping and rapid manufacturing operations?

BIBLIOGRAPHY

Abboud, J.H., Rawlings, R.D. and West, D.R.F. (1995). Functionally graded nickel-aluminide and iron-aluminide coatings produced via laser cladding. *Journal of Materials Science*, **30**, 5931–5938.

Agrawal, G., Kar, A. and Mazumder, J. (1993). Theoretical studies on extended solid solubility and nonequilibrium phase diagram for Nb–Al alloy formed during laser cladding. *Scripta Metallurgica et Materialia*, **28**, 1453–1458.

Aihua, W., Zengyi, T. and Beidi, Z. (1991). Laser beam cladding of seating surfaces on exhaust valves. *Welding Journal*, **70**, (4), 106s–109s.

Alam, N. and Ion, J.C. (2002). Extending the life of components by laser beam cladding. *Australasian Welding Journal*, **47**, (2), 26–27.

Arcella, F.G. and Froes, F.H. (2000). Producing titanium aerospace components from powder using laser forming. *Journal of Materials*, **52** (5), 28–30.

Arnold, J. and Volz, R. (1999). Laser powder technology for cladding and welding. *Journal of Thermal Spray Technology*, **8**, (2), 243–248.

Belforte, D. (1994). Laser cladding: its time has come. *Industrial Laser Review*, **9**, (3), 16–17.

Belmondo, A. and Castagna, M. (1979). Wear resistant coatings by laser processing. *Thin Solid Films*, **64**, 249–256.

Blake, A.G. and Eboo, G.M. (1985). Laser hardfacing turbine blades using dynamic powder feed. SME Technical Paper no. AD85-913. Dearborn: Society of Manufacturing Engineers. pp. 1–14.

Brandt, M., Scott, D.A. and Yellup, J.M. (1995). Fibre optic Nd-YAG laser cladding of preplaced Hastelloy C powder. *Surface Engineering*, **11**, (3), 223–232.

Bruck, G.J. (1987). High power laser beam cladding. *Journal of Metals*, **39**, (2), 10–13.

Capp, M.L. and Rigsbee, J.M. (1984). Laser processing of plasma-sprayed coatings. *Materials Science and Engineering*, **62**, 49–56.

Choi, J., Choudhuri, S.K. and Mazumder, J. (2000). Role of preheating and specific energy input on the evolution of microstructure and wear properties of laser clad Fe–Cr–C–W alloys. *Journal of Materials Science*, **35**, (13), 3213–3219.

Choi, J. and Mazumder, J. (1994). Non-equilibrium synthesis of Fe–Cr–C–W alloy by laser cladding. *Journal of Materials Science*, **29**, 4460–4476.

Com-Nougué, J., Kerrand, E. and Hernandez, J. (1987). Chromium steel laser coating with cobalt base alloy. In: Arata, Y. ed. *Proceedings of the Conference Laser Advanced Materials Processing (LAMP '87)*, 21–23 May 1987, Osaka, Japan. Osaka, Japan: High Temperature Society of Japan and Japan Laser Processing Society. pp. 389–394.

Conde, A., Zubiri, F. and de Damborenea, J. (2002). Cladding of Ni–Cr–B–Si coatings with a high power diode laser. *Materials Science and Engineering*, **A334**, 233–238.

Coquerelle, G., Collin, M. and Fachinetti, J.L. (1986). Laser cladding and alloying. In: Quenzer, A. ed. *Proceedings of the 3rd International Conference on Lasers in Manufacturing (LIM-3)*, 3–5 June 1986, Paris, France. Bedford: IFS Publications. pp. 197–205.

de Damborenea, J.J. and Vázquez, A.J. (1992). Laser processing of NiCrAlFe alloy: microstructural evolution. *Journal of Materials Science*, **27**, 1271–1274.

de Damborenea, J. and Vázquez, A.J. (1993). Laser cladding of high-temperature coatings. *Journal of Materials Science*, **28**, 4775–4780.

De Hosson, J.Th.M. (1995). Fundamental and applied aspects of laser processed ceramic coatings of metals. In: Aliabadi, M.H. and Terranova, A. eds *Proceedings of the Conference Surface Treatment 95*, June 1995, Milan, Italy. Southampton: Computational Mechanics Publications. pp. 1–15.

De Hosson, J.Th.M. and De Mol van Otterloo, L. (1997). Surface engineering with lasers of Co-base materials. In: Aliabadi, M.H. and Brebbia, C.A. eds *Proceedings of the 3rd International Conference Surface Treatment: Computer Methods and Experimental Measurements*, 15–17 July 1997, Oxford, UK. Southampton: WIT Press. pp. 341–359.

Dehm, G. and Bamberger, M. (2002). Laser cladding of Co-based hardfacing on Cu substrate. *Journal of Materials Science*, **37**, 5345–5353.

Duhamel, R.F., Banas C.M. and Kosenski, R.L. (1986). Production laser hardfacing of jet engine turbine blades. In: Cheo, P.K. ed. *Proceedings of the Conference Manufacturing Applications of Lasers.* Vol. 621. Bellingham: SPIE. pp. 31–39.

Eboo, G.M. and Lindemanis, A.E. (1985). Advances in laser cladding process technology. In: Jacobs, R.R. ed. *Proceedings of the Conference Applications of High Power Lasers.* Vol. 527. Bellingham: SPIE. pp. 86–94.

Eiholzer, E., Cusano, C. and Mazumder, J. (1985). Wear properties of laser alloyed and clad Fe–Cr–Mn–C alloys. In: Mazumder, J. ed. *Proceedings of the Materials Processing Symposium (ICALEO '84).* Toledo: Laser Institute of America. pp. 159–167.

Ferraro, F., Nannetti, C.A., Campello, M. and Senin, A. (1986). Aluminium alloy surface hardening. In: Quenzer, A. ed. *Proceedings of the 3rd International Conference Lasers in Manufacturing,* 3–5 June 1986, Paris, France. Bedford: IFS Publications Ltd. pp. 233–244.

Giordano, L., Ramous, E., Ferraro, F., Pasquini, F. and La Rocca, A.V. (1984). Comparison of stellite hardfacing by laser and traditional techniques. *High Temperature Technology,* **2**, (4), 213–216.

Gnanamuthu, D.S. (1979). Laser surface treatment. In: Metzbower, E.A. ed. *Proceedings of the Conference Applications of Lasers in Materials Processing.* Metals Park: American Society for Metals. pp. 177–211.

Grünenwald, B., Shen, J., Dausinger, F. and Nowotny, St. (1992). Laser cladding with a heterogeneous powder mixture of WC/Co and NiCrBSi. In: Mordike, B.L. ed. *Proceedings of the 4th European Conference Laser Treatment of Materials (ECLAT '92),* 12–15 October 1992, Göttingen, Germany. Oberursel: DGM Informationsgesellschaft. pp. 411–416.

Haasch, R.T., Tewari, S.K., Sircar, S., Loxton, C.M. and Mazumder, J. (1992). Nonequilibrium synthesis of $NbAl_3$ and Nb–Al–V alloys by laser cladding: Part II. Oxidation behaviour. *Metallurgical Transactions,* **23A**, (9), 2631–2639.

Haferkamp, H., Schmidt, H., Gerken, J. and Püster, T. (1994). Application of laser powder cladding and the risk of residual powder. In: Mordike, B.L. ed. *Proceedings of the 5th European Conference Laser Treatment of Materials (ECLAT '94),* 26–27 September, 1994, Bremen-Vegesack, Germany. Oberursel: DGM Informationsgesellschaft. pp. 475–483.

Ignat, S., Sallamand, P., Nichici, A., Vannes, A.B., Grevy, D. and Cicală, E. (2001). $MoSi_2$ laser cladding – a comparison between two experimental procedures: Mo–Si online combination and direct use of $MoSi_2$. *Optics and Laser Technology,* **33**, 461–469.

Ion, J.C., Kauppila, J. and Metsola, P. (1999). Laser-based fabrication: opportunities for novel design and manufacturing. In: Kujanpää, V. and Ion, J.C. eds *Proceedings of the 7th Nordic Conference on Laser Materials Processing (NOLAMP-7).* Lappeenranta: Acta Universitatis Lappeenrantaensis. pp. 26–37.

Irving, R. (1991). High-powered lasers gain ground for cladding and hardfacing. *Welding Journal,* **70**, (8), 37–40.

Jansson, A., Ion, J.C. and Kujanpää, V. (2003). CO_2 and Nd:YAG laser cladding using Stellite 6. In: Miyamoto, I., Kobayashi, K.F., Sugioka, K., Poprawe, R. and Helvajian, H. eds *Proceedings of the 1st International Symposium on High-Power Laser Macroprocessing (LAMP 2002).* Vol. 4831. Bellingham: SPIE. pp. 475–480.

Jasim, K.M., Rawlings, R.D. and West, D.R.F. (1990). Thermal barrier coatings produced by laser cladding. *Journal of Materials Science,* **25**, 4943–4948.

Kadolkar, P. and Dahotre, N.B. (2003). Effect of processing parameters on the cohesive strength of laser surface engineered ceramic coatings on aluminium alloys. *Materials Science and Engineering,* **A342**, 183–191.

Kagawa, A. and Ohta, Y. (1995). Wear resistance of laser clad chromium carbide surface layers. *Materials Science and Technology,* **11**, (5), 515–519.

Kaplan, A.F.H. and Groboth, G. (2001). Process analysis of laser beam cladding. *Journal of Manufacturing Science and Engineering,* **123**, 609–614.

Kaul, R., Ganesh, P., Tiwari, M.K., Singh, A.K., Tripathi, P., Gupta, A. and Nath, A.K. (2002). Laser assisted deposition of graded overlay of Stellite 6 on austenitic stainless steel. *Lasers in Engineering,* **12**, (3), 207–225.

Kaul, R., Ganesh, P., Albert, S.K., Jaiswal, A., Lalla, N.P., Gupta, A., Paul, C.P. and Nath, A.K. (2003). Laser cladding of austenitic stainless steel with hardfacing alloy nickel base. *Surface Engineering,* **19**, (4), 269–273.

Kelly, J., Nagarathnam, K. and Mazumder, J. (1998). Laser cladding of cast aluminium–silicon alloys for improved dry sliding wear resistance. *Journal of Laser Applications,* **10**, 45–54.

Koch, J.L. and Mazumder, J. (1994). Rapid prototyping by laser cladding. In: Denney, P., Miyamoto, I. and Mordike, B.L. eds *Proceedings of the Laser Materials Processing Conference (ICALEO '93).* Orlando: Laser Institute of America/Bellingham: SPIE. pp. 556–565.

König, W., Rozsnoki, L. and Kirner, P. (1992). Laser beam surface treatment – is wear no longer the bug bear of old? In: Mordike, B.L. ed. *Proceedings of the 4th European Conference Laser Treatment of Materials (ECLAT '92),* 12–15 October 1992, Göttingen, Germany. Oberursel: DGM Informationsgesellschaft. pp. 217–222.

Kooi, B.J., Pei, Y.T., De Hosson, J.T.M. (2003). The evolution of microstructure in a laser clad TiB–Ti composite coating. *Acta Materialia*, **51**, (3), 831–845.

Koshy, P. (1985). Laser cladding techniques for application to wear and corrosion resistant coatings. In: Jacobs, R.R. ed. *Proceedings of the Conference Applications of High Power Lasers*. Vol. 527. Bellingham: SPIE. pp. 80–85.

Kosuge, S., Ono, M., Nakada, K. and Watanabe, I. (1987). In: Banas, C.M. and Whitney, G.L. eds *Proceedings of the 5th International Congress Applications of Lasers and Electro-optics (ICALEO '86)*. Bedford: IFS (Publications) Ltd/Berlin: Springer-Verlag. pp. 105–112.

Küpper, F., Gasser, A., Kreutz, E.W. and Wissenbach, K. (1990). Cladding of valves with CO_2 laser radiation. In: Bergmann, H.W. and Kupfer, R. eds *Proceedings of the 3rd European Conference Laser Treatment of Materials (ECLAT '90)*, 17–19 September 1990, Erlangen, Germany. Coburg: Sprechsaal Publishing Group. pp. 461–467.

Lei, T.C., Ouyang, J.H., Pei, Y.T. and Zhou, Y. (1995). Microstructure and wear resistance of laser clad TiC particle reinforced coatings. *Materials Science and Technology*, **11**, (5), 520–525.

Li, L.J. and Mazumder, J. (1984). A study of the mechanism of laser cladding processes. In: Mukherjee, K. and Mazumder, J. eds *Proceedings of the Symposium Laser Processing of Materials*, 26 February–1 March 1984, Los Angeles, CA, USA. Warrendale: The Metallurgical Society of AIME. pp. 35–50.

Li, L., Steen, W.M. and Hibberd, R.D. (1990a). Computer aided laser cladding. In: Bergmann, H.W. and Kupfer, R. eds *Proceedings of the 3rd European Conference Laser Treatment of Materials (ECLAT '90)*, 17–19 September 1990, Erlangen, Germany. Coburg: Sprechsaal Publishing Group. pp. 355–369.

Li, L., Steen, W.M., Hibberd, R.D. and Brookfield, D.J. (1990b). In-process clad quality monitoring using optical method. In: Opower, H. ed. *Proceedings of the Conference Laser-Assisted Processing II*. Vol. 1279. Bellingham: SPIE. pp. 89–100.

Li, Y. and Steen, W.M. (1993). Laser cladding of Stellite and silica on aluminium substrate. In: Mei, S.-S. and Zhou, B. eds *Proceedings of the International Conference on Lasers and Optoelectronics*. Vol. 1979. Bellingham: SPIE. pp. 602–608.

Liu, Y., Mazumder, J. and Shibata, K. (1994a). TEM crystal and defect structure study of martensite in laser clad Ni–Al bronze. *Acta Metallurgica et Materialia*, **42**, (5), 1755–1762.

Liu, Y., Mazumder, J. and Shibata, K. (1994b). Transmission electron microscopy study of martensites in laser-clad Ni–Al bronze on

aluminium alloy AA333. *Metallurgical and Materials Transactions*, **25A**, (1), 37–46.

Liu, Y., Mazumder, J. and Shibata, K. (1994c). Laser cladding of Ni–Al bronze on Al alloy AA333. *Metallurgical and Materials Transactions*, **25B**, (10), 749–759.

Lugscheider, E., Bolender, H. and Krappitz, H. (1991). Laser cladding of paste bound hardfacing alloys. *Surface Engineering*, **7**, (4), 341–344.

Macintyre, R.M. (1983a). Laser hardfacing of gas turbine blade shroud interlocks. In: Kimmitt, M.F. ed. *Proceedings of the 1st International Conference Lasers in Manufacturing*. Bedford: IFS Publications. pp. 253–261.

Macintyre, R.M. (1983b). Laser hard-surfacing of turbine blade shroud interlocks. In: Metzbower, E.A. ed. *Proceedings of the 2nd International Conference on Applications of Lasers in Materials Processing*. Metals Park: American Society for Metals. pp. 230–239.

Macintyre, R.M. (1986). The use of lasers in Rolls-Royce. In: Draper, C.W. and Mazzoldi, P. eds *Proceedings of the NATO ASI Laser Surface Treatment of Metals*. Dordrecht: Martinus Nijhoff. pp. 545–549.

Man, H.C., Zhang, S., Cheng, F.T. and Yue, T.M. (2001). Microstructure and formation mechanism of *in situ* synthesized TiC/Ti surface MMC on Ti6–Al–4V by laser cladding. *Scripta Materialia*, **44**, 2801–2807.

Marsden, C.F., Hoadley, A.F.A. and Wagnière, J.D. (1990). In: Bergmann, H.W. and Kupfer, R. eds *Proceedings of the 3rd European Conference Laser Treatment of Materials (ECLAT '90)*, 17–19 September 1990, Erlangen, Germany. Coburg: Sprechsaal Publishing Group. pp. 543–553.

Matthews, S.J. (1983). Laser fusing of hardfacing alloy powders. In: Metzbower, E.A. ed. *Proceedings of the Conference Lasers in Materials Processing*. Metals Park: American Society for Metals. pp. 138–148.

Mehta, P., Cooper, E.B. and Otten, R. (1985). Reverse machining via CO_2 laser. In: Mazumder, J. ed. *Proceedings of the Materials Processing Symposium (ICALEO '84)*. Toledo: Laser Institute of America. pp. 168–176.

Merchant, V. (2002). The slow industrial acceptance of laser cladding. *Industrial Laser Solutions*, **17**, (6), 9–12.

Miyamoto, I., Fujimori, S. and Itakura, K. (1997). Mechanism of dilution in laser cladding with powder feeding. In: Fabbro, R., Kar, A. and Matsunawa, A. eds *Proceedings of the Laser Materials Processing Conference (ICALEO '97)*. Orlando: Laser Institute of America. pp. F1–F10.

Monson, P.J.E. and Steen, W.M. (1990). A comparison of laser hardfacing with conventional processes. In: Johnson, K.I. ed. *Proceedings of the Conference Advances in Joining and Cutting Processes*. Abington: Woodhead Publishing. pp. 274–284.

Morgan, D. (2001). Benefiting from additive metal changes. *Industrial Laser Solutions*, **16**, (8), 6–8.

Nagarathnam, K. and Komvopoulos, K. (1993). Microstructural characterization and *in situ* transmission electron microscopy analysis of laser-processed and thermally treated Fe–Cr–W–C clad coatings. *Metallurgical Transactions*, **24A**, (7), 1621–1629.

Nowotny, S., Richter, A. and Beyer, E. (1998). Laser cladding using high-power diode lasers. In: Beyer, E., Chen, X. and Miyamoto, I. eds *Proceedings of the Laser Materials Processing Conference (ICALEO '98)*. Orlando: Laser Institute of America. pp. G68–G74.

Nurminen, J.I. and Smith, J.E. (1983). Parametric evaluations of laser/clad interactions for hardfacing applications. In: Metzbower, E.A. ed. *Proceedings of the Conference Lasers in Materials Processing*. Metals Park: American Society for Metals. pp. 94–106.

Oberländer, B.C. and Lugscheider, E. (1992). Comparison of properties of coatings produced by laser cladding and conventional methods. *Materials Science and Technology*, **8**, (8), 657–665.

Ono, M., Kosuge, S., Nakada, K. and Watanabe, I. (1987). In: Arata, Y. ed. *Proceedings of the Conference Laser Advanced Materials Processing (LAMP '87)*, 21–23 May 1987, Osaka, Japan. Osaka, Japan: High Temperature Society of Japan and Japan Laser Processing Society. pp. 395–400.

Pei, Y.T., Ouyang, J.H., Lei, T. C. and Zhou, Y. (1995). Laser clad ZrO_2–Y_2O_3 ceramic/Ni-base alloy composite coatings. *Ceramic International*, **21**, 131–136.

Pelletier, J.M., Sahour, M.C., Pilloz, M. and Vannes, A.B. (1993). Influence of processing conditions on geometrical features of laser claddings obtained by powder injection. *Journal of Materials Science*, **28**, 5184–5188.

Pelletier, J.M., Oucherif, F., Sallamand, P. and Vannes, A.B. (1995). Hadfield steel coatings on low carbon steel by laser cladding. *Materials Science and Engineering*, **A202**, 142–147.

Picasso, M., Marsden, C.F., Wagnière, J.-D., Frenk, A. and Rappaz, M. (1994). A simple but realistic model for laser cladding. *Metallurgical and Materials Transactions*, **25B**, (4), 281–291.

Pilloz, M., Pelletier, J.M. and Vannes, A.B. (1992). Residual stresses induced by laser coatings: phenomenological analysis and predictions. *Journal of Materials Science*, **27**, 1240–1244.

Powell, J., Henry, P.S. and Steen, W.M. (1988). Laser cladding with preplaced powder: analysis of thermal cycling and dilution effects. *Surface Engineering*, **4**, (2), 141–149.

Regis, V., Bracchetti, M., Cerri, W., D'Angelo, D. and Mor, G.P. (1990). In: Bachelet, R. *et al.* eds *Proceedings of the Conference High Temperature Materials for Power Engineering*, 24–27 September 1990, Liège, Belgium. Dordrecht: Kluwer. pp. 1747–1756.

Ritter, U., Kahrmann, W., Küpfer, R. and Glardon, R. (1992). Laser coating proven in practice. *Surface Engineering*, **8**, (4), 272–274.

Semionov, S.A. (1989). Laser surfacing of cutting tools with metal powders. *Welding International*, **3**, (5), 435–437.

Singh, J. and Mazumder, J. (1986a). Evolution of microstructure for laser clad Fe–Cr–Mn–C alloys. In: Schuöcker, D. ed. *Proceedings of the Conference High Power Lasers and Their Industrial Applications*. Vol. 650. Bellingham: SPIE. pp. 235–244.

Singh, J. and Mazumder, J. (1986b). Evolution of microstructure in laser clad Fe–Cr–Mn–C alloy. *Materials Science and Technology*, **2**, (7), 709–713.

Singh, J. and Mazumder, J. (1987a). Microstructure and wear properties of laser clad Fe–Cr–Mn–C alloys. *Metallurgical Transactions*, **18A**, (2), 313–321.

Singh, J. and Mazumder, J. (1987b). Effect of extended solid solution of Hf on the micro-structure of the laser clad Ni–Fe–Cr–Al–Hf alloys. *Acta Metallurgica*, **35**, (8), 1995–2003.

Singh, J., Nagarathnam, K. and Mazumder, J. (1987c). Laser cladding of Ni–Cr–Al–Hf on Inconel 718 for improved high temperature oxidation resistance. *High Temperature Technology*, **5**, (3), 131–137.

Singh, J. and Mazumder, J. (1988). Microstructure of laser clad Ni–Cr–Al–Hf alloy on a γ' strengthened Ni-base superalloy. *Metallurgical Transactions*, **19A**, (8), 1981–1990.

Sircar, S., Chattopadhyay, K. and Mazumder, J. (1992). Nonequilibrium synthesis of $NbAl_3$ and Nb–Al–V alloys by laser cladding: Part I. Microstructure evolution. *Metallurgical Transactions*, **23A**, 2419–2429.

Steen, W.M. and Courtney, C.G.H. (1980). Hardfacing of Nimonic 75 using 2kW continuous-wave CO_2 laser. *Metals Technology*, **7**, (6), 232–237.

Subramanian, R., Sircar, S. and Mazumder, J. (1991). Laser cladding of zirconium on magnesium for improved corrosion properties. *Journal of Materials Science*, **26**, 951–956.

Takeda, T., Steen, W.M. and West, D.R.F. (1985). Laser cladding with mixed powder feed. In: Mazumder, J. ed. *Proceedings of the Materials Processing Symposium (ICALEO '84)*. Toledo: Laser Institute of America. pp. 151–158.

Tam, K.F., Cheng, F.T. and Man, H.C. (2002). Laser surfacing of brass with Ni–Cr–Al–Mo–Fe using various laser processing parameters. *Materials Science and Engineering*, **A325**, 365–374.

Tosto, S., Pierdominici, F. and Bianco, M. (1994). Laser cladding and alloying of a Ni-base superalloy on plain carbon steel. *Journal of Materials Science*, **29**, 504–509.

Tuominen, J., Vuoristo, P., Mäntylä, T., Ahmaniemi, S., Vihinen, J. and Andersson, P.H. (2002). Corrosion behaviour of HVOF-sprayed and Nd-YAG laser-remelted high-chromium, nickel–chromium coatings. *Journal of Thermal Spray Technology*, **11**, (2), 233–243.

van de Haar, E. and Molian, P.A. (1987). Effect of process variables on the laser cladding of zirconia. In: Ream, S.L. ed. *Proceedings of the Materials Processing Symposium (ICALEO '87)*. Bedford: IFS Publications/Berlin Springer-Verlag. pp. 189–193.

Walther, H. and Pera, L. (1978). A high power laser metalworking facility and its applications. In: Shull, J.M., Kinney, A.L. and Morse, J.A. eds *Proceedings of the 4th European Electro-optics Conference*. Vol. 164. Bellingham: SPIE. pp. 252–270.

Webber, T. (1992). Application of a hardface coating with CO_2 and CW Nd:YAG lasers. In: Metzbower, E.A., Beyer, E. and Matsunawa, A. eds *Proceedings of the Laser Materials Processing Conference (ICALEO '91)*. Orlando: Laser Institute of America/Bellingham: SPIE. pp. 380–388.

Weerasinghe, V.M., Steen, W.M. and West, D.R.F. (1987). Laser deposited austenitic stainless steel clad layers. *Surface Engineering*, **3**, (2), 147–153.

Wetzig, A., Brenner, B., Fux, V. and Beyer, E. (1998). Induction assisted laser-cladding. A new and effective method for producing high wear resistant coatings on steel components. In: Beyer, E.,

Chen, X. and Miyamoto, I. eds *Proceedings of the Laser Materials Processing Conference (ICALEO '98)*. Orlando: Laser Institute of America. pp. D20–D28.

Wolf, S. and Volz, R. (1995). Application of laser beam cladding when manufacturing plastic processing machines. *Laser und Optoelektronik*, **27**, (2), 47–53. (In German.)

Wolfe, D. and Singh, J. (1998). Functionally gradient ceramic/metallic coatings for gas turbine components by high-energy beams for high-temperature applications. *Journal of Materials Science*, **33**, 3677–3692.

Yang, X., Zhong, M., Zheng, T. and Zhang, N. (1993). Novel reduced-friction materials by laser cladding of copper on cast iron. *Journal of Material Science and Technology (China)*, **9**, 248–252.

Yellup, J.M. (1995). Laser cladding using the powder blowing technique. *Surface and Coatings Technology*, **71**, 121–128.

Yue, T.M., Mei, Z. and Man, H.C. (2001). Improvement of corrosion resistance of magnesium ZM51/SiC composite by laser cladding. *Journal of Materials Science Letters*, **20**, 1479–1482.

Zhou, X.B. and De Hosson, J.Th.M. (1991). Spinel/metal interfaces in laser coated steels: a transmission electron microscopy study. *Acta Metallurgica et Materialia*, **39**, (10), 2267–2273.

CHAPTER 13

CONDUCTION JOINING

INTRODUCTION AND SYNOPSIS

Conduction joining describes a family of processes in which the laser beam is focused to give a power density on the order of 10^3 W mm^{-2}, which is used to fuse material to create a joint without significant vaporization. Conduction welding may be performed in two modes: direct heating; and energy transmission. During *direct heating*, heat flow is governed by classical thermal conduction from a surface heat source and the weld is made by melting portions of the base material. The first conduction welds, made in the early 1960s, used low power pulsed ruby and CO_2 lasers to make joints in wire connectors. Conduction welds can now be made in a wide range of metals and alloys in the form of wires and thin sheets in various configurations using CO_2, Nd:YAG and diode lasers with power levels on the order of tens of watts. Direct heating by a CO_2 laser beam can also be used to make lap and butt welds in polymer sheets. *Transmission welding* is an efficient means of joining polymers that transmit the near infrared radiation of Nd:YAG and diode lasers; energy is absorbed through novel interfacial absorption methods. Composites can be joined providing that the thermal properties of the matrix and reinforcement are similar. In the *laser soldering* and *brazing* processes, the beam is used to melt a filler addition, which wets the edges of the joint without melting the base material. Laser soldering started to gain popularity in the early 1980s for joining the leads of electronic components through holes in printed circuit boards. The process parameters are determined by the material properties: higher power levels are required for materials of high thermal conductivity, high melting temperature and low absorptivity. Soldering uses filler materials that have a relatively low liquidus temperature (below 450°C), such as alloys of lead and tin. Laser brazing uses filler materials with higher liquidus temperatures, such as those based on the aluminium, copper or silver alloy systems.

Laser conduction joining techniques share many of the principles of their conventional counterparts. They also include features of laser surface melting and cladding using filler wire. However, many of the procedures and properties of conduction welded materials are similar to those of *keyhole welding*, which are described in more detail in Chapter 16. Here, we concentrate on the differences between these two modes of welding. In appropriate applications, the precision, speed and low energy input with which joints may be made provide the necessary advantages over conventional techniques to justify the increased investment costs. The principles are relatively well understood, enabling analytical modelling to be applied to joining of metals and alloys, which can be used to construct conduction joining diagrams. Empirical data from trials using polymers indicate that analytical modelling should be possible when the processing mechanisms are understood more fully. The electronics and biomedical industries in particular have recognized the opportunities for miniaturization that precision laser-based microjoining techniques have to offer. Biomedical welding applications range from the construction of precision instruments to the joining of human tissue. Automated laser soldering systems are becoming ever more sophisticated to cater for the increases in component density that surface mount technology enables.

Principles

Conduction Welding

Conduction welding can be carried out in two principal modes: direct heating and energy transmission.

The mechanism of direct heating involves absorption of the beam energy by the material surface and subsequent transfer of energy into the surrounding material by conduction. A hemispherical weld bead and heat affected zone (HAZ) is formed in a similar manner to conventional arc fusion welding processes. Conduction-limited welds, therefore, exhibit a low depth-to-width ratio (aspect ratio), which is often required when limited penetration in the thickness direction is desired. Thermal cycles are rapid, resulting in a fine-grained weld bead with good mechanical properties. Spot welds are made by pulsing the laser beam to melt sufficient material to form the joint. Continuous welds may be made by overlapping pulsed spot welds, or by using a continuous wave beam.

The energy transmission mode of conduction welding is used with materials that transmit near infrared radiation, notably polymers. An absorbing ink is placed at the interface of a lap joint. The ink absorbs the laser beam energy, which is conducted into a limited thickness of surrounding material to form a molten interfacial film that solidifies as the welded joint. Thick section lap joints can thus be made without melting the outer surfaces of the joint. Butt welds can be made by directing the energy towards the joint line at an angle through material at one side of the joint, or from one end if the material is highly transmissive.

Soldering

Soldering is a member of the conduction joining family. The filler material used has a liquidus temperature below 450°C, and which is also below the solidus temperature of the base material. The process uses a precisely controlled laser beam to transfer energy to a soldering location where it is absorbed by the substrate, the solder, or a flux that aids wetting and protects the liquid solder. The absorbed energy heats the solder until it reaches its melting temperature. The solder joint forms itself by the nature of the wetting process, even when the heat and solder are not directed precisely at the location. In contrast to laser welding, no melting of the base materials occurs. The very short heating and cooling cycle results in the formation of a fine solidified solder microstructure, and a reduction in the thickness of the intermetallic films that form at joints, which can act as locations of failure. Soldered joints may perform electrical, thermal or mechanical functions, or combinations of these.

In the process of direct reflow soldering, the components are pretinned. No additional solder is needed for the soldering operation – the beam simply melts the predeposited material, and only a liquid flux is needed to ensure wetting. Solder paste is applied to the components in soldering of untinned components. Fluxless soldering is used when cleanliness is important and flux residues cannot be tolerated, or when post-solder cleaning is impossible.

The localized and well-controlled thermal cycles generated by the laser beam are ideally suited to soldering. Damage to heat-sensitive components and overheating of the solder can be avoided. Since the solder is molten for only a short time, the formation of brittle intermetallic compounds may also be avoided. Finer grain sizes are also formed, which have greater resistance to fatigue cracking.

Brazing

Laser brazing also relies on the addition of a filler material, which is melted by the laser beam and drawn into a narrow joint by capillary action. The principle of laser brazing is similar to that of laser soldering. Brazing is differentiated from soldering by the melting point of the filler material used: 450°C is a commonly used value, brazing being carried out above this temperature and soldering below. Naturally this temperature difference dictates the types of filler material used in the respective processes; aluminium, copper or silver alloy systems are commonly used for brazing operations. The energy input provided by the laser beam is correspondingly higher to achieve melting. Because the brazes used have higher melting temperatures, the strength of a brazed joint is normally higher than that of a soldered joint.

PROCESS SELECTION

CHARACTERISTICS OF CONDUCTION JOINING

Welding

Conduction welding is a stable, low energy input process, which can be performed using a laser that does not need to have a high beam quality. A relatively low power, low cost laser can therefore be used. Providing that fusion can be obtained, the properties of weldable materials are relatively unimportant – hardness does not affect the process, and they do not need to conduct electricity, for example. Materials can be joined in a variety of geometries: wires, strips, thin sheets, or combinations of these. Because of the relatively large beam size, the tolerances for alignment and fit-up can be relatively large. The process is ideally suited to small component geometries, in which the weld bead is on the order of millimetres. Joints can be placed close to heat-sensitive components. Because of the stable welding conditions, the weld bead has a good visual appearance. High temperature strength, reduced manufacturing time and simplified material separation at the end of the life cycle are also important process characteristics.

However, if conduction welding is used to join thicker sections, the energy input per unit length of weld is relatively high, resulting in a large weld bead and high levels of distortion.

Soldering

The most significant characteristic of laser soldering is the ability to localize the low energy input to the joint. Substrate heating is minimized, which reduces differential expansion, and creates joints of low stress. Soldering can be performed close to heat-sensitive devices. The mechanical properties of a laser-soldered joint are good because the rapid thermal cycle creates a fine grain size that is hard and has a long fatigue life.

A wide range of joint types may be soldered because the process is controlled through software, which may be updated easily. Process parameters such as laser power, process time and the geometry of the laser beam can be programmed quickly, enabling increased flexibility in the design of components. Densely packaged devices can be soldered because of the small spot size, and both sides of a circuit board can be used. When joints are soldered sequentially, the propensity for solder to bridge adjacent pads is minimized, enabling solder masks to be dispensed with. Laser soldering is environmentally friendly. Legislation is being introduced to ban the use of lead in solders – alternatives such as tin–silver can be used by simply increasing the laser power. The very high energy efficiency of diode laser systems (>40%) when compared with ovens or flashlamp-pumped solid state lasers leads to a reduction in the use of energy. Because of the difference in melting temperature between the solder and the base materials, components can be disassembled for recycling relatively easily. The process is amenable to on-line quality control by monitoring the infrared solder emission. Precision and repeatability are keys to high productivity.

The main drawback of laser soldering is the relatively low strength of the joint, which results from the inherent low strength of the low melting temperature materials used in solders.

Brazing

Laser brazing is an intermediate process, lying between welding and brazing, and so displays some of the characteristics of both processes. As with laser soldering, many of the process characteristics originate from the precise nature with which the low energy input can be applied. Precise control over the melting of the brazing filler metal and wetting of the base material is possible, enabling complex joint geometries to be processed and the braze properties to be carefully controlled to improve reliability. The process is suitable for sealing joints where uniform heating of the entire joint is impractical. A preplaced braze material or a wire feeder can be used. The high quality that is possible is characterized by a lack of porosity and cracking, and little weld reinforcement. Local heating and the rapid thermal cycle minimize deformation of the base material and reduce diffusion, which could lead to the formation of intermetallic compounds. Brazed joints have a higher tensile strength than soldered joints because of the higher strength of the brazes used. On-line process monitoring and control are also possible.

The main drawback of laser brazing is the high capital cost, and the fact that the process cannot be used in joints where extensive capillary flow is required because of the limited time that the braze material is in the liquid state.

COMPETING METHODS OF CONDUCTION JOINING

Laser-based conduction joining methods compete with a large number of conventional joining techniques. To make a meaningful comparison, here we consider principally those techniques in which joining is achieved in relatively thin materials through the use of a discrete energy source for heating, or discrete mechanical methods for fastening.

Welding

Resistance welding was introduced in the 1930s. It is still the most widely used joining process for fabricating components from formed sheet in the thickness range 0.15–2 mm. Energy is generated by the application of a current between electrodes that clamp the materials in a lap joint. The resistance depends on the properties and surface condition of the materials being welded, as well as the cross-sectional area through which current flows. In spot welding, the current is delivered to the workpiece through shaped electrodes, which concentrate the current through a small area. In seam welding, the electrodes take the form of rollers. Joining is achieved by a combination of resistance heating and pressure. The weld time is normally short, typically 0.2 s per mm of the single sheet thickness being joined. The equipment cost is low, and the process automatically clamps the parts. Resistance welding is only suitable for materials that conduct electricity, and results can be poor for aluminium alloys since the oxide layer makes the process unstable. In addition, electrodes must be replaced periodically, productivity is relatively low, electrode deformation may necessitate a post-weld finishing operation, and access from both sides is required, which limits design possibilities. Resistance welding is probably the technique that has most easily been replaced by laser welding, particularly in the automotive and domestic goods industries. Laser welding is competitive primarily when high productivity, high joint quality and low distortion are required. Additional benefits result from the ability to redesign joints to save material and to improve properties, since access is only required from one side. A continuous laser weld in a three-dimensional framework is stiffer than the equivalent made using resistance spot welds, which enables cheaper or thinner materials to be used.

Ultrasonic welding is used for sheet in the range 0.1–1 mm. Electrical energy is converted into heat by high frequency mechanical vibration. One part of the assembly is vibrated, generating friction heat with the static part. Movement is provided by a sonotrode that is applied at right angles to the surface of the part to be welded. The method is particularly suitable for joining polymers. High production cycles, clean exteriors, and gas-tight assemblies can be produced. Metals can also be joined.

Mechanical fastening covers a family of non-fusion processes. Solidification and HAZ imperfections can therefore be avoided. However, a process such as riveting involves many stages: drilling, reaming, burring, sealing and riveting. It is thus an expensive joining technique with long cycle times. In addition weight penalties are incurred arising from the weight of the rivets themselves, as well as the need for overlapping or strengthened joints. Corrosion properties can suffer if the drilled holes pass through protective layers. Laser welding provides the possibility of a high productivity, single-step process, with the added benefit of potential weight savings of about 10% in comparison with riveting.

Adhesive bonding also avoids distortion and metallurgical changes because of the low levels of energy required compared with fusion welding. Large stress-bearing areas are produced that possess good fatigue strength, stiffness and shock absorption. Dissimilar materials can be joined, as can heat-sensitive materials, using room temperature curing adhesives. However, surfaces generally need to be cleaned and degreased, significant cure times can be required, jigs and fixtures may be required to locate components while the joint cures, and rigid process control is required to obtain consistent results. Laser welding is a fast single-step process that is easily automated, and which produces a joint of higher integrity, particularly in elevated temperature applications.

Hybrid sheet joining techniques involve a conventional joining method and adhesive bonding. Self-piercing riveting and clinching are of increasing interest to manufacturers in the automotive, domestic products and light fabrication industries. A wide range of materials may be joined – uncoated, coated, prepainted, similar and dissimilar metal combinations – without the addition of heat. Hybrid joining is particularly relevant in the joining of lightweight materials, including titanium, magnesium and aluminium alloys, and in obtaining superior joint performance in terms of increased strength in peel and leak tightness. Self-piercing riveting is also an environmentally friendly process that does not produce fume and draws less power than conventional arc or resistance welding processes. Laser welding competes on productivity when used to join materials that are readily fusion weldable.

Soldering

Manual hand-held soldering is the oldest and most versatile soldering technique. The majority of soldering irons used today are electrically heated and have power outputs ranging from a few watts to several hundred watts. Since the process is manual, it is flexible, but requires a high degree of operator skill and is relatively slow. Laser soldering competes when large numbers of identical components are to be soldered with high productivity.

Wave soldering is a method of mass soldering printed circuit boards (PCBs) loaded with through hole components. By pumping the molten solder through a nozzle, a wave is created at the surface of a solder bath over which the PCB is passed. Laser soldering competes when the pitch between joints is very small.

Hot bar soldering systems use an electrically heated bar, or tool, shaped to fit specific components. In hot gas soldering systems, a gas (nitrogen or air) is passed over an electrically heated surface from which it picks up and transfers thermal energy to the joint. Again, laser soldering competes when the scale of joints is small.

Infrared soldering systems are used in the bulk manufacture of surface mounted devices. It is also possible to heat an array of joints by focusing relatively low power infrared radiation onto specific areas in a very controlled manner. In this case laser soldering competes when high precision is required.

Brazing

The brazing technique with which people are most familiar is torch brazing, in which the energy required to melt and flow the filler metal is supplied by a fuel gas flame. The fuel gas can be acetylene, hydrogen or propane, and is combined with oxygen or air to form a flame. The apparatus is simple and inexpensive, and the process is flexible with respect to the sizes and shapes of components that can be joined. Torch brazing is most commonly used for repairs and short production runs. The main drawbacks are the high labour costs since it is a manual operation, the operator skill required, the low production rate, and the difficulty controlling health and safety issues. Laser brazing competes when large numbers of identical components are to be brazed with high productivity.

Electron beam brazing is generally performed under high vacuum ($<10^{-4}$ bar) using a defocused beam to avoid melting of the base metal. Since brazing is performed under vacuum, no flux is required, although the filler metal must be selected such that there is little or no vaporization during brazing. The process is particularly suitable for small assemblies and where components require an internal vacuum. Laser brazing can be carried out without the need for a vacuum, providing greater flexibility and higher productivity.

High frequency induction heating can also be used for brazing. It is clean and rapid, giving accurate control of temperature and location of heat. The principle of induction heating was described in Chapter 9. The coils are designed for individual parts, some of which may be complex, requiring considerable expense. Laser brazing is more flexible – it can be adapted to a wide range of part shapes and sizes.

PRACTICE

The most important process parameters in conduction joining are: the material properties, notably the melting temperature and absorptivity of the base material and the filler; the power, power–time profile, and beam size; and the joint geometry and provision of processing gas.

WELDING

Material Properties

Prior to joining, materials should be cleaned thoroughly. If components contain residues from prior processing, they are often baked to remove moisture and contaminants. Surfaces and abutting edges should be as smooth as possible to avoid weld imperfections. The mechanical properties of the base materials should be measured to establish a baseline against which the properties of the welded joint can be compared.

Beam Characteristics

CO_2, Nd:YAG and diode lasers can be used for direct energy conduction welding of metals and alloys; the absorptivity of metals increases as the wavelength decreases. The absorption bands of most polymers lie in the wavelength range 2–10 μm. CO_2 lasers and those diode lasers that produce an output wavelength greater than 2 μm are therefore suitable for polymer welding. Laser output with a wavelength in the transmission range of most polymers (0.4–1.5 μm) is suitable for transmission welding; Nd:YAG, fibre lasers and many diode lasers are thus suitable sources of energy.

Since conduction welding is normally used with relatively small components, the beam is delivered to the workpiece via a small number of optics. Simple beam defocusing to a projected diameter that corresponds to the size of weld to be made, allowing for any gaps in the joint, is normally sufficient.

Spot welds are made by pulsing the laser beam to melt sufficient material to form the joint. A uniform intensity distribution in the beam profile is desirable to ensure even melting. Continuous welds may be made by overlapping pulsed spot welds, or by using a continuous wave beam.

Processing Parameters

Processing is normally carried out at room temperature in a clean environment. Appropriate fixturing is needed to ensure that parts do not move relative to one another during welding to prevent misalignment and the formation of gaps. Molten weld metal is protected from environmental contamination by a quiescent blanket of inert shielding gas, such as argon.

SOLDERING

Material Properties

Base materials may be similar or dissimilar combinations of metals and alloys, or ceramics and glasses. Laser soldering is well suited to sensitive substrates, such as flex-prints and ceramics. The critical factor in the choice of a solder and base material is that the molten solder wets the base materials forming the joint efficiently.

Lead, tin or indium alloys are popular choices for solder materials. The most common solder is the 63% Sn–37% Pb eutectic alloy, which has been used for decades. However, environmental legislation is forcing industries to adopt lead-free solders. Several alloys are candidate replacements. Characteristics such as availability, melting temperature, shear strength, and solderability must be taken into account. The most promising are based on the Sn/Ag, Sn/Cu, Sn/Ag/Cu, Sn/Bi, Sn/In and Sn/Zn systems. Small alloying additions are made to modify their properties, to improve their wetting ability or to change the melting temperature. Depending on the scale of joining, solder can be dispensed in the form of a wire, paste, preform or screen print.

Beam Characteristics

CO_2, Nd:YAG and diode lasers are primarily used for soldering, although some trials have been performed with excimer lasers. Near infrared radiation from Nd:YAG and diode lasers is generally more suitable for soldering since it is absorbed by the metallic solder more readily than CO_2 laser radiation, and transmitted by laminate substrates and polymer materials. The focused spot size of Nd:YAG and diode laser beams is also smaller and is therefore more suitable for precision soldering. Different solders can be melted by changing the laser power to achieve the necessary temperature.

Processing Parameters

Three principal types of joint are used with laser soldering: lead frame type joints (gull-wing or J type leads), solder pads, and through-hole type joints. The first two are generally pretinned and do not always require additional paste during the reflow operation. In these cases, the joints are small; modern fine pitch devices have leads down 200 μm wide on a 500 μm pitch. Pretinned pads onto which components may be laser soldered have a larger area, but the solder volume is small. The third through-hole joint type always requires filler material. Joint clearances are normally between 0.025 and 0.25 mm.

The filler material may be either solder paste, solder wire or preforms. In most non-telecom applications, a flux is often incorporated into these fillers. Recently there has been a trend towards 'no-clean' fluxes, for environmental reasons. SMT (Surface Mounted Technology) and THT (Through Hole Technology) components can be soldered with the use of solder wire, solder paste, or solder preforms. Single contacts as well as whole components can be soldered in this way without contact or thermal damage to the component, substrate or the surroundings.

BRAZING

Material Properties

In general, materials that can be brazed by conventional methods are suitable for laser brazing. Metals in which imperfections are formed when welded autogenously can often be brazed successfully in thin sections. Laser brazing provides opportunities for joining similar or dissimilar combinations of metals and alloys, and ceramics and glasses.

The selection of the braze material is based on the application of the brazed joint (mechanical, thermal and electrical requirements), compatibility with the base material (the wetting characteristics), and the design of the joint. Braze material may be preplaced in the form of shims, powder, wire or paste, or added during the brazing process in the form of wire. Conventional brazing materials can be used; guidance can be found from the filler materials used in laser keyhole welding listed in Chapter 16. However, the availability of filler in the required form can be a limitation. Most brazing materials are based on the aluminium, copper or silver alloy systems.

Beam Characteristics

The beam characteristics are similar to those used for laser soldering. A higher laser beam power is adopted because of the higher melting temperature of brazes in comparison with solders. Pulsed Nd:YAG lasers have been shown to be suitable for thin materials, less than 0.25 mm in thickness, with the brazing material preplaced on the joint. The hardness of the braze material is increased, providing opportunities for improved joint properties.

Processing Parameters

Specialized joint designs are used in laser brazing to facilitate capillary flow of the braze material. The most common joint designs are the lap, I-butt, T and J configurations.

PROPERTIES OF CONDUCTION-JOINED MATERIALS

The properties of conduction-joined materials depend on the procedures used. General guidelines for the process variables are described above, but it is not possible to prescribe general sets of conditions for specific materials and fillers. The references in Tables 13.1 and 13.2 can be consulted for individual processing advice and the properties of the joints formed.

WELDING

Materials that can be welded using conventional arc fusion processes can normally be laser welded in the conduction-limited mode. References pertaining to some of the earliest work on conduction laser welding are given in Table 13.1. Many of the procedures developed are still in use today.

The low energy input of laser conduction welding results in limited and predictable distortion, which minimizes the need for reworking and reduces material wastage. The rapid thermal cycle reduces the segregation of embrittling elements such as sulphur and phosphorus, and beneficial fine solidification microstructures are formed. Alloys of nickel and tantalum, for example, which are normally susceptible to such imperfections, can therefore be laser welded. Rapid thermal cycles also limit grain growth and the dissolution and coarsening of hardening precipitates, helping to maintain strength in such alloys.

Table 13.1 Sources of data for laser conduction welding: N/S not specified

Grade 1	Grade 2	Laser	Reference
Ag	Al	Ruby	(Platte and Smith, 1964)
Cu, Ag, Au	Cu, Ag, Au	Ruby	(Fairbanks and Adams, 1964)
Ni wire	Ni wire	Ruby	(Anderson and Jackson, 1965)
Ni alloys	Ni alloys	Ruby	(Earvolino and Kennedy, 1966)
Be–Cu wire	Be–Cu wire	Ruby	(Cohen et al., 1969)
AISI 304	AISI 304	CO_2	(Conti, 1969)
C–Mn steel	C–Mn steel	CO_2	(Esposito et al., 1982)
AISI 304	AISI 304	CO_2	(Esposito et al., 1982)
Various	Various	Nd:YAG	(Glasmacher et al., 1996)
Cu alloy	Steel	N/S	(Klages et al., 2003)

Table 13.2 Sources of data for laser soldering: N/S not specified

Grade 1	Grade 2	Solder	Laser	Reference
Various	Various	Various	CO_2	(Chang, 1986)
Various	Various	Various	Various	(Hartmann et al., 1991)
InP laser chip	Submount	80Au–20Sn	Diode	(Lee et al., 1992)
Surface mounted devices	Glass epoxy PCB	Sn–Ag	Nd:YAG	(Yang et al., 1995)
Tissue	Tissue	Albumin protein	Diode	(McNally et al., 1999)
Surface mounted devices	Glass epoxy PCB	Sn–43Bi, Sn–2Ag– 5Bi–0.5Cu	Nd:YAG	(Nakahara et al., 2000)
Leads	PCB	SnAg3, 8CuO, 7Sb0.2	N/S	(Fritz and Lutz, 2002)
Leads	PCB	Lead-free	Diode	(Wang et al., 2003)

Since the laser beam can be positioned with high accuracy, precise control over the weld bead location is possible. Some combinations of dissimilar materials can be joined using conduction welding by positioning the beam within the material of highest melting temperature, such that heat is conducted into the other material to cause limited interfacial melting with only a thin intermetallic interface.

SOLDERING

Procedures for laser soldering and data for the properties of soldered joints are scarce. Many procedures have been developed in-house by manufacturers, particularly those working in the electronics industries. One driving force for research into laser soldering has been the need to find lead-free replacements for the conventional Sn–37Pb solder. The metallic solders listed in Table 13.2 have been shown to exhibit equivalent properties to the conventional solder when fused using a laser beam.

The biomedical sector is also examining the potential for laser soldering as a means of joining tissue without the need for sutures. Biomedical solders are based on albumin proteins that coagulate on interaction with the laser beam to form the joint. A lower energy is used than with laser welding in such procedures, and so the risk of damage to surrounding tissue is reduced.

Table 13.3 Sources of data for laser brazing: N/S not specified

Grade 1	Grade 2	Braze	Laser	Reference
Various	Various	Various	Nd:YAG, ruby, CO_2	(Witherell and Ramos, 1980)
Cu	Cu		CO_2	(Jones and Albright, 1984)
Si_3N_4	Si_3N_4	Y_2O_3, La_2O_3, MgO and Si_3N_4	Nd:YAG	(Miyamoto *et al.*, 1987)
AISI 316	5052	Al–12% Si	N/S	(Park and Na, 1998)
Pin	Plate	–	N/S	(Jeon *et al.*, 1998)
AISI 304	AISI 304	Au–18% Ni	Nd:YAG	(Davé *et al.*, 2001)

BRAZING

Laser brazing is a process that has yet to fulfil its potential. There are few procedures and results available, but a selection of references can be found in Table 13.3.

The properties of laser-brazed joints are reported to be superior to those brazed by conventional techniques because of the lower energy input, which produces a more localized, solidified braze with a finer microstructure.

IMPERFECTIONS

Imperfections found in conduction-joined materials include: lack of fusion; lack of fill; disbonding; cracking; discolouration; distortion; and porosity. A lack of fusion normally results from insufficient energy delivered to the joint. A lack of fill can be avoided by cleaning the components prior to processing, ensuring filler material has the necessary wetting characteristics, controlling the heating cycle to ensure complete melting of the filler material, providing sufficient joint clearance for capillary action to locate the molten filler material, fixturing parts correctly, and ensuring that sufficient filler material is present in the joint. Disbonding can occur when sections of dissimilar thickness are joined, and the thicker section does not attain the temperature required for the filler material to flow and fill the joint. Fissures can arise from solidification cracking within the filler material – eutectic materials, or compositions with a low solidification range are less prone to this type of cracking. Cracking may also arise in base materials of high strength, and is best avoided by using a base metal in the annealed condition, and by proper design of the joint. Discolouration can be caused by overheating during processing, or insufficient cleaning. Distortion arises from localized heating and can be controlled by correct joint design and fixturing. Porosity may occur when gas is entrapped, particularly in blind joints, and can be minimized by designing joints that are able to vent vaporized materials.

LASER CONDUCTION JOINING DIAGRAMS

PROCESS OVERVIEW

The basic thermal conditions required for stable conduction welding are that the surface temperature exceeds the melting temperature, but remains below the vaporization temperature. In Chapter 11 we used mathematical modelling to construct the contours corresponding to these conditions for surface

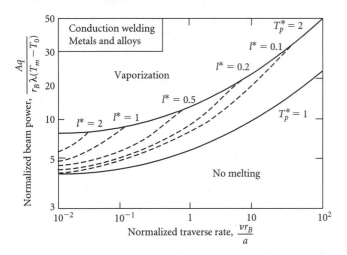

Figure 13.1   Normalized overview of conduction welding of metals and alloys

melting, which enabled an operating window to be presented in a framework of normalized groups of process variables. The same technique is used here to create the operating window for conduction welding shown in Fig. 13.1. The contours of normalized plate thickness are constructed in the same manner as described for surface melting (Chapter 11).

Figure 13.1 may be used to describe a range of conduction welding conditions for a wide range of metals and alloys. The effects of changes in process variables on the depth of penetration, e, can be predicted.

PARAMETER SELECTION

Figure 13.1 may be converted to a material-specific diagram by substituting appropriate thermal properties into the dimensionless variables used as the axes, and by defining a beam radius for processing. By using the material properties listed in Table D.1 (Appendix D) for the austenitic stainless steel AISI 304, and a beam radius of 0.6 mm, Fig. 13.2 is thus constructed.

Figure 13.2 is based on the radial heat flow geometry of a surface energy source. As the penetration of the weld bead increases relative to the plate thickness, lateral heat flow occurs because the plate is no longer semi-infinite with respect to the energy source. In such cases, the power predicted by Fig. 13.2 will be slightly higher than that required in practice. However, Fig. 13.2 can be used to obtain an initial set of processing parameters for the development of a welding procedure.

CALIBRATED PARAMETER SELECTION

We have seen in previous chapters how a practical processing diagram can be constructed using linear axes and calibrated to a known condition. Providing that the mechanism of processing is known, and does not vary over the range of validity of the diagram, the process model can be used to generate a series of contours.

However, the behaviour of some materials is difficult to model. Amorphous polymers, for example, do not exhibit a latent heat of melting – instead the heat capacity continues to rise as the temperature is raised until the glass transition temperature is reached. It is beyond the scope of this book to develop models for processing all classes of engineering material. But it is instructive to examine experimental data to identify trends in the variation of processing parameters. Figure 13.3 is constructed from experimental data obtained during direct heat welding of amorphous polyethylene.

The linear relationship between incident beam power and welding speed shown in Fig. 13.3 for a given material thickness indicates that heat flow could be described by simple analytical relationships, providing the geometry of the molten weld bead can be characterized and the relevant material properties determined. Parameter selection diagrams may then be constructed for a wider range of engineering materials.

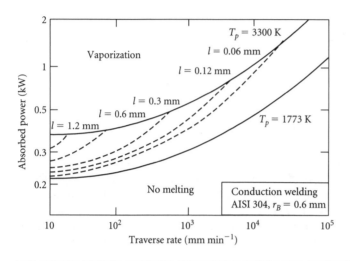

Figure 13.2   Parameter selection diagram for conduction welding of AISI 304

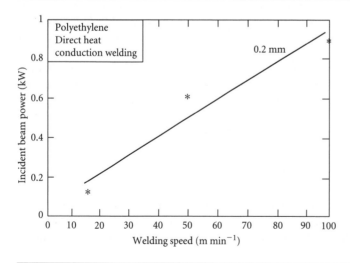

Figure 13.3   Calibrated parameter selection diagram for conduction welding of polyethylene. (Source: calibration data (∗) from Ian Jones, TWI, Abington, UK)

INDUSTRIAL APPLICATION

The most common industrial applications of laser conduction joining are those in which large numbers of identical small parts are joined. The examples shown in Table 13.4 highlight the progress that has been made in the microelectronics and biomedical industries through the use of various microjoining techniques.

Table 13.4   Industrial applications of laser conduction joining

Industry sector	Process	Product	Material(s)	Laser	Reference
Aerospace	Welding	Apollo lunar sample containers	Various	Ruby	(Moorhead and Turner, 1970)
Automotive	Soldering	Ignition module	Brass/alumina	CO_2	(Chang, 1987)
	Brazing	Audi A6	Steel	Nd:YAG	(AudiWorld, 2004)
	Brazing	Audi TT Coupé C-pillar	Steel	Nd:YAG	(Larsson, 1999)
Biomedical	Welding	Cardiac pacemaker batteries	AISI 304L	Nd:YAG	(Fuerschbach and Hinkley, 1997)
	Welding	Dental implants	Titanium	Nd:YAG	(Sjögren *et al.*, 1988; Yamagishi *et al.*, 1993; Wang and Welsch, 1995; Knabe and Hoffmeister, 2003)
	Welding	Dental implants	Ni–Cr–Mo, Cr–Co–Mo	Nd:YAG	(Bertrand *et al.*, 2001)
	Soldering	Middle ear reconstruction	Hydroxyapatite/ albumin solder	Diode	(Ditkoff *et al.*, 2001)
Domestic goods	Welding	Gillette Sensor razor	Stainless steel	Nd:YAG	(Anon., 1990)
Electronics	Welding	Packages	Al alloys	Nd:YAG	(Emerson, 1996)
	Soldering	Flat packs	Cu	CO_2	(Burns and Zyetz, 1981)
	Soldering	Surface mounted devices	Fe–Ni leads/ Sn solders	Various	(Hall *et al.*, 1987; Lea, 1989; Semerad *et al.*, 1993)
	Soldering	Flat panel display	Silica/Ag solder	UV	(Lee *et al.*, 2001)

Figure 13.4   Computer hard drive component joined by pulsed Nd:YAG laser spot welding. (Source: Mo Naeem, GSI Lumonics, Rugby, UK)

ELECTRONIC COMPONENTS

Pulsed lasers are used for a variety of welding applications in the electronics industries, including the attachment of wires to contacts and the sealing of electronic packages. Many hermetic semiconductor packages are made from aluminium alloys with a wall thickness of about 1 mm. The body is typically of 6061, to provide strength, corrosion resistance and machinability, and the lid of 4047 to minimize the likelihood of solidification cracking during welding. The cover is designed with a machined step around its edges, which provides a tight, self-fixturing joint with the body, and a lip into which the laser beam can be directed to make the weld. Prior to welding it is normal to bake the package to remove any trapped moisture from the package cavity and the component mounting epoxies.

Figure 13.4 shows a component from a computer hard drive that has been joined by pulsed Nd:YAG laser spot welds. The main requirements for such parts are good cosmetic appearance and low energy input to eliminate distortion.

CATHETERS

A catheter accurately locates an enteral tube in real time. It thus provides an effective bedside technique for nasoduodenal tube placement to facilitate enteral feeding in critically ill patients. Figure 13.5 shows a novel way to fabricate a catheter from dissimilar materials of small size by using a laser. The microjoining capability of laser spot welding enables the catheter to be designed in a novel way, and produced with high quality, high productivity and low cost.

DENTAL PROSTHESES

It is necessary to use a single metal or alloy for dental restorations to avoid galvanic effects. Based on their physical and chemical properties, titanium and its alloys are particularly suitable for the construction

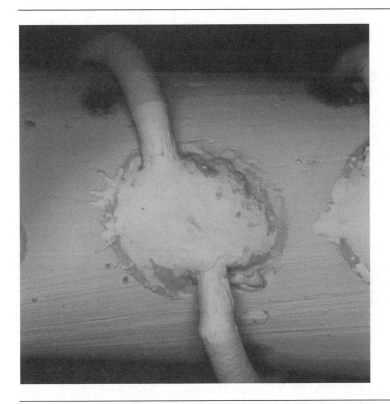

Figure 13.5   Laser spot welding of a 0.05 mm diameter copper wire with polyimide coating to a 0.35 mm AISI 316 stainless steel wire for a catheter. (Source: Simon Doe, CSIRO, Adelaide, Australia)

of dental implants and prostheses such as crowns and bridges. However, processing methods must be selected with care to ensure success. An impression is first made, and a structure fabricated by fusing Ti-cast frameworks to prefabricated titanium copings by laser welding. The precision and low energy input of laser welding enables joints to be placed at selected locations, and minimizes distortion. The process is completed by veneering or by making a removable denture with a titanium reinforcement.

CLEARWELD™

Figure 13.6 shows two pieces of polymethylmethactylate (PMMA) of thickness 3 mm that have been joined using the Clearweld™ process. Pure PMMA is transparent to visible and near infrared radiation. To make the joint, an ink that absorbs such radiation is spread along the facing surfaces of a lap joint. In the example shown in Fig. 13.6, a diode laser of wavelength 940 nm with a beam of diameter 3 mm and a power of 60 W was scanned across the top surface of the joint at a speed of 1.65 m min^{-1}. These parameters are varied for different grades of polymer. Diode lasers of wavelength 940 and 980 nm are normally used in the process. However, Nd:YAG lasers of wavelength 1064 nm are also effective. Beam widths typically range from 2 to 5 mm. Since a weld is only produced where the ink has been applied, the tolerances on the location of beam delivery can be wide – a significant processing advantage. The power level can range from a few watts to a few hundred watts, depending on the process specifications. Speeds also vary depending on the particular application.

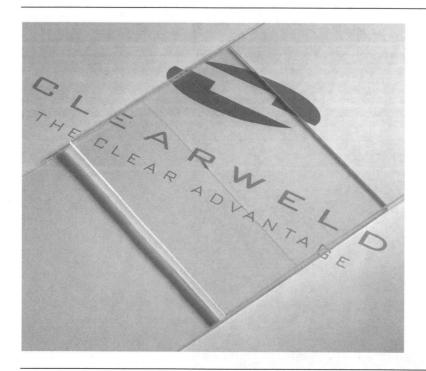

Figure 13.6   A lap joint made in PMMA using the Clearweld™ process. (Source: L.P. Freider III, Gentex Corporation, Carbondale, PA, USA. Clearweld is a registered trademark of TWI)

MERCEDES-BENZ E CLASS CARS

Laser brazing is a preferred joining technique for thin section steels in automotive assembly because of the non-linearity of the joints, the bulkiness of the components, the aesthetic appeal of the finished joint, the ease of automation and repair, the high joining rate, and the reduction in the number of post-braze operations. The hatchback of the T-model for the Mercedes-Benz E class cars consists of two parts. The joining process must be able to cope with small machined radii, gaps, compact three-dimensional structures and soldered joints around the rear lights, and the recess for the licence plate. A fully automatic production laser soldering system has been constructed for this complex component. The system consists of an industrial robot, a clamping jig, a measuring sensor, an Nd:YAG laser and a laser-soldering head. A robot system with an integrated measuring sensor first records the path of the soldering head along the clamped parts. The measuring sensor is then automatically replaced by the soldering head in a docking station. Sensor-assisted laser soldering has been demonstrated to be an excellent solution to the problem of joining complex body-shell components. A major advantage of the system is its very flexible laser soldering head, which is capable of following both small radii and large curves with the same accuracy.

DIODE LASER SOLDERING OF CHIP CARRIERS TO PCBS

Soldering of chip carriers to printed circuit boards (PCBs) has been evaluated for many years using CO_2 and solid state lasers, but never widely implemented by the microelectronics industries. The compactness, efficiency of energy conversion, and ease of use of diode lasers operating in the wavelength range

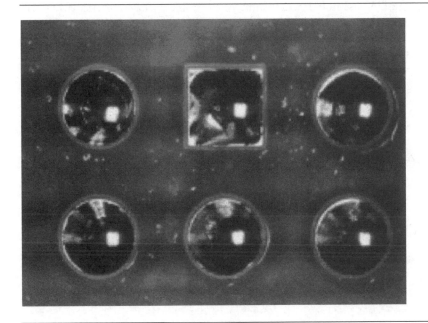

Figure 13.7   Diode laser soldering of six pins of pitch 2.54 mm of a 44 pin chip carrier to a printed circuit board; power 7 W, pulse duration 0.6 s. (Source: Bill Petreski and Neil McMahon, Coherent Inc., Santa Clara, CA, USA)

810–977 nm provide answers to many of the shortcomings of their predecessors. Manufacturers now offer fully integrated diode laser soldering systems, comprising x–y positioning stages or articulated robots, which perform precision, non-contact microsoldering. Applications may be categorized by the area of solder required: small (40–100 µm); medium (100–150 µm); and large (1–3 mm). Small area soldering includes pads used in high density packaging, for which a few watts of power are required, typically delivered via an optical fibre of diameter 200 µm. Medium area applications require about 30 W of power with a dwell time of about 1 second per joint. Large area joints are accommodated by expanding the beam spot size, increasing the power to 80 W, and increasing the dwell time. The objective is to achieve a well-wetted joint, precise location of the solder, and minimal heat transfer in the PCB. Joints are made by positioning solder preforms containing a low amount of flux over the joint region, which are melted in a highly accurate and controllable manner, and do not require post-processing cleaning. Figure 13.7 shows soldered joints made between six pins of pitch 2.54 mm of a standard 44 pin chip carrier to a PCB.

SUMMARY AND CONCLUSIONS

Laser-based conduction joining techniques are used when a well-defined joint is to be made in a precise location with low energy input. Conduction welding is suitable if the base materials can be melted to form the joint, and is applied to precision joining of wires and foils. If the base materials cannot be melted, laser beam soldering and brazing can be used. A filler material is then added, which is melted by the laser beam, wetting the joint faces, and solidifying to form the joint. Laser beam soldering and brazing are stable, reproducible processes for applications requiring a controlled localized energy input and reduced thermal and mechanical stress compared with conventional techniques. Conduction welding of metals and alloys can be modelled, enabling practical conduction welding diagrams to be constructed.

Conduction welding is finding use in hermetic sealing of electronic packages, particularly those containing delicate components that are easily damaged by heating. The increasing use of laser soldering in the microelectronics industries is being driven by the need for a precision joining process capable of handling high densities of fine joints. Its use in biomedical applications is driven by the desire for minimally invasive surgical joining procedures in which surrounding tissue is undamaged. The versatility of laser brazing has not yet been fully exploited; it is a relatively simple process that is suitable for a wide range of materials and can be implemented quickly for mass production.

FURTHER READING

Burns, F. and Zyetz, C. (1981). Laser microsoldering. *Electronic Packaging and Production*, **21**, (5), 109–120.

Messler, R.W. Jr and Millard, D.L. (1994). Laser soldering: new light on an old joining process. *Welding Journal*, **73**, (10), 43–48.

Metals Handbook (1983). 9th ed. Vol. 6. Welding, brazing and soldering. Metals Park: American Society for Metals.

Witherell, C.E. and Ramos, T.J. (1980). Laser brazing. *Welding Journal*, **59**, (10), 267s–277s.

EXERCISES

1. What are the main advantages of laser-based conduction joining techniques in comparison with their conventional counterparts?

2. What are the most important metallurgical properties of solders? What would you suggest as a lead-free replacement for use in laser soldering?

3. An I joint in a plate of thickness 1 mm containing a gap equal to half the plate thickness is to be laser brazed using a continuously fed eutectic aluminium–silicon braze wire of diameter 0.6 mm. The traverse rate of the plate during brazing is 0.3 m min^{-1}. Calculate the wire feed rate and laser power required to fill the gap. Assume that: heat conduction into the base material is negligible; all the laser power is absorbed by the braze wire; and the braze wire has the properties of pure aluminium, given in Table D.1 (Appendix D).

4. By using Fig. 13.1 and the material property data in Table D.1 (Appendix D), calculate the incident laser beam power required to produce a weld of depth 0.5 mm in a carbon–manganese steel sheet with a traverse rate of 0.6 m min^{-1}. Assume that: the focused beam radius is 1 mm; the absorptivity of the steel is 0.5; and the initial temperature of the steel is 25°C.

5. What are the most common imperfections in laser-soldered and laser-brazed joints, and how can they be avoided?

BIBLIOGRAPHY

Anderson, J.E. and Jackson, J.E. (1965). Theory and application of pulsed laser welding. *Welding Journal*, **44**, (12), 1018–1026.

Anon. (1990). Lasers deliver precise welds on new razor. *Welding Journal*, **69**, (10), 47–48.

AudiWorld (2004). *Audi Starts Production of the New A6 Limousine* [Internet]. Available from: <http://www.audiworld.com/news/04/021704/content.shtml> [Accessed 13 April 2004].

Bertrand, C., Le Petitcorps, Y., Albingre, L. and Dupuis, V. (2001). The laser welding technique applied to the non precious dental alloys. Procedure and results. *British Dental Journal*, **190**, (5), 255–257.

Burns, F. and Zyetz, C. (1981). Laser microsoldering. *Electronic Packaging and Production*, **21**, (5), 109–120.

Chang, D.U. (1986). Experimental investigation of laser beam soldering. *Welding Journal*, **65**, (10), 32–41.

Chang, D.U. (1987). An analytical model for laser reflow soldering of an electronic component. *Welding Journal*, **66**, (11), 323s–331s.

Cohen, M.I., Mainwaring, F.J. and Melone, T.G. (1969). Laser interconnection of wires. *Welding Journal*, **48**, (3), 191–197.

Conti, R.J. (1969). Carbon dioxide laser welding. *Welding Journal*, **48**, (10), 800–806.

Davé, V.R., Carpenter, R.W., Christensen, D.T. and Milewski, J.O. (2001). Precision laser brazing utilizing nonimaging optical concentration. *Welding Journal*, **80**, (6), 142s–147s.

Ditkoff, M., Blevins, N.H., Perrault, D. and Shapshay, S.M. (2002). Potential use of diode laser soldering in middle ear reconstruction. *Lasers in Surgery and Medicine*, **31**, 242–246.

Earvolino, L.P. and Kennedy, J.R. (1966). Laser welding of aerospace structural alloys. *Welding Journal*, **45**, (3), 127s–134s.

Emerson, W.F. (1996). Laser-beam welding seals electronic packages. *Welding Design and Fabrication*, (April), 43–44.

Esposito, C., Daurelio, G. and Cingolani, A. (1982). On the conduction welding process of steels with the CO_2 laser. *Optics and Lasers in Engineering*, **3**, 139–151.

Fairbanks, R.H. and Adams, C.M. Jr (1964). Laser beam fusion welding. *Welding Journal*, **43**, (3), 97s–102s.

Fritz, H. and Lutz, D. (2002). Better quality of soldered joints through rapid laser-beam reflowing. In: Courtois, B., Karam, J.M., Markus, K.W., Michel, B., Mukherjee, T. and Walker, J.A. eds *Proceedings of the Conference Design, Test, Integration, and Packaging of MEMS/MOEMS 2002*. Vol. 4755. Bellingham: SPIE. pp. 675–679.

Fuerschbach, P.W. and Hinkley, D.A. (1997). Pulsed Nd:YAG laser welding of cardiac pacemaker batteries with reduced heat input. *Welding Journal*, **76**, (3), 103s–109s.

Glasmacher, M., Pucher, H.-J. and Geiger, M. (1996). Improvement of the reliability of laser beam microwelding as interconnection technique. In: Migliore, L.R., Roychoudhuri, C.S., Schaeffer, R.D., Mazumder, J. and Dubowski, J.J. eds *Proceedings of the Conference Lasers as Tools for Manufacturing of Durable Goods and Microelectronics*. Vol. 2703. Bellingham: SPIE. pp. 411–420.

Hall, D.R., Whitehead, D.G. and Polijanczuk, A.V. (1987). A laser soldering system for surface mounted components. In: Steen, W.M. ed. *Proceedings of the 4th International Conference Lasers in Manufacturing*, 12–14 May 1987, Birmingham, UK. Bedford: IFS Publications. pp. 133–138.

Hartmann, M., Bergmann, H.W. and Kupfer, R. (1991). Experimental investigations in laser microsoldering. In: Braren, B. ed. *Proceedings of the Conference Lasers in Microelectronic Manufacturing*. Vol. 1598. Bellingham: SPIE. pp. 175–185.

Jeon, M.-K., Kim, W.-B., Han, G.-C. and Na, S.-J. (1998). A study on heat flow and temperature monitoring in the laser brazing of a pin-to-plate joint. *Journal of Materials Processing Technology*, **82**, (1–3), 53–60.

Jones, T.A. and Albright, C.E. (1984). Laser beam brazing of small diameter copper wires to laminated copper circuit boards. *Welding Journal*, **63**, (12), 34–47.

Klages, K., Gillner, A., Olowinsky, A.M., Fronczek, S. and Studt, A. (2003). Laser beam micro-welding of dissimilar metals. In: Miyamoto, I., Ostendorf, A., Sugioka, K. and Helvajian, H. eds *Proceedings of the Fourth International Symposium on Laser Precision Microfabrication*. Vol. 5063. Bellingham: SPIE. pp. 303–307.

Knabe, C. and Hoffmeister, B. (2003). Implant-supported titanium prostheses following augmentation procedures: a clinical report. *Australian Dental Journal*, **48**, (1), 55–60.

Larsson, J.K. (1999). Lasers for various materials processing. A review of the latest applications in automotive manufacturing. In: Kujanpää, V. and Ion, J.C. eds *Proceedings of the 7th Nordic Conference on Laser Materials Processing (NOLAMP-7)*. Lappeenranta: Acta Universitatis Lappeenrantaensis. pp. 26–37.

Lea, C. (1989). Laser soldering – production and microstructural benefits for SMT. *Soldering and Surface Mount Technology*, (2), June, 13–21.

Lee, C.H., Wong, Y.M., Doherty, C., Tai, K.L., Lane, E., Bacon, D.D., Baiocchi, F. and Katz, A. (1992). Study of Ni as a barrier metal in AuSn soldering application for laserchip/submount assembly. *Journal of Applied Physics*, **72**, (8), 3808–3815.

Lee, J.-H., Kim, W.-Y., Ahn, D.-H., Lee, Y.-H. and Kim, Y.-S. (2001). Laser soldering for chip-on-glass mounting in flat panel display application. *Journal of Electronic Materials*, **30**, (9), 1255–1261.

McNally, K.M., Sorg, B.S., Welch, A.J., Dawes, J.M. and Owen, E.R. (1999). Photothermal effects of laser tissue soldering. *Physics in Medicine and Biology*, **44**, 983–1002.

Miyamoto, I., Maruo, H., Kuriyama, K. and Horiguchi, Y. (1987). Joining of Si_3N_4 ceramics by laser-activated brazing. In: Arata, Y. ed. *Proceedings of the Conference Laser Advanced Materials Processing (LAMP '87)*, 21–23 May 1987, Osaka, Japan. Osaka: High Temperature Society of Japan and Japan Laser Processing Society. pp. 237–244.

Moorhead, A.J. and Turner, P.W. (1970). Welding a thermocouple gauge to Apollo lunar sample return containers. *Welding Journal*, **49**, (1), 15–21.

Nakahara, S., Kamata, T., Yoneda, N., Hisada, S. and Fujita, T. (2000). Microsoldering using a YAG laser on lead-free solder. In: Miyamoto, I., Sugioka, K. and Sigmon, T.W. eds *Proceedings of the First International Symposium on Laser Precision Microfabrication*. Vol. 4088. Bellingham: SPIE. pp. 276–279.

Park, J.S. and Na, S.J. (1998). Heat transfer in a stud-to-plate laser braze considering filler metal movement. *Welding Journal*, **77**, (4), 155s–163s.

Platte, W.N. and Smith, J.F. (1964). Laser techniques for metal joining. *Welding Journal*, **42**, (11), 481s–489s.

Semerad, E., Musiejovsky, L. and Nicolics, J. (1993). Laser soldering of surface-mounted devices for high-reliability applications. *Journal of Materials Science*, **28**, 5065–5069.

Sjögren, U., Andersson, M. and Bergmann, M. (1988). Laser welding of titanium in dentistry. *Acta Odontologica Scandinavica*, **46**, 247–253.

Wang, J.-B., Watanabe, M., Goto, Y., Fujii, K., Kuriaki, H., Satoh, M., Ikeda, J. and Fujimoto, K.

(2003). Development of high speed laser soldering process for lead-free solder with diode-laser. In: Miyamoto, I., Kobayashi, K.F., Sugioka, K., Poprawe, R. and Helvajian, H. eds *Proceedings of the First International Symposium on High-Power Laser Macroprocessing*. Vol. 4831. Bellingham: SPIE. pp. 26–31.

Wang, R.R. and Welsch, G.E. (1995). Joining titanium materials with tungsten inert gas welding, laser welding, and infrared brazing. *The Journal of Prosthetic Dentistry*, **74**, (6), 521–530.

Witherell, C.E. and Ramos, T.J. (1980). Laser brazing. *Welding Journal*, **59**, (10), 267s–277s.

Yamagishi, T., Ito, M. and Fukimura, Y. (1993). Mechanical properties of laser welds of titanium in dentistry by pulsed Nd:YAG laser apparatus. *The Journal of Prosthetic Dentistry*, **70**, (3), 264–273.

Yang, W., Messler, R.W. Jr and Felton, L.E. (1995). Laser beam soldering behavior of eutectic Sn-Ag solder. *Welding Journal*, **74**, (7), 224s–229s.

CHAPTER 14

CUTTING

INTRODUCTION AND SYNOPSIS

There are few materials that cannot be parted in some way through the action of a focused traversing laser beam. A typical cutting geometry for flat sheet profiling is shown in Fig. 14.1 – the beam is traversed over a programmed path covering several metres of the sheet. Materials that have a distinct molten phase of low viscosity, notably metals and alloys, and thermoplastics, are cut by the heating action of a beam of power density on the order of 10^4 W mm^{-2}. In combination with the melt shearing action of a stream of inert or active assist gas, a molten channel is created through the material, and a kerf (slot) is formed. Materials that are not readily melted (some glasses, ceramics and composites, for example) can be cut by a mechanism involving vaporization that is induced by a higher beam power density, on the order of 10^6 W mm^{-2}. A kerf can be formed in many organic materials by a mechanism involving chemical degradation caused by the heating action of the beam. Hard, brittle

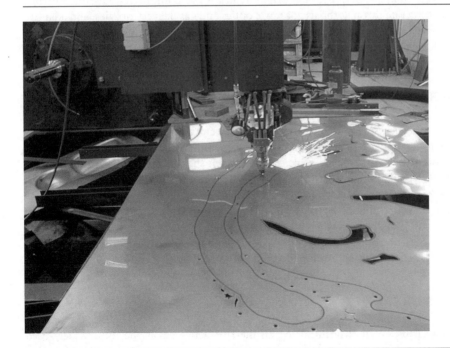

Figure 14.1   Flying optic laser cutting geometry for flat sheets. (Source: Petri Metsola, Lappeenranta University of Technology, Finland)

materials (glasses and domestic ceramics) can be parted by using the laser beam to create notches by vaporization that raise stress locally (scribing), after which fracture is induced by bending. Relatively low power CO_2 lasers combined with a flying optic beam delivery system such as that shown in Fig. 14.1 are the tools of choice for flat sheet cutting. Robot-mounted fibreoptic delivery of the light from an Nd:YAG laser or a portable diode laser head is preferred for three-dimensional cutting. Small complex shapes are cut using variations of robot-mounted light delivery and workpiece manipulation. A useful rule of thumb for CO_2 laser cutting with oxygen is that 1.5 kW of laser power will cut mild steel of thickness 1 mm at 10 m min^{-1}, or a thickness of 10 mm at 1 m min^{-1}. A similar rule of thumb states that the cutting speed of wood is inversely proportional to its density. The objectives of laser cutting are to part materials at the highest rate possible, while producing an edge that can be prescribed in terms of a quality standard that classifies edge roughness and squareness.

Laser cutting shares many principles with conventional fusion cutting methods, but excels in applications requiring high productivity, a high edge quality, and a minimum of waste. If high productivity is the main criterion, then oxygen-assisted melt shearing is the cutting mechanism of choice. If a high edge quality is more important, then the slower high pressure inert gas melt shearing mechanism is used. Laser cutting competes with a range of mechanical techniques. Technical comparisons can be made relatively easily, but economic comparisons are more difficult, particularly when the value of a novel cutting technique must be assessed. In general, laser cutting normally provides the best combination of quality and productivity with homogeneous materials less than about 3 mm in thickness when the equipment is in use for about 16 hours per day. Data show that many engineering materials can be cut at several metres per minute using a relatively cheap 1–2 kW CO_2 laser. Procedures have also been developed to cut sections several tens of millimetres in thickness. Analytical modelling is applied in this chapter to the melt shearing mechanism, for which various forms of laser cutting diagram are constructed. Laser cutting has been adopted as a direct substitute for conventional cutting in many industries because of the similarity in processing geometry. Others have used the process as the basis of innovative procedures for rapid prototyping. The current trend is towards improving edge quality while maintaining a high cutting rate through the use of novel optics and gas delivery nozzles, and adaptive control of processing variables. Hybrid cutting centres, which comprise laser cutting and a conventional parting process, enable manufacturers to select the most appropriate method for a particular job.

PRINCIPLES

A number of different mechanisms may operate during laser cutting. More than one is normally active, but a principal processing mechanism can normally be identified for a particular material, assist gas and laser. There are five distinct mechanisms: inert gas melt shearing; active gas melt shearing; vaporization; chemical degradation; and scribing. The principal cutting mechanisms for the various classes of engineering material are summarized in Table 14.1.

Trepanning is a means of cutting holes using one of the above mechanisms by guiding the beam around a predefined locus. It can be used when the diameter of the hole to be cut is considerably larger than the diameter of the laser beam, and allows bevelled sides to be produced. Material is first pierced at the centre of the hole, and the cutting front guided out to the periphery.

INERT GAS MELT SHEARING

Of the mechanisms available, melt shearing using an inert or active gas is the most widely used industrially. The inert gas melt shearing mechanism is based on the formation of a narrow penetrating cavity that melts surrounding material, which is subsequently removed by the shearing action of a coaxial jet of inert assist gas, as illustrated in Fig. 14.2. The technique is used with materials that melt

Table 14.1 Mechanisms of laser cutting for various engineering materials (a suitable mechanism is indicated by ✓)

Material	Inert gas melt shearing	Reactive gas melt shearing	Vaporization	Chemical degradation	Scribing
Ferrous alloys	✓	✓	–	–	–
Non-ferrous alloys	✓	✓ (Ti)	–	–	–
Polymers	✓ (Thermoplastics)	✓ (Thermosets)	✓ (PMMA)	(Thermosets)	–
Ceramics	✓	–	–	–	✓
Glasses	✓	–	–	–	✓
Elastomers	–	–	–	✓	–
Composites	✓	–	–	✓ (Woods)	✓

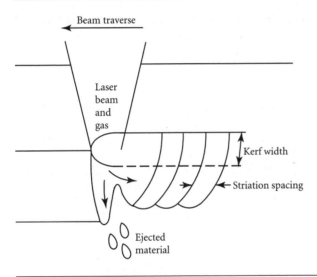

Figure 14.2 Principle of laser cutting by melt shearing. (After Powell, 1998)

readily when heated by a laser beam, such as metals and alloys, thermoplastics including polyethylene, polypropylene, nylon, acrylonitrile-butadiene-styrene (ABS), a limited number of ceramics and glasses, as well as metal matrix and polymer matrix composites. Air is often used for the assist gas because of its availability, but nitrogen is a common choice when no oxidation can be tolerated. Argon or helium may be used for materials such as titanium that form deleterious brittle nitrides. A high gas pressure (above 10 bar) is recommended to remove liquid that can adhere to the underside of the cut and solidify as dross. Inert gas cutting is limited to metal sections approximately 8 mm in thickness because instabilities occur when the cutting speed approaches the rate at which energy is removed from the cutting zone. A high quality cut edge is formed, but cutting speeds are relatively low in comparison with active gas cutting mechanisms. The technique is tolerant to relatively large changes in processing parameters.

The edge of laser-cut material is made up of a regular pattern of striations, which suggests that the cutting process is cyclic in nature. Various theories have been proposed to explain the formation of striations. The most widely accepted is that the laser beam melts a portion of the material at its leading edge, moves forward and is effectively extinguished, and then reignites to continue the procedure

Table 14.2 Energy of exothermic oxidation reaction in a number of metals

Material	Oxidation energy $(kJ\,mol^{-1})$
Iron	822
Aluminium	1670
Copper	160
Titanium	912
Nickel	244
Lead	276

(Miyamoto and Maruo, 1991). Since the molten regions overlap, a continuous cut is produced. However, cyclic variations in the driving force of the oxidation reaction caused by changes in the partial pressure of gas in the cutting zone are also believed to be involved (Ivarson et al., 1994). Viscosity and surface tension effects associated with melt removal also contribute to the mechanism of cutting.

ACTIVE GAS MELT SHEARING

By replacing the inert assist gas with a reactive gas such as oxygen or air, additional process energy may be generated through an exothermic chemical reaction. Cutting speeds can thus be increased in comparison with inert gas melt shearing. The mechanism again relies on the formation of a penetrating cavity, and so the beam must be focused to produce the required power density. Ferrous alloys and some thermoset polymers are cut by active gas melt shearing. Temperatures are higher than in the inert gas process, which can lead to edge charring in carbon-based materials, and a poorer edge quality, particularly in thicker metallic sections. Oxygen is typically used in cutting of mild steel; the principal exothermic reaction is described by the following equation:

$$Fe + \tfrac{1}{2}O_2 \rightarrow FeO.$$

The energy of the oxidation reaction in a number of metals is given in Table 14.2.

Sections of carbon–manganese steel up to 50 mm in thickness have been cut using a power of only 2 kW by maximizing the energy produced by the exothermic reaction, through the use of a short focal length lens that focuses the beam inside the lens housing, while the oxygen is delivered coaxially onto the workpiece surface over an area similar to that of the nozzle exit (O'Neill and Gabzdyl, 2000).

It is important to note that the term 'active gas' is relative to the material: air is considered to be active when cutting aluminium, but inert when cutting alumina.

VAPORIZATION

Vaporization cutting is the mechanism normally used with pulsed lasers, and for continuous wave (CW) cutting of materials that are not easily melted. A high power density is used, on the order of $10^6\,W\,mm^{-2}$, which is around 100 times that used in inert gas melt shearing. Material is heated rapidly to the vaporization temperature before extensive melting though thermal conduction occurs. Material is removed by vaporization and the ejection of liquid by an inert gas jet. Some woods and polymers, notably polymethylmethacrylate (PMMA), are cut using the vaporization mechanism. Since significant melting is avoided, a very high edge quality is produced. Cutting speeds are comparatively low because of the lack of an exothermic cutting reaction.

CHEMICAL DEGRADATION

Chemical degradation relies on the action of the laser beam to break chemical bonds and form new compounds, and is an important mechanism in the cutting of wood, thermoset polymers, elastomers and some composites. Cutting rates are generally lower than melt shearing, and cut edges are of relatively high quality, although residue may require cleaning.

SCRIBING

The objective of scribing is to create a groove or a series of blind holes at the workpiece surface. Low energy, high power density pulses cause vaporization with a restricted heat affected zone (HAZ). The notches serve to raise stress locally such that the material can be fractured along a defined line under subsequent bending. The mechanism is used for some ceramics, notably alumina, as well as some glasses and composites. Very high processing rates are possible.

PROCESS SELECTION

CHARACTERISTICS OF LASER CUTTING

The low energy input of laser cutting results in low deformation, a narrow recast layer, and a narrow HAZ. The high power density enables a high cutting speed to be achieved with a high edge quality. Both characteristics mean that finishing operations are minimal or can be eliminated. The narrow, square kerf that is formed minimizes material wastage by allowing patterns to be nested closely on a sheet, permits sharp corners to be cut, and enables a single cut to form the border of adjacent patterns. Certain fabrics are sealed during cutting, which prevents fraying.

Laser cutting is a high productivity process. A wide range of equipment is commercially available, which can be substituted directly for existing technology. Numerical control provides accuracy (tolerances are typically between 0.05 and 0.1 mm) with little surface disruption, as well as short lead times. Automatic nesting systems can be used to maximize material usage. Labour costs are low, and the machine is often paid for within one year.

The non-contact nature of laser cutting means that there is no tool wear, no tool storage costs, no tool setup time, no deformation of the cut surface and no slippage with only light fixturing.

Laser cutting is flexible. Short (prototype) and long (production) series can be accommodated, as can small and large products with varying shape complexity – there is no minimum internal radius restriction. Many different types of material, irrespective of hardness, can be cut. Thick materials are cut both as monolithic workpieces and multiple layered sheets. Cuts can be started or stopped anywhere. The beam can be switched to different workstations easily to serve several cutting operations in sequence or simultaneously.

The process is also environmentally friendly, since it is quiet, permits the most efficient use of materials, and restricts harmful fumes to a well-defined interaction zone remote from the operator that can be ventilated efficiently.

The main drawbacks are the high initial equipment, running and maintenance costs. Accurate cost analysis is therefore required when laser cutting is proposed as an alternative to a conventional cutting technique. Materials with high reflectivity and high thermal conductivity, such as gold, copper and aluminium, are difficult to cut using an infrared laser beam. The thickness that can be cut is limited by the penetration capability of the beam, and incomplete penetration cuts are difficult to produce. As with all thermal processes involving fusion, an HAZ and a recast layer are formed, which might not be acceptable in certain applications. Inhomogeneous materials containing phases with differing thermal

Table 14.3   Typical technical features of cutting processes

	Laser	Abrasive water jet	Plasma arc	Oxyfuel
Materials	All homogeneous	All	Metallic	Metallic
Max. thickness (steel, mm)	30	100	50	300
Kerf width (mm)	0.1–1.0	0.7–2.5	>1	>2
HAZ width (mm)	0.05	0	>0.4	>0.6
Edge quality (relative)	Square, smooth	Square, smooth	Bevelled ($c.$ 17°)	Square, rough
Edge roughness (R_a, μm)	1–10	2.0–6.5	–	–
Smallest hole diameter (mm)	0.5	>1.5	>1.5	20
Energy input (relative)	Low	Low	High	Medium
Capital cost (relative)	1	1	0.1	0.01
Productivity (relative)	High	Medium–low	Medium	Low

properties may fracture under thermal loading. Charred surfaces are often produced on cut surfaces of carbon-base materials.

COMPETING METHODS OF CUTTING

A comparison of laser cutting with conventional cutting methods requires both the technical and economic characteristics of each process to be considered. Naturally, if the required technical quality can be achieved by using several processes, then the one with the lowest cost is chosen. Similarly, if the cost of using different processes is similar, then the one providing the highest quality is the preferred option. However, it is worth remembering that laser cutting may provide a unique solution to a manufacturing requirement that cannot be met by competing methods, by reconsidering the materials used or by redesigning a product for example. The economic benefit of a novel process can often be difficult to quantify.

Technical Comparison

Competitive cutting processes may be thermal or mechanical. The technical features of cuts made using the principal techniques are given in Table 14.3. A comparison of the operating windows for laser, plasma arc and gas cutting of C–Mn steel is shown in Fig. 14.3.

Plasma arc cutting was developed from the GTA welding method by reducing the nozzle opening to constrict the electric arc and gas, thereby increasing both the arc temperature and voltage. The method came into use in the early 1980s. Gas is delivered at pressures up to 1.4 MPa at a rate of several hundred metres per second. Conventional plasma equipment with air or oxygen is normally used for two-dimensional cutting of sheet materials. Gantries holding multiple torches allow productivity to be increased. Quality can be improved by cutting under water, for example. Plasma arc cutting is used for electrically conductive materials, and is particularly useful for metals of high thermal conductivity because of the concentrated energy input. Thick sections can be cut at a higher rate than by laser cutting. However, laser cutting competes when a square edge of high quality with a narrow kerf width is required in thick materials. In thin materials a higher cutting speed and greater accuracy can be achieved with laser cutting. Non-metals can be laser cut using this method. The HAZ is small. Because

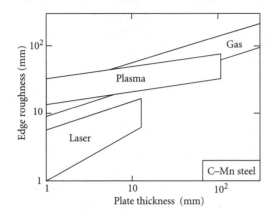

Figure 14.3   Operating windows of edge roughness and plate thickness typical of laser, gas and plasma cutting of carbon–manganese steel

of its non-contact nature, running costs are relatively low – although plasma arc cutting nozzles wear out and must be replaced.

Mechanical cutting of certain materials, such as velour, causes short fibres to be produced, which constitute a health risk. The elastic properties of materials that stretch are an origin of inconsistencies in size when cut using dies. Knives become dull and cannot cope well with sharp profiles requiring rapid changes in direction. Fabrics have a weave, and cutting across the grain produces a ragged edge.

Water jet cutting appeared in the late 1960s. Pressurized water (up to about 400 MPa) is forced through a small diameter sapphire or diamond nozzle (0.1–0.8 mm) at speeds of about 800 m s^{-1}. Abrasive particles such as silica, silicon nitride, garnet or alumina with a diameter in the range 0.1–0.5 mm can be added to enable metals, composites and other hard materials to be cut. The standoff distance is between 0.5 and 1.5 mm. Most systems are designed in a gantry configuration with CNC operation and multiple heads for increased productivity. The cut is made by an erosion process, with no HAZ, which is essential in some applications, especially in the aerospace industry. It is suitable for all engineering materials, especially inhomogeneous materials such as marble and concrete that could fracture on heating, and composites that delaminate when heated. Reflective materials such as copper and aluminium are cut easily, as are heat-sensitive materials such as high alloy steels and titanium alloys. Water jet cutting is particularly suitable for cutting thick materials up to 100 mm. The top of the cut is smooth because of the high energy of the jet, but becomes rougher and striated lower in the workpiece because abrasive particles are scattered. Laser cutting competes for cutting parts thinner than 6 mm, in which a narrow kerf width is required. The laser is also more suitable for three-dimensional out-of-position cutting. PMMA can be laser cut with a flame-polished edge.

Economic Comparison

In Chapter 4, the cost of laser processing was broken down into fixed and variable costs. The total cost of laser cutting can be established by considering the components of these cost factors.

The fixed costs ($ yr^{-1}) are related to the laser and its situation, and comprise: depreciation of capital equipment; maintenance; and floor space. These amounts can be calculated depending on individual circumstances and arrangements made by the facility. The variable costs ($) are a function of the characteristics of the cutting process, and comprise: planning; materials; CAD programming;

laser gas; laser electricity; optical components; other consumables; labour; quality assurance; and post-processing operations. Times are allocated to each component of the variable cost, and summed to produce the total variable cost. Similarly, the proportion of the fixed costs related to the total processing time is calculated. The total job cost is then the sum of the variable and fixed costs.

The total job cost for competitive methods of cutting is calculated in a similar way, enabling an economic comparison to be made.

The analysis described above highlights a few rules of thumb that can be applied when considering process selection in cutting:

- Laser cutting normally provides the best combination of quality and productivity with homogeneous materials less than about 3 mm in thickness, when the equipment is in use for at least 16 hours per day.

- Thicker materials may be cut more quickly with plasma arc methods at the expense of edge quality.

- Oxyfuel cutting is preferred for one-off jobs or short production runs in which quality is not of prime importance.

- Non-thermal cutting methods are more suitable for composite and inhomogeneous materials, but they are relatively slow.

PRACTICE

The cutting head comprises a chamber, sealed at its upper surface by the focusing lens, onto which a nozzle and tip are attached as illustrated in Fig. 14.4. Gas is injected into the chamber at a given pressure, and exits through the tip with a velocity and envelope determined by the nozzle geometry. The cutting head may be traversed over large distances, or fixed above a moving workpiece. Benchmarks for metal cutting using laser processing machines are provided in the standard referenced in Appendix F.

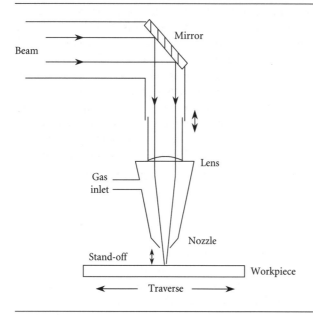

Figure 14.4   Schematic design of a laser cutting head

The laser cutting parameters are dependent on the composition of the material to be cut, its thickness, the beam characteristics, the cutting rate required, and the desired edge quality. The processing parameters described below are general in nature, but primarily relate to cutting of carbon–manganese steel using the melt shearing mechanism. They are provided to establish a baseline. Variations in procedure, cutting productivity, and the properties of laser cuts in other engineering materials are described later.

MATERIAL PROPERTIES

Thermal Properties

The high reflectivity of some materials to infrared laser light can lead to difficulties with initiation and maintenance of the cutting process. A high power density is required to initiate cutting, but once the process has stabilized, the power can be reduced. Such procedures are normally determined through trial and error testing.

Heat is transported rapidly from the cutting front in materials with high thermal conductivity. High power levels or low cutting speeds are therefore required to maintain a cutting front. However, as the cutting speed is reduced, instabilities occur that can result in abnormal molten regions, blowouts and poor edge quality.

Materials with high values of specific heat capacity require greater amounts of energy to raise their temperature to that required for melting. Similarly, materials with high latent heats of melting and vaporization require more energy for the changes in state relevant to the cutting mechanism.

Physical Properties

The presence of grease or paint may interfere with the cutting mechanism, resulting in unpredictable performance. A scale-free surface is desirable for cutting, since changes in scale thickness or the presence of rolled-in defects and grooves are detrimental to edge quality. However, the presence of a thin, uniform oxide layer can facilitate absorption of the laser beam and improve cutting performance. Control of slag formation during cutting is also an important factor in maintaining edge quality.

The surface tension and viscosity of molten material play important roles in determining edge quality and cutting speed. Molten materials with high values of surface tension and low viscosity are more difficult to remove from the cutting front by the assist gas. Dross can therefore adhere to the underside of the cut.

Steel companies now produce special grades for laser cutting, in which particular attention is paid to composition and surface properties. Silicon, for example – used to improve the castability of aluminium alloys by reducing viscosity – has a beneficial effect on edge roughness and cutting speed, at the expense of edge squareness.

The plate thickness determines the relationship between the rate at which cutting can be performed and the incident beam power, for a given set of processing parameters. It is therefore difficult to specify a single plate thickness that distinguishes thin plate cutting from thick plate cutting, without specifying a beam power or cutting speed. For practical purposes we use a value of 4 mm here. Thin section laser cutting is routinely carried out in job shops using a wide range of engineering materials. Conventional thick section laser cutting becomes difficult when the plate thickness exceeds about 15 mm, because the cutting speed is reduced to a level at which the rate of thermal conduction from the cutting front approaches the power input required to maintain the cut. Material ejection at the base of the cut becomes inconsistent and the edge quality deteriorates. Special techniques such as beam spinning, variation of the focal plane, and the use of a high brightness power source are required for very thick section cutting – these are discussed below.

BEAM CHARACTERISTICS

Wavelength

The majority of industrial laser cutting applications use CO_2 lasers, because economic cutting rates can easily be achieved with readily available, modestly priced 1 and 2 kW machines. However, systems incorporating higher power level lasers (up to 6 kW) are becoming popular as the mechanisms of thick section cutting – particularly when an inert assist gas is used – are understood more fully. The CO_2 laser is particularly attractive when high power is needed and when non-metallic materials are to be cut, because of the high efficiency of beam absorption.

The Nd:YAG laser has practically taken over in the field of robotic cutting, because of fibreoptic beam delivery and the availability of higher power Nd:YAG lasers. The primary consideration for selecting the Nd:YAG laser process over competing techniques is edge quality, which is particularly high in thin sheet materials. An Nd:YAG beam can be focused to a smaller diameter than a CO_2 laser beam, providing more accuracy, a narrower kerf width and low edge roughness, which is particularly useful in scribing applications. The Nd:YAG laser can also be operated easily in pulsed mode, which when optimized can give a higher edge quality. The high peak power means that assist gases are often not required. Operation in the second harmonic generation mode produces shorter wavelength light that has better focusability, and results in improved processing properties.

Excimer lasers are particularly effective tools when high quality is required when cutting certain polymers, rubber, ceramics and fabrics. The chemical oxygen iodine (COIL) laser has also been shown to be superior in cutting performance to the CO_2 laser, using both active and inert gases, and has the potential to be scaled to high power levels.

Power

Most flat sheet and thin section cutting is carried out using a laser power level of between 1 and 2 kW. The optimum incident workpiece power is established during procedure development. Excessive power results in a wide kerf width, a thicker recast layer, and an increase in dross. With insufficient power, cutting cannot be initiated. A variation in power around the optimum of $+30\%$ to -10% can normally be tolerated without affecting cutting performance significantly.

Spatial Mode

A TEM_{00} beam mode gives the smallest focused spot and highest power density in comparison with other beam modes, which leads to the highest cutting rate, and lowest roughness. In addition, this beam mode has the largest depth of focus, and therefore will give the best performance when cutting thicker materials. The highest edge quality can be obtained if the Rayleigh length of the focused beam (Chapter 3) is equal to the sheet thickness. Of equal importance is the mode stability, which must be as high as possible in order to achieve high quality cuts.

Temporal Mode

A CW laser beam is preferred for smooth, high cutting rate applications, particularly with thicker sections, whereas a lower energy pulsed beam is preferred for precision cutting of fine components. A pulsed beam of high peak power is also advantageous when processing materials with a high thermal conductivity, and when cutting narrow geometries in complex sections where overheating is a problem. Beam pulsing influences the fluid dynamics of the molten kerf and the temperature of the cutting front, and modulates exothermic reactions. Optimized pulse parameters improve the uniformity of the cut edge profile (Powell *et al.*, 1986). Pulsed processing helps to reduce dross formation in certain metals, e.g. aluminium. Inherent power variations on the millisecond time scale may be present in certain lasers, and can affect cutting performance.

Optic Focal Length

A focusing lens, rather than a mirror, is normally used in laser cutting because of the relatively low power levels used. A lens can be incorporated into a cutting head, acting as the upper wall of the pressure chamber. The focal length of the lens determines the spot size and depth of field of the laser beam. For a TEM_{00} CO_2 laser beam of diameter 15 mm, a 127 mm (5 in) focal length lens produces a spot diameter of about 0.15 mm, with a depth of focus around 1 mm. This is a good combination for metals with a thickness in the range 0.2–8 mm, and is the reason for the popularity of this focal length. A longer focal length is preferred thick section cutting, where the depth of focus should be around half the plate thickness. For thin materials (less than 4 mm in thickness) a shorter focal length, typically 63 mm gives a narrower kerf and smoother edge because of the smaller spot size. Optics can be protected from weld spatter through the use of a cross flow of inert gas (an 'air knife').

Focal Plane Position

When cutting with oxygen, the maximum cutting speed is achieved if the focal plane of the beam is positioned at the plate surface for thin sheets, or about one third of the plate thickness below the surface for thick plates. When using an inert gas, however, the optimum position is closer to the lower surface of the plate, because a wider kerf is produced that allows a larger part of the gas flow to penetrate the kerf and eject molten material. Since the beam diameter at the exit of the nozzle is consequently greater, larger nozzle diameters are used in inert gas cutting. The focal plane may be positioned approximately 1% of the lens focal length above or below the workpiece surface without any appreciable effect on cutting performance. If the focal plane is too high relative to the workpiece surface, the kerf width and recast layer thickness increase to a point at which the power density falls below that required for cutting. Similar changes occur when the focal plane is positioned too far below the surface.

By oscillating the focal plane position vertically within the plate during cutting, the focused spot traces a sinusoidal locus, which enables thick plates to be cut with relatively low power (Geiger *et al.*, 1996). A dual focus optic is available that focuses a part of the beam on the top surface and the remainder at the lower surface. Thicker materials can then be cut more quickly, without subsurface dross, and cost savings in assist gas can be made through the use of a smaller nozzle aperture (Graydon, 1999; Nielsen and Ellis, 2002).

Cutting Speed

The cutting speed must be balanced with the gas flow rate and the power. As cutting speed increases, striations on the cut edge become more prominent, dross is more likely to remain on the underside, and penetration is lost. When the cutting speed is too low, excessive burning of the cut edge occurs, which degrades edge quality and increases the width of the HAZ. In general, the cutting speed for a material is inversely proportional to its thickness. Speed must be decreased when traversing sharp corners, which requires a corresponding reduction in beam power to avoid burning.

Polarization

The high gas flow rate and high speed used in laser cutting mean that little plasma is formed. Energy transfer therefore occurs primarily via Fresnel absorption at the cutting front. Since the beam is at a high angle of incidence to the front (close to grazing) the polarization of the beam affects cutting performance. Beam polarization is of concern for CO_2 laser cutting since the light from a CO_2 laser is linearly polarized. (Light from Nd:YAG lasers is randomly polarized, and so cutting performance is not affected by direction.) For metals, only *p*-polarized light is absorbed significantly at grazing angles of incidence. If the beam is polarized in the direction of travel, *p*-polarized light is absorbed along the front edge of the cut, while *s*-polarized light is absorbed by the sides of the kerf. A narrow kerf with high perpendicularity is then formed and cuts can be made with the highest speeds. If the beam is polarized at an angle to the cut, more absorption takes place at the sides of the cut, producing a wider kerf with a taper that depends on the angle between the cutting direction and the plane of polarization.

A uniform cut can be achieved when the cutting direction changes by using various devices. Optics are available that convert a linearly polarized beam into a circularly polarized beam (Chapter 4), which produces a compromise uniform kerf width and perpendicularity at the expense of cutting speed. Alternatively, an adaptively controlled system can be used to maintain the optimum polarization with respect to the direction of cutting (Keilmann *et al.*, 1992).

Beam Manipulation

By spinning the beam, very thick plates can be cut at low speeds through the formation of a wide kerf (Arai and Riches, 1997). By using two beams separated by several millimetres in the direction of cutting, higher cutting speeds than an equivalent power single beam can be obtained, with enhanced cut thickness (Molian, 1993). The first beam partially penetrates the workpiece, causing melting and vaporization. As the second beam reaches the molten region, it vaporizes some material and superheats the remainder.

PROCESSING PARAMETERS

Process Gas

The process gas has five principal functions during laser cutting. An inert gas, such as nitrogen, must expel molten material without allowing drops to solidify on the underside (dross). An active gas, such as oxygen, participates in an exothermic reaction with the material. When cutting thick sections with high beam intensities, the gas also acts to suppress the formation of plasma. Focusing optics are protected from spatter by the gas flow. Finally, the cut edge is cooled by the gas flow, thus restricting the width of the HAZ. Without an assist gas, it is impossible to use a laser for cutting at high speed with good quality if the thickness is more than a few tenths of a millimetre. The importance of assist gas increases as the material thickness increases.

Gases used in cutting may be delivered in liquid or gaseous states. If cutting gas consumption exceeds about 1000 m^3 each month, then liquid supply in pallet tanks or stationary tanks is more appropriate. This is normally the case when more than one laser is used for low pressure oxygen cutting, or when one or more lasers are used for high pressure cutting. If cutting is primarily carried out using low pressure oxygen on one laser, or the laser is only used for cutting occasionally, gases delivered in cylinder bundles are more economical.

Nitrogen is the most common inert gas because it is relatively inexpensive. The purity is relatively unimportant, provided it is above 99.8%. Nitrogen is often used to produce cuts of high edge quality in ferrous materials such as stainless steels, and nickel-base alloys. Argon is a common choice when cutting titanium, since it prevents the formation of oxides or brittle titanium nitrides. Helium can be used in applications requiring very high quality cut edges, or where contamination with oxygen must be avoided. However, helium is expensive outside the United States. Inert gas pressures at the nozzle are generally higher than those used with oxygen. If the inert gas flow rate is too high, an intermittent cutting effect is produced, whereas insufficient gas flow results in inadequate material removal and consequently the formation of dross on the back of the plate.

Air is often used for cutting aluminium, polymers, wood, composites, alumina, glass and quartz because it is so readily available.

Oxygen is used for cutting mild steel and stainless steels when a high cutting speed is paramount, with edge quality and discolouration of secondary importance. As well as increasing absorption through oxidation in the initial stages of cutting, the presence of oxygen creates an exothermic reaction, effectively increasing the laser power. Around 50% of the total energy available for cutting is supplied by the oxidation reaction, the remainder being provided by the beam. As plate thickness is increased, the oxygen pressure is reduced to avoid burning effects, and the nozzle diameter is increased. Oxygen cutting can be carried out at speeds up to five times those available with air. Gas purity is important – mild steel of thickness 1 mm can be cut up to 30% more quickly using 99.9 or 99.99% purity oxygen, in comparison with the standard oxygen purity of 99.7%.

Excess active gas pressure results in turbulence in the melt and uncontrollable oxidation leading to prominent striations and burning. Shockwaves are formed, causing flow patterns and pressure distributions that are sensitive functions of the nozzle parameters. An increase in kerf width results, with more pronounced drag lines on the cut surface. If the gas pressure is too low, molten material is not ejected efficiently, creating excessive dross and eventually a loss of the cutting front.

Gas mixtures can also be used: 60% He/40% O_2 mixture maintains a good edge quality without sacrificing cutting speed, for example.

Nozzle

The design of the nozzle – and in particular the design of the orifice – determines the shape of the cutting gas jet, and hence the quality of the cut. Several different geometries can be used, depending on the application. Some of the more common designs are illustrated in Fig. 14.5. No single nozzle is superior in all applications. The nozzle has three main functions: to ensure that the gas is coaxial with

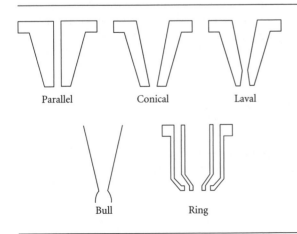

Figure 14.5   Common nozzle designs for laser cutting

the beam; to reduce the pressure to minimize lens movements and misalignments; and to stabilize the pressure on the workpiece surface to minimize turbulence in the met pool. The diameter of the nozzle orifice ranges between 0.8 and 3 mm, and is selected according to the material and plate thickness. A nozzle that is too small creates difficulties in alignment and localizes the gas, resulting in a rough edge. A nozzle that is too large provides insufficient gas flow to expel molten material, and results in high gas consumption. The alignment of the nozzle with the laser beam has a significant effect on the quality of the cut. Misalignment causes gas to flow across the top of the cut zone, which can lead to undesirable burning of the cut edge, and a poor quality cut. Alignment may be ensured by movement of the nozzle holder at the exit, or by movement of the lens holder at the entrance.

Most industrial cutting of flat sheets is carried out using nozzles with a converging profile and a cylindrical orifice. These are simple to manufacture and give good results with stand-off distances up to 1 mm. Convergent–divergent nozzle types, such as the Laval design, avoid the sharp corners at the exit of a cylindrical nozzle, which reduces the divergence of the gas on exit. They deliver higher pressures over larger distances, involving supersonic flow, which produces a favourable shock structure for through-thickness cutting with a clean kerf. A good cut quality is produced with a stand-off distance above 1 mm. They are particularly suitable for three-dimensional cutting of thin materials, in which it may be difficult to maintain a small constant stand-off. Higher cutting speeds can also be used. The major disadvantage is that the smallest cross-section must be adapted to the converging beam, and must therefore be relatively large, leading to higher gas consumption. A ring nozzle enables a chemically reactive gas, such as oxygen, to be delivered to enhance cutting speed, or control burr formation, while an inert gas can be introduced at a different pressure in the outer annulus. In all nozzle designs the stand-off distance between the nozzle exit and the workpiece surface controls the pressure in the kerf. It must be maintained accurately to ensure the correct pressure.

A dual oxygen jet system comprising coaxial gas delivery and an off-axis jet has been found to increase the maximum cutting speed in thick materials substantially (Hsu and Molian, 1992; Ilvarasan and Molian, 1995). Such devices are not yet widely used in industry.

Stand-off

An important practical parameter in laser cutting is the distance between the nozzle and the workpiece – the stand-off distance. This distance influences the flow patterns in the gas, which have a direct bearing on cutting performance and cut quality. If the stand-off distance is greater than about 1 mm, large variations in pressure can occur. The stand-off distance is normally selected in the same range as the diameter of the cutting nozzle – between 0.5 and 1.5 mm – in order to minimize turbulence. A short stand-off distance gives stable cutting conditions, although the risk of damage to the lens from spatter is increased. The stand-off distance is optimized to maximize cutting speed and quality.

Cooling

If the metal surrounding the cut becomes too hot it begins to burn, creating a wide cut with a poor edge quality. To control this and to prevent the burning reaction becoming uncontrollable, a fine mist of water can be applied near the beam, or a water coating applied to the plate surface. Water cooling also improves the surface finish, and spreads the depth of focus over a greater range, widening the operating window, which enables process parameters to be established more quickly.

Piercing

The cut is initiated with a relatively high energy pulse that pierces the material. Piercing is carried out without uncontrolled burning of the material and without causing too much spatter that could

Table 14.4 Methods of monitoring and controlling laser cutting

Source	Sensor	Monitored	Controlled	Reference
Laser	Photodiodes	Reflected light	Stand-off Power, speed Off-line parameters	(Abdullah *et al.*, 1994) (Huang and Chatwin, 1994) (Hansmann *et al.*, 1989)

damage the optics. The laser power can then be increased gradually, the stand-off distance increased, or the cutting gas pressure reduced. These can all be programmed into the CNC system for automatic piercing.

Normally the same gas is used for piercing and cutting. However, helium is sometimes used for piercing before high pressure nitrogen cutting of high alloy steels to avoid plasma absorption of the beam above the workpiece. In the case of aluminium, oxygen is preferred for piercing since oxidized aluminium improves absorption, minimizing problems associated with back reflection.

Hybrid Cutting

The use of an electric arc with a CO_2 laser beam can double the speed at which oxygen-assisted melt shearing of mild steel can be performed (Steen, 1981; Kamalu, 1986). In another hybrid process, a fine laminar water jet directs an Nd:YAG laser beam in the same manner as a fibreoptic cable, enabling advantage to be taken of the most appropriate features of water jet and laser beam cutting (Richerzhagen, 2001).

Process Monitoring

Emissions in various frequency bands are used to monitor the condition of laser cutting. When integrated with a process model or empirical knowledge base, changes to process variables can be made during cutting to ensure optimum cutting characteristics. Light reflected from the cutting zone is one of the most widely used signals, Table 14.4.

PROPERTIES OF LASER-CUT MATERIALS

QUALITY ASSURANCE

Cut quality is assured by developing a qualified cutting procedure that contains the values, or ranges of values, for process parameters that are demonstrated to produce cuts to a quality level prescribed by an end user. The development process may be expedited by basing trials on an existing procedure, or by using predictive methods, some of which are described below.

Imperfections

Imperfections in laser cuts are associated with the edge of the cut and the upper and lower surfaces of the workpiece. Terminology of imperfections in thermal cuts, and the limits that are used to grade cuts are described in standards referenced in Appendix F.

Burrs

A burr is a highly adherent dross (solidified material) or slag (solidified oxides) that forms on the underside of the cut. The burr may take the form of elongated droplets, or a rough, whisker-like layer. Dross can be removed during processing through the use of a gas jet directed from the underside of the workpiece (Birkett *et al.*, 1985). Alternatively, it is removed mechanically after cutting.

Kerf

The kerf is the slot that is formed during through-thickness cutting. It is normally narrower at the bottom of the workpiece than the top. The kerf width is defined as the width at the bottom of the cut, and is typically slightly larger than the focused beam diameter in optimized laser cutting. In contrast, the kerf width of other gas-assisted thermal cutting processes is dependent on the gas jet width.

Pitting

Pits are erosions and craters that form on the cut edge as a result of irregular melt flow during cutting with parameters that have not been optimized. They originate from turbulence in the melt region, particularly when cutting with low speed, and are described by size and number from a visual inspection.

Roughness

The average value of roughness, R_z, is used in quality classification. The length of measurement is 15 mm, which is divided into five partial measuring lengths. The distance between the highest peak and the lowest trough is determined for each partial measuring length. R_z is the average of these five distances, and is measured in micrometres. Quality classes can be defined, based on those for thermal cutting techniques, which express roughness in terms of the material thickness, a, as shown in Fig. 14.6.

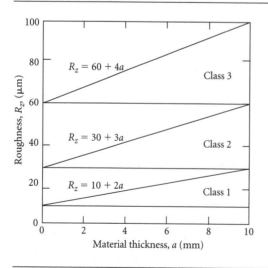

Figure 14.6   Quality classes for thermal cutting based on edge roughness (ISO, 2002)

Groove Lag

Groove lag, or drag, is the deviation of the striations from the perpendicular line in a cut. It is defined as the greatest distance between two adjacent striations. The mechanism of striation formation was described earlier. Groove lag increases as the cutting speed increases, indicating that fluid dynamics play an important role in the formation of striations.

Perpendicularity

The perpendicularity, or squareness and inclination tolerance, u, is the greatest perpendicular distance between the actual surface and the intended surface. It is the sum of the angular deviation and the concavity or convexity of the surface. The units of measurement are mm. Quality classes can be defined for thermal cutting techniques in a similar manner to those describing roughness, in which perpendicularity and angularity are expressed in terms of the material thickness, a, as shown in Fig. 14.7.

HAZ Width

Although not normally included in quality assessment of a laser cut, the width of the HAZ in the parent material adjacent to the cut edge can be measured. The HAZ width increases as the energy input per unit length and cut thickness increases. HAZ width is important if cuts are to be made near heat-sensitive components, which may place a maximum limit on the incident beam power or plate thickness, or a minimum limit on the cutting speed.

Surface Condition

The surface condition after cutting is primarily characterized by the presence of spatter on the upper surface (residual molten material ejected from the cut), and discolouration of the surface. Coatings such as graphite can be used to reduce the amount of spatter.

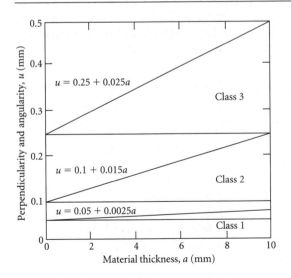

Figure 14.7   Quality classes for thermal cutting based on edge perpendicularity and angularity (ISO, 2002)

PERFORMANCE

Performance can be defined in terms of four main variables: cutting speed, material thickness, a kerf width parameter and a roughness parameter. Gabzdyl (1989) defines a performance factor, *S*:

$$S = \frac{vt}{kR} \tag{14.1}$$

where *v* is the cutting speed, *t* is the material thickness, *k* is the kerf width and *R* is the surface roughness. Equation (14.1) is a useful means of ranking laser cutting processes for various engineering materials.

METALS AND ALLOYS

Carbon–Manganese Steels

Carbon–manganese steels are normally cut using oxygen assist gas. When edge quality is more important than cutting speed, an inert gas such as nitrogen is used. Higher carbon steels are generally more easily cut than low carbon varieties. Cold rolled steels are cut more easily than hot rolled steels, since they are generally made with tighter compositional tolerances. High impurity levels affect the exothermic reaction, causing blowouts and roughness. Steel grades with more than 0.3% C form a noticeable martensitic layer at the cut edge. Low silicon and sulphur concentrations help to ensure the quality and cleanliness of the cut edge; such empirical observations enable steel companies to develop steels with narrow compositional tolerances specifically for laser cutting. References for laser cutting of carbon–manganese steels are given in Table 14.5.

The surface condition of carbon–manganese steels has a large influence on laser cutting performance, particularly when oxygen is used. Rust, which retains moisture, can give rise to dross and notches. Painted surfaces contain hydrocarbon molecules, which readily absorb laser light. Although paints do not normally affect the cutting process, they can become damaged around the cut line. Zinc primers and iron oxide shop primers can lead to dross formation when they face the nozzle. The use of high pressure nitrogen cutting overcomes these problems, although the cutting speed is reduced. Cold rolled steel often has a surface layer of oil to prevent corrosion, which can interfere with the cutting process. Hot rolled steel contains a surface layer of oxides, which is produced during the rolling process. The thickness of the oxide layer may vary considerably, but ideally should be thin, solid and homogeneous. This oxide layer may dilute the chemical reaction of the oxygen, and reduce its effectiveness, leading to

Table 14.5   Sources of data for laser cutting of carbon–manganese steels

Material	Laser	Assist gas	Reference
43A	CO_2	O_2	(O'Neill and Gabzdyl, 2000)
A36	COIL	N_2, O_2	(Carroll *et al.*, 1996; Carroll and Rothenflue, 1997a, b)
AISI 1020	CO	O_2	(Sato *et al.*, 1989)
AISI 1020	CO_2	O_2	(Hsu and Molian, 1994)
C–Mn steel	CO_2	O_2	(Sullivan and Houldcroft, 1967)
C–Mn steel	CO_2	O_2	(Lunau and Paine, 1969)
C–Mn steel	CO_2	O_2	(Adams, 1970)
SS400	Nd:YAG	O_2, air, N_2	(Watanabe *et al.*, 1999)
St14–03	CO_2	O_2	(Olsen *et al.*, 1989)

a reduction in cutting speed. Zinc-coated sheet is normally delivered in two forms: galvanized or hot dipped. A thin galvanized surface presents no problems for laser cutting. Hot-dipped sheet normally contains a thicker zinc coating, which reacts exothermically with oxygen and may ignite, producing harmful zinc fumes. The cutting speed can typically be set anywhere between the maximum and 50% of the maximum before burning becomes evident.

High pressure nitrogen is normally only used when the cutting speed is not a critical factor, or when parts are to be painted in which case an oxide coating on the cut edge is unacceptable. Nitrogen is not completely inert, but is used in preference to argon because of its lower cost. Gas pressures in excess of 1 bar are normally used.

Alloy Steels

Low alloy steels, such as AISI 4140 and 8620 tend to show similar cutting characteristics to carbon–manganese steels. Their lower levels of impurities improve the cut quality, which is similar to cold-rolled low carbon steel. Chromium additions reduce reactivity and produce an adherent scale. Hardenable grades have a martensitic edge. Cutting performance of free cutting steels is similar to that of mild steels, although sulphur can form SO_2, which may reduce the speed and cut quality. Tool steels with high amounts of tungsten cut slowly. References for laser cutting of alloy steels are given in Table 14.6.

Stainless Steels

Stainless steels may be cut by both active gas (oxygen) and inert gas (nitrogen) melt shearing mechanisms. The oxygen assist mechanism comprises exothermic oxidation and the inert melt shearing mechanism. Stainless steels have relatively low thermal conductivity, which enables them to be cut at relatively high rates since energy remains at the cutting front rather than being dissipated into the material ahead of the cutting front. The nickel present in austenitic grades of stainless steel affects energy coupling and heat transfer, which limits the thickness that can be cut with a given laser beam power. Chromium readily forms oxides with a high melting temperature during oxygen-assisted cutting, which do not dissolve in molten material, but which form a seal that limits the exothermic reaction in the upper part of the cutting front. The high surface tension of the oxide leads to dross formation and a rough cut surface. Other alloying elements can also create problems when they react with active gases. A high pressure inert gas such as nitrogen is therefore used when edge quality is of greater importance than cutting speed. Precautions must be taken to avoid inhalation of chromium vapour that is produced when cutting stainless steels. References describing cutting procedures for stainless steels are given in Table 14.7.

Aluminium Alloys

Because of the relatively low melting temperature of aluminium and its highly exothermic oxidation reaction, it would be expected that cutting could be carried out at high speed. However, three properties of aluminium alloys limit the performance and cut quality. High reflectivity means that a high power

Table 14.6 Sources of data for laser cutting of alloy steels

Material	Laser	Assist gas	Reference
High alloy	CO_2	O_2	(Petring *et al.*, 1990)
Tool steel	CO_2	O_2	(Sullivan and Houldcroft, 1967)

Table 14.7 Sources of data for laser cutting of stainless steels

Material	Laser	Assist gas	Reference
AISI 304	CO	O_2	(Sato *et al.*, 1989)
AISI 304L	Nd:YAG	Ar	(Jones and Chen, 1999)
AISI 304L	Nd:YAG	O_2	(Alfillé *et al.*, 1996; Powell, 1990)
AISI 400 series	COIL	He, N_2	(Kar *et al.*, 1996)
Austenitic stainless	CO_2	O_2	(Adams, 1970)
Martensitic stainless	CO_2	O_2	(Adams, 1970)
Stainless	CO_2	O_2	(Sullivan and Houldcroft, 1967; Lunau and Paine, 1969)
SUS304	Nd:YAG	O_2, air, N_2	(Watanabe *et al.*, 1999)

density is required to initiate cutting. Precautions must also be taken to ensure that the beam is not reflected back into the laser cavity; tilting the workpiece be several degrees relative to the beam and protection of surrounding equipment overcomes this problem. High thermal conductivity means that heat is transported from the cutting front rapidly, which can lead to difficulties when initiating the cut. The maximum cutting speed is reduced accordingly. The ease with which aluminium forms a protective oxide layer is the origin of two difficulties. First, variations in the oxide layer thickness along the cutting line result in variations in the absorptivity of the material, which causes changes in the energy absorbed by the material. (An oxide depth in excess of about $4\,\mu$m leads to almost total absorption of a CO_2 laser beam.) It can therefore be difficult to optimize the processing parameters for oxidized sheets. Secondly, oxide films on molten aluminium at the cutting front extinguish the exothermic reaction produced during oxygen-assisted cutting, and increase the tenacity of molten material. The former creates irregularities in the cutting mechanism and results in a rough cut edge; it can be reduced by thorough cleaning of sheets and by minimizing the time between cleaning and cutting. The latter encourages the formation of dross on the underside of the cut, which can be reduced by increasing the assist gas flow rate. A standard oxygen purity of 99.7% is sufficient. Low pressure oxygen cutting (2–6 bar) is frequently used, although this results in cut edges with dross and a rough cut edge. High pressure oxygen cutting (10–15 bar) is preferred with commercial purity aluminium, whereas 7–15 bar can be used for aluminium alloys.

Aluminium can also be cut using nitrogen at high pressure (between 5 and 17.5 bar); a nozzle diameter of 1.4 mm and a stand-off of 0.7 mm are typically selected. A high beam power, above 2 kW, and a low order beam mode enable a high power density to be maintained. A short focal length lens (typically 63 mm or 2.5 in) is advantageous for inert gas cutting of thinner aluminium sheets because of the small spot diameter and high power density that can be produced. The focal plane is located close to the lower surface of the sheet, while maintaining the power density for cutting at the upper surface. An improvement to the edge quality can be achieved at the expense of the cutting speed. A narrow HAZ width can also be obtained because the cut surfaces are cooled. References for laser cutting of aluminium alloys can be found in Table 14.8.

The focal plane is located close to the lower edge of the sheet when cutting with high pressure oxygen. When cutting with low pressure oxygen, the focal plane is located close to the surface of the plate.

Aluminium alloys heat up considerably during cutting, causing distortion, which can negate some of the advantages of laser cutting. Oxygen-assisted laser cutting of aluminium can be carried out at speeds up to three times greater than with air. Low alloy aluminium alloys, such as 1000 and 3000 series, are more difficult to cut than alloys containing manganese (5000 and 7000 series). Aluminium alloys up to about 8 mm in thickness can be laser cut.

Table 14.8   Sources of data for laser cutting of aluminium alloys

Material	Laser	Assist gas	Reference
Aluminium alloy	COIL	N_2, O_2	(Carroll et al., 1996; Carroll and Rothenflue, 1997a, b)
Al–1.5Cu–2.3Mg–5.7Zn	CO_2	He	(Juckenath et al., 1990)

Table 14.9   Sources of data for laser cutting of titanium alloys

Material	Laser	Assist gas	Reference
Ti	Nd:YAG	Ar	(Powell, 1990)
Ti	CO_2	O_2	(Lunau and Paine, 1969)
Ti alloy	CO_2	O_2	(Adams, 1970)
Ti–0.2Fe	CO_2	He	(Juckenath et al., 1990)
Ti–6Al–4V	CO_2	O_2	(Bod et al., 1969)
Ti–36Al intermetallic	CO_2	O_2, Ar	(Shigematsu et al., 1993)

Aluminium alloys and anodized aluminium are normally easier to cut than commercially pure aluminium, because of the enhanced absorption characteristics of the alloy and oxide layer. Laser cut aircraft alloys such as 2024 and 7075 possess a rough edge with considerable microcracking, which prevents the use of laser cutting in structural components.

Titanium Alloys

High pressure inert gas cutting using high purity argon is the preferred method of processing titanium, because of the metallurgical problems associated with the use of oxygen or nitrogen, described below. Flow rates of 30–150 L min^{-1} are typically used. The argon must be of very high purity (99.99–99.999%); traces of oxygen in particular must be avoided. Plasma can form when argon is used, particularly with low cutting speeds. Helium is therefore added to increase the thermal conductivity of the assist gas, reduce the workpiece temperature, and reduce plasma formation. The focal plane of the beam should be positioned below the top surface of the workpiece for high pressure inert gas cutting to achieve the power density required at the surface and to maintain an effective level throughout the plate thickness, because no exothermic reaction takes place to generate additional energy.

Since titanium has a high affinity for oxygen, the removal of oxidized surface layers before cutting is recommended. Titanium has a highly exothermic oxidation reaction, Table 14.2, but the reaction causes uncontrollable burning and the formation of oxides. For rough cutting this might be acceptable, but an inert assist gas is used when high quality edges are required. The oxidation reaction can be reduced by using air instead of pure oxygen, in which case the cutting mechanism changes to a combination of melt shearing and oxidation. Oxygen-assisted cutting of titanium alloys creates an oxide film at the cut edge that is susceptible to cracking, which can act as an initiation site for fatigue cracks. Similarly, nitrogen is absorbed into the surface where a hard, brittle layer of α phase and TiN forms. Titanium alloys are therefore normally cut using the inert gas melt shearing mechanism with argon or helium. References for laser cutting of titanium alloys are given in Table 14.9.

The most pronounced feature of titanium cut by inert gas melt shearing is the recrystallization layer. Any colouration of the cut surface indicates that oxidation has taken place. The edge retains much of the toughness and integrity of the base material but some hardening occurs. Oxidation of the cut edge can also occur after cutting when the edge is still hot.

Nickel Alloys

Nickel has a relatively low thermal conductivity in comparison with other metals, but the reflectivity to CO_2 laser light is still high. It reacts exothermically with oxygen, Table 14.2, and so this gas or air are often used as assist gases. Low pressure oxygen cutting is carried out with a pressure less than 6 bar, but dross is observed on the underside of the cut and the cut surfaces are oxidized. Nitrogen is the preferred assist gas for high pressure inert gas melt shearing. The beam focal plane is positioned below the upper plate surface for the same reasons as in titanium cutting. The cut edges are burr free and oxide free, and possess a similar edge quality to stainless steels, but the cutting speed is reduced in comparison with oxygen cutting. References for laser cutting of nickel alloys are given in Table 14.10.

Pulsing of a high brightness Nd:YAG beam at several hundred hertz enables nickel alloys of thickness up to 38 mm to be cut with 2 kW.

Copper Alloys

The high thermal conductivity and reflectivity of copper lead to similar changes in cutting procedure to those used with aluminium; Nd:YAG laser beams are absorbed more efficiently than CO_2. A short focal length optic, typically 63 mm (2.5 in), is needed to maintain power density. Oxygen-assisted cutting is advantageous when cutting copper alloys since the oxide formed at the cut front improves absorption, although the oxidation reaction does not provide a significant amount of energy. Both low (below 6 bar) and high (up to 20 bar) pressures are used. High pressure nitrogen may also be used. References for laser cutting of copper alloys are given in Table 14.11.

The oxide layer on the surface enhances absorption of the laser beam. Material up to about 5 mm in thickness can be cut. Brass has a lower reflectivity and a lower thermal conductivity than copper, and is therefore easier to cut.

Table 14.10   Sources of data for laser cutting of nickel alloys

Material	Laser	Assist gas	Reference
Hastelloy X	Nd:YAG	Air	(Powell, 1990)
Inconel	Nd:YAG	Air	(Powell, 1990)
Inconel 600	Nd:YAG	Ar	(Jones and Chen, 1999)
Nimonic 75	CO_2	O_2	(Adams, 1970)
Nimonic 80A	CO_2	O_2	(Adams, 1970)
Nimonic 90	CO_2	O_2	(Adams, 1970)

Table 14.11   Sources of data for laser cutting of copper alloys

Material	Laser	Assist gas	Reference
Cu–DHP	CO_2	Various	(Daurelio, 1987)
Cu, brass, 90Cu–10Ni	CO_2, Nd:YAG, excimer	O_2	(Pocklington, 1989)

Table 14.12 Sources of data for laser cutting of Zircaloy

Material	Laser	Assist gas	Reference
Zircaloy-2	Nd:YAG	Ar	(Ghosh *et al.*, 1996)
Zircaloy	Nd:YAG	Air	(Powell, 1990)

Table 14.13 Sources of data for laser cutting of amorphous alloys: N/S not specified

Material	Laser	Assist gas	Reference
Co–Si–Mo–Fe–B	Nd:YAG	N_2	(Dorn *et al.*, 1994)
Metglas 2605CO	N/S	N/S	(Churchill, 1984)
Metglas 2605S-2	Nd:YAG	N/S	(Churchill *et al.*, 1989)

Zirconium

Zircaloy is laser cut using an inert assist gas. Procedures can be found in the references listed in Table 14.12.

Amorphous Metals

Foils of amorphous metals (metallic glasses) can be cut without a change in microstructure by using a low energy input pulsed beam with a short duration. Higher energy input leads to remelting and a crystallized zone around the cut. Procedures can be found in the references given in Table 14.13.

CERAMICS AND GLASSES

Many ceramics are difficult to cut using mechanical means because they are hard and brittle. Many also possess poor thermal shock resistance, which hinders laser cutting. However, some can be laser cut using the mechanisms of melt shearing, vaporization, and scribing by drilling a series of blind holes. Both CO_2 and Nd:YAG lasers are used. The success of cutting depends on the homogeneity of the material; the different phases in inhomogeneous natural ceramics possess different coefficients of thermal expansion, which can lead to undesirable fracture. The formation of a vitreous phase on the cut surface of domestic ceramics, which is susceptible to microcracks, can be avoided more easily using a CO_2 laser. Nd:YAG lasers are favoured when cutting engineering ceramics. The beam is normally pulsed with a frequency of several tens of hundreds of hertz when the vaporization and scribing mechanisms are used, to produce a high peak power that initiates vaporization and material removal with a low energy input that restricts the HAZ, while avoiding melting (which would leave a recast layer). Scribing is a more productive means of parting ceramics; rates on the order of several tens of metres per minute can be achieved. Compressed air is the most common assist gas used when cutting ceramics and glasses. Inert gases such as nitrogen are only used when the material is very flammable. Oxygen can be used to avoid discolouration of the cut surfaces of ceramics. Procedures can be found in the references listed in Table 14.14.

Melt shearing leaves a glassy sealed recast surface on the cut edge. Scribing produces a rough fractured surface. The most serious problem encountered when laser cutting ceramics is cracking, notably on the cut edge. Preheating is the most promising method of avoiding cracking.

Table 14.14 Sources of data for laser cutting of ceramics: N/S not specified

Material	Laser	Assist gas	Reference
Alumina	CO_2	Air	(Powell *et al.*, 1987)
Cement	CO_2	Ar, N_2, O_2	(Lunau and Paine, 1969)
Ceramic tile	CO_2	Ar	(Adams, 1970)
Cordierite ($2MgO$–$2Al_2O_3$–$5SiO_2$)	CO_2	N/S	(Pilletteri *et al.*, 1990)
Silicon nitride	CO_2	Air	(Lei and Li, 1999)
Various	CO	N/S	(Spalding *et al.*, 1993)
Various	CO_2	N/S	(Lumley, 1969)

Table 14.15 Sources of data for laser cutting of glasses: N/S not specified

Material	Laser	Assist gas	Reference
Glass	CO_2	N/S	(Chui, 1975)
Quartz fibre	CO_2	Air	(Dobbs *et al.*, 1994)
Silica	CO_2	Air	(Powell *et al.*, 1987)
Soda glass	CO_2	Air	(Powell *et al.*, 1987)
Soda glass	CO_2	Ar, N_2, O_2	(Lunau and Paine, 1969)

Glasses behave in a similar manner to ceramics during laser cutting. Procedures can be found in the references listed in Table 14.15.

POLYMERS

Most polymers are cut by inert gas melt shearing using air as the assist gas. The fume produced is the main hazard. It is mainly of a particulate nature with the particles about 0.25 μm in diameter. Conventional filtration equipment is normally suitable for removing such pollutants.

Thermoplastics

Thermoplastics are cut by melt shearing, with the exception of PMMA, which is highly crystalline and is cut by air-assisted vaporization, primarily to obtain transparency by allowing a thin molten layer to solidify on the cut edge. Care should be taken when cutting PVC since hydrogen chloride vapour is produced. This can be eliminated by adding ammonia to the assist gas, to form ammonium hydroxide, which is removed by normal means of ventilation. Other by-products are benzene and polycyclic aromatic hydrocarbons, which are suspected carcinogens. Procedures for laser cutting thermoplastics can be found in the references in Table 14.16.

Thermoplastics with a relatively low melting temperature typically display clean cuts with polished edges as a result of resolidified melt. The cutting performance is dependent on the molecular mass and the water content.

Table 14.16   Sources of data for laser cutting of thermoplastic polymers

Material	Laser	Assist gas	Reference
ABS	CO_2	Air	(Powell *et al.*, 1987)
LDPE	CO_2	Air	(Powell *et al.*, 1987)
Nylon	CO_2	Ar, N_2, O_2	(Lunau and Paine, 1969)
PC	CO_2	Air	(Powell *et al.*, 1987)
PP	CO_2	Air	(Powell *et al.*, 1987)
PMMA	CO_2	Air	(Powell *et al.*, 1987)
PMMA	CO_2	Ar	(Adams, 1970)
PMMA	CO_2	Ar, N_2, O_2	(Lunau and Paine, 1969)
PTFE	CO_2	Ar, N_2, O_2	(Lunau and Paine, 1969)
PVC	CO_2	Air	(Powell *et al.*, 1987)

Table 14.17   Sources of data for laser cutting of thermoset polymers

Material	Laser	Assist gas	Reference
Phenolic	CO_2	Air	(Powell *et al.*, 1987)

Table 14.18   Sources of data for laser cutting of elastomers: N/S not specified

Material	Laser	Assist gas	Reference
Natural rubber	CO_2	N/S	(Verbiest, 1973)

Thermosets

Most thermoset polymers are cut by active gas melt shearing and chemical degradation. A surface finish with a roughness of 0.2 µm can typically be achieved. There are few established procedures, but one can be found in Table 14.17.

Elastomers

Most elastomers are cut by chemical degradation using air. Both natural and synthetic rubbers can also be vaporized by a focused laser beam. Cuts can be made without the risk of distortion or stretching. Some edge charring occurs, but this can be removed by wiping. Bevel cuts can be made in seals. A reference is given in Table 14.18.

COMPOSITES

Mechanically cut composites often exhibit a poor edge quality as a result of delamination and the removal of the reinforcing phase from the matrix, leaving voids at the cut edge. The quality of laser cuts, characterized by the uniformity of the surface morphology and the extent of the HAZ, is improved

Table 14.19   Sources of data for laser cutting of metal matrix composites

Material	Laser	Assist gas	Reference
A6061/SiC fibre	CO_2	O_2	(Kagawa *et al.*, 1989)
Steel–polymer laminate	CO_2	O_2	(Haferkamp and Wendorff, 1992)

Table 14.20   Sources of data for laser cutting of ceramic matrix composites: N/S not specified

Material	Laser	Assist gas	Reference
C–SiC fibre	CO_2	Ar	(Rieck, 1990)
Glass ceramic–SiC fibre	CO_2	N/S	(Ridealgh *et al.*, 1990)
SiC–SiC fibre	CO_2	Ar	(Rieck, 1990)
Slate	CO_2	O_2, Ar, N_2	(Boutinguiza *et al.*, 2002)

when the matrix and reinforcing phase have similar thermal properties. The high pulse energy of Nd lasers is advantageous in reducing matrix melt-back, which minimizes fibre exposure.

Metal Matrix

Metal matrix composites (MMCs) are laser cut using procedures that are modified from those used for the matrix material. Thus the additional energy generated by an exothermic reaction can be used to increase cutting speed in metals for which oxidation does not present quality problems. Procedures are specific to particular MMCs; a selection can be found in Table 14.19.

Particulate reinforced MMCs have similar cutting characteristics to the matrix material. In contrast, MMCs reinforced with ceramic fibres such as SiC exhibit separation of the fibre from the matrix because of the difference in melting temperatures and thermal expansion coefficients.

Ceramic Matrix

Inert gases are normally used when laser cutting ceramic matrix composites (CMCs). Procedures are generally similar to those used for the matrix ceramic. A selection can be found in Table 14.20. However, fibre-reinforced CMCs are difficult to cut because of the different thermal and mechanical properties of the matrix and reinforcement. Water jet cutting is often preferred.

Polymer Matrix

Organic/organic polymer matrix composites (PMCs) can be laser cut relatively easily because of the similarities in thermal properties of the matrix and reinforcement. Inorganic/organic combinations are more difficult because of the wide disparity between properties of the components. Graphite-reinforced materials are particularly difficult to cut since graphite must be heated to 3600°C for vaporization, and heat is transferred rapidly to the matrix producing a wide HAZ. Boron–epoxy composites also present problems because the power density required to cut the boron vaporizes the epoxy producing

Table 14.21 Sources of data for laser cutting of polymer matrix composites: N/S not specified

Material	Laser	Assist gas	Reference
Aramid FRP	CO_2	Ar	(Rieck, 1990)
Aramid FRP (Kevlar)	CO_2	N_2	(Jovane *et al.*, 1993)
Aramid FRP	CO_2	Inert	(Di Ilio *et al.*, 1987)
Bond paper	CO_2	Air	(Powell *et al.*, 1987)
CFRP	CO_2	Ar	(Rieck, 1990)
Corrugated cardboard	CO_2	Air	(Powell *et al.*, 1987)
Deal	CO_2	Ar, N_2, O_2	(Lunau and Paine, 1969)
Fibreboard	CO_2	Air	(Powell *et al.*, 1987)
Fibre-reinforced polyesters	CO_2	Inert	(Tagliaferri *et al.*, 1985)
GFRP	CO_2	Ar	(Rieck, 1990)
GFRP	CO_2	N_2	(Di Iorio *et al.*, 1987)
Hardboard	CO_2	Air	(Powell *et al.*, 1987)
Hardwood	CO_2	Ar	(Adams, 1970)
Leather	CO_2	Ar, N_2, O_2	(Lunau and Paine, 1969)
Newsprint	CO_2	Air	(Powell *et al.*, 1987)
Oak	CO_2	Air	(Powell *et al.*, 1987)
Oak	CO_2	Ar, N_2, O_2	(Lunau and Paine, 1969)
Pine	CO_2	Air	(Powell *et al.*, 1987)
Plywood	CO_2	Air	(Powell *et al.*, 1987)
Plywood	CO_2	Ar	(Adams, 1970)
PMC	CO_2	N/S	(Flaum and Karlsson, 1987)
RRIM–polyurethane	CO_2	N/S	(Nuss *et al.*, 1988)
Softwood	CO_2	Ar	(Adams, 1970)
Teak	N/S	Ar, N_2, O_2	(Lunau and Paine, 1969)
Woods	CO_2	Various	(Merchant, 1995)

a frayed, charred edge of unacceptable quality. Fibre/resin separation is a characteristic of laser-cut carbon fibre reinforced composites. Laser cutting may also create hazardous by-products – aromatic compounds and polycyclic hydrocarbons have been identified. Water jet cutting is therefore a more suitable method. Procedures for laser cutting of organic/organic PMCs can be found in Table 14.21.

Wood is cut by chemical degradation, which may produce suspected carcinogens such as acrolein, acetaldehyde, formaldehyde and polycyclic aromatic hydrocarbons. Charring occurs during cutting, particularly in woods with a relatively low water content. The cutting speed is inversely proportional to the density of the wood.

LASER CUTTING DIAGRAMS

PROCESS OVERVIEW

Melt Shearing

The heat source is modelled as a fully penetrating line. Equation (E3.1) in Appendix E describes lateral heat flow from a stationary line heat source within an infinite solid. This solution is fitted to the cutting geometry as follows. The vapour cavity is modelled as a cylinder of diameter equal to the beam radius r_B and length l (the plate thickness), moving at a speed v. The volume of material vaporized in one second is then $L_v r_B l v$. The molten kerf is modelled as a cylinder of length l with an aspect ratio (depth/width) of 12, moving at a speed v. The volume of material melted in one second is then $l^2 v / 12$.

For metals and alloys, the temperature at the fusion line is T_m. The edge of the heat affected zone, at which the temperature has fallen to the initial temperature, T_0, is assumed to extend a radial distance $l/6$ from the heat source. Therefore in equation (E3.1), $T_1 = T_m$ where $r_1 = l/12$, and $T_2 = T_0$ at $r_2 = l/6$. An energy balance between the absorbed power and the power used for melting and heating gives

$$Aq = r_B L_v l v + \frac{l^2}{12} v L_m + \frac{2\pi \lambda l (T_m - T_0)}{\ln(2)} \tag{14.2}$$

where L_v is the volumetric latent heat of vaporization ($J\,m^{-3}$), L_m is the volumetric latent heat of melting ($J\,m^{-3}$), and λ is the thermal conductivity ($J\,s^{-1}\,m^{-1}\,K^{-1}$). In terms of the dimensionless variables described in Chapter 7, this becomes

$$q^* = l^* v^* L_v^* + \frac{l^{*2} v^* L_m^*}{12} + \frac{2\pi l^*}{\ln(2)} \tag{14.3}$$

where $q^* = Aq/[r_B \lambda (T_m - T_0)]$, $l^* = l/r_B$, $v^* = v r_B / a$, $L_m^* = L_m/[\rho c (T_m - T_0)]$ and $a = \lambda/(\rho c)$, where λ is the thermal conductivity of the material ($J\,s^{-1}\,m^{-1}\,K^{-1}$). The value of L_m^* lies around 0.4 for metals and alloys (Richard's rule).

Equation (14.3) gives an explicit relationship between q^* and v^*, for a given normalized plate thickness, l^*, which is illustrated in Fig. 14.8.

Figure 14.8 can be used to describe a range of laser melt shearing conditions for metals and alloys and is useful in understanding and displaying the relationships between the process variables of laser cutting. It can also be converted into a practical diagram with axes of principal process variables by substituting appropriate values for material properties and beam size.

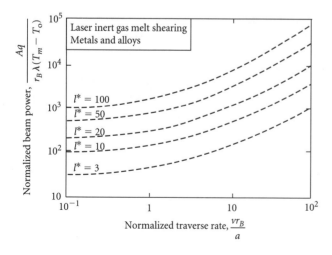

Figure 14.8 Normalized model-based overview of laser inert gas melt shearing of metals and alloys, showing contours of constant normalized plate thickness, l^* (broken lines)

PARAMETER SELECTION

Melt Shearing

For a given material and set of processing parameters, the only unknown in equation (14.2) is the absorptivity, A. This can be calculated from an experimentally determined calibration. Oxygen-assisted melt shearing of a C–Mn steel of thickness 4 mm is found to require 2.5 kW of incident CO_2 laser beam power and a traverse rate of 4 m min^{-1}. Equation (14.2) may then be used to construct a diagram with axes of incident beam power, q, and traverse rate, v, showing contours of constant plate thickness, l, Fig. 14.9. The calibration data point is indicated by a star. The following values are used as material properties for C–Mn steel (Table D.1, Appendix D): $\rho = 7764$ kg m^{-3}; $T_m = 1800$ K; $T_0 = 298$ K; $L_m = 270$ kJ kg^{-1}; and $L_v = 6.05$ MJ kg^{-1}.

Literature data indicate that when oxygen is used for cutting C–Mn steels, the exothermic reaction contributes about the same amount of energy to the process as the laser beam. The left-hand ordinate of Fig. 14.9 refers to the incident power for inert gas melt shearing of C–Mn steels, while the right-hand ordinate is shifted down by 50% and refers to the incident power required for oxygen gas melt shearing.

Figure 14.9 can be used to select an initial set of processing parameters for cutting procedure development. It may be used to extrapolate data to predict cutting parameters for conditions that lie outside current operating windows. The effect of the focused beam size on cutting performance can be established by substituting an appropriate beam radius into equation (14.2). Changes in the geometry of the kerf can be accommodated by modifying the aspect ratio used to establish equation (14.2).

The calibration data point technique can also be used to generate contours of processing conditions for materials other than C–Mn steels by substituting the appropriate material properties (Table D.1, Appendix D) into equation (14.2), as illustrated in Fig. 14.10. The contours for inert gas melt shearing of various materials shown in Fig. 14.10 are generated from a titanium cut data point by assuming that process mechanism and processing parameters are the same, and substituting the appropriate material properties. The calibration data point shown in Fig. 14.10 refers to a cut made in 2 mm titanium using

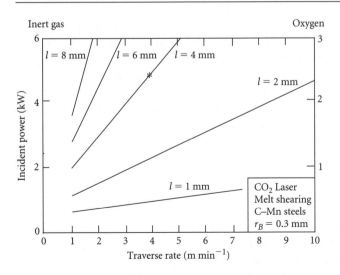

Figure 14.9 Parameter selection diagram for laser melt shearing of C–Mn steels using a focused beam of radius 0.3 mm, showing model-based contours of constant plate thickness calibrated using an experimental data point (*)

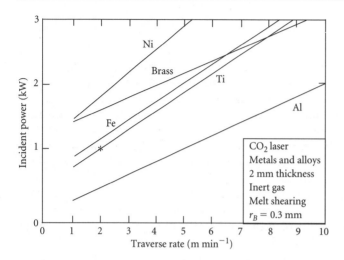

Figure 14.10 Parameter selection diagram for inert gas melt shearing of various metals and alloys of thickness 2 mm, showing model-based contours of constant plate thickness calibrated using an experimental data point established from a laser cut in titanium (*)

1 kW and a speed of $2\,\text{m}\,\text{min}^{-1}$. This may not be valid with all materials and processing geometries, but the information contained in Fig. 14.10 provides a useful initial set of processing parameters for procedure development, which can be refined through experimentation.

INDUSTRIAL APPLICATION

The benefits of laser cutting have been applied in many industrial sectors with a wide range of engineering materials. Increases in productivity are such that the laser normally pays for itself within one year. Improvements to edge quality open up new fields of application – laser welding of laser-cut parts is increasing rapidly. But the process also provides flexibility for prototype production and cutting of custom holes at a late stage in production. It can also be adapted to a method of rapid prototyping that involves cutting sections of an object and bonding them to create a model. Modern laser cutting centres come complete with CAD facilities for optimizing the nesting of parts for maximum material usage. Various industrial applications of laser cutting are given in Table 14.22.

DIEBOARDS

A dieboard is a flat, strong fixture that holds cutting blades used for cutting and creasing of materials for carton boxes and various forms of corrugated containers. The board is typically manufactured from 16 mm thick laminated maple plywood. The dieboard was formerly made by a craftsman who laid out the design and cut slots around 0.7 mm in width with a jigsaw. Sharp cutting blades or blunt creasing blades were then inserted in the appropriate positions, secured by friction.

Laser cutting provides a cost-effective, high quality alternative. Dieboard slotting became the first industrial application of laser cutting, and is described in Chapter 2. The beam produces the uniform, perpendicular, parallel-sided slots required to hold the blades securely. A tolerance of about 0.05 mm can be achieved. The accuracy is reflected in a reduction in the time needed to set the blades, but the

Table 14.22   Industrial applications of laser cutting

Industry sector	Product	Material	Reference
Aerospace	Duct subassemblies	Metals and alloys	(Maher, 1997)
	Engine strut and nacelle tooling	Steels	(Maher, 1997)
Automotive	Volvo V70 and S80 chassis	Steel	(Roos and Johansson, 2000)
	Prototype Audi A8 and A12 bodywork	Al alloys	(Harris, 1999)
	Hydroformed structures	Steel	(Vendramini, 1999)
Domestic goods	Advertising signs	Various	(Powell, 1990)
	Suits	Textiles	(Roux, 1989)
	Appliqués	Embroidery	(Belforte, 1995)
Forestry	Tree trimming	Woods	(Malachowski, 1985)
Machinery	Bushing	Polyurethane	(Belforte, 2002)
	Frames	A36 steel	(Irving, 1994)
Material production	Screen aperture	Sugar	(Atherton *et al.*, 1990)
	Plates	Steel	(Irving, 1994)
Salvaging	Various	Steels	(Hull *et al.*, 2000)
Shipbuilding	Structural sections	C–Mn steels	(Perryman, 1996)
	Lightweight constructions	C–Mn steels	(Anon., 2001)
	Various	C–Mn steels	(Olsen *et al.*, 2003)

main benefit is the repeatability in the positions of cuts and creases, which allows folding and glueing machines to run at rates of more than 400 packs per minute.

PROFILING

Figure 14.11 illustrates the narrow kerf width, flame-cut edge and accuracy with which PMMA can be laser cut. Modern CAD systems enable profiles to be cut efficiently from a wide range of engineering materials.

HYDROFORMED AUTOMOBILE PARTS

The first car ever constructed with prefabricated laser-cut metal parts was the Ford Capri II. Today there is great interest in the use of robot-mounted fibreoptic beam delivery from an Nd:YAG laser for cutting features in hydroformed parts, Fig. 14.12. The hydroforming process, in which tubes and other hollow members are produced by forcing a fluid into a preform, finds many uses in modern automobiles. The first application to raise noticeable interest was the rear axle suspension used in the BMW 5 series, introduced in 1994. Hydroformed members for vehicle frames, engine cradles, roof pillars and suspensions are popular, particularly in trucks and sport utility vehicles (SUVs).

Previously, members were produced from folded and spot-welded assemblies that were punched with conventional tooling before forming. Hydroformed components possess a high stiffness to weight ratio, and can be formed in many geometries. The stand-off distance required is maintained during

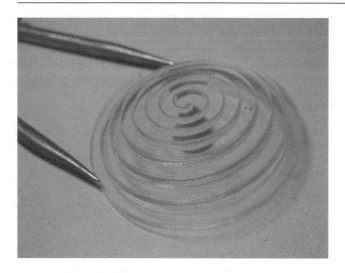

Figure 14.11 Spiral cut in PMMA using a CO_2 laser. (Source: Pertti Kokko, Lappeenranta University of Technology, Finland)

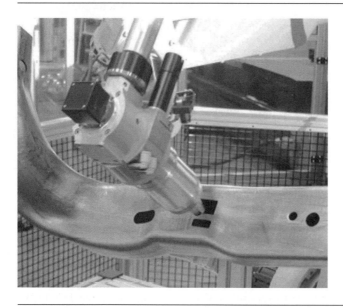

Figure 14.12 Nd:YAG laser cutting of features in a hydroformed tube (Source: Mo Naeem, GSILumonics, Rugby, UK)

the three-dimensional cutting operation through the use of a capacitive height sensor. The cutting process is flexible, and can be programmed to profile tailored option holes, short run production, and prototypes. In addition, access from one side only is required, which enables closed tubes to be cut. In order to maximize the potential benefits of hydroform technology, components must be redesigned for specific applications, and not merely substituted for existing parts.

SHIP PLATES

Vosper Thorneycroft (Southampton, UK) was probably the first shipyard in Europe to install a large CO_2 laser system to accommodate all their steel plate cutting requirements. Two workstations were installed, each capable of accepting a plate measuring 10×2.5 m. Laser cutting brought benefits of increased plate size, thickness and accuracy, which led to improved efficiency and productivity compared with plasma cutting. In addition laser cutting provides clean, accurate cut edges in both a square and bevel configuration. Low distortion results in minimal secondary part processing. All circular holes down to 3 mm in diameter are laser cut, rather than drilled or punched.

The most significant rewards of high accuracy in the cutting and marking processes are gained downstream. Parts fit together more closely, providing good joints that can be welded with the minimum of weld metal, which leads to a reduction in distortion. In addition to cutting, by using 5% of the laser power parts can be marked with part numbers, datums, stiffener positions and flanging lines. A corvette fabricated using laser-cut sections is shown in Fig. 2.6 (Chapter 2).

SUMMARY AND CONCLUSIONS

Five distinct laser cutting mechanisms can be identified: inert gas melt shearing; active gas melt shearing; vaporization; chemical degradation; and scribing. Of these, melt shearing using inert or active gases is the most widely used in industrial production. Costs can be quantified relatively easily, enabling economic and technical comparisons to be made with conventional cutting methods. Practical processing procedures and production data have been generated for many materials, which are available to aid in procedure design. The process can be modelled relatively easily by using simple energy balances that can be calibrated to existing data. The models form the basis of various laser cutting diagrams, which aid in the selection of processing parameters. Industrial applications are found in industry sectors ranging from microsurgery to decommissioning and salvaging.

The productivity and cut quality available from modern industrial laser cutting centres allow a typical job shop to offer economical cutting with a single relatively cheap laser, with short turn-around times, in thickness ranges between a few millimetres in metals and alloys to around 30 mm in polymers and wood. The benefits are many: high quality; flexibility; high productivity; and the favourable ergonomics of the process. Cutting is one of the few laser-based processes that can be substituted directly for a conventional fabrication technique. The main drawback is that laser cutting cannot compete with the cycle times of the CNC turret punch press in notching and hole punching, and does not compete with the cost of blanking parts on a punch press in a production mode. These drawbacks have largely been overcome with the introduction of hybrid cutting centres in which a laser is used for complex profiles or short production runs, while a punch press takes care of repetitive cutting of regular geometries in high volumes. Developments are being made in nozzle design for increased cutting speed and cut thickness. Improved quality control is being achieved through on-line monitoring of process characteristics from emitted signals, which are likely to lead to greater automation and an increase in the range of applications.

FURTHER READING

Caristan, C. (2003). *Laser Cutting Guide for Manufacturing*. Dearborn: Society of Manufacturing Engineers.

Powell, J. (1998). *CO₂ Laser Cutting*. 2nd ed. London: Springer.

Powell, J. (1999). *LIA Guide to Laser Cutting*. Orlando: Laser Institute of America.

VDI Technologiezentrum (1993). *Cutting with CO₂ Lasers*. Düsseldorf: VDI Verlag. (In German.)

EXERCISES

1. Describe the five principal mechanisms of laser cutting, and select the types of engineering material they are most suited to.

2. Describe the effects of: (i) an increase in beam power; (ii) a change from nitrogen to oxygen assist gas; and (iii) a change from a CO_2 laser beam to an Nd:YAG laser beam on the speed and edge quality that can be achieved with laser cutting of C–Mn steels.

3. Estimate the increase in cutting speed that may be achieved in C–Mn steel 4 mm in thickness by using oxygen in preference to nitrogen as the process gas with an incident laser beam power of 2 kW. Disregard any effects on edge quality.

4. A sheet of titanium of thickness 2 mm is cut at $2\,\text{m}\,\text{min}^{-1}$ with an incident beam power of 1 kW and a beam radius of 0.3 mm using an inert gas cutting mechanism. Estimate the incident power required to cut nickel of thickness 1 mm at a speed of $5\,\text{m}\,\text{min}^{-1}$. Assume that the cutting mechanism and all other processing parameters are the same as those used for cutting titanium.

5. Describe the form of the kerf produced when a linearly polarized laser beam is used to cut a circle in a sheet of flat C–Mn sheet. Assume that the direction of polarization is aligned with the beam traverse direction at the start of the cut.

6. Suggest appropriate lasers and cutting mechanisms for the following materials: (a) C–Mn steel; (b) polyethylene; (c) alumina; (d) quartz glass; and (e) a composite of titanium reinforced with silicon carbide particles. Assume that the cutting speed is to be maximized. Justify your choices.

7. Summarize the advantages and limitations of cutting three-dimensional components by using Nd:YAG laser light delivered via a fibre optic to a robot-mounted beam delivery head, in comparison with CO_2 laser cutting.

BIBLIOGRAPHY

Abdullah, H.A., Chatwin, C.R. and Huang, M.-Y. (1994). Performance of an optical stand-off control system for laser materials processing. *Optics and Lasers in Engineering*, **21**, 165–180.

Adams, M.J. (1970). Introduction to gas jet laser cutting. *Metal Construction and British Welding Journal*, **2**, 1–8.

Alfillé, J.P., Pilot, G. and de Prunelé, D. (1996). New pulsed YAG laser performances in cutting thick metallic materials for nuclear applications. In: Osborne, M.R. and Sayegh, G. eds *Proceedings of the Conference High Power Lasers: Applications and Emerging Applications.* Vol. 2789. SPIE: Bellingham. pp. 134–144.

Anon. (2001). Shipbuilding with laser lightweight construction. *Bander Bleche Rohre* (Germany), **42**, (7–8), 12–17.

Arai, T. and Riches, S. (1997). Thick plate cutting with spinning laser beam. In: Fabbro, R., Kar, A. and Matsunawa, A. eds *Proceedings of the Laser Materials Processing Conference (ICALEO '97)*. Orlando: Laser Institute of America. pp. B19–B26.

Atherton, P.G., Brandt, M., Crane, K.C.A. and Noble, A.G. (1990). Laser-cut screens for continuous centrifugals. *International Sugar Journal*, **92**, 243–248.

Belforte, D.A. (1995). Evolution of an embroidery cutter. *Industrial Laser Review*, **10**, (2), 15.

Belforte, D.A. (2002). Cutting large-diameter non-metal parts. *Industrial Laser Solutions*, **17**, (4), 3.

Birkett, F.N., Herbert, D.P. and Powell, J. (1985). Prevention of dross attachment during laser cutting. In: Kimmitt, M.F. ed. *Proceedings of the 2nd International Conference Lasers in Manufacturing (LIM-2)*, 26–28 March 1985, Birmingham, UK. Bedford: IFS (Publications) Ltd. pp. 63–66.

Bod, D., Brasier R.E. and Parks, J. (1969). A powerful CO_2 cutting tool. *Laser Focus*, August, 36–38.

Boutinguiza, M., Pou, J., Lusquiños, F., Quintero, F., Soto, R., Pérez-Amor, M., Watkins, K. and Steen, W.M. (2002) CO_2 laser cutting of slate. *Optics and Lasers in Engineering*, **37**, (1), 15–25.

Carroll, D.L. and Rothenflue, J.A. (1997a). Experimental study of cutting thick aluminum and steel with a chemical oxygen–iodine laser using an N_2 or O_2 gas assist. In: Baker, H.J. ed. *Proceedings of the XI International Symposium on Gas Flow and Chemical Lasers and High Power Laser Conference*, 25–30 August 1996, Edinburgh, UK. Bellingham: SPIE. Vol. 3092. pp. 758–763.

Carroll, D.L. and Rothenflue, J.A. (1997b). Experimental study of cutting thick aluminium and steel with a chemical oxygen–iodine laser using an N_2 or O_2 gas assist. *Journal of Laser Applications*, **9**, 119–128.

Carroll, D.L., Rothenflue, J.A., Kar, A. and Latham, W.L. (1996). Experimental analysis of the materials processing performance of a chemical oxygen-iodine laser (COIL) In: Duley, W., Shibata, K. and Poprawe, R. eds *Proceedings of the Laser Materials Processing Conference (ICALEO '96)*. Orlando: Laser Institute of America. pp. 19–27.

Churchill, R.J. (1983). Laser cutting of amorphous alloys. In: Wright, P.K. ed. *Proceedings of the 11th Conference on Production Research and Technology*, 21–23 May 1984, Pittsburgh, PA, USA. Dearborn: Society of Manufacturing Engineers. pp. 226–232.

Churchill, R.J., Glass, J.M., Zich, R.L. and Groger, H.P. (1989). Nd:YAG laser cutting of metallic glass motor laminations. In Dornfeld, D.A. ed. *Advances in Manufacturing Systems Integration and Processes: Fifteenth Conference on Production Research and Technology*, 9–13 January 1989, Berkeley, CA, USA. Dearborn: Society of Manufacturing Engineers. pp. 631–636.

Chui, G.K. (1975). Laser cutting of hot glass. *American Ceramic Society Bulletin*, **54**, 515–518.

Daurelio, G. (1987). Copper sheets laser cutting: a new goal on laser material processing. In: Arata, Y. ed. *Proceedings of the Conference Laser Advanced Materials Processing (LAMP '87)*, 21–23 May 1987, Osaka, Japan. Osaka, Japan: High Temperature Society of Japan and Japan Laser Processing Society. pp. 261–266.

Di Ilio, A., Tagliaferri, V. and Di Iorio, I. (1987). Laser cutting of aramid FRP. In: Arata, Y. ed. *Proceedings of the Conference Laser Advanced Materials Processing (LAMP '87)*, 21–23 May 1987, Osaka, Japan. Osaka, Japan: High Temperature Society of Japan and Japan Laser Processing Society. pp. 291–296.

Di Iorio, I., Tagliaferri, V. and Di Ilio, A.M. (1987). Cut edge quality of GFRP by pulsed laser: laser–material interaction analysis. In: Arata, Y. ed. *Proceedings of the Conference Laser Advanced Materials Processing (LAMP '87)*, 21–23 May 1987, Osaka, Japan. Osaka, Japan: High Temperature Society of Japan and Japan Laser Processing Society. pp. 279–284.

Dobbs, R., Bishop, P. and Minardi, A. (1994). Laser cutting of fibrous quartz insulation materials. *Journal of Engineering Materials Technology*, **116**, 539–544.

Dorn, L., Lee, K.-L., Munasinghe, N. and Handke, H. (1994). Nd:YAG-laser cutting of amorphous alloys. In: Mordike, B.L. ed. *Proceedings of the 5th European Conference Laser Treatment of Materials (ECLAT '94)*, 26–27 September, 1994, Bremen-Vegesack, Germany. Oberursel: DGM Informationsgesellschaft. pp. 447–455.

Flaum, M. and Karlsson, T. (1987). Cutting of fiber-reinforced polymers with a cw CO_2 laser. In: Kreutz, E.A., Quenzer, A. and Schuöcker, D. eds *Proceedings of the Conference High Power Lasers: Sources, Laser–Material Interactions, High Excitations, and Fast Dynamics*. Vol. 801. Bellingham: SPIE. pp. 142–149.

Gabzdyl, J. (1989). Quality and performance in laser cutting. In: *Proceedings of the 2nd Nordic Laser Materials Processing Conference (NOLAMP 2)*, 30–31 August 1989, Luleå, Sweden. Luleå: Luleå University of Technology. pp. 161–168.

Geiger, M., Schuberth, S. and Hutfless, J. (1996). CO_2 laser beam sawing of thick sheet metal with adaptive optics. *Welding in the World*, **37**, (1), 5–11.

Ghosh, S., Badgujar B.P. and Goswami, G.L. (1996). Parametric studies of cutting zircaloy-2 sheets with a laser beam. *Journal of Laser Applications*, **8**, 143–148.

Graydon, O. (1999). Two foci prove better than one. *Opto and Laser Europe*, Issue 64, July, 20–21.

Haferkamp, H. and Wendorff, H.-P. (1992). Cutting steel–plastic laminate systems with a carbon dioxide laser. *Welding and Cutting*, **44**, (9), 496–499. (In German.)

Hansmann, M., Decker, I. and Ruge, J. (1989). Registration of melt flow during laser beam cutting. In: Schuöcker, D. ed. *Proceedings of the 7th International Symposium Gas Flow and Chemical Lasers*. Vol. 1031. Bellingham: SPIE. pp. 582–585.

Harris, R.A. (1999). Audi prototypes cut with multi-axis laser. *Welding Design and Fabrication*, **72**, (6), 46–47.

Huang, M.Y. and Chatwin, C.R. (1994). A knowledge-based adaptive control environment for an industrial laser cutting system. *Optics and Lasers in Engineering*, **21**, 273–295.

Hsu, M.J. and Molian, P.A. (1992). Dual gas-jet laser cutting of thick stainless steel. In: Matsunawa, A. and Katayama, S. eds *Proceedings of the Conference Laser Advanced Materials Processing (LAMP '92)*, Osaka, Japan: High Temperature Society of Japan and Japan Laser Processing Society. pp. 601–606.

Hsu, M.J. and Molian, P.A. (1994). Thermochemical modelling in CO_2 laser cutting of carbon steel. *Journal of Materials Science*, **29**, 5607–5611.

Hull, R.J., Lander, M.L. and Eric, J.J. (2000). Experiments in laser cutting of thick steel sections using a 100-kW CO_2 laser. In: Hügel, H., Matsunawa, A. and Mazumder, J. eds *Proceedings of* the Laser Materials Processing Conference (ICALEO 2000). Orlando: Laser Institute of America. pp. B78–B86.

Ilvarasan, P.M. and Molian, P.A. (1995). Laser cutting of thick sectioned steels using gas flow impingement on the erosion front. *Journal of Laser Applications*, **7**, (4), 199–209.

Irving, R. (1994). Laser beam cutting of steel plate goes on-stream in Georgia. *Welding Journal*, **63**, (11), 33–36.

ISO (2002). ISO 9013:2002. *Thermal cutting. Classification of thermal cuts. Geometrical product specification and quality tolerances.* Geneva: International Organization for Standardization.

Ivarson, A., Powell, J., Kamalu, J. and Magnusson, C. (1994). The oxidation dynamics of laser cutting of mild steel and the generation of striations on the cut edge. *Journal of Materials Processing Technology*, **40**, 359–374.

Jones, M. and Chen, X. (1999). Thick section cutting with a high brightness solid state laser. In: Denney, P., Miyamoto, I. and Watkins, K. eds *Proceedings of the Laser Materials Processing Conference (ICALEO '99)*. Orlando: Laser Institute of America. pp. A158–A165.

Jovane, F., Di Ilio, A., Tagliaferri, V. and Veniali, F. (1993). Laser machining of composite materials. In: Martellucci, S., Chester, A.N. and Scheggi, A.M. eds *Proceedings of the NATO ASI Laser Applications for Mechanical Industry*, 4–16 April 1992, Erice, Italy. Dordrecht: Kluwer Academic Publishers. pp. 115–129.

Juckenath, B., Bergmann, H.W., Geiger, M. and Kupfer, R. (1990). Cutting of aluminium and titanium alloys by CO_2 lasers. In: Waidelich, W. ed. *Proceedings of the 9th International Congress Laser, Optoelectronics in Engineering* (Laser 89). Berlin: Springer-Verlag. pp. 595–598.

Kagawa, Y., Utsunomiya, S. and Kogo, Y. (1989). Laser cutting of CVD-SiC fibre/A6061 composite. *Journal of Materials Science Letters*, **8**, 681–683.

Kamalu, J.N. (1986). Electric arc augmentation of the laser cutting of mild steel. In: Cheo, P.K. ed. *Proceedings of the Conference Manufacturing Applications of Lasers*. Vol. 621. Bellingham: SPIE. pp. 49–54.

Kar, A., Scott, J.E. and Latham, W.P. (1996). Theoretical and experimental studies of thick-section cutting with a chemical oxygen–iodine laser (COIL). *Journal of Laser Applications*, **8**, 125–133.

Keilmann, F., Hack, R. and Dausinger, F. (1992). Polarisation gives lasers a new cutting edge. *Opto and Laser Europe*, (September), 20–24.

Lei, H., and Li, L. (1999). A study of laser cutting engineering ceramics. *Optics and Laser Technology*, **31**, 531–538.

Lumley, R.M. (1969). Controlled separation of brittle materials using a laser. *American Ceramic Society Bulletin*, **48**, 850–854.

Lunau, F.W. and Paine, E.W. (1969). CO_2 laser cutting. *Welding and Metal Fabrication*, **27**, (1), 3–8.

Maher, W. (1997). Laser applications in manufacturing at the Boeing Company. *Industrial Laser Review*, **12**, (8), 7–9.

Malachowski, M.J. (1985). Trimming trees using a high power CO_2 laser: machining of green and dry wood. In: Mazumder, J. ed. *Proceedings of the Materials Processing Symposium (ICALEO '84)*. Toledo: Laser Institute of America. pp. 185–192.

Merchant, V.E. (1995). The influence of cutting assist gas and pressure on the laser cutting of lumbar products. In: Mazumder, J., Matsunawa, A. and Magnusson, C. eds *Proceedings of the Laser Materials Processing Conference (ICALEO '95)*. Orlando: Laser Institute of America. pp. 128–137.

Miyamoto, I. and Maruo, H. (1991). The mechanism of laser cutting. *Welding in the World*, 29, (9/10), 283–294.

Molian, P.A. (1993). Dual-beam CO_2 laser cutting of thick metallic materials. *Journal of Materials Science*, **28**, 1738–1748.

Nielsen, S.E. and Ellis, N. (2002). Dual-focus laser cutting. *Welding in the World*, **46**, (3/4), 33–40.

Nuss, R., Müller, R. and Geiger, M. (1988). Laser cutting of RRIM-polyurethane components in comparison with other cutting technologies. In: Hügel, H. ed. *Proceedings of the 5th International Conference Lasers in Manufacturing (LIM-5)*, 13–14 September 1988, Stuttgart, Germany. pp. 47–57.

Olsen, F.O., Emmel, A. and Bergmann, H.W. (1989). Contribution to oxygen assisted CO_2 laser cutting. In: Steen, W.M. ed. *Proceedings of the 6th International Conference Lasers in Manufacturing (LIM-6)*. Bedford: IFS Publications/Berlin: Springer-Verlag. pp. 67–79.

Olsen, F., Juhl, T.W. and Nielsen, J.S. (2003). Recent results in high power CO_2-laser cutting for shipbuilding industry. In: Chen, X. and Duley, W. eds *Proceedings of the Laser Materials Processing Conference (ICALEO 2003)*. Orlando: Laser Institute of America. 8p.

O'Neill, W.O. and Gabzdyl, J.T. (2000). New developments in laser-assisted oxygen cutting. *Optics and Lasers in Engineering*, **34**, 355–367.

Perryman, I. (1996). Laser cuts time to completion at Vosper Thorneycroft. *Welding and Metal Fabrication*, **64**, (August/September), 11–12.

Petring, D., Abels, P., Beyer, E., Noldechen, W. and Preissing, K.U. (1990). In: Waidelich, W. ed. *Proceedings of the 9th International Conference Laser Optoelectronics in Engineering (Laser 89)*, Berlin: Springer-Verlag. pp. 599–604.

Pilletteri, V.J., Case, E.D. and Negas, T. (1990). Laser surface melting and cutting of cordierite substrates. *Journal of Materials Science Letters*, **9**, (2), 133–136.

Pocklington, D.N. (1989). Application of lasers to cutting copper and copper alloys. *Materials Science and Technology*, **5**, (1), 77–86.

Powell, J. (1990). CO_2 laser cutting. In: Johnson, K.I. ed. *Proceedings of the Conference Advances in Joining and Cutting Processes*, 30 October–2 November 1989, Harrogate, UK. Abington: Woodhead Publishing. pp. 376–392.

Powell, J., King, T.G., Menzies, I.A. and Frass, K. (1986). Optimisation of pulsed laser cutting of mild steels. In: Quenzer, A. ed. *Proceedings of the 3rd International Conference Lasers in Manufacturing (LIM-3)*, 3–5 June 1986, Paris, France. Bedford: IFS Publications/Berlin: Springer-Verlag. pp. 67–75.

Powell, J., Ellis, G., Menzies, I.A. and Scheyvaerts, P.F. (1987). CO_2 laser cutting of non-metallic materials. In: Steen, W.M. ed. *Proceedings of the 4th International Conference Lasers in Manufacturing (LIM-4)*, 12–14 May 1987, Birmingham, UK. Bedford: IFS Publications/Berlin: Springer-Verlag. pp. 69–82.

Powell, J. (1998). *CO_2 Laser Cutting*. 2nd ed. London: Springer.

Richerzhagen, B. (2001). The best of both worlds – laser and water jet combined in a new process: the water jet guided laser. In: Yao, Y.L., Kreutz, E.W. and Lim, G.C. eds *Proceedings of the Laser Materials Processing Conference (ICALEO 2001)*. Orlando: Laser Institute of America. pp. 1815–1824.

Ridealgh, J.A., Rawlings, R.D. and West, D.R.F. (1990). Laser cutting of glass ceramic matrix composite. *Materials Science and Technology*, **6**, (4), 395–398.

Rieck, K. (1990). CO_2 laser cutting of fibre-reinforced materials. In: Bergmann, H.W. and Kupfer, R. eds *Proceedings of the 3rd European Conference Laser Treatment of Materials (ECLAT '90)*, 17–19 September 1990, Erlangen, Germany. Coburg: Sprechsaal Publishing Group. pp. 777–788.

Roos, S.-E. and Johansson, G. (2000). Laser cutting on the production line in Volvo's chassis production. *Verkstäderna*, (6), 6–8. (In Swedish.)

Roux, R. (1989). The laser in the French clothing industry. In: Belforte, D. and Levitt, M. eds *The Industrial Laser Annual Handbook*. Tulsa: PennWell Books. pp. 129–138.

Sato, S., Takahashi, K., Saito, S., Fujioka, T., Noda, O., Kuribayashi, S., Imatake S. and Kondo, M. (1989). In: Steen, W.M. ed. *Proceedings of the 6th International Conference Lasers in Manufacturing (LIM-6)*, 10–11 May 1989, Birmingham, UK. Bedford: IFS Publications/Berlin: Springer-Verlag. pp. 31–40.

Shigematsu, I., Kozuka, T., Kanayama, K., Hirai, Y. and Nakamura, M. (1993). Cutting of TiAl intermetallic compound by CO_2 laser. *Journal of Materials Science Letters*, **12**, 1404–1407.

Spalding, I.J., Stamatakis, T., Wlodarczyk, G. and Megaw, J.H.P.C. (1993). Industrial laser developments in the UK. In: *Proceedings of the 24th Conference Plasmadynamics and Lasers*, 6–9 July, 1993, Orlando, FL, USA. Washington: American Institute of Aeronautics and Astronautics. Paper 93–3152.

Steen, W.M. (1981). Arc augmented laser processing of materials. *Journal of Applied Physics*, **51**, (11), 5636–5641.

Sullivan, A.B.J. and Houldcroft, P.T. (1967). Gas-jet laser cutting. *British Welding Journal*, **14**, (8), 443–445.

Tagliaferri, V., Di Ilio, A. and Crivelli-Visconti, I. (1985). Laser cutting of fibre-reinforced polyesters. *Composites*, **16**, (4), 317–325.

Vendramini, A. (1999). VW laser cuts hydroformed parts. *Industrial Laser Solutions*, **14**, (5), 9–11.

Verbiest, R. (1973). Laser cutting and welding. *Revue de la Soudure*, **29**, (4), 230–236. (In Flemish.)

Watanabe, T., Kobayashi, H., Suzuki, K. and Beppu, S. (1999). Cutting of thick steel with fiber-delivered high power Nd:YAG laser beam. In: Chen, X., Fujioka, T. and Matsunawa, A. eds *Proceedings of the Conference High-Power Lasers in Manufacturing*. Vol. 3888. Bellingham: SPIE. pp. 635–642.

MARKING

INTRODUCTION AND SYNOPSIS

Laser marking is a rapid, non-contact means of producing permanent high resolution images on the surface of most engineering materials. The most popular sources are pulsed CO_2, Nd:YAG and excimer lasers with average power levels of several tens of watts. The beam may be scanned or rastered over the material using computer-controlled mirrors oscillating along orthogonal axes, or projected through a mask or stencil to generate the image. Scanning provides flexibility in character generation, and allows the entire source output to be used. Mask marking is a more rapid method of producing a series of identical marks since the complete image is produced by a single pulse, at the expense of quality and versatility. Dot matrix marking, in which the beam is formed into a matrix of dots that can be turned on or off, allows characters to be generated 'on-the-fly'. Marks are made using several different thermal mechanisms, each associated with a characteristic temperature rise, which depends on the properties of the laser beam and material, and the nature of the mark required. As a rule of thumb, the CO_2 laser is generally (but not always) the more appropriate source for organic materials such as paper, wood and some polymers containing additives. The Nd:YAG laser produces a superior result on metals and alloys. Excimer lasers are used when high precision micromarking is required on brittle materials such as ceramics and glasses. An increase in the use of laser marking has been driven by the need for superior labelling for product inventories, improved consumer information on products, and the flexibility to mark a wide range of products that the laser provides.

Since various thermal mechanisms are utilized, marking cannot be categorized as a process that depends purely on heating, melting or vaporization; all three mechanisms may be used in some applications. Laser marking has little in common with conventional methods of labelling, except for the techniques that are used to manipulate the beam into the desired marking pattern. The advantages of laser marking lie in the range of materials that can be marked, the ease with which markings may be changed through changes in software, the speed of the process, the quality of the mark produced, and the low environmental impact. Five distinct mechanisms of marking can be identified, but none is understood well enough to be modelled accurately. Trial and error testing is therefore the most common means of establishing processing parameters. Around 90% of industrial laser marking is carried out for product identification in industries including domestic goods, microelectronics, automotive and aerospace – the remainder being for aesthetic applications such as decorative patterns and company logos. As the scale of marking decreases through the use of short wavelength lasers, opportunities for micromarking will grow; security marking and anti-counterfeiting features are large potential markets.

PRINCIPLES

Five principal mechanisms of marking are used in practical applications, each associated with a particular group of materials and the wavelength of the laser beam. The key to selecting the most appropriate method is to match the material with a laser beam wavelength that will be absorbed only in surface regions.

FOAMING AND MICROCRACKING

Foaming involves the formation of gas bubbles in surface regions at low temperatures. It is a common marking method for polymers. The bubbles are surrounded by molten material, which may provide sufficient contrast for the mark to be formed. If the temperature is raised slightly, the bubbles may break the surface, thus improving contrast. The laser intensity is low, since no thermal degradation or ablation is required; only the microstructure of the surface is changed.

Microcracking results from shallow surface melting (to about $20\,\mu m$). Resolidified material contains many microcracks that scatter light and provide contrast with the substrate. Glass is marked using the CO_2 laser in this manner – soda glass marks more easily than thermally resistant types. If fine detail is required, a shorter wavelength excimer laser is used.

DISCOLOURATION

Discolouration occurs when the absorbed beam energy heats the material to a temperature at which thermal degradation is induced. In polymers and woods, the mechanism is referred to as charring, and normally results in a black mark. This method does not affect the surface typography, and is mainly used on specially pigmented plastics and high alloy steel or brass. It is used when the surface must remain flat.

BLEACHING

Coloured pigments in special polymers can be changed by a photochemical reaction. Radiation of sufficiently short wavelength, such as that from an excimer laser, dissociates molecules to cause a colour change and a permanent mark.

ENGRAVING

Engraving involves localized melting or vaporization of the surface, typically to a depth of about 0.1 mm. Contrast is provided by the different optical properties of the resolidified/vaporized regions and the substrate. This mechanism is commonly used to mark thermoplastics with CO_2 lasers. Marks may be produced by rastering the beam or by direct vector engraving. Engraving is desirable when an abrasion-resistant mark is required.

The image is converted into a matrix of dots in rows in raster marking. Each dot is assigned a value, e.g. 0 representing white and 1 representing black. As the beam is scanned, it is pulsed when it encounters a dot assigned 1, creating the image line by line. The resolution is determined by the number of dots per unit length.

Vector engraving involves tracing a line around an object – the beam acts as a pen. The image is converted into a series of coordinates to direct the beam. The conversion can be done electronically, thus saving time and increasing quality.

ABLATION

Marking by ablating surface regions involves the highest temperature rise, since material must be raised above the vaporization temperature. Material is thus removed, often with little disruption to the surrounding surface. The technique may be used on homogeneous materials, in which case the mark arises as a result of the contrast between the substrate and the ablated regions. Alternatively, it may be applied to layered substrates; the ablated region reveals an underlying region that provides high contrast – a popular technique for marking polymers.

PROCESS SELECTION

CHARACTERISTICS OF LASER MARKING

Laser marking is flexible; a wide range of materials and components can be marked by selecting an appropriate laser beam. Numerically controlled beam motion provides opportunities to create marks directly from computer-generated images in an unlimited variety of patterns and fonts. Recessed locations with difficult access can be marked, as can moving objects. Laser-marked characters have uniform contrast, good resistance, high resolution and good durability.

Laser marking is a clean process. No paints, inks or acids are used, which could contaminate the product. This also avoids the need to dispose of toxic solvents. Labels are not needed – the mark is made directly on the product. There is no material distortion and no tool wear since it is a non-contact process involving low levels of power.

Productivity is high in comparison with conventional processes, especially for short production runs; changes in characters and fonts can be made through changes in software. A laser marker is readily integrated into high volume, high speed process lines.

The initial investment cost is high, but in comparison with conventional marking operations, running costs are low.

COMPETING METHODS OF MARKING

Ink jet printing is a strong competitor to laser marking, particularly when organic products that cannot be damaged by heat are involved. Dot matrix delivery of ink provides a flexible system of marking alphanumeric characters with high throughput. However, the permanence and legibility of the mark is dependent on the nature and topography of the substrate. The investment cost of ink jet printing is typically one quarter that of a transversely excited atmospheric (TEA) CO_2 laser, but the running and maintenance costs of ink jet printing can be ten times higher because of the consumable costs. Laser printing competes when high volumes are to be marked on substrates that can vary in topography and properties.

Metal stamping produces marks with high permanence that are difficult to remove without producing evidence of tampering. Dies are required, which limit the flexibility of the process if a large range of characters is to be marked. The cost of consumables is high because dies become worn. Worn dies can also cause problems for automatic vision systems. The mark is produced by plastic deformation of the substrate, which presents no problems for labels that are to be applied to components, but which may be unsuitable for direct marking. Laser marking produces less deformation.

Preprinted labels provide clear marks, but large quantities must normally be produced, resulting in high waste. Labelling is not a very flexible process if codes are changed regularly. Label making also involves an added production stage.

Chemical etching produces marks of high permanence, but relatively slowly. Since masking is often required, the flexibility of the process is also low. Of increasing concern is the environmental impact of chemicals.

Processes such as embossing, silkscreen printing and pantograph printing all suffer from high consumable costs, and are less clean operations than laser marking.

PRACTICE

The choice of a laser marking system depends on many factors: the materials to be marked, the throughput, the permanence, the flexibility and the tolerance of the substrate to marking. There are no simple rules that can be applied to all cases; each must be considered on its own merit. Additional complications are introduced by the interactions between process variables.

MATERIAL PROPERTIES

Metals and alloys are good reflectors of the far infrared light of the CO_2 laser, and so Nd:YAG or excimer lasers are normally the sources of choice for marking bare metals. The mechanism of marking is normally ablation or discolouration resulting from the deposition of metal compounds from the process gas. Alternatively, a coating can be applied to the surface, which is fused by the laser beam, leaving a permanent mark; the remaining coating is washed away. Aluminium can be anodized and coloured – marks can then be produced by ablating the anodized layer (which absorbs infrared radiation well) to reveal the substrate. The gold-coated lids of electronics packages are marked using Nd:YAG lasers. The gold is present as a thin-plated layer on nickel, which is used to coat a Kovar substrate. A black mark, with good contrast against the gold background, is produced.

Many polymers, notably polyethylene and polypropylene, absorb the infrared radiation of CO_2 and Nd:YAG lasers poorly. The absorptivity of polyolefins can be increased by mixing in pigments, typically in quantities around 0.1 vol.% Pigments such as metal oxides and mica in particle sizes about 2–3 μm in diameter improve absorptivity without degrading the quality of the mark. Suitable pigments should not change the colour or mechanical properties of the polymer, and be non-toxic if used in applications involving food packaging.

Glasses, which are used in transmissive optics with Nd:YAG laser output, cannot normally be marked using this laser. Pulsed CO_2 laser output is suitable for etching characters by ablating or microcracking the substrate. For high precision markings, the excimer laser can be used to generate microcracks in the surface without thermally disrupting the surrounding material.

Ceramics absorb light of a wide range of wavelength well, and so a variety of lasers can be used for thermal marking. However, the brittle nature of ceramics, particularly in the fused state, places limitations on the mechanisms of marking that can be used. Silicon wafers can be marked by the Nd:YAG laser at the beginning of the semiconductor manufacturing process by melting a thin surface layer without thermally affecting the substrate.

BEAM CHARACTERISTICS

The pulsed TEA CO_2 laser is the most common source for marking non-metallic consumer products. A typical set of processing parameters would be: 50–100 W average power; 1–2 μs pulse duration; and 1–20 J cm^{-2}. Slow axial flow and sealed CO_2 lasers are also used, typically with power levels in the range 10–25 W in CW mode, and 3–5 J per pulse in pulsed mode. A CO_2 laser beam can be delivered using a mask, beam manipulation, and dot matrix rastering, discussed below. Optics are relatively expensive because of the far infrared wavelength of the beam. Marks are produced principally by a thermochemical reaction (melting or vaporization). Far infrared light is particularly suitable for organic materials such as paper and other wood products, many plastics and ceramics, and for removing thin layers of paint or ink from a substrate. Metallic surfaces can be marked using absorptive coatings, and polymers marked by colour changes induced by heating.

The Nd:YAG laser can also be used in CW or pulsed operation. The average power level normally lies between 10 and 120 W, while pulses of peak energy 10 mJ and peak power 60 kW are produced in pulsed mode. Q-switching is used to develop high peak pulse energies. Marks are produced by a thermochemical reaction in a similar manner to the CO_2 laser, but because of the shorter wavelength of Nd:YAG laser light, finer detail can be marked. The beam is normally delivered via a scanning optical system. Nd:YAG laser marking is suitable for the same materials as the CO_2 laser, but the light is absorbed more strongly by metals. Frequency-doubled output at 532 nm can be produced, which is suitable for marking by photochemical reactions such as bleaching.

The excimer laser is well absorbed by polymers and glasses because of its short wavelength. The output is pulsed, with an energy in the range 150 to 400 mJ, and marking is normally achieved using masks. The longer wavelength excimer lasers (xenon chloride at 308 nm and xenon fluoride at 351 nm) are used for surface marking of plastics because most polymers absorb well at these wavelengths. The short wavelength of the argon fluoride laser (193 nm) cleanly ablates glass by a photochemical mechanism with no peripheral damage.

PROCESSING PARAMETERS

Masks

The mark may be formed by illuminating a mask, or reticule, that contains the desired pattern, as illustrated in Fig. 15.1. The mask may be a metal stencil, which contains connecting strips, or glass substrates covered with dielectrics. It may be positioned close to the workpiece, or the transmitted pattern may be focused to the desired size on the object using a lens. Mask marking is rapid if the mask does not need to be changed. Different masks can be mounted on a revolving wheel and positioned accordingly, although this reduces the marking rate. Since only a part of the beam is transmitted to make the mark, fairly high power lasers are needed. TEA CO_2 and excimer lasers are normally used in

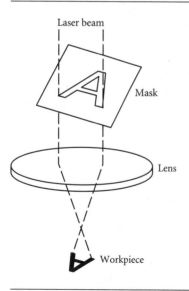

Figure 15.1   Marking using a mask

mask marking. This technique is popular for sequential coding, batch coding, and open or closed date coding, with a wide range of materials including paper, glass, plastics, coated materials and ceramics, which are marked by discolouration. Masking is also the method of choice for complex patterns that do not change – it is considerably faster than beam scanning.

Rastering

Alternatively, the laser beam can be guided using computer numerically controlled mirrors over a marking area, which is typically 100×100 mm. The beam is first expanded, to reduce divergence, and directed towards the galvanometer mirrors. It passes though a lens and is focused onto the workpiece, as illustrated in Fig. 15.2. Any desired image may be drawn using appropriate programming. Software is available in control systems, providing a flexible system capable of marking with a variety of fonts. The mark is of very high quality, but the process is slower than single mask marking. The complex optics and the necessary shielding increase the system cost. An Nd:YAG laser is normally used in CW or Q-switched mode since the narrower lines produced are suited to this technique. Raster marking is a popular replacement for acid and electro-etch marking, stamping and punching, ink jet and other printing systems. Slow axial flow and sealed CO_2 lasers can give higher throughput on non-metals, and these systems are generally cheaper than Nd:YAG systems. Scanning is typically used for serialization of plastic and ceramic products, and products requiring high quality graphics. One advantage is that the entire power of the laser beam is used for marking.

Dot Matrix

Dot matrix marking systems produce characters by producing small dots in given patterns to generate the characters required. The beam is scanned over the matrix, and pulsed when a dot is required. The pulse is generated by a rotating polygon mirror, acousto-optical device, or piezoelectric scanning. Dot matrix printing is particularly suitable for producing characters in a well-defined series, such as alphanumeric codes.

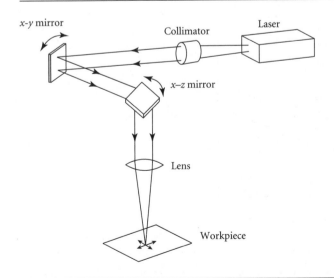

Figure 15.2   Marking using rastering mirrors

LASER MARKING DIAGRAMS

Data for the various mechanisms of modelling can be displayed on the normalized overview chart shown in Fig. 15.3. The process parameters are normalized with the thermal properties of the materials (Appendix D).

Data for the various thermal marking mechanisms cluster into appropriate regions of Fig. 15.3 with respect to the normalized surface peak temperature attained. Bleaching data (a photochemical mechanism) fall in regions of low surface temperature rise. The normalized traverse rate of all mechanisms is high – marking is a procedure that is carried out as rapidly as possible.

INDUSTRIAL APPLICATION

Marking is now the major laser application in the electronics industries. Wires, substrates, devices, boards, packages and assembled instruments are all laser marked. Parts can be marked with logos and part numbers. Barcodes can be marked onto edge connectors, silicon wafers or boards. Diamond scribing – the previous process – was less flexible and sometimes resulted in wafers breaking. Rapid marking, and soft tooling, which can readily handle ever decreasing package sizes, are the principal attractions of laser marking. Few industrial applications appear in the literature; Table 15.1 contains a selection from industry sources.

PRODUCT IDENTIFICATION CODES

Allied Distillers, who manufacture such brands as Beefeater Gin, Ballantines Whisky and Kahlua, have installed a number of pulsed CO_2 laser systems for marking manufacturing information on bottles. Between 30 and 280 bottles are marked per minute, depending on the product and bottle size. Mumm also uses laser marking on its champagne bottles.

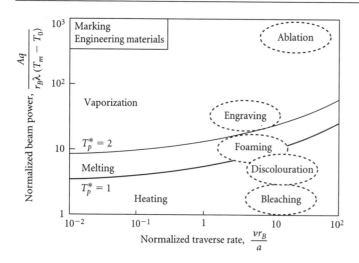

Figure 15.3   Normalized model-based diagram showing contours for the onset of surface melting and vaporization constructed from a surface heat flow model (solid lines); experimental data for various mechanisms of marking cluster into regions bounded by broken lines

Table 15.1   Industrial applications of laser marking

Industry sector	Product	Laser	Material
Aerospace	Control cables	Excimer	Polymer
Automotive	Windscreens	CO_2	Glass
	Vehicle identification numbers	CO_2	Metals
	Dashboard displays	Nd:YAG	Polymers
	Windscreen edging	CO_2	Glass
Domestic goods	Product information	CO_2	Glass
	Etching	CO_2	Granite plaques
	Packaging	CO_2	Coated paper
Electronics	Tubes and panels	Pulsed CO_2	Glass
	Integrated circuits	CO_2	Semiconductors
	Keyboards	Nd:YAG	Pigmented polymers
Medical	Syringes	Nd:YAG	Polymers
	Containers	Excimer	Glass
Shipbuilding	Identification of sections	CO_2	Structural steel

Clairol (Stamford, CT, USA), a major producer of hair care products, uses pulsed CO_2 lasers to mark painted aluminium tubes and paperboard cartons with product information. Laser marking dispensed with the large inventory of adhesive labels required in the previous process. It was also a superior process to the competing ink jet method.

At the Kraft Jacobs Suchard plant in Berlin, dot matrix pulsed CO_2 laser marking is used to code the polyethylene wrappers of chocolate bars at rates of up to 650 per minute.

TAPS

Laser marking of taps by Oras (Rauma, Finland) was described in Chapter 1; an illustration is shown in Fig. 1.2. The polymer taps are electroplated with layers of chromium and nickel. Part of this layer thickness is ablated by the beam from an 85 watt Nd:YAG laser, enabling the product logo to be written. The lasers are fully integrated into the production line, working 24 hr per day. The levers and thermostatic shower mixer covers come off the automatic chromium plating line in frames, and are then visually inspected and placed into pallets. Each pallet carries a memory chip that stores information on both the product and the pallet. When the laser marking system receives the pallet, it reads the chip and checks the production database for the graphics required for that product. As well as handling all communication with the laser, this customized software then either loads the required graphic or uses a file already held in the laser. The software also controls the position of the laser marks via the pallet positioning system and generates production statistics. The typical marking time is between 2 and 5 seconds for each item, depending on the product being marked. When the pallet is finished, a robot places the marked products in product-specific transfer boxes for removal to final assembly.

Since the introduction of laser marking, the use of conventional printing techniques has fallen from 100% to about 1% of Oras marking applications. Only the colour-printed levers are still tampo-printed, and that method will be replaced once a laser marking system is developed.

JEWELLERY

Laser-based surface ornamentation is actively being developed. Marks are produced by growing discrete oxide layers; the thickness of the oxide layer determines the characteristics of the mark. By controlling

Figure 15.4 Barcode and text marked on 'Hysol' epoxy ink (ink area 32×6 mm), using a CO_2 laser with 9 W at 1.27 m s^{-1} and a spot size of 180 microns; Hysol inks supplied by Dexter Electronic Materials Division. (Source: Geoff Shannon, Synrad Inc., Mukilteo, WA, USA)

the processing parameters carefully, a variety of colours can be obtained. An Nd:YAG laser and a marking head with galvanometer mirror beam delivery is a suitable system for such an application. A laser is also used to optically transfer two-dimensional images to three-dimensional objects. By defocusing the beam, novel aesthetic effects can be produced. The laser system parameters for marking include pulsing at 5 kHz, a traverse rate of 100 mm s^{-1}, and a 60% track overlap.

PRINTED CIRCUIT BOARD BARCODES

Printed circuit boards (PCBs) need to be identified by a code that can be read by machines, humans, or both. The reduction in size of electronic technology and the need for marking flexibility offers an excellent opportunity for laser markers. Printed circuit boards are made up from a number of layers, with the central copper layer defining the boards' conducting paths. When marking the board, this layer must not be exposed, otherwise board integrity is lost. Since this covering layer is only tenths of millimetres in thickness, controlling input energy and thus mark penetration is critical.

The contrast available by direct writing onto the board is relatively low, therefore inks are used to enhance the contrast of the marked barcodes. These inks are typically white or yellow. The laser may completely remove or darken the ink to provide contrast, as shown in Fig. 15.4.

PRINTED CIRCUIT BOARD DATA MATRIX CODES

Laser dot matrix marking combines the reliability of lasers with the flexibility of dot matrix markers. The process is cost effective and maintenance free. Like ink jet systems, it relies on product motion

Figure 15.5   A 2 mm square data matrix code containing 26 alphanumeric characters directly marked onto a printed circuit board marked with a 5 W CO_2 laser beam and a 115 micrometre spot size at 0.38 m s^{-1}. (Source: Geoff Shannon, Synrad Inc., Mukilteo, WA, USA)

to provide one axis of mark positioning, with the pulsed laser being scanned vertically to produce a column of dots. Masks are not needed for imaging code information. Multiple lines of marked information can be produced. Direct marking on to PCBs is considerably more convenient, and with the introduction of the data matrix (DM) code, this is possible. The key difference between barcodes and the data matrix code is the level of contrast required for reading. Barcodes require at least 80% contrast, hence the need for ink, whereas the level of contrast can be as low as 20% with DM codes. In addition, an order of magnitude reduction in the space and marking time of the DM code can be achieved in comparison with the barcode. In the direct marking process the laser effectively thermally bleaches the surface of the board, with negligible material removal, as illustrated in Fig. 15.5.

SUMMARY AND CONCLUSIONS

Laser marks can be produced using both thermal and athermal modes of beam–material interaction. Each thermal mechanism is characterized by a temperature rise that ranges from heating (in the case of discolouration), melting (foaming and microcracking), and vaporization (engraving). Both CO_2 and Nd:YAG lasers can be used to induce thermal modes of marking. Athermal marking involves chemical change on the level of atomic bonds, for which excimer lasers are more suited. Since the capital cost of laser marking equipment is high in comparison with conventional methods, the technique is most suitable for high productions runs, applications in which flexibility is required, and where quality must be high. The capital cost can then be offset against the relatively low running costs and the durability of the equipment. The number of industrial sectors that have adopted laser marking is testament to the advantages that it provides.

Few materials cannot be laser marked. The CO_2 laser is generally the more appropriate source for organic materials, the Nd:YAG laser produces better results with metals and alloys, and excimer lasers are used when high precision micromarking is required. Materials can be modified, e.g. by adding fillers that absorb light, to overcome problems of transparency that hinder marking. The lasers themselves may be modified to be more suitable for various classes of engineering material; the frequency of the Nd:YAG laser can be multiplied to improve its absorptivity, or the pulse energy increased through the use of Q-switching. The ability to make such modifications to materials and lasers will ensure that laser marking remains a competitive manufacturing technique.

EXERCISES

1. Name five methods of laser marking, and for each give the principal mechanism of marking, the type of laser most likely to be used, and the class of engineering material for which the method is most suitable.

2. Which type of laser and marking procedure would you choose to label varieties of citrus fruit with alphanumeric codes, without damaging the fruit?

CHAPTER 16

KEYHOLE WELDING

INTRODUCTION AND SYNOPSIS

The *conduction-limited* mode of laser welding, based on melting, was described in Chapter 13. In the *keyhole* mode of laser welding, the beam is focused to produce an incident power density at the workpiece surface that is sufficient to initiate vaporization. A narrow, deeply penetrating vapour cavity, or keyhole, is then formed by multiple internal reflection of the beam. The keyhole is surrounded by molten material. By traversing the beam relative to the workpiece, a narrow weld bead is formed with a high aspect ratio (depth/width), illustrated in Fig. 16.1. The keyhole is maintained during welding by equilibrium between the forces created by vapour pressure and those exerted by the surrounding molten material. In fully penetrating welds, the heat affected zone (HAZ) is narrow and parallel with the weld bead; in partially penetrating welds it resembles that of a conduction-limited weld.

The multikilowatt CO_2 laser is the most economical source for welding linear and rotationally symmetric joints since the beam can be manipulated relatively easily using mirrors. Complex three-dimensional parts are more easily welded using Nd:YAG laser light delivered via a robot-mounted fibreoptic cable. A compact multikilowatt diode laser head with a suitable focusing optic can be mounted on an articulated robot to create a highly flexible welding tool.

Figure 16.1 Transverse section through a full penetration keyhole weld made in a carbon–manganese steel of thickness 6 mm using an Nd:YAG laser beam

Structural and stainless steels are particularly suitable for laser keyhole welding; their physical and chemical properties favour keyhole formation and dynamic weld bead stability. Laser keyhole welding is one of the few fusion joining processes able to deliver sufficient power to overcome rapid heat flow in alloys of high thermal conductivity. Light alloys of aluminium, magnesium and titanium are also laser welded, but measures must be taken to offset their high reflectivity to infrared radiation and the vaporization of volatile alloying elements. Ceramics and glasses are poor candidates for laser welding because they are susceptible to cracking in high temperature gradients. Polymers absorb far infrared radiation well, but a stable keyhole is difficult to maintain – they are more suitable for conduction joining methods. Composite materials can be joined provided that steps are taken to avoid damage to the reinforcement, which is often non-metallic and absorbs the laser beam more readily than the matrix.

The data in Fig. 6.3 (Chapter 6) indicate that laser keyhole welding requires a power density in excess of 10^4 W mm^{-2} (corresponding to the onset of surface vaporization) with a beam interaction time between 0.1 and 0.01 s (a compromise between maintaining the keyhole and an economical rate of welding). Simple relationships between incident beam power, traverse rate and penetration exist. A common rule of thumb for structural steels is that 1 kW of power is needed per 1.5 mm of steel to be penetrated at a welding speed of 1 m min^{-1}. Analogous relationships are found with other materials. The objective of keyhole laser welding is to join materials as rapidly as possible, while meeting specified quality criteria.

In this chapter we consider the reasons for the remarkable growth in the industrial application of laser beam keyhole welding. The principles of keyhole formation, weld bead solidification, and phase transformations in the HAZ of weldable engineering materials are described. The process characteristics are compared with those of conventional joining techniques, in order to identify materials and products for which laser welding has a competitive edge. Practical procedures for laser welding of carbon–manganese steels are described, together with the most common imperfections. These are used as a baseline against which welding procedures for other engineering materials are contrasted. Laser welding diagrams are developed; these provide an overview of the process and can also be used to select processing parameters. The extent of industrial application is illustrated in a number of cases that highlight the advantages of keyhole welding and the opportunities that the process offers for novel design.

PRINCIPLES

Laser beam keyhole welding is referred to below as laser welding for conciseness.

THE KEYHOLE

The power density available from an industrial laser beam spans many orders of magnitude, attaining 10^6 W mm^{-2} in a high quality focused beam. However, such a high power density is difficult to control, and keyhole welding is normally carried out with a power density on the order of 10^4 W mm^{-2}. The surface of the material vaporizes at the point of interaction. The recoil force of vaporization from the liquid surface causes a surface depression, which develops into a deeply penetrating vapour cavity by multiple internal reflection of the beam, as illustrated in Fig. 16.2. The diameter of the keyhole is approximately that of the beam diameter.

Energy is absorbed by the material through two mechanisms, which determine the overall energy transfer efficiency. Inverse *Bremsstrahlung* absorption (transfer of energy from photons to electrons) takes place in the partially ionized plasma formed in and above the keyhole; it is the dominant mechanism at low welding speeds. Fresnel absorption by multiple reflections at the walls of the keyhole dominates at high welding speeds, and is dependent on the polarization of the beam. Plasma (ionized vapour) and plume (vaporized material) facilitate energy transfer from the beam to the material, but they also defocus the laser beam, reducing its power density.

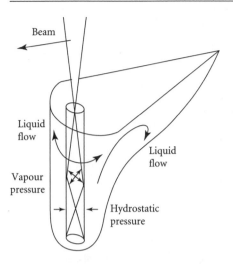

Figure 16.2   Principle of laser beam keyhole welding

The vapour cavity is surrounded by molten material. The cavity is maintained through equilibrium between opening forces arising from material ablation and plasma formation, and forces caused by the surface tension and hydrostatic pressure of the molten pool, which act to close it. As the beam and material move relative to one another, material is progressively melted at the leading edge of the molten pool and flows around the deep penetration cavity to the rear of the pool where it solidifies in a characteristic chevron pattern. The requirement to maintain this balance leads to practical minimum and maximum traverse rates for keyhole welding – excessive speed causes the keyhole to collapse, whereas insufficient speed results in a wide weld bead that sags. The shape and size of the keyhole fluctuate during welding. The molten pool temperature is considerably higher than that of a conventional arc weld. Heat is conducted into the surrounding material to produce a narrow HAZ. When the laser beam is turned off, several processes occur: the plasma inside the keyhole is extinguished; vaporization pressure decays; and the keyhole collapses through the effects of surface tension and gravity.

REGIONS OF THE WELDED ZONE

Keyhole welds exhibit two distinct regions: a fusion zone; and an HAZ. Within the HAZ, subregions can be identified; their extent depends on the material composition and the peak temperature attained during welding. The relationship between microstructure and peak temperature is illustrated for a structural steel weld in Fig. 16.3. (In some materials, notably alloys of aluminium, a third partially melted region is observed.) Each region has a specific composition, microstructure and set of properties.

In the fusion zone, material is melted and solidified rapidly. Grains grow epitaxially with the adjacent HAZ grains in a columnar morphology. Equiaxed grains may also be formed in the centre of the weld bead. The grain morphology depends on the welding speed: a high speed results in abrupt changes in grain orientation, causing parallel elongated grains to form along the weld centreline that are susceptible to solidification cracking. The weld bead microstructure and properties are essentially those of rapidly cooled cast material. Depending on the nature of the alloy and the composition of the weld metal, it might be possible to regain the properties of the base material through post-weld heat treatment.

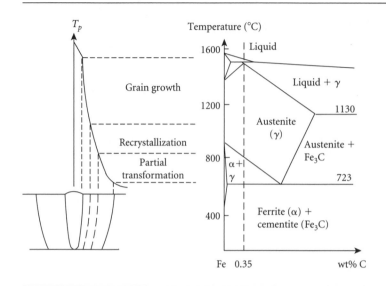

Figure 16.3 Schematic illustration of the regions of the heat affected zone in a laser beam keyhole weld in a 0.35 wt% C structural steel

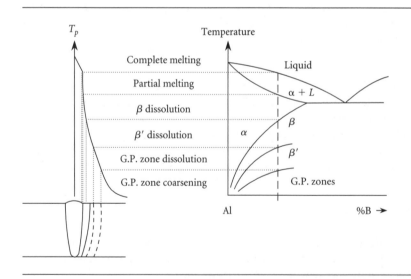

Figure 16.4 Schematic illustration of the regions of the heat affected zone in a laser beam keyhole weld in a precipitation-hardened aluminium alloy

The partially melted zone is defined as the region in which a peak temperature between the liquidus and solidus was attained during welding. Localized melting occurs, accompanied by segregation at grain boundaries. A microstructure is produced that is unable to withstand the contraction stresses generated when the weld metal solidifies, rendering it susceptible to solidification cracking. Heat treatable aluminium alloys of the 6000 series, for example, are susceptible to this type of cracking.

The microstructure at a location within the HAZ is determined by solid state reactions induced by the thermal cycle experienced. In the case of precipitation-hardened alloys, coarsening and dissolution of strengthening precipitates can occur; this is illustrated schematically in Fig. 16.4 for heat treatable aluminium alloys.

Precipitate dissolution is dominant in regions of high peak temperature; particle coarsening is observed in regions of relatively low peak temperature. Precipitate dissolution and coarsening, and the formation of incoherent particles, result in a loss of strength. Reprecipitation is possible close to the fusion zone on cooling, but a post-weld heat treatment is normally necessary to fully regain the strength of the base material. In work-hardenable alloys, strength is lost through annealing of the dislocation substructure.

Continuous cooling transformation (CCT) diagrams provide some information about the microstructure of ferrous alloys after laser welding. However, rapid laser thermal cycles may result in incomplete transformation on heating, and non-equilibrium distributions of alloying elements, which should be taken into account. Coarse phases in steels, particularly carbides, and phases that require long range diffusion for transformation, may be retained during rapid laser processing.

PROCESS SELECTION

CHARACTERISTICS OF LASER KEYHOLE WELDING

The small focused spot has a high power density, which enables a deeply penetrating weld pool to be created and maintained during welding. Through-thickness welds can be made rapidly in a single pass. However, small fitup tolerances are required to avoid sagging and other imperfections in the weld bead. These may represent a high initial setup cost, but closer production tolerances have a positive effect on the complete fabrication philosophy: the requirements for modular assembly can be achieved more easily.

The energy input during laser welding is low. A narrow weld bead is produced, with a narrow HAZ. Distortion is limited and predictable, which minimizes the need for reworking and reduces material wastage. Rapid solidification provides metallurgical advantages: segregation of embrittling elements such as sulphur and phosphorus can be reduced; and beneficial fine solidification microstructures are formed. Alloys of nickel and tantalum, which are normally susceptible to such imperfections, can therefore be laser welded. In the HAZ, the rapid thermal cycles limit grain growth and the dissolution and coarsening of strengthening precipitates – an advantage for heat treatable aluminium alloys. However, high cooling rates can result in unacceptably high hardness values in hardenable structural steels. The effect of energy input on the hardness developed in the HAZ of C–Mn steels is shown in Fig. 16.5. (The method for calculating HAZ hardness is discussed later.) For a given maximum HAZ hardness, e.g. 350 HV, the data in Fig. 16.5 indicate that the acceptable range of the steel carbon equivalent is narrower, and has a lower maximum value for laser welding than arc welding processes, which means that greater control of the processing parameters is required for laser welding.

Ranges of power density and energy for various fusion welding processes are shown in Fig. 16.6. The power density of laser welding is seen to be similar to electron beam welding, and is higher than gas and arc fusion processes (in which a keyhole is not formed). The energy input of laser welding is lower than conventional fusion processes, which is the origin of many of the advantages of the process.

A laser beam can be positioned accurately, giving precise control over weld bead location. This enables dissimilar materials to be joined, since the beam can be positioned within one material and the energy transferred to the other by conduction. Welds can also be placed close to heat-sensitive components because of the limited HAZ. Parts can be redesigned – located closer to one another, for example – since the beam can be guided into locations that would be inaccessible for a conventional welding torch.

The capital cost of a laser is several orders of magnitude higher than arc fusion welding equipment, and the running costs typically a factor of 5–10 higher. This cost penalty is offset by increased

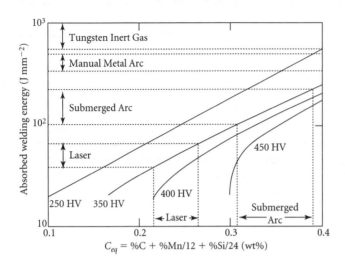

Figure 16.5 Welding energy plotted against a simple carbon equivalent for laser and arc fusion welding processes. Theoretical hardness contours are shown as solid lines. Practical limits of process operation are shown as dotted lines. The carbon equivalent that results in a maximum heat affected zone hardness below a certain value, e.g. 350 HV, has a narrower range and a lower maximum value for laser welding than arc processes.

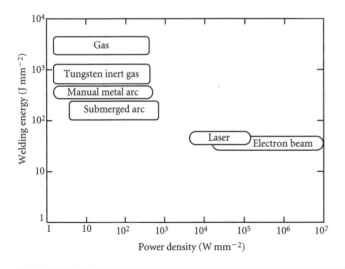

Figure 16.6 Comparison of welding energy per unit length of weld and plate thickness, and power density for various fusion welding processes

productivity, product quality and flexibility. Beam sharing or switching between workstations is often designed into a production line to maximize the beam-on time.

Opportunities also exist to reduce the total cost of a component by redesign. The component geometry can be designed based on its function rather than the limitations imposed by a conventional joining technique (such as the limited accessibility of a bulky welding torch). Cheaper or higher performance materials can be selected, which may be laser welded. Improvements can be made in the preparation and fixturing of joints. The benefits of such changes become apparent in the long term, but they require an increase in investment. Similarly, the investment in procedure development for a new process is often regarded as a drawback: quality assurance is considerably more demanding and certification of laser welding can be expensive. Approved welding procedures are available, but because of the high number of process variables they are normally restricted to limited ranges of process parameters. The methods developed later in this chapter aid in procedure development.

Laser welding can be automated easily, and is suitable for computer control, giving freedom from dependence on welder skill. Different welding configurations may be accommodated through changes in software, providing flexibility. The ease of automation enables adaptive control systems to be integrated into closed-loop production equipment.

Keyhole welding is a relatively clean process. Ionizing radiation is not produced by the interaction of a laser beam and engineering materials, and so screening for such is not required. The production of particulate fume and gases can be monitored and controlled. The laser beam is not deflected by magnetic fields. Processing can be carried out under normal atmospheric conditions.

COMPETING METHODS OF JOINING

Many conventional joining methods can be considered as competitors to laser keyhole welding. When selecting a process, a variety of factors must be taken into account: the thickness of the plate; the welding position required; the total cost of joining; the compatibility of the process with the materials to be joined; the required production rate; the ability to automate the process; the quality required; the properties of the weld; the knowledge and skill available; and the environmental impact.

Electron beam welding was developed after the Second World War, initially to weld refractory metals such as zirconium for the nuclear industry. Industrial production started in 1951. In the early 1960s most applications were found in the nuclear and aerospace industries where low distortion and the ability to weld reactive metals were crucial. In the 1970s, electron beam welding became a popular method of joining components for jet engines, later migrating into the automotive industry for gear welding. Of all the competing joining processes, the principle of electron beam welding is closest to that of laser keyhole welding; the energy source is a focused high energy beam of electrons (rather than photons), which forms the keyhole. The process is denoted high (>80 kV) or low (<80 kV) voltage. Welding takes place in a chamber under vacuum; contaminants are therefore excluded. By relaxing the degree of the vacuum, flexibility can be improved at the expense of weld bead penetration and weld quality. Electron beam welding is particularly suitable for joining very thick section parts, up to 100 mm. Since reflective alloys such as aluminium and copper contain many free electrons, they are more amenable to welding using an electron beam than a laser beam. Low distortion and high penetration can be achieved with high welding speeds. Laser keyhole welding becomes competitive under a number of conditions. The chamber size limits the part size that can be welded by an electron beam. Loading and pumping, which may take up to 5 minutes for a chamber with dimensions $1 \times 1 \times 2$ m, limits productivity – a laser beam can be switched between workstations to eliminate down time. The electron beam is deflected by magnetic and electric fields; magnetic fixturing is therefore not possible, and joints between metals with dissimilar magnetic and electrical properties can be difficult to produce. The part geometry may not allow an electron beam unit to move close enough to the joint line for welding, whereas a long focal length optic can be used to focus a laser beam, providing a large stand-off distance.

The manual metal arc process (MMA) – referred to as shielded metal arc (SMA) welding in the United States – is probably the most familiar technique of fusion welding. Most of the weld metal is provided by a flux-coated metal stick electrode, whose composition can be controlled accurately. The flux melts to protect the molten weld pool from atmospheric contamination. The length of the arc is controlled manually by a skilled operator; complex joints can therefore be welded. Common applications include welding of structural steelwork, and repair and hardfacing of constructional plant. MMA welding is suitable for joining plates between about 1.5 and 10 mm in thickness. Thick joints must be prepared in V, X or Y geometries to accommodate filler material. The equipment is relatively inexpensive. Laser welding is competitive for long welds with a regular geometry, or when the plate thickness is too large to be welded manually without joint preparation – particularly when high productivity is required.

The terms MIG (metal inert gas), MAG (metal active gas) and GMAW (gas metal arc welding) all describe the process of fusion welding using an electric arc and a shielding gas. The term GMAW is used in the United States while MIG and MAG are used in Europe and Japan. The inert shielding gas used in MIG is argon or helium, or a mixture. MIG is particularly suitable for welding aluminium and its alloys. MAG uses an active shielding gas, e.g. CO_2, or a mixture of an inert and an active gas. MAG is primarily used for welding carbon steels, low alloy steels and stainless steels. MIG and MAG both use solid filler wire, which is fed continuously into the weld pool during welding. Tubular filler wires are also manufactured, containing a mixture of flux compounds, alloying elements and metallic particles. This process, flux-cored arc welding (FCAW), provides deeper penetration and a higher deposition rate (about three times that of MMA). Arc-gas welding processes are characterized by a number of favourable properties: they can be used in manual or automatic modes; they are easy to learn; they can be used for a range of plate thicknesses by using multiple passes; they can be used with common metals and alloys; they are suitable for welding in all positions; they do not leave a flux; and productivity is high (typically four times that of MMA). However, since a hemispherical or elliptical weld bead section is formed, the penetration of a single pass is limited. Thick joints must therefore be prepared prior to welding. The energy input is relatively high, which results in distortion. Laser welding becomes competitive when a single pass, fully penetrating weld of low distortion is required.

Gas tungsten arc welding (GTAW) is the term used in the United States to describe the fusion joining process in which energy is provided by an electric arc formed between the workpiece and a non-consumable tungsten electrode. Outside the United States the process is known as tungsten inert gas (TIG) welding. Filler material may be fed into the weld pool if required. It is a cheap, well-known, high quality process for joining sheets in the range 0.1–3 mm. The main disadvantages of TIG welding are: low penetration; low productivity; a large weld bead size; and thermal distortion. Laser keyhole welding is competitive when a narrow, fully penetrating weld bead is required in unprepared sheets.

Plasma arc welding (PAW) is similar to TIG welding – the main difference is that argon or helium is used to produce a plasma arc, which is separated from the shielding gas by locating the pointed tungsten electrode in the body of the torch. The plasma is forced through a fine-bore copper nozzle, which constricts the arc. Three principal modes of operation are possible, depending on the bore diameter and the plasma gas flow rate. Microplasma welding is performed with very low welding currents, in the range 0.1 to 15 A. At higher currents, from 15 to 200 A, the characteristics of the plasma arc are similar to those of the TIG arc. However, since the plasma is constricted, the arc is 'stiffer', and can be directed with high precision, giving deeper penetration and greater tolerance to surface contamination. Above about 100 A, a penetrating plasma beam is produced with high gas flow rates. A deeply penetrating keyhole is then formed by a mechanism similar to laser and electron beam welding, which can penetrate about 10 mm of steel. A filler material is added to ensure a smooth weld bead profile. PAW is performed in a mechanized system. The main drawback is the bulkiness of the torch, which restricts access – a laser beam can be directed into locations with poor accessibility.

Submerged arc welding (SAW) is an automatic process that is used for larger-scale constructions such as pressure vessels, shipbuilding and structural engineering. The arc melts a bare wire electrode that is constantly fed by a set of rolls that keep the arc length constant. The weld pool is protected by granular flux that melts to provide a protective blanket. SAW is suitable for plates in the thickness range 3–100 mm. Laser welding competes on productivity and through a reduction in post-welding finishing operations because of the lower energy input.

Friction stir welding (FSW), in common with other types of friction welding, involves heating the workpiece to a temperature at which it undergoes plastic flow – joining is performed in the solid state, rather than via a liquid phase. The workpiece is placed on a backing plate and clamped to prevent abutting joint faces being forced apart. A cylindrical, shouldered tool with a profiled projecting pin is rotated and slowly plunged into the joint line. The pin length is similar to the penetration required. When the pin contacts the work surface the material is heated rapidly by friction, which lowers the mechanical strength. Under an applied force the pin forges and extrudes the material until the shoulder of the pin is in contact with the workpiece surface. Friction heating then produces substantial plastic deformation. When the workpiece is moved, material is forced from the leading edge to the trailing edge by mechanical stirring and forging. Oxide layers are broken, and a solid phase weld is formed. Friction stir welding is clean, does not require a protective gas, and weld fume, smoke and ozone production can be eliminated. Flat, high quality joints with good mechanical properties can be made quickly. Apart from degreasing, no special material preparation methods are required. It is particularly suitable for light alloys that contain volatile alloying elements that cause difficulties when welding in the liquid phase. However, the process is currently limited to flat plates with straight seams. The process parameters must be controlled accurately, and the underside of the weld must be accessible for full penetration. Laser welding is a more suitable choice for materials with high melting temperatures, for complex three-dimensional joints, and joints in which access is available from one side only.

PRACTICE

Practical procedures for laser welding are described in this section. Unlike many other laser-based processes, the procedure for laser welding is heavily dependent on the material because the process mechanisms are complex and the material properties play an important role in determining suitable process variables. Generic procedures are described first, to illustrate the effect of process variables on weld quality and to set a baseline against which variations in procedure for other materials can be compared. Procedural variations are described in the section that follows, which also contains the properties typical of laser welds, in terms of workmanship standards and the imperfections that they classify.

MATERIAL PROPERTIES

A full description of the material includes its designation, trade name, manufacturer, batch number and composition – all important properties for traceability. The delivery condition comprises the surface condition (residue from rolling operations, coatings such as primers, markings and surface roughness), and the heat treatment, which determines the microstructure and mechanical properties such as hardness, yield strength and ultimate tensile strength. Such mechanical properties act as a baseline against which the properties of the weld can be compared. The orientation of the rolling direction of the plate relative to the welding direction is important; there can be small differences in properties parallel and perpendicular to the rolling direction.

The preparation of the workpiece prior to welding includes surface cleaning such as degreasing, pickling and wire brushing. Consistent welding results are obtained most easily if the surface condition of the workpiece does not vary along the joint line. Oxide layers and coatings such as primers act to increase the absorptivity of the beam, but introduce instabilities if their properties change during welding.

The material thickness plays a significant role in determining the most efficient welding process for a given joint, and dictates the processing parameters required. As mentioned earlier, a common rule of thumb for structural steels is that 1 kW of power is needed per 1.5 mm of steel to be penetrated at a welding speed of 1 m min^{-1}. Analogous relationships are found with other materials. Thick sections may be partially laser welded from both sides, either simultaneously or sequentially, such that the weld bead roots overlap. The required laser power can thus be reduced for a given welding speed, or the welding speed can be increased with a given laser power. The material thickness also defines any necessary joint preparation. With sufficient laser beam power it can be possible to produce a full penetration weld in a single pass. Thicker sections may be joined by machining grooves to form a Y-joint; the root pass may be completed autogenously and the groove filled using an alloy addition, for example. The method of edge preparation is important; a guillotined edge (which has a certain amount of deformation) behaves in a different manner during welding to an edge machined to a qualified specification. Ideally, machined edges with gaps no larger than 5% of the material thickness should be used. Abutting edges should also be cleaned prior to welding.

Prewelding treatments include preheating to reduce the cooling rate after welding, and mechanical prebending to compensate for distortion introduced during welding. The time permitted between such treatments and the welding process itself should be specified to ensure that required conditions are met.

BEAM CHARACTERISTICS

The standard information describing the laser includes its designation, the nature of the active medium, the geometry of the optical cavity (which affects the beam quality), the method of excitation, the number of lasers used (if more than one is involved in welding), and the method by which the beams are combined.

The wavelength of the beam influences a number of factors that determine the absorptivity of the beam by the material. Benefits can be gained by using shorter wavelength light because it is absorbed more efficiently by materials and the size of the focused spot is reduced, which results in an increase in power density, aiding initial beam–material coupling. There is also a decreased tendency for plasma to absorb a short wavelength beam. Multikilowatt CO_2 lasers are used predominantly for welding long straight seams in large structures, for which simple mirror-based flying optics can be used. Multikilowatt Nd:YAG sources with robot-mounted fibreoptic beam delivery are used for more complex three-dimensional components. When combined with suitable focusing optics, multikilowatt diode lasers can also be mounted directly on articulated robots to weld complex components.

Power may be delivered to the workpiece in the form of a continuous wave (CW), or as a series of pulses. For CW welding, the bead penetration increases with beam power for a given welding speed. A power density on the order of $10^4 \, \text{W mm}^{-2}$ is required to initiate and maintain the keyhole mode of welding. Welding is normally performed with a power level close to 90% of the maximum to ensure high productivity, with a margin of about 10% to allow for potential variations in workpiece geometry and laser output during welding. The power level may be increased or decreased during welding to maintain full penetration in plates of varying thickness. Transient effects are associated with the beginning and end of welds, which can lead to imperfections such as weld crater cracking; these can be eliminated by ramping the power up at the start of the weld, ramping down as the beam approaches the end of the joint, and overlapping the start and end points of the weld.

The most appropriate transverse spatial beam mode for keyhole welding is one that contains a central intensity peak. TEM_{00}, TEM_{10} and TEM_{20} modes are commonly used with stable cavity lasers. An annular beam intensity distribution is used with some unstable cavity lasers; deeper penetration and a higher weld bead aspect ratio is obtained in such a beam if it possesses a large magnification number (the ratio of the external to internal beam diameters). The annular intensity distribution may be transformed into a distribution with a central intensity peak by focusing.

The temporal beam mode determines the appearance and energy input of the weld. A pulsed beam is used to produce either spot welds, stitch welds, or a continuous weld made from overlapping spot welds. Pulsed welding is appropriate when precision and low energy input are required. Beam pulsing can be used to increase the available peak power density, thus increasing the penetration achievable.

Plasma formation can also be reduced by pulsing the beam. A CW beam is specified when welding speed is the main criterion. Pulse shaping (profiling the power with time) to incorporate a power tail that reduces the cooling rate can prevent cracking and porosity in susceptible materials.

The beam transmission system, or workstation, is characterized by its geometry (e.g. a gantry or articulated robot) and the method of transmission (turning mirrors or fibreoptic beam delivery). The essential properties of the workstation are the size of its working envelope, its positioning accuracy (typically ± 0.1 mm), and its positioning repeatability. The speed limits and accelerations of the axes are important in planning a procedure for complex geometry welding. The different types of workstations for various beam–component geometries were described in Chapter 4. The path length between the laser output window and the workpiece is particularly important in mirror-based transmission systems since each mirror can absorb about 1% of the incident radiation, and the divergence of the beam means that its diameter changes significantly depending on its location within the working envelope, which affects the focusing properties of the beam. Adaptive mirrors may be installed to compensate for such changes. The beam may be split and shared simultaneously or sequentially between workstations to increase productivity.

The beam delivery system itself is often referred to as the head (although this term is sometimes used to describe the laser source). It comprises: beam manipulation optics such as focusing lenses, mirrors or integrators; rotational or Cartesian axes to position the beam; equipment required to deliver process gases and filler materials; and process monitoring and control instrumentation. Very small spot sizes, obtained with small f-number optics, are ideal for high speed welding of thin materials, but vulnerable to matter ejected during welding. As the plate thickness increases, improved performance is obtained if an optic with a larger f-number is used, in order to maintain a high power density throughout the plate thickness. In practice, welding speed can be maximized by using an optic with a depth of focus that matches the plate thickness. Other factors must also be considered when selecting the depth of focus of the optic. For example, if the workpiece surface is uneven, penetration can easily be lost when a small f-number optic is used. A larger f-number optic provides greater tolerance to variations in the workpiece, and enables a larger stand-off distance to be used, which reduces the likelihood of damage. Broadening the beam helps to overcome problems caused by the presence of an air gap, but the welding speed is reduced accordingly.

There is still some debate about optimum position of the beam focal plane in relation to the surface of the workpiece. When welding relatively thin materials (<5 mm), the focal plane is commonly located at the workpiece surface. When welding thicker materials, the focal plane position has a significant effect on the maximum welding speed, as well as the shape of the weld bead. Much experimental data point to the fact that penetration is slightly higher, and the weld narrower, if the focal plane is positioned slightly below the surface of the workpiece in such cases. A displacement of 0.5–1 mm is normal, although the exact distance is dependent on the material thickness and the beam power. No clear explanation has been put forward, but keyhole formation by multiple internal reflection might be aided if the beam converges at the keyhole wall as it impinges the surface of the workpiece. If the focal plane is positioned significantly below the workpiece surface, a wide cap is formed, although the root may be narrower, and it may still be possible to obtain full penetration. If the focal plane is positioned significantly above the workpiece surface, a wide cap is formed, and penetration may be lost.

The polarization of a beam describes the orientation of the electric field in the direction of propagation of the wave. Linearly polarized beams produce different bead profiles in different directions, which can result in an unacceptable loss of fusion, for example. Hence it is important to know the polarization of the beam. The welding speed can be maximized when the direction of polarization is parallel with the abutting surfaces of the joint. Perpendicular polarization can reduce the welding speed to half the maximum, and circular polarization to about 75%. Above a threshold welding speed (determined by the various beam parameters), absorption of the beam is predominantly through Fresnel absorption at the keyhole wall, and plasma absorption is reduced. In this region there is a strong effect of polarization on welding speed.

By splitting the beam in the direction of welding, an elongated weld pool shape is created, which results in improved degasification and a reduction in the sensitivity to porosity, particularly when welding aluminium and coated materials such as primed plates. If the beams are not aligned in the welding direction, but displaced slightly transverse to the welding direction, the gap-bridging ability of the weld bead can be improved at a low power. However, the welding speed is affected: an angular misalignment of about 20° results in a reduction in weld speed of about 30%.

The beam can also be oscillated in the direction of welding, perpendicular to it, or with a combination of both (beam spinning). Air gaps along the joint line can then be bridged, and differences in sheet thickness can be accommodated, again at the expense of welding speed.

PROCESSING PARAMETERS

The main processing parameters of keyhole welding are the welding position, fixturing, the process gases, the filler materials, and the welding environment. The welding speed is normally considered to be a constant of the process (except when it is varied to overcome transient welding effects). The highest possible welding speed is normally used to maximize productivity.

Welding Position

Most laser welding is performed in the flat position (downhand, 1G) because this position is normally the most convenient for accommodating auxiliary equipment such as a plasma control jet, filler wire feeder or process monitoring equipment. However, welds can also be produced in the vertical up and down (3G) and overhead (4G) positions. Circumferential welding of a stationary pipe with a horizontal axis (5G) is also possible. Welding limits are determined by the need for surface tension to support the weld pool. It is more difficult to weld overhead because the unsupported volume of material is larger.

Process Gases

Gases have three main functions in laser welding: shielding; plasma suppression; and protection of the optics.

Shielding of the molten weld pool prevents oxidation and contamination, which could lead to porosity and embrittlement. Both the weld bead face and root require shielding. The compromise between the quiescent gas curtain required for shielding and the flow in a gas jet for plasma suppression has led to unique nozzle designs, some of which are illustrated in Fig. 16.7. The simplest is a side tube with a diameter between 2 and 9 mm, which blows gas across the weld pool. A coaxial flow is a simple, compact design that can be used with a laser power up to about 5 kW when plasma formation is not a serious problem. The exit diameter is large (6–20 mm) to enable large areas to be shielded with low gas flow rates that do not disturb the melt pool. Typical stand-off distances lie between 5 and 8 mm. They allow access in limited spaces, and are suitable for industrial production, especially in three-dimensional operation. A trailing shield is a rectangular shroud which extends some distance behind the weld, often several centimetres, Fig. 16.7. It is used when welding metals that are sensitive to oxidation, e.g. titanium, to shield the heated material. It can also be used to protect the molten weld metal when welding at high speeds. Its bulk makes it unsuitable for applications with limited access, or with robotic multi-axis beam delivery. A backing gas supports the root, which is particularly useful when welding materials of low viscosity in the molten state.

Plasma is caused by ionization of the ejected metal vapour and of the shielding gas. Plasma formation is particularly noticeable when welding with high power levels and at speeds below about $1 \, \text{m} \, \text{min}^{-1}$. The effect is to defocus the beam, reduce the energy absorbed by the weld, and to produce

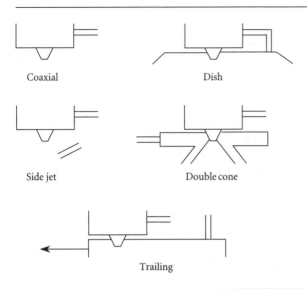

Figure 16.7   Gas delivery nozzles used in laser welding

Table 16.1   Properties of gases used in laser welding

Gas	Molecular weight	Thermal conductivity @1200 K ($W\,m^{-1}\,K^{-1}$)	1st ionization potential (eV)	Density ($g\,L^{-1}$)	Relative cost (Europe)
He	4	0.405	24.46	0.1769	1
Ar	40	0.049	15.68	1.7828	0.4
N_2	28	0.076	15.65	1.2507	0.1
O_2	16	0.082	12.50	1.4289	0.1
CO_2	44	0.080	14.41	1.9768	0.1

a distinctive 'nail-head' appearance in the weld bead section. Plasma is removed from the interaction zone by a gas jet that may trail or lead the beam. A typical gas jet diameter is 1.5 mm, which is used with a gas flow rate of about 10 L min^{-1} (helium). The jet is aligned in the plane of the joint, impinging about 1 mm ahead of the beam–material interaction point in the direction of welding, and crossing the beam about 1 mm above the interaction point. The angle of the jet with the workpiece may need to be increased for thicker plates, e.g. 25° with 6 mm plate, up to 45° with 13 mm plate.

Gases also protect optics from weld spatter. An air knife of rapidly flowing compressed air is located directly below the optic, in order to protect it by deflecting ejections from the weld.

The most important properties of common process gases used in keyhole welding are given in Table 16.1.

Helium is a popular choice for plasma suppression, because of its high ionization potential (resistance to plasma formation), its low molecular weight (which aids recombination of electrons and ions in the plasma), and its high thermal conductivity (which removes energy from the interaction zone). However, it is expensive outside the United States, and has a low density, which reduces its ability to blanket and protect the weld, and so high flow rates are required. In comparison with other process gases, helium produces a narrow weld bead profile, high penetration, a narrow HAZ, low porosity, and is tolerant to changes in procedure. It is particularly suitable for thick section welding using far infrared radiation – a low welding speed promotes plasma build-up in the interaction zone. The addition of about 4% oxygen increases the formation of acicular ferrite on solidification of a structural steel weld bead, resulting in improved weld ductility.

Argon is often substituted for helium, particularly when welding with near infrared laser beam, since plasma is more transparent to shorter wavelength radiation. Its high density assists in removing plasma. It has a lower ionization potential than helium, and so is not as effective for plasma control, but with a density that is ten times greater than helium, it shields the weld pool more effectively. Argon may promote undercut in conventional welding of steels, discussed later but this can be reduced or eliminated by adding 1–5% O_2 or 3–25% CO_2, which increases wetting, enabling the undercut to be filled. Argon is relatively cheap, except in Japan. The addition of argon to helium, in amounts up to 50%, may improve the economics of welding, without sacrificing plasma control.

Nitrogen is also an economic replacement for helium, particularly in mass-production applications such as joining of steel automobile bodies. However, high flow rates are required, which may result in entrainment of air. Keyhole stability is then reduced and coarse porosity can form if the shielding geometry is inadequate. Fine porosity can arise from degassing on solidification – nitrogen is considerably less soluble in molten than solid steel. Under optimum conditions, a welding speed and bead appearance comparable with those obtained with helium are possible. In structural steels, nitrogen absorbed in the weld metal stabilizes austenite, resulting in the formation of more martensite and bainite on cooling, increasing the hardness of the weld bead, which increases its susceptibility to cracking. Nitrides may also form; if present at grain boundaries they can lead to embrittlement. Nitrogen absorption in the weld metal of austenitic stainless steels can result in an improved solidified microstructure with superior corrosion characteristics.

Carbon dioxide is also relatively cheap. It blankets the interaction zone effectively because of its high density. An exothermic reaction can be supported: CO_2 can dissociate to provide oxygen that increases the effective energy input of the weld. However, a CO_2 laser beam is absorbed by carbon dioxide, and so the gas is not as suitable as for conventional welding. The exothermic reaction widens the weld bead, which can be desirable in rapid sheet welding where small gaps exist along long joint lines. Similar welding speeds to those obtained with helium are possible in sheet materials because of the exothermic reaction. Porosity can be high unless the weld pool contains sufficient deoxidizing elements to control the formation of carbon monoxide. Carbon absorbed in the weld metal increases the amount of martensite formed on cooling, which reduces the formability of sheet steels.

Joints

Close-fitting joints designed for conventional welding are suitable for laser welding. However, novel joints can be designed because of the accessibility of the beam. Groove preparation is not required unless the section thickness is greater than twice the maximum bead penetration. Figure 16.8 shows a selection of joint designs used in laser welding.

The square butt or I-joint is ideal for laser welding. Strength is generated from the complete weld bead penetration. However, it is the least forgiving. Air gaps arise from poor fitup of parts, or from the roughness of cut plate edges. Air gaps must be less than about 5% of the plate thickness to avoid bead concavity and sagging. The beam must be aligned with the joint line over its entire length.

T-joints can be welded in various configurations. The web is the vertical member of the joint and the flange is the horizontal member, as illustrated in Fig. 16.9. A butt weld is produced either with a fully penetrating weld bead from one side only, or by overlapping weld beads made from two sides, either sequentially using a single beam or simultaneously using two beams. The weld bead tends to

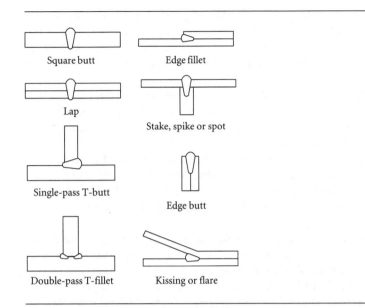

Figure 16.8   Selection of joint designs and weld bead shapes for keyhole welding

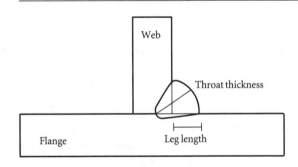

Figure 16.9   Nomenclature used in T-joints

follow the line of the joint. Single pass welds require access from only one side, but the root profile is important – a large radius of curvature is desirable in order to minimize stress concentration. Also, a single weld is asymmetric, which may lead to the build-up of stresses. Double pass welding is therefore often recommended. The leg length is the distance from the web to the edge of the fillet, as measured along the flange. The throat thickness is the distance between the bottom and the face of the fillet, as illustrated in Fig. 16.9. The use of a single pass butt weld provides a means of visually assessing the weld since the weld root is visible. The joint can also be welded by making two partially penetrating fillet welds, in which case weld integrity must be tested by radiography.

T-joints are used to fabricate stiffeners in ships and other constructions, accounting typically for around 80% of the total number of joints. In comparison with arc welds, laser welds have a small leg length, although the throat thickness is of a similar size. Weld toes in T-joints are susceptible locations for fatigue crack initiation, because of the stress concentration. Conventionally, the toes are ground, in order to reduce the discontinuity. Laser remelting of weld toes achieves the same result.

Alternatively, the weld may be made by fully penetrating the flange and partially penetrating the web to produce a stake, spike or spot weld (Fig. 16.8). This is a relatively easy weld to make since the tolerance over which the beam can be positioned is wide and access is required from one side only. It is often used to weld a thin upper sheet to a thicker lower sheet. However, the unwelded portions can act as crack initiation sites under transverse loading.

The lap joint is the most common laser-welded sheet metal joint because it offers the highest flexibility from the design viewpoint with respect to ease of manufacture under production conditions. It is also easy to make because there are no beam–joint alignment problems, and joint preparation, location and access are easy. The main drawback is that energy is used to penetrate the upper plate before the joint is achieved, and there is no visual evidence of penetration unless full penetration is achieved. Good contact between clean faces is required. Gaps of about 10% of one plate thickness can be tolerated, while the beam focal plane is maintained. Joint strength is determined by the weld width at the interface. A general requirement is that the weld width should equal at least the thickness of one plate.

Edge butt welds can be made rapidly, typically at twice the rate of stake welding. However, the beam must be oriented and positioned accurately relative to the joint line. Accuracy requirements may result in a considerable increase in the time required for fixturing.

The kissing or flare joint is particularly suitable for power beam welding. Energy is reflected into the base of the joint, increasing absorptivity. The highest welding speeds can be obtained if the plane of polarization is normal to the plane of the joint line. This technique has been investigated for welding the longitudinal seams of pipelines.

Fixturing

Accurate fixturing is necessary in laser welding because gaps along the joint line cannot be tolerated by the small focused beam. Fixturing is a time-consuming and expensive manufacturing phase, but is compensated for by a higher quality product and a reduced need for post-weld reworking (which can represent a third of the total welding cost).

Joint parts may be fixtured in a frame to avoid angular and bending shrinkage. Clamping prevents opening in the horizontal and vertical planes caused by shrinkage of the weld bead. Arching – a common feature of welds made in thin sheet metal plates – is minimized by fixturing close to the joint line. Large forces induced by transverse and longitudinal shrinkage may exceed frictional fixturing forces; such displacements are minimized more effectively by using groove gap inserts. In some joints, the weight of the parts themselves may provide sufficient force for effective fixturing, obviating the need for external elements. However, only a small amount of springback shrinkage can be accommodated when the component is released.

Tack welds may also be used to hold parts during welding. Their length should be two to three times the required penetration, and they should be spaced with an interval of about 150 mm. The first tack weld is placed in the centre of the joint line, with subsequent tacks placed either side alternately. The cooling rate associated with a tack weld is high because its size is small in comparison with the joint, and so hardening and cracking may ensue. The section must then be preheated locally prior to tack welding. Cracks and defects can also form when a tack weld is welded over during the welding pass. This can be avoided by preheating, and by removing slag and other artefacts prior to the welding pass.

Parts can be prebent, or presprung prior to welding to prestress the assembly, which counteracts shrinkage during welding. Prebending through the use of wedges can be used to compensate for angular distortion; parts regain correct alignment when the wedges are removed after welding.

Hollow structures containing single-sided longitudinal welds are braced to compensate for shrinkage induced by bending. The magnitude and direction of the bending deflection are selected such that the component springs back into the correct geometry after welding. The weld is located on the convex side of the component because deflection occurs 'towards the weld' during welding. The degree of bending is set to be slightly larger than the assumed bending shrinkage induced when the component is not braced.

Misalignment up to about 10% of the plate thickness and less than 1 mm can normally be tolerated in a butt joint before an unacceptable weld geometry results. In lap welding the gap between the two parts should not exceed 10% of the material thickness. An angular mismatch up to 5° can be tolerated before filler material becomes necessary. Practical techniques for increasing tolerances include: using an optic with a large depth of field; defocusing the beam to increase its projected size; oscillating or spinning the beam; and using a filler wire.

Filler material

Autogenous laser welding allows high quality welds to be made quickly, and provides opportunities for products to be designed more efficiently. However, strict fitup tolerances must be imposed because of the small size of the beam. In some materials the high cooling rates induced may lead to problems associated with high weld metal hardness. Filler materials can be used to relax fitup tolerances and control weld metallurgy. Air gaps up to 2 mm can typically be filled in material up to 12 mm in thickness. By controlling weld pool chemistry, improvements in weld properties can be achieved, and dissimilar metals can be joined with an appropriate material. Multi-pass welding enable sections several times thicker than the nominal single pass autogenous capability to be joined. The root pass in a thick Y-joint can be laser welded autogenously, while filler material can be added to fill the groove.

Filler material may be introduced in the form of a wire, insert or powder. Wire feed is currently considered to be the most practical and flexible method, illustrated in Fig. 16.10. Wire has the advantage that contaminant pickup is low, productivity is high, and a wide range of filler materials are available as MIG welding wire, typically of diameter 1.2 mm. The wire may be fed into the beam above the workpiece, into the leading edge of the weld pool, or into the trailing edge of the weld pool. Adequate gas shielding is important, as the weld pool is generally wider and longer than an autogenous weld pool. Gas may be delivered coaxially with the wire, or from a separate nozzle that is normally located

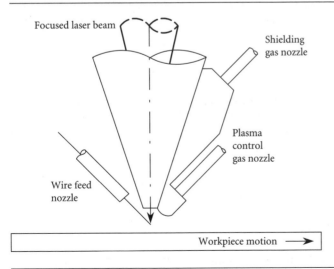

Figure 16.10   Geometry of leading edge filler wire laser welding.

on the opposite side of the beam to the wire feed nozzle. Coaxial gas delivery affords protection to the heated parts of the wire, and provides more space for sensor equipment, but the amount of protruding wire is longer, which may lead to inconsistencies in positioning.

When wire is fed into the laser beam, part of the beam energy is used to melt the wire, part is reflected, and part passes through a keyhole generated in the wire or passes around the wire and is available to form the weld bead. The beam may be reflected in many different directions by the wire, producing a variety of intensity distributions, depending on the beam power and mode, focal point size and wire feed rate. An increase in the fraction of the beam reflected by the wire is observed with an increase in wire feed rate and a decrease in laser power. The polarization of the beam also has an effect on the reflection of the beam from the wire. For a linearly polarized beam, a larger portion of the beam is reflected when the wire is fed transverse to the angle of polarization. Such wire feeding has the greatest tolerance to variations in gap width, but requires very precise alignment of the equipment. If the wire intersects the beam at the surface of the workpiece, or slightly below it, a uniform distribution of alloying elements is more likely to be produced. Intersection above the surface is more likely to distribute the alloying addition in the upper region of the weld.

Leading edge wire feeding is used in the majority of practical welds. The process efficiency is increased through beam reflection into the interaction zone, which contributes to weld pool formation. Mixing in the weld pool is improved, although practical setup and control may be difficult. The filler wire must be positioned very accurately. Deviation of little more than a millimetre around the keyhole can result in poor welds. The wire may be projected into the beam path and cause excessive spatter, or vaporize and form a plasma cloud that defocuses the laser beam, reradiating the energy elsewhere.

Trailing edge feeding has a number of characteristics that must be understood for successful application. If the distance between the wire and the beam in the welding direction is too small the wire is partially vaporized, and cannot be incorporated into the weld pool. If the distance is too large the wire can stick to the solidifying weld metal. Uneven mixing of the weld pool is more likely, particularly with high welding speeds. However, with high welding speeds, trailing wire feed can be a more robust technique.

By heating the filler wire, the process efficiency can be increased through two effects: an increase in the absorptivity of the laser beam; and a reduction in the energy required for melting the wire. A 30% increase in wire feed rate is typical when the wire is heated to 250°C. Preheating the wire also allows greater control over the weld metal cooling rate, allowing for improvements in microstructure and toughness. Resistance heating is a suitable method in practice. However, the possibility of the formation of weld bead defects such as porosity and inadequate geometry is increased, and precise control of the process is required.

Heat Treatment

Because of the low energy input of laser welding, preheating is often necessary to avoid cracking. In a 0.4% C steel, a preheating temperature of 500°C is not unusual. Small components can be preheated in a furnace and transferred to the workstation. Localized regions of larger components can be preheated using induction pads and sleeves. By splitting the laser beam, one portion can be used to preheat the region ahead of the keyhole.

Post-weld heat treatment is performed at relatively low temperatures to improve toughness by tempering the microstructures of the weld bead and HAZ.

Environment

As ambient pressure is decreased, plasma formation is reduced, and the bead profile changes from a 'nail-head' to a parallel-sided bead. Penetration increases to a limiting value, which corresponds approximately to that obtained during electron beam welding. The minimum ambient pressure required to attain the penetration limit varies with gas type; values for helium, nitrogen and argon of 100, 10 and 0.1 mbar, respectively, are typical. An additional effect is the reduction in vaporization temperature with increasing ambient pressure, which facilitates keyhole formation and stability, thereby reducing the level of coarse porosity.

Underwater welding can be simulated in a pressurized helium environment. A pressure of 50 bar represents an underwater depth of about 500 m. Plasma production is greater than that produced at atmospheric pressure, which results in a wider, shallower weld bead. Underwater welding can be carried out practically, although water strongly absorbs far infrared radiation. An interaction with a workpiece results from the formation of a vaporized channel, through which the beam is directed to the workpiece surface.

Process Monitoring

A primary concern in high volume automated laser welding is the detection of imperfections in real time, and their correction at an early stage in the manufacturing process. Various sensing techniques have been developed, based on acoustic emission, audible sound, infrared radiation and ultraviolet radiation emitted from the weld zone. Such signals can be correlated with the properties of the weld, e.g. the achievement of full penetration, or the following of a joint line. Table 16.2 lists a number of such systems.

As laser welding becomes more automated, the use of such sensoring techniques will increase.

Hybrid Welding Processes

The combination of a laser beam and an electrical arc in a hybrid welding process was first investigated in the late 1970s. The thermally affected zone produced by a focused carbon dioxide laser beam was found to root a gas tungsten arc (GTA). Not only can welds then be made at several times the rate possible with the laser beam alone, but penetration can also be increased. GTA sources have been successfully combined with CO_2 and Nd:YAG laser beams. Similarly, laser beams have been combined with gas metal arc (GMA) sources. Coaxial laser–GTA and laser–GMA sources are highly versatile. The beneficial effects of combining plasma arc and high frequency (HF) induction energy sources with

Table 16.2   Methods of monitoring and controlling laser welding

Source	Sensor	Monitored	Controlled	Reference
Optical scanner	CCD camera	Joint line and geometry	Beam position, traverse rate, wire feed rate	(Andersen and Holm, 2002)
Vapour plasma	CCD camera	Vapour plume	–	(Bagger et al., 1991)
	Monochromator, acoustic	Plasma	Welding speed	(Hoffman et al., 2002)
Optical scanner	CCD/CMOS camera	Joint geometry, weld bead	Beam focus position, traverse rate, power	(Boillot and Noruk, 2001)
Optical scanner	CCD camera	Joint line and geometry	Beam oscillation amplitude, traverse rate	(Coste et al., 1999)
Laser line	CCD camera	Joint line	Beam position	(Drews and Willms, 1995)

Table 16.3 Hybrid laser welding techniques.

Laser	Auxiliary	Reference
CO_2	TIG	(Eboo *et al.*, 1978; Steen and Eboo, 1979; Alexander and Steen, 1980, 1981; Diebold and Albright, 1984; Abe *et al.*, 1998; Hamatani *et al.*, 2001; Ishide *et al.*, 2001; Makino *et al.*, 2002)
CO_2	MIG	(Dilthey *et al.*, 1998a; Abe *et al.*, 1996; Shida *et al.*, 1997, Ishide *et al.*, 2001)
Nd:YAG	TIG	(Beyer *et al.*, 1994; Ishide *et al.*, 1997)
Nd:YAG	MIG	(Missori *et al.*, 1997; Abe *et al.*, 1998; Jokinen *et al.*, 2000; Ono *et al.*, 2002)
CO_2, Nd:YAG	PAW	(Walduck and Biffin 1994; Biffin, 1997; Biffin *et al.*, 1998; Blundell, 1998; Page *et al.*, 2002)
CO_2	Arc	(Tse *et al.*, 2000)
CO_2	Multiple arc	(Dilthey and Wieschemann, 2000)
CO_2	HF preheating	(Inaba and Shintani, 1993)

laser beams have also been demonstrated for welding. Invariably the performance of the hybrid system is greater than the sum of the individual components, implying that synergistic effects on processing capability are produced. Examples of hybrid welding techniques are given in Table 16.3.

In recent years, there has been a renewed interest in hybrid welding technology, motivated by the demands of the automotive industry, notably in the production of tailored blanks. Commercial systems based on laser–TIG, laser–MIG (Fig. 16.11), laser–multiple arc, and laser–plasma systems may now be purchased. In addition to the improvements in productivity and penetration outlined above, tolerances to joint gaps and misalignments may be increased from about 10% to 30% and 60% of plate thickness, respectively. A superior weld face profile, suppression of zinc spatter in coated steels and expulsion in aluminium alloys, and a reduction in weld hardness may also be achieved. In addition to such technical advantages, the cost of a hybrid system can be a fraction of the cost of an equivalent laser-only solution, particularly for welding thick sections.

Despite the advances made in equipment and process design, the fundamental process principles of hybrid welding are not well understood, in particular the interaction between the laser beam and the arc, and its effect on weld pool geometry. As a result, it is difficult to establish practical operating windows of processing parameters from the large number of variables, thereby necessitating trial-and-error testing for procedure development.

PROPERTIES OF LASER-WELDED MATERIALS

The properties of a laser weld are determined by the interaction of the material properties, the beam characteristics and the processing parameters. Properties include those of the weld metal, the HAZ and the parent material, all of which contribute to the overall weld quality. Weld requirements are specified in terms of the properties and performance of the weld, and the levels of imperfections that can be tolerated. The most common imperfections – those that can arise in welds in all types of material – are described below. Modifications to procedures for specific materials to minimize the levels of imperfections are discussed later.

With the exception of relatively thin structural steels and some aluminium alloys used in aerospace applications, standard welding procedures are few. It is difficult to provide general advice – indeed, contradictory findings are sometimes reported in the literature. For this reason, references are given to

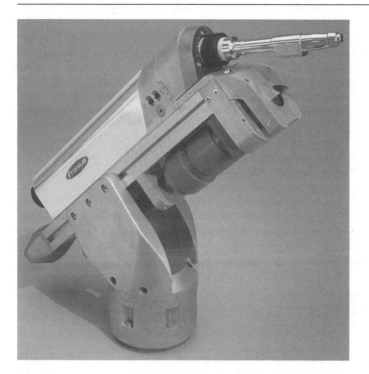

Figure 16.11   Hybrid Nd:YAG laser–MIG welding head. (Source: Herbert Staufer, Fronius, Wels-Thalheim, Austria)

allow the reader to trace original work and examine the procedures used and the reasons for the choice of processing parameters.

QUALITY ASSURANCE

Weld quality is assured by developing a qualified welding procedure that meets standard conditions, or requirements provided by an end user. An approved welding procedure must first be developed that contains the values, or ranges of values, for the process parameters that are demonstrated to produce welds to a prescribed quality level. This involves trial-and-error welding and testing. The development process may be expedited by basing trials on an existing procedure, or by using predictive methods, some of which are described later.

Welding Procedure Development

The development of an approved welding procedure begins with the design of a welding procedure specification (WPS), which describes the relevant process parameters for a particular weld. The contents are defined in a standard, listed in Appendix F. The WPS is submitted for acceptance testing (again performed in accordance with a standard given in Appendix F), and the results registered in a welding procedure qualification test results record (WPQ-Test). If the WPS and the WPQ-Test are approved by the relevant authority then the welding procedure is accepted and the data in the WPS record can be copied into the welding procedure qualification record (WPQ). If the welding procedure is not approved, then the WPS must be redesigned, and the testing procedure repeated for approval.

The design of a welding procedure for full penetration butt joints in a structural steel in the downhand position is now described.

A preliminary set of basic welding parameters is first established. Melt runs are made by systematically varying the traverse rate of the laser beam with a given power level. This procedure is repeated for different power levels such that a 'weldability lobe' is constructed that shows the conditions of power and traverse rate required for full penetration. These parameters are then used to produce welds in abutting plates.

Plates are cut to a standard length, such that the joint line is not less than 150 mm in length. The abutting edges of the plates are degreased with acetone and are clamped in an I-joint configuration. Five welds are typically made for each combination of processing parameters obtained from the weldability lobe to provide sufficient specimens for destructive and non-destructive testing. All the processing parameters are recorded.

Distortion, weld spatter and macroscopic imperfections are identified by visual examination. Welds are radiographed perpendicular to the plate surface to reveal the presence of porosity and cracking. Liquid penetrant testing is used to detect geometrical weld bead imperfections and cracks that break the surface. Welds that do not meet the requirements are discarded.

The welds are sectioned transverse to the welding direction, ground, polished and etched to reveal the weld bead and HAZ by optical metallography. The weld bead is assessed by measuring the following features: undercut, excess weld metal, excess penetration, linear misalignment, sagging, incomplete groove filling, and root concavity (discussed below). Micrographs are made in the weld bead, at the fusion line, in the coarse-grained HAZ, the fine-grained HAZ, and the parent material to identify the microstructural phases.

Harness measurements are taken from a transverse metallographic sample of each of the welded plates, traversing the parent material, HAZ and weld bead on both sides of the weld centreline. The traverses are made 2 mm from the upper and lower plate surfaces, parallel with the surfaces. Any welds in which the maximum allowable hardness is exceeded are discarded.

Tensile testing specimens are machined from the welded plates such that the weld, of length 25 mm, is positioned in the centre of the specimen, oriented transverse to the specimen axis. The ultimate tensile stress, the 0.2% proof stress and the elongation at failure of a 50 mm gauge length are noted, as is the location of fracture.

Impact testing is carried out at specified temperatures by machining standard Charpy V specimens of length 55 mm and width 10 mm, transverse to the welding direction. The samples are positioned such that the notches, of depth 2 mm, are located in the weld metal, the coarse-grained HAZ, and the fine-grained HAZ, in the different specimens. The notch is oriented such that the fracture path runs parallel with the direction of welding during testing. The energy absorbed and the fraction of ductile failure on each fracture surface are recorded after each test. A reference set of impact properties is obtained from tests performed on the parent material. Care should be taken when interpreting impact test results since fracture path deviation from one zone to a neighbouring zone is common, which yields a false result.

Three-point bend testing is performed using a standard apparatus. A former of diameter 24 mm and supporting rolls of diameter 60 mm with a surface-to-surface separation of 24 mm are typically used. One specimen is tested such that the weld cap forms the outer fibre in tension, the other having the weld root outermost. Two longitudinal specimens are also tested; one in root bending and the other in face bending. The specimens are bent through an angle of 180°. The weld reinforcements are machined flat prior to testing. The specimens are examined for imperfections after testing.

Imperfections

An imperfection in a weld is a fault or blemish associated with the weld metal, the HAZ or the surrounding parent material; it becomes a defect when it is considered detrimental to weld properties.

Imperfections arise from deviations from a qualified welding procedure with respect to the material properties, laser beam characteristics or processing parameters. Some imperfections are associated with the geometry of the weld bead, and can be assessed from a visual inspection, while others are concealed in the weld bead or HAZ, and can only be detected by non-destructive methods if they do not break the surface. The most common imperfections are described below, and in the workmanship standards listed in Appendix F. Levels of imperfections that qualify welds in steels and aluminium alloys into one of three classes – moderate (D), intermediate (C) and stringent (B) – are also described in these standards.

Excess Weld Metal

Excess weld metal (also referred to as overfill, reinforcement and cap/root height) refers to excessive convexity of the face or root of the weld bead, i.e. weld metal lying outside the plane containing the weld toes. It appears in autogenous welds in response to four conditions: insufficient welding speed; contraction of the plates; excess shielding gas pressure; and longitudinal flow in the weld bead, which draws material into the welding region. An excessive filler material feed rate may also create this imperfection. The workmanship standard for steels defines quality classes in terms of the excess weld metal height, h, and the plate thickness, d, such that $h <= 0.1 + 0.2d$, $h <= 0.1 + 0.15d$ and $h <= 0.1 + 0.1d$ (with a maximum limit of 5 mm) for classes D, C and B, respectively. The limits for aluminium alloys are slightly less rigorous. Excess weld metal adds no value to the weld; it lowers fatigue strength because stress concentrations occur at weld toes that contain a high contact angle (the angle between the parent material and the adjacent weld bead).

In contrast, incomplete groove filling is caused by poor fitup, which results in an excessive joint air gap that cannot be filled by autogenous welding. The static and dynamic strength of the weld is lower than that of the parent materials. This imperfection can be corrected by using a filler material.

Sagging

Sagging describes a collapse of the weld metal as a result of excessive fusion and gravity. It increases the risk of cracking because of the increase in the density of shrinkage stresses in the weld bead. It can normally be rectified by increasing the welding speed.

Undercut

Undercut refers to a groove melted into the base metal adjacent to the weld bead, which remains unfilled by weld metal. It is formed if the molten weld metal begins to wet the gouged region, but solidifies before the groove can be filled completely. Several forces play a role in the distribution of molten weld metal: surface tension; gravity; frictional drag forces; and vapour pressure. Undercut can be avoided by reducing the welding speed. It does not significantly lower the static strength of the joint, but it can degrade fatigue properties by acting as a site for crack initiation.

Lack of Fusion

Weld bead imperfections such as a lack of fusion or penetration may be caused by poor joint preparation, poor positioning of the beam, or insufficient welding energy. Incomplete penetration creates a notch that can act as a site for fatigue crack initiation. A lack of penetration results from an incorrect choice of processing parameters, e.g. insufficient beam power, excessive welding speed, or incorrect focus position.

Humping

Humping is the term used to denote a series of periodic longitudinal swellings in the weld bead, normally accompanied by undercut and a loss of penetration. It originates from instabilities in the fluid dynamics of the molten weld bead, and is more likely to occur with a high welding speed. It can be avoided if the welding speed is chosen such that the Froude number (Chapter 11), which is based on the weld pool length, lies below 2.0. This is likely when the weld pool is long. In practice, a twin-beam welding technique, in which the beams are displaced in the direction of welding, or a beam elongated in the direction of welding, can be used to increase the threshold welding speed for humping.

Deformation

Structural deformation (displacement and rotation of the welded plate) is caused by the presence of residual stresses after welding. It may be temporary or permanent, and is referred to as shrinkage, distortion or warping. The four basic types of welding deformation are longitudinal, transverse, angular and bending shrinkage. Distortion may be minimized or eliminated by careful control of design parameters and welding variables. Part fitup tolerances should be minimized, preferably to less than 5% of the section thickness. The transverse joint area should be minimized to reduce the energy input required – a close-fitting I-joint is preferable to a V-joint – and should be optimized to use the beam energy in the most efficient manner (a kissing joint maximizes energy absorption, for example). The geometry, location and sequence of weld passes may be designed to balance distortion about a neutral axis. A joint can be preset at an angle in an elastic fixture to compensate for distortion introduced by welding. Static fixturing, such as clamping and tack welding, restrains the joint during welding. Tandem welding, in which two or more beams are oriented relative to the workpiece to create the joint, enables distortion to be balanced statically (by using sequential passes) or dynamically (by using simultaneous passes). Energy input may be minimized by using a hybrid process, for example one that combines a laser beam and a MIG arc. Joints may then be welded using sequential passes with a single welding head; the land of a Y-joint can be fully penetrated in a single pass with the laser beam, and the groove filled using the MIG arc. Alternatively, the penetration capability of the laser beam may be combined simultaneously with the gap bridging capability of the MIG process to enable a thicker section joint with air gaps along its length to be welded in a single pass. Distortion may be reduced by using techniques such as back-step and skip welding. Back-step welding involves making short adjacent weld lengths out of sequence in the opposite direction to the net welding direction. Skip welding involves laying short weld lengths in a predetermined, evenly spaced, sequence along the seam.

Distortion can be longitudinal (bowing) or transverse (winging). It is caused by shrinkage following solidification of the weld metal and rapid cooling of the HAZ, which result in the formation of residual stresses in welds. In a typical double-V butt weld, the upper weld pass creates a tensile stress state in regions adjacent to the weld bead.

Cracking

Solidification cracking occurs during the solidification phase of the weld bead in susceptible alloys. Solid forms in the weld metal as interlocking dendrites during cooling, accompanied by the rejection of alloying elements and impurities at the solid–liquid interface. A solute-rich interdendritic liquid of low solidification temperature is formed, which remains at the grain boundaries at the end of primary solidification, lowering the cohesion of the solidified grains. The rejected solute subsequently solidifies, reducing the strength of the weld bead in the grain boundary region. Cracking may then occur in a low ductility trough below the solidus temperature. In both cases, the presence of tensile stresses can lead to solidification cracks in these regions. Solidification cracks in the weld bead can take three characteristic forms: intergranular cracks in the centre of the weld bead, which are associated with equiaxed crystallization; cracks at the edges of columnar grains; and intergranular cracks extending along the fusion line, which are associated with epitaxial solidification. The following *material* factors promote solidification cracking: a coarse microstructure; the presence of elements such as sulphur, phosphorus, niobium, nickel and boron, which promote the formation of low strength grain boundary films; and a wide solidus–liquidus temperature range, which enables high contraction strains to be developed during solidification. The following *process* factors promote solidification cracking: a high energy input, which produces a coarse solidification structure and allows time for diffusion of elements to the moving solid–liquid interface; a high welding speed, which tends to produce a centreline region at which dendrites form in an orientation perpendicular to the welding direction, which cannot easily accommodate solidification strains; a non parallel-sided weld bead, containing bulges; the presence of tensile stresses; and rapid solidification, which can lead to an inhomogeneous distribution of alloying elements, particularly in welds made with filler material.

Cold cracking, or hydrogen-induced cracking, is the most serious weldability problem with materials that have low room temperature ductility, such as martensitic steels. It is transgranular in nature, and can occur in the HAZ or the weld metal. Cracks may be longitudinal or transverse, with respect to the weld bead. Cracks develop through an incubation and propagation mechanism, and appear towards the end of cooling, at temperatures around 150°C, or in the case of delayed cracking, several hours after the weld has been made. The following factors increase the likelihood of cold cracking: a hardenable material; microstructural inhomogeneities; the presence of hydrogen; and restraint. The likelihood of cold cracking can be reduced by various means: machining a stress relieving groove adjacent to the weld to restrict heat flow in partial penetration welds; welding the material in a ductile condition; preheating; thorough cleaning; and welding in an inert atmosphere.

Porosity

Porosity in welds arises from gas bubbles that become trapped on solidification of the weld metal. Porosity may be classified into two broad types, fine and coarse, which are normally differentiated by an average pore diameter of 0.5 mm.

Fine porosity often appears as a regular distribution of spherical pores. It originates mainly from dissolved gases that have a lower solubility in solid than liquid metal, and which are rejected at the melt boundary on solidification. A regular line of fine porosity can sometimes be observed at or parallel with the fusion line. Hydrogen, oxygen and nitrogen result in significant outgassing. The major sources of these are the base material, filler additions, surface contaminants and process gases.

Coarse porosity is characterized by larger, more irregularly shaped voids that are distributed randomly throughout the weld bead, but which are often found along the weld centreline, in the lower half of the weld bead. They arise from instabilities in the keyhole, which create voids in the molten weld metal that become trapped during solidification. Such instabilities are often associated with the root of the keyhole, and are frequently observed in welds made at speeds close to the penetration limit. They have many origins: multiple internal reflections of the beam; high metal vapour pressure; disturbances at the trailing edge of the keyhole; fluid dynamics in the surrounding molten metal; plasma formation at the mouth of the keyhole; entrainment of gases; and small temporal and spatial fluctuations in beam intensity and mode. Coarse porosity is often found in the termination region of circumferential welds in pipes.

Fine-scale porosity, evenly distributed throughout the weld, degrades load-bearing properties significantly when present in amounts above 6 vol%, resulting in a reduction in static strength. Pores containing hydrogen can be particularly dangerous; their presence in a susceptible microstructure can lead to cold cracking. Root porosity can affect root bend test results, by creating nucleation sites for crack formation. This is particularly likely when pores are present in elongated form with a sharp tail. Pores may also act as nucleation sites for solidification cracks, and can increase the likelihood of brittle fracture. Porosity is particularly detrimental when present in clusters, which degrade the formability of sheet steel welds. A potentially dangerous effect is that dense porosity may mask other more serious defects in a radiograph. In a given material, porosity can be reduced or eliminated by a number of means: creating a wider bead by reducing the welding speed, typically by about 20% of that required to achieve full penetration; by moving the focal plane relative to the workpiece; by beam spinning or oscillation to widen the weld bead; by pulsing the beam with an appropriate shape such that the keyhole is withdrawn gradually from the bottom to the top; by ensuring that abutting edges are free of oxides; and by using joints in which gases can escape through the root.

Inclusions may originate from the cleaning process, filler material additions or from a backing plate. They act as sites for crack initiation, or sources of hydrogen. Metallic inclusions may be brittle, providing sites for corrosion.

Misalignment

Linear misalignment (also known as high–low in the United States) describes the condition when the surfaces of the plates are not at the same level. It is a measure of the difference in the locations of the plates, and is caused by incorrect plate fixturing, or the failure of tack welds during welding. The resistance to service loads may be reduced by this type of defect.

Angular misalignment is measured in degrees and refers to the deviation from 180° of the welded plates.

Spatter

Spatter is the term used to describe liquid particles that are expelled during welding, which adhere to the surface of the base metal or weld bead. Spatter can arise from the volatility of alloying elements or instabilities in the keyhole during welding. It is normally removed mechanically after welding, and does not affect the mechanical properties of the weld.

Vaporization of volatile alloying elements weakens the weld bead in alloys that rely on solid solution strengthening. For example, the loss of magnesium reduces the weld bead strength of 5000 series aluminium alloys.

CARBON–MANGANESE STEELS

One measure of the effect of composition on the ease with which a material can be welded is its *weldability*. This is referred to given specifications, which may include limits on distortion, hardness or cracking, and can be defined in a standard or prescribed by a contractor. In addition to the material composition, weldability depends on: the welding process (the energy input); the environment; the type of joint (the degree of restraint); and the material thickness. For example, the composition of a carbon–manganese steel determines its hardenability (the tendency to form martensite); and an increase in the content of carbon and other austenite-stabilizing elements increases hardenability. The composition of a structural steel is therefore often expressed in terms of a *carbon equivalent* formula, which sums the contributions of elements to increasing hardenability with that of carbon. Various carbon equivalent formulae are available, depending on the criterion chosen to define weldability; they are discussed later. In addition to the amount of martensite formed, its hardness determines the average hardness of the weld. The hardness of martensite is dependent principally on carbon content, increasing linearly up to a maximum of about 900 HV at about 0.5 wt%. A maximum HAZ hardness is often stipulated as a measure of susceptibility to cracking; this determines the composition and process parameters that can be used. Steel manufacturers produce grades suitable for the relatively low energy input of laser welding. Weldability may also be characterized in terms of susceptibility to liquid state cracking; various parameters, similar to carbon equivalent formulae, have been proposed to quantify susceptibility to solidification cracking, and are also discussed later.

Laser welding procedures have been developed and certified for relatively thin section C–Mn steels, of low carbon content, such as those used in blanks for pressing. Thick section welding requires careful control of the process parameters because the low energy input per unit length results in high cooling rates. Higher carbon steels have high hardenability, and so procedure development is based on a criterion such as an upper hardness limit, typically 350 HV. Various empirical formulae (carbon equivalents), have been proposed to express the effect of alloying elements, relative to that of carbon, on the hardenability, cold cracking susceptibility, or a mechanical property of welded steel. They differ in the range of composition and cooling rate over which they were established, and for which they are valid. The most common is the International Institute of Welding formula (Svetskommissionen, 1971):

$$CE = C + \frac{Mn}{6} + \frac{Cr + Mo + V}{5} + \frac{Ni + Cu}{15}$$

where element symbols refer to nominal steel composition in wt%. According to this formula, a maximum carbon equivalent, CE, of 0.38 wt% is considered acceptable for welding of carbon–manganese steels. An alternative formula that defines the equivalent P_{cm} has been developed in Japan for welding of high strength steels (Ito and Bessyo, 1968):

$$P_{cm} = C + \frac{Mn}{20} + \frac{Si}{30} + \frac{Cu}{20} + \frac{Ni}{60} + \frac{Cr}{20} + \frac{Nb}{15} + \frac{V}{10} + 5B$$

where element symbols refer to nominal steel composition in wt%. According to this formula, a maximum carbon equivalent, P_{cm}, of 0.22 wt% is considered acceptable for welding of carbon–manganese steels.

Guidelines have also been produced for composition limits, based on acceptable mechanical properties, when laser beam welding hull structural steels, Table 16.4.

Empirical relationships also relate the solidification crack length, L (mm), and chemical composition (wt%), for steels containing 0.09–0.16 wt% C, laser welded using 8.5 kW and a welding speed of 1 m min^{-1} (Russell, 1999):

$$L = -2846C + 9810C^2 + 426S + 200$$

where element symbols refer to composition (wt%). This formula can be used to rank materials in order of crack susceptibility. A similar method defines an empirical UCS number (units of crack susceptibility) (CEN, 2001):

$$UCS = 230C + 190Si + 74P + 45Nb - 12.3Si - 5.4Mn$$

where element symbols refer to composition (wt%). A high number denotes a high susceptibility. In structural steels, the level of impurities such as sulphur and phosphorus must be reduced to below those acceptable for arc welding.

Table 16.4 Upper limits of chemical composition of laser weldable hull structural steels (wt%) (Lloyd's Register, 1997)

C	Mn	Si	S	P	Al	Nb	V	Ti	Cu	Cr	Ni	Mo	N	CE	P_{cm}
0.12	0.9–1.6	0.1–0.5	0.005	0.010	0.015	0.05	0.10	0.02	0.35	0.20	0.40	0.08	0.012	0.38	0.22

Table 16.5 Sources of data for laser welding of carbon–manganese steels

Grade	Laser	Reference
AISI 1008	CO_2	(Albright *et al.*, 1991)
SAE 1005	CO_2	(Wang, 1995)
Deep drawing	CO_2	(Dawes and Watson, 1984)
AMS 5525, AMS 5544	CO_2	(Alwang *et al.*, 1969)
StE 355, StE 690	CO_2	(Winderlich *et al.*, 1994)
Various	Nd:YAG	(Moore *et al.*, 2002)
A36	CO_2	(Gill *et al.*, 1986; Moon and Metzbower, 1988)

Low Carbon Steels

References for procedures for laser keyhole welding of low carbon steels, such as those used in automobile bodies, domestic appliances and light structures are given in Table 16.5.

The most common problems when welding low carbon steels are porosity and hot cracking. In rimmed steels, the reduction of ferrous oxides by carbon produces carbon monoxide, which can become trapped in the solidifying weld bead. Rimmed steels can be laser welded by ensuring that sufficient aluminium is added to deoxidize the weld pool. Improved fatigue properties, in comparison with the base material, are often reported in laser welds in low carbon steels, attributed to the generation of compressive residual stresses.

Primed carbon–manganese steels are coated to protect the steel from corrosion and oxidation before fabrication. The primer should be removed from the weld line in lap joints to avoid contamination and imperfections in the weld metal. Primer normally has little effect on weld quality in butt joints because it can be vented easily. Galvanized steels possess a coating of zinc, which has a vaporization temperature of 900°C. Zinc is therefore present at both the surface of the workpiece and the interface of joints – a particular problem in lap joints. Vaporized zinc readily forms plasma above the workpiece, which defocuses the laser beam resulting in a wider weld bead, a loss of penetration, a loss of zinc adjacent to the bead, and distortion in sheet materials. Zinc vapour becomes trapped at joint interfaces, forming vapour that cannot escape from the weld metal before solidification. Porosity and an uneven weld bead appearance, resulting in poor mechanical properties, are the two main consequences. By careful fixturing or the use of pressure devices, a controlled gap of around 0.1 mm at the interface can be created, which allows zinc vapour to escape from the weld region. Similar precautions are taken when welding sheets coated in other materials. Alternatively, a projection may be pressed into the top sheet prior to welding to maintain the desired gap (Graham *et al.*, 1996).

Medium Carbon Steels

Medium carbon steels can be laser welded relatively easily, providing impurities such as sulphur and phosphorus are not too high. Procedures are referenced in Table 16.5. The microstructure and hardness developed depend on the process variables; a medium energy input often produces microstructures of ferrite and martensite, while a low energy input gives more martensite, and higher hardness. Acicular ferrite is desirable in the weld metal microstructure since it reduces the likelihood of cracking.

Sulphur increases the weld bead aspect ratio, providing that welding is performed under conditions in which convective heat transfer is important, i.e. with high Peclet numbers. For pure metals, the temperature coefficient of surface tension, $d\gamma/dT$, is negative; the flow of molten metal is radially outward, i.e. from regions of high temperature in the centre of the weld pool to regions of low temperature at the solid–liquid interface. The presence of surface active elements such as sulphur and oxygen significantly affects $d\gamma/dT$, such that it can become positive. Radially inward fluid flow is then possible, which transports heat downwards, thus increasing penetration.

The maximum hardness developed in welded high strength steels is an important measure for evaluating weld zone ductility and the susceptibility to cold cracking and solidification cracking. Acceptable levels vary. Hardness levels of 350 or 375 HV are often used in offshore applications with conventional welds. 400 HV is proposed as an acceptable criterion for laser welding because careful control of the hydrogen present in the environment can be exercised. A hardness level below 400 HV can be achieved in steels with a carbon content below 0.12 wt% over a range of welding conditions. For welding speeds greater than 1 m min^{-1}, the maximum hardness is generally observed at the weld centreline, but may occur in other locations with lower welding speeds. Welds made in T-butt joints are generally harder than those made in I-butt joints using the same parameters, because of the larger heat sink and consequently higher cooling rate. However, if the microstructure of the I-joint is fully martensitic, a similar microstructure is produced with the same welding parameters in a T-butt weld, and so the difference in hardness is small.

Current rules specify a minimum weld ductility of 22%, which simulates the damage likely to be experienced in a ship. Based on the acceptability of 400 HV as an upper hardness level for laser beam welds, longitudinal bend tests show that structural steels can withstand this elongation without the formation of defects. Ductility decreases with increasing hardness in laser welds in structural steels. The microstructure of the top of the weld bead can be influenced by the shielding gas; nitrogen, for example, promotes martensite formation, which is a factor to take into account when bend testing the weld bead face. The root of a weld is generally more susceptible to defects, and lies on the outside of a ship structure – a root bend test is therefore a more stringent test than a face bend in ship welds.

The fatigue properties of laser-welded medium carbon steels are often reported to be superior to those of arc-welded steels. This may arise from the greater attention paid to fitup prior to welding, which reduces the likelihood of stress-raising notches being formed, or from the lower contact angle between the weld face and the base material.

High Carbon Steels

High carbon steel welds are susceptible to high hardness with high cooling rates. A low carbon filler material can be used to reduce the hardness of the weld metal. Preheating is normally applied to reduce the cooling rate of the weld and minimize the hardness developed in the weld bead and HAZ.

STAINLESS STEELS

Keyhole welding procedures for austenitic stainless steels are referenced in Table 16.6.

Austenitic Stainless Steels

Stainless steels that solidify as primary austenite are normally more susceptible to solidification cracking than those which solidify in the primary ferrite mode, since harmful elements such as sulphur and phosphorus are less soluble in austenite than ferrite, and segregate more easily.

Constitution diagrams have been constructed to describe the mode of solidification and microstructure in terms of composition indices. The Schaeffler diagram (Schaeffler, 1949) displays

Table 16.6 Sources of data for laser welding of stainless steels: N/S not specified

Type	Grade	Laser	Reference
Austenitic	AISI 304	CO_2	(Bonollo *et al.*, 1993a; Fedrizzi *et al.*, 1988; Zambon and Bonollo, 1994; Locke *et al.*, 1972)
	AISI 304	Nd:YAG	(Bransch *et al.*, 1994)
	AISI 304	CO_2	(Zacharia *et al.*, 1989a, b)
	AISI 316	Nd:YAG	(Goodwin, 1988)
	AISI 316	CO_2	(Zambon and Bonollo, 1994; Fedrizzi *et al.*, 1988)
	Various	CO_2	(David *et al.*, 1987; Nakao, 1989)
	Various	Nd:YAG, CO_2	(Chehaibou *et al.*, 1993; Lippold, 1994)
Ferritic	AISI 409		(Tullmin *et al.*, 1989)
	AISI 409	Nd:YAG	(Liu *et al.*, 1993)
Precipitation hardening	15-5 PH		(Ozbaysal and Inal, 1990)
Duplex	UNS 31803	CO_2	(Bonollo *et al.*, 1993b, c, 1996; Zambon and Bonollo, 1994)

the primary solidification mode of stainless steels using axes of chromium equivalent, Cr_{eq} (abscissa), and nickel equivalent, Ni_{eq} (ordinate), when welded using arc fusion processes. Its construction was described in Chapter 5. The abscissa represents elements that promote the formation of ferrite, and is defined as

$$Cr_{eq} = Cr + Mo + 1.5Si + 0.5Nb$$

whereas the ordinate represents elements that promote austenite formation, defined by

$$Ni_{eq} = Ni + 30C + 0.5Mn$$

where element symbols represent concentration (wt%). The diagram can be used to identify ranges of composition in which welds are susceptible to solidification cracking; for arc welding, a ferrite content of 3–8% gives a reasonable assurance of crack-free welds. A similar ferrite content can be expected to be desirable in the microstructure of laser welds, but the higher cooling rate of laser welds in comparison with arc welds influences the solidification mode, and hence the room temperature microstructure of stainless steel welds. The duplex region of low crack susceptibility in the Schaeffler diagram contracts, in comparison with arc fusion welds, as shown in Fig. 16.12. Thus a steel composition designed to produce a microstructure containing 3–8% ferrite after arc welding may solidify as a single phase during laser welding. Care must therefore be taken when using such diagrams with laser welding.

The Suutala diagram (Kujanpää *et al.*, 1979) is constructed using an ordinate that represents the ratio of ferrite to austenite stabilizing elements, Cr_{eq}/Ni_{eq}, and an abscissa that indicates the concentration of elements associated with harmful grain boundary segregation (sulphur and phosphorus). Crack-free welds are associated with a combined $S + P$ concentration below 0.02 wt%, or at higher concentrations a Cr_{eq}/Ni_{eq} ratio greater than 1.5. The solidification cracking susceptibility of laser welds extends to a wider range of composition than arc welds, and so a higher concentration of ferrite-stabilizing elements is required to prevent cracking in laser welds in stainless steels. Inhomogeneous mixing resulting from the high solidification rate may also contribute to the effect.

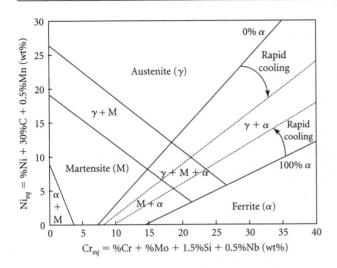

Figure 16.12 Contraction of the austenite–ferrite duplex field of the Schaeffler diagram caused by rapid cooling

Laser welding may reduce intermetallic phase formation because of the high cooling rate. Austenitic stainless steels can be welded by using a nickel-containing filler wire and post-weld heat treatment at 1100°C to obtain the desired balance of phases in the weld metal to avoid solidification cracking.

Ferritic Stainless Steels

Ferritic stainless steels possess good laser weldability. Welds have good impact properties, and the low energy input results in limited grain growth, and no weld sensitization. Procedures developed for welding medium carbon steels may be applied to welding ferritic stainless steels.

Martensitic Stainless Steels

Martensitic stainless steels should be tempered after welding, or preheated to 260–315°C to avoid cracking. On welding, high strength lower ductility microstructures containing bainite and carbides may form. In higher Cr grades the weld metal consists of tempered martensite. Cracking may be a problem, particularly if more than 0.1 wt% V and residual elements such as copper and antimony are present. Reheat cracking during post-weld heat treatment and elevated temperature service may occur, associated with embrittlement of grain boundaries, particularly in the coarse-grained HAZ.

Duplex Stainless Steels

Duplex stainless steels exhibit good laser weldability because of the low carbon content and high nitrogen content. In general, the weld bead contains more delta ferrite than predicted by the Schaeffler diagram. Typically the weld bead microstructure contains about 20% austenite. Rapid solidification after welding does not affect ferrite formation, but rapid cooling suppresses the solid state austenite phase transformation. Post-weld heat treatment in the range 1050–1100°C enables the parent material microstructural balance (50% austenite) to be regained, and good stress corrosion properties are obtained. Annealing and quenching encourages austenite formation in autogenous welds. High nickel filler wire may also be used to increase the austenite content. The hardness after welding typically increases up to 350 HV.

Precipitation-Hardened Stainless Steels

Precipitation-hardened stainless steels are welded in the solution annealed or overaged condition. In order to achieve full properties, the weld must be heat treated after welding. If these alloys are welded in the hardened condition, a narrow region is formed in the HAZ where the precipitate is dissolved. Heat treatment is required to regain full properties.

ALUMINIUM ALLOYS

Fusion welding of aluminium alloys results in the formation of three regions, which depend on the peak temperature attained: the fusion zone; the partially melted zone; and the HAZ. Each zone has a characteristic composition, microstructure and properties. In the as-welded condition, the principal strengthening mechanism is solid solution hardening, which endows relatively low strength but good ductility. In the fusion zone, material is melted and solidified rapidly with a microstructure characterized by epitaxial grain growth, columnar grains, and in some cases equiaxed grains in the centre of the weld bead. In the partially melted zone, peak temperatures between the liquidus and solidus are attained. The main effects in the HAZ are solid state reactions involving the coarsening and dissolution of strengthening precipitates during the thermal cycle. Dissolution is dominant in regions of relatively high peak temperature, and coarsening in regions of relatively low peak temperature. Strength is lost through both precipitate dissolution and coarsening, and the formation of incoherent particles. Reprecipitation during cooling is possible close to the fusion zone in some alloys. The most common defects in laser welds in aluminium alloys have been reviewed by Dubiel (1987).

Material Properties

The weldability of aluminium alloys is influenced by the thermal properties of aluminium and the alloying additions present. The high value of thermal conductivity means that heat is conducted away from the beam–material interaction zone easily; it is then difficult to maintain the power density required for keyhole formation, and heat is able to build up ahead of the weld, thus increasing absorption of the beam, which can result in process instability. The high coefficient of thermal expansion results in greater weld distortion than ferrous alloys. The high density of free electrons in aluminium makes it one of the best reflectors of infrared light, and measures must be taken to avoid back reflection of the beam. Common alloying additions in aluminium alloys, such as zinc and magnesium, have low vaporization temperatures, and so vaporize readily to reduce the threshold power density for keyhole and plasma formation, which can also lead to welding instabilities. Loss of material through vaporization can result in concavities of the root and face. The low viscosity of molten aluminium can cause the weld bead to sag; in extreme circumstances holes can form in the seam. Aluminium oxidizes readily, and so precautions must be taken to remove the oxide coating prior to welding, and to avoid oxide formation during welding. Individual welding procedures are therefore needed for specific aluminium alloys.

Beam Characteristics

The focal plane should be placed at or slightly below the workpiece surface to improve the stability of the keyhole during welding of aluminium alloys. By splitting the beam into two energy sources, aligned in the direction of welding and displaced by a fraction of a millimetre, the weld pool is elongated, enabling gas pockets to escape before solidification. Improvements to the weld bead can also be made by using two Nd:YAG beams with different focal planes (Ga-Orza, 1999).

When a pulsed beam is used for welding, a high duty cycle helps to prevent cracking.

Processing Parameters

A reflected perpendicular beam of light can damage the optics, or even enter the cavity to amplify the laser power in an uncontrollable manner. A set of polarizing mirrors may be used to eliminate back reflection (Chapter 4). Alternatively the beam or workpiece may be tilted.

Weld bead porosity is the main imperfection found in laser welds in aluminium alloys. It can be reduced by adopting special procedures for material handling and by modifying the processing parameters. Oxide films and other contaminants on surfaces and edges in the beam–material interaction zone are removed by chemical etching, laser ablation, machining or wire brushing immediately prior to welding. Filler materials should be subjected to similar treatment. High purity, low dew-point helium process gas is preferred, delivered through a system that prevents condensation. Porosity can be reduced by using a high welding speed to reduce the nucleation and growth of bubbles with an increase in solidification rate. Similarly, by defocusing the beam by up to 3 mm above or below the workpiece, a wider bead is formed, which enables bubbles to escape. By using helium alone as the shielding and plasma formation gas, the interaction region becomes unstable, causing fluctuations in the keyhole and coarse porosity. Argon–helium mixtures are therefore preferred because argon promotes keyhole stability, reducing coarse porosity. Brittle nitrides can be formed in aluminium alloys if nitrogen gas is used.

Segregation of the eutectic phase at grain boundaries may also be observed, leading to low ductility and liquation cracking next to the fusion zone. Solidification cracking can occur along the centreline because of the segregation of the eutectic phase. Such problems may be alleviated through the use of filler materials.

The purpose of using filler materials in welding of aluminium is to reduce the solidification temperature interval during which the alloy is susceptible to solidification cracking. Thus 2319 filler wire is used to avoid solidification cracking when welding 2024; the former has a higher copper content, which reduces the solidification range between the liquidus and solidus. Similarly, 4043 or 4047 filler wire, with silicon contents of 5 and 12 wt% (eutectic composition), respectively, reduce the solidification range, and are used when welding 5000 and 6000 series alloys. A summary of recommended filler materials for MIG welding of different alloys is given in Table 16.7, which can be used as a guide to selecting appropriate filler materials for laser welding.

Laser beam welding of aluminium alloys has been researched extensively, primarily because of the importance of aluminium in the transport industries. Examples of procedures developed for various alloys can be found in the references listed in Table 16.8.

Arc welding practice recommends that the silicon content of the weld metal in Al–Mg–Si alloys is kept above 2% to prevent solidification cracking. (This corresponds to a concentration of Mg_2Si of around 1%.) Empirical solidification crack indices (SCIs) have been developed for binary and ternary alloys, based on laser welds:

$$SCI_{binary} = 15.32 - (\%Si - 0.7)^2 - 0.25(\%Cu - 2.0)^2 - 0.04(\%Mg - 1.5)^2$$
$$SCI_{ternary} = 16.00 - (\%Si - 0.7)^2 - 0.25(\%Cu - 2.0)^2 - 0.04(\%Mg - 1.5)^2 - 0.0143(\%Zn - 8.3)^2.$$

The maximum crack length observed increases with an increase in SCI, from 0.4 mm to 0.8 mm with an increase in SCI from 13.7 to 14.5. Material with a low SCI is therefore preferred.

Table 16.7 Selection of filler materials for MIG welding of aluminium alloys; the upper, middle and lower designations at intersections of alloy rows and columns refer to selections for maximum strength, corrosion resistance and freedom from cracking, respectively (Dudas and Collins, 1966)

	Al–Si casting	Al–Mg casting	1xxx	2xxx	3xxx	5xxx	5005	5083	6xxx	7020
7020	NR	5	5	NR	5	5	5	5556A	5	5556A
		5	5		5	5	5		5	5
		5	5		5	5	5		5	5
6xxx	4	5	4	NR	4	5	5	55	4/5	
	4	5	4		4	5	4	5556A	4	
	4	5	4		4	5	4	5	4	
5083	NR	5	5	NR	5	5	5	5556A		
		5	5		5	5	5	5		
		5	5		5	5	5	5		
5005	4	5	5	NR	5	5	5			
	4	5	4		4	5	5			
	4	5	4		4	5	5			
5xxx	NR	5	5	NR	5	5				
		5	5		5	5				
		5	5		5	5				
3xxx	4	5	4	NR	3					
	4	5	3/4		3					
	4	5	4		3					
2xxx	NR	NR	NR	2						
				4						
				4						
1xxx	4	5	1							
	4	5	1							
	4	5	1							
Al–Mg casting	NR	5								
		5								
		5								
Al–Si casting	4									
	4									
	4									

1xxx = 1080A, 1050A, 1200, 1350; 2xxx = 2219, 2014, 2024; 3xxx = 3103, 3105; 5xxx = 5251, 5454, 5154A; 6xxx = 6063, 6061, 6082, 6101A; 1 = 1080A, 1050A; 2 = 2319; 3 = 3103; 4 = 4043A, 4047A; 5 = 5056A, 5356, 5556A, 5183; NR = Not recommended

MAGNESIUM ALLOYS

Procedures for laser welding magnesium alloys have been developed based on those for aluminium alloys. Many of the difficulties in welding magnesium have similar origins to those in aluminium, and the measures taken to reduce their effects on weld properties are therefore similar. It is not possible to provide generalized procedures because of the individual nature of each alloy; the parameters used for some of the more popular alloys can be found in the references listed in Table 16.9.

Table 16.8   Sources of data for laser welding of aluminium alloys

Grade	Laser	Reference
2024	CO_2	(Kokkkonen and Ion, 1999)
2024	Nd:YAG	(Milewski et al., 1993)
2090	CO_2	(Marsico and Kossowsky, 1989)
2090-T8E41	CO_2	(Molian and Srivatsan, 1990)
2195	CO_2	(Jan et al., 1995; Hou and Baeslack, 1996)
2196-T8	CO_2	(Martukanitz and Howell, 1995)
2219, 5083, 6063	CO_2	(Kutsuna et al., 1993)
2219 T3	CO_2	(Kimura et al., 1991)
2219/6061	CO_2	(Gopinathan et al., 1994)
5xxx	Diode	(Herfurth et al., 2001)
5083	CO_2	(Blake and Mazumder, 1985)
5083	Nd:YAG	(Junai et al., 1990)
5086-H32	Nd:YAG	(Cieslak and Fuerschbach, 1988)
5182-O	CO_2	(Leong et al., 1999)
5456-H116	CO_2	(Katayama and Lundin, 1992)
5456-H116	Nd:YAG	(Cieslak and Fuerschbach, 1988)
5456-H116	CO_2	(Moon and Metzbower, 1983)
5754	CO_2	(Duley, 1994; Smith et al., 1993)
5754	Nd:YAG	(Junai et al., 1990)
5754-O	Nd:YAG	(Martukanitz and Altshuller, 1996)
5754-O	Nd:YAG, CO_2	(Ramasamy and Albright, 2001)
6013-T6	CO_2	(Guitterez et al., 1996; Douglass et al., 1996;
6061-T6	CO_2	Hirose et al., 1997, 2000)
6061-T6	Nd:YAG	(Cieslak and Fuerschbach, 1988)
6063	CO_2	(Yamaoka et al., 1992)
6082 (cellular)	CO_2, Nd:YAG	(Haferkamp et al., 2004)
G-AlSi9Cu3	CO_2	(Chen and Roth, 1993)
AMg6	CO_2	(Avramchenko and Molchan, 1983)
7075	CO_2	(Adair, 1994; Kimura et al., 1991)
7475	CO_2	(Gnanamuthu and Moores, 1987)
8090	CO_2	(Lee et al., 1996; Whitaker et al., 1993; Whitaker and McCartney, 1994)
8090 T3	CO_2	(Gnanamuthu and Moores, 1988)
Al–8.5Fe–1.2V–1.7Si	CO_2	(Whitaker and McCartney, 1995)
Al–8Fe–2Mo	Nd:YAG	(Sampath and Baeslack, 1993)

TITANIUM ALLOYS

Good results are generally obtained when laser welding is applied to titanium alloys. They are not prone to solidification cracking, except under conditions of considerable restraint. Titanium alloys are susceptible to oxidation, and so a trailing inert gas shield should be used to protect the weld bead and the HAZ. Titanium alloys are also susceptible to hydrogen embrittlement, but maintenance of clean conditions and the use of shielding gas is normally sufficient to prevent problems arising. The short thermal cycle of laser welding restricts grain growth, and as a result laser-welded materials have

Table 16.9 Sources of data for laser welding of magnesium alloys

Grade	Laser	Reference
AZ31B-H24	CO_2	(Sun *et al.*, 2002)
AZ, AM series	Nd:YAG	(Draugelates *et al.*, 2000)
AZ31B	Nd:YAG	(Haferkamp *et al.*, 1997)
AZ31B-F, AZ61A-F, AZ91HP-F, AM50HP-F	Nd:YAG, CO_2	(Haferkamp *et al.*, 2000)
AZ91, Mg Y4 Se3	CO_2	(Maisenhälder *et al.*, 1993)
WE54X	Nd:YAG	(Baeslack *et al.*, 1986)
AZ91, AM60, ZC63, ZE41, QE22, WE54, AZ31, AZ61, ZW3, ZC71	CO_2	(Weisheit *et al.*, 1998)

Table 16.10 Sources of data for laser welding of titanium alloys

Type	Grade	Laser	Reference
α	CP Ti	CO_2	(Juckenath *et al.*, 1988; Metzbower, 1990)
	Ti–14Al–21Nb	CO_2	(Martin *et al.*, 1995)
$\alpha + \beta$	Ti–6Al–4V	CO_2	(Mazumder and Steen, 1980, 1982; Metzbower, 1990; Juckenath *et al.*, 1988)
	Ti–6Al–2Sn–4Zr–2Mo, Ti–5Al–5Sn–2Zr–4Mo		(Baeslack and Banas, 1981)
β	Ti–22V–4Al	CO_2	(Shinoda *et al.*, 1991)

superior ductility to their arc-welded equivalents. Procedures can be found in the references listed in Table 16.10.

NICKEL ALLOYS

The normal recommendation is to weld nickel alloys in the solution-treated state and then to carry out solution and ageing treatments to facilitate precipitation hardening. Laser welding results in superior properties to TIG welds because of reduced agglomeration of dispersoids. The rapid thermal cycles in laser welds have also been found to reduce microfissuring, probably as a result of the lack of time for segregation of embrittling elements to grain boundaries. The reflectivity of nickel alloys to infrared light is high, therefore a high power density is required. References for keyhole welding of nickel alloys can be found in Table 16.11.

By mechanically alloying the nickel-based Inconel alloy MA 754, oxide dispersion strengthening (ODS) low melting temperature elements such as Al and high melting temperature elements such as Cr and Ni can be alloyed with yttrium oxide (Y_2O_3) in order to extend the creep, oxidation and rupture performance of the alloy in turbine vane applications in advanced gas turbine engines. These alloys require joining, but excessive heating of the ODS alloy can cause oxide coalescence leading to severe agglomeration, reducing the effectiveness of the dispersoid in pinning dislocations. Melting produces an equiaxed grain structure with a grain boundary orientation, giving a plane of weakness in the weld. As a result, diffusion bonding and hot isostatic pressing (HIP) are used to join components. Laser

Table 16.11 Sources of data for laser welding of nickel alloys; Inconel is a registered trade mark of Special Metals Corporation

Grade	Laser	Reference
Inconel® 718	Nd:YAG	(Cornu *et al.*, 1995)
MA 754	CO_2	(Molian *et al.*, 1992a)
Nickel aluminide	CO_2	(Molian *et al.*, 1992b)

Table 16.12 Sources of data for laser welding of copper alloys

Grade	Laser	Reference
C1020 1/2H (Oxygen-free)	CO_2	(Kimura *et al.*, 1991)
Cu	Nd:YAG	(Ramos *et al.*, 2001; Biro *et al.*, 2002; Hashimoto *et al.*, 1991)
Cu–9Al–4Fe–4Ni	CO_2	(Petrolonis, 1993)
Brass	Nd:YAG	(Ramos *et al.*, 2001)
Bronze	Nd:YAG	(Ramos *et al.*, 2001)

welding melts a minimum of base material with a low energy input, and has been found to produce weldments with good yield strength, tensile strength, ductility and hot corrosion resistance.

Ductile nickel aluminide, $Ni_3Al + B$, is an intermetallic alloy with high strength and ductility, and has potential for both elevated temperature and cryogenic use. The alloy has been laser welded, although cracking in the fusion zone is reported.

Laser welding of nickel aluminium bronze is effective in producing primarily martensitic weld metal and a narrow HAZ, which reduce selective phase corrosion.

COPPER ALLOYS

The high reflectivity of copper to infrared radiation is the main problem to overcome when welding its alloys. The workpiece, beam, or both may be inclined to prevent back reflection into the laser cavity. The use of a polarizing lens (Chapter 4) also prevents back reflection. By adding oxygen in amounts up to about 50% in an inert process gas, absorptivity can be improved, without the formation of embrittling oxides in the weld pool. The formation of plasma with CO_2 laser light has been found to be beneficial in coupling the beam. Preheating, nickel plating and welding in an oxidizing atmosphere are also methods of improving absorptivity. Keyhole welding references are listed in Table 16.12.

Laser welding of brass is difficult: porosity arises from vaporization of alloyed zinc.

OTHER NON-FERROUS ALLOYS

Limited trials have been conducted on laser welding of other metallic alloys. A list of references is given in Table 16.13.

Laser welds in molybdenum are usually brittle, but may be acceptable where high strength is not required. Ductile welds may be produced in tantalum, but precautions must be taken to prevent

Table 16.13 Sources of data for laser welding of various non-ferrous metallic alloys

Grade	Laser	Reference
Ir–0.3% W	CO_2	(David and Liu, 1982)
Zircaloy 2	CO_2	(Ram *et al.*, 1986)

oxidation. A high power input is required to weld tungsten because of its high melting temperature and high thermal conductivity.

DISSIMILAR METALS AND ALLOYS

The following features of laser welding provide advantages over other processes for joining dissimilar metal combinations. The high power density of the focused beam is sufficient to fuse the most important engineering metals and alloys, and overcomes many of the problems associated with large differences in thermal conductivity. The small size of the focused beam and the accuracy with which it can be positioned allow the fusion ratio of the base materials to be controlled, thus providing control over the formation of sensitive microstructures. The beam is normally positioned in the material with the higher conductivity. The low energy input of the process results in rapid solidification and cooling rates, reducing the extent of segregation in some alloys, and providing the opportunity for the formation of novel microstructures. Filler materials can be used to join plain carbon steel to austenitic stainless steels, for example. The nickel addition ensures an austenitic weld metal microstructure, rather than the predominantly martensitic microstructure produced by autogenous welding. References for a selection of laser welding procedures for dissimilar metal combinations are given in Table 16.14.

Electron beam welding (EBW) also uses the deep penetration keyhole mechanism, and is therefore a candidate process for joining dissimilar metal joints. Since deep penetration EBW is performed in vacuum, greater penetration is possible in comparison with a laser of similar power. However, the size of the vacuum chamber limits the size of component that can be welded, and the vacuum pumping time may be a critical factor in high production rate applications for all but the very thickest parts. Differences in the magnetic properties of the joint components can cause deflections of the beam, to the extent that the joint line may be missed. The energy input of gas and arc processes such as TIG, MIG, MMA, PAW and SA can be an order of magnitude greater than that of laser welding. The cooling and solidification rates are therefore lower, increasing the risk of segregation and the formation of brittle intergranular phases. The fusion and heat affected zones are larger, producing more distortion and larger residual stresses. Alignment of the heat source relative to the joint line, and the subsequent control of dilution is more difficult. Brazing and soldering are suitable joining processes for a wide range of dissimilar metal combinations, in particular those in which metallurgical incompatibility results in undesirable microstructures. The energy input is low, but joints often have a limited operational temperature range, and the possibilities for galvanic corrosion must be appreciated. Resistance welding is suitable for many dissimilar metal combinations, but is particularly amenable to automation, and avoids many fusion-related problems because of the very limited area of melting. Spot, projection, seam and upset welding are all variations of the process. However, dissimilar metal combinations having large differences in electrical conductivity may give problems, and the clamp produced by the electrodes requires access from both sides of the joint.

Table 16.14 Sources of data for laser welding of dissimilar metals and alloys

Grade 1	Grade 2	Laser	Reference
Cu	AISI 300 series	Nd:YAG	(Shen and Gupta, 2004)
ASTM A 387	17740 X2 CrNi 1911		(Missori and Koerber, 1997)
GGG-40, GGG-60	20 MnCr 5, 42 CrMo 4	CO_2	(Dilthey et al., 1998b)
Ti–6Al–4V/SiC fibre	Ti–6Al–4V	CO_2	(Fukumoto et al., 1993)

Table 16.15 Sources of data for laser welding of ceramics

Grade	Laser	Reference
Alumina	CO_2	(Exner, 1994)
Alumina	CO_2, Nd:YAG	(Reinecke and Exner, 2001)
Various	Various	(Maruo et al., 1986; Santella, 1992)

CERAMICS

The availability of effective joining techniques for ceramics could broaden their use in mass-produced components. However, the methods used for ceramic processing and the materials themselves present serious challenges. Deformation of densified ceramics to form complex shapes is practically impossible because most ceramic materials are brittle even at elevated temperatures. One of the main problems encountered when fusion welding ceramics is the control of cracking caused by thermal stresses, which are induced in both the HAZ and solidifying weld bead. Several techniques have been developed to join ceramics for structural application: brazing with filler metals; diffusion bonding; microwave joining; and the use of interfacial layers designed to form a thin transient liquid phase at a relatively low bonding temperature.

Despite these difficulties, attempts to develop laser welding procedures for ceramics have been made; Table 16.15 lists some references.

Alumina has been joined successfully to itself and silica by preheating to 1600°C using a scanning CO_2 laser beam, followed by welding using an Nd:YAG laser beam. Preheating reduces the thermal shock to the ceramic, which would cause cracks to form.

POLYMERS

There are about 15 different techniques used to join polymers. One class of techniques generates heat internally, and includes friction and ultrasonic welding. A second uses an external heat source such as a hot bar or a gas. A third class uses the effects of electromagnetic waves; these include induction, microwave and laser welding.

Polymers absorb far infrared radiation well, and so CO_2 laser welding procedures have been developed, particularly for thin sheet in a lap joint configuration. However, a novel interfacial method of joining transparent materials, which avoids problems associated with through-thickness welding, was described in Chapter 13; absorbing dyes are placed at the interface of the lap joint, and welding is achieved using a diode laser.

References for through-thickness laser welding of various polymers are given in Table 16.16.

Table 16.16   Sources of data for laser welding of polymers

Grade	Laser	Reference
LDPE	CO_2	(Ruffler and Gürs, 1972)
LDPE, PP	CO_2	(Duley and Gonsalves, 1972)
PE, PP	CO_2	(Jones and Taylor, 1994)
PE, PP	Nd:YAG	(Jones and Taylor, 1994)
ABS	CO_2, Nd:YAG	(Joma-Plastik, 1987)

Table 16.17   Sources of data for laser welding of composites

Class	Grade	Laser	Reference
Metal matrix	SiC particle–Al alloy	CO_2	(Dahotre *et al.*, 1990)
	SiC particle–Al alloy	CO_2, Nd:YAG	(Cola *et al.*, 1990)
	SiC particle reinforced aluminium A-356	CO_2	(Lienert *et al.*, 1993)

Low power (100 W) CO_2 laser welding is suitable for thin (<0.2 mm) lap joints in polyethylene and polypropylene, which can be welded at rates on the order of 100 m s^{-1}. An Nd:YAG laser beam is suitable for lap and butt joints 0.2–2 mm in thickness.

COMPOSITES

Little has been reported on laser welding of composites. The majority of work has involved the metal matrix class. Problems associated with laser welding composite materials can arise from a number of sources. Preferential absorption of the beam by one component (normally the non-metallic) can lead to overheating and dissolution, with possible reprecipitation of deleterious phases. Differences in the thermal properties can lead to cracking. Keyhole welding references are given in Table 16.17.

Preferential absorption by particulate SiC in the composite A-356 has been reported to lead to dissolution of the particles, and subsequent precipitation of deleterious Al_4C_3. However, low energy input levels and the rapid thermal cycles have been reported to preclude the formation of Al_4C_3. The duty cycle of a pulsed laser beam can be used to vary the degree of SiC dissolution.

Silicon carbide fibre reinforced Ti–6Al–4V composites have been joined by inserting a Ti–6Al–4V filler material into the joint. By using a wide bead, and by positioning the beam in the unreinforced plate, successful welds have been made autogenously between the composite and Ti–6Al–4V. A post-weld heat treatment improves properties. Full penetration welds with no apparent fibre damage can be made under optimum conditions, using welding directions both parallel and perpendicular to the fibre direction.

A potential application of laser welding polymer matrix composites is joining of skin tissue. Traditional methods of closing a wound are stitching and stapling. Both can allow puckering around the wound and seepage of fluids, and neither can be used in inaccesible locations. Laser welding would be faster than traditional methods, could be used in difficult locations, would result in better joints, more rapid healing and decrease the risk of complications.

LASER KEYHOLE WELDING DIAGRAMS

Because of the large number of variables involved in laser welding, it is unlikely that a complete process model could ever be developed. A small change in, for example, the shielding gas flow rate is sufficient to produce a dramatic effect on the geometry of the weld bead. The cause of this – changes in the plasma density above the keyhole – is difficult to model accurately, because of the complexity of the physical mechanisms involved and the large number of poorly known physical constants. However, individual components of the process have been modelled successfully.

Here we do not attempt to model all the features of keyhole welding. Under certain conditions, e.g. full penetration welding, simplifications can be made that allow analytical expressions to be obtained between the process variables. Thus the keyhole can be modelled as a moving through-thickness line heat source (Rosenthal, 1941, 1946), which allows the widths of the fusion zone and the HAZ to be established directly (Adams, 1958; Swift-Hook and Gick, 1973). Such methods can be used to construct a variety of laser welding diagrams.

OVERVIEW

Equation (E3.1) in Appendix E describes lateral heat flow from a stationary line heat source within an infinite solid. This solution is fitted to the welding geometry as follows. The vapour cavity is modelled as a cylinder of length l (the plate thickness) with a diameter equal to the beam radius r_B, moving at a speed v. The aspect ratio of the molten zone (depth/width) is assumed to take a value of 4, such that the zone extends a radial distance $l/8$ from the heat source, at which point the temperature is T_m. (Variations in the aspect ratio may be reflected in the model if necessary.) The cross-sectional area of the molten zone is then $l^2/4$. The edge of the heat affected zone, at which the temperature has fallen to the initial temperature, T_0, is assumed to extend a radial distance l from the heat source. Therefore in equation (E3.1), $T_1 = T_m$ where $r_1 = l/8$, and $T_2 = T_0$ at $r_2 = l$. An energy balance between the absorbed power and the power used for vaporization, melting and heating gives

$$Aq = r_B l v L_v + \frac{l^2}{4} v L_m + \frac{2\pi \lambda l (T_m - T_0)}{\ln(8)} \tag{16.1}$$

where L_v is the volumetric latent heat of vaporization ($\mathrm{J\,m^{-3}}$), L_m is the volumetric latent heat of melting ($\mathrm{J\,m^{-3}}$), and λ is the thermal conductivity ($\mathrm{J\,s^{-1}\,m^{-1}\,K^{-1}}$). In terms of dimensionless variables this becomes

$$q^* = l^* v^* L_v^* + \frac{l^{*2} v^* L_m^*}{4} + \frac{2\pi l^*}{\ln(8)} \tag{16.2}$$

where $L_m^* = L_m/[\rho c(T_m - T_0)]$ and $L_v^* = L_v/[\rho c(T_m - T_0)]$. The values of L_m^* and L_v^* lie around 0.4 and 4, respectively for metals and alloys (Richard's and Trouton's rules).

Equation (16.2) is an explicit relationship between q^* and v^*, for a given normalized plate thickness, l^*. The values of L_m^* and L_v^* that are used determine the class of materials for which the diagram is valid −0.4 and 4, respectively are appropriate for metals and alloys. Figure 16.13 is constructed by using equation (16.2) with values of L_m^* and L_v^* representative of metals and alloys, and shows contours of l^*.

In its general form, Fig. 16.13 may be used to describe a range of keyhole welding conditions for engineering materials within a particular class. But it can also be converted into a practical diagram with axes of principal process variables by substituting appropriate values for material properties (Appendix D) and beam size.

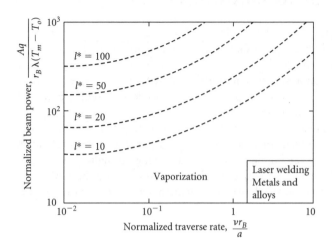

Figure 16.13 Model-based overview diagram for laser keyhole welding of metals and alloys, showing contours of constant normalized plate thickness, l^* (broken lines)

PARAMETER SELECTION

Figure 16.13 is converted to a material-specific diagram by substituting appropriate thermal properties into the variables used as the axes. Figure 16.14 is thus constructed for commercial purity aluminium. A focused beam radius of 0.3 mm is assumed. The positions of the contours are fixed through a single calibration, indicated by a star. Figure 16.14 provides an overview of keyhole welding of the material, and is useful when a rough estimate of process parameters is required, e.g. when planning the purchase of a laser and beam handling system. It is constructed using a simple model, which is valid only for the processing setup that was used to establish the data point. When a welding procedure is developed, a more accurate calibration is required, and it is then more useful to present the model results using linear axes.

The keyhole is treated as a through-thickness line heat source, from which heat flows laterally in the workpiece. Equation (16.3) gives an explicit relationship between beam power, q, welding speed, v, and plate thickness, d. However, it contains a number of poorly known constants, such as the latent heats of transformation. This difficulty may be overcome by calibrating the equation to a single experimental data point, thus eliminating the constants while maintaining the essential relationships.

Consider the case of full penetration welding of a material. A small set of experiments is conducted to establish that full penetration is just achieved for the conditions: $l = l_{cal}$, $q = q_{cal}$, $v = v_{cal}$. Equation (16.1) then becomes

$$A q_{cal} = v_{cal}(r_B l_{cal} L_v + \frac{l^2}{4} L_m) + \frac{2\pi \lambda l_{cal}(T_m - T_0)}{\ln(8)}. \tag{16.3}$$

For a given material and weld, the only unknown in equation (16.3) is the absorptivity, A. A can be calculated from a calibration. Equation (16.3) may then be used to construct a diagram with axes of incident beam power, q, and traverse rate, v.

Provided that the mechanism does not vary, the data point technique can be used with an analytical model to establish the *variation* about the known condition with changes in process variables. Note that the calibration reduces the generality of the diagram, but increases its range of practical use.

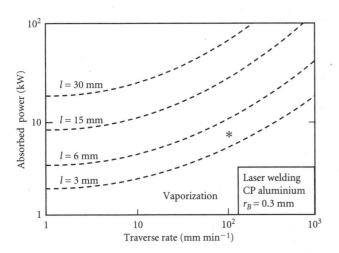

Figure 16.14   Calibrated model-based parameter selection diagram for laser keyhole welding of commercial purity aluminium using a focused beam of diameter 0.6 mm. The axes are the principal process variables power (kW) and traverse rate (mm min^{-1}). Contours for full penetration welding to various depths (mm) are shown as broken lines. The contours are calibrated to an experimental measurment, which is indicated by a star

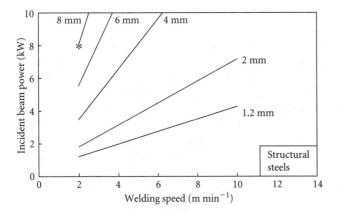

Figure 16.15   Laser keyhole welding diagram, showing conditions for full penetration; the model is calibrated using an experimental measurement, which is indicated by a star

Figure 16.15 is based on the following calibration: a beam power of 8 kW is required to fully penetrate a plate of thickness 8 mm with a welding speed of 2 m min^{-1}. By using this data point in equation (16.3), the other contours of penetration shown in Fig. 16.15 may be constructed. Note, however, that the model is unable to predict the loss of penetration with low beam power (irrespective of the welding speed), because this represents a change in the processing mechanism.

WELD PROPERTIES

Figure 16.15 provides a framework in which the properties of completed welds may be presented. Modelling is used to describe the constitutive relationships between processing parameters, e.g. for full

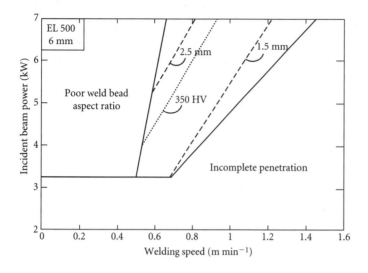

Figure 16.16 Laser weldability diagram for a 6 mm plate of the structural steel EL 500 showing the empirically established region of acceptable weld bead geometry (solid boundaries), which contains model contours of HAZ width (dashed) and maximum HAZ hardness (dotted)

penetration welding, but empirical data are used to establish the precise functional relationships. Such a diagram is shown in Fig. 16.16.

Figure 16.16 shows the weld penetration limits (the weldability lobe), together with the properties of the HAZ, for a 6 mm plate of the structural steel EL 500. The limit of penetration is established by using equation (16.3) in conjunction with a data point. The theoretical contours of constant HAZ width are established using equation (E3.8) (Appendix E). The A_{c1} and T_m isotherms denote the HAZ boundaries, and are calculated using empirical equations based on chemical composition (Equations E5.1 and E5.3 (Appendix E)). The contour of constant hardness is calculated by first establishing the cooling time required using equation (E3.5) (Appendix E). The resulting phase transformation products and phase hardness are then calculated by using the method described for laser hardening in Chapter 9. Welding parameters to the right of this hardness contour are predicted to produce an HAZ hardness greater than 350 HV. The diagram may be used to estimate processing parameters according to the need for full penetration, or criteria such as a maximum HAZ hardness, or a maximum HAZ width.

Model-based Hardenability Overview

The methods described above may be used to construct the diagram shown in Fig. 16.17. This is an overview of the effects of structural steel composition and welding energy on the maximum HAZ hardness developed. It enables either composition, or welding parameters, to be selected for a criterion of a given tolerable HAZ hardness. Thus materials and process parameter combinations may be sifted quickly, in order to expedite the development of a welding procedure specification for qualification testing.

SOFTWARE FOR PARAMETER SELECTION

A PC-based system for procedure development in laser welding using the methods described above has been described (Ion and Anisdahl, 1995). The software prompts the user for known data, such as

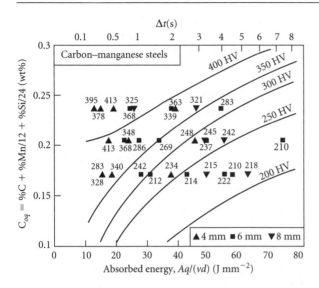

Figure 16.17   Hardenability overview for laser welding of carbon–manganese steels; the boundaries are model predictions of maximum HAZ hardness. Experimental measurements of maximum HAZ hardness (VPN), established by laser welding steels of various thickness, are included

the material composition, beam power and size, as well as user requirements, for example the depth of hardening. Model-based calculations are then used to establish unknown parameters, e.g. the traverse rate. The effects of variations in material composition, as well as changes in processing parameters, on the properties of the hardened region can be established rapidly. The system also includes a facility for archiving and retrieving procedures for future use.

INDUSTRIAL APPLICATION

The dominant role that the automotive industry has played in the development of industrial applications can be appreciated from the data given in Table 16.18.

Many of the applications developed for the automotive industry were adopted by other users of steel sheet in the late 1980s and 1990s, notably in the manufacture of domestic goods. The CO_2 laser beam continues to be the source of choice for welding of long linear and rotationally symmetric parts. However, the development of multikilowatt Nd:YAG lasers towards the end of the 1980s, with the advantage of fibreoptic beam delivery, enables laser welding to be applied to more complex three-dimensional parts.

The shipbuilding industry played a central role in the 1990s in obtaining approval for laser welding of structural steel in the range 6–25 mm, which resulted in guidelines covering steel composition and laser welding parameters that were drawn up by the Classification Societies (Howarth, 1999). The increasing use of aluminium in the transportation sector stimulates research into laser welding of various industrial alloys, which are now common, and are being put into production in the aerospace industries.

Today, applications of laser welding are growing rapidly, because of a better understanding of the joint requirements, the implications and opportunities for product redesign, and their acceptance through codes produced by industry bodies. In addition to promoting growth in established procedures, the availability of diode lasers provides solutions in both microjoining applications and

Table 16.18   Industrial applications of laser keyhole welding: N/S not specified

Industry sector	Product	Material	Laser	Reference
Aerospace	Airbus airframe skin stringers	AA 6013	CO_2	(Eritt, 1995)
	Aircraft structures	AA 2195	Nd:YAG	(Lang *et al.*, 2000)
	Jet engine casing flanges	N/S	N/S	(Kunzig, 1994)
	Space shuttle engines	Inconel	N/S	(Irving, 1992)
Automotive	Door panels	N/S		(Bransch, 1996)
	Audi A2 panels and extrusions	Aluminium	N/S	(Larsson, 1999a)
	Audi A8 bodywork	Aluminium	Nd:YAG–MIG	(Anon, 2002)
	Audi floorpan blanks	Low C steel		(Uddin, 1991)
	BMW 5-series bodywork	Low C steel	CO_2, Nd:YAG	(Hornig, 1997)
	Cadillac Seville centre pillar	Low C steel	CO_2	(Neiheisel and Cary, 1991)
	Chrysler Ultra-Drive transmission	C–Mn steel	CO_2	(Ogle, 1991)
	Exhaust systems	AISI 409	CO_2	(Irving, 1993; Anon., 1994)
	Ford estate roof sections	Low C steel		(Breuer, 1994)
	Fuel injectors	Stainless and C–Mn steels	Nd:YAG	(Anon., 1999)
	GM Cadillac rear shelf	Low C steel	CO_2	(Forbis-Parrott, 1991)
	Mercedes key system	Polymer	Diode	(Larsson, 1999b)
	Porsche Carrera side doors	Aluminium	N/S	(Larsson, 1999a)
	Tailored blanks	Coated steel	Various	(Auty, 1998; Aristotile and Fersini, 1999; Azuma and Ikemoto, 1993; Baysore *et al.*, 1995)
	Volvo S70 and C70 bodywork	Low C steel		(Larsson *et al.*, 2000)
	Volvo 960 bonnet	AlMg2Mn0.3, AlMg2.5	Nd:YAG	(Carlsson, 1997)
	VW Phaeton door	Al cast alloy– 4047 filler	Nd:YAG–MIG	(Graf and Staufer, 2003)
Biomedical	Dental prostheses	Ni–Cr–Mo, Cr–Co–Mo	Nd:YAG	(Bertrand *et al.*, 2001)
Construction	Box beams	C–Mn steel	CO_2	(Ion, 1992)
	Tailored blanks	Various	Various	(Pallett and Lark, 2001)
Domestic goods	Beverage cans	Steel	CO_2	(Mazumder and Steen, 1981; Sharp and Nielsen, 1988)
	Filter cannisters	Steel	CO_2	(Weeter, 1998)
	Water tanks	N/S	N/S	(Garrison, 1993)

(Contd)

Table 16.18 *(Contd)*

Industry sector	Product	Material	Laser	Reference
Domestic goods	Window frames	Aluminium	N/S	(Weeter, 1998)
	Window spacers	Aluminium alloy	N/S	(Eckersley, 1982)
	Packaging	Nylon 66 substrate	CO_2	(Brown *et al.*, 2000)
Materials	Steel coil production	C–Mn steel	CO_2	(Kawai *et al.*, 1984)
	Synchrotron chamber	A6063EX/A3003		(Yamaoka *et al.*, 2000)
Power generation	Pipes	Ferritic stainless steel		(Cerri *et al.*, 1989)
	Steam generator tubes			(Ishide *et al.*, 1992)
	Tubular components	Inconel 600, Inconel 900	Nd:YAG	(Kim *et al.*, 2001)
Railway	Carriage bodies	Various	Various	(Tamaschke, 1996; Oikawa *et al.*, 1993)
	Modular components	AISI 304	CO_2	(Daurelio *et al.*, 1999)
Shipbuilding	Decking	Steel	CO_2	(Roland *et al.*, 2002)
	T-stiffeners	Steel	CO_2	(Brooke, 1988)
Yellow goods	Abrams tank engine recuperator	Inconel 625	CO_2	(Miller and Chevalier, 1983)

keyhole welding. Laser welding provides opportunities for more efficient joint designs, novel procedures, improved health and safety factors, easy automation, as well as the traditional benefits of high productivity, high quality and flexibility.

AUTOMOBILE BODYWORK

Virtually every car manufacturer now uses laser welding in bodywork construction. Continuous seam welds not only improve stiffness, handling, road noise and crashworthiness, but also enable new joint geometries to be used, leading to material savings. The first applications involved simple butt and lap joints between steel sheets of similar thickness. Today, mass produced cars contain several tens of metres of such welded seams. Trends in recent years have included: an increasing use of lightweight materials, notably aluminium alloys in chassis members and bodywork; the replacement of electron beam welding for powertrain joints requiring penetration less than about 5 mm; the use of Nd:YAG lasers for bodywork welding in three dimensions; the development of automatic process monitoring and control systems; the use of hybrid welding systems comprising a laser beam and an electric arc; and an increase in the use of laser-welded tailored blanks, discussed below.

Two classes of product have been developed for laser welding of automobile bodies: the CO_2 laser with articulated mirror delivery, and the Nd:YAG laser with fibreoptic beam delivery. The latter product has a number of important benefits over the former. When setting up, fibreoptic beam delivery is a relatively simple plug-in accessory with no mirrors to align. It can be used with standard relatively cheap welding robots, which are often already being used in a factory. From an application point of view, welding of mild steel or zinc coated steel can be achieved without the need for shielding gas – only a cross jet of air is required to protect the optics. Compressed air is normally installed in the plant, and so no bottles of gas or high pressure gas lines need to be provided. The focused spot is quite large, at least 0.5 mm, which provides increased tolerance for weld fitup and reduces the required accuracy of the robot. In comparison with conventional resistance spot welding, laser spot welding is a

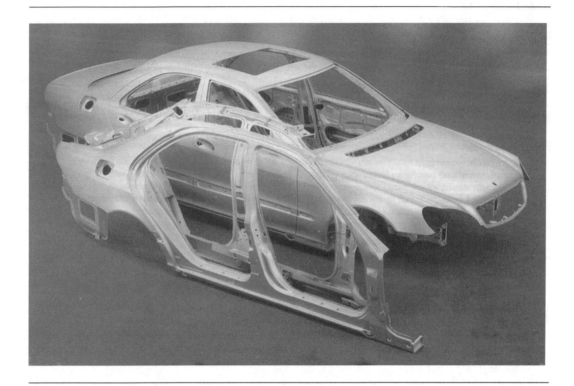

Figure 16.18   Side body ring of a Mercedes C class automobile, constructed by pressing a laser-welded tailored blank. (Source: Klaus Ziegler, DaimlerChrysler, Sindelfingen, Germany)

non-contact process, which eliminates the feature of electrode wear. Continuous seam welds not only improve performance, but also provide opportunities for new joint geometries to be used, which can lead to material savings as well as the use of novel materials. Recent years have seen a noticeable trend towards greater use of Nd:YAG laser welding, particularly for complex geometry joints.

Tailored blanks are made by butt welding steel sheets of differing quality, thickness and/or coating, to form a composite section. The first tailored blanks were made by EB welding in the 1960s. Toyota introduced the concept of laser-welded tailored blanks in 1985. In August 1985, Thyssen Stahl AG began butt welding two pieces of hot-dipped galvanized steel to make a blank wide enough for the floorpan of the Audi 80. Today, resistance mash seam welding and laser welding are the main fabrication techniques, and the expression 'tailor-welded blanks' is now common terminology.

Laser welding has a number of advantages over other joining techniques for sheet materials: a narrow, even and aesthetic weld is produced, which is necessary for subsequent drawing operations, and there are no restrictions on the size of the blanks. They are typically drawn to depths of 16 cm. For example, a 0.8 mm thick galvanized section for the base of a car door can be joined to a section 1.8 mm in thickness for hinge supports, and completed with uncoated steel with a large draw depth to follow the contour of the vehicle. These customized blanks are then ready for drawing or stamping, and contain the desired properties exactly where they are needed. An example is shown in Fig. 16.18.

Laser-welded tailored blanks provide many advantages over bodywork components manufactured in the traditional way, i.e. separate stampings that are subsequently joined using, for example, resistance spot welding. Greater flexibility is offered at the design stage; the designer can bypass limitations normally imposed on part width by available coil widths. In the various manufacturing phases, fewer parts and stamping dies are needed, scrap can typically be reduced from around 50% to less than 15%, the total number of joining phases is reduced, 100% visual inspection is possible, and reductions in finishing operations can be achieved. Since the cost of material for the bodywork is approximately 50% of the total manufacturing cost, the potential for cost savings through a reduction in material usage is high. Butt welding enables overlap flanges to be eliminated, giving a reduction in weight of around 10 kg per car, and a reduction in the use of sealing operations. Such factors translate into higher productivity and a lower overall cost. Improvements in quality include a flush surface which is aesthetically pleasing, and increased part stiffness, which allows tighter tolerances to be used, resulting in improved fit. In terms of performance, the weight reduction decreases fuel consumption. Improved crash performance, fatigue properties and corrosion properties can also be obtained. The full penetration weld bead is clearly visible on both sides of the sheet for inspection. The only requirement is high quality edge preparation, typically to within 0.05 mm over the length of the weld.

Tailored blanks are normally welded using CO_2 lasers, which are good for two-dimensional applications. Nd:YAG lasers and fibreoptic beam delivery allow designers to specify blanks of any shape, with three-dimensional curved welds. Laser-welded aluminium blanks are now to be found in cars such as the Lamborghini Gallardo (the front wheel arch), and are finding popularity with many automobile manufacturers.

AIRCRAFT SKIN STRINGERS

Many aerospace products are characterized by relatively short runs of high quality, customized components. However, there are applications in which sections are joined in long production lengths. Although the first aircraft made by Boeing was welded (after wooden structures were replaced), mechanical fastening methods such as riveting have dominated because of the difficulties associated with fusion welding of the common aluminium aerospace alloys 2024 and 7075. The riveting process involves many stages: drilling, reaming, burring, sealing and placing the rivets. It is an expensive joining technique, which in addition incurs weight penalties arising from the weight of the rivets themselves, as well as the need for overlapping or strengthened joints. Rivet holes can also act as sources of cracking and corrosion. Welding provides the possibility of a high productivity, single-step process, with the added benefit of potential weight savings of about 10% in comparison with riveting. Laser welding can reduce joining costs relative to adhesive bonding and riveting by about 20%. The aviation authorities demand extensive testing, and so any changes in manufacturing procedures mean that a new process, especially welding, is normally only introduced when a new aircraft is designed.

EADS Airbus (Bremen, Germany) has invested heavily in laser welding as a replacement for riveting in non-critical applications. One application involves joining stiffening stringers to the skin of the fuselage. It took almost 10 years for laser beam welding to be qualified for this application, in which the damage-tolerant alloy 6013 is the base material and 4047 the filler material. The stringers are welded from two sides at 10 m min^{-1}, using two 2.5 kW CO_2 laser beams. A welded lower fuselage section is shown in Fig. 16.19. The joint is designed such that the HAZ is contained in the stringer, and does not impinge on the skin. The process is used in series production of the Airbus 318, and will be implemented in future aircraft types.

LIGHTWEIGHT STRUCTURES

Modular lightweight structures that possess high stiffness and low weight are attractive to many sectors of industry, particularly in transportation and construction. Structural panels, which can be manufactured from flat sheets joined to pressed sheet cores, ribs or hollow beams, are characterized

Figure 16.19 Laser-welded skin stringers on a lower aircraft fuselage section. (Source: Karl-Heinz Rendigs, EADS Airbus, Bremen, Germany)

by good crush performance, a high ratio of stiffness to weight, high strength, good heat resistance and the accommodation of point and distributed loads. Lightweight structures can be made by a variety of conventional joining technologies, such as resistance welding, arc welding and adhesive bonding. Laser stake welding provides a number of advantages over conventional fabrication techniques:

- Access from only one side is required, compared with two-sided access when using resistance welding.

- The energy input of the laser joining process is around one-tenth that of conventional arc welding processes, which results in a reduction in distortion, and a significant saving in time and cost through the reduction in reworking necessary before the panels can be used.

- The high penetration allows the bead to fuse both the sheet and the core if necessary, which would be difficult to achieve using, for example, TIG welding.

- Laser-welded panels possess superior high temperature strength to adhesively bonded panels.

- Increases in productivity by a factor of 10 can be gained because of the high welding speed possible.

- The manufacturing method is simpler and more easily automated than arc fusion welding and adhesive bonding.

- Product flexibility is improved – sheet assembly components can be obtained in a wide range of sizes, and formed into a range of core shapes, or tubes can be obtained in a range of geometries.

- Product quality, product safety, and quality assurance can be improved, since visual inspection techniques, such as the observation of heat marks from welding, can be applied.

Figure 16.20   Example of a laser-welded lightweight structural panel. (Source: Petri Metsola, Lappeenranta University of Technology, Finland)

CO_2 laser welding is ideally suited to long, linear welds. A curved panel manufactured this way is shown in Fig. 16.20. The lap weld in a panel made between the flat sheet and corrugated pressed sheet steel core is shown in Fig. 16.21. Nd:YAG laser welding is more suited to smaller panels of more complex design, because of the ease with which the beam can be delivered via a fibre optic. Fibreoptic beam delivery also provides an opportunity for the beam to be delivered from inside the panel, producing a weld that fully penetrates the core, but only partially penetrates the skin, leaving a flat outer surface that can be painted easily. Lightweight panels are used for decking, hatches, partitions and walkways, and are ideal for transportable structures. On a smaller scale, such panel geometries might be suitable for crash-resistant structures that absorb energy on impact. They would be considerably cheaper than alternatives such as metal foams.

SHIP HULL T-STIFFENERS

Large container ships are still the most economical and environmentally friendly means of transporting goods over long distances. Japan is the largest producer of ships by tonnage, followed by South Korea, Germany, Taiwan and Denmark. A large luxury cruise liner costs about US$400 million, of which the cost of materials accounts for about 60%, with labour and overhead charges sharing the remainder in equal proportions. There is pressure to reduce fabrication costs, with a strong interest in reducing the amount of corrective work needed. Shipbuilders estimate that rectifying the distortions produced in 6–8 mm plate accounts for up to 33% of the labour costs of a passenger ship, or fast naval craft. A large part of the existing tonnage is constantly being replaced for a number of reasons: simple wear and tear, uneconomical use, and the need for technical modifications to meet safety and environmental standards. Of particular current interest is the use of an Nd:YAG laser for autogenous welding of the root pass in a Y-joint, followed by laser welding with a filler wire to fill the groove. Thus thick sections may be joined using several passes.

Figure 16.21   Laser lap weld in a lightweight structure with a pressed sheet steel corrugated core. (Source: Petri Metsola, Lappeenranta University of Technology, Finland)

SUMMARY AND CONCLUSIONS

Keyhole laser welding is now an established joining technique in many sectors of industry, and is being investigated actively in others. Most of the principles of keyhole formation are understood, and the formation and properties of the weld bead and HAZ can be established by adapting predictive methods that have been developed for conventional fusion welding processes. Most materials that can be welded by conventional fusion techniques are suitable for keyhole welding. Low alloy steels are particularly suitable – welds can be made rapidly in thin sections, and full penetration can be achieved in thicker sections in a single pass. Many variations of keyhole welding have been developed – hybrid arc–beam processes; multiple spot sources; filler wire welding – all of which exploit the advantages inherent in the individual techniques. The process can be modelled by using relatively simple analytical methods to obtain initial estimates for parameter selection in procedure development. To achieve the highest level of detail for practical use, models may be calibrated to known data points to produce a practical processing diagram.

The main benefits of laser keyhole welding become evident when it is used at the limits of its capability: high speed welding of sheet materials, or deep penetration joining in thick sections. As with all laser-based processes, the low energy input is the origin of many of the advantages over competing fusion joining processes, since distortion is low, productivity is high, quality is superior, and new materials may be considered. The accessibility of the beam enables new joints to be designed into products. The controllability of the beam allows dissimilar materials to be joined. Low distortion reduces the total number of manufacturing steps by removing post-weld straightening. However, the process is often not used to its fullest advantage because of a lack of understanding of the opportunities available at the design stage of a component. In addition, procedures for keyhole welding and quality standards must be generated and approved by classification societies, particularly for thick section materials. Both are currently obstacles to greater application. Equipment for on-line process monitoring and quality control is continually being sophisticated. This will enable rugged systems to be made, which will increase process automation, and make the process more attractive to a wider range of industry sectors.

FURTHER READING

Beyer, E. (1987). *Welding with High Power CO$_2$ Lasers*. Düsseldorf: VDI Verlag. (In German.)

Dawes, C. (1992). *Laser Welding*. Abington: Abington Publishing.

Duley, W.W. (1999). *Laser Welding*. New York: John Wiley & Sons.

Easterling, K.E. (1992). *Introduction to the Physical Metallurgy of Welding*. 2nd ed. Oxford: Butterworth-Heinemann.

Ion, J.C. (2000). Laser beam welding of wrought aluminium alloys. *Science and Technology of Welding and Joining*, **5**, (5), 265–276.

Kou, S. (1987). *Welding Metallurgy*. New York: John Wiley & Sons Inc.

Mathers, G. (2002). *The Welding of Aluminium and its Alloys*. Cambridge: Woodhead Publishing Ltd.

Mazumder, J. (1982). Laser Welding: State of the Art Review. *Journal of Metals*, **34**, 16–24.

Radaj, D. (1992). *Heat Effects of Welding*. Berlin: Springer.

Rykalin, N., Uglov, A. and Kokora, A. (1978). *Laser Machining and Welding*. Oxford: Pergamon Press.

Schwartz, M.M. (1981). *Source Book on Electron Beam and Laser Welding*. Metals Park: American Society for Metals.

Seyffarth, P. and Krivtsun, I. (2002). *Laser-arc Processes and their Applications in Welding and Material Treatment*. London: Taylor & Francis.

Zhao, H., White, D.R. and DebRoy, T. (1999). Current issues and problems in laser welding of automotive aluminium alloys. *International Materials Reviews*, **44**, (6), 238–266.

Sun, Z. and Ion, J.C. (1995). Review: laser welding of dissimilar metal combinations. *Journal of Materials Science*, **30**, 4205–4214.

EXERCISES

1. Explain the principles underlying the formation of the keyhole, weld pool and HAZ in laser keyhole welding.

2. Why are transverse sections of arc fusion weld beads and keyhole laser weld beads in structural steel different?

3. List the advantages and drawbacks of laser welding in comparison with electron beam welding, gas metal arc welding, and resistance welding.

4. Describe the effect of carbon on the hardenability, microstructure and hardness developed in the heat affected zone of keyhole laser welds in structural steels.

5. List the most common imperfections and their origins in keyhole laser welds in steels.

6. What are the main differences between the material properties of structural steels and aluminium alloys, and what effects do these have on laser beam weldability?

7. Describe the effects of silicon and magnesium on the microstructure and properties of 6000 series alloys. What effects do these alloying elements have on the laser weldability of these alloys?

8. What are the main problems when laser welding aluminium alloys, and how can they be addressed?

9. The bodywork and power train components of an automobile are to be laser welded. Give examples of parts that could be joined using a CO$_2$ laser beam delivered by mirrors, and parts that are more suited to joining using an Nd:YAG laser beam delivered via a robot-mounted fibreoptic. Discuss the advantages of each type of welding system.

10. It is often noted that penetration during laser keyhole welding can be increased by placing the focal plane a small distance below the surface of the workpiece, rather than at the surface or with the same displacement above the surface. Suggest reasons for this.

BIBLIOGRAPHY

Abe, N., Agano, Y., Tsukamoto, M., Makino, T., Hayashi, M. and Kurosawa, T. (1996). Effect of CO_2 laser irradiation on arc welding. In: Duley, W., Shibata, K. and Poprawe, R. eds *Proceedings of the Laser Materials Processing Conference (ICALEO '96)*. Orlando: Laser Institute of America. pp. 64–73.

Abe, N., Kunugita, Y., Miyake, S., Hayashi, M., Tsuchitani, Y. and Mihara, T. (1998). The mechanism of high speed leading path laser–arc combination welding. In: Beyer, E., Chen, X. and Miyamoto, I. eds *Proceedings of the Laser Materials Processing Conference (ICALEO '98)*. Orlando: Laser Institute of America. pp. F37–F45.

Adair, R. Jr (1994). Welding aluminum alloys with CO_2 lasers. *Fabricator*, **24**, (6), 96–98.

Adams, C.M. Jr (1958). Cooling rates and peak temperatures in fusion welding. *Welding Journal*, **37**, (5), 210s–215s.

Albright, C.E., Hsu, C. and Lund, R.O. (1991). Fatigue strength of laser-welded lap joints. In: Ream, S.L., Dausinger, F. and Fujioka T. eds *Proceedings of the Laser Materials Processing Conference ICALEO '90*, Orlando: Laser Institute of America. pp. 357–363.

Alexander, J. and Steen, W.M. (1980). Penetration studies on arc augmented laser welding. In: *Proceedings of the International Conference Welding Research in the 1980s*, 27–29 October 1980, Osaka, Japan. Osaka: Osaka University Welding Research Institute. pp. 121–125.

Alexander, J. and Steen, W.M. (1981). Arc augmented laser welding – process variables, structure and properties. In: *Proceedings of the Conference the Joining of Metals: Practice and Performance*, 10–12 April 1981, Coventry, UK. London: The Institution of Metallurgists. pp. 155–160.

Alwang, W.G., Cavanaugh, L.A. and Sammartino, E. (1969). Continuous butt welding using a carbon dioxide laser. *Welding Journal*, **48**, (3), 110s–115s.

Andersen, H.J. and Holm, H. (2002). Extending the tolerances for laser beam welding in heavy industry. *Journal of Advanced Materials*, **34**, (1), 42–46.

Anon. (1994). Laser welding systems muffle quality concerns. *Industrial Laser Review*, **9**, (2), 11–12.

Anon. (1999). Laser welding helps standardize automobile fuel injectors. *Welding Journal*, **78**, (4), 51–52.

Anon. (2002). Manufacturing the new Audi A8. *AutoTechnology*, No. 6, pp. 36–39.

Aristotile, R. and Fersini, M. (1999). 'Tailored blanks' for automotive components. Evaluation of mechanical and metallurgical properties and corrosion resistance of laser-welded joints. *Welding International*, **13**, (3), 194–203.

Auty, T. (1998). Laser welded tailored blanks – a practical guide. *Engineering Lasers & Power Beam Processing*, (May), 12–18.

Avramchenko, P.F. and Molchan, I.V. (1983). Continuous laser beam welding AMg6 alloy. *Automatic Welding*, **36**, (5), 61–62.

Azuma, K. and Ikemoto, K. (1993). Laser welding technology for joining different sheet metals for one piece stamping. In: Martellucci, S. *et al.* eds *Proceedings of the Conference Laser Applications for Mechanical Industry*. Kluwer: Dordrecht. pp. 219–223.

Baeslack, W.A. III and Banas, C.M. (1981). A comparative evaluation of laser and gas tungsten arc weldments in high-temperature titanium alloys. *Welding Journal*, **60**, (7), 121s–130s.

Baeslack, W.A. III, Savage, S.J. and Froes, F.H. (1986). Laser-weld heat-affected zone liquation and cracking in a high-strength Mg-based alloy. *Journal of Materials Science Letters*, **5**, 935–939.

Baeslack, W.A. III, Chiang, S. and Albright, C.A. (1990a). Laser welding of an advanced rapidly-solidified titanium alloy. *Journal of Materials Science Letters*, **9**, 698–702.

Baeslack, W.A. III, Cieslak, M.J. and Headley, T.J. (1990b). Heat treatment of pulsed Nd:YAG laser welds in a Ti–14.8 wt%Al–21.3 wt% Nb titanium aluminide. In: David, S.A. and Vitek, J.M. eds *Proceedings of the 2nd International Conference Trends in Welding Research*. Materials Park: ASM International. pp. 211–216.

Bagger, C., Miyamoto, I., Olsen F. and Maruo, H. (1991). On-line control of the CO_2 laser welding process. In: *Proceedings of the 3rd International Beam Technology Conference*, 13–14 March 1991, Karlsruhe, Germany. DVS-Berichte 135. Düsseldorf: Deutscher Verband für Schweißtechnik. pp. 1–6.

Baysore, J.K., Williamson, M.S., Adonyi, Y. and Milian, J.L. (1995). Laser beam welding and formability of tailored blanks. *Welding Journal*, **74**, (10), 345s–352s.

Bertrand, C., Le Petitcorps, Y., Albingre, L. and Dupuis, V. (2001). The laser welding technique applied to the non precious dental alloys procedure and results. *British Dental Journal*, **190**, (5), 255–257.

Beyer, E., Dilthey, U., Imhoff, R., Meyer, C., Neuenhahn, J. and Behler, K. (1994). New aspects in laser welding with an increased efficiency. In: McCay, T.D., Matsunawa, A. and Hügel, H. eds *Proceedings of the Laser Materials Processing Conference ICALEO '94*, 17–20 October 1994. Orlando: Laser Institute of America. pp. 183–192.

Biffin, J. (1997). Plasma augmented lasers shine ahead in auto body application. *Welding and Metal Fabrication*, **56**, (4), 19–21.

Biffin, J., Blundell, N., Johnson, T. and Page, C. (1998). Enhancing the performance of industrial lasers with a plasma arc. In: David, S.A., Johnson, J.A., Smartt, H.B., DebRoy, T. and Vitek, J.M. eds

Proceedings of the 5th International Conference on Trends in Welding Research, 1–5 June 1998, Pine Mountain, GA, USA. Metals Park: ASM International. pp. 492–495.

Biro, E., Weckman, D.C. and Zhou, Y. (2002). Pulsed Nd:YAG laser welding of copper using oxygenated assist gases. *Metallurgical and Materials Transactions*, **33A**, 2019–2030.

Blake, A. and Mazumder, J. (1985). Control of magnesium loss during laser welding of Al-5083 using a plasma suppression technique. *Journal of Engineering for Industry*, **107**, 275–280.

Blundell, N. (1998). Arc takes laser welding into new territory. *Materials World*, **6**, (9), 537–538.

Boillot, J.-P. and Noruk, J. (2001). Benefits of vision in laser welding. *Industrial Laser Solutions*, **16**, (11), 14–18.

Bonollo, F., Tiziani, A. and Zambon, A. (1993a). Model for CO_2 laser welding of stainless steel, titanium and nickel: parametric study. *Materials Science and Technology*, **9**, 1137–1144.

Bonollo, F., Giordano, L., Tiziani, A. and Zambon, A. (1993b). Microstructural optimization of laser beam welded duplex stainless steels. In: Nicodemi, W. ed. *Proceedings of the Conference Innovation Stainless Steel*, Vol. 3, 11–14 October 1993, Florence, Italy. Associazone Italiana di Metallurgia. pp. 251–257.

Bonollo, F., Brunoro, G., Tiziani, A. and Zucchi, F. (1993c). Laser beam welded duplex stainless steels: a study on corrosion behaviour. In: Nicodemi, W. ed. *Proceedings of the Conference Innovation Stainless Steel*, Vol. 3, 11–14 October 1993, Florence, Italy. Associazone Italiana di Metallurgia. pp. 367–372.

Bonollo, F., Tiziani, A., Brunoro, G. and Zucchi, F. (1996). Study of stress corrosion behaviour of laser welded superduplex stainless steels. *Welding International*, **10**, 124–127.

Bransch, H. (1996). Nd:YAG lasers – economic benefits in sheet metal welding. *Metal Forming*, **30**, (4), 20–28.

Bransch, H.N., Weckman, D.C. and Kerr, H.W. (1994). Effects of pulse shaping on Nd:YAG spot welds in austenitic stainless steel. *Welding Journal*, **73**, (6), 141s–151s.

Breuer, P. (1994). Nd:YAG laser welding used in car body manufacturing. *Industrial Laser Review*, **9**, (8), 13–15.

Brooke, S.J. (1988). High power laser welding of T joints in ship production. *The North East Coast Institution of Engineers and Shipbuilders*, **104**, (3), 93–103.

Brown, N., Kerr, D., Jackson, M.R. and Parkin, R.M. (2000). Laser welding of thin polymer films to container substrates for aseptic packaging. *Optics and Laser Technology*, **32**, 139–146.

Carlsson, T. (1997). Production test of aluminium welding with Nd-YAG laser at Volvo Olofstrom plants. In: Roller, D. ed. *Proceedings of the 30th International Symposium on Automotive Technology and Automation (ISATA)*, 14–19 June 1997, Florence, Italy. Croydon: Automotive Automation Ltd. pp. 305–314.

CEN (2001). EN 1011-2:2001. *Welding. Recommendations for welding of metallic materials. Arc welding of ferritic steels*. Brussels: European Committee for Standardization.

Chehaibou, A., Goussain, J.-C. and Barouillet, M. (1993). Study of the susceptibility to hot cracking in austenitic stainless steels welded by laser beam. In: Charissoux, C. ed. *Proceedings of the 5th International Conference on Welding and Melting by Electron and Laser Beams (CISFFEL-5)*. Saclay: SDEM. pp. 299–306. (In French.)

Chen, G. and Roth, G. (1993). Comparison of the seam geometry and pore formation during laser beam welding of aluminium by the pulse and continuous methods. *Schweißen und Schneiden*, **45**, (8), 419–423. (In German.)

Cieslak, M.J. and Fuerschbach, P.W. (1988). On the weldability, composition and hardness of pulsed and continuous Nd:YAG laser welds in aluminum alloys 6061, 5456, and 5086. *Metallurgical Transactions*, **19B**, (4), 319–329.

Cola, M.J., Lienert, T.L., Gould, J.E. and Hurley, J.P. (1990). Laser welding of a SiC particulate reinforced aluminum metal matrix composite. In: Patterson, R.A. and Mahin, K.W. eds *Proceedings of the Conference on Weldability of Materials*. Materials Park: American Society for Metals. pp. 297–303.

Cornu, D., Gouhier, D., Richard, I., Bobin, V., Boudot, C., Gaudin, J.-P., Andrzejewski, H., Grevey, D. and Portrat, J. (1995). Weldability of superalloys by Nd:YAG laser. *Welding International*, **9**, (10), 802–811.

Coste, F., Fabbro, R. and Sabatier, L. (1999). Adaptive control of high-thickness laser welding. *Welding International*, **13**, (6), 465–469.

Dahotre, N.B., McCay, M.H., McCay, T.D., Sharp, C.M., Gopinathan, S. and Allard, L.F. (1990). Laser welding of a SiC/Al-alloy metal matrix composite. In: Ream, S.L., Dausinger, F. and Fujioka, T. eds *Proceedings of the Laser Materials Processing Conference (ICALEO '90)*. Orlando: Laser Institute of America/Bellingham: SPIE. pp. 343–356.

Daurelio, G., Ludovico, A.D., Nenci, F. and Zucchini, A. (1999). Production of modular structural components for railway vehicle platforms using CO_2 laser technology. *Welding International*, **13**, (1), 49–59.

David, S.A. and Liu, C.T. (1982). High-power laser and arc welding of thorium-doped iridium alloys. *Welding Journal*, **61**, (5), 157s–163s.

David, S.A. Vitek, J.M. and Hebble, T.L. (1987). Effect of rapid solidification on stainless steel weld metal and its implications on the Schaeffler diagram. *Welding Journal*, **66**, (10), 289s–300s.

Dawes, C.J. and Watson, M.N. (1984). CO_2 laser welding of deep drawing steel sheet and microalloyed steel plate. In: Metzbower, E.A. ed. *Proceedings of the Materials Processing Symposium (ICALEO '83)*. Toledo: Laser Institute of America. pp. 73–79.

Denney, P.E. and Metzbower, E.A. (1989). Laser beam welding of titanium. *Welding Journal*, **68**, (8), 342s–346s.

De Paris, A., Robin, M. and Fantozzi, I.G. (1992). CO_2 laser welding of SiO_2–Al_2O_3 ceramic tubes. In: Mordike, B.L. ed. *Proceedings of the European Conference on Laser Treatment of Materials (ECLAT '92)*, 12–15 October 1992, Göttingen, Germany. Oberursel: DGM Informationsgesellschaft. pp. 131–136.

Diebold, T.P. and Albright, C.E. (1984). 'Laser-GTA' welding of aluminium alloy 5052. *Welding Journal*, **63**, (6), 18–24.

Dilthey, U., Lüder, F. and Wieschemann, A. (1998a). Process-technical investigations on hybrid technology laser-beam-arc-welding. In: *Proceedings of the 6th International Conference on Welding and Melting by Electron and Laser Beams*, 15–19 June 1998, Toulon, France. Roissy CDG Cedex: Institute de Soudure. pp. 417–424.

Dilthey, U., Böhm, S., Träger, G. and Ghandehari, A. (1998b). Laser-beam and electron-beam welding of material combinations consisting of cast iron and case-hardening or heat-treatable steels. *Welding and Cutting*, **50**, (11), 718–723. (In German.)

Dilthey, U. and Wieschemann, A. (2000). Prospects by combining and coupling laser beam and arc welding processes. *Welding in the World*, **44**, (3), 37–46.

Douglass, D.M., Mazumder, J. and Nagarathnam, K. (1996). Laser welding of Al 6061-T6. In: Smartt, H.B., Johnson, J.A. and David, S.A. eds *Proceedings of the 4th International Conference Trends in Welding Research*, 5–8 June 1995, Gatlinburg, TN, USA. Materials Park: ASM International. pp. 467–478.

Draugelates, U., Schram, A. and Kettler, C. (2000). Welding of magnesium alloys. In: Aghion, E. and Eliezer, D. eds *Proceedings of the Second Israeli International Conference on Magnesium Science and Technology*. pp. 441–448.

Drews, P. and Willms, K. (1995). Sensor system for high precision detection of welding grooves in laser beam welding. *Schweißen und Schneiden*, **47**, (11), 924–927. (In German.)

Dubiel, D. (1987). Investigations on the causing mechanism of defects in laser weldings. *Practical Metallography*, **24**, 457–467.

Dudas, J.H. and Collins, F.R. (1966). Preventing weld cracks in high-strength aluminum alloys. *Welding Journal*, **45**, (6), 241s–249s.

Duley, W.W. and Gonsalves, J.N. (1972). Industrial applications of carbon dioxide lasers. *Canadian Research and Development*. (January), 25–29.

Duley, W.W. and Mueller, R.E. (1992). CO_2 laser welding of polymers. *Polymer Engineering and Science*, **32**, (9), 582–585.

Duley, W.W. (1994). CO_2 laser welding of 5000 series aluminium alloys for tailor blanking. In: Uddin, M.N. ed. *Proceedings of the Conference on Automotive Body Materials (IBEC '94)*. Warren: IBEC Ltd. pp. 49–51.

Eboo, M., Steen, W.M. and Clarke, J. (1978). Arc-augmented laser welding. In: Needham, J.C. ed. *Proceedings of the 4th International Conference on Advances in Welding Processes*, 9–11 May 1978, Harrogate, UK. Abington: The Welding Institute. pp. 257–265.

Eckersley, J.S. (1982). CO_2 laser welding of aluminium alloy air spacers for insulated windows. In: Bass, M. ed. *Proceedings of the Materials Processing Symposium (ICALEO '82)*. Toledo: Laser Institute of America. pp. 61–64.

Eritt, J. (1995). Laser welding of high strength aluminium alloys in aircraft construction. *VDI-Z*, **137**, (6), 34–38. (In German.)

Exner, H. (1994). Laser welding of ceramics. In: *Proceedings of the 5th European Conference on Laser Treatment of Materials (ECLAT '94)*, 26–27 September 1994, Bremen-Vegesack, Germany. Düsseldorf: Deutscher Verband für Schweißtechnik. pp. 290–293.

Fedrizzi, L., Molinari, A., Tiziani A. and Magrini, M. (1998). Microstructural characterization and corrosion behaviour of laser welded austenitic stainless steel. In: Contré, M. and Kuncevic, M. eds *Proceedings of the 4th International Colloquium on Welding and Melting by Electron and Laser Beams (CISFFEL-4)*, 26–30 September 1988, Cannes, France. Saclay: CEN. pp. 417–424.

Flavenot, J.-F., Deville, J.-P., Diboine, A., Cantello, M. and Gobbi, S.-L. (1993). Fatigue resistance of laser welded lap joints of steel sheets. *Welding in the World*, **31**, (5), 358–361.

Forbis-Parrott, C.A. (1991). Laser beam welding is ready to go to work at Cadillac. *Welding Journal*, **70**, (7), 37–42.

Fukumoto, S., Hirose, A. and Kobayashi, K.F. (1993). Application of laser beam welding to joining of continuous fibre reinforced composite to metal. *Materials Science and Technology*, **9**, (3) 264–271.

Ga-Orza, J.A. (1999). Welding of aluminium alloys with high-power Nd:YAG lasers. *Welding International*, **13**, (4), 282–284.

Garrison, M. (1993). Laser beam welding goes into high-speed production of home hot water tanks. *Welding Journal*, **72**, (12), 53–57.

Gill, S.J., Moon, D.W., Metzbower, E.A. and Crooker, T.W. (1986). Fatigue of A36 steel laser beam weldments. *Welding Journal*, **65**, (2), 48s–50s.

Gnanamuthu, D.S. and Moores, R.J. (1987). Laser welding of 7475 aluminium alloy. In: Bakish, R. ed. *Proceedings of the Conference the Laser vs the Electron Beam in Welding, Cutting and Surface Treatment – State of the Art 1987.* Englewood: Bakish Materials Corporation. pp. 295–300.

Gnanamuthu, D.S. and Moores, R.J. (1988). Laser welding of 8090 aluminium–lithium alloy. In: Metzbower, E.A. and Hauser, D. eds *Proceedings of the International Power Beam Conference*, 2–4 May 1988, San Diego, CA, USA. Metals Park: ASM International. pp. 181–183.

Goodwin, G.M. (1988). The effects of heat input and weld process on hot cracking in stainless steel. *Welding Journal*, **67**, (4), 88s–94s.

Gopinathan, S., Murthy, J., McCay, T.D., McCay, M.H. and Spiegel, L. (1994). The autogenous laser welding of Al2219 to Al6061. In: Denney, P., Miyamoto, I. and Mordike, B.L. eds *Proceedings of the Laser Materials Processing Conference (ICALEO '93).* Orlando: Laser Institute of America. pp. 794–804.

Gouveia, H., Norrish, J. and Quintino, L. (1994). CO_2 laser welding of copper. *Welding Review International*, **13**, 266–273.

Graf, T. and Staufer, H. (2003). Laser welding drives VW improvements. *Welding Journal*, **82**, (1), 43–48.

Graham, M.P., Weckman, D.C., Kerr, H.W and Hirak, D.M. (1996). Nd:YAG laser beam welding of coated steels using a modified lap geometry. *Welding Journal*, **75**, (5), 162s–170s.

Guitterez, L.A., Neye, G. and Zschech, E. (1996). Microstructure, hardness profile and tensile strength in welds of AA6013 T6 extrusions. *Welding Journal*, **75**, (4), 115s–121s.

Haferkamp, H., Niemeyer, M., Dilthey, U. and Träger, G. (2000). Laser and electron beam welding of magnesium materials. *Welding and Cutting*, **52**, (8), E178–E180.

Haferkamp, H., Burmester, I., Niemeyer, M., Doege, E. and Dröder, K (1997). Innovative production technologies for magnesium light-weight constructions – laser beam welding and sheet metal forming. In: Roller, D. ed. *Proceedings of the 30th International Symposium on Automotive Technology and Automation (ISATA '97)*, 16–19 June 1997, Florence, Italy. Croydon: Automotive Automation Ltd. pp. 247–258.

Haferkamp, H., Bunte, J., Herzog, D. and Ostendorf, A. (2004). Laser based welding of cellular aluminium. *Science and Technology of Welding and Joining*, **9**, (1), 65–71.

Hamatani, H., Miyazaki, Y., Ohara, M. and Tanaka, T. (2001). Experimental study of laser welding with applied electrical potential. *Science and Technology of Welding and Joining*, **6**, (4), 197–202.

Hamazaki M. *et al.* (1983). Effect of TIG current in combined welding of TIG arc and laser welding.

Journal of High Temperature Society, **9**, (2), 79–83.

Hashimoto, K., Sato, T. and Niwa, K. (1991). Laser welding copper and copper alloys. *Journal of Laser Applications*, **3**, (1), 21–25.

Herfurth, H.J., Cryderman, M. and Clarke, J.A. (2001). Diode laser welding of aluminium. *Industrial Laser Solutions*, **16**, (4), 29–31.

Hirose, A., Matsuhiro, Y., Kotoh, M., Fukumoto, S. and Kobayashi, K.F. (1993). Laser beam welding of SiC fiber-reinforced Ti–6Al–4V composite. *Journal of Materials Science*, **28**, 349–355.

Hirose, A., Todaka, H. and Kobayashi, K.F. (1997). CO_2 laser beam welding of 6061-T6 aluminium alloy thin plate. *Metallurgical and Materials Transactions*, **28A**, 2657–2662.

Hirose, A., Kobayashi, K.F., Yamaoka, Y. and Kurosawa, N. (2000). Evaluation of properties in laser welds of A6061-T6 aluminium alloy. *Welding International*, **14**, (6), 431–438.

Hoffman, J., Szymanski, Z., Jakubowski, J. and Kolasa, A. (2002). Analysis of acoustic and optical signals used as a basis for controlling laser-welding processes. *Welding International*, **16**, (1), 18–25.

Hornig, J. (1997). Laser welding processing on the new BMW 5-series. In: *Proceedings of the Automotive Laser Applications Workshop*, March 1997, Ann Arbor, Michigan, USA. Michigan: University of Michigan. 29pp.

Hou, K.H. and Baeslack, W.A. III (1996). Effect of solute segregation on the weld fusion zone microstructure in CO_2 laser beam and gas tungsten arc welds in Al–Li–Cu alloy 2195. *Journal of Materials Science Letters*, **15**, 208–213.

Howarth, D.J. (1999). Laser welding in ship building. The role of classification. In: *Proceedings of the European Symposium Assessment of Power Beam Welds*, 4–5 February 1999, Geesthacht, Germany. Geesthacht: GKSS-Forschungszentrum. Paper 13.

Inaba, Y. and Shintani, S. (1993). The present status and applications of laser processing: a hybrid manufacturing process for laser welding of stainless steel pipes. *Welding International*, **7**, (6), 487–492.

Ion, J.C. (1992). High power CO_2 lasers for cost-effective materials processing. *Metals and Materials*, **8**, (9), 485–489.

Ion, J.C. and Anisdahl, L.M. (1995). Development of procedures for laser welding of carbon manganese steels using a PC-based system. In: Mazumder, J., Matsunawa, A. and Magnusson, C. eds *Proceedings of the Laser Materials Processing Conference (ICALEO '95).* Orlando: Laser Institute of America. pp. 524–533.

Irving, R. (1992). Lasers: made in the USA. *Welding Journal*, **71**, (6), 67–73.

Irving, R. (1993). Laser beam and GMA welding lines go on-stream at Arvin Industries. *Welding Journal*, **72**, (7), 52–56.

Ishide, T., Nagura, Y., Matsumoto, O., Nagashima, T., Kidera, T. and Yokoyama, A. (1992). High power YAG laser welded sleeving technology for steam generator tubes in nuclear power plants. In: Matsunawa, A. and Katayama, S. eds *Proceedings of the Conference Laser Advanced Materials Processing (LAMP '92)*, Osaka, Japan: High Temperature Society of Japan and Japan Laser Processing Society. pp. 957–962.

Ishide, T., Hashimoto, Y., Akada, T., Nagashima T. and Hamada, S. (1997). The latest YAG laser welding system – development of hybrid YAG laser welding technology. In: Fabbro, R., Kar, A. and Matsunawa, A. eds *Proceedings of the Laser Materials Processing Conference (ICALEO '97)*. Orlando: Laser Institute of America. pp. A149–A156.

Ishide, T., Nayama, M., Watanabe, M. and Nagashima, T. (2001). Coaxial TIG-YAG and MIG-YAG welding methods. *Journal of the Japan Welding Society*, **70**, 12–17.

Ito, Y. and Bessyo, K. (1968). *Weldability formula of high strength steels*. IIW Doc. IX-576-68. Roissy CDG Cedex: International Institute of Welding.

Jan, R., Howell, P.R. and Martukanitz, R.P. (1995). Optimizing the parameters for laser beam welding of aluminium–lithium alloy 2195. In: Smartt, H.B., Johnson, J.A. and David, S.A. eds *Proceedings of the 4th International Conference Trends in Welding Research*, 5–8 June 1995, Gatlinburg, TN, USA. Materials Park: ASM International. pp. 329–334.

Jokinen, T., Vihervä, T., Riikonen, H. and Kujanpää, V. (2000). Welding of ship structural steel A36 using a Nd:YAG laser and gas–metal arc welding. *Journal of Laser Applications*, **12**, (5), 185–188.

Joma-Plastik (1987). Welding of plastics with the aid of a laser. *Plastverarbeiter*, **38**, (10), 110–120. (In German.)

Jones, I.A. and Taylor, N.S. (1994). High speed welding of plastics using lasers. In: *Proceedings of the Conference ANTEC '94*. pp. 1360–1363.

Juckenath, B., Cantello, M., Breme, J. and Bergmann, H.W. (1988). Laser welding of titanium and titanium alloys. In: *Proceedings of the 6th World Conference on Titanium*, 6–9 June 1988, Cannes, France. Les Ulis Cedex: Les Editions de Physique. pp. 1397–1402.

Junai, A.A., van Dijk, M., Hiensch, M., Rijnders, A., Notenboom, G. and Jelmorini, G. (1990). Pulsed Nd:YAG welding of crack sensitive aluminium–magnesium alloys. In: Ireland, C.L.M. ed. *Proceedings of the Conference High-Power Solid State Lasers and Applications (ECO3)*. Vol. 1277. Bellingham: SPIE. pp. 217–231.

Katayama, S. and Lundin, C.D. (1992). Laser welding of aluminium alloy 5456. *Welding International*, **6**, (6), 425–435.

Kawai, Y., Aihara, M., Ishii, K., Tabuchi, M. and Sasaki, H. (1984). Development of laser welder for strip processing line. *Kawasaki Steel Technical Report*, No. 10, December 1984.

Kim, J.-D., Kim, C.-J. and Chung, C.-M (2001). Repair welding of etched tubular components of nuclear power plant by Nd:YAG laser. *Journal of Materials Processing Technology*, **114**, 51–56.

Kimura, S., Kubo, T., Makino, Y., Honda, K. and Sugiyama, S. (1991). *CO_2 laser welding of copper and aluminium alloys*. IIW Doc. IV-565-91. Roissy CDG Cedex: International Institute of Welding.

Klemens, P.G. (1976). Heat balance and flow conditions for electron beam and laser welding. *Journal of Applied Physics*, **47**, (5), 2165–2174.

Kokkonen, J. and Ion, J.C. (1999). CO_2 laser welding of the aluminium aerospace alloy AA2024. In: Kujanpää V. and Ion, J.C. eds *Proceedings of the 7th Nordic Conference on Laser Materials Processing NOLAMP-7*. Lappeenranta: Acta Universitatis Lappeenrantaensis. pp. 418–429.

Kujanpää, V., Suutala, N., Takalo, T. and Moisio, T. (1979). Correlation between solidification cracking and microstructure in austenitic–ferritic stainless steel welds. *Welding Research International*, **9**, (2), 55–76.

Kunzig, L. (1994). Laser welding of automotive and aero components. *Welding and Metal Fabrication*, **62**, (1), 14–16.

Kutsuna, M., Suzuki, J., Kimura, S., Sugiyama, S., Yuhki, M. and Yamaoka, H. (1993). CO_2 laser welding of A2219, A5083 and A6063 aluminium alloys. *Welding in the World*, **31**, (2), 126–135.

Lang, R., Kullick, M. and Muller-Hummel, P. (2000). Laser beam welding of AlLi2195 (aluminium alloy) high strength aerospace structures. *DVS-Berichte*, **208**, 25–29. (In German.)

Larsson, J.K. (1999a). Laser welding. A mature process technology with various application fields. *Svetsaren*, Issue 1/1999, 43–50.

Larsson, J.K. (1999b). Lasers for various materials processing. A review of the latest applications in automotive manufacturing. In: Kujanpää, V. and Ion, J.C. eds *Proceedings of the 7th Nordic Conference on Laser Materials Processing NOLAMP-7*. Lappeenranta: Acta Universitatis Lappeenrantaensis. pp. 26–37.

Larsson, L.-O., Palmquist, N. and Larsson, J.K. (2000). High quality aluminium welding – a key factor in future car body production. *Svetsaren*, Issue 2/2000, 43–50.

Lee, M.F., Huang, J.C. and Ho, N.J. (1996). Microstructural and mechanical characterization of laser-beam welding of a 8090 Al–Li thin sheet. *Journal of Materials Science*, **31**, 1455–1468.

Lehner, C. (1998). Welding of die-casted magnesium alloys on production machines. In: Beyer, E., Chen, X. and Miyamoto, I. eds *Proceedings of the Laser Materials Processing Conference (ICALEO '98)*.

Orlando: Laser Institute of America. pp. F18–F27.

Leong, K.H., Sabo, K.R., Altshuller, B., Wilkinson, T.L. and Albright, C.E. (1999). Laser beam welding of 5182 aluminium alloy sheet. *Journal of Laser Applications*, **11**, (3), 109–118.

Lienert, T.J., Brandon, E.D. and Lippold, J.C. (1993). Laser and electron beam welding of SiC$_p$ reinforced aluminum A-356 metal matrix composite. *Scripta Metallurgica et Materialia*, **28**, 1341–1346.

Lippold, J.C. (1994). Solidification behaviour and cracking susceptibility of pulsed-laser welds in austenitic stainless steels. *Welding Journal*, **73**, (6), 129s–139s.

Liu, J.T., Weckman, D.C. and Kerr, H.W. (1993). The effects of process variables on pulsed Nd:YAG laser spot welds: Part 1. AISI 409 stainless steel. *Metallurgical Transactions*, **24B**, 1065–1076.

Lloyd's Register (1997). *Guidelines for Approval of CO$_2$-laser Welding*. London: Lloyd's Register of Shipping.

Locke, E., Hoag, E. and Hella, R. (1972). Deep penetration welding with high power CO$_2$ lasers. *Welding Journal*, **51**, (5), 245s–249s.

Maisenhälder, F., Chen, G. and Roth, G. (1993). Laser beam cutting and welding of magnesium-based alloys. In: Roller, D. ed. *Proceedings of the Conference Laser Applications in the Automotive Industries (ISATA '93)*, 14–19 June 1997, Florence, Italy. Croydon: Automotive Automation Ltd. pp. 335–342.

Makino, Y., Shiihara, K. and Asai, S. (2002). Combination between CO$_2$ laser beam and MIG arc. *Welding International*, **16**, (2), 99–103.

Maruo, H., Miyamoto, I. and Arata, Y. (1986). CO$_2$ laser welding of ceramics. In: Arata, Y. ed. *Plasma, Electron, and Laser Beam Technology – Development and Use in Materials Processing*. Metals Park: American Society of Metals.

Marsico, T.A. and Kossowsky, R. (1989). Physical properties of laser-welded aluminium–lithium alloy 2090. In: *Proceedings of the Conference Aluminium–Lithium Alloys*. Birmingham: Materials and Component Engineering Publications Ltd. pp. 1447–1456.

Martin, G.S., Albright, C.E and Jones, T.A. (1995). An evaluation of CO$_2$ laser beam welding on a Ti$_3$Al–Nb alloy. *Welding Journal*, **74**, (2), 77s–82s.

Martukanitz, R.P. and Altshuller, B. (1996). Laser beam welding of aluminum alloy 5754-O using a 3 kW Nd:YAG laser beam and fiber optic beam delivery. In: Duley, W., Shibata, K. and Poprawe, R. eds *Proceedings of the Laser Materials Processing Conference (ICALEO '96)*. Orlando: Laser Institute of America. pp. 39–44.

Martukanitz, R.P. and Howell, P.R. (1995). Relationships involving process, microstructure, and properties of weldments of Al–Cu and Al–Cu–Li alloys. In: Smartt, H.B., Johnson, J.A. and David, S.A. eds *Proceedings of the 4th International Conference Trends in Welding Research*, 5–8 June 1995, Gatlinburg, TN, USA. Materials Park: ASM International. pp. 553–560.

Mazumder, J. and Steen, W.M. (1980). Welding of Ti–6Al–4V by a continuous wave CO$_2$ laser. *Metal Construction*, **12**, 423–427.

Mazumder, J. and Steen, W.M. (1981). The laser welding of steels used in can making. *Welding Journal*, **60**, (6), 19–25.

Mazumder, J. and Steen, W.M. (1982). Microstructure and mechanical properties of laser welded titanium 6Al–4V. *Metallurgical Transactions*, **13A**, 865–871.

Metzbower, E.A. (1990). Laser beam welding of titanium. In: Patterson, R.A. and Mahin, K.W. eds *Proceedings of the Conference Weldability of Materials*. Materials Park: American Society for Metals. pp. 311–318.

Milewski, J.O., Lewis, G.K. and Wittig, J.E. (1993). Microstructural evaluation of low and high duty cycle Nd:YAG laser beam welds in 2024-T3 aluminum. *Welding Journal*, **72**, (7), 341s–346s.

Miller, J.A. and Chevalier, J. (1983). Development of an automated laser beam welding facility for high volume production. *Welding Journal*, **62**, (7), 49–54.

Missori, S. and Koerber, C. (1997). Laser beam welding of austenitic–ferritic transition joints. *Welding Journal*, **76**, (3), 125s–133s.

Missori, S., Koerber, C., Neuenhahn, J.C. and Tosto, S. (1997). Process combination of CO$_2$ laser-beam welding and metal–inert gas welding in order to join clad plates. *Welding and Cutting*, **49**, (11), E165–E168.

Molian, P.A. and Srivatsan, T.S. (1990). Weldability of aluminium–lithium alloy 2090 using laser welding. *Journal of Materials Science*, **25**, 3347–3358.

Molian, P.A., Yang, Y.M. and Patnaik, P.C. (1992a). Laser welding of oxide dispersion-strengthened alloy MA 754. *Journal of Materials Science*, **27**, 2687–2694.

Molian, P.A., Yang, Y.M. and Srivatsan, T.S. (1992b). Laser-welding behaviour of cast Ni$_3$Al intermetallic alloy. *Journal of Materials Science*, **27**, 1857–1868.

Moon, D.W. and Metzbower, E.A. (1983). Laser beam welding of aluminium alloy 5456. *Welding Journal*, **62**, 53s–58s.

Moon, D.W. and Metzbower, E.A. (1988). Temperature measurements in a mid-plane of a laser beam weldments in A36 steel. In: Metzbower, E.A. and Hauser, D. eds *Proceedings of the International Power Beam Conference*, 2–4 May 1988, San Diego, CA, USA. Metals Park: ASM International. pp. 125–130.

Moore, P.L., Howse, D.S. and Wallach, E.R. (2002). Microstructure and properties of autogenous high-power Nd:YAG laser welds in C–Mn steels. In: David, S.A., DebRoy, T., Lippold, J.C., Smartt, H.B.

and Vitek, J.M. eds *Proceedings of the 6th International Conference Trends in Welding Research*, 15–19 April 2002, Pine Mountain, GA, USA. Materials Park: ASM International. pp. 748–753.

Nakao, Y. (1989). Surface treatment by laser quenching. *Welding International*, **3**, (7), 619–623.

Neiheisel, G. and Cary, R. (1991). From scrap reclamation to tailored blanks. *Industrial Laser Review*, **6**, (6), 5–9.

Ogle, M. (1991). Lasers geared to transmissions. *Welding Design and Fabrication*, (September), 49–50.

Oikawa, M., Minamida, K., Goto, N. and Tendo, M. (1993). Development of all-laser-welded honeycomb structures for high speed civil transports. In: Denney, P., Miyamoto, I. and Mordike, B.L. eds *Proceedings of the Laser Materials Processing Conference (ICALEO '93)*. Orlando: Laser Institute of America/Bellingham: SPIE. pp. 453–462.

Ono, M., Shinbo, Y., Yoshitake, A. and Ohmu, M. (2002). Development of laser–arc hybrid welding. *NKK Technical Review*, No. **86**, 8–12.

Ozbaysal, K. and Inal, O.T. (1990). Thermodynamics and structure of solidification in the fusion zone of CO_2 laser welds of 15-5 PH stainless steels. *Materials Science and Engineering*, **A130**, 205–217.

Page, C.J., Devermann, T., Biffin, J. and Blundell, N. (2002). Plasma augmented laser welding and its applications. *Science and Technology of Welding and Joining*, **7**, 1–10.

Pallett, R.J. and Lark, R.J. (2001). The use of tailored blanks in the manufacture of construction components. *Journal of Materials Processing Technology*, **117**, 249–254.

Petrolonis, K. (1993). Laser beam welded nickel aluminum bronze (Cu–9Al–4Fe–4Ni). *Welding Journal*, **72**, (7), 301s–306s.

Ram, V., Kohn, G. and Stern, A. (1986). CO_2 laser beam weldability of zircaloy 2. *Welding Journal*, **65**, (7), 33–37.

Ramasamy, S. and Albright, C.E. (2001). CO_2 and Nd-YAG laser beam welding of 5754-O aluminium alloy for automotive applications. *Science and Technology of Welding and Joining*, **6**, (3), 182–190.

Ramos, J.A., Tyszka, E. and Lechuga, F. (2001). Welding red metal. *Industrial Laser Solutions*, **16**, (4), 25–28.

Reinecke, A.-M. and Exner, H. (2001). A new promising joining technology. *Journal of Ceramic Processing Research*, **2**, (2), 45–50.

Roland, F., Reinert, T. and Pethan, G. (2002). Laser welding in shipbuilding – an overview of the activities at Meyer Werft. In: Kristensen, J.K. ed. *Proceedings of the Conference Advanced Processes and Technologies on Welding and Allied Processes*. Roissy CDG Cedex: International Institute of Welding.

Rosenthal, D. (1941). Mathematical theory of heat distribution during welding and cutting. *Welding Journal*, **20**, 220s–234s.

Rosenthal, D. (1946). The theory of moving sources of heat and its application to metal treatments. *Transactions of the American Society of Mechanical Engineers*, **68**, 849–866.

Ruffler, C. and Gürs, K. (1972). *Optics and Laser Technology*, **4**, 265.

Russell, J.D. (1999). Laser weldability of C–Mn steels. In: Kocakdos, M. and Santos, J.F. eds *Proceedings of the European Symposium Assessment of Power Beam Welds*, 4–5 February 1999, Geesthacht, Paper 2. 14pp.

Sampath, K. and Baeslack, W.A. III (1993). Weldability of RS–PM Al–8Fe–2Mo alloy. *Welding Journal*, **72**, (8), 416s–427s.

Santella, M.L. (1992). A review of techniques for joining advanced ceramics. *Ceramic Bulletin*, **71**, (6), 947–954.

Schaeffler, A.L. (1949). Constitution diagram for stainless steel weld metal. *Metal Progress*, **56**, (11), 680–680B.

Sharp, C.M. and Nielsen, C.J. (1988). The development and implementation of high-speed laser beam welding in the can-making industry. *Welding Journal*, **67**, (1), 25–28.

Shen, H. and Gupta, M.C. (2004). Nd:YAG laser welding of copper to stainless steel. *Journal of Laser Applications*, **16**, 2–8.

Shida, T., Hirokawa, M. and Sato, S. (1997). CO_2 laser welding of aluminium alloys. Report 1: welding of aluminium alloys using CO_2 laser beam in combination with MIG arc. *Quarterly Journal of the Japan Welding Society*, **15**, (1), 18–23.

Shinoda, T., Matsunaga, K. and Shinhara, M. (1991). Laser welding of titanium alloy. *Welding International*, **5**, (5), 346–351.

Smith, D.J., Martukanitz, R.P. and Howell, P.R. (1993). Laser beam welding of aluminum alloy 5754 for automotive applications. In: Bickert, C. ed. *Proceedings of the Conference Light Metals Processing and Applications*. Montreal: Canadian Institute of Mining, Metallurgy and Petroleum. pp. 581–588.

Steen, W.M. (1980). Arc augmented laser processing of materials. *Journal of Applied Physics*, **51**, (11), 5683–5641.

Steen, W.M. and Eboo, M. (1979). Arc augmented laser welding. *Metal Construction*, (July), 332–335.

Sun, Z., Pan, D. and Wei, J. (2002). Inert gas and laser welding of AZ31 magnesium alloy. *Science and Technology of Welding and Joining*, **7**, 343–351.

Svetskommissionen (1971). *Guide to the welding and weldability of C–Mn steels and C–Mn microalloyed steels*. IIW Doc. IIW-382-71. Stockholm: Svetskommissionen.

Swift-Hook, D.T. and Gick, A.E.F. (1973). Penetration welding with lasers. *Welding Journal*, **52**, (11), 492s–499s.

Tamaschke, W. (1996). Cutting and welding large sheets. *Industrial Laser Review*, **11**, (1), 12–13.

Tse, H.C., Man, H.C. and Yue, T.M. (2000). Effect of electric field on plasma control during CO_2 laser welding. *Optics and Lasers in Engineering*, **33**, 181–189.

Tullmin, M., Robinson, F.P.A., Henning, C.A.O., Strauss, A. and Le Grange, J. (1989). Properties of laser-welded and electron-beam-welded ferritic stainless steel. *Journal of the South African Institute of Mining and Metallurgy*, **89**, (8), 243–249.

Uddin, M.N. (1991). How laser welding is changing the auto industry. *Industrial Laser Review*, **6** (2), 11–14.

Walduck, R.P. and Biffin, J. (1994). Plasma arc augmented laser welding. *Welding and Metal Fabrication*, **62**, (4), 172–176.

Wang, P.C. (1995). Fracture mechanics parameter for the fatigue resistance of laser welds. *International Journal of Fatigue*, **17**, 25–34.

Weeter, L. (1998). Technological advances in aluminum laser welding. *Practical Welding Today*, **2**, (1), 56–57.

Weisheit, A., Galun, R. and Mordike, B.L. (1998). CO_2 laser beam welding of magnesium-based alloys. *Welding Journal*, **77**, (4), 149s–154s.

Whitaker, I.R. and McCartney, D.G. (1994). Fracture of bead-on-plate CO_2 laser welds in the Al–Li alloy 8090. *Scripta Metallurgica et Materialia*, **31**, (12), 1717–1722.

Whitaker, I.R. and McCartney, D.G. (1995).The microstructure of CO_2 laser welds in an Al–Fe–V–Si alloy. *Materials Science and Engineering*, **A196**, 155–163.

Whitaker, I.R., McCartney, D.G., Calder, N. and Steen, W.M. (1993). Microstructural characterization of CO_2 laser welds in the Al–Li based alloy 8090. *Journal of Materials Science*, **28**, 5469–5478.

Winderlich, B., Heidel, M., Füssel, U., Behler, K. and Kalla, G. (1994). Fatigue strength of laser welded structural steels. In: *Proceedings of the 5th European Conference on Laser Treatment of Materials (ECLAT '94)*, 26–27 September 1994, Bremen-Vegesack, Germany. Düsseldorf: Deutscher Verband für Schweißtechnik. pp. 498–503.

Yamaoka, H., Yuki, M., Murayama, T., Tsuchiya, K. and Irisawa, T. (1992). CO_2 laser welding of aluminium A6063 alloy. *Welding International*, **6**, (10), 766–773.

Yamaoka, H., Murayama, T., Tsuchiya, K., Nishidono, T. and Ohkuma, H. (2000). Application of laser welding to the aluminium alloy chamber for a huge synchrotron radiation facility (SPring-8). *Welding International*, **14**, (8), 606–613.

Zacharia, T., David, S.A., Vitek, J.M. and DebRoy, T. (1989a). Weld pool development during GTA and laser beam welding of type 304 stainless steel, part 1 – theoretical analysis. *Welding Journal*, **68**, (12), 499s–509s.

Zacharia, T., David, S.A., Vitek, J.M. and DebRoy, T. (1989b). Weld pool development during GTA and laser beam welding of type 304 stainless steel, part 2 – experimental correlation. *Welding Journal*, **68**, (12), 510s–519s.

Zambon, A. and Bonollo, F. (1994). Rapid solidification in laser welding of stainless steels. *Materials Science and Engineering*, **A178**, 203–207.

CHAPTER 17

THERMAL MACHINING

INTRODUCTION AND SYNOPSIS

Laser machining covers a range of scribing and drilling techniques that are based on both thermal and athermal modes of processing. Athermal mechanisms involve breaking chemical bonds using ultraviolet or ultrashort pulsed laser beams, and were described in Chapter 7. Here we consider thermal laser machining, which is based on conventional mechanisms of laser beam heating including melting and vaporization. Most classes of engineering material can be machined using an appropriate laser beam: high energy, short wavelength visible and ultraviolet output from pulsed copper vapour and frequency-multiplied Nd:YAG lasers for metals, alloys and many ceramics; or the lower energy, longer wavelength radiation of pulsed red ruby and far infrared CO_2 lasers for some ceramics, glasses and polymers. The power density required for processing typically varies between 10^4 W mm^{-2} for mechanisms in which melting is dominant and 10^6 W mm^{-2} for vaporization-dominated mechanisms; the corresponding beam interaction times lie around 10^{-6} and 10^{-8} s, respectively. The objective of thermal machining is to remove material as quickly as possible within specified dimensional tolerances, with minimal effect on the parent material.

In this chapter we consider the principles, procedure and industrial application of thermal laser machining. The precision that can be achieved is several orders of magnitude lower than athermal techniques (which can achieve an accuracy on the scale of nanometres), but the rate of machining is several orders of magnitude higher (typically on the scale of cubic millimetres per minute). The process therefore competes with conventional methods of mechanical and thermal machining capable of operating with tolerances on the order of micrometres. Procedures and performance data are provided for a range of engineering materials. Analytical process modelling is applied to illustrate the effects of changes in processing variables on performance, and to enable practical processing parameters to be selected. The first industrial application of thermal laser drilling (and of laser material processing) involved punching holes in diamond for wire-drawing dies (Chapter 2). Since then, the process has evolved into a precision technique of microfabrication that has the potential to contribute to the exponential growth expected in micromachining applications, notably in the fields of electronics and biomedicine.

PRINCIPLES

Thermal laser machining is referred to hereafter as laser machining for expediency. It should not be confused with the athermal mode of material removal that was described in Chapter 7.

SCRIBING

Thermal machining processes dominate when the interaction time of the laser beam with the material is on the order of microseconds and the power density is sufficient to initiate melting or vaporization.

Laser energy is absorbed in different classes of engineering material by the classical mechanisms described in Chapter 5. Excited species transfer their energy into surrounding material on a time scale of picoseconds (10^{-12} s), causing the material to melt, and in most practical cases, vaporize. The mechanism of thermal machining depends on the required geometry, the material type, the rate of removal, and the surface quality required.

The melt fusion mechanism is similar to that used in cutting, in which material is expelled by the kinetic energy of a process gas, which may be inert or active, delivered coaxially or at an angle to the beam. A relatively rough surface is produced with material removal rates on the order of a few cubic millimetres per minute.

With the controlled oxidation mechanism, material is melted and reacted with oxygen to form a brittle oxide, which breaks away. Material removal rates up to $100 \, mm^3 \, min^{-1}$ are then possible (Haferkamp *et al.*, 1995).

Scribing can be performed by increasing the power density to about $10^6 \, W \, mm^{-2}$, causing material to be removed principally by vaporization. A slot of high aspect ratio (depth/width) is produced with a surface of high quality and accuracy, at the expense of a lower material removal rate.

DRILLING

Drilling is a continuation of scribing, in which the beam is normally stationary with respect to the workpiece. The aim is to produce a cavity with a very high aspect ratio (typically up to 30). The beam heats the material to the vaporization temperature, after which it penetrates to form a cavity in a similar manner to that of keyhole welding. The pressure induced by vaporized material, together with any assist gas, forces molten material at the cavity wall to its outer rim where it is expelled. Radiation becomes trapped in the keyhole, inducing plasma formation. A portion of the energy may then be absorbed by the plasma and reradiated to the cavity wall, increasing the process efficiency. However, if plasma formation is excessive, energy can be scattered from the keyhole, reducing the process efficiency. Efficient drilling therefore requires plasma formation to be limited through careful control of an assist gas. Holes can be formed by three practical techniques: direct drilling, percussion drilling and trepanning.

The earliest laser drilling technique was direct single-shot drilling. Each hole is drilled with a single pulse, typically with an energy of tens of joules and a duration on the order of 1 ms. A large quantity of molten and vaporized material is produced, which creates difficulties in maintaining a small dimensional tolerance (typically $\pm 10\%$) and limited recast layer (about 0.02 mm). The hole diameter is determined by the temporal and spatial modes of the laser pulse, the beam spot size, and the material. It is normally slightly less than the beam diameter since only the central part of the beam is used in processing.

In the more common multipulse percussion drilling technique, several lower energy pulses are used to drill each hole, thus limiting the amount of molten and gaseous material produced, enabling tighter tolerances to be met. The diameter of the hole (20–800 μm) is again normally smaller than that of the focused beam, the recast layer is typically less than 0.1 mm, the hole diameter tolerance is about $\pm 5\%$, and the hole aspect ratio can attain a value up to 50.

Larger holes may be drilled by trepanning (described for cutting in Chapter 14), in which the focused beam is moved over the workpiece by a rotating lens offset from the axis of the beam to produce a circular contour.

PROCESS SELECTION

Laser machining competes with both thermal and mechanical methods of material removal, but to provide a meaningful comparison, we consider predominantly thermal techniques here.

CHARACTERISTICS OF LASER MACHINING

The most notable characteristic of laser machining is its ability to process materials independent of their physical properties. Hard, brittle materials can be machined as easily as soft, deformable materials. Machining is not limited by the need for the material to conduct electricity. The low energy input results in machined elements of high accuracy with minimal distortion of surrounding material, eliminating the need for corrective work or the disposal of machining chips. The high power density is sufficient to induce vaporization and a machining mechanism that produces a high quality machined edge. The controllable focused beam diameter enables elements 0.02–1.5 mm in depth to be machined directly. Both short and long production runs are possible, in a wide range of shapes, with flexible machining orientations and geometries. The non-contact nature of laser machining eliminates tool wear and contamination. Automation is easy, leading to higher processing rates than conventional methods of thermal machining. Percussive laser drilling can be used to produce holes up to 25 mm in depth, with an aspect ratio approaching 50. Large diameter holes can be produced by trepanning.

The negative aspects of laser machining are associated with the high equipment cost, the thermal mode of processing, and the nature of a focused laser beam. A recast layer on the machined edge can act as a location for crack initiation. Holes with an aspect ratio greater than 50 are difficult to drill with high quality and low taper because of the cone angle of the focused beam. The quality of holes in thick materials of high thermal conductivity is often inferior to that produced by electrodischarge and electrochemical machining.

COMPETING METHODS OF MACHINING

Technical Comparison

Electrodischarge machining (EDM) is a spark erosion process used to create complex two- and three-dimensional shapes in electrically conductive workpieces. A thin wire 0.05–0.30 mm in diameter is used as an electrode. A DC power supply delivers high frequency pulses to the electrode and workpiece. The gap between the electrode and the workpiece is flooded with deionized water, which acts as a dielectric. Material is eroded by spark discharges. Multiple electrodes can be used to drill holes. This non-contact process is easily automated and used to machine deep grooves with a good surface finish and high accuracy. However, productivity is relatively low, the capital cost is high, and a recast layer is formed. The hole quality of percussion laser drilling competes with that produced by EDM, and the production rate when using a high quality beam with large depth of focus is normally higher. Laser-trepanned

Table 17.1 Characteristics of thermal machining processes (EDM: electrodischarge machining; ECM: electrochemical machining; ECD: electrochemical drilling)

	Laser	EDM	ECM	ECD
Minimum hole diameter (μm)	20	50	750	500
Minimum taper (μm)	10	0.5	25	1
Minimum recast layer (μm)	40	25	–	–
Minimum angle to surface (°)	10	20	–	15
Minimum surface roughness (μm)	20	6	2	6
Maximum aspect ratio	50	200	200	300
Maximum material removal rate ($mm^3\ min^{-1}$)	100	70	1600	1000

holes can exceed EDM hole quality. In general, the laser process competes for hole diameters below about 1 mm, and for depths less than about 15 mm.

In the electrochemical machining (ECM) process, metal is removed by a chemical reaction rather than the electrical action of the EDM arc. The workpiece is the positive anode of an electrochemical cell. The cathode is the 'tool', which is normally formed in an inverse of the shape to be produced. The cathode and the workpiece are brought together in a high pressure circulation of salt solution electrolyte in a DC electrolytic cell. The anode dissolves, with a rate that is inversely proportional to the separation of the electrodes, through the formation of positive ions that break free and are removed by the flowing electrolyte. The finished component thus has the inverse form of the tool, with high dimensional accuracy and a good surface finish. Laser machining competes when products with a range of sizes and shapes are to be machined, and when cavities of small diameter are required.

Electrochemical drilling (ECD) is also an electrolytic process, which uses a chemical 'drill'. The drill is a conducting cylinder with an insulating coating on the outside. The electrolyte is pumped down through the centre of the tube. The tool is moved towards the workpiece, and material at the base of the electrode is removed. Since the outer surface of the tool is insulated, the surface of the hole is not eroded by the process. The process is also referred to as shaped tube electrolytic machining (STEM), and is a common technique for drilling cooling channels of high aspect ratio in turbine blades. Laser drilling competes under similar conditions to those favouring laser machining.

Ultrasonic drilling is based on mechanical vibrations – typically 20 kHz in frequency – that are conducted into an abrasive medium via a cutting tool. Small diameter holes (0.4–1 mm) can be made in hard materials with a depth up to 5 mm.

Focused ion beam machining was developed as a fabrication tool for semiconductor devices. The equipment cost is prohibitively high, but the technique is capable of producing submicrometre details in a vacuum chamber through the use of assist gases.

Economic Comparison

An economic analysis of laser machining and conventional techniques of material removal can be carried out by considering the fixed and variable costs associated with each process. This method was described for laser cutting in Chapter 14.

Comparison Summary

Nd:YAG laser machining using the Q-switched fundamental wavelength or frequency-multiplied output is a cost-effective means of machining features greater than about 50 μm in size. Output from the copper vapour laser provides greater flexibility and better quality in many materials, but the higher cost of ownership must be taken into consideration. Electrodischarge machining also has a relatively high cost of ownership, but provides the highest quality in electrically conductive materials. Electrochemical methods of machining and drilling are relatively inexpensive means of removing material at high rates, with high aspect ratios, but are limited to about 0.5 mm in the minimum size of feature that can be machined.

PRACTICE

Laser machining is normally performed in air by focusing the laser beam onto the workpiece. An assist gas is used principally to increase the rate of material removal using the melt shearing mechanism. Machining may be blind (partial penetration) or through-thickness (slotting and hole punching). Scribing is carried out by moving a continuous wave (CW) or pulsed beam relative to the workpiece.

Drilling is performed using a pulsed beam with no relative lateral movement, but movement in the direction of drilling as the channel is formed. Practical laser machining parameters depend on the composition of the material, the beam characteristics, the machining rate required, and the properties of the machined region (its aspect ratio and edge quality). Processing parameters are readily available for metals, alloys and ceramics, to which the procedures described below relate.

MATERIAL PROPERTIES

The laser beam wavelength is chosen to match the absorptive characteristics of the material, i.e. short wavelength light is used to machine metals, alloys and some ceramics, whereas organic materials absorb relatively long wavelength far infrared light sufficiently for efficient machining.

Materials with high values of specific heat capacity require larger amounts of energy to raise their temperature to that required for melting. Similarly, materials with high latent heats of melting and vaporization require more energy for the changes in state relevant to the machining mechanism.

BEAM CHARACTERISTICS

Wavelength

Green light emitted from the copper vapour laser can be focused to a small spot size, and delivered in pulses of high peak power and energy with high repetition rates. Such output is suitable for machining a wide range of engineering materials with high material removal rates in comparison with other laser sources. The equipment costs more than an Nd:YAG laser, but is similar in price to an excimer source.

Pulsed Nd:YAG lasers, up to 500 W average power, have been used as a cost-effective source of energy for drilling small holes in ceramics for many years. Diode-pumped, frequency-multiplied, Q-switched Nd:YAG lasers now provide a feasible alternative to the copper vapour laser for some applications: the focused spot size is reduced (to about 40 μm), pulses are highly controllable, a high peak power can be generated, and plasma absorption at short wavelengths is relatively low. Frequency-tripled Nd:YAG laser output at 355 nm is the wavelength of choice for machining polymers. Frequency quadrupled output lies in the ultraviolet, and is particularly effective for polymer drilling, where a high repetition rate results in high removal rates and high quality (Glover *et al.*, 1995). This type of laser is the most cost effective choice for simple slotting of grooves over 50 μm in width. Nd:glass lasers are not capable of rapid repetition, but produce good quality machining with low material removal rates.

The ruby laser was used in the first drilling applications, which included diamond wire-drawing dies (Chapter 2). The ruby laser is hampered by the limited repetition rate possible because of the inherent nature of pulsed ruby laser emission, but good quality machining and drilling can be produced with low rates of material removal.

Low and medium power CO_2 lasers are used in drilling of non-metals. The focused spot size is relatively large, and therefore less easily located than other lasers. However, a CO_2 laser beam of high frequency and short pulse length is able to remove polymers from an electronic substrate more rapidly than its ultraviolet counterparts, and would be the source of choice when the precision required is relatively low.

Other infrared lasers, such as the carbon monoxide (CO), chemical oxygen iodine laser (COIL), and chemical lasers developed for military use (e.g. hydrogen fluoride), are absorbed well by natural ceramics such as granite and sandstone. Their performance for underground drilling is being evaluated. The low machining accuracy associated with thermally induced stresses in such inhomogeneous materials is not a concern when the basic requirement is to remove as much material as possible in the shortest possible time.

Spatial Mode

A high quality beam mode is essential for precision machining. The edges of a poor quality beam contain insufficient energy to vaporize, but enough energy to heat and damage peripheral material. Copper vapour lasers produce a beam of high quality that can be focused to a small spot size, typically $10\,\mu m$ in diameter. A fundamental wavelength Nd:YAG laser beam can be focused to a spot around $40\,\mu m$ in diameter, with a corresponding reduction for frequency multiplied output.

Temporal Mode

The pulse profile should be as square as possible to avoid heating effects from the 'tails' of the profile. It is more efficient to drill deep holes using the percussive drilling mechanism than by using a single high energy pulse because plasma shielding effects are reduced, which results in squarer holes with less taper, although the overall material removal rate may be lower. Short pulses cause the surface to reach the vaporization temperature rapidly, restricting the amount of molten material generated, which could form a recast layer.

The high peak power and high repetition rate of copper vapour laser pulses are particularly suitable for machining metals and semiconductor materials. The main advantages over other laser sources are accuracy, a thinner recast layer, the ability to drill holes at shallow angles, and the absence of a heat affected zone. Typical drilling parameters are: repetition rate 7 kHz; pulse energy 6 mJ; and pulse duration 35 ns. The material removal rate is typically 10^{-6} mm^3 per pulse, which is higher than that characteristic of Nd:YAG or excimer laser pulses.

Typical parameters for Q-switched Nd:YAG drilling are: repetition rate 10 Hz; pulse energy 5–10 J; and pulse duration 0.5–1.5 ms. Around 0.1 mm^3 of material is removed per pulse.

Optical Focal Length

An optic of long focal length produces a less tapered groove or hole with a higher aspect ratio because the beam convergence and divergence angles above and below the focal plane are low, which enables a high power density to be maintained throughout the workpiece thickness. However, the diameter of a spot focused by a long focal length optic is greater than that produced by one of shorter focal length, which results in a reduction in the power density. The focal length is therefore selected based on the material thickness and the aspect ratio of the groove or hole required.

Focal Plane Position

The focal plane is normally positioned at the workpiece surface to ensure the maximum power density at the start of machining. Shallow, tapered holes of large diameter are produced if the plane lies above the surface. Shallower, conical holes are produced if it lies below the surface. If the focal plane of a short focal length optic is placed at the surface, it is possible to produce holes of negative taper (wider at the bottom than the top). As with the focal length of the optic, the focal plane position is determined by the material thickness, the aspect ratio of the groove or hole, and the cone angle of the focused beam. The focal plane is moved in the direction of machining to maintain a high power density at the material boundary, within the limits specified by the groove aspect ratio and beam cone angle.

PROCESSING PARAMETERS

Gases

Gases are used to protect the focusing optics, and to improve the processing rate in two ways: by blowing melt out of the hole, and preventing the vapour expanding in the direction of the beam to reducing blocking. An inert gas such as nitrogen or argon prevents reaction of molten material with the atmosphere. Compressed air or oxygen is used to assist through exothermic reaction.

The gas jet is generated in a nozzle. The velocity of the jet may be subsonic, or supersonic if the gas pressure is above about 2.2 bar. Gas pressures between 4 and 6 bar are used in practical drilling nozzles.

PROPERTIES OF LASER-MACHINED MATERIALS

Quality standards for laser machining are not yet available. However, those developed for thermal cutting, described in Chapter 14, provide a means of assessing edge quality in terms of depth and plate thickness.

IMPERFECTIONS

The main imperfections in laser-machined parts relate to the geometry of the cavity. A recast layer, if present, is a source of imperfections related to integrity and metallurgy.

The geometry is defined as the depth, aspect ratio, roughness and squareness of the cavity. A low aspect ratio is a result of insufficient power density to initiate vaporization, or insufficient assist gas flow with the melt shearing mechanism. A non-square cavity with parallel sides indicates deviation of the beam axis from the normal to the workpiece surface. Cavity taper is caused by a deviation of the focal plane position from the optimum, which is at the surface for a cavity of low aspect ratio, or within the workpiece for a cavity of high aspect ratio. A burr may form at the entrance to a blind hole if insufficient beam pulse energy or assist gas is available to eject molten material completely.

A recast layer on the sides of the cavity is also a sign of insufficient beam pulse energy or assist gas. Cracking on the macro scale is associated with high solidification and cooling rates. The recast layer can contain microcracks, which are initiation sites for fatigue cracking.

METALS AND ALLOYS

Laser machining is a well-established technique of scribing and drilling holes and channels in hard metals and alloys, and can be achieved using most of the lasers described above. Major fields of interest include drilling cooling channels in jet engine components, feeder channels in fuel injection systems and ink delivery nozzles in printers, all of which are made from metals and alloys. Selective micromachining of nickel layers on ceramics provides a means of writing and patterning microcircuits. Procedures can be found in the references in Table 17.2.

The dominance of the copper vapour laser for machining metals and alloys, and the emergence of the frequency-multiplied Nd:YAG laser, is demonstrated by their appearance in the procedures listed in Table 17.2.

CERAMICS AND GLASSES

Ceramics and glasses include some of the most difficult materials to process with lasers. The main problems are the build-up of a glass phase at the machined edge, microcracking, and the formation of

Table 17.2 Sources of data for laser machining of metals and alloys

Material	Laser	Reference
Al	Cu vapour	(Lash and Gilgenbach, 1993)
C–Mn steel	CO_2	(O'Neill et al., 1995)
C–Mn steel	Cu vapour	(Chang and Warner, 1996)
Co-based Haynes 188	Cu vapour	(Knowles et al., 1995)
Cu	Nd:YAG ($\lambda/2$, $\lambda/3$, Q-switched)	(Tunna et al., 2001)
Fe	Cu vapour	(Lash and Gilgenbach, 1993)
Ni on ceramic substrate	Cu vapour	(Knowles, 2000)
Nimonic	Nd:YAG	(Kamalu et al., 2002)
Ti	Cu vapour	(Lash and Gilgenbach, 1993)
Various	Cu vapour	(Chang et al., 1998)

Table 17.3 Sources of data for laser machining of ceramics and glasses

Material	Laser	Reference
Alumina	CO_2	(Olson and Swope, 1992)
Ceramics	Nd:YAG	(Pfefferkorn et al., 2003)
Cordierite ($2MgO–2Al_2O_3–5SiO_2$)	Nd:YAG	(Kirby and Jankiewicz, 1998)
Diamond	Cu vapour	(Barnes, 2001)
	Nd:YAG	(Miyazawa et al., 1994)
	Ruby	(Anon., 1966)
Diamond (CVD)	Cu vapour	(Knowles, 2000)
Diamond (CVD)	Nd:YLF	(Schaeffer, 1995)
Engineering ceramics	Nd:YAG	(Harryson and Herbertson, 1987)
Glass	CO_2 (Q-switched), Nd:YAG ($\lambda/4$)	(Schaeffer et al., 2002)
	Nd:YAG	(Harryson and Herbertson, 1987)
	Various	(Atanasov and Gendjov, 1987)
Rocks	CO, CO_2	(Graves et al., 2000)
	COIL	(Hallada et al., 2000)
Sandstone	HF	(Graves and O'Brien, 1998)
Si_3N_4	CO_2	(Bang et al., 1993; Roy and Modest, 1993; Takeno et al., 1992; Wallace and Copley, 1986)
Si_3N_4	Nd:YAG	(Rozzi et al., 2000)
Si_3N_4, SiC	CO_2	(Kitagawa and Matsunawa, 1990)

a bell-mouth to the drilled holes. However, in the absence of conventional techniques of machining hard ceramics such as diamond, laser-based procedures have gained a foothold, some of which are referenced in Table 17.3.

The information in Table 17.4 illustrates the popularity of machining engineering ceramics such as silicon nitride (Si_3N_4) using a variety of laser sources. In general, a short pulse duration, a high peak power and a low repetition rate produce superior results when drilling Si_3N_4 with a CO_2 laser beam,

Table 17.4 Sources of data for laser machining of polymers

Material	Laser	Reference
Paint	CO_2 (TEA)	(Cottam *et al.*, 1998; Schweizer, 1995)
	CO_2 (pulsed)	(Foley, 1991; Head, 1991)
	Nd:YAG	(Forrest, 1992)
	Excimer	(Lovoi, 1994)
PMMA	CO_2	(Berrie and Birkett, 1980; Takeno *et al.*, 1992)
Polyimide	Cu vapour (λ-mixed)	(Knowles, 2000)
Polymers	CO_2 (Q-switched), Nd:YAG ($\lambda/3$)	(Schaeffer and O'Connell, 2002)

since plasma interaction effects are minimized. Three-dimensional machining of engineering ceramics has been achieved using a CO_2 laser beam of relatively low power (below 1 kW), and Nd:YAG lasers in combination with accurate preheating. Micromachining of diamond, particularly that deposited as films for heat sinks, has many potential applications in the microelectronics industry.

POLYMERS

Interest in laser machining of polymers has been driven by niche applications in various industry sectors. The microelectronics industry requires a tool for ablating insulating polymers from copper substrates to create channels tens of micrometres in width on printed circuit boards. Many medical instruments can be used only once – the opportunities for manufacturing instruments from polymers that can be micromachined with laser beams open up many potential fields of application. Methods of removing polymer-based coatings in an environmentally friendly way have received attention from airlines that need to strip and repaint aircraft regularly. (Other methods of laser cleaning are discussed in Chapter 10.) Procedures can be found in the references in Table 17.4.

COMPOSITES

Composite materials present many problems for both mechanical and thermal cutting techniques. Most arise from the difference in mechanical and thermal properties of the matrix and reinforcement – the very properties that endow composites with their unique characteristics. Many industries are looking for methods of machining man-made composites with improved quality and productivity. A field of particular interest is the use of laser machining in biomedical applications: both as a technique of manufacturing medical devices, and as a method of machining the natural composites of which soft and hard tissue are made. Opportunities for replacing the conventional mechanical dentist's drill have driven much of the research into the use of lasers for removing dental caries without pain. A selection of procedures can be found in Table 17.5.

In general, particulate-reinforced metal matrix composites (MMCs) have similar machining characteristics to the matrix material. The main imperfection is the irregular nature of the machined edge, where the reinforcement may protrude or have been melted back, depending on its melting temperature. In contrast, MMCs reinforced with ceramic *fibres* such as SiC exhibit separation of the fibre from the matrix because of the difference in melting temperatures and thermal expansion coefficients.

Table 17.5 Sources of data for laser machining of composites: N/S not specified

Matrix	Reinforcement	Laser	Reference
Alloy	Ceramic coating	N/S	(Forget *et al.*, 1989)
Epoxy	Carbon fibre	CO_2	(Crane and Brown, 1981)
	Glass fibre	CO_2	(Crane and Brown, 1981)
	Kevlar fibre	CO_2	(Crane and Brown, 1981)
Polyester	Glass fibre	CO_2	(Chryssolouris *et al.*, 1990)
Polymer	Glass fibre	Nd:YAG ($\lambda/4$)	(Yung *et al.*, 2002)
PTFE	Carbon whiskers	CO_2	(Chryssolouris *et al.*, 1990)
Soft gum tissue	–	Nd:YAG	(Myers, 1993)
Soft tissue	–	IR free electron	(Benson *et al.*, 1989)
Wood	–	CO_2	(Barnekov *et al.*, 1989)

Of the fibre-reinforced polymer matrix composites (PMCs), those containing carbon are the most difficult to machine because the fibre separates from the matrix on heating as a result of the large differences in melting temperature. Less damage is observed when laser machining glass and aramid reinforced composites, although carbon deposits are observed on the machined faces. In addition, potentially hazardous fumes are generated by the decomposition of organic compounds. Water jet machining appears to be a more satisfactory means of machining such materials.

LASER MACHINING DIAGRAMS

The geometry of the machined element is modelled as follows. Machining is assumed to be carried out using a stationary pulsed laser beam or a moving CW beam. In the former case, the beam interaction time is the pulse duration. In the latter case, the beam interaction time, τ, is given by the formula $\tau = 2r_B/v$ where r_B is the beam radius and v is the traverse rate. The volume of material removed by a single pulse, or in a given beam interaction time, is equal to the volume melted in that time, which is modelled as a cylinder of length l with an aspect ratio (depth/width) of 2, as illustrated schematically in Fig. 17.1. Penetration is assumed to occur via the formation of a cylindrical vapour cavity of length l and aspect ratio 8. The heat affected zone is defined as a region in the parent material bounded by the machined edge (located at a radial distance $l/4$ from the heat source where the temperature is equal to the material melting temperature, T_m), and which extends to a radial distance $l/2$ (where the temperature has fallen to the initial temperature, T_0). The cylinder moves at a speed v. An analytical solution for lateral heat flow from a stationary line heat source within an infinite solid – equation (E3.1) in Appendix E – is now used to construct two laser machining diagrams.

MACHINING OVERVIEW

The cross-sectional area of the vaporized region is $l^2/8$. The cross-sectional area of the molten region is $l^2/2$. To describe the HAZ, the following are substituted in equation (E3.1) (Appendix E): $T_1 = T_m$ where $r_1 = l/4$; and $T_2 = T_0$ at $r_2 = l/2$. An energy balance between the absorbed power and the power used for vaporization, melting and heating gives

$$Aq = \frac{l^2}{8}vL_v + \frac{l^2}{2}vL_m + \frac{2\pi\lambda l(T_m - T_0)}{\ln(2)} \tag{17.1}$$

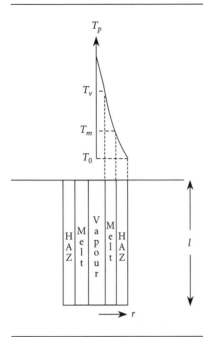

Figure 17.1 Model used for laser machining (not to scale)

where L_v is the volumetric latent heat of vaporization (J m^{-3}), L_m is the volumetric latent heat of melting (J m^{-3}), and λ is the thermal conductivity $(\text{J s}^{-1}\,\text{m}^{-1}\,\text{K}^{-1})$. In terms of the dimensionless variables defined in Chapter 6, equation (17.1) can be written:

$$q^* = \frac{l^{*2}v^*L_v^*}{8} + \frac{l^{*2}v^*L_m^*}{2} + \frac{2\pi l^*}{\ln(2)} \tag{17.2}$$

where $q^* = Aq/[r_B\lambda(T_m - T_0)]$, $l^* = l/r_B$, $v^* = vr_B/a$, $L_m^* = L_m/[\rho c\,(T_m - T_0)]$, $L_v^* = L_v/[\rho c(T_m - T_0)]$, and $a = \lambda/(\rho c)$, where ρ is the material density (kg m^{-3}) and c is the specific heat capacity $(\text{J kg}^{-1}\,\text{K}^{-1})$. The values of L_m^* and L_v^* lie around 0.4 and 4, respectively, for metals and alloys (Richard's and Trouton's rules). Equation (17.2) gives an explicit relationship between q^* and v^*, for a given normalized plate thickness, l^*, and is used to construct the contours of constant machined depth shown in Fig. 17.2. The boundary for machining is defined as the condition required to initiate surface melting from a surface point source of energy, which was established for metals and alloys in Chapter 6.

Although machining is normally performed using a stationary pulsed laser beam, Fig. 17.2 is constructed with the dimensionless groups used in the other normalized laser processing overview diagrams described in the book so that comparisons between processes can be made: a pulse duration is converted into an effective traverse rate using the technique described above. This effective traverse rate is then used to establish values for process variable groups from Fig. 17.2. The normalized machined depth for multiple pulse machining is obtained by summing the contributions from each pulse.

Figure 17.2 provides an overview of machining metals and alloys, and illustrates the relationships between the groups of process variables, which helps in understanding the process mechanisms. The feasibility of laser machining with available equipment can be established. The effects of changes in process variables on the geometry of the machined cavity can be calculated. Variations in the geometry

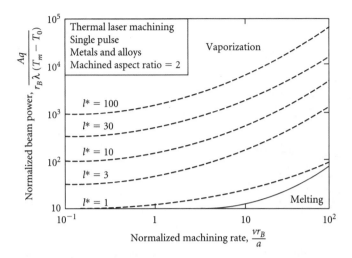

Normalized beam power, $\dfrac{Aq}{r_B\lambda(T_m-T_0)}$

10^5

10^4

10^3

10^2

10

Thermal laser machining
Single pulse
Metals and alloys
Machined aspect ratio = 2

Vaporization

$l^* = 100$

$l^* = 30$

$l^* = 10$

$l^* = 3$

$l^* = 1$

Melting

10^{-1} 1 10 10^2

Normalized machining rate, $\dfrac{vr_B}{a}$

Figure 17.2   Normalized model-based overview of thermal laser machining of metals and alloys, showing vaporization contour (solid line) and contours of constant normalized machining depth, l^* (broken lines)

of the machined element may be taken into account in the model by changing the aspect ratio of the molten section.

Figure 17.2 also provides a basis for constructing diagrams of process parameter selection.

PARAMETER SELECTION

For a given material and set of processing parameters, the only unknown in equation (17.2) is the absorptivity, A, which is contained in the dimensionless variable q^*. An effective value is established from an experimentally determined calibration. A groove of depth 0.5 mm is measured after machining nickel using an incident Nd:YAG laser beam of power 1.0 kW with a traverse rate of 1 m min^{-1}. By substituting these parameters and the material properties for nickel into equation (17.2), a value of 0.5 is obtained for the effective absorptivity. By assuming that the absorptivity and aspect ratio of the machined groove remain constant with changes in beam power and machining rate, equation (17.2) may then be used to construct a diagram with axes of incident beam power, q, and traverse rate, v, showing contours of constant plate thickness, l, Fig. 17.3. The calibration data point is indicated by a star. The following values are used as material properties for nickel: (Table D.1, Appendix D): $\rho = 8900 \text{ kg m}^{-3}$; $T_m = 1700 \text{ K}$; $T_0 = 298 \text{ K}$; $L_m = 270 \text{ kJ kg}^{-1}$; and $L_v = 6.05 \text{ MJ kg}^{-1}$.

Figure 17.3 may be used to expedite machining procedure development by enabling candidate procedures to be sifted rapidly, allowing an initial set of processing parameters to be selected. The depth produced by multiple pulse machining is obtained by summing the contributions from each pulse. The figure may be used to extrapolate data to predict cutting parameters for conditions that lie outside current operating windows.

The calibration data point can also be used to generate contours of machining depth for materials other than nickel by substituting the appropriate material properties (Table D.1, Appendix D) into equation (17.2), and assuming that the mechanism, absorptivity and aspect ratio of the machined element remain constant. Changes in the geometry of the machined element may be accommodated by modifying the aspect ratio of the molten region.

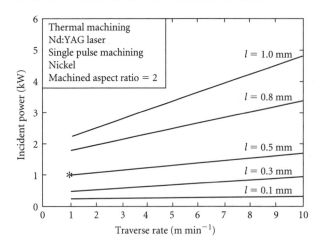

Figure 17.3 Parameter selection diagram for thermal laser machining of nickel, showing model-based contours of constant machining depth for a machined aspect ratio of 2 that are calibrated to an experimental data point ($*$)

Table 17.6 Industrial applications of laser scribing and ablation

Industry sector	Application	Material	Laser	Reference
Aerospace	Machinery	Hard metals	CO_2	(Maher, 1995)
	Paint removal	Al alloys	CO_2	(Anon., 1995)
Biomedical	Ablation	Tissue	Free electron	(Benson *et al.*, 1989)
	Atherectomy	Polymer	Ho:YAG	(Aretz *et al.*, 1991)
	Disposable medical devices	Polymers	Nd:YAG ($\lambda/3$)	(Schaeffer, 1996)
	Fragmentation	Gall stone	Ho:YAG	(Blomley *et al.*, 1995)
	Stents	AISI 316L	Yb fibre	(Kleine *et al.*, 2002)
Machinery	Relief pattern plate engraving	Rubber	CO_2	(Wakabayashi and Sugishima, 1989)
	Steel sheet roller texturing	Steel	CO_2	(Ujihara and Cooke, 1992; Crahay *et al.*, 1986)

INDUSTRIAL APPLICATION

Applications of laser machining are found in industrial sectors comparable with those that use laser cutting on a microscopic scale. The flexibility of the process enables it to be applied to such diverse procedures as keyhole atherectomy (arterial plaque removal), drilling cooling channels in jet engine components, and producing energy by inertial confinement fusion. Table 17.6 illustrates some of the more common applications of laser scribing and ablation, while Table 17.7 gives examples of applications of laser drilling.

Retinal Attachment

The first medical use of the laser in the eye – and the application that did most to raise the public profile of laser material processing – was retinal attachment. The retina of the eye can tear in some regions,

Table 17.7   Industrial applications of laser drilling: N/S not specified

Industry sector	Product	Material	Laser	Reference
Aerospace	Engine components	Ni alloy, Co alloy	Nd:YAG	(van Dijk *et al.*, 1989)
	Prototype wing surface	Ti alloy	Nd:YAG	(Fishlock, 1992)
	Turbine airfoils	Ni-based	Nd:YAG	(Rockstroh *et al.*, 2002)
	Turbine blades and vanes	Ni-based	Nd:YAG	(Heglin, 1979)
Automotive	Fuel injectors	Steels	Cu vapour	(Knowles, 2000)
Biomedical	Dental caries	Tooth enamel	Nd:YAG	(Petty and Krevsky, 1990)
	Soft tissue	N/S	Various	(Petty and Krevsky, 1990)
Domestic goods	Baby bottle nipples	Polymer	N/S	(Anon., 1969)
	Compact disc masters	N/S	Ar ion	(Darby, 1991)
	Ink-jet printer nozzle orifice	Stainless steel	Cu vapour	(Knowles, 2000)
Electrical	PCB vias	Polyimide	Cu vapour	(Knowles, 2000)
Material production	Wire drawing dies	Diamond	Ruby	(Anon., 1966)

eventually leading to complete detachment from the back of the eye. Light is focused to drill hundreds of small holes in the region surrounding the lesion, thus sealing it off. The burned spots develop tough scar tissue, which helps hold the weakened retina in place. Red ruby laser light was used in the first trials; blue argon ion laser light is now more commonly used.

Glaucoma

Glaucoma encompasses a group of conditions that involve damage to the optic nerve of the eye. It is caused by an imbalance in internal pressure within the eye. In the case of open angle glaucoma, peripheral vision gradually becomes blurred. Narrow angle glaucoma causes halos to appear around lights, and can be painful. The blue light of an argon ion laser may be used in the treatment of both types of glaucoma.

The treatment for open angle glaucoma is known as laser trabeculoplasty. The laser beam is focused on the filtration ring (the boundary between the coloured and white parts of the eye). Heat is generated in the microscopic filtration holes and meshwork that make up the filter, causing changes in the shape of these holes that facilitate increased drainage. Treatment typically reduces pressure by about 80%.

The treatment for narrow angle glaucoma is known as laser peripheral iridotomy. A small hole is made in the iris, through which excess fluid is vented.

Catheter Drilling

The precision and control of laser drilling and machining enables intravenous catheters of superior quality to be produced, Figure 17.4. Liquid flow is precise and insertion is smooth and easy. The holes in this example were drilled with a pulsed Nd:YAG laser.

Figure 17.4   Intravenous catheter drilled using a pulsed Nd:YAG laser. (Source: Mo Naeem, GSI Lumonics, Rugby, UK)

Resistor Trimming

Thick film resistor trimming was the first application of laser processing in microelectronics. Lasers are now candidates for any machining operation requiring high precision, such as trimming, repair, exposure, drilling or refinement.

Film circuits consist of films of insulating dielectric layers and conductive tracks, which are deposited on a substrate to form electrical circuits. *Thin* film circuits are made by vacuum deposition followed by etching, and are very accurate, but the manufacturing process is time consuming. *Thick* film circuits are made by a screen printing process, in which the conductor tracks are printed in position – an economic mass-production process suited to both long and short series. Flat resistors are printed as resistive pastes in a similar way. The resistors are printed larger (less resistive) than required, and are laser trimmed to the correct value.

The main benefits of laser trimming are the high production rate, accuracy, ease of automation, and cleanliness. Trimming can be carried out by air blasting with an abrasive jet, but laser micromachining provides two distinct advantages: high accuracy and a high quality sealed edge. Pulsed Nd:YAG lasers are often used, with powers in the range 0.25 W at 1 kHz to 3 W at 4 kHz. Trimming takes the form of narrow channels (vias) micromachined into the resistor. Thus the process enjoys the benefits of easier and cheaper manufacture than thin film technology, while achieving the desired accuracy.

Jet Engine Components

Combustor

In 1981, Pratt & Whitney (Middletown, CT, USA) redesigned the combustor of the JT9D engine to produce a more uniform distribution of cooling air. The new design called for numerous holes with a variable section thickness to be produced in Hasteloy X nickel alloy sheet. The largest rings contain up to 600 cooling holes, each typically 2 mm in diameter in material about 3 mm in thickness. The holes are drilled by mechanically shuttering the beam from a multikilowatt CO_2 laser. The production rate of one hole per second is between 10 and 100 times higher than pulsed low power laser drilling, and around 20 times higher than the alternative process of EDM. The holes are slightly bell-mouthed, which provides a higher flow coefficient than laser trepanned, twist drill or EDM-formed holes. (A similar laser is used to make an adjacent seam weld without intruding on the holes.) The system pays for itself every five months. The new combustor design has significantly lengthened the time between hot section inspections, resulting in a reduction in engine maintenance costs. A laser is also used to drill over 10 000 holes in the Inconel 625 combustor of the Pratt & Whitney 4000 series engines used in the Boeing 777.

Turbine Blades

The temperature of combustion gases in aircraft jet engines can attain 2000°C. This is higher than the melting temperature of the nickel-base alloys used in the engine. To overcome this problem, cooling holes are drilled in the turbine blade with specific orientations to the surface, which produce films of insulating air over the surface. Each turbine blade typically requires 20 000 holes. This is one of the most successful metalworking applications today, and is performed using Nd:YAG lasers.

National Ignition Facility

Energy is produced by inertial confinement fusion (ICF) at the National Ignition Facility (NIF), located at the Lawrence Livermore National Laboratory (Livermore, CA, USA). Unlike magnetic fusion energy designs such as the Joint European Torus (JET), which hold fuel in a magnetic field, ICF relies on isentropic compression of fusion fuel. Fuel is ignited by high energy beams that are formed from a variety of lasers.

The first step in the process uses a nanojoule energy pulse from the master oscillator (a diode-pumped fibre laser), which is transported to preamplifier modules (PAMs) for beam shaping and amplification: first to about 1 mJ, and then to 22 J in a flashlamp-pumped amplifier. A total of 48 PAMs feed a quad of four laser beams. The beams are then amplified in slabs of Nd:glass (phosphate) excited by flashlamps, creating powerful infrared radiation. Further beam manipulation, including the use of a Pockels cell to convert infrared light to an ultraviolet wavelength (351 nm), eventually enables 1.8 MJ pulses of energy of duration 3 ns to be produced. The pulses are focused onto small metallic fusion targets that are compressed under conditions in which they ignite and burn, liberating sufficient energy to initiate fusion reactions. Thermonuclear ignition and burning can also be used to test nuclear weapons without underground trials.

SUMMARY AND CONCLUSIONS

The two principal forms of thermal laser machining are scribing and drilling. They compete with athermal methods of laser machining when relatively large material removal rates are required (cubic millimetres per minute), and dimensional tolerances on the order of micrometres are specified. Laser machining competes with conventional machining methods in applications involving rapid processing with simple fixturing and flexible geometries (e.g. shallow angle drilling). By using an appropriate wavelength of laser light, all classes of engineering material may be machined, irrespective of their physical and thermal properties. The copper vapour laser has gained a reputation for macromachining hard materials, producing cavities with a high aspect ratio (up to 30), because of its ability to produce high energy pulses with high repetition rates. Laser machining is also popular for processing hard ceramics, some of which are too brittle to machine using conventional methods, because ceramics absorb visible and ultraviolet light efficiently. Analytical modelling can be used to construct laser machining diagrams, which are applicable to both scribing and drilling. The models are simple enough to be modified to suit a variety of processing geometries (cavity aspect ratios), and are flexible enough to be applied to a range of engineering materials. The aerospace industry initially developed techniques for scribing and drilling, which have been extended to micromachining applications, particularly in the biomedical field, as shorter wavelength laser sources have been developed.

Laser machining is ideally suited to all classes of hard engineering material, in which high aspect ratio cavities are required in flexible locations and orientations, with high material removal rates and minimal effects on the parent material. The low process energy input ensures accurate machining, minimal reworking, and low scrap rates. As sources of even shorter wavelength are developed, capable of producing high energy pulses with rapid repetition rates that permit high rates of material removal, the thermal mode of laser machining will make a significant contribution (along with athermal laser machining), to the exponential growth in micromachining applications expected in the decades to come.

FURTHER READING

Ahlers, R.-J., Hoffman, P., Lindl, H. and Rothe, R. eds (1994). *Laser Materials Processing and Machining*. Bellingham: SPIE.

Berns, M.W. (1990). Laser surgery: organs to organelles. *Journal of Laser Applications*, **1**, (Summer/Fall), 58–60.

Chryssolouris, G. (1991). *Laser Machining: Theory and Practice*. New York: Springer.

Gaillard, M. and Quenzer, A. eds (1989). *High Power Lasers and Laser Machining Technology*. Bellingham: SPIE.

Gower, M.C., Helvajian, H., Sugioka, K. and Dubowski, J.J. eds (2001). *Laser Applications in Microelectronic and Optoelectronic Manufacturing VI*. Bellingham: SPIE.

Heglin, L.M. (1986). Introduction to laser drilling. In: Belforte, D.A. and Levitt, M. eds *The Industrial Laser Annual Handbook*. Tulsa: PennWell. pp. 116–120.

Miyamoto, I., Ostendorf, A., Sugioka, K. and Helvajian, H. eds (2003). *Fourth International Symposium on Laser Precision Microfabrication*. Bellingham: SPIE.

Phipps, C.R. ed. (1998). *Proceedings of the Conference High-Power Laser Ablation*. Vol. 3343. Bellingham: SPIE.

Phipps, C.R. and Niino, M. eds (2000). *Proceedings of the Conference High-Power Laser Ablation II*. Vol. 3885. Bellingham: SPIE.

Phipps, C.R. ed. (2002). *Proceedings of the Conference High-Power Laser Ablation III*. Vol. 4065. Bellingham: SPIE.

Phipps, C.R. ed. (2004). *Proceedings of the Conference High-Power Laser Ablation IV*. Vol. 4760. Bellingham: SPIE.

Russo, R.E. (1995). Laser ablation. *Applied Spectroscopy*, **49**, (9), 14A–28A.

Rykalin, N., Uglov, A. and Kokora, A. (1978). *Laser Machining and Welding*. Oxford: Pergamon Press.

EXERCISES

1. Discuss the main differences between laser machining using thermal and athermal techniques in terms of the processing mechanisms and the properties of machined material.

2. Discuss the role of absorptivity in the selection of lasers and materials for thermal machining.

3. By considering the mechanism by which laser light is generated (Chapter 3), explain why the ruby laser is an appropriate source of light for thermal drilling applications.

4. If nickel can be thermally machined to a depth of 0.5 mm with an incident laser beam power of 1.28 kW and a traverse rate of $5\,\text{m}\,\text{min}^{-1}$, calculate an effective absorptivity for the process. Estimate the laser power required to machine nickel to a depth of 0.1 mm at a traverse rate of $1\,\text{m}\,\text{min}^{-1}$. Assume that the machining mechanism, machined aspect ratio and absorptivity remain constant.

5. If nickel can be machined to a depth of 0.8 mm with an incident power of 2.39 kW at a rate of 5 m min^{-1}, calculate the incident power required to machine titanium to a depth of 0.5 mm at a traverse rate of 1 m min^{-1}. Assume that the machining mechanism, machined aspect ratio and absorptivity remain constant.

BIBLIOGRAPHY

Anon. (1966). Laser 'punches' holes in diamond wire-drawing dies. *Laser Focus*, **2**, (1), 4–7.

Anon. (1969). CO_2 beam perforates baby bottle nipples. *Electrotechnology*, **84**, (1), 46.

Anon. (1995). Laser-based paint removal system. *Industrial Laser Review*, **10**, (4), 6.

Aretz, H.T., Gregory, K.W., Martinelli, M.A., Gregg, R.E., LeDet, E.G., Hatch, G.F., Sedlacek, T. and Haase, W.C. (1991). Ultrasound guidance of laser atherectomy. *International Journal of Cardiac Imaging*, **6**, (3–4), 231–237.

Atanasov, P.A. and Gendjov, S.I. (1987). Laser cutting of glass tubing – a theoretical model. *Journal of Physics D, Applied Physics*, **20**, (5), 597–601.

Bang, S.Y., Roy, S. and Modest, M.F. (1993). CW laser machining of hard ceramics – II. Effects of multiple reflections. *International Journal of Heat and Mass Transfer*, **36**, (14), 3529–3540.

Barnekov, V., Huber, H.A. and McMillin, C.W. (1989). Laser machining wood composites. *Forest Products Journal*, **39**, (10), 76–78.

Barnes, C. (2001). Copper at the cutting edge: the copper vapor laser [Internet]. Available from: <http://www.copper.org/innovations/2001/06/cutting_edge.html> [Accessed 29 July 2004].

Benson, S., Madey, J., Straight, R. and Hooper, B. (1989). Performance of a broadband free-electron laser and preliminary studies on its application to biology and medicine. *Journal of Laser Applications*, **2**, (July), 49–58.

Berrie, P.G. and Birkett, F.N. (1980). The drilling and cutting of polymethyl methacrylate (perspex) by CO_2 laser. *Optics and Lasers in Engineering*, **1**, 107–129.

Blomley, M.J., Nicholson, D.A., Bartal, G., Foster, C., Bradley, A., Myers, M., Man, W., Li, S. and Banks, L.M. (1995). Holmium-YAG laser for gall stone fragmentation: an endoscopic tool. *Gut*, **36**, 442–445.

Chang, J.J. and Warner, B.E. (1996). Laser–plasma interaction during visible-laser ablation of methods. *Applied Physics Letters*, **69**, (4), 473–475.

Chang, J.J., Warner, B.E., Dragon, E.P. and Martinez, M.W. (1998). Precision micromachining with pulsed green lasers. *Journal of Laser Applications*, **10**, (6), 285–290.

Chryssolouris, G., Sheng, P. and Choi, W.C. (1990). Three-dimensional laser machining of composite materials. *Journal of Engineering Materials and Technology*, **112**, 387–392.

Cottam, C.A., Emmony, D.C., Cuesta, A. and Bradley, R.H. (1998). XPS monitoring of the removal of an aged polymer coating from a metal substrate by TEA-CO_2 laser ablation. *Journal of Materials Science*, **33**, 3245–3249.

Crahay, J., Renauld, Y., Monfort, G. and Bragard, A. (1986). Present state of development of the 'Lasertex' process. In: Quenzer, A. ed. *Proceedings of the 3rd International Conference Lasers in Manufacturing (LIM-3)*, 3–5 June 1986, Paris, France. IFS Publications: Bedford/Springer-Verlag: Berlin. pp. 245–259.

Crane, K.C.A. and Brown, J.R. (1981). Laser-induced ablation of fibre/epoxy composites. *Journal of Physics D: Applied Physics*, **14**, 2341–2349.

Darby, M.J. (1991). Compact video disc plant employs ion lasers. *Laser Magazine*, (6), 25–27.

Fishlock, D. (1992). Loads of time and money. *Financial Times*, 2 September 1992.

Foley, J.S. (1991). Laser paint stripping: an automated solution. *Industrial Laser Review*, **6**, (3), 4–9.

Forget, P., Jeandin, M., Lechervy, P. and Varela, D. (1989). Laser drilling application to a ceramic-coated alloy. *Materials and Manufacturing Processes*, **4**, (2), 263–272.

Forrest, G.T. (1992). Nd:YAG lasers strip paint effectively. *Industrial Laser Review*, **7**, (12), 12–14.

Glover, A.C.J., Withford M.J., Illy, E.K. and Piper, J.A. (1995). Ablation threshold and etch rate measurements in high-speed ultraviolet (uv)-micromachining of polymers with uv-copper vapour lasers. In: Mazumder, J., Matsunawa, A. and Magnusson, C. eds *Proceedings of the Laser Materials Processing Conference (ICALEO '95)*. Orlando: Laser Institute of America. pp. 361–370.

Graves, R.M. and O'Brien, D.G. (1998). *Targeted literature review: determining the benefits of StarWars laser technology for drilling and completing natural gas wells: topical report number 1.* Report 98/0163. Chicago: Gas Research Institute.

Graves, R.M., Ionin, A.A., Klimachev, Yu.M., Mukhammedgalieva, A.F., O'Brien, D.F., Sinitsyn, D.V. and Zvorykin, V.D. (2000). Interaction of pulsed CO and CO_2 laser radiation with rocks. In: Phipps, C.R. ed. *Proceedings of the Conference High-Power Laser Ablation III*. Vol. 4065. Bellingham: SPIE. pp. 602–613.

Haferkamp, H., Bach, Fr.-W., von Alvensleben, F. and Seebaum, D. (1995). Material removal using high power lasers and its industrial potential. In: Mazumder, J., Matsunawa, A. and Magnusson, C. eds *Proceedings of the Laser Materials Processing Conference (ICALEO '95)*. Orlando: Laser Institute of America. pp. 148–157.

Hallada, M.R., Walter, R.F. and Seiffert, S.L. (2000). High-power laser rock cutting and drilling in mining operations: initial feasibility tests. In: Phipps, C.R. ed. *Proceedings of the Conference High-Power Laser Ablation III*. Vol. 4065. Bellingham: SPIE. pp. 614–620.

Harryson, R. and Herbertson, H. (1987). Machining of high performance ceramics and thermal etching of glass by laser. In: Steen, W.M. ed. *Proceedings of the 4th International Conference Lasers in Manufacturing (LIM-4)*, 12–14 May 1987, Birmingham, UK. Bedford: IFS Publications/ Berlin: Springer-Verlag. pp. 211–219.

Head, J.D. (1991). New CO_2 laser cuts paint-stripping damage. *Industrial Laser Review*, **6**, (3), 11–14.

Heglin, L.M. (1979). Laser drilling for materials fabrication. In: Metzbower, E.A. ed. *Proceedings of the Conference Applications of Lasers in Materials Processing*. Metals Park: American Society for Metals. pp. 129–147.

Kamalu, J., Byrd, P. and Pitman, A. (2002). Variable angle laser drilling of thermal barrier coated nimonic. *Journal of Materials Processing Technology*, **122**, 355–362.

Kirby, K.W. and Jankiewicz, A.T. (1998). Laser machining of glass forming ceramics. *Journal of Laser Applications*, **10**, (1), 1–10.

Kitagawa, A. and Matsunawa, A. (1990). Three dimensional shaping of ceramics by using CO_2 laser and its optimum processing condition. In: Ream, S.L., Dausinger, F. and Fujioka, T. eds *Proceedings of the Laser Materials Processing Conference (ICALEO '90)*. Orlando: Laser Institute of America/Bellingham: SPIE. pp. 294–301.

Kleine, K.F., Whitney, B. and Watkins, K.G. (2002). Use of fiber lasers for micro cutting applications in the medical device industry. In: *Proceedings of the Laser Materials Processing Conference (ICALEO 2002)*. Orlando: Laser Institute of America. 11p.

Knowles, M.R.H. (2000). Micro-ablation with high power pulsed copper vapour lasers. *Optics Express*, **7**, (2), 50–55.

Knowles, M.R.H., Foster-Turner, R., Bell, A.I., Kearsley, A.J., Hoult, A.P., Lim, S.W. and Bisset, H. (1995). Drilling of shallow angled holes in aerospace alloys using a copper laser. In: Mazumder, J., Matsunawa, A. and Magnusson, C. eds *Proceedings of the Laser Materials Processing Conference (ICALEO '95)*. Orlando: Laser Institute of America. pp. 321–330.

Lash, J.S. and Gilgenbach, R.M. (1993). Copper vapor laser drilling of copper, iron, and titanium foils in atmospheric pressure air and argon. *Review of Scientific Instruments*, **64**, (11), 3308–3313.

Lovoi, P. (1994). Laser paint stripping offers control and flexibility. *Industrial Laser Review*, **10**, (3), 15–18.

Maher, W. (1997). Laser applications in manufacturing at the Boeing Company. *Industrial Laser Review*, **12**, (8), 7–9.

Miyazawa, H., Miyake, S., Watanabe, S., Murakawa, M. and Miyazaki, T. (1994). Laser-assisted thermochemical processing of diamond. *Applied Physics Letters*, **64**, (3), 387–389.

Myers, T. (1993). Laser in dentistry. *Lasers and Optronics*, **12**, (5), 23–26.

Olson, R.W. and Swope, W.C. (1992). Laser drilling with focused Gaussian beams. *Journal of Applied Physics*, **72**, (8), 3686–3696.

O'Neill, W., Volgsanger, M., Elboughey, A. and Steen, W.M. (1995). Selective removal of steel by a laser slotting process. In: Mazumder, J., Matsunawa, A. and Magnusson, C. eds *Proceedings of the Laser Materials Processing Conference (ICALEO '95)*. Orlando: Laser Institute of America. pp. 158–167.

Petty, H.R. and Krevsky, B. (1990). A survey of laser applications in biomedicine. *Lasers and Optoelectronics*, **9**, (4), 63–79.

Pfefferkorn, F.E., Incropera, F.P. and Shin, Y.C. (2003). Surface temperature measurement of semi-transparent ceramics by long-wavelength pyrometry. *Journal of Heat Transfer*, **125**, (1), 48–56.

Rockstroh, T.J., Scheidt, D. and Ash, C. (2002). Advances in laser drilling of turbine airfoils. *Industrial Laser Solutions*, **17**, (8), 15–20.

Roy, S. and Modest, M.F. (1993). CW laser machining of hard ceramics – I. Effects of three-dimensional conduction, variable properties and various laser parameters. *International Journal of Heat and Mass Transfer*, **36**, (14), 3515–3528.

Rozzi, J.C., Pfefferkorn, F.E., Shin, Y.C. and Incropera, F.P. (2000). Experimental evaluation of the laser-assisted machining of silicon nitride ceramics. *Journal of Manufacturing Science and Engineering*, **122**, 666–670.

Schaeffer, R. (1995). Novel high-power Nd:YLF laser for CVD-diamond micromachining. In: Markus, K.W. ed. *Proceedings of the Conference Micromachining and Microfabrication Process Technology*. Vol. 2639. Bellingham: SPIE. pp. 325–334.

Schaeffer, R. (1996). Laser micromachining of medical devices. *Medical Plastics and Biomaterials*, **3**, (3) 32–38, 1996.

Schaeffer, R.D., Kardos, G. and Derkach, O. (2002). Laser processing of glass. *Industrial Laser Solutions*, **17**, (9), 11–13.

Schaeffer, R.D. and O'Connell, J. (2002). Comparing lasers for micromachining plastics. *Industrial Laser Solutions*, **17**, (5), 16–19.

Schweitzer, G. (1995). CO_2 lasers reduce waste from aircraft paint removal. *Opto and Laser Europe*, **18**, (March), 42–43.

Takeno, S., Moriyasu, M. and Hiramoto, S. (1992). Study on material processing by high peak pulse CO_2 laser – effects of beam–plasma interaction and heat accumulation on characteristics of drilling. In: Matsunawa, A. and Katayama, S. eds *Proceedings of the Conference Laser Advanced Materials Processing (LAMP '92)*. Osaka: High Temperature Society of Japan and Japan Laser Processing Society. pp. 329–333.

Tunna, L., Kearns, A., O'Neill, W. and Sutcliffe, C.J. (2001). Micromachining of copper using Nd:YAG laser radiation at 1064, 532, and 355 nm wavelengths. *Optics and Laser Technology*, **33**, 135–143.

Ujihara, S. and Cooke, D.W. (1992). Application of laser textured dull steel to automobile panels. *Surface Engineering*, **8**, (2), 124–130.

van Dijk, M.H.H., de Vlieger, G. and Brouwer, J.E. (1989). Laser precision hole drilling in aero-engine components. In: Steen, W.M. ed. *Proceedings of the 6th International Conference Lasers in Manufacturing (LIM-6)*, 10–11 May 1989, Birmingham, UK. Bedford: IFS Publications/Berlin: Springer-Verlag. 11p.

Wakabayashi, K. and Sugishima, N. (1989). CO_2 laser engraving system for relief plate to print corrugated cardboard. *Journal of Laser Applications*, **1**, (3), 26–30.

Wallace, R.J. and Copley, S.M. (1986). Laser machining of silicon nitride: energetics. *Advances in Ceramic Materials*, **1**, 277–283.

Yung, K.C., Mei, S.M. and Yue, T.M. (2002). A study of the heat-affected zone in the UV YAG laser drilling of GFRP materials. *Journal of Materials Processing Technology*, **122**, 278–285.

CHAPTER 18

OPPORTUNITIES

INTRODUCTION AND SYNOPSIS

The aim of the book has been four-fold:

- to provide an understanding of the principles of the most common methods of laser material processing;
- to describe the practice adopted for successful implementation;
- to use analytical process modelling to clarify the effect of processing variables on the properties of processed materials; and
- to examine the technical and economic criteria that underlie industrial application.

Opportunities for sustainable growth in the field of laser material processing can be identified by considering each of these objectives in turn.

The principles of most of the processes described are based on the heating effect of a laser beam – the thermal mode of material processing – for which transformations can be explained in terms of the underlying mechanisms of heating, melting and vaporization. We noted that laser processing can also be carried out in an athermal mode, involving for example, photochemical and photophysical mechanisms to make and break chemical bonds – mechanisms that cannot be described using classical heat transfer. Opportunities can be identified by understanding the underlying principles of both modes of beam–material interaction, and comparing them with the properties of engineering materials.

Practice makes perfect, it is said. The difference between success and failure can lie in the choice of a single parameter. Process principles guide the engineer, but only trial and error testing can validate selections. Some results come as a complete surprise, and have formed a basis for novel methods of processing. Others enable promising ideas to be developed gradually into productive applications.

Laser processing is amenable to mathematical modelling because of the precise control over the process variables that is possible. Modelling has many powerful uses, which have been demonstrated through the use of *Laser Processing Diagrams*. The response of metals and alloys to laser heating can be predicted well since many of the relevant thermal properties may be combined into dimensionless constants that simplify analysis. The methods work because metals and alloys obey laws of structure and properties that are based on their regular arrangements of atoms. By identifying analogous methods for non-metals, there are opportunities to extend modelling techniques to other classes of engineering material.

Industrial practice has developed to such an extent that many processes are now carried out in turnkey systems, thanks to the use of online monitoring and process control systems, many of which incorporate mathematical models to predict material response during processing. Turnkey systems *may* eliminate the need for experts, providing opportunities for laser processing to be adopted by more industries, particularly small and medium size enterprises that have limited budgets for research and training. Some processes are simple replacements for existing processes, in which the increase in productivity or product quality, or both, have provided the economic advantage. But we have also seen how the unique properties of the laser beam have been extended to create novel manufacturing operations. Parts may then be redesigned based on their function instead of the limitations imposed by conventional methods of fabrication.

The purpose of this concluding chapter is to provide means of identifying opportunities for future growth in the industry. The characteristics of laser processing are summarized, to form the basis of a check list that can be used in discussions when a laser-based method of fabrication is first compared with a conventional method. Developments in laser systems, materials and processes are described to provide insight into market trends. The role of process modelling in the construction of processing procedures and the operation of intelligent processing systems is discussed; the power of modelling for novel procedure development should never be underestimated. The driving forces for improved efficiency and sustainable growth in various industry sectors are considered, and matched with the opportunities that laser processing provides. The chapter concludes by comparing the spectacular growth in the industry with the relatively slow development of education and training in the subject – areas that need to be addressed if the momentum of past decades is to be sustained.

CHARACTERISTICS OF LASER MATERIAL PROCESSING

Many opportunities arise from the inherent characteristics of laser processing. A benefit might be possible without a change in technical specifications – a simple substitution of a laser cutting system for a bandsaw may provide the necessary improvements to process economics to justify the change. Alternatively, technical improvements in product quality often more than compensate for increased production costs. The most profitable scenario involves improvements in both technical and economic factors. The characteristics of laser processing are summarized below to enable the most significant components of an initial comparison between a conventional process and a laser-based alternative to be identified and compared.

Higher productivity is possible in laser processing through: a reduction in processing time because of the low inertia of laser processing; a reduction in the use of materials; the use of cheaper materials; a reduction in labour costs; a reduction in equipment maintenance costs because of the absence of tool wear; a reduction in scrap; a reduction in process cycle time through the minimization of post-treatment working; a reduction in response time to orders; the high equipment availability time; and a reduction in energy costs.

Higher product quality, in terms of improved in-service properties, can be achieved through: improved tolerances; accurate control of process parameters; selection of new materials; and product redesign.

New opportunities to manufacture arise from: the nature of laser processing – it is a non-contact process that causes no tool wear; the ability to process conventionally untreatable materials; the ability to treat discrete portions of large components; the precise treatment of small components; the treatment of selected areas; and the accessibility of a laser beam in concealed locations.

Laser processing is environmentally friendly; it uses clean energy and few contaminating materials. There is very little environmental disturbance (sound, external electrical and magnetic fields) in delivering optical energy, which enables processing to be carried out in ergonomic surroundings.

Increased flexibility arises from: the wide ranges of available laser wavelength, power and energy, which enable processing to be carried out in thermal and athermal modes; the variety of mechanisms that can be used – photoelectrical, photochemical and photophysical (athermal), and heating, melting and vaporization (thermal); the wide range of processes available, which can be used for both surface and penetration treatments; the adaptability to automation and flexible manufacturing; the different methods of optical beam delivery; the ability to switch rapidly between different treatment geometries; and the ability to process a wide range of shapes through the ease of beam shaping. As a wider range of engineering materials are taken into use, and the cost of lasers decreases, opportunities that once were either technically or economically unfeasible will become attractive.

These advantages have enabled laser processing to gain footholds in a variety of conventional fields of treatment.

Against these advantages must be balanced a number of negative characteristics, the main ones being: the high equipment cost; a lack of knowledge of the potential benefits; management roadblocks to implementation following negative past experiences with early unreliable lasers; a lack of acceptance of potential reduction in manpower; a need for retraining; and a scarcity of codes of practice for some processes.

The characteristics described above can be used as the basis for a check list to assess the suitability of laser-based manufacturing for a particular operation. Consultants are available who can quantify much of the data required for decision making. Manufacturers are keen to highlight the benefits of their lasers and systems, and are also able to provide technical and economic comparisons with conventional processing. Research organizations and many manufacturers will carry out preproduction trials to examine the feasibility of an operation. However, each has its own agenda. The information in this book is intended to provide an unbiased means of identifying the critical factors in individual applications so that appropriate questions can be asked at an early stage in the selection process.

Developments in Laser Systems, Materials and Processes

As with all manufacturing processes, opportunities arise through developments in the equipment and materials used. Some lead to a reduction in cost, while others provide a step-change in quality.

Laser Systems

The functionality of laser systems is constantly being improved. From the user's point of view, the time and expertise needed for operation and servicing are being reduced, which enables extended warranties to be offered. Smaller laser footprints mean that processing can be carried out in smaller premises. Improvements to laser efficiency, for example through the use of diode pumping in solid state lasers, and diffusion cooling in CO_2 'slab' lasers, lead to lower operating costs. New designs, such as slab, tube, fibre and disc geometries in solid state lasers, also provide improvements to running efficiency, which offset the high initial cost of a laser system.

Laser manufacturers seek means to reduce manufacturing costs and improve performance. As the popularity of laser-based manufacturing grows, so opportunities to implement techniques of mass production arise. Lasers can then be built on production lines, using techniques found in the automotive industry for large volumes, or methods favoured by the aerospace industry for niche products. The unit cost of lasers will then fall significantly, which will feed expansion of the market. The use of lasers in the manufacture of other lasers is just one example of this trend.

Significant improvements can be made through developments in optical components. Lenses capable of focusing the low quality beam from diode lasers provide opportunities for penetration processing. Novel designs, such as twin focus optics, lead to improvements in cutting. Work is being done in the field of diffractive optics, which enable a low quality beam to be transformed into an appropriate tool for a variety of processing mechanisms. Adaptive optics compensate for irregularities in beam propagation, improving processing performance.

Light is easily transmitted over large distances in air. Workstations can therefore be located remote from the laser. The beam can be switched between workstations quickly, or divided to serve many workstations simultaneously. This aspect of *remote processing* provides advantages in terms of productivity, flexibility and production engineering. Individual production cells can then be constructed, in which process monitoring and adaptive control enable manufacturing to continue with little human intervention. Systems are becoming more user friendly, thanks to advances in software and control systems. Components for fabrication are now routinely created as digital files, which can be transferred via the Internet to the processing site – another meaning of the term 'remote processing'.

Light, compact lasers are ideal tools for autonomous, flexible, robotic systems for material processing, which enable small batches of tailored components to be fabricated quickly. Discerning customers expect customized products. Manufacturers endeavour to reduce the lead time from concept to manufacturing. Current dedicated automatic processes are unable to respond with sufficient agility. Robot-mounted lasers provide solutions to such problems, and can be expected to be found in an increasing number of manufacturing sites in the future.

Industrial micromachining was revolutionized when ultrashort (femtosecond) pulsed lasers became commercially available in highly automated systems. Infrared diode lasers are used to write information on CDs and DVDs; now shorter wavelength blue–violet diode lasers enable optical discs to store larger amounts of data. Such applications represent two extremes of the laser processing industry: the *technology push* of ultrashort pulse lasers into systems into micromachining (after a few years); and the *industry pull* that drove the development of blue–violet gallium nitride diode lasers for high density optical disc storage. 'Technology push' and 'industry pull' will continue to present opportunities for new applications of existing lasers, and the development of new devices.

MATERIALS

By considering the interaction of the thermal profiles induced during laser processing with the physical and thermal properties of engineering materials, opportunities for novel treatments and material design can be realized.

The thermal properties of engineering materials were illustrated in Chapter 5 with charts that reveal discrete ranges associated with each material class. Within these regions, subranges can be identified for individual materials. The charts reveal similarities in properties between certain materials in different classes; the values for thermal conductivity and diffusivity of metals and alloys are similar to engineering ceramics such as silicon nitride and boron carbide. (This is not surprising since many engineering ceramics have been designed to respond to heating in the same manner as metals, while being considerably lighter.) The effect of thermal fatigue on mechanical properties can be simulated in metals by accelerated testing using thermal histories generated by laser heating. The potential for testing the response of engineering ceramics, glasses and composites to steady state and transient laser heating in a similar manner is therefore large, and largely unexplored. Model-based laser heating diagrams can be constructed by using the models and properties described, which not only predict response, but also provide information that assists in the design of engineering materials for thermal applications.

Laser-based processes that exploit the mechanism of melting are used widely in surface engineering. Rapid, highly controllable thermal profiles are used to create surfaces with properties tailored to particular applications. Rapid solidification can be induced, resulting in alloys that possess enhanced solid solubility compared with the equilibrium phase diagram. The amorphous structure of the liquid phase can be retained at room temperature, resulting in the production of metallic glasses. Not only do such microstructures enhance surface properties, but when processed in an appropriate manner they can be used as materials in their own right. Laser beams may thus be used to *synthesize* new materials as well as to process existing materials. A good example is the *in-situ* production of functional materials containing gradients in properties. Variations can be produced by constantly changing processing parameters for a given material in a controlled manner, or by introducing new materials during processing. The opportunities for producing multipurpose components that incorporate materials whose function depends on their location in the component provide means of reducing assembly times, discussed below. Biomaterials in particular require specific properties at discrete locations; the formation of *in-situ* composite surface structures by laser deposition provides a means of attaching prostheses to bone via an intermediate biocompatible layer. Further processing of this layer, e.g. by drilling microscopic bulbous holes in the surface, may aid in the growth of bone and improve the integrity of the joint.

PROCESSES

An approximate distribution of laser-based processes at the turn of the millennium is given in Fig. 18.1. The relative importance of processes changes, sometimes dramatically: demands for fine microstructuring in the microelectronics sector drives exponential growth in the area of scribing and optical lithography, for example. However, progress continues to be made in all areas of processing; some involve process refinement, while others involve the introduction of novel processes. Current trends help to identify future opportunities.

Although laser cladding continues to be an efficient means of surfacing discrete parts of components, the basic processing geometry has undergone modifications to create an industry in which parts can be manufactured rapidly and flexibly. Other processes have joined the family of rapid manufacturing: laminated object manufacturing, selective laser sintering, stereolithography, to name just three. All are based on innovative applications of a basic process. In the field of 'traditional' laser thermal processing, more such developments can be expected, driven by the availability of even smaller, cheaper lasers.

As an example of the development of traditional thermal laser processing, welding is normally thought of as a through-thickness method of joining. However, the optical nature of the laser beam provides possibilities for novel welding geometries and techniques. The beam may be directed into a highly absorptive opening between plates (a kissing joint) to join thick section materials rapidly at their surface. It can also be used when access to a joint is available from only one side (a stake weld), which provides the designer with a greater range of joint types. A laser beam can pass through joint members to be absorbed by interfacial films, thus producing a joint line in a range of shapes, with no external

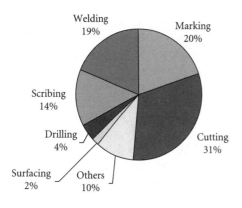

Figure 18.1 Industrial application of laser-based processes at the turn of the millennium, based on volume

markings. Hybrid welding processes combine the characteristics of their constituent techniques: a laser beam is highly penetrating, but intolerant to variations along the joint line; these can be accommodated by more forgiving arc fusion processes. Hybrid processing enables technical problems to be overcome as well as reducing the requirements placed on the laser, leading to a reduction in capital investment. This type of reasoning can be applied to most thermal mechanisms of laser processing, presenting opportunities for novel application of an existing technology.

Novel applications can also arise as a result of the introduction of a new laser system. Photorefractive surgery and other cosmetic procedures have been revolutionized as user-friendly turnkey excimer laser systems found their way into consulting rooms. Similarly, femtosecond pulsed lasers in user-friendly workstations caused a revolution in athermal micromachining. The search for sources of even shorter wavelength laser light continues in order to meet increasing industry demands.

One aspect of laser processing that has not been developed extensively is the ability to use a single beam for a number of processes. This provides opportunities for redesign and material selection. Figure 18.2 shows a gear–hub assembly. The gear wheel is joined to the hub by laser welding. The flanks of the gear teeth are hardened using the same laser. This case study highlights opportunities for redesign.

The conventional method of gear–hub fabrication starts by machining the assembly in a single unit from a block of 20 MnCr 5 steel. The gear is then treated by gas carburization to produce a surface hardness of 750 HV, and a profile in which the hardness lies above 550 HV to a depth of 0.6 mm. The carburizing process causes the gear to distort. It must be machined again to meet dimensional tolerances. The alternative laser-based fabrication process uses a two-piece gear–hub assembly. The hub is made from a relatively cheap structural steel, such as Fe52, which can be laser welded without exceeding a hardness of 400 HV, above which cracking could be initiated. The material for the gear wheel is selected on the basis of two criteria: the hardness profile of the tooth flank must be equivalent to that produced by gas carburizing; and the hardness developed during laser welding must remain below 400 HV. The minimum carbon equivalent and the processing parameters are found using the

Figure 18.2   Gear–hub assembly made using laser welding and laser hardening

model-based methods described in Chapter 9 for surface hardening. The maximum carbon equivalent and the welding parameters are determined using the methods for penetration processing described in Chapter 16. Thus we are able to select materials for the gear and hub that satisfy performance requirements, while minimizing fabrication time and material cost.

THE POWER OF PROCESS MODELLING

Simple analytical models have been used in the book to couple the laser beam (the energy source) to the material. The beam is idealized as a finite, distributed, surface energy source to model techniques of surface processing, or a through-thickness line source for penetration processes. Temperature fields in the material have thus been generated. By incorporating temperature fields with models for phase transformations, the properties of processed materials can be predicted with an accuracy that can be used for practical purposes.

The models described are valid for thermal processing of metals and alloys, since the response of such materials to laser heating is characterized well by dimensionless groups of material properties. For example, the latent heat of melting for metals and alloys, when normalized with the melting temperature and the volumetric heat capacity, lies around 0.4. Dimensionless groups then become the true variables of processing, and can be expressed in terms of explicit relationships. Conditions (group relationships) required for particular thermal process mechanisms – heating, melting and vaporization – may then be established and applied to different processes. Suitable dimensionless groups could also be derived for other classes of engineering material (ceramics and glasses, polymers and composites), enabling the use of analytical modelling to be extended to laser processing these materials.

A number of well-established processes have been modelled, in which the underlying physical and chemical principles are understood. There are opportunities to extend the application of modelling. Selective Laser Sintering (SLS), for example, is based on well-established diffusion-controlled mechanisms of sintering. The kinetics of diffusion might therefore be modelled and applied to laser-induced thermal cycles, providing opportunities to examine innovative sintering techniques and assess novel materials prior to trial and error testing. Similarly, as the mechanisms of athermal processing are elucidated, modelling techniques might be developed to describe photoablative processes such as femtosecond pulse machining.

Material selection charts are valuable tools for assessing the suitability of engineering materials for particular applications. However, they also provide information that is of value when designing new methods of laser-based processing. For example, by constructing a chart that displays mechanical strain as a function of thermal strain, concrete is seen to fail mechanically under low thermal loading in comparison with other engineering materials. By modelling the temperature profiles developed during laser heating of concrete, beam parameters can then be established for selective removal of surface layers over given areas and to specified depths, which are governed by the stresses required for failure. Thus the process of laser scabbling may be modelled. Other candidate materials for this process can then be identified. An analogous technique could be used to characterize solidification cracking: the thermal stresses that are tolerated before solidified grains are separated could be established, and used as the criterion to predict appropriate laser welding parameters. Similarly, material selection charts can be used to identify materials for given processes, processes for given materials, or novel materials for innovative processes.

Models have been used to construct laser processing diagrams. In their most general form they provide an overview of a process for a given class of engineering material by using dimensionless groups of processing variables and material properties. They provide a greater understanding of the role of the essential process variables, and are a useful teaching tool. In practice they can be used to sift candidate materials and processing parameters rapidly to reduce the amount of costly trial and error testing involved in developing processing procedures. When greater accuracy is required, the models may be calibrated to a known condition (the data point technique). The effects of variations in process parameters about the data point on the properties of processed materials can then be established. A significant benefit of using calibrated *analytical* modelling is that results with sufficient accuracy for practical application can be obtained by using a PC, in only a few seconds. By identifying the relevant processing mechanisms, diagrams could be constructed for laser treatment of other classes of engineering material. We saw that laser-induced interstitial thermotherapy (LITT) involves heating tissue to about 60° to destroy diseased regions; models of heat flow from a surface source could be combined with thermal constants for tissue, and calibrated using an experimental observation, to develop a processing diagram to aid in the selection of parameters for different types of treatment. In addition, when the mechanisms involved in athermal processing (making and breaking bonds) can be modelled, results could be presented in a similar graphical form within the frameworks provided.

Process models can also be incorporated into adaptive control systems, in which established relationships between process variables are used to adjust the processing parameters online in response to changes in processing conditions or requirements – *intelligent processing*. This has led to the introduction of turnkey systems for some processes. The setup time is often considerably longer than the actual processing time, and so techniques for automatic process setup and control provide significant savings in time and money. In addition, an expert operator is then not required – laser processing becomes simply another production line tool.

Codes of practice are common in conventional material processing. In contrast, codes are only now being developed for laser-based fabrication. In many cases these are merely workmanship standards, which cannot be related to in-service performance. Process modelling could aid in two fields: expediting procedure development by sifting candidate materials and processing parameters; and providing a link between existing workmanship standards and in-service performance through an understanding of the underlying principles. Thus laser processing diagrams could be complemented with regions that indicate predicted quality classes, and hence performance. Materials are already being designed specifically for laser processing. It is not unreasonable to imagine that in future they will be delivered with model-based laser processing guidelines to enable manufacturing quality classes to be achieved.

Process modelling provides opportunities to extend some of the fundamental predictive tools used in thermal treatment. We have seen how continuous cooling transformation (CCT) diagrams can be used to predict the response of steels to laser heating (surface hardening and the heat affected zone of welds). Laser-induced thermal histories can be used to generate time-temperature-austenitization (TTA) diagrams for rapid heating, to predict degrees of superheating. Modifications to solidification diagrams, such as the Schaeffler diagram that predicts the microstructure of the solidified weld metal of stainless steels, can be generated either empirically (by carrying out large numbers of accurately controlled laser-based experiments quickly), or predicted by combining models of phase transformation with laser-induced cooling profiles. Opportunities also exist for producing phase diagrams for ceramic systems, and predicting transformation temperatures in polymers in a similar manner. At a fundamental level, the controlled heating effect of a laser beam is also an ideal tool for determining thermal properties of materials by using calorimetric techniques.

DRIVERS FOR INDUSTRIAL APPLICATION

A well-known rule in conventional manufacturing is that it takes about three seconds to add a part (Boothroyd and Dewhurst, 1989). Automobiles can be assembled in hours, whereas commercial aircraft, with several million parts, take months. A strong driving force for all sectors of industry is therefore to minimize the number of parts to be assembled by making each perform several functions. Structures can then be designed in a modular fashion, by making joints self-fixturing, adopting parallel processing procedures, and eliminating fasteners. Laser-based fabrication provides such solutions in many sectors, some of which we have seen already. Industry sectors have characteristic driving forces for improving competitiveness, as well as obstacles that hinder the implementation of new technology. However, most sectors provide opportunities for the introduction, or expansion, of laser-based fabrication.

AEROSPACE

The major driving forces for change in the aerospace industry are: improved safety; improved structural monitoring; lower costs of manufacturing and maintenance; ease of fabrication; and a reduction in operating costs. The growth in air travel has led to new designs for larger aircraft with improved fuel consumption and increased payloads. Airlines would like to be able to fly sectors of more than 16 hours over distances greater than 15 000 km. Weight reduction is critical; a reduction of 1% in the weight of a large aircraft saves about 150 000 litres of fuel each year (every kilogram saved is worth about $50 per year).

Laser processing has been introduced in the manufacture and repair of airframe, control system and engine components. Aerospace products are characterized by relatively short runs of very high quality, customized components. Rapid prototyping and manufacturing techniques have provided solutions to fabrication problems in these areas, particularly with materials that are difficult to forge, such as titanium alloys. The aviation authorities demand extensive testing before certification, and so any changes in manufacturing procedures mean that a new process, especially welding, is normally only introduced when a new aircraft is designed. Fuselage stiffeners are now laser welded, air conditioning ducting is laser cut, and aircraft control cables are laser marked.

Growth in a number of areas can be expected: rapid manufacturing and repair of components; high quality cutting, welding and marking; and the migration of laser-based techniques from tertiary to secondary structures. The replacement of riveting alone results in significant weight savings, simplifications to fabrication sequences, and higher joint integrity.

AUTOMOTIVE

The automotive sector accounts for about 11% of global Gross Domestic Product, and provides directly or indirectly about one job in nine in industrialized countries. Of all the industrial sectors, this industry has done the most to develop and implement large-scale laser processing in a production environment. Automobile manufacturers are constantly seeking to increase market share in arguably the most competitive of manufacturing sectors. Weight reduction translates into reductions in fuel consumption and emissions. Stiffer, tailor-made structural members improve quality and crash worthiness. Product development and manufacturing costs are constantly being reduced (typically by about 5% per year). The customer demands a wider, more tailored product range.

Laser processing provides solutions in five principal component areas: bodywork; chassis members; engine; powertrain and accessories. Pressures are also imposed by legislative organizations empowered to reduce carbon dioxide emission, increase fuel economy, improve crashworthiness, and maximize recycling. In the early 1990s, it was estimated that North American automobile manufacturers could save one thousand million dollars annually by using laser processing, with savings of at least $100 per car through changes to the bodywork alone. In Japan, more than one quarter of all CO_2 lasers delivered are destined for the automotive industry. The growth of laser welding in the automotive industry has been spectacular, to the extent that virtually every automobile manufactured in high volumes now contains laser-welded parts in all the product categories.

There is a trend towards the use of light materials, particularly aluminium alloys and polymers, for which techniques of laser processing, including cutting, welding, marking, hardening and cladding, are being developed. Recycling has led to demands for methods of disassembly and material identification – potential areas for laser processing.

BIOMEDICAL

Laser applications in the biomedical sector can be divided into those in which the laser beam is used as the tool for treatment of a particular condition, and those in which it is used in the manufacture of devices.

In the field of medical treatments, dedicated laser systems have been developed to cater for a growing number of applications. The procedure determines the most suitable laser wavelength. Red ruby laser light find uses in a variety of thermal cosmetic applications, including the removal of tattoos, birthmarks and acne. Ultraviolet excimer lasers ablate corneal tissue athermally in a variety of ophthalmological procedures. Dye and diode lasers can treat activated cells by photochemical means to respond to certain treatments. Tuneable dye lasers are also attractive sources because the wavelength of light can be selected for a range of applications. The tuneability and pulse structure of the free electron laser are important characteristics in the ablation of both soft and hard tissue, and maintain interest in the development of this laser. User-friendly systems incorporating such new sources provide opportunities for a range of procedures to be performed with a single unit. Public demand for laser-based medical procedures will ensure the growth of laser-based healthcare.

Conduction joining, marking and drilling are just three processes used in the manufacture of medical devices such as catheters, heart pacemakers and scalpels. Standard pulsed neodymium lasers are used in microjoining techniques including welding, soldering and brazing. Frequency-multiplied Nd:YAG units and copper vapour lasers provide the accuracy required with thermal machining techniques. Micromachining using ultrashort pulse Ti:sapphire laser systems enables the scale of details to be reduced by orders of magnitude in comparison with thermal processing. The potential market for devices fabricated by laser processing is enormous.

CONSTRUCTION

The construction industry has been relatively slow to adopt laser processing. The look of a car is important – often the deciding factor in a purchase – but the construction of a building less so in the eyes of many. The cost of modern office buildings is around the same as fresh fruit and vegetables. In comparison, a modern airliner costs about $450 per kilogram. Construction is not an industry with a high intrinsic cost, and there is little incentive for innovation. But there are opportunities.

There is a potential market for portable high power lasers for on-site welding of beams and sections, which saves time and increases design flexibility. There is scope for cutting holes of various sizes in beams and welding sections such as box girders. Computer-aided designs can be downloaded directly to a robot-delivered beam of light to cut and weld complex three-dimensional structures. Modular sections can then be completed more easily, ready for integration at the construction site.

Lightweight structures, constructed from skins welded to cores with various geometries are ideal for transportable structures in a range of industry sectors; the geometries are limited only by the imagination of the designer, and the scale can encompass primary structural members as well as secondary structures, which may be on the order of millimetres or even smaller.

DOMESTIC GOODS

The use of laser cutting, marking and welding is increasing in the domestic goods industries. Consumer legislation throughout the world calls for stringent marking of products, including information such as batch identification, manufacturing date and latest consumption date. Mask marking is most prevalent in the consumer-related industries that require marking on non-metallic labels. Metals and alloys are more often marked using CNC mirror systems. Money is saved in fabric cutting by reducing scrap. Computer-generated nesting patterns allow this to be achieved for products ranging from suites to kites. It is an industry in which novel products and processes provide a competitive edge, particularly in marketing. Manufacturers are therefore keen to continue to use innovative processing methods, notably laser welding.

ELECTRICAL AND ELECTRONICS

Miniaturization, cost reduction, and reduced environmental impact are the main drivers for change in the electrical and electronics industries. The requirements of industries involved with mobile telephony have driven changes in the market. The relatively high cost of laser systems has hindered acceptance in the past. Compact solid state lasers, based on diode laser pumping, are now available for producing short wavelength light that is ideal for processing novel materials and implementing innovative designs on the microscopic scale. As short wavelength lasers are developed that extend further into the ultraviolet, the precision and scale of lithography decrease. Moore's law (the doubling of microelectronic component density every 18 months) looks set to continue to be valid more than 40 years after its proposition, thanks largely to laser lithography.

Electronic packages are becoming increasingly sophisticated, requiring high densities of connection points to printed wiring boards. Boards are becoming thinner, and require the use of high performance materials that can often only be processed with light. Holes must be smaller – conventional drilling techniques are reaching their limits of operation – and laser cutting and drilling have much to offer.

The electrical and electronics industries use more resistance welding systems than any other industry. As this industry continues its growth, so the need for dedicated laser-based precision microjoining and microprocessing will grow. Packages must be sealed, often hermetically (air tight); precision laser welding provides solutions, particularly when joining light materials.

Constant demands for greater data storage have driven the development of milliwatt class diode lasers into the blue range of the electromagnetic spectrum, where light can be focused to a smaller spot, enabling the density of information stored on optical disc media to be increased. The benefits provided by the compact disc for sound reproduction led to the demise of vinyl in the 1980s. The opportunities presented by the digital versatile disc (DVD) did the same to the market for video tape. The availability of recordable high capacity DVDs will ensure growth in this medium.

MACHINERY

The machinery industry is wide ranging, and includes the construction of machine tools, yellow goods (e.g. excavators) and armoured vehicles. The main opportunities for laser processing lie in welding and cutting of thick section structures. The main obstacles are those common to all heavy industries: codes of practice. Many are being addressed by the shipbuilding industry, and are described below.

PETROCHEMICAL

The drivers for change in the petrochemical industry sector are related to reducing operating costs, applying novel processing techniques in deep environments, improving reliability, corrosion management and a reduction in the time required for fabrication. So far this sector has limited its use of laser processing to cladding and welding.

Cladding is becoming the surfacing process of choice for life extension in components with load-bearing surfaces. As welding procedures for thick section pipeline steel become available, the laser provides joining solutions, particularly in remote locations with difficult access. Lightweight structures are potential solutions to the problem of transporting accommodation units for oil exploration.

POWER GENERATION

The main driving forces for change in the conventional power generation sector are: life extension for existing equipment through repair and redesign; cost reduction for existing fabrication procedures; increased safety requirements; reduced environmental impact; higher efficiency; and the development of renewable sources of energy.

Remote processing by fibreoptic beam delivery provides possibilities for decommissioning and cleaning of nuclear power stations. Novel techniques, such as scabbling, have been developed to remove contaminated concrete through controlled fracture. Laser welding of dissimilar metal combinations enables materials to be chosen for plant based on their function, thus reducing overall cost. Encapsulation techniques for waste materials, particularly from nuclear power generation, are strong candidates for laser welding.

However, there is an increasing demand for power generation that is not based on fossil fuels – the hydrogen economy. Carbon nanotubes, produced by laser ablation, provide means of constructing the membranes and nanostructures needed in efficient fuel cells. Gratings for power conversion can be etched on the microscopic scale by short wavelength laser machining. Semiconductor manufacture can be made more efficient through small-scale laser annealing and other procedures. In contrast to the use of large, high power lasers in the manufacture of plant for conventional power generation, the development of the hydrogen economy will be catalysed by the use of compact, highly efficient lasers in microprocessing applications.

SHIPBUILDING

Shipping is still the most economical and environmentally friendly means of transporting goods over long distances. There is pressure to reduce fabrication costs, with a strong interest in reducing the amount of corrective work needed. Shipbuilders estimate that rectifying the distortions produced in 6–8 mm plate accounts for up to 33% of labour costs. The shipbuilding industry has been at the forefront of procedure development for cutting and welding of thick section materials. Plates of thickness 5–25 mm are used in cruise ships, 10–40 mm in tankers, and plate lengths normally vary up to 22 m. Reduction in deformation, rather than an increase in the welding speed, is the key to improved economics.

Significant savings are possible by making improvements to accepted fabrication tolerances. The tolerances associated with laser cutting are about ±0.5 mm, i.e. around one fifth of those of conventional cutting techniques. Improvements in the quality of laser welds arise from the increased demands placed on plate edge preparation and fitup. A typical 300 000 tonne tanker contains around 280 km of joints that show potential for laser welding, comprising T-stiffeners (10–15 mm in thickness), fillet welds in section assemblies, butt welds in the shell structure (18–25 mm), and two-sided fillet welds on transverse elements (5–6 mm). These comprise about 40% of the total weld length. However, benefits can only be realized if laser welds can be produced in appropriate steel grades in a shipyard environment that meet the quality requirements of the Classification Societies. Standards for laser welding and material composition are therefore now being produced through extensive international testing programmes. Craft have already been manufactured with the laser used exclusively for cutting and marking panels.

EDUCATION AND TRAINING

Sales of laser systems increased spectacularly during the last four decades of the twentieth century. Annual double-digit growth was common during the 1980s and 1990s, tempered only by the economic

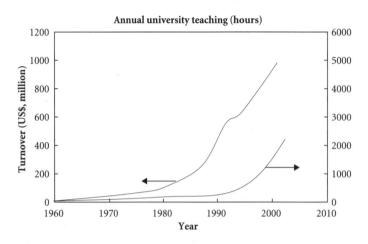

Figure 18.3 Comparison of annual turnover in laser-based manufacturing and approximate number of university teaching hours devoted to laser processing

recession of the early 1990s. At the turn of the millennium, about 125 000 industrial lasers had been sold since the business started at the beginning of the 1960s. If sales continue to rise at the same rate in the coming years – and they are expected to – the number of systems will more than double every five years. This is remarkable growth, which is illustrated in Fig. 18.3.

In contrast, the importance of teaching the subject of laser material processing has only recently been acknowledged. The number of hours devoted to university teaching in the subject is relatively small in comparison with conventional manufacturing technologies, and its growth has lagged behind the increase in industrial application. Engineering undergraduates might receive a few hours of instruction in the use of laser-based fabrication and its *technical* advantages. But few courses are able to cover the vital roles that opportunities for design, material selection, production engineering and novel manufacturing play in the *economic* equation. There are opportunities for industries and centres of continuing education to come together to produce relevant courses to meet future needs. The rapid rate of change in laser-based industries means that qualifications will require constant updating.

Conferences and industrial workshops play a vital role in spreading the capabilities of laser-based fabrication among existing practitioners and interested parties, to meet the demand for innovative and sustainable solutions to manufacturing problems.

SUMMARY AND CONCLUSIONS

Opportunities exist in many areas for growing the business of industrial laser material processing. Developments in the fields of laser systems, materials and processes generate conditions for innovation: the availability of a new laser source can underpin a novel process; a new material might lead to an original application of an existing source; or an innovative process might demand the development of a new material and laser source. Industry sectors have characteristic drivers for change, most resulting from the need to sustain growth and maintain competitiveness. The benefits of laser processing complement many of the requirements of manufacturing.

Inventions often become seminal when they facilitate a transformation; the printing press enabled thoughts to be put into words and disseminated widely for the first time, the automobile provided personal freedom to travel, and the Internet is an efficient means of instantly accessing a wealth of knowledge. They are all inventions that were used by select groups of people initially, and which did not need to be understood fully to be used successfully. When their potential was realized, their growth into people's lives gathered pace; over centuries, decades and years in the cases of printing, the automobile and the Internet, respectively. The laser is the device that facilitates the application of photonics to manufacturing. Its potential is just being realized. Significant growth can be expected year by year. Similarly, advances in the design and production of engineering materials facilitate their migration from the laboratory into everyday life. Laser processing of engineering materials is an outstanding example of an industry capable of sustainable growth – a subject that provides practical opportunities for lifelong learning.

FURTHER READING

LASERS IN MATERIAL PROCESSING

Bass, M. ed. (1983). *Laser Materials Processing*. Amsterdam: North-Holland.

Bimberg, D. (1977). *Laser in Industrie und Technik*. Grafenau: Lexica. (In German.)

Charschan, S.S. ed. (1972). *Lasers in Industry*. Toledo: Laser Institute of America.

Charschan, S.S. ed. (1993). *Guide to Laser Materials Processing*. Orlando: Laser Institute of America.

Crafer, R.C. and Oakley, P.J. eds (1993). *Laser Processing in Manufacturing*. London: Chapman & Hall.

Duley, W.W. (1982). *Laser Processing and Materials Analysis*. New York: Plenum Press.

Grigoryants, A.G. (1994). *Basics of Laser Material Processing*. Boca Raton: CRC Press.

Hallmark, C.L. and Horn, D.T. (1987). *Lasers – The Light Fantastic*, 2nd ed. Blue Ridge Summit: TAB Books.

Harry, J.E. (1974). *Industrial Lasers and their Applications*. Maidenhead: McGraw-Hill.

Hecht, J. and Teresi, D. (1998). *Laser: Light of a Million Uses*. Mineola: Dover.

Koebner, H. ed. (1984). *Industrial Applications of Lasers*. Chichester: John Wiley & Sons.

Lasers in Materials Processing (1989). Lake Geneva, WI: Tech Tran Consultants Inc.

Lugomer, S. (1990). *Laser Technology: Laser Driven Processes*. Englewood Cliffs: Prentice Hall.

Luxon, J.T. and Parker, D.E. (1992). *Industrial Lasers and Their Applications*. 2nd ed. Englewood Cliffs: Prentice Hall.

Luxon, J.T., Parker, D.E. and Plotkowski, P.D. (1987). *Lasers in Manufacturing*. Bedford: IFS Publications.

Metzbower, E.A. ed. (1981). *Source Book on Applications of the Laser in Metalworking*. Metals Park: American Society for Metals.

Migliore, L. ed. (1996). *Laser Materials Processing*. New York: Marcel Dekker.

Perera, J. (1989). *Lasers – The Success Story: The Industry in the UK*. London: Financial Times Business Information.

Ready, J.F. (1997). *Industrial Applications of Lasers*. 2nd ed. San Diego: Academic Press.

Ready, J.F. ed. (2001). *LIA Handbook of Laser Materials Processing*. Berlin: Springer.

Rubahn, H.-G. (1999). *Laser Applications in Surface Science and Technology*. New York: John Wiley & Sons.

Rykalin, N., Uglov, A. and Kokora, A. (1978). *Laser Machining and Welding*. Oxford: Pergamon Press.

Schuöcker, D. (1999). *High Power Lasers in Production Engineering*. London: Imperial College Press/World Scientific.

Soares, O.D.D. and Perez-Amor, M. eds (1987). *Applied Laser Tooling*. Dordrecht: Martinus Nijhoff.

Steen, W.M. (2003). *Laser Material Processing*. 3rd ed. London: Springer.

White, C.W. and Peercy, P.S. (1980). *Laser and Electron Beam Processing of Materials*. New York: Academic Press.

DEVELOPMENTS IN LASER MATERIAL PROCESSING

Industrial Laser Review. Tulsa: PennWell. Published monthly 1985–1997.

Industrial Laser Solutions in Manufacturing. Tulsa: PennWell. Published monthly since 1998.

Proceedings of the International Congress on Applications of Lasers and Electro-Optics (ICALEO). Orlando: Laser Institute of America. Published annually since 1982.

Journal of Laser Applications. Orlando: Laser Institute of America. Published four times a year.

Proceedings of the Conference Laser Advanced Materials Processing (LAMP). Osaka: High Temperature Society of Japan and Japan Laser Processing Society. Published every five years since 1987.

Laser Focus World. Tulsa: PennWell. Published monthly since 1964.

Proceedings of the Conference Lasers in Manufacturing (LIM). Bedford: IFS Publications. Published annually 1983–1989.

Opto and Laser Europe. Bristol: IoP Publishing. First published September 1992, and published monthly since August 1994.

Proceedings of the Nordic Laser Materials Processing Conference (NOLAMP). Published biennially since 1987.

Proceedings of the Symposium on Automotive Technology and Automation. Croydon: Automotive Automation Ltd. Published annually since 1967.

Volumes of SPIE. Bellingham: The International Society for Optical Engineering (SPIE).

BIBLIOGRAPHY

Boothroyd, G. and Dewhurst, P. (1989). *Product Design for Assembly*. Wakefield: Boothroyd Dewhurst Inc.

GLOSSARY

Aberration Deviation from ideal behaviour by an optical component. May be caused by light being focused to different positions on the *optical axis* because of variations in *wavelength* (chromatic aberration), or because of differences in the position that the rays impinge on the optical component (spherical aberration).

Abrasive wear Wear caused by hard, sharp particles.

Absorption coefficient The factor β in the *Beer–Lambert law*, which relates the intensity, I, at a depth, z, to the incident intensity, I_0, in an absorbing medium: $I = I_0 \exp(-\beta z)$.

Absorptivity The ratio of the *power* absorbed by the workpiece to the *power* of the incident *radiation*. (Normally defined in photochemistry as the ratio of the *power* transmitted by a medium to the *power* of the incident *radiation*.)

Active medium Collection of *species* capable of undergoing *stimulated emission* at a given *wavelength*. Normally used to describe the type of a *laser*.

Adaptive control Variation of one or more *process variables* during processing in order to maintain a given process condition.

Air knife A restricted rapidly moving flow of air designed to protect optics from damage, normally from weld *spatter*.

Allotrope Each of two different morphologies in which an element or compound may exist.

Alloy A mixture of a *metal* with other *metals* or non-metals.

Amorphous A disordered non-*crystalline* atomic arrangement in *metals* and *ceramics*. A random arrangement of chains in a *polymer*.

Amplification A measure of the intensification of the *radiation* field in the *laser resonator* as light is reflected between the cavity mirrors.

Amplifier A device that amplifies light produced by an external *laser*, but which does not contain mirrors capable of sustaining oscillation to independently produce a *laser beam*.

Anisotropy The change in physical properties with direction.

Annealing Heating to a specific temperature followed by slow cooling to a specific final temperature. Used to change the microstructure to improve *ductility* and *toughness*.

Aperture An opening through which light passes in an optical device.

AR coating Anti-reflection coating. A coating applied to an *optical device* in order to increase transmission of the *beam*.

Atactic structure A *polymer* structure in which the side groups are arranged randomly on either side of the molecular chain.

Ausforming Mechanical working of *metastable austenite* in a ferrous *alloy* to produce a fine uniform dispersion of strengthening *carbides*.

Austenite A *solid solution* of carbon in iron possessing a face-centred cubic structure. Named after Sir William Chandler Roberts-Austen (1843–1902) of the Royal School of Mines, London, UK.

Austenitizing Heating of a ferrous *alloy* to a temperature at which *austenite* forms.

Autogenous processing Processing without the addition of alloying elements.

Axial flow laser A *laser* in which the *active medium* flows continuously along the *optical axis*.

Bainite A *constituent* comprising fine aggregates of *ferrite* plates (or laths) and *cementite*, which forms in the temperature range intermediate between diffusional transformation of *austenite* to *pearlite* and the displacive transformation to *martensite*. Its *morphology* is dependent on the rate of cooling and its composition, and may appear in upper or lower forms. Named in 1934 by the US Steel Corporation (New Jersey, USA), after the American chemist and metallurgist Edgar C. Bain (1891–1971).

Bandwidth The difference between *wavelengths* emitted about a peak intensity at which the intensities are half of the peak value.

Bead on plate weld See *melt run*.

Beam A collection of rays that may be parallel, convergent, or divergent.

Beam axis The straight line connecting the centroids defined by the first spatial moment of the cross-sectional profile of *power* at successive positions in the direction of propagation in a homogeneous medium.

Beam diameter The diameter of an *aperture* in plane perpendicular to the *beam axis* that contains a given fraction of the total *beam power*. The points are commonly defined as the positions at which the intensity has fallen to $1/e$ (0.368), or $1/e^2$ (0.135) of the peak level.

Beam radius The radius of an *aperture* in plane perpendicular to the *beam axis* that contains a given fraction of the total *beam power*. The location is defined in a similar manner to the *beam diameter*.

Beam width The size of the smallest *aperture* perpendicular to the direction of motion of the *beam* (when moving) that transmits a given fraction of the total *beam power*.

Beam length The size of the smallest *aperture* parallel to the direction of motion of the *beam* (when moving) that transmits a given fraction of the total *beam power*.

Beam parameter product The product of the *beam* waist radius and the half *divergence* angle.

Beam waist The location of the minimum *beam diameter* or *beam width*.

Beam waist radius The radius of the *beam* at the location of the *beam waist*.

Beer–Lambert law The *absorbance* of a *beam* of collimated monochromatic *radiation* in a homogeneous isotropic medium is proportional to the absorption path length and to the concentration of the absorbing *species*. Also known as the Beer–Lambert–Bouguer law. Formulated independently by the French mathematician, physicist and astronomer Johann Lambert (1728–1777), the German mathematician, chemist and physicist August Beer (1825–1863), and the French mathematician Pierre Bouguer (1698–1758).

Black body A body that absorbs 100% of the total incident *radiation*.

Blank A sheet of *metal* that is prepared for a subsequent forming operation. See also *tailored blanks*.

Blind cut A cut of incomplete penetration.

Blow hole A through-thickness hole in the weld *metal* caused by ejection of material during welding.

Bremsstrahlung **effect** The emission of *photons* from excited electrons. *Photons* are absorbed by electrons by the inverse *bremsstrahlung* effect. *Bremsstrahlung* is German for braking *radiation*.

Brewster angle The angle of incidence at which light polarized in a given plane undergoes no loss through reflection. Named after the Scottish physicist Sir David Brewster (1781–1868).

Brightness The perception of the luminous *power* of the *beam*.

Brinell hardness number A measure of the resistance of a material to local *plastic deformation* under a given load imposed by a tungsten *carbide* ball. Named after the Swedish engineer Johan August Brinell (1849–1925).

Brittle fracture Failure of a material by a mechanism that does not involve significant *plastic deformation* adjacent to the crack. Elucidated in 1920 by the British materials scientist and aeronautical engineer Alan Arnold (A.A.) Griffith (1893–1963), who also designed gas turbine engines for aircraft.

Calorimeter A device that absorbs *energy*, which can be used to measure the *power* of an incident *laser beam*.

Carbide A compound of carbon and one or more other elements.

Carbon dioxide laser A *laser* in which the *active medium* is gaseous carbon dioxide. The most common *laser* used for cutting.

Carbon equivalent A measure of the *weldability* of structural steels. Obtained by expressing the effectiveness, in comparison with carbon, of elements to stabilize *austenite* by attributing a dividing factor to the amount of each element present in the steel. The carbon equivalent is the sum of the carbon content and the quotients. An equivalent of 0.4 wt% is considered to be an upper limit to avoid *cold cracking* after welding.

Carburizing Diffusion of carbon into a surface through heating in carbon-rich atmosphere.

Case depth The depth to which the concentration of a diffusing element has risen to a given value, as in *carburizing*. Also used to describe the depth at which a given hardness is achieved in transformation hardening (although no diffusing species is involved).

Cast iron A range of ferrous *alloys* containing between 2 and 4.5 wt% carbon.

Cavitation erosion Removal of material by the action of vapour bubbles in a turbulent liquid.

Cavity dumping Periodic removal of coherent *radiation* from an *optical cavity*.

Cementite Iron *carbide*, Fe_3C.

Ceramic Class of *crystalline*, inorganic non-*metals*.

Cermet A composite of *metal* and *ceramic*.

Charpy test A means of measuring the *energy* absorbed by a material at a given temperature by impacting a specimen 10 mm square by 60 mm in length, containing a sharp notch (2 mm in depth of radius 0.015 mm) to localize the stress.

Coherence The degree to which light waves are in phase in both time and space.

Coherence length The distance in the direction of *beam* propagation within which *radiation* maintains a significant phase relationship.

Coherence (spatial) The correlation of the phases of a wave at different locations at a given time.

Coherence (temporal) The correlation of the phases of a wave for different times at a given location.

Coherence time The time within which *radiation* maintains a significant phase relationship.

Cold cracking Cracking that occurs at low temperatures, usually below 300°C, in the coarse-*grained HAZ* of steels. It is time dependent and propagates in bursts. It requires a susceptible microstructure (hardened), and the presence of hydrogen and stress. Often referred to as *hydrogen cracking*.

Collimation The process of making light rays as parallel as possible such that they have low *divergence*.

Component An element that is present in an *alloy*.

Composite A class of material that comprises two or more classes of material in the form of a matrix and a distinct second phase addition.

Constituent A clearly identifiable feature in a microstructure, comprising *phases*, e.g. *pearlite*, which is a lamellar mixture of *ferrite* and *cementite*.

Constitution diagram See *phase diagram*.

Continuous wave (CW) operation The continuous-emission *mode* of a *laser*, in which *radiation* is emitted continuously over a time greater than 0.25 s.

Copolymer A *polymer* produced by combining two or more *monomers* in a single chain.

Covalent bond A bond formed by the sharing of electrons to achieve full outer electron shells in both elements. Commonly found in organic materials. Also known as a dative bond.

Creep *Plastic deformation* occurring over a long period of time at a relatively high fraction of the melting temperature.

Crystalline Denoting a regular ordered arrangement of atoms or molecules.

Defect A *imperfection* that is sufficiently deleterious to affect the performance of a sample.

Dendrite A branched structure formed when liquids solidify into *crystalline* structures.

Depth of field See *depth of focus*.

Depth of focus The distance over which the *beam diameter* does not vary by more than a specified fraction of the focused *beam diameter*, e.g. 5% or $\sqrt{2}$.

Diode laser A *laser* in which the *active medium* is a *p–n* junction in a *semiconductor* material.

Dislocation A line *imperfection* in the *lattice*, caused by the addition of a plane of atoms (edge *dislocation*) or shear of part of the *lattice* (screw *dislocation*).

Dissociation The splitting of molecules into smaller molecules or atoms.

Divergence Angle of *beam* spread measured in radians. For small angles where the cord is approximately equal to the arc, the *beam divergence* can be closely approximated by the ratio of the cord length (*beam diameter*) to the distance (range) from the *laser aperture*.

Ductile fracture Failure of a material by a mechanism that involves significant *plastic deformation* adjacent to the crack.

Ductility The ability withstand *plastic deformation* without failing.

Duty cycle The ratio of the *beam* on time to the total time of operation in pulsed *laser* output. Also the length of time that the *laser* is used for welding compared with the time that it is in operation.

Efficiency (laser) The quotient of the total *power* in the *laser beam* and the total input *power* to the *laser*. Often referred to as the wall plug efficiency.

Efficiency (quantum) The quotient of the *energy* of a single *laser photon* and the *energy* of a single *pumping photon* that causes a *population inversion* in an optically pumped *laser*.

Elastic deformation Strain that is recovered on unloading.

Elastic limit The maximum stress at which strain is recovered on unloading.

Elastomer A class of material that exhibits a low *Young's modulus*, high tensile strength and large elongation to failure, e.g. rubbers.

Electromagnetic wave The simultaneous periodic variation of electric and magnetic fields.

Emissivity The ratio of the radiant *energy* emitted by a source to that emitted by a blackbody at the same temperature.

Energy The ability to do work.

Equilibrium diagram See *phase diagram*.

Erosion Removal of material by the flow of particles within a liquid or gas.

Etalon Two parallel, flat, partially reflecting mirrors with a spacing on the order of millimetres or centimetres. Used in the *optical cavity* of a *laser* to select a *mode*.

Eutectic A mixture of *constituents* that has the lowest solidification temperature.

Eutectoid The *solid state* equivalent of a *eutectic*. Often used with reference to steels, describing an *alloy* containing 0.80% carbon, which solidifies with a microstructure of *pearlite* on slow cooling.

Excimer A *species* that can only exist in an *excited state*.

Excimer laser A *laser* in which the *active medium* is a diatomic *excimer* molecule.

Excitation The process of raising the *energy* of a *species* from a low state to a higher state. See *pumping*.

Excited state A state of higher *energy* than the *ground state* of a *species*.

f-number The focal length of a lens divided by its effective diameter, i.e. the diameter of the *laser beam* or an *aperture* which restricts a *laser beam*, whichever is the smaller.

Fabry–Pérot interferometer A container, based on the design of an *etalon*, bounded by two parallel plane mirrors that can be used as an *optical cavity*. Invented by the French physicists Charles Fabry (1867–1945) (discoverer of the ozone layer) and Alfred Pérot (1863–1925).

Far-field The field of *radiation* at a distance in the direction of *beam* propagation greater than 50 times the *Rayleigh length*.

Far infrared radiation Electromagnetic *radiation* with a *wavelength* between $10\,\mu m$ and 1 mm.

Fatigue Crack growth caused by repeated applications of stress below the ultimate tensile strength of a material.

Ferrite A *solid solution* of carbon in iron possessing a body-centred cubic structure.

Fibreoptic See *optical fibre*.

Flashlamp A gas-filled lamp that is excited by en electrical pulse to produce a short, bright flash of light. Used for optically *pumping* lasers.

Fluorescence The emission of light of a particular *wavelength* resulting from absorption of *energy* from light of shorter *wavelengths*.

Focal length The distance between the centre of an optical device and the point on the *optical axis* to which parallel rays of light are brought to focus. It is assigned a negative value if rays diverge from an apparent source behind the optical device.

Fracture toughness A measure of the probability of crack growth in a material.

Free electron laser A *laser* in which the *active medium* is a *beam* of free electrons that passes through a magnetic field that periodically changes its polarity with respect to distance.

Frequency multiplication The generation of a *harmonic* of a fundamental frequency through a *non-linear optical effect* in certain materials.

Fresnel number A measure of the tendency for a stable laser cavity to sustain low order *modes*. Defined as the square of the cavity radius divided by the product of the *radiation wavelength* and the length of the *optical cavity*. A small *Fresnel number* favours low order *mode* generation. Developed by the French physicist Augustin Fresnel (1788–1827), whose name is also associated with a mechanism of light absorption.

Fretting Surface damage caused by small relative motion between two surfaces under high loading.

FWHM (Full width at half maximum) The interval between values about a peak at which the values are half of the peak value.

Gain See *amplification*.

Galling Surface damage caused by localized welding of materials sliding in contact with each other.

Galvanizing A hot dip process for depositing zinc for galvanic corrosion protection.

Gas dynamic laser A *laser* in which the *active medium* is a hot, dense gas that is expanded into a low pressure region to create a *population inversion*.

Gaussian beam A *beam* in which the intensity in a given plane varies about a central peak as a classical Gaussian distribution. Also known as a TEM_{00} *mode beam*.

Glass Class of non-*crystalline* (*amorphous*) solids.

Glass transition temperature The temperature at which the behaviour of a *polymer* changes from rigid to flexible, i.e. at which the *Young's modulus* changes from that of a *glass* to that of an *elastomer*.

Graded index optical fibre An *optical fibre* in which the *refractive index* varies monotonically between the centre and periphery of the fibre, in order to refract light such that it remains in the fibre.

Grain A regular three-dimensional crystal arrangement delineated by *grain* boundaries that denote a change in crystal orientation.

Ground state The lowest *energy* state of a *species*.

Guinier–Preston zones Clusters of solute atoms that form coherently with the matrix *lattice*. The precursor to precipitate formation during *precipitation hardening*. Discovered independently in 1938 by the French physicist and crystallographer André Guinier (1911–2000) and the British crystallographer George Dawson Preston (1896–1972).

Hardenability The ability of a material to be hardened, normally by heat treatment.

Harmonic An integral multiple of a fundamental frequency.

HAZ Heat affected zone. The region in the base material that has experienced a thermal transformation, without melting, caused by heat conducted from the weld bead.

Homopolymer A *polymer* that comprises molecules of a single *monomer*.

Hydrogen cracking See *cold cracking*.

Illuminance The quantity of light visible per unit area.

Imperfection A fault or blemish in a sample that distinguishes it from perfection, but which may not be sufficiently deleterious to be classed as a *defect*.

Interlock A switch that when defeated causes a *laser* system to operate in a safe mode.

Interstitial site A region in a *lattice* that is not occupied by an atom.

Ionic bond A bond formed by the donation of electrons from one element and the acceptance of electrons by another to achieve full outer electron shells in both elements.

Ionization The process by which a neutral atom or molecule loses or gains electrons, thereby acquiring a net charge.

Irradiance The *power* incident on an element divided by the area of that element.

Isotactic structure A *polymer* structure in which the side groups are arranged on one side of the molecular chain only.

Izod test A means of measuring the toughness of a material, similar in principle to the *Charpy test*.

Joint efficiency The ratio of the strength of a welded joint to the strength of the base material.

Kerf width The width of a cut.

Kerr cell An optical device consisting of a transparent cell with two electrodes between two polarizing media that passes light only when the two planes of *polarization* are the same. Used in a *Q-switch* to produce short pulses of light. Named after the Scottish minister, mathematician and physicist the Reverend John Kerr (1824–1907).

Keyhole A cavity formed in a material through vaporization. The mechanism for deep penetration welding.

Knoop microhardness A measure of the resistance of a material to local *plastic deformation* under a given low load imposed by an elongated diamond pyramid with an axis ratio of 7:1. Devised for thin sheets and brittle materials in 1939 by Frederick Knoop and colleagues at the National Bureau of Standards in the United States.

Laser A source of coherent ultraviolet, visible or infrared *radiation* produced by means of *stimulated emission*. The acronym for *light amplification* by *stimulated emission* of *radiation*.

Lattice A regular periodic arrangement of atoms, ions or molecules in a *crystalline* solid.

Ledeburite A fine mixture of *pearlite* and *cementite* formed by rapidly solidifying a *eutectic* composition of *cast iron* to prevent the formation of grey iron. Named after the German metallurgist Adolf Ledebur (1837–1916).

Liquidus The boundary on a *phase diagram* separating a liquid *phase* and a mixture of solid and liquid *phases*.

Maraging *Precipitation hardening* by ageing of *martensite* below its decomposition temperature.

Martensite A body-centred tetragonal *phase* formed, in ferrous *alloys*, by rapidly cooling *austenite*. Named after the German microscopist Adolf Martens (1850–1914). Also formed in *alloy* systems other than iron–carbon.

Maser A device capable of emitting coherent microwave *radiation* by means of *stimulated emission*. The predecessor of the *laser*. The acronym for *microwave amplification* by *stimulated emission* of *radiation*.

Melt run A weld made in a plate in which no joint is present. Often used when establishing welding parameters.

Metal Class of material in which atoms are arranged in the sold state in an ordered *lattice*, which are in general lustrous, malleable, fusible, ductile, and good conductors of heat and electricity.

Metastable A non-equilibrium condition under which a state can exist for a limited time. Diamond and graphite at room temperature and pressure are examples of metastable and stable forms of carbon.

Mode A stable condition of oscillation in a *laser*.

Mode locking A method of producing short *laser* pulses (less than 1 ps in certain *lasers*) in bursts or a continuous train.

Mode (longitudinal) The electric field distribution within an *optical cavity* in the direction of propagation of the *beam*.

Mode (spatial) The distribution of *energy* in a specified plane of the *beam*.

Mode (temporal) The profile of intensity versus time at a point in the *beam*.

Mode (transverse) The electric field distribution within an *optical cavity* perpendicular to the direction of *beam* propagation.

Monomer A building unit comprising relatively few atoms that repeats to form a *polymer*.

Moore Sitterly notation Shorthand notation for describing the electronic structure of a *species* written as $^{2S+1}L_J$. Developed by the American astrophysicist Charlotte Moore Sitterly (1898–1990).

Morphology The form in which a *phase* appears.

Multiplexing The transmission of two or more signals on the same frequency during the time normally required for the transmission of a single signal.

Near-field The field of *radiation* at a distance in the direction of *beam* propagation less than 50 times the *Rayleigh length*.

NDT Non-destructive testing. Testing in which the sample is not damaged, e.g. radiography.

Near infrared radiation Electromagnetic *radiation* with a *wavelength* between 700 and 1000 nm.

Nitriding Diffusion of nitrogen into a surface through heating in a nitrogen-rich atmosphere.

Non-linear optical effect An effect induced by electromagnetic *radiation* in which the magnitude is not proportional to the *irradiance*. Examples include *frequency multiplication* and *Raman shifting*.

Normalizing *Annealing* in which cooling to room temperature takes place in air.

Numerical aperture The product of the sine of the angle of rays that can enter or leave an optical element and the *refractive index* of the optical medium.

Optical axis The line passing through the centres of curvature of the optical elements in an *optical cavity*.

Optical cavity The assembly of the cavity containing the *active medium* and the mirrors.

Optical device A device, such as a mirror or lens, that is used to manipulate the *beam*.

Optical fibre A fibre that is able to transmit light by internal reflection.

Paschen notation Electron *energy* level notation based on the arrangement of electrons in quantum shells. Developed by the German physicist Louis Paschen (1865–1947).

Pauli exclusion principle The principle that no two electrons in an atom can be at the same time in the same state or configuration. Proposed in 1925 by the Austrian physicist Wolfgang Pauli (1900–1958) to explain the observed patterns of light emission from atoms.

Pearlite *Constituent* comprising *ferrite* and *cementite* in a lamellar *morphology*, whose appearance in reflected light resembles that of a pearl.

Phase A physically distinct homogeneous body of matter. In physics, the solid, liquid and gas states are referred to as phases. In materials science, the definition is extended to describe a region in a material with a distinct crystal structure, a single composition, and constant physical and chemical properties. See *phase diagram, ferrite, austenite, martensite*.

Phase diagram A diagram of composition (abscissa) versus temperature (ordinate) that shows the regions of stability of various *phases* that can occur in the system at a given pressure (normally atmospheric). Also known as a *constitution diagram* or *equilibrium diagram*.

Photon The quantum of electromagnetic *radiation*.

Piezoelectric The process of generating a voltage when a mechanical force is applied, or vice versa.

Plasma Ionized gas comprising process gas and *metal* vapour, often observed during *laser* welding.

Plastic deformation Strain that is not recovered on unloading.

Plume Non-ionized gas, comprising process gas and *metal* vapour, often observed during *laser* welding.

Pockels cell An electro-optic device in which birefringence (*anisotropy* of the *refractive index*, which varies as a function of polarization and orientation with respect to the incident ray) is modified under the influence of an applied voltage. Used in *Q-switches* to produce short pulses of light. Named after the German scientist Agnes Pockels (1862–1935).

Polarization The orientation of the plane containing the oscillation of the electric field of an *electromagnetic wave*.

Polymer Class of material composed of one or more large molecules that are formed from repeated units of smaller molecules.

Population inversion *Excitation* of a *species*, such that a higher *energy* state is more populated than a lower *energy* state. A prerequisite for *laser* action.

Porosity *Imperfections* in the form of coarse (>0.5 mm diameter) or fine (<0.5 mm diameter) holes in solidified material.

Power The rate of doing work.

Power density The quotient of the *beam power* and the cross-sectional area of the *beam*.

Precipitation hardening A strengthening mechanism that relies on the precipitation of a small uniform second *phase*, which obstructs *dislocation* movement.

Process variables The material properties, *beam* characteristics and processing conditions that may be varied during laser processing.

Pulsed operation The pulsed-emission *temporal mode* of a *laser*, in which *radiation* is emitted in the form of a single pulse or a train of pulses with a duration less than 0.25 s.

Pumping See *excitation*.

Q-factor The quality (Q) factor of an *optical cavity*, which is a measure of the efficiency of the cavity for storing *energy*.

Q-switch An *optical device* that changes the *Q-factor* of an *optical cavity*.

Radiance See *brightness*.

Radiation The emission of *energy* in the form of *electromagnetic waves* or moving particles.

Raman shifting The change in frequency of *radiation* because of *non-linear optical effects*, such as inelastic scattering by molecules. Named after the Indian physicist Sir Venkata Raman (1888–1970).

Rayleigh length The distance from the *beam waist* in the direction of propagation at which the *beam* cross-sectional area is twice that at the *beam waist*. Named after the English physicist the third Baron Rayleigh, born John William Strutt (1842–1919).

Recrystallization Production of a strain-free *grain* structure by heating.

Reflectance The ratio of reflected *power* to incident *power* at a surface.

Refractive index The ratio of the speed of light in a vacuum to the speed of light in a material.

Refractory A material, normally a *ceramic*, which can be used in high temperature applications.

Relaxation Return to a lower *energy* state in a *species* accompanied by the release of heat (non-radiative) or *radiation* (radiative).

Residual stress A state of internal stress normally resulting from the effects of a thermal cycle.

Resonator The assembly of the *optical cavity* and the means of *excitation*.

Rockwell hardness number A measure of the resistance of a material to local *plastic deformation* under a given load imposed by a ball (B scale) or diamond (C scale).

Semiconductor A material with a resistivity between that of insulators and conductors.

Shore hardness number A measure of the resistance of rubber or plastic materials to local *plastic deformation* under a given load imposed by a hardened steel rod with a truncated 35° cone (Type A) or a 30° conical point (Type D).

Solid solution A crystal containing two or more types of atoms in the same *lattice*.

Solid solution hardening A strengthening mechanism based on strain induced by the difference in atomic sizes of atoms in a *solid solution*.

Solid state Used to denote that the *active medium* of a *laser* is in the solid form.

Solidus The boundary on a *phase diagram* separating a solid *phase* and a mixture of solid and liquid *phases*.

Solution treatment Heat treatment to form a *solid solution*.

Solvus The boundary on a *phase diagram* separating a region of solid solubility and a region in which multiple solid *phases* exist at equilibrium.

Sorbite Spheroidized microstructure resulting from the tempering of *martensite* in steels. Named after the English geologist and microscopist Henry Clifton Sorby (1826–1908).

Spalling Lifting or detachment of a coating from a substrate.

Spatter Material ejected from the weld pool during welding.

Species An ion, molecule or atom in which a *laser* transition occurs.

Spontaneous emission The emission of a *photon* from an excited *species* as a result of its natural decay to a lower *energy* state.

Stimulated emission The emission of a *photon* from an excited *species* caused by interaction with another *photon* of the same frequency.

Substitutional site A site on a *lattice* that is occupied by an atom. Used in relation to replacement of *lattice* atoms by *alloy* atoms of different atomic diameter, which strain and strengthen the *lattice*.

Syndiotactic structure A *polymer* structure in which the side groups are arranged alternately on either side of the molecular chain.

Tailored blank A *blank* comprising joined areas of materials of differing grade, thickness or coating.

TEA laser Transversely excited, atmospheric pressure *laser*.

TEM Transverse electromagnetic *mode*. The oscillation *mode* of a *laser*, defined by the distribution of *energy* in a plane transverse to the *optical axis*.

Temper As a verb, used to describe the practice of heat or work treatment applied to a material to improve its properties. As an adjective, used to denote a particular heat or work treatment.

Tempering Heating to, and holding at, a relatively low temperature for the purpose of altering the microstructure of an as-quenched *alloy*.

Thyratron A hot-cathode gas tube in which one or more control electrodes initiate, but do not limit the anode current except under certain operating conditions. Used in *excimer lasers*.

Toughness A measure of the resistance to crack growth in a material.

Transition temperature The temperature at which the mechanism of material failure changes from *ductile fracture* to *brittle fracture*.

Transverse flow laser A *laser* in which the *active medium* flows transverse to both the axis of *excitation* and the *optical axis*.

Ultraviolet radiation Electromagnetic *radiation* with a *wavelength* between 100 and 300 nm.

Van der Waals forces Weak intermolecular forces arising from temporal and spatial variations of charge within neighbouring atoms and molecules. Named after the Dutch physicist Johannes Diderik van der Waals (1837–1923).

Vickers hardness number A measure of the resistance of a material to local *plastic deformation* under a given load imposed by square-based diamond pyramid with an angle of 136° between the opposite faces at the vertex. Devised in the 1920s by engineers at Vickers Ltd in the United Kingdom.

Visible radiation Electromagnetic *radiation* with a *wavelength* between 300 and 700 nm.

Wavelength The distance between successive crests of a wave.

Wear Removal of material from a surface. May be abrasive (caused by harder particles) or adhesive (adhesion of minute contact points that pulls material from surface).

Weld bead The region of fused *metal* in a weld.

Weldability A measure of how easily a material can be welded to given requirements. These may include: weld bead geometry, level of *porosity*, absence of cracks, adequate strength, adequate *fatigue* properties, adequate corrosion properties. Often assessed for a steel by using a *carbon equivalent*.

Work hardening A strengthening mechanism based on an increase in *dislocation* density by *plastic deformation* at temperatures below the *recrystallization* temperature.

YAG laser A *solid state laser* in which the host for the *active medium* is the crystal yttrium aluminium garnet.

Young's modulus The ratio of tensile stress to tensile strain. Named after the English physicist Thomas Young (1773–1829).

APPENDIX

PROPERTIES OF LASERS FOR MATERIAL PROCESSING

INTRODUCTION AND SYNOPSIS

About 45 different lasers were described in Chapter 3. In the early days of laser material processing, processes were based on the thermal effect of infrared laser beams to heat, melt and vaporize materials. As shorter wavelength ultraviolet sources were discovered, new modes of processing became available, involving athermal mechanisms of making and breaking bonds. As techniques of ultrafast pulsing were developed, interactions between the beam and materials were able to be generated that are shorter than the mean free time between collisions in atoms and molecules, leading to processing mechanisms that do not obey laws of classical heat conduction. Such developments have broadened the scope of laser material processing; lasers that were once thought to have insufficient power to be used for material processing are now finding niche uses in a variety of processes on the microscopic scale. Conversely, as the mechanisms of such athermal processes are understood, materials may be selected to take advantage of the properties of the laser beam.

This appendix lists the principal properties of commercial lasers. The data may be used with those of engineering materials, listed in Appendix D, to select appropriate lasers for different processes. For the thermal mode of material processing, power and wavelength are the dominant parameters in continuous wave operation, with pulse duration, repetition rate and energy having complementary significance in pulsed operation. The wavelength and pulse duration are of significance when selecting lasers for athermal mechanisms involving the making and breaking of chemical bonds. (A selection of bond energies can be found in Table 5.1.)

Table B.1 Properties of commercial lasers for material processing in free running operation (note that the frequency and temporal characteristics of the beam can be modified using optical devices; * depends on pumping; + depends on array

Active medium	Type of active medium	Laser transition states (upper → lower)	Wavelength(s) of principal lines (nm)	Photon energy ($J \times 10^{-19}$)	Continuous operation Power (W)	Pulsed operation Duration (ns)	Repetition rate (Hz)	Energy (J)	Average power (W)	Wall plug efficiency (%)
F_2	Excimer gas	${}^3\Pi_{2g} \rightarrow {}^3\Pi_{2u}$	157	12.65	–	10	250–1000	1–10	2–7.5	1–2.5
ArF	Excimer gas	${}^2\Sigma^+_{1/2} \rightarrow {}^2\Sigma^+_{1/2}$	193	10.3	–	5–25	1–1000	0.5	70	1
KrF	Excimer gas	${}^2\Sigma^+_{1/2} \rightarrow {}^2\Sigma^+_{1/2}$	248	8.01	–	2–60	1–500	2	500	2
XeCl	Excimer gas	${}^2\Sigma^+_{1/2} \rightarrow {}^2\Sigma^+_{1/2}$	308	6.45	–	1–250	1–500	2	1–200	2.5
He–Cd	Ionic gas	${}^2D_{3/2} \rightarrow {}^2P_{1/2}$, ${}^2D_{5/2} \rightarrow {}^2P_{3/2}$	325, 442	6.11, 4.50	to 0.035	–	–	–	–	1
N_2	Molecular gas	$C^3\Pi_u \rightarrow B^3\Pi_g$	337	5.89	–	0.3–10	1–1000	1×10^{-6}–0.01	0.4	0.1–5
XeF	Excimer gas	${}^2\Sigma_{1/2} \rightarrow {}^2\Sigma_{1/2}$	353	5.68	–	0.3–35	1–1000	0.5	1–70	2
Dye	Liquid	$S_1 \rightarrow S_0$	370–1000	5.36–1.98	140	3–50	0–10000	0–10	0.05–15	0.1
Ti:sapphire	Insulating solid	${}^2T_2 \rightarrow {}^2E$	470–1650	3.01–1.75	5	1.5×10^{-4}	250–1000	1×10^{-6}	0.005	1
Ar^+	Ionic gas	$4p^2D^0_{5/2} \rightarrow 4s^2P_{3/2}$, $4p^4D^0 \rightarrow 4s^2P$	488, 515	4.07, 3.86	20	–	–	–	–	0.1
Cu	Ionic gas	${}^2P^0_{3/2} \rightarrow {}^2D_{5/2}$, ${}^2P^0_{1/2} \rightarrow {}^2D_{3/2}$	511, 578	3.44, 3.89	–	15–60	2000–30000	20	120	1

Laser	Type	Transition	Wavelength (nm)							
He–Ne	Atomic gas	$^3S_2 \rightarrow {}^2P_4$	633	3.14	0.05	–	–	–	–	0.01
Ruby	Insulating solid	$^2E \rightarrow {}^4A_2$	694	2.86	–	0.3–3	<4	1–100	0.005–100	0.1–1
Alexandrite	Insulating solid	$^4T_2$ and $^2E \rightarrow$ vibrational levels	720–800	2.76–2.48	to 100	to 0.02	2	40	0.005–20	1
AlGaAs	Semiconductor	–	808	2.50	+	+	+	+	+	40
InGaAs	Semiconductor	–	940	2.12	+	+	+	+	+	30
Nd:YLF	Insulating solid	$^4F_{3/2} \rightarrow {}^4I_{11/2}$	1053	1.89	*	*	*	*	*	*
Nd:glass	Insulating solid	$^4F_{3/2} \rightarrow {}^4I_{11/2}$	1053	1.89	–	10	0.1–1	0.004	0.1–100	1–5
Nd:YAG	Insulating solid	$^4F_{3/2} \rightarrow {}^4I_{11/2}$	1064	1.87	10000	500–1500	10–50000	2–6	0.1–10000	3–8
InGaAsP	Semiconductor	–	1100–1600	1.80–1.24	+	+	+	+	+	20
Iodine	Atomic gas	$^2P_{1/2} \rightarrow {}^2P_{3/2}$	1315	1.51	10^6	–	–	–	10^{12}	0.3
Colour centre	Insulating solid	–	1430, 2300, 3500	1.39, 0.864, 0.568	1	–	–	–	–	10
HF	Molecular gas	–	2600, 3000	0.764, 0.662	>150	50–200	0.5–20	>3	500	1
Er:YAG	Insulating solid	$^4I_{11/2} \rightarrow {}^4I_{13/2}$	2937	0.676	*	*	*	*	*	*
DF	Molecular gas	–	3600	0.552	>100	50–200	0.5–20	>3	0.01–0.1	1
CO	Molecular gas	$X^1\Sigma^+$ transitions	5430	0.366	to 35000	–	–	–	–	40
CO2	Molecular gas	$(00^01) \rightarrow (10^00)$	10600	0.187	to 60000 (TF), to 30000 (FAF), to 600 (sealed)	mechanical (TF), 20×10^3 (FAF), 1×10^6 (TEA)	mechanical (TF), 20000 (FAF), 1000 (TEA)	mechanical (TF), 20000 (FAF), 1000 (TEA)	mechanical (TF), 100000 (FAF), 500 (TEA)	5–10

DESIGNATIONS FOR METALS AND ALLOYS

INTRODUCTION AND SYNOPSIS

Engineering materials are named using a variety of conventions. Detailed codes have been drawn up to specify the composition and properties of metals and alloys. Polymers, ceramics, glasses and composites are more commonly referred to by proprietary names, many of which have become generic – it is beyond the scope of this book to describe the thousands of such materials; their specifications and properties can be found in manufacturers' handbooks. Each process chapter of the book includes tables of processing data and sources, to enable the results of processing and the individual procedures used to be located. The materials are identified using the naming conventions of the country in which the data were generated. This appendix provides details of the conventions and standards relevant to engineering metals and alloys.

Engineering metals and alloys are identified by reference to a *standard* and a *designation*. The former refers to the organization that has produced the standard, its number, and year of publication, and describes features such as appropriate terminology, the range of application, and the units used. It is often omitted for brevity, particularly with familiar materials. The designation identifies the grade, type or class of material with a number, letter or symbol, or a combination of these. Designations are generally based on chemical composition, mechanical properties or application.

Much of the data concerning laser-processed materials has been generated in the United States, the United Kingdom, Germany, France and Japan. Each country has developed a designation defined in a national standard, and data are normally presented in that format. Although designations in Europe are being harmonized through the use of Euronorm (EN) standards, and worldwide by the International Organization for Standardization (ISO), familiar national codes remain in use. There is therefore much variation in reporting metals and alloys, which can lead to confusion when comparing data from different sources.

The information provided in this appendix can be used in a number of ways: the codings used in different countries (and even in the same country at different times) can be interpreted; data produced by different sources can be compared; and the salient material characteristics can be extracted from the designation, providing a means of understanding trends and developing new materials for novel laser-based applications.

STANDARDIZATION BODIES

Standards used to characterize materials in the United States of America are defined by bodies such as the American Society for Testing and Materials (ASTM), the American Iron and Steel Institute (AISI), the American Society of Mechanical Engineers (ASME), the Aluminum Association (AA), and the Society of Automotive Engineers (SAE). A Unified Numbering System (UNS) has been developed for steels by ASTM and SAE in collaboration with technical societies, trade associations and government agencies. The UNS number is a designation of chemical composition. It comprises a letter and five numerals. The letter indicates the broad class of alloy; the numerals define specific alloys within the class. The numerals are derived from the SAE–AISI designation.

The British Standards Institution (BSI) develops British standards. The designation for metals and alloys includes the standard defining the product form and an alloy code, prefixed with BS. Many of the British standards have been incorporated into Euronorm standards that are published by the European Committee for Standardization (CEN) in Brussels, Belgium.

Standards and designations developed by *Deutsches Institut für Normung e.V.* (DIN) are used in Germany. A single system applies to all classes of material. Materials are given an abbreviation (*Kurzzeichen* or *Kurznamen*) or a material number (*Werkstoff Nummer*), or both. Designations are prefixed with DIN, although this is often omitted for brevity. The abbreviation may be based either on composition or mechanical properties. The alphanumeric composition designation contains symbols and numerals representing the principal alloying additions and their amounts. Multiplication factors are defined for certain groups of elements in order to produce integer values for the designation. Cast materials are indicated by the prefix G. The designation based on mechanical properties refers to the yield strength of the material. The material number comprises five numerals in the form x.xxxx. It defines a unique material composition range for a comprehensive range of engineering materials. The first numeral denotes the major material characteristic: 0 for pig iron and ferrous alloys; 1 for steels; 2 for heavy non-ferrous alloys; 3 for light metals; 4–8 for non-metallic materials; and 9 for other materials. The first two numerals following the decimal point denote the class of material. The final two numerals represent a running number. A further two numerals may be added following a further decimal point to give additional information regarding the manufacturing process and heat treatment. A key is required to interpret the number.

National standards produced by the member states of the European Union are gradually being incorporated into Euronorm standards in order to provide common technical specifications throughout Europe. These standards cover, among others properties, material grades, product forms, dimensional tolerances, delivery conditions, testing and certification, designation and quality assurance. Conventional designations for metallic materials are described in the standard EN 2032-1:2001 (CEN, 2001).

The Japanese Industrial Standards (JIS) committee produces Japanese standards. The designation prefix is JIS followed by an upper case letter that designates the division, or product form, of the standard. This letter is followed by a series of numerals and letters that identify the specific steel, for which a key is required.

In France the *Association Française de Normalization* (AFNOR) designates steel standards. The designation is similar to the DIN abbreviation system, but with single letters used to denote alloying elements rather than elemental symbols. The prefix is NF and an alphanumeric code represents the relevant standard.

In Italy UNI standards are produced by the *Ente Nazionale Italiano di Unificazione*. The prefix is UNI, which is followed by a four-digit product form code, and an alphanumeric alloy identification similar to the DIN abbreviation system.

Classification Societies, based in different countries, have also produced codes for designating materials, which are described below.

Carbon–Manganese Steels

United States of America

The most widely used specifications for structural steels in the United States are those published by ASTM. ASME has adopted many of the ASTM specifications with little or no modification. The SAE–AISI system is the most commonly used designation, normally appearing simply as AISI. It comprises four numerals: the first two define the alloy type; the final two denote the carbon composition in wt% multiplied by 100. AISI 1040 is therefore an untreated structural steel containing 0.4% C. Examples are given in Table C.1. The UNS number for structural steels comprises the letter G followed by the SAE–AISI number appended with the numeral 0; G10430 refers to AISI 1043.

United Kingdom

A familiar British standard for structural steels is BS 4360:1990 (BSI, 1990). Although it is being replaced with European versions, it remains a popular reference. It describes four basic grades of steel: 40, 43, 50 and 55; the numbers denoting the minimum ultimate tensile strength (UTS, $N\,mm^{-2}$), divided by 10. The former two grades are based on alloying with carbon and manganese, whereas the latter two achieve higher strength values through microalloying. Within each grade, a number of subdivisions denote increasing levels of impact strength (a minimum Charpy V value of 27 J) at various temperatures: A indicates no requirement; B refers to 20°C; C refers to 0°C; D refers to −20°C; DD refers to −30°C; E refers to −40°C; and EE refers to −50°C and −60°C.

Grade 43D therefore refers to a steel with a minimum tensile strength of $430\,N\,mm^{-2}$ and a minimum Charpy V impact energy value of 27 J at −20°C. An alternative designation defined in BS 970-1:1996 (BSI, 1996) comprises three numerals, a letter, and two further numerals. The first three numerals designate the type of steel: 000 to 199 referring to carbon and carbon–manganese types, the number being the manganese content multiplied by 100. The letter indicates whether the steel is supplied to a chemical composition (A), a hardenability specification (H), or a mechanical property specification (M). The final two numerals represent the carbon content in wt% multiplied by 100; 080M18 refers to a steel containing 0.8 wt% Mn and 0.18 wt% C that has been supplied to a mechanical property specification. A third designation, EN 10025:1993 (CEN, 1993a), defines steel grades according to mechanical properties. Fe 510C has a minimum UTS of $490\,N\,mm^{-2}$, with the Grade C toughness requirement described above, and is equivalent to BS 4360 Grade 50C. A previous system of designation, replaced in the early 1970s, but still found in some literature, used the prefix En followed by a number and an optional letter; En8 is equivalent to 080M40.

Germany

The DIN abbreviation based on composition for structural steels comprises the symbol C, followed by the carbon index, which is the composition in wt% multiplied by 100. Additional symbols denote specific subgroups: f for flame and induction hardening steels; k for steels with particularly low phosphorus and sulphur contents; m for steels with a guaranteed range of sulphur content (not just an upper limit); and q for case hardening and heat-treatable steels suitable for cold forming.

Ck 45 is a steel containing 0.45% C with low P and S contents. The abbreviation based on material properties includes the yield strength; St E 255 has a yield strength of $255\,N\,mm^{-2}$. Previously the yield strength was given in $kg\,mm^{-2}$, in which case the E was omitted, e.g. St 37. A letter that indicates a particular steel type or heat treatment may precede this designation: Q for cold-forming steels; R for killed and semi-killed steels; and U for rimming steels.

The DIN material numbers describing steels begin with 1.

Table C.1 AISI designation of major groups of carbon and alloy steels. xx indicates carbon content (wt%) multiplied by 100 (ASM, 1990)

Series	Steel type	Sub-division	Description
1000	Carbon–manganese	10xx	Carbon steel with 1% max. Mn
		11xx	Resulphurized
		12xx	Resulphurized and rephosphorized
		13xx	1.7% Mn
		15xx	1.00 to 1.65% Mn
2000	Nickel	23xx	3.5% Ni
		25xx	5.0% Ni
3000	Nickel–chromium	31xx	1.25% Ni; 0.65 or 0.80% Cr
		32xx	1.75% Ni; 1.07% Cr
		33xx	3.5% Ni; 1.5 or 1.57% Cr
		34xx	3% Ni; 0.77% Cr
4000	Molybdenum	40xx	0.20 or 0.25% Mo
		44xx	0.40 or 0.52% Mo
	Chromium–molybdenum	41xx	0.50, 0.80 or 0.95% Cr; 0.12, 0.20, 0.25 or 0.30% Mo
	Nickel–chromium–molybdenum	43xx	1.82% Ni; 0.50 or 0.80% Cr; 0.25% Mo
	Nickel–molybdenum	46xx	0.85 or 1.82% Ni; 0.20 or 0.25% Mo
5000	Chromium	50xx	0.27, 0.40, 0.50 or 0.65% Cr
		51xx	0.80, 0.87, 0.92, 0.95, 1.00 or 1.05% Cr
6000	Chromium–vanadium	61xx	0.60, 0.80 or 0.95% Cr; 0.10 or 0.15% min. V
7000	Tungsten–chromium	72xx	1.75% W; 0.75% Cr
8000	Nickel–chromium–molybdenum	81xx	0.3% Ni; 0.40% Cr; 0.12% Mo
		86xx	0.55% Ni; 0.50% Cr; 0.20% Mo
		87xx	0.55% Ni; 0.50% Cr; 0.25% Mo
		88xx	0.55% Ni; 0.50% Cr; 0.35% Mo
9000	Silicon–manganese	92xx	1.40 or 2.00% Si; 0.65, 0.82 or 0.85% Mn; 0 or 0.65% Cr
	Nickel–chromium–molybdenum	93xx	3.25% Ni; 1.20% Cr; 0.12% Mo
		94xx	0.45% Ni; 0.40% Cr; 0.12% Mo
		97xx	0.55% Ni; 0.20% Cr; 0.20% Mo

EUROPEAN UNION

Euronorm steel designation names, symbols and numbers are described in the standards EN 10027-1:1992 (CEN, 1992) and EN 10027-2:1991 (CEN, 1991). The general delivery conditions for hot rolled products in weldable fine grain structural steels are described in the standard EN 10113-1:1993 (CEN, 1993b), and those for normalized and thermomechanical rolled steels described in the standards EN 10113-2:1993 (CEN, 1993c) and EN 10113-3:1993 (CEN, 1993d), respectively. Designations may be based on application and mechanical or physical properties (group 1), or chemical composition (group 2). The designation is alphanumeric, starting with the letter S or E to denote structural or engineering steels, respectively. The minimum yield strength of a 16 mm section is then given in $N\,mm^{-2}$. The characters that follow indicate the temperature of the 27 J impact toughness requirement: JR, JO, and J2 representing 25, 0 and $-20°C$, respectively, with K2 representing a 40 J requirement at $-20°C$. Subsequent characters provide further identification of the type and delivery condition of the steel. The letters W, P and H denote steels with improved weather resistance, a greater phosphorus content, or a hollow section product, respectively. S355JO is the European designation of BS 4360 Grade 50C.

JAPAN

JIS steel specifications begin with the letter G to denote the division (product form) of the standard followed by a series of numbers and letters that identify the specific steel, for example: 4051 for carbon steels; and 4052 for structural steels with specified hardenability bands.

An alphanumeric code similar to the DIN abbreviation is then used; G4051 S 43 C denotes a steel containing 0.43% C.

CLASSIFICATION SOCIETIES

Specifications for shipbuilding steels have been produced by the Classification Societies (American Bureau of Shipping, Bureau Veritas, *Det Norske Veritas*, *Germanischer Lloyd*, Lloyd's Register of Shipping, *Nippon Kaiji Kyokai* and *Registro Italiano Navalo*). The societies have collaborated since the 1950s when there was an urgent need to harmonize the approaches taken in formulating steel specifications with improved resistance to brittle fracture. The American Bureau of Shipping (ABS) favoured specifications based on deoxidization practice, composition and heat treatment, whereas the European Societies preferred specifications based primarily on mechanical properties. Five specifications for the three basic types of steel have been produced: Grade A – ordinary shipbuilding steel; Grade B – an intermediate grade based on the ABS approach; Grade C – the highest grade based on the ABS approach; Grade D – an intermediate grade based on specified impact strength at 0°C (European approach); and Grade E – the highest grade based on specified impact strength at $-10°C$ (European approach).

Note that although the letters A to E are used to denote increasing levels of toughness, there are differences in their meaning. Lloyd's higher strength steel AH36 must absorb 34 J when impact tested at 0°C, whereas no toughness requirements are associated with Grade A steels defined in BS 4360:1990 (BSI, 1990).

ALLOY STEELS

The first two numerals of the AISI designation indicate the major alloying additions. The remaining numerals indicate the carbon content in wt% multiplied by 100; AISI 4340 contains 0.4 wt% C, 0.7 wt% Mn, 0.8 wt% Cr, 0.25 wt% Mo and 1.83 wt% Ni.

The BSI system for alloy steels is an extension of that used for carbon–manganese steels. The first three digits cover the range 500–999, and refer to the main alloying elements; e.g. the range 800–839 refers to steels containing nickel–chromium and molybdenum. The letter following refers to the supply specification, as for C–Mn steels, and the final two digits correspond to the carbon content multiplied by 100; BS 817M40 is equivalent to AISI 4340.

The DIN designation for alloy steels uses element symbols to represent the alloying additions that distinguish the steel from other grades, and numerals to quantify the amount of the respective alloying additions. The symbol for carbon is omitted for brevity, but the first number represents the carbon content. The symbol(s) that follow are arranged in order of decreasing content. The number(s) that follow indicate the weight percentage of significant elements, in their respective order, multiplied by an alloy-dependent factor. The following multiplication factors are used to produce integers (values of 0.5 or higher being rounded upwards): 4 for Cr, Co, Mn, Ni, Si and W; 10 for Al, Be, Pb, B, Cu, Mo, Nb, Ta, Ti, V and Zr; 100 for C, P, S, N and Ce; and 1000 for B. A steel containing 0.34% C, 1.5% Cr, 1.5% Ni and 0.2% Mo is designated 34 CrNiMo 6. The nickel and molybdenum contents are not included since they are not necessary for identification. Steels containing more than 5% alloying addition are designated in a similar manner, but preceded by X. Carbon retains its multiplier of 100, but the multiplier for all other elements is 1; X 10 CrNi 18 8 refers to a highly alloyed steel containing 0.1% C, 18% Cr and 8% Ni.

STAINLESS STEELS

AISI uses a three-digit code, based on alloying addition. Steels containing chromium, with nickel as the austenite stabilizing element, are referred to as 300 series steels. Steels containing chromium, nickel and manganese, with nitrogen as the austenite stabilizing element, are referred to as 200 series steels. The 400 series steels are ferritic and martensitic stainless steels. Precipitation-hardened steels are classified according to their chromium and nickel contents, and are designated PH. Duplex stainless steels can be divided into three categories, according to their composition. The most common grades are 22% Cr and 25% Cr, which typically contain 4–7% Ni and up to 4% Mo, and other minor elements such as nitrogen. The third category, superduplex stainless steels, contains alloys with more than 25% Cr and additional alloying elements.

The BSI system follows that for carbon–manganese steels, using the first three digits of the AISI designation, which lie in the range 300–499, followed by S, and two more digits between 11 and 99 that represent a variant of the steel; BS 316S16 is equivalent to AISI 316.

EN 10088-1 (CEN, 1995a) lists grades of stainless steel using the material number and abbreviation, with the analysis and corresponding grade according to BS 970-1:1996 (BSI, 1996).

TOOL STEELS

AISI uses a letter to indicate the method of hardening or the main alloying element, followed by a number that denotes a particular composition, Table C.1. The letters are explained in Table C.2.

The BSI system is identical to that of AISI; the letter is simply prefixed with B. A summary is given in Table C.2.

Table C.2 British and American designation systems for tool steels

BSI	AISI	Steel
BW	W	Water hardening
BO	O	Oil hardening for cold work
BA	A	Medium alloy hardening for cold work
BD	D	High carbon and high chromium for cold work
BH	H	Chromium or tungsten alloyed for hot work
BM	M	Molybdenum alloyed high speed steel
BT	T	Tungsten alloyed high speed steel
BSI	S	Shock resisting
BP	P	Mould steel
BL	L	Low alloy tool steel for special applications
BF	F	Carbon tungsten steels

The DIN compositional abbreviation for tool steels follows that described for carbon–manganese steels, with the grade being indicated by a symbol after the carbon index: W for a normal tool steel; W1 for a first class grade; and W2 for a second class grade.

CAST IRONS

The terminology used to classify cast irons depends on the context in which the material is used. The carbon composition determines the mode of solidification, with respect to the eutectic composition, and provides a metallurgical distinction: *hypoeutectic* – a carbon content less than 4.3%; *eutectic* – a carbon content of 4.3%, which produces a microstructure of ledeburite; and *hypereutectic* – a carbon content greater than 4.3%.

The colour and appearance of the fracture surface provide a visual distinction: *grey* – flake graphite; *white* – cementite; *blackheart* – heat treated cementite; and *whiteheart* – heat treated cementite.

The morphology of the graphite is an unambiguous discriminator. *Nodular* (spheroidal graphite) cast irons are classified into types: type I – nodular; type II – nodular, imperfectly formed; type III – aggregate, or temper carbon; type IV – quasi-flake graphite; type V – crab-form graphite; type VI – irregular, or open type nodules; and type VII – flake graphite. *Flake* (lamellar graphite) cast irons are also denoted by types: type A – uniform distribution, random orientation; type B – rosette grouping, random orientation; type C – variable flake size, random orientation; type D – flakes between dendrites, random orientation; type E – flakes between dendrites, aligned orientation.

The mechanical properties provide a guide to their use: *ductile* irons contain spheroidal graphite; and *malleable* irons contain heat treated cementite.

ASTM standards are based on mechanical properties expressed in Imperial units. Grey irons are designated according to minimum ultimate tensile strength (UTS) measured in ksi ($klb\ in^{-2}$), ranging from class 20 (20 ksi or $140\ N\ mm^{-2}$) to class 60 (60 ksi or $410\ N\ mm^{-2}$). The designation 100-70-03 refers to a pearlitic nodular cast iron with a minimum ultimate tensile strength (UTS) of 100 ksi, a yield stress of 70 ksi and an elongation at failure of 3%. Two numbers, the yield stress in 10^2 psi and the elongation, are used to designate malleable irons.

BSI standards designate grey irons into seven grades: 150, 180, 220, 260, 300, 340 and 400, which represent the minimum UTS in $N\ mm^{-2}$ of a 30 mm diameter test bar. Nodular irons are also designated according to mechanical properties: 420-12 denotes an iron with a minimum UTS of $420\ N\ mm^{-2}$ and an elongation of 12%. The designation for malleable irons contains the letter B, P or A, to indicate a blackheart, pearlitic or whiteheart iron, respectively. UTS ($N\ mm^{-2}$) and elongation (%) are included in the same manner as with nodular irons. Alloy cast irons are specified for graphite-free white irons by a number 1, 2 or 3, which indicates whether the alloy is low alloy, nickel–chromium or high chromium, respectively. The letter that follows indicates the specific alloy. For graphite-containing alloy irons, austenitic irons containing lamellar or spheroidal graphite are designated L or S, respectively, followed by symbols and numbers representing the alloy additions. Ferritic high silicon, graphite-containing alloys are designated by the main alloying elements and their percentages, e.g. Si10.

The DIN abbreviation for a cast iron is the capital letter G, followed by a number that represents the minimum UTS ($kg\ mm^{-2}$). Grey irons are denoted by GG, nodular irons by GGG; GGG 50 is a nodular iron with a minimum UTS of $50\ kg\ mm^{-2}$. Alloy irons are described using a similar designation to alloy steels; G-X 300 CrMoNi 15 2 1 contains 3% C, 15% Cr, 2% Mo and 1% Ni. The material number for cast irons begins with 0, the grade numbers being 60 and 70 for grey and nodular irons, respectively.

The European designations for cast irons are related to the BSI and DIN systems; EN-GJS-350-22 denotes a nodular iron of minimum UTS $350\ N\ mm^{-2}$ and elongation 22%.

The Japanese JIS designation is similar to the BSI coding; FCD 450-10 is a nodular iron with a minimum UTS of $450\ N\ mm^{-2}$, and an elongation of 10%.

ALUMINIUM ALLOYS

The two most common systems used to identify wrought aluminium alloys are the four-digit International Alloy Register (IAR) number, prefixed AA (Aluminum Association) and an alphanumeric system used by the International Organization for Standardization (ISO). The IAR number is most commonly used in the United States. It comprises four digits, the first representing the principal alloying addition, which identifies the alloy series, Table C.3. The second digit indicates modifications to the original alloy or impurity limits. The last two digits denote the individual alloy. In the case of 1000 series alloys, the last two digits represent the aluminium content above 99.00% in hundredths. A prefix, e.g. P-, indicates a preliminary designation. A suffix, e.g. A, indicates variations in the allowable amount of impurities.

The complete specification of a wrought aluminium alloy requires an indication of its temper, i.e. the degree of cold work or heat treatment. Some of the more common temper codings are given in Table C.4. Thus 6061-T4 is an aluminium–magnesium–silicon alloy that has been solution heat treated and aged naturally.

The most common coding system for casting alloys is that of the Aluminum Association, which uses four digits, with the last separated by a full stop. The first digit represents the principal alloying element, denoting the series, as shown in Table C.3. The second and third digits identify specific alloys. The fourth digit represents the product form: 0 for a casting and 1 for an ingot. A letter before the numerical code indicates a modification of the original composition.

The Euronorm systems based on numerical and chemical designations for the chemical composition and form of wrought products are described in the standards EN 573-1:1995 (CEN, 1995b) and EN 573-2:1995 (CEN, 1995c), respectively.

The ISO designation is most commonly used in Germany. It is based on the content of the major alloying additions. AlCuMg2 describes an aluminium–copper alloy containing around 2% magnesium.

Table C.3   Aluminum Association designation system for aluminium alloys

Type	Designation	Principal alloying addition(s)
Wrought	1xxx	Aluminium of 99.00% minimum purity
	2xxx	Copper
	3xxx	Manganese
	4xxx	Silicon
	5xxx	Magnesium
	6xxx	Magnesium and silicon
	7xxx	Zinc
	8xxx	Other
	9xxx	Unused
Cast	1xx.x	Aluminium of 99.00% minimum purity
	2xx.x	Copper
	3xx.x	Silicon with copper or magnesium
	4xx.x	Silicon
	5xx.x	Magnesium
	6xx.x	Unused
	7xx.x	Zinc
	8xx.x	Tin
	9xx.x	Other

Table C.4   Aluminum Association and British designations for temper codes for aluminium alloys

AA	BSI	Temper
F	M	As fabricated
O	O	Annealed
H1	H1	Strain hardened only. Degree of hardening indicated by a second digit between 1 and 8. 2, 4, 6 and 8 represent quarter, half, three quarters and full hardness, respectively.
H2	H2	Strain hardened and partially annealed. Degree of hardening after annealing indicated in same manner as H1.
H3	H3	Strain hardened and stabilized by a low temperature heat treatment. Degree of hardening, before stabilization, indicated in same manner as H1.
T1		Cooled from an elevated temperature forming process, and naturally aged.
T2		Cooled from an elevated temperature forming process, cold worked, and naturally aged.
T3	TD	Solution heat treated, cold worked and naturally aged.
T4	TB	Solution heat treated and naturally aged.
T42		Solution treated from the O or F temper.
T5	TE	Cooled from an elevated temperature forming process, and then artificially aged.
T51		Stress relieved by tension, after solution treatment or cooling from an elevated temperature forming process.
T52		Stress relieved by compression, after solution treatment or cooling from an elevated temperature forming process.
T54		Stress relieved by combined tension and compression.
T6	TF	Solution heat treated and then artificially aged.
T62		Solution treated from the O or F temper.
T7		Solution heat treated and stabilized.
T73		Step ageing for improved stress corrosion properties.
T7351		T73 with controlled stretching before ageing.
T8	TH	Solution heat treated, cold worked and artificially aged.
T9		Solution heat treated, artificially aged and then cold worked.
T10	TE	Cooled from an elevated temperature forming process, cold worked, and then artificially aged.
Tx51		Stress relieved by controlled stretching, x = temper.
W		Solution heat treatment for alloys that age harden naturally.

MAGNESIUM ALLOYS

ASTM uses two letters to indicate the major alloying elements in magnesium alloys, Table C.5, followed by two numbers that denote the nominal amounts (wt%) of the two alloying elements. Two alloys with the same nominal composition are differentiated by a third letter following the numbers. The designation is completed with the temper, expressed using the Aluminum Association system described above for aluminium alloys. Thus AZ31B-H24 denotes a type B wrought alloy containing 3 wt% Al and 1 wt% Zn that has undergone a strengthening treatment comprising half strain hardening and partial annealing to produce recrystallization without grain growth.

The BSI system uses the notation MAG followed by an arbitrary number to indicate the specific alloy, for which a key is required. MAG 1 refers to the alloy AZ81A. Wrought alloys are also designated

Table C.5 ASTM designation system for additions in magnesium alloys

Letter	Element
A	Aluminium
B	Bismuth
C	Copper
D	Cadmium
E	Rare earth
F	Iron
H	Thorium
K	Zirconium
L	Beryllium
M	Manganese
N	Nickel
P	Lead
Q	Arsenic
R	Chromium
S	Silicon
T	Tin
W	Silver
Z	Zinc

MAG, followed by a dash and a letter to denote the delivery form: S refers to plate, sheet and strip; and E refers to bars, sections and tubes including extrusions. A further dash is followed by a key number, starting from 100 denotes a specific alloy. The designation is completed with the temper, using the BSI system for aluminium alloys.

The Euronorm designation system for magnesium and its alloys is described in the standard EN 1754:1997 (CEN, 1997).

TITANIUM ALLOYS

Commercially pure (CP) titanium alpha alloys are referred to by their ASTM 'grade' designation number. The commonly used CP alloys have oxygen as their primary interstitial element. They are known as: grade 1, which has a maximum of 0.18% oxygen; grade 2 (0.25%); grade 3 (0.35%); and grade 4 (0.40%).

The most common system of designating titanium alloys is based on the crystal structure(s) present, and the amount of principal alloying additions expressed using the ISO system for aluminium alloys, but without the use of multipliers. Alpha (α) alloys denote commercially pure titanium and those alloys containing strong alpha-stabilizing elements. A beta (β) alloy, e.g. Ti11.5Mo6Zr, contains the respective weight percentages of molybdenum (a beta stabilizer) and zirconium (for solid solution strengthening). Alpha–beta ($\alpha + \beta$) alloys contain significant amounts of beta phase at room temperature; Ti6Al4V denotes principal alloying additions of 6 wt% Al and 4 wt% V. Heat treatment is described by the solution treatment temperature and time, method of cooling, and the precipitation hardening temperature and time.

NICKEL ALLOYS

Nickel alloys are most commonly referred to by well-known registered trade marks, such as Inconel® and Nimonic®, followed by a key number to identify a specific alloy. There is a BSI system, in which alloys are referred to by NA followed by a key number to indicate the specific alloy; superalloys are denoted by the letters HR followed by an identifying number. ASTM and ASME identify nickel alloys by using the letters B and BS, respectively, followed by a key three-digit number, and a dash with a further identification number if necessary. The nickel–chromium alloy Incoloy® 825 is referred to as NA 16, B163 and SB 163 in the BSI, ASTM and ASME systems, respectively. (Inconel, Nimonic and Incoloy are registered trademarks of Special Metals Corporation, Huntington, USA.)

Table C.6 CDA and BSI designations for copper alloys, and some common alloy names

Form	CDA	BSI	Alloying elements	Common name
Wrought	C1xx	C	Minimum 99.3 wt% Cu, and alloys with 96–99.3 wt% Cu	
	C2xx	CZ	Zinc	Brasses
	C3xx		Zinc and lead	Leaded brasses
	C4xx		Zinc–tin	Tin brasses
	C5xx	PB	Tin and phosphorus	Bronzes and phosphor bronzes
	C6xx	CA	Aluminium	Aluminium bronzes
		CS	Silicon	Silicon bronzes
		CB	Beryllium	Beryllium bronzes
	C7xx	CN	Nickel	Cupro-nickels
		NS	Nickel and zinc	Nickel silvers
Cast	C8xx	HCC1	High copper alloys	
		DCBx, HCBx, PCBx, SCBx	Brasses	
	C862		Manganese bronzes	
	C863			
		CMA1	Manganese and aluminium Zinc and silicon	
	C9xx	CT1, G1	Tin	Tin bronzes, gunmetals
		LBx	Tin and lead	Leaded bronzes
		CT2, G3	Tin and nickel	Nickel gunmetals
		LGx	Tin, nickel and lead	Leaded gunmetals
		PBx	Tin and phosphorus	Phosphor bronzes
			Lead, tin and phosphorus	Leaded phosphor bronzes
		ABx	Aluminium	Aluminium bronzes
			Aluminium and iron	Aluminium bronzes
		CN1	Nickel and chromium	
			Nickel and iron	
		CN2	Nickel and niobium	
			Nickel and zinc	Nickel silvers

COPPER ALLOYS

The Copper Development Association (CDA) system of coding is common in the United States. The letter C (which is often omitted) is followed by three digits. The first denotes a specific group of alloys, and the remaining two key digits identify a specific alloy, Table C.6.

The BSI system uses letters to denote a group of alloys, followed by a number to indicate a particular property of the alloy: SCB1 represents a general purpose sand casting brass; and SCB6 represents a brazable quality sand casting brass, Table C.6. Euronorm designations are defined in the standard EN 1173:1995 (CEN, 1995d).

Temper designations are based on the AA and BSI systems for aluminium alloys. The BSI designation follows the aluminium codes more closely, with small variations, e.g. 1/2H denotes half hard. The AA system requires a key for interpreting the designations.

BIBLIOGRAPHY

ASM (1990). *Metals Handbook*. Metals Park: ASM International.

BSI (1990). BS 4360:1990. *Specification for weldable structural steels*. Milton Keynes: British Standards Institution.

BSI (1996). BS 970-1:1996. *Specification for wrought steels for mechanical and engineering purposes. General inspection and testing procedures and specific requirements for carbon, carbon manganese, alloy and stainless steels*. Milton Keynes: British Standards Institution.

CEN (1991). EN 10027-2:1991. *Designation systems for steels. Steel numbers*. Brussels: European Committee for Standardization.

CEN (1992). EN 10027-1:1992. *Designation systems for steel. Steel names. Principal symbols*. Brussels: European Committee for Standardization.

CEN (1993a). EN 10025:1993. *Hot rolled products of non-alloy structural steels. Technical delivery conditions*. Brussels: European Committee for Standardization.

CEN (1993b). EN 10113-1:1993. *Hot rolled products in weldable fine grain structural steels. General delivery conditions*. Brussels: European Committee for Standardization.

CEN (1993c). EN 10113-2:1993. *Hot rolled products in weldable fine grain structural steels. Delivery conditions for normalized/normalized rolled steels*.

Brussels: European Committee for Standardization.

CEN (1993d). EN 10113-3:1993. *Hot rolled products in weldable fine grain structural steels. Delivery conditions for thermomechanical rolled steels*. Brussels: European Committee for Standardization.

CEN (1995a). EN 10088-1:1995. *Stainless Steels. List of Stainless Steels*. Brussels: European Committee for Standardization.

CEN (1995b). EN 573-1. *Aluminium and aluminium alloys. Chemical composition and form of wrought products. Numerical designation system*. Brussels: European Committee for Standardization.

CEN (1995c). EN 573-2. *Aluminium and aluminium alloys. Chemical composition and form of wrought products. Chemical symbol based designation system*. Brussels: European Committee for Standardization.

CEN (1995d). EN 1173. *Copper and copper alloys. Material condition or temper designation*. Brussels: European Committee for Standardization.

CEN (1997). EN 1754. *Magnesium and magnesium alloys. Magnesium and magnesium alloy anodes, ingots and castings. Designation system*. Brussels: European Committee for Standardization.

CEN (2001). EN 2032-1:2001. *Metallic materials. Conventional designation*. Brussels: European Committee for Standardization.

PROPERTIES OF ENGINEERING MATERIALS

INTRODUCTION AND SYNOPSIS

The thermal and mechanical properties of engineering materials play the greatest role in determining the response of components to laser processing. Collections of property data can be found in books (such as those mentioned in Chapter 5) and by electronic means. However, data vary – often considerably. In some cases, variations result from the conditions under which materials are tested – many properties vary with temperature, for example. In other cases, anisotropy in the material produces variations in results, as is the case when wood is tested parallel with or perpendicular to the grain. However, by careful choice of data, bearing in mind the application in which they will be used, representative values can be found to both characterize and differentiate materials during laser processing.

The variation of the thermal properties of metals and alloys with temperature has often led to the use of sophisticated numerical methods for establishing temperature fields during laser processing. For example, the thermal cycles produced during laser heating have been obtained by numerical means using appropriate values for thermal properties at each temperature in the cycle. However, the thermal cycle can be reproduced with an accuracy within the tolerances used in this book (±5%) by using the analytical formulae described in Appendix E, provided that average values for thermal properties corresponding to 60% of the melting temperature are used (Kou *et al.*, 1983). Thus the computation of temperature fields during laser heating may be simplified considerably. Similar principles can be applied to other properties to enable simple analytical models to be developed.

The data contained in this appendix have been selected to be representative of conditions pertaining to laser material processing; thermal properties corresponding to 60% of the melting temperature of metals and alloys are quoted. The data have many uses: they may be substituted into the formulae given in Appendix E to make rapid calculations of the temperature profiles generated during laser processing; they may be used with the material selection charts in Chapter 5 to choose materials for laser processing; and they can be used to design novel materials and laser-based procedures using the process principles outlined in the book.

Table D.1 Mechanical and thermal properties of some engineering metals and alloys: ρ is density; E is Young's modulus; σ_y is yield stress; σ_{TS} is ultimate tensile stress; ε_f is elongation at failure; λ is thermal conductivity; a is thermal diffusivity; c_p is specific heat capacity at constant pressure; T_m is melting temperature; T_v is vaporization temperature; L_m is latent heat of melting; L_v is latent heat of vaporization; α is coefficient of linear expansivity (the quantities λ, c_p and a refer to values at 60% of the melting temperature)

Material	Grade	Mechanical properties					Thermal properties							
		ρ $(\mathrm{kg\,m^{-3}})$	E $(\mathrm{GN\,m^{-2}})$	σ_y $(\mathrm{MN\,m^{-2}})$	σ_{TS} $(\mathrm{MN\,m^{-2}})$	ε_f (%)	λ $(\mathrm{J\,s^{-1}\,m^{-1}\,K^{-1}})$	a $(\mathrm{m^2\,s^{-1}} \times 10^{-6})$	c_p $(\mathrm{J\,kg^{-1}\,K^{-1}})$	T_m (K)	T_v (K)	L_m $(\mathrm{kJ\,kg^{-1}})$	L_v $(\mathrm{MJ\,kg^{-1}})$	α $(\mathrm{K^{-1}} \times 10^{-6})$
Aluminium	CP	2704	71	50	200	43	238	88	1000	932	2740	388	10.79	23
Copper	CP	8930	117	75	400	45	375	112	471	1356	2868	205	4.79	17
Copper	Brass	8500	115	450	550	8	110	35	370	1300	1180	150	3.75	18
Copper	Bronze	8800	110	140	260	10	180	57	360	1300	2540	105	3.50	17
Gold	CP	19 300	71	40	220	40	296	116	132	1340	3239	65	1.28	14
Iron	CP	7790	206	165	300	45	32.5	7.5	560	1810	3300	272	6.10	12
Iron	C–Mn	7764	210	300	460	35	30	9.2	420	1800	3100	270	6.05	15
Iron	AISI 304	7870	195	230	600	60	25.5	7.2	450	1773	3300	280	6.20	16
Iron	Grey	7142	110	180	275	0.5	75	21	500	1500	2900	100	5.80	11
Iron	Nodular	7891	197	275	415	18	25	6.6	480	1810	2900	100	5.80	12
Iron	Malleable	7700	180	190	230	0	75	22	440	1420	2900	140	5.80	11
Lead	CP	11 538	16.1	12	168	50	33	22	130	600	2017	24	0.863	27
Magnesium	CP	1740	44	95	190	5	100	70	821	924	1380	358	5.51	25
Nickel	CP	8900	207	60	300	30	72	14	560	1726	3005	302	6.38	13
Nickel	Monel	8800	150	240	520	40	360	97	420	1600	2800	250	5.20	14
Silver	CP	10 500	70	150	180	45	419	160	235	1230	2485	106	2.33	19
Tin	CP	7300	40	26	30	0.5	68	39	264	505	2540	60	2.50	23
Titanium	CP	4500	116	480	620	20	23	9.8	523	1950	3530	392	10.67	9
Zinc	CP	7140	110	260	150	50	111	40	420	693	1180	110	1.76	31

Table D.2 Mechanical and thermal properties of some engineering glasses: ρ is density; E is Young's modulus; σ_{TS} is ultimate tensile stress; λ is thermal conductivity; c_p is specific heat capacity at constant pressure; a is thermal diffusivity; T_g is glass transition temperature; T_m is melting temperature; α is coefficient of linear expansivity

Material	Grade	Mechanical properties			Thermal properties					
		ρ (kg m^{-3})	E (GN m^{-2})	σ_{TS} (MN m^{-2})	λ (J s^{-1} m^{-1} K^{-1})	c_p (J kg^{-1} K^{-1})	a (m^2 s^{-1} × 10^{-6})	T_g (K)	T_m (K)	α (K^{-1} × 10^{-6})
Alumino-silicate	Fibre glass	2700	85	60	1.0	800	0.46	863	1500	5.2
Boro-silicate	Pyrex	2300	65	55	1.0	753	0.57	743	1400	3.2
Glass	Silica	2200	75	70	1.4	840	0.76	1253	1715	0.6
Glass	Soda	2500	74	50	1.0	503	0.80	793	1996	7.8
Glass	Lead	3000	60	65	0.8	670	0.40	663	1996	9.4

Table D.3 Mechanical and thermal properties of some engineering ceramics: ρ is density; E is Young's modulus; σ_y is yield stress; σ_{TS} is ultimate tensile stress; ε_f is elongation at failure; λ is thermal conductivity; c_p is specific heat capacity at constant pressure; a is thermal diffusivity; T_m is melting temperature; α is coefficient of linear expansivity (*four point bending; + compression)

Material	Grade	Mechanical properties					Thermal properties				
		ρ $(\mathrm{kg\,m^{-3}})$	E $(\mathrm{GN\,m^{-2}})$	$\sigma_y^\star$ $(\mathrm{MN\,m^{-2}})$	σ_{TS} $(\mathrm{MN\,m^{-2}})$	ε_f (%)	λ $(\mathrm{J\,s^{-1}\,m^{-1}\,K^{-1}})$	c_p $(\mathrm{J\,kg^{-1}\,K^{-1}})$	a $(\mathrm{m^2\,s^{-1}} \times 10^{-6})$	T_m (K)	α $(\mathrm{K^{-1}} \times 10^{-6})$
Alumina	Al_2O_3	3970	345	455	250$^+$	–	7.80	800	9.54	2303	8.5
Aluminium nitride	AlN	2700	320	441	425	–	180	700	95.2	2173	4.4
Beryllia	BeO	3100	400	246	250	–	210	1080	62.7	2803	7.4
Boron carbide	B_4C	2520	450	350	350	–	37.0	950	15.5	2623	4.3
Brick	Red	2300	17	500	36$^+$	–	0.60	440	59.3	2980	9.0
Diamond		3510	500	1500	10 000$^+$	–	2000	472	1207	4300	1.2
Granite		2750	70$^+$	–	175$^+$	23	1.5	800	0.68	7000	8.0
Limestone	$CaCO_3$	2000	63$^+$	–	110$^+$	20	1	1000	0.50	2750	7.0
Magnesia	MgO	3580	395	280	–	–	62	940	18.4	3073	12
Marble		2850	75$^+$	–	115$^+$	–	3	880	1.20	–	6.5
Porcelain		2400	70	–	–	45	5	1070	1.95	1823	3.0
Sialon	Si_3N_4–Al_2O_3	3200	300	425	400	–	22	750	9.2	1600	3.1
Silicon carbide	SiC	3170	400	140	450	0.6	100	1240	0.16	2500	4.5
Silicon nitride	Si_3N_4	3100	175	210	275	–	16	750	6.9	2023	2.6
Tungsten carbide	WC	15 800	700	–	344	–	70	340	13	3073	7.3
Zirconia	ZrO_2	5600	138	175	325	–	19	460	0.6	2770	11

Table D.4 Mechanical and thermal properties of some engineering polymers; ρ is density; E is Young's modulus; σ_y is yield stress; σ_{TS} is ultimate tensile stress; ε_f is elongation at failure; λ is thermal conductivity; c_p is specific heat capacity at constant pressure; a is thermal diffusivity; T_g is glass transition temperature; T_m is melting temperature; α is coefficient of linear expansivity (Kevlar is a registered trademark of E.I. du Pont de Nemours and Company)

Material	Grade	Mechanical properties					Thermal properties					
		ρ (kg m^{-3})	E (GN m^{-2})	σ_y (MN m^{-2})	σ_{TS} (MN m^{-2})	ε_f (%)	λ (J s^{-1} m^{-1} K^{-1})	c_p (J kg^{-1} K^{-1})	a (m^2 s^{-1} × 10^{-6})	T_g (K)	T_m (K)	α (K^{-1} × 10^{-6})
Acrylic	Moulded	1150	3.1	74	69	6	0.20	1500	0.12	373	403	73
Epoxy	Cast	1120	3	80	90	3	0.17	1400	0.11	380	420	5
Melamine	Phenolic	1630	10.7	94	70	80	0.56	1700	0.20	–	360	4
Nylon	6	1130	3	79	100	70	0.23	700	0.29	340	480	94
Polyamide	Kevlar®	1450	110	120	3200	6.6	0.37	1100	0.23	553	1273	28
Polycarbonate		1200	2.38	62	66	80	0.19	1300	0.12	573	783	75
Polyetheretherketone		1330	4.5	99	110	37	0.25	2000	0.01	413	613	39
Polyethylene	Low density	920	0.18	20	13	600	0.25	2300	0.12	270	360	200
Polyethylene	High density	930	0.43	25	30	200	0.52	2300	0.24	300	383	200
Polymethylmethacrylate	Atactic	1190	3.18	80	76	5	0.20	1500	0.11	378	400	250
Polypropylene	Atactic	900	1	25	35	220	0.16	2100	0.01	253	310	62
Polystyrene	Amorphous	1050	3.1	40	50	2	0.13	1300	0.01	373	373	70
Polyurethane		1100	0.44	25	27	250	0.19	1500	0.12	–	358	280
Polyvinylchloride		1352	0.01	28	37	300	0.13	1800	0.05	350	370	150
Rubber	Polyisoprene	916	0.02	–	17	500	0.13	1905	0.07	220	350	660

Table D.5 Mechanical and thermal properties of some engineering composites: ρ is density; E is Young's modulus; σ_y is yield stress; σ_{TS} is ultimate tensile stress; λ is thermal conductivity; c_p is specific heat capacity at constant pressure; a is thermal diffusivity; T_m is melting temperature; α is coefficient of linear expansivity (*bending; + compression; || parallel to grain/reinforcement; ⊥ perpendicular to grain/reinforcement)

Material	Grade	Mechanical properties				Thermal properties				
		ρ (kg m^{-3})	E (GN m^{-2})	σ_y (MN m^{-2})	σ_{TS} (MN m^{-2})	λ (J s^{-1} m^{-1} K^{-1})	c_p (J kg^{-1} K^{-1})	a (m^2 s^{-1} ×10^{-6})	T_m (K)	α (K^{-1} ×10^{-6})
Aluminium alloy–SiC	Particulate	3050	200	300	350	180	850	69.4	823	16.5
Bone	In vivo	2100	20	–	150	0.38	440	0.29	–	1.46
Carbon fibre reinforced polymer	CFRP	1700	100	–	250	0.75	1373	0.30	500	8
Concrete		2400	30$^+$	400	40$^+$	0.10	3350	0.01	1600	12
Glass fibre reinforced polymer	GFRP	2400	80	–	220	0.60	3350	0.01	1600	10
Human tissue	Muscle	1010	–	–	–	0.46	3662	12.4	–	–
Titanium alloy–SiC	Fibre	4100	320	170	450	23	523	9.8	1950	6.8 (||), 8.0 (⊥)
Wood	Ash	600	10.5*	55*	75*	1.0 (||), 0.4 (⊥)	1800	0.65	–	50 (||), 27 (⊥)
	Fir	520	12.0*	40*	65*	1.0 (||), 0.4 (⊥)	1700	0.79	–	55 (||), 25 (⊥)
	Oak	680	11.5*	50*	75*	1.5 (||), 0.6 (⊥)	1900	0.77	–	70 (||), 30 (⊥)
	Pine	600	12.5*	50	75*	1.0 (||), 0.4 (⊥)	1700	0.69	–	60 (||), 25 (⊥)

BIBLIOGRAPHY

Kou, S., Sun, D.K. and Le, Y.P. (1983). A fundamental study of laser transformation hardening. *Metallurgical Transactions*, **14A**, (4), 643–653.

ANALYTICAL EQUATIONS

INTRODUCTION AND SYNOPSIS

There is much commonality in the principles underlying the different methods of thermal laser material processing. The temperature field generated during vaporization cutting is analogous to that found in keyhole welding – both involve lateral heat flow from a through-thickness energy source. The phase transformations responsible for surface hardening in ferrous alloys are the same as those that determine the heat affected zone hardness of a weld – only the thermal cycle is different. Physical processes are therefore presented here as modular components of heat flow and structural change. The modules can be used interchangeably, and applied across a range of processes.

This appendix contains *analytical* solutions for the temperature fields around a variety of energy sources and the structural transformations that they induce in materials. Analytical equations describe explicit relationships between the process variables. Thus the effects of changes in process variables on the structure and properties of materials can be estimated quickly, or, if a particular result is desired, combinations of processing parameters can be assessed rapidly. The notation used is defined in Table E.1.

Table E.1 Notation

Symbol	Definition	Units
A	Absorptivity	–
A	Constant of precipitate solubility product	–
A_B	Beam area	m^2
B	Constant of precipitate solubility product	K
C_{eq}	Carbon equivalent	wt%
C_m	Concentration of metal species of precipitate M_aC_b in solution	wt%
C_c	Concentration of non-metal species of precipitate M_aC_b in solution	wt%
D	Diffusion coefficient, $D = D_0 \exp -(Q/RT)$	$m^2\,s^{-1}$
D_0	Pre-exponential of diffusion coefficient	$m^2\,s^{-1}$
E	Beam power density	$J\,s^{-1}\,m^{-2}$
F	Heat flux	$J\,m^{-2}\,s^{-1}$
H	Vickers hardness number	VPN
H_b	Vickers hardness number of bainite	VPN

(Contd)

Table E.1 (*Contd*)

Symbol	Definition	Units
H_{fp}	Vickers hardness number of ferrite–pearlite mixture	VPN
H_m	Vickers hardness number of martensite	VPN
I	Kinetic strength of a thermal cycle	–
L_m	Volumetric latent heat of melting	$J\,m^{-3}$
L_m^*	Normalized latent heat of melting, $L_m^* = L_m/[\rho c(T_m - T_0)]$	–
L_v	Volumetric latent heat of vaporization	$J\,m^{-3}$
L_v^*	Normalized latent heat of vaporization, $L_v^* = L_v/[\rho c(T_m - T_0)]$	–
M_f	Martensite finish temperature	K
M_s	Martensite start temperature	K
Q	Activation energy	$J\,mol^{-1}$
R	Gas constant, 8.314	$J\,mol^{-1}\,K^{-1}$
T	Temperature	K
T_0	Initial temperature	K
T_{Ac1}	Temperature at which pearlite transforms to austenite on heating	K
T_{Ac3}	Temperature at which ferrite transforms to austenite on heating	K
T_{Ms}	Temperature at which martensite starts to form on cooling	K
T_{M50}	Temperature at which martensite formation is 50% complete on cooling	K
T_{Mf}	Temperature at which martensite formation is complete on cooling	K
T_m	Melting temperature	K
T_s	Dissolution temperature of a precipitate	K
T_v	Vaporization temperature	K
V	Volume fraction	–
V_b	Volume fraction of bainite	–
V_{fp}	Volume fraction of ferrite–pearlite mixture	–
V_m	Volume fraction of martensite	–
V'	Cooling rate at 923 K	$K\,h^{-1}$
a	Thermal diffusivity, $\lambda/(\rho c)$	$m^2\,s^{-1}$
c	Specific heat capacity	$J\,kg^{-1}\,K^{-1}$
d	Plate thickness	m
e	Base of natural logarithms, 2.718	–
f	Matrix volume fraction available for precipitate dissolution	–
g	Grain size	m
g_0	Original grain size	m
k	Kinetic constant	Various
l	Depth of treatment	m
p	Precipitate size	m
p_0	Original precipitate size	m

Table E.1   (*Contd*)

Symbol	Definition	Units
q	Beam power	$J\,s^{-1}$
q_{max}	Peak beam power in a Gaussian distribution	$J\,s^{-1}$
q_v	Volumetric rate of heat generation	$J\,s^{-1}\,m^{-3}$
r	Radial distance from centre of a surface heat source	m
r	Lateral distance from centre of a through-thickness heat source	m
r_B	Beam radius defined where $q = q_{max}/e$, or beam half-width	m
t	Time	s
t_0	Time for heat to diffuse over beam radius, $r_B^2/(4a)$	s
t_p	Time taken to attain peak temperature	s
v	Beam traverse rate	$m\,s^{-1}$
w	Width (spacing of two isotherms)	m
z	Depth	m
z_0	Model displacement of workpiece surface	m
z_t	Thermal penetration depth	m
$\Delta t_{2/1}$	Time to cool from T_2 to T_1	s
$\Delta t_{8/5}$	Time to cool from 800 to 500°C	s
Δt_m^{100}	Characteristic $\Delta t_{8/5}$ for 100% martensite formation	s
Δt_m^{50}	Characteristic $\Delta t_{8/5}$ for 50% martensite formation	s
Δt_m^{0}	Characteristic $\Delta t_{8/5}$ for 0% martensite formation	s
Δt_f^{0}	Characteristic $\Delta t_{8/5}$ for 0% ferrite formation	s
Δt_p^{0}	Characteristic $\Delta t_{8/5}$ for 0% pearlite formation	s
Δt_b^{50}	Characteristic $\Delta t_{8/5}$ for 50% bainite formation	s
Δt_b^{0}	Characteristic $\Delta t_{8/5}$ for 0% bainite formation	s
λ	Thermal conductivity	$J\,s^{-1}\,m^{-1}\,K^{-1}$
ρ	Density	$kg\,m^{-3}$
τ	Beam interaction time	s

E1   EQUATIONS OF HEAT FLOW

Heat flow in laser processing can be complex. However, for many processes it may be approximated to three fundamental conditions: steady state, transient, or quasi-steady state. Fourier's first law describes steady state conditions:

$$F = -\lambda \nabla T \tag{E1.1}$$

where F is the heat flux ($J\,m^{-2}\,s^{-1}$), ∇T is the thermal gradient ($K\,m^{-1}$), and λ is the thermal conductivity ($J\,s^{-1}\,m^{-1}\,K^{-1}$). In this state, the temperature field does not change with time at a location in the material.

Fourier's second law describes transient conditions:

$$\frac{q_v}{\rho c} = \frac{\partial T}{\partial t} - a \nabla^2 T \tag{E1.2}$$

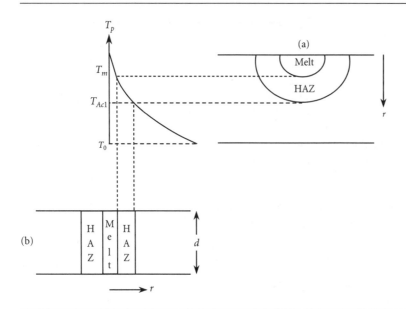

Figure E1.1 Schematic illustration of limiting geometries of heat flow: (a) radial from a point source, (b) lateral from a line source; note that a single temperature profile is used for clarity – in practice the peak temperature varies as $1/r^2$ and $1/r$ for radial and lateral heat flow, respectively

where q_v is the energy generation per unit time and volume ($J\,s^{-1}\,m^{-3}$), ρ is the density ($kg\,m^{-3}$), c is the specific heat capacity ($J\,kg^{-1}\,K^{-1}$), T is the temperature (K), a is the thermal diffusivity ($m^2\,s^{-1}$), and t is the time (s). In this state, a thermal cycle is experienced at a location in the material.

Quasi-steady state heat flow describes a condition in which the temperature field observed from a *moving* energy source remains constant. The variable $\xi = x - vt$ is defined, where ξ is the distance of the point of interest from the source, measured along the x axis (the direction of source motion) and v is velocity. If no energy is generated, equation (E1.2) becomes

$$\frac{\partial^2 T}{\partial \xi^2} + \frac{\partial^2 T}{\partial y^2} + \frac{\partial^2 T}{\partial z^2} = -\frac{v}{a}\frac{\partial T}{\partial \xi}. \tag{E1.3}$$

In a state of quasi-steady state heat flow, transient effects can be ignored, which enables analytical expressions for temperature fields to be derived.

In addition, heat flow in most laser-based processes may be considered to be limited to one of two types of source: a surface source, from which heat is conducted radially into the material (three-dimensional or 'thick plate' heat flow); or a through-thickness source, from which heat is conducted laterally into the material (two-dimensional or 'thin plate' heat flow). Radial and lateral heat flow are illustrated schematically for a structural steel in Figs E1.1(a) and (b), respectively.

Analytical solutions to equations (E1.1)–(E1.3) can be derived by making assumptions, such as: the material is homogeneous and isotropic; heat flow occurs exclusively by conduction; and no energy is generated by material transformations. Under conditions typical of thermally induced change during laser processing, these are justifiable for the level of accuracy sought in this treatment ($\pm 5\%$).

E2 TEMPERATURE FIELD AROUND A SURFACE ENERGY SOURCE

Table E2.1 Analytical equations for the temperature fields around a surface energy source

Type	Shape	Energy distribution	Equation	Eqn no.	Source
Stationary	Area A_B	Uniform	$T_{(0,0,z,t)} - T_0 = \dfrac{2Aq}{A_B}\left[\left(\dfrac{at}{\pi}\right)^{1/2}\exp-\left(\dfrac{z^2}{4at}\right) - \dfrac{z}{2}erfc\dfrac{z}{\sqrt{4at}}\right]$	E2.1	(Carslaw and Jaeger, 1959)
Stationary			$T_{(0,0,0,t)} - T_0 = \dfrac{Aq}{\lambda A_B}\left(\dfrac{4at}{\pi}\right)^{1/2}$	E2.2	(Carslaw and Jaeger, 1959)
Stationary	Circular	Uniform	$T_{(0,0,0,t)} - T_0 = \dfrac{2Aq}{\pi^{3/2}\lambda r_B^2}(at)^{1/2}$	E2.3	(Bass, 1983a)
			$T_{(0,0,0,\infty)} - T_0 = \dfrac{Aq}{\pi\lambda r_B}$	E2.4	(Bass, 1983b)
Stationary	Rectangular	Uniform	$T_{(0,0,0,\infty)} - T_0 = \dfrac{Aq}{2\pi\lambda r_x r_y}\left(r_x\sinh^{-1}\dfrac{r_y}{r_x} + r_y\sinh^{-1}\dfrac{r_x}{r_y}\right)$	E2.5	(Bass, 1983c)
Stationary	Square	Uniform	$T_{(0,0,0,\infty)} - T_0 = \dfrac{\sinh^{-1}(1)Aq}{\pi\lambda r_B} = \dfrac{0.881Aq}{\pi\lambda r_B}$	E2.6	(Carslaw and Jaeger, 1959)
Stationary	Circular	Gaussian	$T_{(x,y,z,t)} - T_0 = \dfrac{Aq}{4\lambda c(at)^{1/2}(t+t_0)}$ $\times \exp-\dfrac{1}{4a}\left(\dfrac{z^2}{t} + \dfrac{x^2+y^2}{t+t_0}\right)$	E2.7	(Ready, 1971)

(Contd)

Table E2.1 (*Contd*)

Type	Shape	Energy distribution	Equation	Eqn no.	Source
			$T_{(0,0,0,t)} - T_0 = \dfrac{Aq}{\pi^{3/2}\lambda r_B}\tan^{-1}\left(\dfrac{4at}{r_B^2}\right)^{1/2}$	E2.8	(Ready, 1971)
			$T_{0,0,0,\infty} = \dfrac{\pi^{1/2}}{2}\dfrac{Aq}{\pi\lambda r_B} = 0.866\dfrac{Aq}{\pi\lambda r_B}$	E2.9	(Ready, 1971)
Moving	Point		$T_{(r,0,0,t)} - T_0 = \dfrac{Aq}{2\pi\lambda vt}\exp\left(\dfrac{r^2}{4at}\right)$	E2.10	(Rosenthal, 1946; Ashby and Easterling, 1982)
			$\left(\dfrac{dT}{dt}\right)_{T>T_p} = -2\pi\lambda\dfrac{v}{Aq}(T-T_0)^2$	E2.11	
			$T_p - T_0 = \dfrac{2}{\pi e}\dfrac{Aq}{\rho c v r^2}$	E2.12	
			$(\Delta t_{T_2-T_1})_{T>T_p} = \dfrac{Aq}{v}\dfrac{1}{2\pi\lambda}\dfrac{1}{\theta_1}$	E2.13	
			$t_p = \dfrac{Aq}{v}\dfrac{1}{2\pi\lambda e}\dfrac{1}{T_p-T_0} = \dfrac{\Delta t_{T_2-T_1}\,\theta_1}{e(T_p-T_0)} = \dfrac{r^2}{4a}$	E2.14	
			$T - T_0 = \theta_1\dfrac{\Delta t}{t}\exp-\left[\dfrac{\Delta t}{et}\left(\dfrac{\theta_1}{(T_p-T_0)}\right)\right]$ where $\dfrac{1}{\theta_1} = \dfrac{1}{T_1-T_0} - \dfrac{1}{T_2-T_0}$ and $T_2 > T_1$	E2.15	
			$w_{T_{p_1}-T_{p_2}} = \left\{\dfrac{Aq}{v}\dfrac{2}{\pi e}\dfrac{1}{\rho c}\left[\dfrac{1}{(T_{p_1}-T_0)} - \dfrac{1}{(T_{p_2}-T_0)}\right]\right\}^{1/2}$	E2.16	

Fast moving   Circular   Gaussian

$$T_{(0,y,z,t)} - T_0 = \frac{Aq}{2\pi\lambda v[t(t+t_0)]^{1/2}} \times \exp{-\frac{1}{4a}\left(\frac{z^2}{t} + \frac{y^2}{t+t_0}\right)}$$

E2.17   (Rykalin et al., 1978)

Moving   Circular   Gaussian

$$T_{(0,y,z,t)} - T_0 = \frac{Aq}{2\pi\lambda v[t(t+t_0)]^{1/2}} \times \exp{-\frac{1}{4a}\left(\frac{(z+z_0)^2}{t} + \frac{y^2}{t+t_0}\right)}$$

E2.18   (Ashby and Easterling, 1984)

$$T_{(0,0,z,t)} - T_0 = \frac{Aq}{2\pi\lambda v[t(t+t_0)]^{1/2}} \times \exp{-\frac{1}{4a}\left(\frac{(z+z_0)^2}{t}\right)}$$

E2.19

$$\frac{dT}{dt} = \frac{(T-T_0)}{t}\left(\frac{(z+z_0)^2}{4at} - \frac{(2t+t_0)}{(2t+2t_0)}\right)$$

E2.20

$$t_p = \frac{t_0}{4}\left(\frac{2(z+z_0)^2}{r_B^2} - 1 + \left(\frac{4(z+z_0)^4}{r_B^4} + \frac{12(z+z_0)^2}{r_B^2} + 1\right)^{1/2}\right)$$

E2.21

E3   TEMPERATURE FIELD AROUND A THROUGH-THICKNESS ENERGY SOURCE

Table E3.1   Analytical equations for the temperature field around a through-thickness energy source

Type	Shape	Equation	Eqn no.	Source
Stationary	Line	$$\frac{Aq}{d} = \frac{2\pi\lambda\,(T_1 - T_2)}{\ln\left(\dfrac{r_2}{r_1}\right)}$$	E3.1	(Klemens, 1976)
Moving	Line	$$T_{(r,t)} - T_0 = \frac{Aq}{vd}\frac{1}{(4\pi\lambda\rho ct)^{1/2}}\exp-\frac{r^2}{4at}$$	E3.2	(Rosenthal, 1941; Ashby and Easterling, 1982)
		$$\frac{dT}{dt} = -\left(\frac{vd}{Aq}\right)^{1/2}\frac{1}{4(\pi\lambda\rho c)^{1/2}}\,(T - T_0)^{3/2}$$	E3.3	
		$$T_p - T_0 = \frac{Aq}{vd}\left(\frac{1}{2\pi e}\right)^{1/2}\frac{1}{\rho cr}$$	E3.4	
		$$\left(\Delta t_{T_2 - T_1}\right)_{T > T_p} = \left(\frac{Aq}{vd}\right)^2\frac{1}{4\pi\lambda\rho c}\frac{1}{\theta_2^2}$$	E3.5	
		$$t_p = \left(\frac{Aq}{vd}\right)^2\frac{1}{4\pi\lambda\rho ce}\frac{1}{(T_p - T_0)^2}$$ $$= \frac{\Delta t\theta_2^2}{e(T_p - T_0)^2} = \frac{r^2}{2a}$$	E3.6	
		$$T - T_0 = \theta_2\left(\frac{\Delta t}{t}\right)^{1/2}\exp-\left(\frac{\theta_2^2\Delta t}{2et(T_p - T_0)^2}\right)$$ where $\dfrac{1}{\theta_2^2} = \dfrac{1}{(T_1 - T_0)^2} - \dfrac{1}{(T_2 - T_0)^2}$ and $T_2 > T_1$	E3.7	
		$$w_{T_{p_1} - T_{p_2}} = \frac{Aq}{vd}\left(\frac{1}{2\pi e}\right)^{1/2}$$ $$\times\frac{1}{\rho c}\left[\frac{1}{(T_{p_1} - T_0)} - \frac{1}{(T_{p_2} - T_0)}\right]$$	E3.8	

E4   KINETIC EFFECT OF A THERMAL CYCLE AND DIFFUSION-CONTROLLED STRUCTURAL CHANGES

Table E4.1   Analytical equations for the kinetic effect of a thermal cycle and diffusion-controlled structural changes

	Equation	Eqn no.	Source
Kinetic effect	$I = \alpha t_p \exp -\dfrac{Q}{RT_p}$ where $\alpha = \sqrt{\dfrac{2\pi RT_p}{Q}}$ and $t_p = \dfrac{Aq}{v}\dfrac{1}{2\pi\lambda e}\dfrac{1}{T_p - T_0}$ for radial heat flow, and $\alpha = 2\sqrt{\dfrac{\pi RT_p}{Q}}$ and $t_p = \left(\dfrac{Aq}{vd}\right)^2 \dfrac{1}{4\pi\lambda\rho ce}\dfrac{1}{(T_p - T_0)^2}$ for lateral heat flow	E4.1	(Ion *et al.*, 1984)
Grain growth	$g^2 - g_0^2 = k\alpha t_p \exp -\dfrac{Q}{RT_p}$	E4.2	(Avrami, 1939; Johnson and Mehl, 1939)
Precipitate coarsening	$p^3 - p_0^3 = \dfrac{k\alpha t_p}{T_p} \exp -\dfrac{Q}{RT_p}$	E4.3	(Lifshitz and Slyozov, 1961; Wagner, 1961)
Precipitate dissolution	$T_s = \dfrac{B}{A - \log\left[\dfrac{C_m^a C_c^b}{f^{a+b}}\right]}$ where $f = 1 - \exp -\left\{\dfrac{\alpha t_p}{\alpha^* t_p^*}\exp -\dfrac{Q}{R}\left(\dfrac{1}{T_p} - \dfrac{1}{T_p^*}\right)\right\}^{3/2}$ and starred variables denote calibration conditions	E4.4	(Ion *et al.*, 1984)

E5 PHASE TRANSFORMATIONS AND HARDNESS IN CARBON–MANGANESE STEELS

Because of their importance in engineering, analytical equations describing phase transformations and hardness in plain carbon–manganese steels are provided in Table E5.1. Figure E5.1 illustrates the critical cooling times for phase formation. Note that the carbon equivalent C_{eq} used here to calculate critical cooling times is that defined in equation (E5.7).

Table E5.1 Empirical and theoretical equations for phase transformations in steels: alloying additions in wt%

	Equation	Eqn no.	Source
Phase transformation temperatures	T_{Ac1} (K) $= 996 - 30\text{Ni} - 25\text{Mn} - 5\text{Co} + 25\text{Si}$ $+ 30\text{Al} + 25\text{Mo} + 50\text{V}$	E5.1	(Andrews, 1965)
	T_{Ac3}(K) $= 1183 - 416\text{C} + 228\text{C}^2 - 40\text{Cr}$ $- 50\text{Mn} - 40\text{Ni} + 800\text{P} + 60\text{V} + 130\text{Mo}$ $+ 50\text{W} + 50\text{S}$	E5.2	(Kumar, 1968)
	T_m(K) $= 1810 - 90\text{C}$	E5.3	(Hansen, 1958)
	T_{Ms}(K) $= 812 - 423\text{C} - 30.4\text{Mn}$ $- 17.7\text{Ni} - 12.1\text{Cr} - 7.5\text{Mo}$	E5.4	(Andrews, 1965)
	T_{M50} (K) $= T_{Ms} - 47$	E5.5	(Steven and Haynes, 1956)
	T_{Mf} (K) $= T_{Ms} - 215$	E5.6	(Steven and Haynes, 1956)
Carbon equivalent	$C_{eq} = C + \text{Mn}/12 + \text{Si}/24$	E5.7	(Inagaki and Sekiguchi, 1960)
Critical cooling times	$\Delta t_m^{50} = \exp\,(17.724 C_{eq} - 2.926)$	E5.8	(Ion *et al.*, 1984)
	$\Delta t_f^0 = \exp\,(19.954 C_{eq} - 3.944)$	E5.9	(Ion *et al.*, 1984)
	$\Delta t_b^0 = \exp(16.929 C_{eq} + 1.453)$	E5.10	(Ion *et al.*, 1984)
	$\Delta t_b^{50} = \exp(\ln(\Delta t_b^0 \cdot \Delta t_f^0)/2)$	E5.11	(Ion *et al.*, 1984)
Phase volume fractions	$V_m = \exp\{\ln(0.5) \cdot (\Delta t/\Delta t_m^{50})^2\}$	E5.12	(Ion *et al.*, 1984)
	$V_b = \exp\{\ln(0.5) \cdot (\Delta t/\Delta t_b^{50})^2\} - V_m$	E5.13	(Ion *et al.*, 1984)
	$V_{fp} = 1 - (V_m + V_b)$	E5.14	(Ion *et al.*, 1984)
Phase hardness	$H_m = 295 + 515\,C_{eq}$	E5.15	(Ion *et al.*, 1984)
	$H_b = 223 + 147\,C_{eq}$	E5.16	(Ion *et al.*, 1984)
	$H_{fp} = 140 + 139\,C_{eq}$	E5.17	(Ion *et al.*, 1984)
Average hardness	$H_{max} = V_m H_m + V_b H_b + V_{fp} H_{fp}$	E5.18	

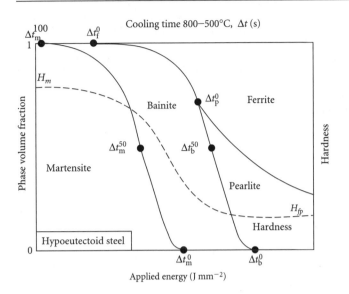

Figure E5.1 Critical cooling times for the formation of various microstructures in hypoeutectoid carbon–manganese steels, with a schematic hardness variation superimposed (broken line)

BIBLIOGRAPHY

Andrews, K.W. (1965). Empirical formulae for the calculation of some transformation temperatures – *Journal of the Iron and Steel Institute*, **203**, 721–727.

Ashby, M.F. and Easterling, K.E. (1982). A first report on diagrams for grain growth in welds. *Acta Metallurgica*, **30**, (11), 1969–1978.

Ashby, M.F. and Easterling, K.E. (1984). The transformation hardening of steel surfaces by laser beams – 1. Hypoeutectoid steels. *Acta Metallurgica*, **32**, (11), 1935–1948.

Avrami, M. (1939). Kinetics of phase change I. *Journal of Chemical Physics*, **7**, 1103–1112.

Bass, M. (1983a). Laser heating of solids. In: Bertolotti, M. ed. *Physical Processes in Laser–material Interactions*. New York: Plenum Press. p. 83.

Bass, M. (1983b). Laser heating of solids. In: Bertolotti, M. ed. *Physical Processes in Laser–material Interactions*. New York: Plenum Press. p. 86.

Bass, M. (1983c). Laser heating of solids. In: Bertolotti, M. ed. *Physical Processes in Laser–material Interactions*. New York: Plenum Press. p. 87.

Carslaw, H.S. and Jaeger, J.C. (1959). *Conduction of Heat in Solids.* 2nd ed. Oxford: Clarendon Press.

Hansen, M. (1958). *Constitution of Binary Alloys*. New York: McGraw-Hill.

Inagaki, M. and Sekiguchi, H. (1960). Continuous cooling transformation diagrams of steels for welding and their applications. *Transactions of the National Institute for Metals*, Japan, **2**, 102–125.

Ion, J.C., Easterling, K.E. and Ashby, M.F. (1984). A second report on diagrams of microstructure and hardness for heat-affected zones in welds. *Acta Metallurgica*, **32**, (11), 1949–1962.

Johnson, W.A. and Mehl, R.F. (1939). Reaction kinetics in processes of nucleation and growth. *Transactions of the American Institute of Mining and Metallurgical Engineering, Iron Steel Division*, **135**, 416–458.

Klemens, P.G. (1976). Heat balance and flow conditions for electron beam and laser welding. *Journal of Applied Physics*, **47**, (5), 2165–2174.

Kumar, R. (1968). *Physical Metallurgy of Iron and Steel.* London: Asia.

Lifshitz, I.M. and Slyozov, V.V. (1961). The kinetics of precipitation from supersaturated solid solutions. *Journal of the Physics and Chemistry of Solids*, **19**, 35–50.

Ready, J.F. (1971). *Effects of High-Power Laser Radiation*. New York: Academic Press.

Rosenthal, D. (1941). Mathematical theory of heat distribution during welding and cutting. *Welding Journal*, **20**, 220s–234s.

Rosenthal, D. (1946). The theory of moving sources of heat and its application to metal treatments. *Transactions of the American Society of Mechanical Engineers*, **68**, 849–866.

Rykalin, N., Uglov, A. and Kokora, A. (1978). *Laser Machining and Welding*. Oxford: Pergamon Press.

Steven, W. and Haynes, A.G. (1956). The temperature of formation of martensite and bainite in low-alloy steels. *Journal of the Iron and Steel Institute*, **183**, 349–359.

Wagner, C. (1961). Theory of precipitate ageing during re-solution (Ostwald ripening). *Zeitschrift der Elektrochemie*, **65**, 581–591. (In German.)

STANDARDS AND GUIDELINES

INTRODUCTION AND SYNOPSIS

The standards and guidelines given in this appendix can be used in a number of ways. Some provide information regarding the safety requirements for personnel and machinery – the regulations they describe are mandatory. They are to be used when designing or modifying a laser or material processing system to meet the demands of a safety audit.

Others provide valuable guidance, and their contents will be used increasingly to harmonize the recording of procedures and results. Since reports of laser material processing procedures began to appear in the early 1960s, the parameters used – notably the beam characteristics – have been measured in different ways; this has led to difficulties in comparing data from different sources. Standards now exist by which such variables can be reported in a consistent manner. Similarly, the procedures for testing and the records of results can now be described by referring to appropriate standards, e.g. those describing the edge quality of laser cuts, or the levels of imperfections in laser welds.

One of the drawbacks facing designers and production engineers in selecting laser processing as a fabrication route has been the difficulty in comparing the properties of a laser-processed component with those of a component manufactured using a traditional technique. Standardized quality levels of workmanship provide means of establishing in-service properties, enabling comparisons to be made between laser-processed and traditionally fabricated parts.

Standards and guidelines are continually being prepared and updated. Information about current documents can be obtained from the bodies listed in the bibliography.

Table F.1 Standards and guidelines for laser material processing

Topic	Subject	Number	Reference
Lasers	Safe use of lasers	ANSI Z136.1-2000	(ANSI, 2000)
Lasers	Safe use of lasers in the healthcare environment	ANSI Z136.3-1996	(ANSI, 1996)
Lasers	Laser diode modules for fibreoptic systems	IEC 60747-12-2:1995	(IEC, 1995)
Lasers	Spatial characterization of laser beam	ISO 11146:1999	(ISO, 1999)
Lasers	Spatial characterization of laser beam	ISO/TR 11146-3:2004	(ISO, 2004a)
Lasers	Components for UV, visible and near infrared ranges	ISO 11151-1:2000	(ISO, 2000a)

(Contd)

Table F.1 *(Contd)*

Topic	Subject	Number	Reference
Lasers	Components for infrared spectral range	ISO 11151-2:2000	(ISO, 2000b)
Lasers	Minimum requirements for documentation	ISO 11252:1993	(ISO, 1993a)
Lasers	Mechanical interfaces	ISO 11253:1993	(ISO, 1993b)
Lasers	Laser-induced damage of optical surfaces – 1-on-1 test	ISO 11254-1:2000	(ISO, 2000c)
Lasers	Laser-induced damage of optical surfaces – S-on-1 test	ISO 11254-2:2001	(ISO, 2001a)
Lasers	Laser beam positional stability	ISO 11670:2003	(ISO, 2003a)
Lasers	Laser beam polarization	ISO 12005:2003	(ISO, 2003b)
Lasers	Shape of a laser beam wavefront	ISO 15367-1:2003	(ISO, 2003c)
Lasers	Optical instruments – vocabulary and symbols	ISO 11145:2001	(ISO, 2001b)
Lasers	Fibreoptic connectors for laser applications	ISO 11149:1997	(ISO, 1997a)
Lasers	Absorptance of optical laser components	ISO 11551:2003	(ISO, 2003d)
Lasers	Laser beam power, energy and temporal characteristics	ISO 11554:2003	(ISO, 2003e)
Lasers	Laser beam power (energy) density distribution	ISO 13694:2000	(ISO, 2000d)
Lasers	Lifetime of lasers	ISO 17526:2003	(ISO, 2003f)
Engineering materials	Determination of thermal diffusivity	EN 821-2:1997	(CEN, 1997a)
Engineering materials	Micrographic determination of grain size of steels	ISO 643:2003	(ISO, 2003g)
Engineering materials	Determination of grain size of copper alloys	ISO 2624:1995	(ISO, 1995a)
Engineering materials	Tensile testing of metals at ambient temperature	EN 10002-1:2001	(CEN, 2001a)
Engineering materials	Tensile testing of metals at elevated temperatures	EN 10002-5:1992	(CEN, 1992a)
Systems	Optical elements and systems	ISO 10110-17:2004	(ISO, 2004b)
Systems	CO_2-laser beam machines – principles, acceptance	ISO 15616-1:2003	(ISO, 2003h)
Systems	CO_2-laser beam machines – static and dynamic accuracy	ISO 15616-2:2003	(ISO, 2003i)
Systems	CO_2-laser beam machines – gas flow and pressure	ISO 15616-3:2003	(ISO, 2003j)
Systems	CO_2-laser beam machines using 2D moving optics	ISO/TS 17477:2003	(ISO, 2003k)
Systems	Filters and eye-protectors against laser radiation	ISO 6161:1981	(ISO, 1981)
Systems	Filters and eye-protectors against laser radiation	EN 207:1998	(CEN, 1998a)
Systems	Laser adjustment eye-protectors	EN 208:1998	(CEN, 1998b)

Table F.1 (*Contd*)

Topic	Subject	Number	Reference
Systems	Safety – diagnostic and therapeutic laser equipment	EN 60601-2-22:1996	(CEN, 1996a)
Systems	Safety – equipment, requirements and user's guide	IEC 60825-1:2001	(IEC, 2001)
Systems	Safety of optical fibre communication systems	EN 60825-2:2000	(CEN, 2000a)
Systems	Safety of laser products – laser guards	EN 60825-4:1997	(CEN, 1997b)
Systems	Safety – manufacturer's check list for IEC 60825-1	IEC/TR 60825-5:2003	(IEC, 2003)
Systems	Safety – products for visible information transmission	IEC TS 60825-6:1999	(IEC, 1999a)
Systems	Guidelines for the safe use of medical laser equipment	IEC 60825-8:1999	(IEC, 1999b)
Systems	Maximum permissible exposure to incoherent optical radiation	IEC 60825-9:1999	(IEC, 1999c)
Systems	Safety – guidelines and notes for IEC 60825-1	IEC 60825-10:2002	(IEC, 2002)
Systems	Safety of laser products – a user's guide	IEC 60825-14:2004	(IEC, 2004a)
Systems	Safety of laser products	IEC 60825-SER:2004	(IEC, 2004b)
Systems	Interlocking devices associated with guards	EN 1088:1995	(CEN, 1995a)
Systems	Safety requirements for laser processing machines	EN 12626:1997	(CEN, 1997c)
Systems	Safety requirements for laser processing machines	ISO 11553:1996	(ISO, 1996a)
Systems	Power and energy measuring equipment	EN 61040:1992	(CEN, 1992b)
Hardening	Depth of hardening after flame or induction hardening	DIN 50190-2:1979	(DIN, 1979a)
Cutting	Terminology of imperfections in thermal cuts	EN 12584:1999	(CEN, 1999)
Cutting	Metal cutting benchmarks for laser processing machines	ISO TR 11552:1997	(ISO, 1997b)
Cutting	Classification of thermal cuts	ISO 9013:2002	(ISO, 2002a)
Melting	Specification for fusion hardening	DIN 30960:1999	(DIN, 1999a)
Melting	Depth of hardening after nitriding	DIN 50190-3:1979	(DIN, 1979b)
Melting	Fusion hardening depth and fusion depth	DIN 50190-4:1999	(DIN, 1999b)
Melting	Laser surface treatment specification using consumables	DIN 2311-1:1995	(DIN, 1995)
Marking	CO_2-laser printable aerospace cables	EN 2265-004:1995	(CEN, 1995b)
Marking	UV laser printable aerospace cables	EN 2265-005:1995	(CEN, 1995c)
Marking	YAG X3 laser printable aerospace cables	EN 2265-006:1995	(CEN, 1995d)
Welding	Guidelines for approval of CO_2-laser welding	–	(Lloyd's, 1997)
Welding	Nomenclature of welding and allied processes	ISO 4063:1998	(ISO, 1998a)
Welding	Classification of imperfections in fusion welds in metals	ISO 6520-1:1998	(ISO, 1998b)

(*Contd*)

Table F.1   (*Contd*)

Topic	Subject	Number	Reference
Welding	Joint preparation for beam welding of steels	ISO 9692-1:2003	(ISO, 2003l)
Welding	Angles of slope and rotation during welding	ISO 6947:1990	(ISO, 1990)
Welding	Shielding gases for arc welding and cutting	EN 439:1994	(CEN, 1994)
Welding	General guidance for arc welding	EN 1011-1:1998	(CEN, 1998c)
Welding	Arc welding of ferritic steels	EN 1011-2:2001	(CEN 2001b)
Welding	Arc welding of stainless steels	EN 1011-3:2000	(CEN, 2000b)
Welding	Arc welding of aluminium and aluminium alloys	EN 1011-4:2000	(CEN, 2000c)
Welding	Welding of clad steel	EN 1011-5:2003	(CEN, 2003)
Welding	Fusion welding procedures for metals	ISO 9956-1:1995/ Amd 1:1998	(ISO, 1995b)
Welding	Procedure specification for laser beam welding of metals	ISO 9956-11:1996	(ISO, 1996b)
Welding	Specification and qualification of procedures for metals	ISO 15607:2003	(ISO, 2003m)
Welding	Electron and laser beam welding procedure test	ISO 15614-11:2001	(ISO, 2001c)
Welding	Radiographic procedures for welds	ISO 2504:1973	(ISO, 1973)
Welding	Visual examination of welds	EN 970:1997	(CEN, 1997d)
Welding	Magnetic particle examination of welds	EN 1290:1998	(CEN, 1998d)
Welding	Radiographic examination of welded joints	EN 1435:1997	(CEN, 1997e)
Welding	Ultrasonic examination of welded joints	EN 1714:1997	(CEN, 1997f)
Welding	Penetrant testing	EN 571-1:1997	(CEN, 1997g)
Welding	General principles of magnetic particle testing	ISO 9934-1:2001	(ISO, 2001d)
Welding	Detection media for magnetic particle testing	ISO 9934-2:2002	(ISO, 2002b)
Welding	Equipment for magnetic particle testing	ISO 9934-3:2002	(ISO, 2002c)
Welding	Search unit and sound field for ultrasonic inspection	ISO 10375:1997	(ISO, 1997c)
Welding	Hardness test on arc welded joints	EN 1043-1:1995	(CEN, 1995e)
Welding	Micro hardness testing on welded joints	EN 1043-2:1997	(CEN, 1997h)
Welding	Macroscopic and microscopic examination of welds	EN 1321:1996	(CEN, 1996b)
Welding	Fracture tests on welds	EN 1320:1997	(CEN, 1997i)
Welding	Impact testing of welds	EN 875:1995	(CEN, 1995f)
Welding	Transverse tensile testing of welds	ISO 4136:2001	(ISO, 2001e)
Welding	Longitudinal tensile testing of welds	ISO 5178:2001	(ISO, 2001f)
Welding	Tensile test on cruciform and lapped weld joints	ISO 9018:2003	(ISO, 2003n)
Welding	Bend testing of welds	EN 910:1996	(CEN, 1996c)
Welding	Imperfections in beam-welded steel joints	ISO 13919-1:1996	(ISO, 1996c)

(Contd)

Table F.1 (*Contd*)

Topic	Subject	Number	Reference
Welding	Imperfections in beam-welded aluminium joints	ISO 13919-2:2001	(ISO, 2001g)
Welding	Terminology of imperfections in laser beam cuts	ISO 17658:2002	(ISO, 2002d)
Welding	Fusion welding for aerospace applications	AWS D17.1:2001	(AWS, 2001)
Machining	Laser beam ablation	DIN V 32540:1997	(DIN, 1997)

BIBLIOGRAPHY

ANSI (1996). ANSI Z136.3-1996. *American national standard for safe use of lasers in the healthcare environment.* Orlando: Laser Institute of America.

ANSI (2000). ANSI Z136.1-2000. *Safe use of lasers.* Orlando: Laser Institute of America.

AWS (2001). AWS D17.1:2001. *Specification for fusion welding for aerospace applications.* Miami: American Welding Society.

CEN (1992a). EN 10002-5:1992. *Tensile testing of metallic materials. Method of test at elevated temperatures.* Brussels: European Committee for Standardization.

CEN (1992b). EN 61040:1992. *Specification for power and energy measuring detectors, instruments and equipment for laser radiation.* Brussels: European Committee for Standardization.

CEN (1994). EN 439:1994. *Welding consumables. Shielding gases for arc welding and cutting.* Brussels: European Committee for Standardization.

CEN (1995a). EN 1088:1995. *Safety of machinery. Interlocking devices associated with guards. Principles for design and selection.* Brussels: European Committee for Standardization.

CEN (1995b). EN 2265-004:1995. *Aerospace series. Cables, electrical, for general purpose; operating temperatures between $-55°$ C and $150°$ C. Part 004: CO_2-laser printable; product standard.* Brussels: European Committee for Standardization.

CEN (1995c). EN 2265-005:1995. *Aerospace series. Cables, electrical, for general purpose; operating temperatures between $-55°$ C and $150°$ C. Part 005: UV laser printable; product standard.* Brussels: European Committee for Standardization.

CEN (1995d). EN 2265-006:1995. *Aerospace series. Cables, electrical, for general purpose; operating temperatures between $-55°$ C and $150°$ C. Part 006: YAG X3 laser printable; product standard.* Brussels: European Committee for Standardization.

CEN (1995e). EN 1043-1:1995. *Destructive tests on welds in metallic materials. Hardness testing. Hardness test on arc welded joints.* Brussels: European Committee for Standardization.

CEN (1995f). EN 875:1995. *Destructive tests on welds in metallic materials. Impact tests. Test specimen location, notch orientation and examination.* Brussels: European Committee for Standardization.

CEN (1996a). EN 60601-2-22:1996. *Medical electrical equipment. Particular requirements for safety. Specification for diagnostic and therapeutic laser equipment.* Brussels: European Committee for Standardization.

CEN (1996b). EN 1321:1996. *Destructive test on welds in metallic materials. Macroscopic and microscopic examination of welds.* Brussels: European Committee for Standardization.

CEN (1996c). EN 910:1996. *Destructive tests on welds in metallic materials. Bend tests.* Brussels: European Committee for Standardization.

CEN (1997a). EN 821-2:1997. *Advanced technical ceramics. Monolithic ceramics. Thermo-physical properties. Determination of thermal diffusivity by the laser flash (or heat pulse) method.* Brussels: European Committee for Standardization.

CEN (1997b). EN 60825-4:1997. *Safety of laser products. Laser guards.* Brussels: European Committee for Standardization.

CEN (1997c). EN 12626:1997. *Safety of machinery. Laser processing machines. Safety requirements.* Brussels: European Committee for Standardization.

CEN (1997d). EN 970:1997. *Non-destructive examination of fusion welds. Visual examination.* Brussels: European Committee for Standardization.

CEN (1997e). EN 1435:1997. *Non-destructive examination of welds. Radiographic examination of welded joints.* Brussels: European Committee for Standardization.

CEN (1997f). EN 1714:1997. *Non-destructive examination of welded joints. Ultrasonic examination of welded joints.* Brussels: European Committee for Standardization.

CEN (1997g). EN 571-1:1997. *Non-destructive testing. Penetrant testing. General principles.* Brussels: European Committee for Standardization.

CEN (1997h). EN 1043-2:1997. *Destructive tests on welds in metallic materials. Hardness testing. Micro hardness testing on welded joints.* Brussels: European Committee for Standardization.

CEN (1997i). EN 1320:1997. *Destructive tests on welds in metallic materials. Fracture tests.* Brussels: European Committee for Standardization.

CEN (1998a). EN 207:1998. *Personal eye-protection. Filters and eye-protectors against laser radiation (laser eye-protectors).* Brussels: European Committee for Standardization.

CEN (1998b). EN 208:1998. *Personal eye-protection. Eye-protectors for adjustment work on lasers and laser systems (laser adjustment eye-protectors).* Brussels: European Committee for Standardization.

CEN (1998c). EN 1011-1:1998. *Welding. Recommendations for welding of metallic materials. General guidance for arc welding.* Brussels: European Committee for Standardization.

CEN (1998d). EN 1290:1998. *Non-destructive examination of welds. Magnetic particle examination of welds.* Brussels: European Committee for Standardization.

CEN (1999). EN 12584:1999. *Imperfections in oxyfuel flame cuts, laser beam cuts and plasma cuts. Terminology.* Brussels: European Committee for Standardization.

CEN (2000a). EN 60825-2:2000. *Safety of laser products – Part 2: Safety of optical fibre communication systems.* Brussels: European Committee for Standardization.

CEN (2000b). EN 1011-3:2000. *Welding. Recommendations for welding of metallic materials. Arc welding of stainless steels.* Brussels: European Committee for Standardization.

CEN (2000c). EN 1011-4:2000. *Welding. Recommendations for welding of metallic materials. Arc welding of aluminium and aluminium alloys.* Brussels: European Committee for Standardization.

CEN (2001a). EN 10002-1:2001. *Tensile testing of metallic materials. Method of test at ambient temperature.* Brussels: European Committee for Standardization.

CEN (2001b). EN 1011-2:2001. *Welding. Recommendations for welding of metallic materials. Arc welding of ferritic steels.* Brussels: European Committee for Standardization.

CEN (2003). EN 1011-5:2003. *Welding. Recommendations for welding of metallic materials. Welding of clad steel.* Brussels: European Committee for Standardization.

DIN (1979a). DIN 50190-2:1979. *Hardness depth of heat-treated parts; determination of the effective depth of hardening after flame or induction hardening.* Berlin: Deutsches Institut für Normung e.V.

DIN (1979b). DIN 50190-3:1979. *Hardness depth of heat-treated parts; determination of the effective depth of hardening after nitriding.* Berlin: Deutsches Institut für Normung e.V.

DIN (1995). DIN 2311-1:1995. *Specification and approval of laser beam surface treatment using additional materials – Part 1: Laser surface treatment specification using consumables.* Berlin: Deutsches Institut für Normung e.V.

DIN (1997). DIN V 32540:1997. *Laser beam removing. Thermal removing with laser beam. Definitions, influence factors, procedure.* Berlin: Deutsches Institut für Normung e.V.

DIN (1999a). DIN 30960:1999. *Fusion hardening with additional materials. Specification for fusion hardening (SHA). Form sheet.* Berlin: Deutsches Institut für Normung e.V.

DIN (1999b). DIN 50190-4:1999. *Hardness depth of heat-treated parts. Part 4: Determination of the fusion hardening depth and the fusion depth.* Berlin: Deutsches Institut für Normung e.V.

IEC (1995). IEC 60747-12-2:1995. *Blank detail specification for laser diode modules with pigtail for fibre optic systems and subsystems.* Geneva: International Electrotechnical Commission.

IEC (1999a). IEC 60825-6:1999. *Safety of laser products – Part 6: Safety of products with optical sources, exclusively used for visible information transmission to the human eye.* Geneva: International Electrotechnical Commission.

IEC (1999b). IEC 60825-8:1999. *Safety of laser products – Part 8: Guidelines for the safe use of medical laser equipment.* Geneva: International Electrotechnical Commission.

IEC (1999c). IEC 60825-9:1999. *Safety of laser products – Part 9: Compilation of maximum permissible exposure to incoherent optical radiation.* Geneva: International Electrotechnical Commission.

IEC (2001). IEC 60825-1:2001. *Safety of laser products – Part 1: Equipment classification, requirements and user's guide.* Geneva: International Electrotechnical Commission.

IEC (2002). IEC 60825-10:2002. *Safety of laser products – Part 10: Application guidelines and explanatory notes to IEC 60825-1.* Geneva: International Electrotechnical Commission.

IEC (2003). IEC/TR 60825-5:2003. *Safety of laser products – Part 5: Manufacturer's checklist for IEC 60825-1.* Geneva: International Electrotechnical Commission.

IEC (2004a). IEC 60825-14:2004. *Safety of laser products – Part 14: A user's guide.* Geneva: International Electrotechnical Commission.

IEC (2004b). IEC 60825-SER:2004. *Safety of laser products – ALL PARTS.* Geneva: International Electrotechnical Commission.

ISO (1973). ISO 2504:1973. *Radiography of welds and viewing conditions for films. Utilization of recommended patterns of image quality indicators (I.Q.I.).* Geneva: International Organization for Standardization.

ISO (1981). ISO 6161:1981. *Personal eye-protectors. Filters and eye-protectors against laser radiation.* Geneva: International Organization for Standardization.

ISO (1990). ISO 6947:1990. *Welds. Working positions. Definitions of angles of slope and rotation.* Geneva: International Organization for Standardization.

ISO (1993a). ISO 11252:1993. *Lasers and laser-related equipment. Laser device. Minimum requirements for documentation.* Geneva: International Organization for Standardization.

ISO (1993b). ISO 11253:1993. *Lasers and laser-related equipment. Laser device. Mechanical interfaces.* Geneva: International Organization for Standardization.

ISO (1995a). ISO 2624:1995. *Copper and copper alloys. Estimation of average grain size.* Geneva: International Organization for Standardization.

ISO (1995b). ISO 9956-1:1995/Amd 1:1998. *Specification and approval of welding procedures for metallic materials. General rules for fusion welding.* Geneva: International Organization for Standardization.

ISO (1996a). ISO 11553:1996. *Safety of machinery. Laser processing machines. Safety requirements.* Geneva: International Organization for Standardization.

ISO (1996b). ISO 9956-11:1996. *Specification and approval of welding procedures for metallic materials. Welding procedure specification for laser beam welding.* Geneva: International Organization for Standardization.

ISO (1996c). ISO 13919-1:1996. *Welding. Electron and laser beam welded joints. Guidance on quality levels for imperfections. Steel.* Geneva: International Organization for Standardization.

ISO (1997a). ISO 11149:1997. *Optics and optical instruments. Lasers and laser related equipment. Fibre optic connectors for non-telecommunication laser applications.* Geneva: International Organization for Standardization.

ISO (1997b). ISO TR 11552:1997. *Lasers and laser-related equipment. Laser materials-processing machines. Performance specifications and benchmarks for cutting of metals.* Geneva: International Organization for Standardization.

ISO (1997c). ISO 10375:1997. *Non-destructive testing. Ultrasonic inspection. Characterization of search unit and sound field.* Geneva: International Organization for Standardization.

ISO (1998a). ISO 4063:1998. *Welding and allied processes. Nomenclature of processes and reference numbers.* Geneva: International Organization for Standardization.

ISO (1998b). ISO 6520-1:1998. *Welding and allied processes. Classification of geometric imperfections in metallic materials. Part 1: Fusion welding.* Geneva: International Organization for Standardization.

ISO (1999). ISO 11146:1999. *Lasers and laser-related equipment. Test methods for laser beam parameters. Beam widths, divergence angle and beam propagation factor.* Geneva: International Organization for Standardization.

ISO (2000a). ISO 11151-1:2000. *Lasers and laser-related equipment. Standard optical components. Components for the UV, visible and near-infrared spectral ranges.* Geneva: International Organization for Standardization.

ISO (2000b). ISO 11151-2:2000. *Lasers and laser-related equipment. Standard optical components. Components for the infrared spectral range.* Geneva: International Organization for Standardization.

ISO (2000c). ISO 11254-1:2000. *Lasers and laser-related equipment. Determination of laser-induced damage threshold of optical surfaces. 1-on-1 test.* Geneva: International Organization for Standardization.

ISO (2000d). ISO 13694:2000. *Optics and optical instruments. Lasers and laser-related equipment. Test methods for laser beam power (energy) density distribution.* Geneva: International Organization for Standardization.

ISO (2001a). ISO 11254-2:2001. *Lasers and laser-related equipment. Determination of laser-induced damage threshold of optical surfaces. S-on-1 test.* Geneva: International Organization for Standardization.

ISO (2001b). ISO 11145:2001. *Optics and optical instruments. Lasers and laser-related equipment. Vocabulary and symbols.* Geneva: International Organization for Standardization.

ISO (2001c). ISO 15614-11:2001. *Specification and qualification of welding procedures for metallic materials. Welding procedure test. Electron and laser beam welding.* Geneva: International Organization for Standardization.

ISO (2001d). ISO 9934-1:2001. *Non-destructive testing. Magnetic particle testing. Part 1: General principles.* Geneva: International Organization for Standardization.

ISO (2001e). ISO 4136:2001. *Destructive tests on welds in metallic materials. Transverse tensile test.* Geneva: International Organization for Standardization.

ISO (2001f). ISO 5178:2001. *Destructive tests on welds in metallic materials. Longitudinal tensile test on weld metal in fusion welded joints.* Geneva: International Organization for Standardization.

ISO (2001g). ISO 13919-2:2001. *Welding. Electron and laser beam welded joints. Guidance on quality levels for imperfections. Aluminium and its weldable alloys.* Geneva: International Organization for Standardization.

ISO (2002a). ISO 9013:2002. *Thermal cutting. Classification of thermal cuts. Geometrical product specification and quality tolerances.* Geneva: International Organization for Standardization.

ISO (2002b). ISO 9934-2:2002. *Non-destructive testing. Magnetic particle testing. Part 2: Detection media.* Geneva: International Organization for Standardization.

ISO (2002c). ISO 9934-3:2002. *Non-destructive testing. Magnetic particle testing. Part 3: Equipment.* Geneva: International Organization for Standardization.

ISO (2002d). ISO 17658:2002. *Welding. Imperfections in oxyfuel flame cuts, laser beam cuts and plasma cuts. Terminology.* Geneva: International Organization for Standardization.

ISO (2003a). ISO 11670:2003. *Lasers and laser-related equipment. Test methods for laser beam parameters. Beam positional stability.* Geneva: International Organization for Standardization.

ISO (2003b). ISO 12005:2003. *Lasers and laser-related equipment. Test methods for laser beam parameters. Polarization.* Geneva: International Organization for Standardization.

ISO (2003c). ISO 15367-1:2003. *Lasers and laser-related equipment. Test methods for determination of the shape of a laser beam wavefront. Terminology and fundamental aspects.* Geneva: International Organization for Standardization.

ISO (2003d). ISO 11551:2003. *Optics and optical instruments. Lasers and laser-related equipment. Test method for absorbance of optical laser components.* Geneva: International Organization for Standardization.

ISO (2003e). ISO 11554:2003. *Optics and optical instruments. Lasers and laser-related equipment. Test methods for laser beam power, energy and temporal characteristics.* Geneva: International Organization for Standardization.

ISO (2003f). ISO 17526:2003. *Optics and optical instruments. Lasers and laser-related equipment. Lifetime of lasers.* Geneva: International Organization for Standardization.

ISO (2003g). ISO 643:2003. *Steels. Micrographic determination of the apparent grain size.* Geneva: International Organization for Standardization.

ISO (2003h). ISO 15616-1:2003. *Acceptance tests for CO_2-laser beam machines for high quality welding and cutting. General principles, acceptance conditions.* Geneva: International Organization for Standardization.

ISO (2003i). ISO 15616-2:2003. *Acceptance tests for CO_2-laser beam machines for high quality welding and cutting. Measurement of static and dynamic accuracy.* Geneva: International Organization for Standardization.

ISO (2003j). ISO 15616-3:2003. *Acceptance tests for CO_2-laser beam machines for high quality welding and cutting. Calibration of instruments for measurement of gas flow and pressure.* Geneva: International Organization for Standardization.

ISO (2003k). ISO/TS 17477:2003. *Acceptance tests for CO_2-laser beam machines for welding and cutting using 2D moving optics type.* Geneva: International Organization for Standardization.

ISO (2003l). ISO 9692-1:2003. *Welding and allied processes. Recommendations for joint preparation. Part 1: Manual metal-arc welding, gas-shielded metal-arc welding, gas welding, TIG welding and beam welding of steels.* Geneva: International Organization for Standardization.

ISO (2003m). ISO 15607:2003. *Specification and qualification of welding procedures for metallic materials. General rules.* Geneva: International Organization for Standardization.

ISO (2003n). ISO 9018:2003. *Destructive tests on welds in metallic materials. Tensile test on cruciform and lapped joints.* Geneva: International Organization for Standardization.

ISO (2004a). ISO/TR 11146-3:2004. *Lasers and laser-related equipment. Test methods for laser beam widths, divergence angles and beam propagation ratios. Part 3: Intrinsic and geometrical laser beam classification, propagation and details of test methods.* Geneva: International Organization for Standardization.

ISO (2004b). ISO 10110-17:2004. *Optics and photonics. Preparation of drawings for optical elements and systems. Part 17: Laser irradiation damage threshold.* Geneva: International Organization for Standardization.

Lloyd's (1997). *Guidelines for approval of CO_2-laser welding.* London: Lloyd's Register of Shipping.